Advances in Groundwater Governance

Advances in Groundwater Governance

Editors

Karen G. Villholth

International Water Management Institute (IWMI), Pretoria, South Africa

Elena López-Gunn

ICATALIST, Madrid, Spain & University of Leeds, Leeds, UK

Kirstin I. Conti

IGRAC, Delft, The Netherlands

Alberto Garrido

Universidad Politécnica de Madrid (UPM), Madrid, Spain & Water Observatory, Botín Foundation, Spain

Jac van der Gun

Van der Gun Hydro-Consulting, Schalkhaar, The Netherlands

CRC Press is an imprint of the
Taylor & Francis Group, an **informa** business
A BALKEMA BOOK

Sources of front cover images:

Upper row, from left to right: Iran Front Page; Watershed Organisation Trust WOTR; The Water Channel
Second row, from left to right: World Bank; Arjen Y. Hoekstra; Geomol.

CRC Press
Taylor & Francis Group
6000 Broken Sound Parkway NW, Suite 300
Boca Raton, FL 33487-2742

First issued in paperback 2019

Typeset by MPS Limited, Chennai, India

ISBN-13: 978-1-138-02980-4 (hbk)
ISBN-13: 978-0-367-89010-0 (pbk)

Library of Congress Cataloging-in-Publication Data

Applied for

Visit the Taylor & Francis Web site at
http://www.taylorandfrancis.com

and the CRC Press Web site at
http://www.crcpress.com

Table of contents

Foreword

First of all, I would like to thank the editors for this timely book, for which I have been invited to write a short prologue. As you know, the chapters focus on the many facets and challenges of groundwater governance. Groundwater governance is highly multi- and interdisciplinary, and thus it would be impossible to attempt to synthesize all these different facets and challenges into this short space. The prologue includes reflections from my own experience of more than half a century working on matters related to groundwater and groundwater governance.

What springs to mind is a common trajectory of anthropogenic groundwater development in semi-arid and arid areas: first, the traditional neglect of groundwater (also known as 'hydro-schizophrenia'), second, the 'silent revolution' in the intensive use of groundwater, and third, emerging risks and tensions as evidenced around the world today. I conclude with some reflections on the concept of sustainability in the context of groundwater and the key role that groundwater governance will need to play.

Nace[1] in a brief note defined 'hydro-schizophrenia' as the mental attitude of water managers who deal primarily with surface water, forgetting or neglecting groundwater. In Llamas (1975 and 1985), this "illness" was analysed for the case of Spain[2]. The identified diagnosis related to general aesthetic and ethical factors related to groundwater resources. A key element was the fact that groundwater is invisible, and that the engineering structures to take advantage of groundwater have little aesthetic beauty. For example, the head of a well that may be 1000 meters deep is practically invisible on the surface. This contrasts with the usual aesthetic splendour of hydraulic works for the use of surface water. To illustrate this, a new well field has much less visibility and aesthetic beauty during the media coverage of an inauguration than the opening of a surface water reservoir. This fact is well-known by politicians, who attach great importance to the simple visibility of their achievements. It may be argued that other activities related to groundwater are also not very visible, still groundwater works have a great social importance and political relevance.

[1]Nace, R. L. (1973) On a 1972 American Water Resources Association Meeting, *Ground Water*, Vol. 11, No. 1, pp. 48–49.
[2]Llamas, M.R. (1975) Hydroschizophrenia, *Ground Water*, 13 – no. 3, pp. 296–300.
Llamas, M. R. (1985) Spanish Water Resources Policy: The Illogical Influence of Certain Physical and Administrative Factors, *Mem. of the 18th International Congress of the International Association of Hydrogeologists*, Vol. XVIII, part. 2, pp. 160–168.

Moving to the reality of 2017, and also widening the geographical scope, the increased interest in groundwater governance represents a crucial step. During this recent period, certainly the problem of hydro-schizophrenia has diminished considerably, as a result of groundwater governance, seen as the conglomerate of increased knowledge on groundwater, the wide adoption of 'integrated water resources management' approaches, awareness raising and lobbying for political support, development of groundwater management institutions, etc. Equally, the visibility and recognition of groundwater and its use is expected to increase further as more knowledge and outreach activities are undertaken on the significance of this resource.

Llamas and Martinez-Santos (2005)[3] argued that the intensification of groundwater use was driven by the concurrence of three factors: the first was the invention of the turbine pump, which allowed the extraction of water from a well of half a metre of diameter, if the geology was suitable. The water flow obtained was, in favourable cases, sufficient to irrigate 100 hectares of an average water-consuming crop or to supply a population of 50,000 inhabitants. The second factor was the substantial advance in water well drilling technology as a by-product of oil well technology. The third factor was increased underground hydrological science, or hydrogeology, now considered a solid science respected by almost everyone. To illustrate, a legal case frequently quoted refers to a lawsuit in a Kansas State Court in the late nineteenth century. The judge in his ruling stated that the subject of groundwater is so complex that a cause-and-effect relationship between groundwater abstraction and its alleged impacts could not be established. Interestingly, a century later, in the same court, in a similar case – although evidently not with the same judge – the sentence commented how hydrogeological science had allowed to establish a relationship between cause and effect and therefore to give a court ruling based on scientific evidence.

The intensive exploitation of groundwater began in many semiarid regions of the world only a little more than half a century ago. Typically, exploitation was promoted, *e.g.* through subsidies, by the departments or ministries of agriculture to establish new irrigation. In general, the departments of agriculture operated independently of the departments responsible for water management. As a result, the latter departments did not take an active role in these new developments. What is interesting is that the three drivers mentioned above were also a catalyst for the activity of many private farmers who started to establish groundwater-based irrigation on their own account, without subsidies. This private move to developing groundwater, which for the most part was inconspicuous to the public sector, gave rise to the name 'the silent revolution' of groundwater. Interestingly, in many countries where the 'silent revolution' was intensive, it did not only produce economic gains; it also at later stages triggered (a) the awareness of the need for good governance in order to ensure sustainability of the groundwater resources; and (b) a wide range of actions to initiate and support groundwater management.

While the 'silent revolution' has produced many social and economic benefits, it has also given rise to both socio-economic and environmental problems in certain regions. Thus, it is not surprising that in more developed countries, social and environmentally

[3]Llamas M.R., Martínez-Santos P. (2005) Intensive groundwater use: silent revolution and potential source of social conflicts. *J Water Resources Planning and Management*, 131: 337–341.

conscious groups emerged that were opposed to this intensive groundwater use, mainly alluding to the fact that such use is not sustainable and often damages terrestrial and aquatic ecosystems related to groundwater. In other parts of the world, impacts of intensive groundwater use is increasingly turning to social disbenefits and unrest, partly due to the inequity in access to the resource that keeps disappearing below the bottom of many wells due to increased pumping.

Already in 1989 I highlighted how some of the conflicts were related to attitudes of ignorance, arrogance, negligence and corruption. Oftentimes, however, conflicts simply arise due to diverging stakes or preferences of diverse stakeholders. Hence, the importance and timeliness of this edited volume, giving considerations on groundwater governance and how and when it needs to be advocated and practised to facilitate the reversal of current trends.

Regarding intensive exploitation of groundwater there are different views and positions. Some people consider that the exploitation of groundwater in semi-arid zones is almost always unsustainable, in the sense that reserves of groundwater stored in previous centuries or millennia are consumed that could not be replenished for future generations. This is the case of mining groundwater, which is currently occurring in many parts of the world for both fossil and renewable aquifers. From this point of view we have to be very careful on promoting or allowing the use of groundwater. Other people argue that this could be a simplistic view that in many cases can be meaningless or misleading. In reality, the conflicts regarding groundwater intensive use often ultimately relate to who gets the water, and who controls access, who benefits (economic interests, social welfare), etc. This points to the urgency and much needed work and expertise on groundwater governance, the lynchpin that provides the needed lenses, approaches and perspectives to help shed light on this debate, but more important on the reality of groundwater management decisions on the ground.

It is very difficult, if not impossible, to predict the future for the next 20 or 30 years and beyond. In the coming years, the prevailing paradigm of climate change will greatly enhance the knowledge of and focus on groundwater as one of the main resources for adaptation to increasing periods of drought. However, it seems necessary to avoid simplistic interpretations of the concept of sustainability, because these are not only meaningless but also potentially misleading. Indeed, much work is needed on the constructive role a broader and more integrated concept of sustainability may play in groundwater governance. Not only in relation to the 'silent revolution' but also on how sustainability relates to many other key issues in groundwater management, in particular groundwater pollution.

Groundwater governance has many more facets and linkages with both natural phenomena and human activities than covered in this short foreword. A rich selection of these facets and linkages is included in the chapters of this book. I congratulate the authors and editors on this much needed book that can help us understand what groundwater governance can offer in terms of getting the sustainability of groundwater use right, from a quantitative and qualitative as well as a socio-economic perspective, to enhance human well-being while protecting critical groundwater functions.

M. Ramón Llamas Madurga
Member Spanish Royal Academy of Sciences;
Honorary Director Botín Water Observatory;
Emeritus Professor Universidad Complutense of Madrid (Spain)

Preface

As the saying goes: "out of sight, out of mind". This book tackles this saying as it relates to groundwater, focusing on what is the Achilles heel of groundwater: groundwater governance. Groundwater governance is a complex concept, rather difficult to define satisfactorily in plain language. Some like to define it as "*the process by which groundwater resources are managed through the application of responsibility, participation, information availability, transparency, custom, and rule of law. It is the art of coordinating administrative actions and decision making between and among different jurisdictional levels – one of which may be global*" (adapted after Saunier and Meganck, 2007)[1]. Others prefer "*Groundwater governance comprises the promotion of collective action to ensure control, protection and socially-sustainable utilisation of groundwater resources for the benefit of humankind and dependent ecosystems. This action is facilitated by an enabling framework and guiding principles*" (adapted from a definition by Foster and Garduño, 2013)[2]. Yet, and despite this ambiguity, it is increasingly acknowledged that groundwater governance is crucial to sustaining the multitude of benefits accruing to humankind from groundwater.

Groundwater, as part of the hydrological system, is known to almost everybody, but few people are aware of its enormous importance in daily life and even fewer are well-informed about it. Nevertheless, it is fundamental for the development of emerging economies, as well as for supporting welfare, prosperity or survival everywhere, in rich and poor countries alike. By volume, it is the most massively abstracted subsurface natural resource, indispensable for agricultural development, the sole or primary source of drinking water in many parts of the world, and highly relevant for oil and gas extraction or mining operations. Equally, groundwater is an important element of the environment, since it feeds springs, supplies baseflow to streams and forms a fundamental ingredient to many aquatic ecosystems like groundwater-dependent wetlands. In addition, land use in low-lying areas and a variety of uses of the subsurface space may require groundwater levels to be strictly controlled, sometimes even by intensive drainage.

The rationale and idea for producing this book comes from the editors' awareness of the very limited attention to groundwater governance and consequent lack

[1]Saunier, R.E., and R.A. Meganck, 2007. *Dictionary and Introduction to Global Environmental Governance*. London: Earthscan, 410 p.

[2]Foster and Garduño, 2013. Groundwater resource governance: are governments and stakeholders responding to the challenge? *Hydrogeology J* 21: 317–320.

of dissemination of available knowledge and experiences in relation to groundwater governance. Population growth and socio-economic development, combined with the climate changing at a rate unprecedented in history, means that groundwater is becoming an increasingly valuable and indispensable resource, due to its general pervasiveness and reliability. However, this is not generally commensurate with the political attention accorded to groundwater, which is further reflected in poor technical and management capacities, obviously to the detriment of good groundwater governance. Widely proliferating pollution of water resources and steady degradation of water-related environments, although not yet recognized in all countries, are additional challenges. 'Water crises' have been ranked recently among the highest global risks (Global Risks 2015 Report, World Economic Forum)[3] and the international community of water professionals widely adheres to the Global Water Partnership's statement at the Second World Water Forum in 2000 that "the water crisis is often a crisis of governance". Thus, interest in governing groundwater – which represents by far the largest part of the liquid freshwater volume on our planet – is slowly but surely coming to the forefront. This is reflected by the recently finalised Groundwater Governance project carried out jointly by GEF, FAO, UNESCO, the World Bank and IAH (http://www.groundwatergovernance.org/). It is very clear that the world has to face up to the challenge of groundwater governance without delay; otherwise the huge potentials and benefits derived from groundwater will definitely decrease and eventually be lost.

The aim of our book is to contribute to a wider understanding and enhancement of groundwater governance. The book presents in 28 chapters an authoritative and contemporary overview of groundwater governance advances in the world, written by leading experts from all continents with a diversity of professional backgrounds and views. It covers the multiple and complex components of groundwater governance (*e.g.* responsibility, participation, information and knowledge, transparency, custom and the rule of law), as well as the roles of stakeholders, challenges, current practices and options for improvement in groundwater governance. Although there are numerous books on groundwater, we believe that this book covers a gap, because to our knowledge a book exclusively dedicated to groundwater governance does not yet exist. We hope it will trigger a fruitful debate and discussion on relevant aspects of groundwater governance, grounded in robust analysis and experience, in order to gain inspiration, promote further debate, and – as far as possible – achieve consensus on approaches that improve groundwater governance. This book is aimed at all people interested in groundwater in a professional capacity or due to their concern over this crucial resource. It is written in a rigorous, yet accessible manner based on the best scientific evidence available to date, while also 'bringing the elephants into the room', *i.e.* addressing often overlooked or inconvenient topics and truths related to groundwater governance.

Our book looks at groundwater and more broadly at the subsurface environment from the perspective of the many different goals and aspirations of human society. It is not uncommon that some of these are conflicting in a specific local setting: water and

[3]World Economic Forum, *Global Risk Report 2015* (http://reports.weforum.org/global-risks-2015/).

food security, energy provision, climate change adaptation, preservation of ecosystems and the environment, as well as multiple uses of the subsurface. Important trade-offs will have to be made clear and balanced decisions have to be taken. The book addresses new issues and aspects that in current groundwater management and governance practices are often ignored, such as groundwater ecosystem services, increased activities in the subsurface space, impacts of land use practices and land use change, co-governance of groundwater and surface water systems, and transboundary issues. It also highlights the significant role of capacity development, political will and awareness raising for enhancing groundwater governance.

The different chapters have been assigned to individual authors in an attempt to achieve an optimal match between available expertise and the topics to be addressed. The authors were given overall guidelines, yet with freedom to explore their assigned topic as they saw fit. As editors, we wanted to give as much space as possible to freedom of thought and innovative ideas to make this book attractive and pertinent to you, the readers. As a result, several authors surprised us by producing a chapter surpassing our expectations in terms of scope, information or vision. On the other hand, some of the planned chapters could not be produced within the time frame set, and thus had to be cancelled, unfortunately. This is why a number of subjects – such as the links between groundwater and ecosystems, energy production or climate change – are receiving less attention in the book than we originally had in mind.

The book has been structured into four sections. The first section introduces the concept and sets the scene for understanding groundwater governance, looking at both groundwater in its diversity and how people interact with it, *e.g.* how groundwater governance can be analysed from an evolutionary perspective as part of a complex socio-ecological and adaptive system. The next section looks at key features of groundwater governance and discusses important elements and aspects of groundwater governance separately. The third section puts groundwater governance into a broader context and zooms into the increasingly important linkages between groundwater and other resources and sectors, and between local groundwater systems and phenomena or action at the international or even global level. Finally, the fourth section presents a number of case studies in order to illustrate, albeit not exhaustively, groundwater governance in practice. The book does not present final conclusions or recommendations as no silver bullets exist for groundwater governance. Rather, the editors want the book to speak for itself in its entirety and intentionally keep the thinking open on ways forward. This may in fact be the conclusion to bring out, if any.

This book has come into being thanks to the contributions and cooperation provided by many persons. In the first place we like to thank our 55 authors, for their dedication and time spent on writing the chapters, for their patience to wait for feed-back from editors and reviewers, and for the flexible way they incorporated the reviewers' suggestions into a consolidated version of their chapter. A second group we are indebted to are our professional colleagues who carefully reviewed the first drafts of the chapters and provided valuable suggestions on how to improve. They form a rather invisible group, but their impact on the quality of the book is significant. Thanks are also extended to M. Ramón Llamas, our groundwater science Nestor, who kindly wrote a Foreword to this book; to those who granted permission for using illustrations subject to copyright; and to IWMI, IGRAC and the Botín Foundation for providing financial support. Finally, we thank our publisher CRC Press/Balkema, in particular Alistair

Bright, for the interest and confidence shown from the onset in our book project, and for the excellent guidance given throughout the entire process.

We hope that you will enjoy this book. We also hope that the book contributes to bringing groundwater governance clearly into the sight and mind of decision-makers, and also of scientists, planners, stakeholders and the general public alike.

The Editors

Authors

Josef Adler
International Center for Research on the Environment and the Economy (ICRE8), Athens, Greece
Josef.adler@icre8.eu

Aziza Akhmouch
OECD Water Governance Programme, Paris, France
aziza.akhmouch@oecd.org

Ebun Akinsete
International Center for Research on the Environment and the Economy (ICRE8), Athens, Greece; School of Applied Social Studies, Robert Gordon University, Aberdeen, UK
ebun.akinsete@icre8.eu

Waleed Al-Zubari
Water Resources Management Program, College of Graduate Studies, Arabian Gulf University, Manama, Kingdom of Bahrain
waleed@agu.edu.bh

Luiz Amore
Foreign Affairs Chief Adviser, National Water Agency – ANA, Brasilia, DF, Brazil
luiz.amore@gmail.com

Stefano Burchi
International Association for Water Law – AIDA, Rome, Italy
stefano.burchi@gmail.com

John Chilton
International Association of Hydrogeologists – Past Executive Manager, UK
jchilton@iah.org

Delphine Clavreul
OECD Water Governance Programme, Paris, France
delphine.clavreul@oecd.org

Alvar Closas
International Water Management Institute (IWMI), Cairo, Egypt
a.closas@cgiar.org

Kirstin I. Conti
IGRAC, Delft, The Netherlands; University of Amsterdam, The Netherlands
kirstin.conti@gmail.com

Emilio Custodio
Technical University of Catalonia (UPC), Barcelona, Spain
emilio.custodio@upc.edu

Nathaniel Delano
U.S. Environmental Protection Agency Region 2, New York, NY, USA
delano.nathaniel@epa.gov

Peter Dillon
CSIRO Land and Water, and NCGRT Flinders University, Adelaide, Australia
pdillon500@gmail.com

Alfred M. Duda
Retired, World Bank Group and Global Environment Facility Secretariat, Washington DC, US
alfredduda@gmail.com

Nikolaos Englezos
International Center for Research on the Environment and the Economy (ICRE8), Athens, Greece;
Department of Banking and Financial Management, University of Piraeus, Greece
Nikolaos.englezos@icre8.eu

Oscar Escolero
Investigador titular, Departamento de Geología Regional, Instituto de Geología, Universidad Nacional Autónoma de México (UNAM)
escolero@geologia.unam.mx

Richard S. Evans
Infrastructure and Environment, Jacobs, Melbourne, Australia
richard.evans2@jacobs.com

Stephen Foster
International Association of Hydrogeologists–Past President; University College London—Visiting Professor of Groundwater Management, UK
drstephenfoster@aol.com

Jean Fried
Urban Planning and Public Policy, School of Social Ecology, University of California, Irvine
jfried@uci.edu

Sean G. Furey
Rural Water Supply Network (RWSN), Skat Consulting Ltd., St. Gallen, Switzerland
sean.furey@skat.ch

Marco García
WaterFocus, Bunnik, The Netherlands
marcotulio.garcia@gmail.com

Andrea K. Gerlak
Udall Center for Studies in Public Policy, The University of Arizona,
Tucson, AZ, USA
agerlak@email.arizona.edu

Jac van der Gun
Van der Gun Hydro-Consulting, Schalkhaar, The Netherlands
j.vandergun@home.nl

Rien A. Habermehl
(Previously) Geoscience Australia and Bureau of Rural Sciences, Canberra, ACT,
Australia
rien.habermehl@grapevine.com.au

Ricardo Hirata
Vice Director of Centro de Pesquisas de Águas Subterrâneas (CEPAS|USP), Instituto
de Geociências, Universidade de São Paulo (Brazil)
rhirata@usp.br

Arjen Y. Hoekstra
Twente Water Centre, University of Twente, Enschede, The Netherlands;
Institute of Water Policy, Lee Kuan Yew School of Public Policy, National University
of Singapore, Singapore
a.y.hoekstra@utwente.nl

Xanthi I. Kartala
School of Economics, Athens University of Economics and Business, Greece;
International Center for Research on the Environment and the Economy (ICRE8),
Athens, Greece
Xanthi.kartala@icre8.eu

Phoebe Koundouri
School of Economics, Athens University of Economics and Business, Greece;
Grantham Research Institute, London School of Economics and Political Science, UK;
International Center for Research on the Environment and the Economy (ICRE8),
Athens, Greece
pkoundouri@aueb.gr

M. Dinesh Kumar
Executive Director, Institute for Resource Analysis and Policy,
Hyderabad 82, India
dinesh@irapindia.org

Elena López-Gunn
ICATALIST, Madrid, Spain, and University of Leeds, UK
elopezgunn@gmail.com

Sharon B. Megdal
Water Resources Research Center, The University of Arizona, Tucson, AZ, USA
smegdal@email.arizona.edu

Bruce Misstear
School of Engineering, Trinity College Dublin, Ireland
bmisster@tcd.ie

François Molle
Institut de Recherche pour le Développement (IRD), France; and International Water Management Institute (IWMI), Cairo, Egypt
f.molle@cgiar.org

Aditi Mukherji
Theme leader, Water and Air, International Centre for Integrated Mountain Development, Kathmandu, Nepal
aditi.mukherji@icimod.org

Olivier Petit
Université d'Artois, Arras, France
olivier.petit@univ-artois.fr

Shaminder Puri
Former Secretary General, International Association of Hydrogeologists and Chair of IAH Commission Transboundary Aquifers
shammypuri@aol.com

Philippe Quevauviller
Vrije Universiteit Brussel, Dept. of Hydrology and Hydrological Engineering, Brussels, Belgium
Philippe.Quevauviller@ec.europa.eu

Viviana Re
Department of Earth and Environmental Sciences, University of Pavia, Italy
viviana.re@unipv.it

Marta Rica
I-CATALIST, Madrid, Spain
mrica@icatalist.eu

Edella Schlager
School of Government and Public Policy, University of Arizona, Tucson, AZ, US
schlager@email.arizona.edu

Paul Seward
Independent Groundwater Consultant, Cape Town, South Africa
sewardp@vodamail.co.za

Ebel Smidt
SG Consultancy and Mediation, Engelen, The Netherlands and WaterFocus, Bunnik, The Netherlands
esmidt@sgmediation.nl

Ioannis Souliotis
International Center for Research on the Environment and the Economy (ICRE8), Athens, Greece; Centre for Environmental Policy, Imperial College London,

United Kingdom
Ioannis.souliotis@icre8.eu

Frank van Steenbergen
MetaMeta Research, 's-Hertogenbosch, The Netherlands
fvansteenbergen@metameta.nl

Andrew Stone
Executive Director, American Ground Water Trust, Concord, New Hampshire, USA
astone@agwt.org

Zachary Sugg
Martin Daniel Gould Center for Conflict Resolution, Stanford Law School;
Water in the West Program, Stanford University, Stanford, CA, USA
zachary.sugg@gmail.com

W. Todd Jarvis
Institute for Water and Watersheds and College of Earth, Ocean and Atmospheric Sciences, Oregon State University, Corvallis, OR, USA
todd.jarvis@oregonstate.edu

Gabriel du Toit van Dyk
Department of Water and Sanitation, Kimberly, South Africa
vandykg@dws.gov.za

Robert G. Varady
Udall Center for Studies in Public Policy, The University of Arizona, Tucson, AZ, USA
rvarady@email.arizona.edu

Elisa Vargas Amelin
European Commission, DG Environment, Brussels, Belgium
Elisa.VARGAS-AMELIN@ec.europa.eu

Karen G. Villholth
International Water Management Institute (IWMI), Pretoria, South Africa
k.villholth@cgiar.org

Ethan T. Vimont
Water Resources Research Center, The University of Arizona, Tucson, AZ, USA
vimont@email.arizona.edu

Jacobus J. de Vries
Emeritus professor of Hydrogeology, VU, Amsterdam, The Netherlands
devries.selle@kpnmail.nl

Daniel A. Wiegant
MetaMeta Research, 's-Hertogenbosch, The Netherlands
dwiegant@metameta.nl

Adriana Zuniga Teran
Udall Center for Studies in Public Policy, The University of Arizona, Tucson, AZ, USA
aazuniga@email.arizona.edu

About the Editors

Karen G. Villholth has more than 25 years of experience in groundwater resources assessment and management. She deals with research, policy advice, and capacity development related to groundwater irrigation for smallholders, transboundary aquifers, groundwater resources assessment and modelling, climate change and groundwater, adaptation through underground solutions, role of depleting aquifers in global food production, groundwater and ecosystem services, and groundwater management and governance for institutions at various levels, from local to global. She engages with multidisciplinary teams and stakeholders in co-developing tools, approaches, and policies for a more sustainable use of groundwater for livelihoods, food security, and environmental integrity. Karen is a Principal Researcher and a Research Group Leader at IWMI, International Water Management Institute at the Southern Africa regional office. She is coordinating the global partnership initiative on Groundwater Solutions for Policy and Practice (GRIPP).[1] Karen holds a PhD in Groundwater Assessment and a MSc in Chemical Engineering from the Technical University of Denmark and a MSc in Civil Engineering from the University of Washington. She previously worked for DHI-Water and Environment and the Geological Survey of Denmark and Greenland. She is co-author of three books related to groundwater and more than 50 peer-reviewed journal papers.

Elena López Gunn is the Founder and Director of ICATALIST and a Cheney Fellow at University of Leeds in the United Kingdom. Elena finished her PhD at King's College, London. She also holds a Masters from the University of Cambridge. She was an Associate Professor at IE Business School and a Visiting Senior Fellow at the London School of Economics as an Alcoa Research Fellow. Professionally, Elena has collaborated with a number of organizations including UNESCO, FAO, UNDP, EU DG Research and Innovation, universities (Spanish and Dutch) and river basin agencies, the England and Wales Environment Agency, as well as the private sector like Repsol, and NGOs like Transparency International-Spanish Chapter. She has published on a range of topics mainly related to water security, social innovation, collaborative decision making, water governance, evaluation of public policy, knowledge management and transfer.

[1]The Groundwater Solutions Initiative for Policy and Practice (GRIPP) is an IWMI-led partnership initiative designed to co-develop solutions to the challenges of sustainable groundwater development (http://gripp.iwmi.org).

Her current main focus is climate change adaptation, disaster risk reduction and socio-ecological systems. She has recently worked as Technical Expert for the International Atomic Agency on projects related to groundwater governance and capacity building. Elena lives in Madrid with her husband, Dave, and her three fabulous children.

Kirstin I. Conti has 10 years' experience in the environmental and water policy sectors. As a child, Kirstin's favourite gardening project was creating a compost bin – placing worms in a compost box and feeding them produce scraps. Years later her love for nature led her to her major in Earth Systems at Stanford University, which is an interdisciplinary program focused on environmental problem solving. At that time, her experiences in the classroom and conducting field research in the Southern Africa region sparked her passion for a career in water management. Her professional experiences revealed the complexity of managing shared water resources. As an environmental consultant in San Francisco, she witnessed stakeholder conflicts prevent the effective management of California's water. She completed a Masters of Laws at University of Dundee and recently received a PhD for her doctoral research at the University of Amsterdam and the International Groundwater Resources Assessment Centre (IGRAC). Her focus is improving groundwater governance and conflict prevention mechanisms for the purpose of ensuring that our most abundant fresh water resource, groundwater, is managed sustainably, equitably, and cooperatively.

Alberto Garrido is Spanish Professor of Agricultural and Natural Resource Economics and Vice-Rector of Quality and Efficiency at the Universidad Politécnica de Madrid (UPM), Director of the Water Observatory of the Botín Foundation and Researcher of CEIGRAM (UPM). He has a Bachelor Degree and MSc in Agricultural Engineering; a MSc in Agricultural and Natural Resource Economics from the University of California, Davis (1992) and a Doctoral Degree in Agricultural Economics from the UPM (1995). He has supervised 17 Doctoral Dissertations, published 195 references, of which 70 are academic articles and 15 books. He has 25 years of experience in leading research projects and grants. His research focuses on agricultural risks and insurance, natural resource economics and policy and water policy. He has consulted for the main international organisations (FAO, BID, The World Bank, IFAD), several national governments of Europe, Asia and America, and numerous private companies and foundations. He has coordinated 55 research projects, eight of which with international consortia. Since 1996, he is a member of the Advisory Board of the Rosenberg Forum of International Policy.

Jac van der Gun is a Dutch groundwater hydrologist and water resources specialist (MSc, Wageningen University). He has been employed successively by a Dutch water supply company (WMG, Velp), UN-OTC (New York) and the R&D organisation TNO (Delft/Utrecht). His professional career spans almost half a century and has focused on (1) water resources exploration, assessment and monitoring (including significant fieldwork campaigns); (2) hydrogeological mapping; (3) water resources planning and management; and (4) training, capacity building and institutional development. He has been involved actively in the Groundwater Reconnaissance of The Netherlands (entrusted with the overall responsibility for this programme). His long-term assignments abroad include positions as a resident hydrologist in Bolivia and as a resident water resources assessment project manager in Yemen and in Paraguay. In addition, he carried out numerous short missions in Asia, Latin America, Africa and Europe for various international and national organisations, providing scientific-technical input, supervising projects, and formulating or evaluating projects and programmes. He has also lectured at UNESCO-IHE on groundwater for more than thirty years. In 2003 he became the founding director of the International Groundwater Resources Assessment Centre (IGRAC) and since then he is mainly active in groundwater-related projects of international organisations such as UNESCO, FAO and GEF (*e.g.* the World Water Assessment Programme (WWAP) and the Groundwater Governance Project). Together with Jean Margat, he wrote the book "Groundwater around the World: A Geographic Synopsis", published in 2013.

Part 1

Setting the scene

Chapter 1

Groundwater governance: rationale, definition, current state and heuristic framework

Karen G. Villholth[1] & *Kirstin I. Conti*[2,3]

[1]International Water Management Institute (IWMI), South Africa
[2]International Groundwater Resources Assessment Centre (IGRAC), The Netherlands
[3]Governance and Inclusive Development, Amsterdam Institute of Social Sciences Research (AISSR), University of Amsterdam, The Netherlands

ABSTRACT

Groundwater governance has emerged as a relatively new concept in water resources discourses. This opening chapter sets the scene for the book in terms of an introduction to the concept and definition of groundwater governance, its critical and distinctive features, and the rationale for bringing groundwater governance into the equation of broader water and natural resources governance. Building on the recent global Groundwater Governance project, it takes a brief stock of empirical advances in groundwater governance, from the local to the global level. Finally, a heuristic framework for matching the inherent ISD (invisible, slow, distributed) signature of groundwater with governance tenets as a tool to embracing the concept of groundwater governance as well as its expression in practice is proposed. A key conclusion is that groundwater governance provides a comprehensive overarching framework that may accommodate and support more concerted and conscious approaches to targeted, while integrated, management of increasingly at-risk groundwater resources globally.

1.1 RATIONALE FOR GROUNDWATER GOVERNANCE

Human population growth as well as lifestyle changes or lifestyle aspirations associated with economic growth and development are increasingly exerting pressure on water resources globally through increasing water demands. With water seemingly becoming a progressively limiting factor in human development, whether physically or economically and whether directly for consumption or for indirect use via food and other water-dependent products, the need for governance support structures arises. As such, water governance has been framed as a tool to address these needs. Critically, water governance is raised as a means to overcome deficiencies in historical, more engineering and linear approaches to water management. It has also been advocated as a superior approach than the more recent paradigms of sustainable water management and integrated water resources management, which to some extent have been disqualified as too difficult to implement (Biswas & Tortajada, 2011). Water governance is seen as the necessary and effective instrument to address increasingly complex issues of 'water scarcity'[1], which are rooted in the interphase between the resource and human relationships. Water governance is, in fact, seen as the key instrument to ensure water security for all, as well as tackling water scarcity, which is merely the result of, and expression of, poor governance in the first place, rather than a physical condition. In other words, with good water governance, which necessarily encompasses scales beyond the local, where physical water scarcity may be overriding, proper institutions as well as technologies for engineered water and water solutions[2], it is contended that there is enough water for all – the challenge is to properly govern the resource, its use and other related human interactions and impacts (Biswas & Tortajada, 2011). Critically, the simple 'water scarcity'-governance nexus portrays a simplified but strong argument for a change in mind-set, from focus on the physical expression and reasons for water problems to the more human and political aspects. However, it is clear that groundwater governance needs to encompass broader issues than 'water scarcity', *e.g.* water-related environmental issues (ecosystems, land stability, conditions for and interactions with land use, use of subsurface space, mining, etc.) as well as the socio-economic aspects (lack of or inequity in access).

The rationale for specifically focusing on groundwater governance, as a subset of water governance, stems from a number of factors:

1 Groundwater, as the largest store of freshwater on earth, has been developed at unprecedented rates over the past half century – for agriculture and domestic as well as industrial use. Rates of abstraction have exceeded natural replenishment rates over extended periods in many parts of the world, and environmental signs of unsustainable use and negative socio-economic impacts are increasingly evident (Famiglietti, 2014; Foster & Chilton, 2003). In many areas, groundwater is the water resource mostly relied on, either historically (especially in arid and semi-arid regions), or progressively as surface water resources either deplete, become contaminated, or become excessively variable to satisfy never-ceasing needs. Hence, there is an urgent need to address unsustainable trends of groundwater development and use.
2 Groundwater possesses natural distinctive characteristics that inherently complicate its effective and efficient management. These relate to three factors: 1. It is, in effect, an invisible underground resource; 2. It has relatively slow flow rates, and 3. It has a distributed occurrence, with open access opportunities to all stakeholders, at least in principle[3]. In short, this invisible-slow-distributed (ISD) signature unique to groundwater entails that the resource is susceptible to short-sighted and unaccountable exploitation under a first-come, first-served setting as well as to contamination from various indiscriminate or uninformed land uses and waste handling practices under a reactive management setting, rather than pro-active planning and governance.

[1]Water scarcity broadly perceived as the lack of water availability and access to water.
[2]E.g. desalinisation, condensation, wastewater treatment, rainfall enhancement, iceberg melting, virtual water trade – also called unconventional water sources back in 2001 (Smakhtin *et al.*, 2001).
[3]This is also referred to as a common pool resource (Ostrom, 1990).

3 Groundwater has been developed under generally very favourable policy conditions, especially in agriculture, the largest user of groundwater. These policies have provided subsidies for input costs, *e.g.* to well installation and energy consumption and guaranteed output (*e.g.* crop) prices. The policies have been justified in the need to support livelihoods, ensure food security, and enhance rural and economic development, but with often intentional or unintentional consequences of elite capture[4] and skewed access to the resource (Closas & Molle, 2016). In addition, these policies and practices have proven difficult to change, due to vested interests and political gains for strong minorities, which calls for specific governance attention.

These three factors – the reaching of critical environmental or socio-economic tipping points[5] due to unsustainability in use, the inherent ISD-signature, and the historic path dependence, which are somewhat related – all point to the need for concerted groundwater governance. On the other hand, these complexities also explain why groundwater has not received the required attention. It is basically not simple. Global institutions normally outside the community of groundwater management are increasingly arguing for better approaches, seeing groundwater depletion as a geopolitical challenge to sustainable growth (Earth Security Group, 2016). Clearly, better understanding of these challenges and developing governance schemes to address them are needed. This is also the justification of this chapter and the book it introduces.

Recognising that groundwater is part of the larger hydrological system, it is however important to stress that groundwater governance in isolation may not prove effective. Linkages across various water sources critically determine the physical status of groundwater. Conversely, groundwater status is underpinning a vast number of terrestrial, freshwater aquatic and near-shore marine systems, which implies that finding solutions necessarily will have to involve this broader perspective (Chapter 17 in this book).

1.2 DEFINING AND CONCEPTUALIZING GROUNDWATER GOVERNANCE

1.2.1 Background, justification and approach

The term 'governance' has been in use since the 1600s, but has mostly been used in its current conception since the 1980s. Contemporary conceptualizations of governance capture the increasing number of interactions, organizations and activities occurring outside of centralized state government (*e.g.* non-governmental and civil society-based), which, in effect coordinates and adds rules and structure to society

[4]The phenomenon that resources transferred for the benefit of the larger population (usually for the poorest) are usurped by a few individuals – be it economic, political, education ethnic or otherwise.

[5]An environmental tipping point could be the permanent drying out of an important groundwater-fed wetland. A socio-economic tipping point could be the reverting back to rain-fed agriculture or the emigration of farmers away from rural areas previously dependent on groundwater for irrigated farming.

through various mechanisms such as public participation and cooperation between a wide range of actors (Bevir, 2011; Chandhoke, 2003). Since then, many forms of governance have been conceptualized to understand governance in various locations and geographic levels (*e.g.* global governance, EU governance, multilevel governance), of various sectors (*e.g.* corporate governance), of various resources (*e.g.* fisheries governance, forests governance, groundwater governance) as well as different modes of governance (*e.g.* interactive governance, network governance, adaptive governance). These forms of governance have their own bodies of literature and empirical cases, which vary in their levels of development and advancement. The purpose of this section is to discuss the concept of groundwater governance as it is currently defined, understood, and practised.

We conducted a comparative analysis of definitions related to groundwater governance found in the literature in order to understand how the concept originated and evolved over time, what its key attributes are, where it currently stands in terms of its development as well as identifying some key areas where the concept's definition could be refined going forwards. In terms of the latter, we focused on three aspects: (1) key elements of the groundwater governance concept; (2) how groundwater governance differs from groundwater management; (3) issues of vertical and horizontal integration of these frameworks and principles within and across actors, processes and geographic levels. Finally, we will explore whether a new definition of groundwater governance is warranted.

In order to frame the analysis and understand the origin of groundwater governance, we looked at a hierarchical order of increasingly specific definitions: governance, environmental governance, water governance, and finally groundwater governance.

1.2.2 Origins of the concept: from governance to groundwater governance

1.2.2.1 Governance

As the use of the term governance began to increase through the 1980s and 1990s, the number of ways the term was defined also increased. Definitions highlight management as the central purpose of governance and positioned governance as state power *e.g.* "the exercise of political power to manage a nation's affairs (World Bank, 1989:60)." However, by the mid-to-late 1990s and early 2000s, conceptualization of governance had shifted in three key ways:

- First, governance was now conceived as a process *e.g.* "the process whereby societies or organizations make their important decisions, determine who has voice, who is engaged in the process and how account is rendered" (Institute on Governance, 2006); a relationship, *e.g.* "changing relationships between State and society and a growing reliance on less coercive policy instruments" (Pierre and Peters, 2000:12); an interaction, *e.g.* "the interactions among structures, processes and traditions that determine how power and responsibilities are exercised, how decisions are taken, and how citizens or other stakeholders have their say. Fundamentally, it is about power, relationships and accountability: who has influence, who decides, and how decision-makers are held accountable" (Graham, 2003: ii); and/or a framework, *e.g.* "creating an effective political framework conducive to private economic action: stable regimes, the rule of law, efficient State administration adapted to the roles that Governments can actually perform and a strong civil society independent of the State (Hirst, 2000:14)".

- Second, the geographic scope of governance moved beyond the national to include the international and global: "Global governance is the sum of the many ways individuals and institutions, public and private, manage their common affairs" (Commission on Global Governance, 1995:4).
- Third, 'good governance' emerged as a spin-off concept in the governance discourse. Typically, groundwater governance was defined in accordance with the United Nation's (UN) set of eight core tenets (in no particular order): (1) responsibility, (2) accountability, (3) transparency, (4) efficiency, (5) legitimacy, (6) participation, (7) equity and inclusiveness, and (8) rule of law (UN ESCAP, 2006). The good governance concept is primarily used in the international development community – especially, development agencies such as the World Bank and United Nations Development Program (UNDP).

By the late 2000s, framing governance in terms of centralized power or authority had mostly fallen out of vogue. Further, the definitions of governance had become more specific with regard to the scope of what is being governed (the 'object' of governance) and who is governing and being governed (the governance 'actors'). The objects quickly moved from focusing on state issues (*e.g.* 'national affairs') to issues common to broader groups of individuals (*e.g.* 'common affairs'), including the transboundary and transnational. More recent definitions also explicitly positioned non-state actors (*e.g.* society, civil society, individuals, citizens, stakeholders) as key participants in governance. Nevertheless, the definition of governance continues to be contested, perhaps because of the myriad perspectives and disciplines from which people approach the concept (Green, 2007).

1.2.2.2 Environmental governance

The environmental governance concept co-evolved with the concept of 'global governance' in the early 1990s in response to the international legal frameworks addressing Chlorofluorocarbon (CFC) emissions, climate change and sustainable development. It was generally used as a conceptual umbrella for themes such as forest governance, fisheries governance, Antarctic governance, climate governance, land use governance, water governance, etc. However by the 2000s, conceptualizations of 'global environmental governance' included:

- "the sum of the overlapping networks of inter-state regimes on environmental issues" (Patterson *et al.*, 2003:3); and
- "the protection of the Earth's ecosystems under conditions in which human actions have become fundamental driving forces" (Young 2008:24).

To this day, a large proportion of environmental governance literature is focused on the global level *i.e.* 'global environmental governance' (Morin and Orsini, 2013;

Young, 2002, 2008). But like 'governance', the conceptual development of 'environmental governance' also experienced three key shifts. The first is the inclusion of other geographic levels: the regional, transboundary, national, and multi-level (IUCN, 2010; Meadowcroft, 2002). The second is the inclusion of private entities and corporations in environmental governance (Clapp, 2014; Falkner, 2014). The third is the adoption of the concept of 'earth system governance' (Biermann *et al.*, 2009; Dryzek & Stevenson, 2011; Spagnuolo, 2011), which could be considered an expanded global environmental governance, because it integrates sustainable development as a key norm as well as multilevel aspects of governance.

It appears that the concept of environmental governance is defined using its own terminology – distinct from the four framings of 'governance' identified above – process, relationship, interaction, framework. In contrast, 'environmental governance' is considered systems/networks and/or a set of rules/strategies *e.g.* "the rules, practices, policies and institutions that shape how humans interact with the environment." (UNEP 2009:2) or "the sum of the formal and informal rule systems and actor-networks at all levels of human society that are set up in order to influence the coevolution of human and natural systems in a way that secures the sustainable development of human society" (Biermann, 2007:326).

Through environmental governance's conceptual advancements, the object and actors identified remained relatively consistent. Most definitions indicate that all humans or 'humanity' are the key actors in environmental governance and the environment is the object of governance. This relative consistency in framing, object and actors is, in part, attributable to the global-level origins of the concept, which initially focused on legal systems and regimes created to address various environmental challenges to which all humans could potentially be contributing.

1.2.2.3 Water governance

Water governance could be considered a type or sub-theme of environmental governance. Yet, its development as a self-standing concept occurred principally without the involvement of environmental governance academics and practitioners. Rather, it emerged in the early 2000s largely from the development agencies and practitioners primarily concerned with water infrastructure, services and supply. This is observable from the framing of the two early definitions by the Inter-American Development Bank (IADB) and the Global Water Partnership (GWP):

- "the range of political, social, economic and administrative systems that are in place to develop and manage water resources, and the delivery of water services, at different levels of society" (Rogers & Hall, 2003:7).
- "a sub-set of the more general issue of society's creation of physical and institutional infrastructure, and of the still more general issue of social cooperation, which reminds us of the problems of defining who are the stakeholders, communication among stakeholders, the allocating of contributions and outputs, and the creation of institutions" (Rogers, 2002:1).

The GWP publication by Rogers and Hall (2003) not only presented one of the most frequently cited definitions of water governance; it simultaneously introduced the concept of 'effective' water governance, which applied the eight tenets of good governance to water governance (Rogers & Hall, 2003). Many scholars use the term good water governance interchangeably with effective water governance. Good water governance was quickly contested, calling into question whether it is good for everyone or leaves behind the poor (Cleaver, 2006). Subsequently, social scientists, particularly political ecologists, moved away from technically driven (or 'technocratic') approaches towards socio-political approaches in order to emphasize the societal relationships and institutions (including beliefs, traditions, laws) and influence water governance practices.[6] For example, Cleaver states "water governance is conducted through formal and informal institutions, social relationships and more specifically through the 'rules in practice' of everyday water use" (Cleaver, 2007:301).

It is also worth noting that throughout the evolution of the concept of water governance, it has been applied at a range of geographic levels, including 'transboundary', 'regional' (especially the EU) and across multiple levels (*i.e.* 'multilevel') (Pahl-Wostl *et al.*, 2013). However, these applications have not yet moved towards self-standing types of water governance and are not considered an explicit shift.

1.2.2.4 Groundwater governance

The concept of groundwater governance was introduced in the late 2000s, shortly after that of water governance. In 2009, Foster *et al.* (2009:3) – under the World Bank's Groundwater MATE program (GW-MATE) – defined groundwater governance as "*the exercise of appropriate authority and promotion of responsible collective action to ensure sustainable and efficient utilization of groundwater resources for the benefit of humankind and dependent ecosystems.*" (Table 1.1, definition no. 1) The emphasis on 'authority' in this definition harkens back to early definitions of groundwater governance also put forth by the World Bank.

In 2013, the fifth thematic paper on groundwater policy and governance for the Global Environment Facilities' (GEF) Groundwater Governance project (FAO, 2013), defined groundwater governance as "*the process by which groundwater is managed through the application of responsibility, participation, information availability, transparency, custom, and rule of law. It is the art of coordinating administrative actions and decision making between and among different jurisdictional levels – one of which may be global*" (FAO 2013:7) (Table 1.1, definition no. 2). This definition built upon a definition of governance development by Saunier and Meganck (2007:159).[7] The digest of the fifth thematic paper, written by many of the same authors, expanded upon the 2013 FAO definition, saying: "*In practice, groundwater governance is the complex and overarching framework that determines the management of groundwater resources and the use of the aquifers. The local, regional or national governance framework establishes "who" participates in formulating strategies and is responsible*

[6] See generally the works of Rhodante Ahlers, Karen Bakker, Francis Cleaver and Jeroen Warner.
[7] Saunier and Meganck define 'governance' as a "concept describing the way power is exercised in the management of a country's economic and social resources through application of responsibility, participation, information availability, transparency and the rule of law. Governance is not equal to government, which is the art of administration at a given level of power. Rather, it is the art of coordinating administration actions between different territorial levels – one of which may be global." They go on to cite the Commission on Global Governance definition (1995:4).

Table 1.1 Definitions of groundwater governance over time.

No.	*Year*	*Definition*	*Source*
1	2009	The exercise of appropriate authority and promotion of responsible collective action to ensure sustainable and efficient utilization of groundwater resources for the benefit of humankind and dependent ecosystems.	Foster *et al.* (2009:1), *Groundwater governance: conceptual framework for assessment of provisions and needs;* Foster *et al.* (2013:691). *Groundwater—a global focus on the "local resource"*
2	2013	Groundwater governance is the process by which groundwater is managed through the application of responsibility, participation, information availability, transparency, custom, and rule of law. It is the art of coordinating administrative actions and decision making between and among different jurisdictional levels–one of which may be global.	FAO (2013:7), *Groundwater Governance project*[a] *Thematic Paper No. 5, Groundwater Policy and Governance*
3	2015	The overarching framework of groundwater use laws, regulations, and customs, as well as the processes of engaging the public sector, the private sector, and civil society [that] shapes how groundwater resources are managed and how aquifers are used.	Megdal *et al.* (2015:2), *Groundwater governance in the United States: Common priorities and challenges*
4	2016	Groundwater governance comprises the enabling framework and guiding principles for responsible collective action to ensure control, protection and socially-sustainable utilisation of groundwater resources for the benefit of humankind and dependent ecosystems.	FAO (2016a:37), *Global diagnostic on groundwater governance*
5	2016	Groundwater governance comprises the promotion of responsible collective action to ensure control, protection and socially-sustainable utilisation of groundwater resources and aquifer systems for the benefit of humankind and dependent ecosystems. This action is facilitated by an enabling framework and guiding principles.	FAO (2016b:16), *Global framework for action to achieve the vision on groundwater governance*
6	2016	Effective groundwater governance comprises the promotion of responsible action to ensure the protection and sustainable use of groundwater resources and long term management of aquifer systems.	FAO (2016c:5), *Shared global vision for groundwater governance 2030 and a call-for-action*

[a]GEF project 'Groundwater Governance – A Global Framework for Action' (2011–2015), http://www.groundwatergovernance.org/

for their execution and "how" the different actors (governmental, public sector, non-governmental, private sector, and civil society) interact" (FAO, n.d.). This proposed definition and its expansion represented a shift, in which the elements from definitions of 'governance' (*e.g.* governance as a process and a framework), 'good governance' (*e.g.* the tenets of responsibility, participation, etc.), 'environmental governance' (*e.g.* governance as multi-level), and 'water governance' (*e.g.* the relationships and rules in practice) are explicitly incorporated in the concept of groundwater governance.

A few years later, this same group of researchers who developed the 2013 FAO definition, Megdal *et al.* (2015:2), further refined their conceptualization of groundwater governance to be "*the overarching framework of groundwater use laws, regulations, and customs, as well as the processes of engaging the public sector, the private sector, and civil society*" that "*shapes how groundwater resources are managed and how aquifers are used.*" (Table 1.1, definition no. 3). This definition removed the normative elements; emphasized governance as both as a framework and a process; and integrated the who and how these authors previously articulated in the fifth thematic paper for the Groundwater Governance project.

In the course of the execution of the Groundwater Governance project, the definition posed by the project contributors was exchanged for an adapted version of the Foster *et al.* (2009) definition. These adaptations were eventually published in three final project documents – the *Global Diagnostic on Groundwater Governance* (FAO, 2016a:37); the *Global Framework for Action to Achieve the Vision on Groundwater Governance* (FAO, 2016b: 16); and the *Shared Global Vision for Groundwater Governance 2030 and A Call for Action* (FAO, 2016c:5) (Table 1.1, definition no. 4–6). The three documents present slightly varied versions of the same definition. In essence, the practitioners involved removed 'authority' from the definition and instead framed governance as 'an enabling framework and guiding principles' and the 'promotion of responsible collective action'.

One of the reasons for adopting another definition was that the definition presented in the fifth thematic paper (Table 1.1, definition no. 2) was considered normative by focusing on 'good groundwater governance' (according to the criteria of the authors) and thus would exclude forms of groundwater governance, which may not meet this criterion of goodness (van der Gun, pers. comm.). The removal of normative elements in the 2015 definition by Megdal *et al.* (Table 1.1, definition no. 2) indicates some consensus between researchers and practitioners that groundwater governance is not inherently good or bad. Nevertheless, the final definitions presented by the Groundwater Governance project include elements such as socially sustainable utilization, benefit to human kind, and ecosystems protection. These elements could be viewed as normative and/or aspirational/goal-oriented.

1.2.3 Carrying the concept forward

This analysis indicates that the concept of groundwater governance is in its adolescent stage and there is room for it to further mature and develop. As such, the following section highlights key elements that need further consideration and refinement and, on this basis, makes a proposal for carrying the groundwater governance concept forward.

1.2.3.1 Key conceptual elements

The evolutionary account of groundwater governance and the comparative analysis of the definitions of 'governance,' 'environmental governance,' 'water governance,' and 'groundwater governance' elucidated several common conceptual elements:

- Object – The object is *what* is being governed. In our case, groundwater resources are clearly being governed.

- Mode – Mode is the *how* of groundwater governance. While a vast majority of definitions analysed include one or more modes of governance (*e.g.* processes, relationships, interactions, frameworks, systems, networks, formal and informal rules, and/or strategies), the specific mode(s) included vary greatly. This is likely the result of the variety of disciplines and backgrounds of those engaging with the governance concept. Further, it implies that the definition of groundwater governance does not have to focus on one of these modes, to the exclusion of others.
- Actors – Actors are *who* is governing and *who* are being governed. These are not mutually exclusive. People may be simultaneously governing groundwater through every day practices and also being governed by an overarching governance framework. Not all definitions explicitly named a who. Those that did, sometimes, did so broadly (*e.g.* institutions, stakeholders, society, decision-makers) or other times specifically (*e.g.* government, civil society, NGOs, communities, corporations). While a more specific list has the potential to unintentionally exclude key actors, it also emphasises that governance does not purely reside in the domain of the government/state or a specific group.
- Geography – Geography is the *where* of groundwater governance. Over time, many of the conceptualizations of groundwater governance grew to include geographic elements. Groundwater is typically considered a local or national resource. However, groundwater governance is increasingly being addressed from global and regional/transboundary perspectives (Conti & Gupta, 2015). Further, the definition by Megdal *et al.* (2015) already highlighted the multi-jurisdictional nature of groundwater governance.

It also revealed two debated elements:

- Normativity – Normativity refers to whether groundwater governance is regarded as either *good or bad*, in accordance with a specific set of governance criteria (*e.g.* transparency, accountability, responsibility, etc.). Being good or bad is not inherent to the concept of governance. Rather, it is an additional, value-laden element. As discussed in 1.2.2, normative framings are contested because the difference between poor/inadequate governance and good governance may depend on stakeholder view, and could be subject to various value systems.
- Prescription – Prescription refers to a specified *goal, outcome or aspiration* for groundwater governance. Groundwater governance is not inherently prescriptive, however common concerns such as sustainability or equity may be considered as additional elements. The challenge is defining what sustainability might be, for example, in the context of non-renewable (fossil) groundwater resources or what equity looks like in a given society context.

1.2.3.2 Linkages with the practice of groundwater governance

The concepts and definitions discussed above have come into our common usage relatively recently. However, the practice of these forms of governance have been active for many years. For example, a surge in environmental governance practices was triggered at the global level by Hardin's essay "The Tragedy of the Commons"

(Hardin, 1968).[8] This led to an increase in multilateral initiatives and legal frameworks designed to combat transboundary pollution in the 1970s, nearly two decades before the 'environmental governance' concept was first defined (see French, 1992 for an overview of global environmental governance practices in this period). Similarly, groundwater governance has likely occurred in local contexts for thousands of years (see Chapter 2 of this book). Yet, the concept was only defined in 2007 (Table 1.1). Further, groundwater governance cannot be fully understood in a vacuum but must rather be understood as embedded in everyday practices of actors. Consequently, evaluating how the current conceptualization of groundwater governance may or may not align with practice is key to refining its conceptualization.

1.2.3.3 The distinction between groundwater management and governance

Groundwater governance and groundwater management can be difficult to distinguish. We understand management as specific day-to-day actions taken to ensure the strategic use/and or protection of groundwater resources. In practice, the range of actors that participate in management and the scope of activities involved in management are often far narrower than those involved in governance. The practice of groundwater governance is the decision of which management actions should be taken, when, by whom, and for what purpose.

We offer here two characterizations of the distinction between groundwater management and governance. (1) Muhkerji & Shah (2005) say that management comprises activities such as monitoring, model building, and implementation of groundwater laws, while governance is more holistic and inclusive, taking into consideration the concerns of scientists, policy makers and groundwater users. (2) The Groundwater Governance project's *Global Diagnostic on Groundwater Governance* (FAO 2016a) says that groundwater governance establishes the governance actors and determines how they interact; drives decision-making with information, knowledge and science; and creates policies and plans that define why activities are needed and when they should occur. Groundwater management is what activities the actors do within the governance framework related to the development and protection of groundwater. In the practice of groundwater governance, there are many locations with limited or no access to groundwater data, models and no formal groundwater laws implemented, which would imply that day-to-day management is absent or severely limited. Yet, governance can still manifest through the broader planning concerns of the community, historical knowledge of trends in groundwater levels, and informal rules of conduct or procedures. As such, the definition of governance should capture these instances.

1.2.3.4 Coordinating and/or integrating groundwater governance practices

The governance of a groundwater resource or a particular set of groundwater resources may simultaneously occur within and across geographic and institutional levels. There may be multiple governance frameworks for a single resource at the national level that

[8]Note: Multilateral environmental agreements are documented as early as the 1920s.

span different sectors – necessitating some degree of horizontal coordination or integration. There may also be multiple frameworks governing a single resource, which span vertically from global conventions, regional/transboundary treaties, to national policies.

This phenomenon is articulated when Chandhoke (2003:2957) defines governance as the "de-centering and the pluralisation of the state into a number of levels that stretch horizontally from civil society and market organisations on the one hand and vertically from the transnational to local self-government institutions on the other." However, this definition implies that there is a pre-existing centralized governance mechanism. The phenomenon of horizontal and vertical stretch can also occur organically in the absence of a central framework.

The practice of coordinating and/or integrating these groundwater governance frameworks is a challenge facing an increasing number of actors in the form of legal pluralism (Conti & Gupta, 2014). As groundwater governance types of actors expand, along with their scopes and mandates, and the number of groundwater laws and policies increase, the concept of groundwater governance will need to include multilevel issues and address integrating and coordinating governance practices. This could potentially include cross-sectoral coordination and integration with other closely related areas of governance, such as surface water, land use, waste, wetlands and the subsurface.

1.2.3.5 Proposing a step forward

Overall, this analysis presented agreed-upon and contested elements of a definition of groundwater governance, it distinguished groundwater governance from groundwater management and presented the merits of coordination and/or integration across geographical levels and sectors. Based on this analysis, crafting a new definition could be a step forward in refining the concept of groundwater governance. The proposal is not intended to be a one-size-fits-all approach to governance. As such, it does not commend specific governance goals, and normative and prescriptive elements are not included. Rather, we want to articulate the key elements of the who, what, where, and how of groundwater governance. In this way, the basic concept of groundwater governance is clearer. Also, practical challenges of groundwater governance can be taken into account and can be seen as distinct from the practice of groundwater management.

As such, we offer the following definition:

> *Groundwater governance is the framework encompassing the processes, interactions, and institutions, in which actors (i.e. government, private sector, civil society, academia, etc.) participate and decide on management of groundwater within and across multiple geographic (i.e. sub-national, national, transboundary, and global) and institutional/sectoral levels, as applicable.*

1.3 THE CURRENT STATE OF GROUNDWATER GOVERNANCE

As described in the previous section, groundwater governance may be viewed as acting at various levels, defined by informal institutions, politically determined jurisdictions and administrative levels, often linked to various geographic scales. It is interesting to understand the ways groundwater governance plays out at the various levels, the interactions between levels, and variability across the globe. In the following, we present overarching findings at national, transboundary, and international/global level from the global diagnostic analysis of groundwater governance conducted as a critical part of the Groundwater Governance project. This project was first-of-its-kind and instrumental in bringing the issue and challenges of groundwater governance into mainstream international discourse on water management and sustainable development. The project collected information on groundwater governance across the globe, tapping into and compiling information and views from hundreds of professionals from the water sector as well as outside the sector[9]. It used a consultative open process, collecting mostly 'soft', yet valuable feedback, especially on perceived shortcomings. This effort included, among others, five regional consultations[10], which had not only the purpose of collecting information and views, as mentioned above, but also of raising awareness on groundwater governance.

1.3.1 The domestic level

1.3.1.1 Geographic diversity

The results of the comprehensive survey of the Groundwater Governance project revealed considerable variations in groundwater governance between countries, and between regions of countries. The variability may be expected, but the survey marked a huge step forward in describing them and identifying how they can be attributed to differences in factors related to the country-specific context, such as hydrogeological and climatic conditions, socio-economic, cultural, and political settings, as well as the history of local and national groundwater development and management. If groundwater governance is seen as an evolving process, various stages of groundwater governance may be observed. Hence, groundwater management and governance may have reached a more or less advanced stage in some countries, whereas groundwater governance in other countries is in a relatively immature and initial stage, or even still virtually non-existent (pre-management phase). One key conclusion from the analysis is that there is a close relationship between stage of groundwater governance and wealth of countries. Industrialised countries generally work under far more advanced groundwater governance, while pre-management prevails in the majority of poor countries.

Another key conclusion is that context matters. Countries tend to follow different paths, defined by the country-specific context and challenges. Consequently, the focus and characteristics of groundwater governance vary with the local needs and local conditions. One example of that is the USA. Here, groundwater issues related to water quantity are governed by the states, reflecting the individualized, state-wise approach and consequent diversity in historic development of water allocation and water rights systems between states. In contrast, issues related to water quality are delineated by

[9] http://www.groundwatergovernance.org/regional-consultations/consultations/en/

[10] The regions covered by the consultations were: Latin America and the Caribbean, Sub-Saharan Africa, Asia and the Pacific, the Arab region, and the UNECE region.

Table 1.2 Perceived differences regarding groundwater governance between global regions *(Source: FAO, 2016)*.

	Region				
	LAC *Latin America & Caribbean*	**SSA** *Sub-Saharan Africa*	**A&P** *Asia & the Pacific*	**AR** *Arab Region*	**UNECE** *UNECE Region*
Predominant stage of groundwater management					
Pre-development		x			
Initial management	x		x	x	
Advanced management					x
Society's dependence on groundwater					
Moderate	x				
High		x	x		x
Very high				x	
Key management issues currently driving governance					
Improving domestic/public water supply	++[a]	+++	++	++	+
Improving sanitation and wastewater treatment	++	++	++	++	+
Groundwater use for irrigation	+	+	+++	++	+
Impact of rapid urbanisation	++	+	+++	+	+
Groundwater pollution from agricultural land-use	++	+	++	+	+++
Impact of industrial activities	++	+	++	+	++
Environmental control and ecosystem protection	++	+	++	++	+++
Constraints to good groundwater governance					
Lack of awareness and knowledge of groundwater	++	+++	++	+++	+
Insufficient political commitment	++	+++	++	++	+
Poverty and lack of funds	++	+++	++	+	+
Weak institutions	++	+++	++	+++	+

[a]The number of +'es indicate the applicability of proposed management issues and constraints, ranging from comparatively low (+) to very high (+++).

the federal government and implemented and/or augmented at state level (Conti & Gupta, 2014). This probably reflects the fact that groundwater quality arose as an issue much later in history, and at that point, was viewed and dealt with in more general and coordinated fashion. However, redefining and harmonizing water rights systems across the states and at the federal level have not been pursued consistently, possibly because of path dependence (see also Chapter 24 of this book). Another example is the European Water Framework Directive, which is relatively advanced in terms of regional cooperation and harmonisation of approaches to water management and where groundwater has received explicit attention. Here, focus is on environmental quality, reflecting the fact that in most (not all) EU countries, groundwater quality is a greater issue than water quantity (see Chapter 23 of this book).

A representation of geographic diversity is presented in Table 1.2. It summarises some of the perceived differences between the five major regions of the world with regard to groundwater governance-related features and as articulated in the regional consultations. The ratings are basically subjective and qualitative, but nevertheless represent the general perception over these vast regions. Developing more detailed information at national and local scale and more quantitative assessment would require further research, in terms of setting up indicator-based systems and systematic surveys across and within individual countries. However, this was outside the scope of the Groundwater Governance project. The challenges related to monitoring groundwater governance are further elaborated in Chapter 13 of this book. Table 1.2 confirms the clear link between stage of groundwater governance and strength of institutions, *i.e.* weak institutions correspond to no or incipient groundwater governance.

1.3.1.2 Perceived and reported gaps and deficiencies

The Groundwater Governance project made a more detailed assessment of the perceived gaps and deficiencies in groundwater governance. Table 1.3 lists the main gaps and deficiencies in groundwater governance as reported during the regional consultation meetings. According to the assessment, many of the deficiencies are common, while some may be more relevant for specific regions or categories of countries. Lack of awareness and understanding of groundwater issues leads to absence of a sense of urgency and forms in many countries a key obstacle to the development of leadership and commitment to effective groundwater governance. Legal systems may not be perfect, but poor implementation of the law and regulations is a more pervasive and critical problem in most countries. Policy and planning require vision and a good knowledge of the local context and all relevant interdependencies. Often this information is limited, which, combined with the ubiquitous lack of groundwater monitoring data, may result in poor policies and plans with relevance to groundwater.

1.3.1.3 Addressing the gaps and deficiencies

Around the world, purposeful efforts have been made to address gaps and deficiencies in groundwater governance under variable circumstances. A modest but impressive selection of examples, mostly at national but also at regional level, can be found in Part 4 of this book ('Cases'). Table 1.4 presents identified options that have been applied in practice for improving groundwater governance in various parts of the world. Reviewing these approaches, it is clear that focus is still quite biased towards the more functional and practical, project-based aspects of groundwater management. Aspects of core dimensions of good governance, such as decision-making, transparency, accountability and responsibility are receiving limited coverage, and limited guidelines in that respect have so far been developed.

1.3.1.4 Sub-national groundwater governance

At local level, groundwater is often the primary source of water supply available for communities and even single households, most often through individual wells or smaller reticulated systems and for multiple uses. This is especially the case in rural areas, but also increasingly in urban areas in developing countries, where the public

Table 1.3 Perceived and reported main deficiencies in groundwater governance at the national level (*Source: FAO, 2016*).

Component	*Common or occasional deficiencies*
Actors	• Lack of awareness/understanding of groundwater and its role, problems and opportunities (potentially applicable to all categories of stakeholders) • No sense of urgency for governing groundwater properly • Low political commitment related to groundwater issues • Reactive rather than proactive attitudes • Poorly defined mandates or responsibilities of government agencies • Insufficient capacity of government agencies • Poor budgets of government agencies, or dependency on foreign parties • Lack of initiative and commitment of mandated government organisations • Poor accountability and transparency of mandated government organisations • Lack of cooperation between involved government agencies (or even rivalries) • Poor law enforcement or implementation of certain instruments (e.g. licensing) • Poor stakeholder involvement in groundwater governance • Lack of trust between the different categories of actors • Lack of adequate communication between all relevant partners • No balanced and smooth cooperation between all relevant partners
Legal frameworks	• Fragmentation and inconsistencies in legislation • Old groundwater legislation out of line with current views • Groundwater quantity and quality in separate laws • Groundwater law separate from laws governing surface water, land use, mining, subsurface use, environment, etc. • Institutional mandates and responsibilities not clearly defined • Overlapping institutional mandates and responsibilities • Laws ignoring customary rights • Laws inconsistent with realities on the ground (e.g. institutional capacity or perceptions of local groundwater users) • Draft Articles on the Law on Transboundary Aquifers not yet endorsed by countries • Legal instruments existing for very few TBAs only
Policies and management planning	• Limited scope (single use sector and/or neglecting obvious linkages) • Inconsistencies with policies of related domains • Potentially vital role of groundwater overlooked or undervalued • Waste of money due to pursuing unrealistic goals • Short-sightedness (due to time mismatch between political and hydrological/environmental cycles, or ignorance) • Overlooking the importance of involving stakeholders • Lack of practical instruments and approaches for transboundary aquifer management • Wrong 'solutions' due to insufficient knowledge of human behaviour • Negative impacts of some categories of incentives • Inadequate design of certain types of instruments (e.g. licensing systems, pollution fines) • Lack of regular systematic planning for groundwater management and protection
Data, information and knowledge	• Lack of sufficiently detailed groundwater assessments (especially in Africa and in Latin America & the Caribbean) • Monitoring of time-dependent variables is rare and often only fragmentary • Sharing data and information is still in its infancy • Presentation of information not tailor-made for the different categories of actors

Table 1.4 Options and opportunities for improving groundwater governance *(Source: FAO, 2016).*

Category	*Opportunities*
Information, knowledge and awareness	• Structural provisions for data and information • Modern technologies for data acquisition and information management • National or international projects and programmes • Cooperation with the private sector • Making groundwater information available • Awareness raising and lobbying
Legal frameworks	• Legal reforms • Bringing groundwater resources under public control • Legally enforceable regulations • Legal instruments for transboundary aquifers
Policy and planning	• Aligning groundwater management with macro-policies • Adopting suitable principles and approaches • Adopting IWRM and related approaches (conjunctive management, MAR) • Establishing policy and planning linkages with interrelated sectors • Introducing periodic and coherent groundwater management planning
Actors	• Enhancing political commitment • Creating and developing leadership • Institutional reforms • Involvement of the private sector • Involvement of local stakeholders • Improving cooperation by accountability and transparency • Capacity building • Funding and financing • The role of international organisations and partnerships

water supply is falling short (Foster *et al.*, 2016). This fact inevitably brings to the fore the need to bring in local stakeholders/users as legitimate and key players in groundwater governance to ensure sustainable and equitable outcomes, *e.g.* in terms of access, livelihoods, and health. These systems are often shallow and vulnerable to depletion and pollution, especially with intensified human development, and under such scenarios, the poorest could be disproportionately impacted, as they do not have the means to improve their source. Hence, these are critical local issues of groundwater governance. They also need to be linked to governance of public services in water supply, irrigation, health, etc. (Chapter 14 of this book). This opens both important pathways for participatory processes to address the common pool resource challenge, but also predictably complicates the process (see Chapter 7 in this book). While significant progress and experience in this field is developing, it is also clear that there is no common or universal model for such participatory approaches. So far, the experience is that to achieve long-term effective governance and acceptable outcomes, transaction costs to set up functioning and lasting systems are often high, driven either by communities themselves, or in facilitated models with external support. Further work on participatory processes supporting groundwater governance is needed by joining forces across communities and building on cross-contextual experiences as appropriate. Importantly, the challenges are also linked to reconciling top-down and bottom-up approaches to groundwater governance (Varady *et al.*, 2016), formal vs. customary law in controlling water access and allocation, etc.

1.3.2 Transboundary aquifers and regional groundwater governance

At the regional and transboundary level, groundwater is rising on the agenda, related to the discourse on and challenges related to the governance of transboundary aquifers[11] (Chapter 19 in this book). This transboundary focus complements the more established international emphasis on governing transboundary river basins, and is evidenced through emerging frameworks and principles for governance of aquifer systems at this level, mostly in the context of international water law. The emphasis is also evident from increasing numbers of particular aquifers being subjected to transboundary efforts in terms of joint management and governance. This is due to the critical importance of these resources for the states involved and because of advocacy from international organisations who want to advance an agenda on international cooperation in order to improve *e.g.* regional integration and national security for the countries involved.

The governance of transboundary aquifer resources adds significant complexity when compared to the governance of aquifers located within single national territories. Firstly, the fact that aquifers may be shared across political boundaries, subdividing the aquifer's territory into zones under different national jurisdictions, makes it pertinent to clearly delineate the aquifer (system) in question, which is not always a straightforward task. Furthermore, the complexity is considerably amplified by foreign policy considerations and potential contrasts and misunderstandings related to variable groundwater management approaches as well as cultural and language differences (Fried & Ganoulis, 2016).

Nevertheless, substantial experience is developing from these efforts on the scope for international groundwater cooperation, on institutionalisation of shared governance mechanisms and on the more practical aspects of managing the resources, monitoring and sharing information, etc. A potential virtue of these efforts and with particular relevance to groundwater governance is the opportunity that such programmes and lessons-learned offer in terms of developing and testing various higher-level thinking on groundwater governance at the global level (Villholth, 2015), which are presently lacking. Groundwater governance at the global level entails looking at the inter-regional sharing of benefits from larger shared aquifer resources, which are globally very unevenly distributed, yet large compared to surface water resources. These resources are of critical importance in the future, including offshore transboundary aquifers (Martin-Nagle, 2016).

[11] A transboundary aquifer or aquifer system is defined as 'an aquifer or aquifer system, parts of which are situated in different States' (Stephan, 2009).

1.3.3 Global dimensions of groundwater governance

1.3.3.1 Regional and global programmes and initiatives of international agencies

Besides the global Groundwater Governance project, which was supported by the Global Environment Facility (GEF) and implemented by the Food and Agriculture Organization of the United Nations (FAO) together with United Nations Education, Scientific and Cultural Organisation's International Hydrological Programme (UNESCO-IHP), the International Association of Hydrogeologists (IAH), and the World Bank, a host of international organisations is involved in groundwater and water governance. These include, but are not limited to, the United Nations International Law Commission (UNILC), the International Groundwater Resources Assessment Centre (IGRAC), the International Water Management Institute (IWMI), the International Union for Conservation of Nature (IUCN), various geological surveys, training and capacity building organisations and networks, and water supply development organizations. Programmes and networks of importance are further the OECD Water Governance Initiative (OECD-WGI), the Water Integrity Network (WIN), the Groundwater Initiative for Policy and Practice (GRIPP), the Transboundary Waters Assessment Program (TWAP), the Internationally Shared Aquifer Resources Management Programme (ISARM), the Worldwide Hydrogeological Mapping and Assessment Programme (WHYMAP), the Global Water Partnership (GWP), and the World Water Assessment Programme (WWAP). At the regional level, the following organizations are involved in groundwater governance: EU, UNECE, OAS/OEA, OSS, ACSAD, SADC, CCOP, UNESCAP, and AMCOW[12], to mention key ones. Together, these organisations and initiatives play a significant role in promoting and catalysing effective groundwater management and governance. While there is still some disconnect between groundwater-focused agencies, the broader water initiatives, and the governance initiatives, there is definitely an increasing merging of concerns and joint activities, which in combination are raising the issue of groundwater governance on the global agenda.

1.3.3.2 Sustainable Development Goals (SDGs)

The Sustainable Development Goals, ratified in 2015, signaled a new approach and thinking in terms of addressing global imbalances and deficits in development gains globally. Importantly, the approach is integrated, cross-sectoral, embracing both developing and developed countries, and entails improved approaches to track progress (UN, 2015) as compared to their predecessors, the Millennium Development Goals (MDGs). The thinking on aspects related to water has advanced since the MDGs in the sense that the new goals address broader issues of access to water and sanitation.

[12]EU: The European Union; UNECE: The United Nations Economic Commission for Europe; OAS/OEA: The Organization of American States; OSS: Sahara and Sahel Observatory; ACSAD: Arab Center for the Studies of Arid Zones and Dry Lands; SADC: Southern African Development Community; CCOP: The Coordinating Committee for Geoscience Programmes in East and Southeast Asia; UNESCAP: The United Nations Economic and Social Commission for Asia and the Pacific; and AMCOW: African Ministers' Council on Water.

Importantly, they include considerations of the sustainability of the water resources to underpin secure and safe access for human populations, partly reflecting the increasing recognition of the global risks associated with increasing pressure on water resources (World Economic Forum, 2017). Groundwater, as a key resource to achieve SDG 6 on 'Clean Water and Sanitation', though implicitly included, does not receive specific attention. In particular, little preparatory work has been devoted to treat some of the key challenges related to the operational definition and delineation of the resource and the assessment of sustainable use of groundwater, in particular related to systems with limited renewability. Also, equitable access becomes very relevant in groundwater-dependent societies, especially if resources are depleting, because groundwater levels drop out of reach of poor people.

1.3.3.3 The human right to water

The human right to water was affirmed in 2010 with the UN General Assembly Resolution (UN, 2010), as a global subscription to the universal human right to basic and safe water and sanitation (Chapter 14 of this book). This of course aligns with the SDGs, but brings the issue to a more prominent rights-based level, reflecting the international community's legal obligations to address basic needs and dignity. However, the concept has received some comments of not being workable, because the governance structures required to put the principle and required infrastructure in place are fragmented (Gupta *et al.*, 2010). Also, conflicts exist in interpreting and reconciling the human right to water and sanitation with the concept of water as an economic good – the latter representing the principle of achieving cost-recovery in water and sanitation provision, basically the user-pays principle, despite the poorest people generally lacking means to cover these costs (Conti, 2017). In conclusion, groundwater governance, as part of broader water governance, is facing intractable problems related to overarching principles of equity, which need to be tackled conceptually as well as concretely in order to support goals not only in terms of health and workforce considerations, but also in terms of social stability and political security at various scales, including the international.

1.3.3.4 Climate change

The climate change discourse, as expressed in the 2015 Paris Pact on water and adaptation to climate change[13] has been instrumental in raising the issues around groundwater to a higher level of attention globally. Or maybe more correctly, the climate discourse has been applied to sell the importance of groundwater as the ultimate buffer of water that can support resilience of landscapes, populations and socio-economies during drought and increasing climate variability. While it is clear that groundwater is already playing this role and de facto is substituting dwindling surface water resources, either because of climate change-related, increasing water variability or because of increasing water demands, what is missed is that this natural buffer property of groundwater is progressively being undermined in critical areas by excessive abstraction or pollution (Taylor, 2014). Hence, there is a strong need to stress groundwater governance

[13] http://www.riob.org/eletter/COP21-Signatures-Pacte-EN.html

approaches that realistically identify the role that groundwater can play in enhancing resilience of groundwater-dependent communities and societies in stressed regions. The resource needs pro-active management and protection that strike the balance between development for economic growth, while ensuring the role of buffer and water security source.

1.3.3.5 Global implications of groundwater

Despite its documented significance, and irreplaceable character in many parts of the world, as an element in sustainable development, groundwater still receives limited dedicated focus. Emerging research shows the significant interlinkages of groundwater with issues of regional and global water scarcity and depletion (Aeschbach-Hertig & Gleeson, 2012), environmental water requirements (Sood *et al.*, 2017), environmental degradation (van der Gun, 2012), climate change (Taylor *et al.*, 2012), human migration (Kelly *et al.*, 2014), and virtual water flows through commodity trade (mostly food) (CGIAR/WLE, 2017). Especially the linkage to food trade and embedded groundwater depletion is grossly neglected. About 14–17% of all global food produced by groundwater is derived from groundwater that will essentially not be replenished (CGIAR/WLE, 2017) and some parts of this derives from poor nations with increasingly limiting water resources exporting to richer countries (Chapter 18 in this book). Hence, groundwater is no longer a local phenomenon and a local resource as it is often portrayed. It is moved around the world virtually in massive amounts and as such has linkages to the global economy and geopolitics through trade, economics, climate, environment, etc. Bringing these aspects on the global development agenda as a governance issue may still be far fledged, but clearly of critical and increasing importance. Importantly, many of these issues cannot be solved isolated and at a local and national level, and are governed by drivers blind to groundwater. There is an increasing need to bring nations together to seriously consider implications of the reliance on essentially non-replaceable resources and the impending groundwater crisis and develop high-level governance mechanisms that can curb further global-level impacts, and discuss various options to address them for sustainable outcomes. This also implies linkages to global consumerism and consumer patterns, groundwater footprint of products and lifestyles, and inherent and economic values of groundwater. Besides the governance implications, this is a significant evolving research field.

1.3.3.6 Groundwater governance at integrated scales

Having analysed the scale issues related to groundwater governance above, it should be clear that all levels are important, and that they are interrelated and interdependent through the resource aspects, the institutions involved, technology, trade and also through legislations, policies and other decision-making happening at various levels that may have repercussions at other levels. One example of this could be how reduced import restrictions on Chinese pumps in Malawi, coupled with government subsidies, could help poor farmers get access to groundwater for (improved) irrigation. Yet, local factors including poor traffic infrastructure, bureaucracy at national level and high transaction costs for the farmer may make it an infeasible proposition. This may, in effect, entail that rather than benefitting the poor farmers, these technologies and support policies benefitted the richer, more peri-urban farmers (Colenbrander & van Koppen, 2013). To overcome such intricate issues and to implement solutions that would benefit poor farmers and increase their income sustainably would require system and value chain thinking in governance structures well beyond groundwater, well beyond a single scale/level, as well as well beyond the common practice today.

1.4 A HEURISTIC FRAMEWORK FOR UNDERSTANDING AND OPERATIONALISING GROUNDWATER GOVERNANCE

Because groundwater has been a source of water for humans since time immemorial, groundwater governance has always existed in some form or another (Chapter 2 of this book). The scope, issues and scale have changed with time, however, to the point where stakes today of a conscious, explicit and embracing groundwater governance, or lack of the same, are much higher than just a generation ago. Hence, coming to grips with groundwater governance and how to influence it in a direction that may generate more desirable long-term outcomes in terms of adopted societal goals (such as equity and sustainability), is ever more critical now.

So, the question naturally is, how can governance make a difference in terms of addressing mounting groundwater problems and ensuring long-term sustainability and equity, if it does not necessarily have an agreed or overriding sustainability/equity goal nor necessarily works according to aforementioned governance tenets (Section 1.2)?

First, it is important not to assume that governance automatically will solve groundwater unsustainability problems. As already made clear, governance is not inherently good (or bad), nor does it have an inherent humanitarian or sustainability vision. There are many interests at play at various levels, which may have priority, and which may not be compatible with adopted higher goals of sustainability and sharing a common good. Also, due to groundwater invisibility, stakeholders may be unaware of the consequences of their behaviour, and so ignorance about groundwater is prevalent, and maybe more so for a large majority of people who either do not personally deal with groundwater or do not know that their water source or other assets or benefits are derived from groundwater. Farmers and other people directly abstracting or interacting with groundwater will naturally be more aware and knowledgeable about it and hence have a direct stake in it.

Yet, governance is the key mechanism, through which a more pro-active, conscious, long-term and informed groundwater governance and management can be fostered and institutionalised. Furthermore, adopting tenets of governance that are widely considered as positive (responsibility, accountability, transparency, efficiency, legitimacy, participation, equity and inclusiveness, and rule of law) is assumed to fundamentally support the process of contextualised, acceptable and agreed decisions on groundwater.

Building on this assumption and assuming that challenges of groundwater governance are closely linked to the ISD signature of groundwater as mentioned at the outset of this chapter, a conceptual heuristic framework is proposed, which in turn helps identify key action areas required to improve the situation. Hence, Table 1.5 shows the cross-section between the components of the ISD-signature of groundwater

Table 1.5 A conceptual diagram to understand governance requirements of groundwater, as affected by its inherent properties and governance challenges.

	Governance tenets[b]							
ISD-signature component[a]	*Responsibility*	*Accountability*	*Transparency*	*Efficiency*	*Legitimacy*	*Participation*	*Equity and inclusiveness*	*Rule of law*
Invisible	x	x	x	x				
Slow			x	x		x	x	
Distributed	x	x	x		x	x	x	x

[a]See Section 1.1.
[b]From UN ESCAP (2006).

and the key tenets of good governance. The crosses indicate where these intersections are critically and ideally important, according to the authors[14].

Hence, Table 1.5 indicates that transparency is critical for addressing the I-signature. I.e., there is a strong need to generate and expand the knowledge and understanding of groundwater resources to all stakeholders, whether direct or indirect. But it is even more important to share the information openly and efficiently among all interested stakeholders. And linked to this, it is critical to focus on the information that is often less well available, but very critical for the management, namely the human interaction with it (like abstraction, pollution activities), the granted rights/allocations and other regulations' impacts on human behaviour and their compliance. This also links to the tenets of accountability to address the I-signature. Furthermore, it is important to keep track of the trends in these parameters, changes, and possible explanations and drivers (socio-economic, cultural, political, etc.) for them. Finally, subsequent to policy and regulation implementation, it also becomes important to monitor the impacts of these policies, on the resource and feedback to human behaviour, which again links to 'responsibility'.

The S-signature of groundwater needs to be tackled concertedly through transparency (to understand long-term impacts and the sustainability of human interaction with the resource). In addition, participation and equity and inclusiveness is required

[14]Recognising that some countries in reality opt for not optimizing some of these tenets, with some reason. For instance, participation is useless in the absence of a certain basic level of understanding and knowledge on the subject at hand; participation is also sometimes a real cause of delay – compare China (with comparatively pragmatic top-down decision-making) versus European countries (traditionally more inclined to participatory decision-making, which is time-consuming). Equity is in several cases formally not pursued (*e.g.* West Bank).

to discuss and decide on acceptable trade-offs (*e.g.* between short and long-term goals, between inter-sectoral goals, or between economic and social/environmental goals) and the share of burden as well as benefits from the resource and its potential (partial) degradation over time. Finally, efficiency is also important as reversal of impacts is slow, and so ensuring efficiency in remediation helps address long-term negative impacts.

Finally, who determines the regulations, the parameters to monitor and the sanctions is also a fundamental governance issue, which links very much to the tenets of participation, equity and inclusiveness, and rule of law. These factors are strong in groundwater governance because of the D-signature. It needs to be governed at various levels, but with some coherence between them. Responsibility under the D-signature relates to the need for clearly defined roles in a semi-decentralized (integrated top-down and bottom-up) approach required due to the distributed resource.

Table 1.5 gives a quick overview of the interlinkages between groundwater governance challenges and the needed tenets to tackle them. The highlighting of certain tenets, however, does not indicate that other tenets are not important to address the challenges.

What permeates from the framework and how to apply it, is the principal need for transparency, which encompasses both knowledge and awareness creation, but also the efficient sharing of information related to the resource. Knowledge generation may seem 'apolitical', but is affected by decisions and desires of whether to create information in the first place and hence is part of governance. Only by creating and sharing knowledge will a governance model be effective, whatever its goal is. A second key proposition is that a leadership that comprehends these challenges and how to operationalise them is a prerequisite to effective groundwater governance (Chapter 5 of this book). It is not one of the key tenets of governance, as in fact governance is posited on leadership (to govern derives from the Greek verb κυβερνάω [kubernáo], meaning to steer[15]). As such, leadership is the overarching (implicit) foundation of groundwater governance. Again, as the governance is viewed as a multi-tiered and multi-level governance (Section 1.2), coherent leadership at all levels is required.

A key part of this is to raise public awareness and create interest, ownership, and trust among multiple stakeholders, which may foster common vision. There is also a strong need for building the capacity to link detailed knowledge of groundwater physical science with the knowledge of socio-political science and processes. Public awareness may create demand for responsible development of groundwater and related resources (Chapter 12 of this book), enhancing effective response to crises and conflicts. Fostering such processes requires tenacity, dedication, collaboration, negotiation skills, and vision.

Finally, to support the conceptual framework in terms of the existing groundwater-related risks, some persistent, contemporary, and emerging management challenges that require concerted governance processes are listed in Table 1.6. These are further discussed as part of this book.

[15]https://en.wikipedia.org/wiki/Governance

Table 1.6 Lingering, contemporary or emerging groundwater management challenges.

Category of challenge	*Type of challenge*	*Chapter in this book*
Persistent challenges	• Managing groundwater for equitable water access and health • Managing land use for groundwater sustainability • Governing non-renewable aquifers • Managing saltwater intrusion in coastal and agricultural areas	• Chapter 14 • Chapter 16 • Chapter 26
Contemporary challenges	• Managing the groundwater-energy nexus • Managing groundwater and surface water in conjunction • Managing transboundary aquifers • Managing the subsurface space and its resources • Understanding impacts of climate change on groundwater	• Chapter 15 • Chapter 17 • Chapter 19 • Chapter 20
Emerging challenges	• Governing groundwater depletion through virtual groundwater flows • Governing groundwater in large cities • Governing emerging complex and persistent organic contaminants • Global sharing of large transboundary aquifers	• Chapter 18 • Chapter 28

1.5 CONCLUSION

Groundwater governance is an evolving concept. It remains somewhat intangible and not consciously and pro-actively pursued by many stakeholders and managers, yet is paramount to proper groundwater development and management. Ideally, as awareness increases, groundwater governance becomes a deliberate, conscious, explicit and targeted concept, typically associated with aspirational goals (such as sustainability and equity) and core guiding governance tenets (transparency, accountability, integrity, fairness, etc.). It is distinct from groundwater management but the two are intricately interrelated. It also relates very much to overall governance, as groundwater governance may reflect the general status of governance. Hence, broader governance approaches need to be enhanced as prerequisite to good groundwater governance. This chapter has explored the groundwater governance concept systematically, based on a slight revision and rephrasing of previous definitions, to distinguish it as the overarching framework for human interaction with and decision around groundwater resources. Though the governance system may not necessarily and automatically reflect the degree of effectiveness of the implementation (*i.e.* the management) and the other way around (Lautze *et al.*, 2014), it is evident that a well-founded governance will stand a better chance of achieving its goals through support from groundwater management. We conclude that it is important to keep exploring the concept of groundwater governance, defining its dimensions, operationalisation and effectiveness. Such discussions are key to moving forward on the achievement and collaboration on goals associated with groundwater at all levels. Empirical understanding is also growing, but further frameworks for analysing and supporting practical approaches are needed (Megdal *et al.*, 2017).

REFERENCES

Aeschbach-Hertig, W. & Gleeson, T. (2012) Regional strategies for the accelerating global problem of groundwater depletion. *Nature Geoscience*, 5, Dec 2012. www.nature.com/naturegeoscience.

Bevir, M. (2011) Governance as theory, practice, and dilemma. In M. Bevir (Ed.), *The SAGE handbook of governance* (pp. 1–16). London, UK: Sage London.

Biermann, F. (2007) "Earth system governance" as a crosscutting theme of global change research. *Global Environmental Change*, 17(3–4), 326–337. doi:10.1016/j.gloenvcha.2006.11.010

Biermann, F., Betsill, M.M., Gupta, J., Kanie, N., Lebel, L., Liverman, D., *et al.* (2009) *Earth System Governance: People, Places and the Planet. Science and Implementation Plan of the Earth System Governance Project.* doi:10.1787/9789264203419-101-en.

Biswas, A.K. & Tortajada, C. (2011) Future water governance – Problems and perspectives. In: Biswas, A.K. & Tortajada, C. (Eds.): Improving Water Policy and Governance. Routledge, Taylor & Francis Group, London and New York. ISBN: 978-0-415-60628-8.

Chandhoke, N. (2003) Governance and the pluralisation of the state: implications for democratic citizenship. *Economic and Political Weekly*, 38(28), 2957–2968.

CGIAR Research Program on Water, Land and Ecosystems (WLE) (2017) Building resilience through sustainable groundwater use. Colombo, Sri Lanka: International Water Management Institute (IWMI). 12p. doi:10.5337/2017.208.

Clapp, J. (2014) Global Environmental Governance for Corporate Responsibility and Accountability. *Global Environmental Politics*, 5(3), 23–34.

Cleaver, F., Franks, T., Boesten, J. & Kiire, A. (2006) Water governance and poverty: What works for the poor? DFID Research Report, University of Bradford.

Closas, A. & Molle, F. (2016) *Groundwater Governance in the Middle East and North Africa.* IWMI Project Report no 1. Groundwater Governance in the Arab World. http://gw-mena.iwmi.org/wp-content/uploads/sites/3/2017/04/Rep.1-Groundwater-Governance-in-MENA.pdf.

Colenbrander, W. & van Koppen, B. (2013) Improving the supply chain of motor pumps to accelerate mechanized small-scale private irrigation in Zambia, *Water Intern.*, 38(4), 493-503, DOI: 10.1080/02508060.2013.819602.

Commission on Global Governance. (1995) *Our global neighbourhood: the report.* Oxford, UK: Oxford University Press.

Conti, K.I. (2017) *Norms in Multilevel Groundwater Governance and Sustainable Development.* Unpublished Ph.D. thesis. University of Amsterdam, The Netherlands.

Conti, K.I. & Gupta, J. (2015) Global governance principles for the sustainable development of groundwater resources. *International Environmental Agreements: Politics, Law and Economics*, 1–23.

Conti, K.I. & Gupta, J. (2014) Protected by pluralism? Grappling with multiple legal frameworks in groundwater governance. *Current Opinion in Environmental Sustainability*, 11, 39–47.

Dryzek, J.S. & Stevenson, H. (2011) Global democracy and earth system governance. *Ecological Economics*, 70(11), 1865–1874. doi:10.1016/j.ecolecon.2011.01.021.

Earth Security Group, 2016. *CEO Briefing: Global Depletion of Aquifers. Global companies must take an active role in groundwater governance to avoid existential risks.* www.earthsecuritygroup.com.

Falkner, R. (2014) Private Environmental Governance and International Relations: Exploring the Links. *Global Environmental Politics*, 3(2), 72–87.

Famiglietti, J.S. (2014) The global groundwater crisis. *Nat. Clim. Change*, 4, November 2014, pp. 945–948.

Food and Agriculture Organization of the United Nations (FAO) (2016a) *Global Diagnostic on Groundwater Governance.*

Food and Agriculture Organization of the United Nations (FAO) (2016b) Global Framework for Action to achieve the Vision on Groundwater Governance. Rome, Italy.

Food and Agriculture Organization of the United Nations (FAO) (2016c) Shared global vision for Groundwater Governance 2030 and A call-for-action. Rome, Italy.

Food and Agriculture Organization of the United Nations (FAO) (2013) *GEF-FAO Groundwater Governance Project A Global Framework for Country Action Thematic Paper 5: Groundwater Policy and Governance*. Rome, Italy.

Food and Agriculture Organization of the United Nations (FAO) (n.d.) *GEF-FAO Groundwater Governance Project A Global Framework for Country Action Digest of Thematic Paper 5: Groundwater Policy and Governance*. Rome, Italy.

Foster, S., Tyson, G., Howard, K., Hirata, R., Shivakoti, B.R., Warner, K., Gogu, R. & Nkhuwa, D. (2015) *Resilient Cities and Groundwater*. International Association of Hydrogeologists (IAH) Strategic Overview Series.

Foster, S., Chilton, J., Nijsten, G. & Richts, A. (2013) Groundwater—a global focus on the "local resource." *Current Opinion in Environmental Sustainability*, 5(6), 685–695. doi:10.1016/j.cosust.2013.10.010

Foster, S., Garduño, H., Tuinhof, A. & Tovey, C. (2009) *Groundwater Governance*. Strategic Overview Series No. 1. Washington, D.C.: World Bank.

Foster, S.S.D. & Chilton, P.J. (2003) Groundwater: the processes and global significance of aquifer degradation. *Phil. Trans. R. Soc. Lond.* B 358, 1957–1972. doi: 10.1098/rstb.2003.1380.

Franks, T. & Cleaver, F. (2007) Water governance and poverty: a framework for analysis. *Progress in Development Studies*, 7(4), 291–306.

Fried, J. & Ganoulis, J. (Eds.) (2016) *Transboundary Groundwater Resources: Sustainable Management and Conflict Resolution*. Lambert Academic Publishing. ISBN:978-3-659-96418-3.

Graham, J., Amos, B., & Plumptre, T. W. (2003) *Governance principles for protected areas in the 21st century*. Ottawa, Ontario: Institute on Governance Canada.

Green, C. (2007) *Mapping the field: The landscapes of Governance*. EU-funded SWITCH Project, accessed 24 August 2017, available at http://www.switchurbanwater.eu/outputs/pdfs/W6-1_GEN_RPT_D6.1.1b_Mapping_Landscapes_of_Governance.pdf.

Gupta, J., Ahlers, R. & Ahmed, A. (2010) The human right to water: Moving towards consensus in a fragmented world. *Rev. Eur. Commun. & Intern. Environ. Law*, 19: 294–305. doi:10.1111/j.1467-9388.2010.00688.x.

Hardin, G. (1968) Tragedy of the commons. *Science*, 162, 1243–1248.

Hirst, P. (2000) *Democracy and Governance* in Pierre, J. (ed.), Debating Governance: Authority, Steering, and Democracy. Oxford, UK: Oxford University Press.

Institute on Governance (IOG) (Canada) (2006) accessed 24 August 2017, available at http://www.iog.ca/.

IUCN (2010) 'Environmental Governance' accessed 29 February 2016, available at www.iucn.org/about/work/programmes/environmental_law/elp_work/elp_work_issues/elp_work_governance.

Kelly, C.P., Mohtadi, S., Cane, M.A., Seager, R. & Kushnir Y. (2014). Climate change in the Fertile Crescent and implications of the recent Syrian drought. *PNAS*, 112 (11), 3241–3246, doi: 10.1073/pnas.1421533112.

Lautze, J., de Silva, S., Giordano, M. & Sanford, L. (2014) Water Governance In: Lautze, J. (Ed.) (2012) *Key Concepts in Water Resource Management: A Review and Critical Evaluation*. Routledge, New York. ISBN. 978-0-515-71172-2.

Martin-Nagle, R. (2016) Transboundary offshore aquifers – A search for a governance regime. *Int. Water Law*, 1.2, 1–79. doi: 10.1163/23529369-12340002.

Meadowcroft, J. (2002) Politics and scale: Some implications for environmental governance. *Landscape and Urban Planning*, 61(2–4), 169–179. doi:10.1016/S0169-2046(02)00111-1.

Megdal, S.B., Eden, S. & Shamir, E. (2017) Water governance, stakeholder engagement, and sustainable water resources management. *Water*, 9(190). doi:10.3390/w9030190.

Megdal, S.B., Gerlak, A.K., Varady, R.G. & Huang, L.Y. (2015) Groundwater Governance in the United States: Common Priorities and Challenges. *Groundwater*, 53(5), 677–684. doi:10.1111/gwat.12294.

Morin, J.-F. & Orsini, A. (2013) Insights from Global Environmental Governance. *International Studies Review*, 15(4), 562–589. doi:10.1111/misr.12070

Mukherji, A. & Shah, T. (2005) Groundwater socio-ecology and governance: A review of institutions and policies in selected countries. *Hydrogeology Journal*. doi:10.1007/s10040-005-0434-9.

Ostrom, E. (1990) *Governing the Commons. The evolution of institutions for collective action*. Cambridge, Cambridge University Press.

Pahl-Wostl, C., Conca, K., Kramer, A., Maestu, J. & Schmidt, F. (2013) Missing Links in Global Water Governance: a Processes-Oriented. *Ecology and Society*, 18(2), 33. doi:10.5751/ES-05554-180233.

Paterson, M., Humphreys, D. & Pettiford, L. (2003) Conceptualizing Global Environmental Governance: From Interstate regimes to Counter-Hegemonic struggles. *Global Environmental Politics*, 3(2), 1–10. doi:10.1162/152638003322068173.

Pierre, J. and Peters, G.B. (2000) Governance, Politics and the State. New York: St. Martin's Press.

Rogers, P. (2002) *Water Governance in Latin America and the Caribbean Water*. Washington DC, USA: Inter-American Development Bank.

Rogers, P. and Hall, A.W. (2003) *Effective Water Governance*, TEC Background Paper No. 7. Novum, Sweden: Global Water Partnership.

Smakhtin, V.Y., Ashton, P.J., Batchelor, A., Meyer, R., Murray, R., Barta, B., Bauer, N., Naidoo, D., Olivier, J. & Terblanche, D. (2001) Unconventional Water Supply Options in South Africa. *Water Int.*, 26: 3, 314–334. doi: 10.1080/02508060108686924.

Sood, A., Smakhtin, V., Eriyagama, N., Villholth, K.G., Liyanage, N., Wada, Y., Ebrahim, G. & Dickens, C. (2017) *Global environmental flow information for the sustainable development goals*. Colombo, Sri Lanka: International Water Management Institute (IWMI). 37p. (IWMI Research Report 168). doi: 10.5337/2017.201.

Spagnuolo, F. (2011) Diversity and pluralism in earth system governance: Contemplating the role for global administrative law. *Ecological Economics*, 70(11), 1875–1881. doi:10.1016/j.ecolecon.2011.01.024.

Taylor, R. (2014) When wells run dry. *Nature*, 616.

Taylor, R.G., Scanlon, B., Döll, P., Rodell, M., Beek, R. van, Wada, Y., Longuevergne, L., Leblanc, M., Famiglietti, J.S., Edmunds, M., Konikow, L., Green, T.R., Chen, J., Taniguchi, M., Bierkens, M.F.P, MacDonald, A., Fan, Y., Maxwell, R.M., Yechieli, Y., Gurdak, J.J., Allen, D.M., Shamsudduha, M., Hiscock, K., Yeh, P.J.-F., Holman, I. & Treidel, H. (2012) Groundwater and Climate change. *Nature Clim. Change* http://dx.doi.org/10.1038/nclimate1744.

United Nations (UN) (2015) Transforming our world: The 2030 Agenda for Sustainable Development. A/RES/70/1. http://www.un.org/ga/search/view_doc.asp?symbol=A/RES/70/1&Lang=E.

United Nations (UN) (2010) Resolution adopted by the General Assembly on 28 July 2010: 64/292. The human right to water and sanitation, United Nations General Assembly. http://www.un.org/es/comun/docs/?symbol=A/RES/64/292&lang=E.

United Nations Economic and Social Commission for Asia and the Pacific (ESCAP) (2006) *What is good governance?* Accessed 24 August 2017, available at http://www.unescap.org/sites/default/files/good-governance.pdf.

United Nations Environmental Programme (2009) *Environmental Governance.* Accessed on 24 August 2017, available at http://staging.unep.org/pdf/brochures/EnvironmentalGovernance.pdf.

van der Gun, J. (2012) *Groundwater and Global Change: Trends, Opportunities and Challenges.* UN World Water Assessment Programme. World Water Development Report. 38 pp. ISBN 978-92-3-001049-2.

Varady, R.G., Zuniga-Teran, A.A., Gerlak, A.K. & Megdal, S.B. (2016) Modes and approaches of groundwater governance: A survey of lessons learned from selected cases across the globe. *Water*, 8, 417. doi:10.3390/w8100417.

Villholth, K.G. (2015) Reconciling climate change and transboundary groundwater management for sustainable agricultural production. In: Hoanh C.T., V. Smakhtin, and R. Johnston (Eds.): Climate Change and Agricultural Water Management in Developing Countries. CABI. ISBN 978-1-78064-366-3.

World Bank (1989) *Sub-Saharan Africa: From Crisis to Growth.* Washington, DC: World Bank.

World Bank (1992) *Governance and Development.* Washington, DC: World Bank.

World Economic Forum (2017) *The Global Risks Report 2017.* 12th Edition.

Young, O.R. (2002) *The institutional dimensions of environmental change: fit, interplay, and scale.* Cambridge, MA: MIT press.

Chapter 2

Emergence and evolution of groundwater management and governance

Marco García[1], *Ebel Smidt*[2,3] *& Jacobus J. de Vries*[4]
[1]*Hydrogeologist, WaterFocus, Bunnik, The Netherlands*
[2]*SG Consultancy and Mediation, Engelen, The Netherlands*
[3]*WaterFocus, Bunnik, The Netherlands*
[4]*Emeritus Professor of Hydrogeology, VU, Amsterdam, The Netherlands*

ABSTRACT

The evolution of groundwater management and governance are slow processes. On the one hand, this is due to a poor understanding of the subsurface processes, which slowly impact the state of groundwater systems. On the other hand, the implementation of groundwater management measures is hampered by conflict of interests among stakeholders. In this chapter, we give an overview of the emergence and evolution of groundwater management and groundwater governance, with attempts to clarify the corresponding mechanisms by using the Driver-Pressure-State-Impact-Response (DPSIR) framework of analysis. Examples from various historical and geographical situations are presented to illustrate the variety of conditions, solutions and trends. We conclude that the increasing pressure and its negative impacts on the state of groundwater systems have triggered better management and governance. The long-term importance of groundwater to overcome short-term societal challenges has been better understood the last fifty years resulting in more pro-active groundwater governance serving sustainable development goals.

2.1 INTRODUCTION

Groundwater has played an important role in human development since its early footprints. The use of springs marked the first direct interactions between humans and groundwater. From there, the exploitation of groundwater evolved for centuries under the influence of a variety of agricultural, demographic and socio-economic factors, but also in relation with the development of new technologies. These technologies went from groundwater exploitation through dug wells and qanats (Box 2.1), to mechanical drilling and pumping; while other technological efforts focused on achieving sustainable water use through groundwater conservation and replenishment by terraces and dams.

For this chapter, it is important to distinguish between groundwater exploitation, management and governance. Exploitation of groundwater deals specifically with the extraction of the resource for individual use or for the benefits of a selected group. Management and governance serve collective goals, among others achieving sustainability

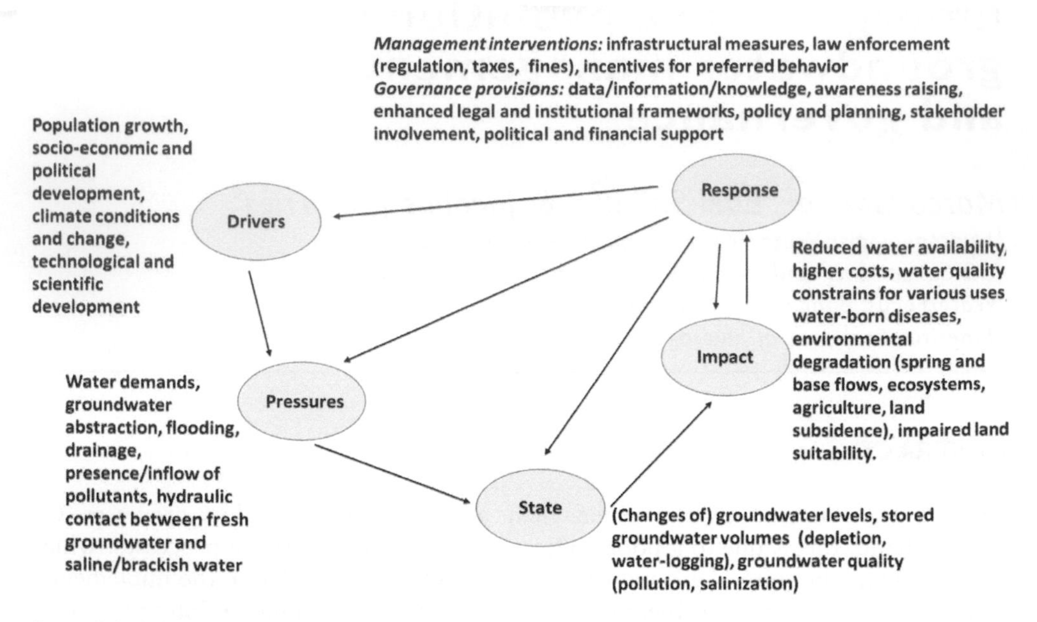

Figure 2.1 DPSIR assessment framework for groundwater management and governance (based on Kristensen, 2004).

for the benefit of the entire society. In practice, groundwater management and groundwater governance are sometimes interchangeable as there is a quite large area in which both overlap. More specifically, groundwater management deals with practical interventions, while groundwater governance focuses on provisions (legal, institutional, information systems, policy and planning, finances, among others) and processes that enable the interventions to take place.

The purpose of this chapter is to present a general picture of the development of groundwater management and governance in a variety of geographical, socio-economic and political environments and to clarify the underlying mechanisms of these processes. The Driver, Pressure, State, Impact and Response (DPSIR) model is used to bring order in the vast diversity of specific management and governance practices of dealing with groundwater problems under various conditions (Fig. 2.1).

The chapter provides examples of groundwater management and governance to elucidate general trends without pretending to provide a complete and coherent picture.

2.2 EXAMPLES OF EARLY EMERGING GROUNDWATER MANAGEMENT AND GOVERNANCE

The exploitation of natural springs and shallow wells goes back to thousands of years in different regions around the world. The use of springs and dug wells as a permanent source of water established the first individual and/or communal ownership of groundwater. An example is the town of Jericho (nowadays in Palestine), where a large

Figure 2.2 Access points of the vertical shafts in a Qanat system (Source: Iran Front Page).

Box 2.1: Qanats

A qanat is a gently sloping tunnel to collect groundwater from an aquifer and transport it to areas downslope for irrigation and public supply. The technique is particularly suited to arid and semi-arid environments, notably where groundwater-rich foothills border dry plains. It was constructed via access shafts (see Fig. 2.2).

spring has been used for irrigation for more than 9000 years. General pressures on the resource such as the increasing water demand at these early sites forced the users or local societies to think and agree about rights to access and use the resource. Over time, the continuous use of groundwater brought about the development of rules regarding allocation in cases of water scarcity, maintenance of the infrastructure, and protection of the abstraction site. Another example of early groundwater management is the construction and collective use of *qanats* (Box 2.1, Fig. 2.2).

Qanats originated in North-western Persia in the early 1st millennium BC. The technology quickly spread first over what is now the Middle East. Later in history, through the expansion and conquests of Islam, the technology reached farther west to southern Spain (Martínez-Santos and Martínez-Alfaro, 2012). The sustainable use of qanats requires good cooperation among the users regarding the allocation of water and also for proper protection of the resource and maintenance of the technical infrastructure.

The *Water Authorities* in the Netherlands are an interesting example of early groundwater governance and stakeholder cooperation. From the 10th century marshlands have been reclaimed and made suitable for agriculture through groundwater drainage by groups of farmers. Subsequently, the local communities organized the required management and maintenance of the drainage and discharge systems infrastructure and flood protection works. Because the drained area enclosed many towns or villages, representatives were elected or appointed to discuss by whom and how the works would be financed and supervised. These technical, administrative and legal requirements spurred the establishment of water authorities responsible for areas of different size. Cities and feudal lords were involved in issues of regional interest (NHV, 1998). This example shows how (ground)water and land management issues can successfully be addressed in the development of communities and their institutional structures. The water authorities still exist in the Netherlands as a special layer in the democratic system.

The concept of environment and its protection, specifically of springs and wells, was well represented in the Islamic Golden Age. There is evidence that both in the Ottoman states and in Muslim India, the government oversaw the use of springs and wells, as well as building water reservoirs and lakes. Islamic law provided an environmental system which included protecting water from misuse and pollution, also known as 'harim', which literally means protected or banned zone (Izzi Dien, 2000). The concept was derived from the experiences of the Prophet Muhammad. He cursed a person for soiling three places; a high road, the shade of a tree, and a river bank (Izzi Dien, 2000). The protected zone was based on ancient units (cubits) for wells (about 20 meters' diameter) and for springs. The practices observed in implementing the concept of 'harim' reflect the Islamic attitude of protecting public environmental interests.

These examples of early groundwater management and governance reflect the diversity of pressures and drivers that triggered new developments. Important elements of the latter were the response or adaptation to climate variability, stakeholder cooperation for land reclamation and religious aspects with regard to the protection of a resource. However, major groundwater management interventions and governance provisions only evolved since the industrial revolution, in response to increasing water demands for industrial, irrigation and domestic uses; facilitated by the improvement of technologies.

2.3 MAIN GROUNDWATER MANAGEMENT CHALLENGES AND OBSERVED RESPONSES

Groundwater management is usually emerging in response to observed groundwater-related problems. Such problems (acting as triggers) may be related to internal factors, like stress on the state of the resource by overexploitation and pollution, or to external factors like conflicting interests and water-borne diseases (*e.g.* cholera outbreaks). This type of reaction or response to what actually happens is known as 'reactive mode'. Responses observed over time are in general influenced by gradually progressing activities and features such as: land reclamation and its linked water level control; climate variability (nowadays linked and accentuated by climate change); the constant progress of technology since the industrial revolution; and in some cases, the acknowledgment of the importance of groundwater for different domains. Many of the observed management responses throughout history can be classified as 'reactive'. Nevertheless 'pro-active' steps in groundwater management (not triggered by what actually happened, but anticipating on what may happen) have also been encountered. The

Box 2.2: Groundwater management stages

Countries (or areas within countries) have different groundwater resources development and management stories, but their evolution can generally be categorized in three stages:

- Pre-management stage – with groundwater being abstracted for local use without people having any notion that its control, management or protection were possible or desirable.
- Initial management stage – in response to emerging problems, steps towards management or protection are taken, according to the problems on an essentially 'single-issue oriented' basis.
- Advanced management stage – some countries have subsequently moved towards more comprehensive and integrated approaches to groundwater administration and protection.

Source: Global Diagnostic on Groundwater Governance, 2016

proactive approach is often motivated by bad experiences in the past and requires a good understanding of the relevant groundwater-related processes, mechanism and interdependencies in the local context. The Netherlands and its long tradition of water management (both surface and groundwater) can serve as a good illustration. Primarily driven by increasing water demands since the late 19th century, groundwater management and governance interventions took place over time. Management and governance alternated in reactive and proactive approaches within the constraints of competing users and a shallow aquifer environment that is vulnerable to groundwater level decline and salinization.

For each other country or region, the type of response to its specific groundwater issues depends in general on their management stage (Box 2.2), which includes – among others – the technical and administrative capacity available to react or anticipate on stresses. The following paragraphs present a selection of 'main' challenges observed in groundwater management, often presented in relation to a specific local context.

2.3.1 Promoting groundwater use for public water supply

Outbreak of water-borne diseases, famine, water constrained agriculture, needs for stimulating rural economic development, and lack of safe drinking water supplies are among the main factors that stimulated the beneficial use of groundwater resources since the 19th century. Notably, the outbreaks of water-borne diseases such as the cholera in Europe in the early 19th century triggered the use of groundwater instead of surface water as a safer source of drinking water.

The Indian sub-continent suffered from many famines throughout its history; some 90 famines in 2,500 years of history (Murton, 2000). In addition, poor service delivery from public water supply systems acted as pressure and stimulated many farmers, and rural and urban households to organize their own private supply, contributing to the remarkable expansion of groundwater use of the last five decades (World Bank, 2010). This change in roles triggered groundwater resources development programs aiming to support the agricultural sector in countering food scarcity and to improve poor public supply services. During the last 50 years, these programs in India brought along a huge infrastructure for groundwater abstraction and distribution: the construction of millions of wells. The programs also introduced management practices and governance provisions such as technical assistance, training, energy subsidies, budget allocations, political support and programs for international cooperation.

2.3.2 Control of declining groundwater levels

Since the early 20th century many cases of groundwater level decline have been observed. Increasing exploitation often has resulted in lowering groundwater levels and loss of artesian pressure, with as consequence higher energy requirements to bring groundwater to the surface. Aquifer degradation (storage depletion, exhaustion, salinization due to sea water intrusion,) within areas of intensive groundwater abstraction is an associated change of state (Fig. 2.1), as well as environmental impacts such as land subsidence and drying up of springs and wetlands.

The evolution of groundwater abstraction in selected countries (Fig. 2.3) illustrates the boom in groundwater development during the second half of the 20th century; with India, U.S.A. and China as the major abstractors followed by countries like Pakistan, Iran and Mexico. The strong increase in ground water abstraction is largely the result of numerous individual and collective initiatives (farmers and decision makers) without centralized planning or coordination for many decades. Access to modern well drilling and pump technology has triggered and catalysed intensive 'wild cat' groundwater exploitation in many parts of the world. The ultimate negative impacts caused by the lowering of groundwater levels are the depletion of exploitable groundwater volumes, both from renewable and non-renewable aquifers, and the irreversible impacts on ecosystems and the environment in general. It should be noticed, however, that several countries (in particular the wealthier industrialized countries) have become successful in controlling groundwater abstraction and thus in reducing or eliminating groundwater level declines and their negative impacts.

In many of the water-poor (semi-)arid zones around the world there has been a lack of adequate governance to control groundwater level declines in renewable and non-renewable aquifers. Monitoring the groundwater quality, groundwater levels and the volumes abstracted, which forms the basis of groundwater management, is absent in most Arab countries (Al-Zubari, 2013). Required management interventions have not been implemented, notably because of water rights, political preferences, lack of awareness, ignorance, and stakeholder opposition. Only in a few parts of the (semi-)arid zones, groundwater management practices have developed, including licensing groundwater abstractions, providing incentives for desired behaviour (subsidies, power supply, etc.) and managed aquifer recharge (MAR). Among the successful examples of the Middle East region are the recharge dams, among others in Israel, Jordan and Oman. These dams intercept and retain flash floods produced by erratic rainfall in order to enhance groundwater recharge and at the same time to reduce damage caused by flooding. This type of water conservation scheme, using intermittent

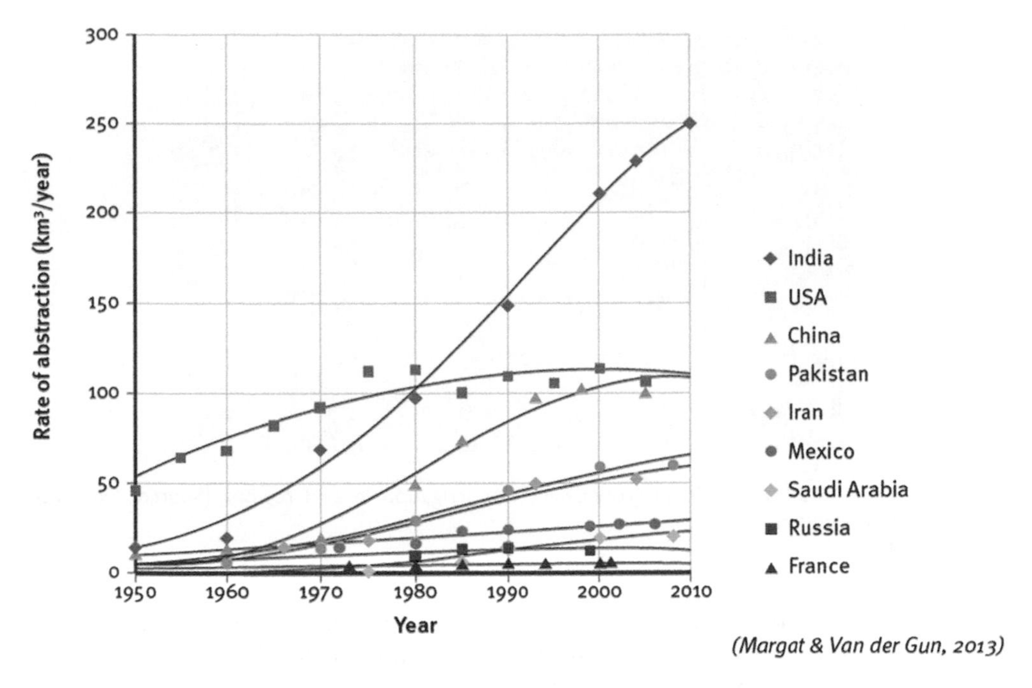

Figure 2.3 Evolution of groundwater abstraction in selected countries.

flood regimes of wadis, may well contribute to the alleviation of water scarcity in many arid and semi-arid regions of the world (Petry *et al.*, 1998).

Groundwater level declines in fossil aquifers are certainly more critical because of their irreversible character. Aquifer systems with recognized non-renewable resources, where significant groundwater mining has taken place, are located principally in North Africa and the Arabian Peninsula (Foster & Loucks, 2006) (Fig. 2.4). These aquifers have been exploited in an unplanned manner and with uncertain trajectory, which threatens the future availability of groundwater for the socio-economic development of the region. In contrast to inactivity elsewhere, impressive efforts have been made to reduce the rate of decline of groundwater levels of the Australian Great Artesian Basin. The Great Artesian Basin Sustainability Initiative (GABSI), that involves the participation of Australian state and territory governments as well as landholders, has proven to be effective and successful in this respect. Its technical components include the rehabilitation of uncontrolled artesian bores and replacement of open earthen bore drains by piped water reticulation systems (MSK, 2013; GAB Coordinating Committee, 2017; Queensland Government, 2017).

2.3.3 Land reclamation and water-logging

In the coastal lowlands of The Netherlands, land reclamation has taken place since about 900 AD. Artificial drainage of peat marshes between 900 AD to 1500 AD resulted in new agriculture land, but at the same time it caused new impacts such as

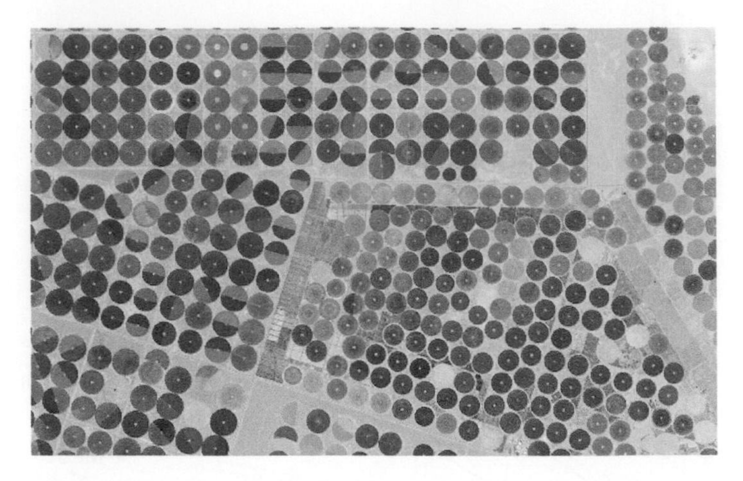

Figure 2.4 Large scale use of fossil groundwater for irrigation in arid regions (Source: Axelspace Corporation).

land subsidence and increasing problems with the evacuation of excess water. In the 15th century, the innovative development of the windmill, which lifted and discharged water, helped to successfully combat this problem. Moreover, the windmill made possible the reclamation of lakes and sea embayments. This latter type of expensive land reclamation was stimulated by rich merchants in the prosperous cities, looking for profitable investments. Over time, roughly 600,000 hectares have been reclaimed in The Netherlands (Van de Ven, 1993; NHV, 1998) (Fig. 2.5).

The reclamation of land worldwide has taken place particularly in low-lands adjacent to the ocean, river beds or lake beds. Examples can be found in many parts of the world, including China, Canada, and South Korea among others (Xue *et al.*, 2016; Glaeser *et al.*, 2016; Lee *et al.*, 2014). Lack of adequate groundwater drainage in those shallow-water-table areas can result in water-logging.

Flooding and associated water-logging is part of life in countries like Egypt, India and Bangladesh, located on the extensive flood plains of major rivers. This includes many urban areas of Bangladesh without adequate drainage, leaving the areas inundated for days (Anisha & Hossain, 2014).

2.3.4 Control of saline/brackish water intrusion

Intensive exploitation of many coastal aquifers around the world has caused deterioration of groundwater quality due to saline/brackish water intrusion. Sea-water intrusion into aquifers hydraulically connected to the sea and the up-coning of salt water in aquifers with underlying saline/brackish groundwater have been studied since the late-19th century. In Bahrain for instance, the heavy reliance on groundwater and excessive abstraction rates since the 1970s has caused the loss of most of the fresh groundwater reserves due to salinization (Al-Zubari, 2013). Adequate control of abstraction, artificial recharge, and hydraulic or physical barriers (underground dams) are among the

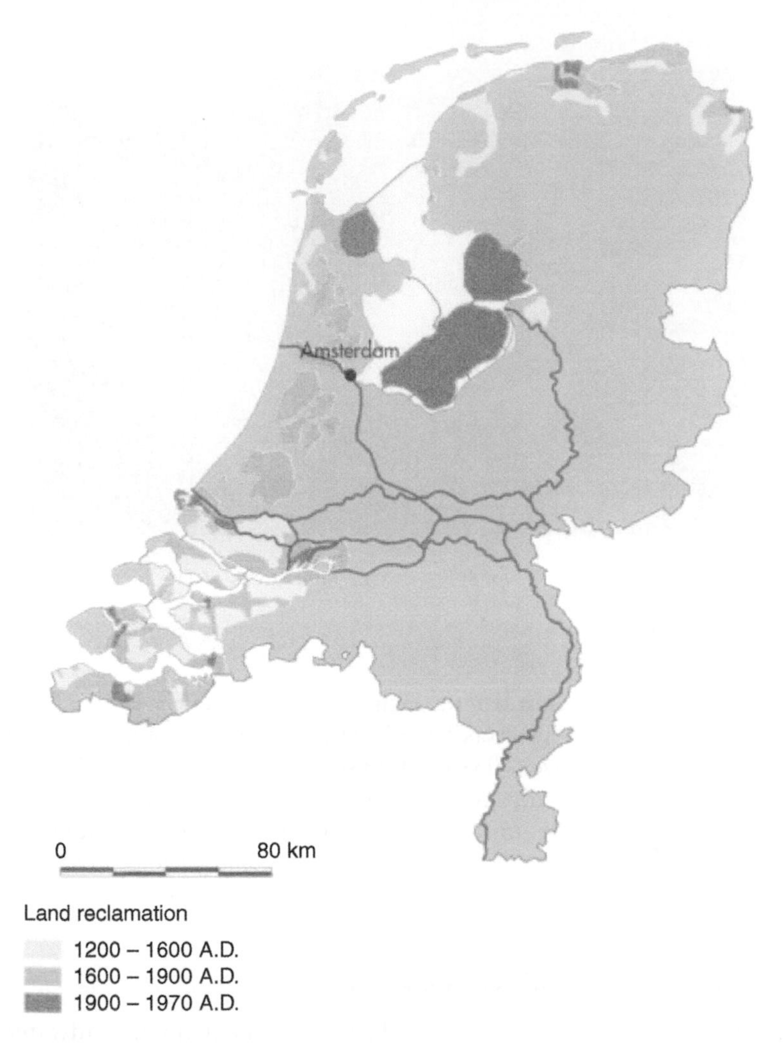

Figure 2.5 Overview of land reclamation in The Netherlands over time (Source: Nederlandse Hydrologische Vereniging).

management responses that might stop or reverse the intrusion. A successful example of reversing the depletion is found in The Netherlands, where since the 1940s artificial recharge takes place in the dune areas as a means to reduce up-coning of saline water and to use the aquifer for groundwater storage. Initially, open recharge systems such as basins, ponds and canals were used (Fig. 2.6), but since 1990 also injection wells have been applied (Margat & Van der Gun, 2013).

Another successful application of artificial recharge is the use of hydrodynamic barriers in the Central and West Coast Basins of Los Angeles, California. Between the 1950s and 1960s the L.A. Flood Control District implemented three barrier projects,

Figure 2.6 Artificial recharge in the North Sea coastal dune areas of The Netherlands (Source: The Water Channel).

together stretching over 17 km length, including 290 injection wells and 773 observation wells. These barrier projects have been adequately protecting the freshwater aquifers of Los Angeles for over 50 years (Johnson, 2007; Barlow and Reichard, 2010). More recently, artificial recharge systems have been constructed in the Lower Llobregat aquifers in Barcelona, Spain (Martin-Alonso, 2016). These systems are used to increase the water reserves, improve the water quality, and to act as hydraulic barriers against sea water intrusion.

2.3.5 Pollution and pollution control

Pollution and pollution control is perceived as one of the main groundwater management issues in many industrialized countries around the world. Apart from current pollution threats, many countries experience widespread and serious groundwater quality impairment from the legacy of industrial and mining contamination in the past (Global Diagnostic on Groundwater Governance, 2016). Groundwater quality has been degraded and is at risk of being degraded by a large variety of anthropogenic pollution types.

Arsenic contamination of groundwater occurs where high natural concentrations of arsenic in deeper parts of the aquifers have been reached by excessive extraction and depletion. A recent example is the Indo-Gangetic Basin in India. Groundwater abstraction from the transboundary Indo-Gangetic Basin comprises 25% of the global groundwater withdrawals, sustaining agricultural productivity in Pakistan, India, Nepal and Bangladesh. Excessive abstraction has meanwhile caused 60% of the aquifer to become unsuitable for drinking purposes due to high levels of salinity and arsenic (MacDonald *et al.*, 2016).

Figure 2.7 Well field protection zone in the Netherlands showing a telephone number to be called to report any detected case of pollution (Photo: Jac van der Gun).

Well field protection is the restriction on activities which generate or apply polluting materials in zones of a water supply catchment close to a well field. Control is often executed by delineating 'protection zones' (Schmoll *et al.*, 2006) (Fig. 2.7). These zones are delineated by various distances between potential sources of pollution and the point of abstraction. These distances correspond to different ranges of groundwater travel time between the pollution source and the object to be protected. Unfortunately, protection zones are not applied in all countries (Bannerman, 2000). The extent to which the protection zones are monitored and the corresponding restrictions are enforced depends on the management stage of the country and the environmental policy frameworks (if available).

The problem of *diffuse pollution* has been a concern since the mid-20th century with the increase in agricultural activity and the use of pesticides and fertilizers. Steps forward in control are usually triggered by accidents or detection of previously unknown subsurface pollution. Urban development, forestry and land use practices, atmospheric deposition, industry, modern transport and rural dwellings can also be the origin of serious contamination (EEA, 2007).

Point-source pollution control has often been triggered by accidents or by detection of previously unknown subsurface pollution. The Lekkerkerk pollution case in The Netherlands is a good example. New residential quarters were built in the late 1960s and early 1970s on former agricultural land where ditches had been filled with domestic and industrial waste. Soil pollution (by aromatic hydrocarbons) was discovered after almost a decade (1979). Consequently, remediation works began in 1980, starting with the evacuation of the area. Subsurface remediation and decontamination involved dig-and-dump of large volumes of soil, and pump-and-treatment of the groundwater, which became a very expensive process. This was the beginning of a country-wide inventory and investigation of the numerous waste dumps and landfills, resulting in a variety of measures, like dig-and-dump, site isolation, and in situ remediation by stimulated microbial degradation and purification processes. These activities are still going on.

Anthropogenic *underground activities* polluting groundwater such as oil and gas extraction, mining, subsurface storage, etc., must be controlled to minimize their negative impact on groundwater. In practice this is often not directly within the mandate and power of groundwater management agencies, thus this requires strong coordination and enforcement by higher government institutions. However, failing to comply with the rules is often not sanctioned because of the influential power of the industries involved. Thanks to the increasingly critical attitude of the general public regarding environmental topics (opposition to fracking, CO_2 storage, storage of nuclear waste, etc.) the pressure to monitor and sanction non-compliance in these sectors is becoming gradually stronger. As a result, many companies' Corporate Social Responsibility (CSR) statements include environmental protection, showing a change from reactive to pro-active attitudes.

2.4 TRENDS AND DIVERSITY IN GROUNDWATER GOVERNANCE

The next four paragraphs are arranged under the four basic components of groundwater governance: (i) data, information and knowledge; (ii) actors; (iii) legal and institutional frameworks; and (iv) policies (Global Diagnostic on Groundwater Governance, 2016). They are followed by paragraphs on additional groundwater governance aspects.

2.4.1 Data, information and knowledge

The statement '*If you cannot measure it, you cannot manage it, and thus cannot govern it*' is of special relevance within groundwater governance due to the hidden nature of the resource. The acquisition of scientific data and knowledge since the industrial revolution helped to base groundwater management systems on scientific knowledge rather than on tradition, superstition or customary law. In recent times, an impressive amount of information and knowledge has been accumulated, especially since the middle of the twentieth century. Numerous hydrogeological maps are available at different scales for a large number of countries (or parts of countries), for continents and even for the entire world (Global Diagnostic on Groundwater Governance, 2016). However, there is still more to be done. In developing countries, the information is limited and basic data is missing in many regions. Observed deficiencies include the scarcity of data in general and the lack of uniformity and synchronization of the data and thus an inherent poor quality of the processed information. There is notably a lack of groundwater monitoring data, which implies that changes in groundwater conditions are very poorly known, precluding rational and efficient actions to exploit, manage and protect the groundwater resources properly (Van der Gun, 2007). This deficit of data, information and knowledge is partly due to the lack of financial resources and partly due to policy makers not recognizing the urgency of groundwater resources assessment and monitoring. This translates into a lack of triggers for adequate groundwater management and governance responses, leading to only few management actions and governance provisions; reducing the flow of the cyclic process illustrated in the DPSIR framework (Fig. 2.1). In addition to the deficit of data, in many countries in a pre- and initial management stage, the lack of sufficient scientific knowledge in general, and a lack of understanding the nature of groundwater has been one of the big gaps that need to be bridged to achieve adequate groundwater governance.

The progress in data acquisition through satellite-based monitoring and other remote sensing techniques has proven to be very helpful in reducing this gap. Recent projects like the Gravity Recovery and Climate Experiment (GRACE), have made it possible to assess groundwater storage variations in some of the world's major aquifer systems via satellites – highlighting the trend towards using innovative techniques in support of hydrogeological investigations. Global simulation models linking the terrestrial and atmospheric components of the hydrological cycle contribute to this goal as well. A limitation of these programs and models is that their applicability to management and governance at a local level is hampered by their low spatial resolution.

During the last decades several programs and organizations have emerged based on the idea of exchanging, sharing, compiling and analysing area-specific information on groundwater, as a contribution to the dissemination of knowledge, *e.g.* the World-wide Hydrogeological Mapping and Assessment Programme (WHYMAP), the International Groundwater Resources Assessment Centre (IGRAC), the Groundwater Management Advisory Team (GW-MATE), the International Waters Learning Exchange and Resource Network (IWLEARN), and the Internationally Shared Aquifers Resources Management Programme (ISARM). The contribution of these programs and organizations to knowledge on groundwater has been of great importance. However, outside the community of hydrologists and hydrogeologists, general knowledge and public awareness about groundwater systems is still rather limited.

Groundwater is often not yet sufficiently prominent on the national water agenda and in international budgets, while decision-makers lack vision to make proper decisions on water and the general public fails to adopt water-friendly behaviour (Van der Gun, 2007). Efforts have been made worldwide to raise public awareness. Additionally, there have been training and education programs on groundwater resources conservation, its potential benefits, and the spread of knowledge on how to explore, develop and use them properly. However, this knowledge stills need to be put into practice worldwide to help achieve adequate management and governance of the resource.

Figure 2.8 Stakeholder involvement and interaction among groundwater actors (Source: Watershed Organization Trust WOTR).

2.4.2 Actors and their roles

The number of actors and their roles has seen an increase over time with the evolution of groundwater resource governance and the development of interlinkages with other environmental and human sectors. Water is no longer the almost exclusive domain of hydrogeologists and engineers; as the knowledge and awareness of groundwater increase, its importance reaches other professions. Over the last decades, groundwater has also received a good deal of attention from economists, sociologists, ecologists, climatologists, lawyers, institutional experts, communication specialists and others (Van der Gun, 2012).

The 19th and 20th century saw the tendency in many countries to shift from traditional instruments and informal institutions, such as customary law, towards formal institutions and special groundwater legislation. In countries where groundwater resources are public or state owned, government organizations have taken or should take the lead (regardless of their capacity or performance) as the main actor. The increasing pressures, and sometimes the inadequate management performance of governmental agencies, have paved the road for community organizations, non-governmental organizations and sometimes the private sector to become important actors within the groundwater governance of their region or country. Less authoritarian governments (as compared with those in previous centuries) have also supported and stimulated the role of stakeholders. Stakeholders are no longer exclusively users or consumers; often, they also have the opportunity to play a distinctive role in the management and governance of their groundwater. They thus may play an important role in putting pressure on governments in policy-making and the implementation of management plans (Fig. 2.8).

In some areas, farmers have expanded their roles from users and/or potential polluters to become also pressure groups, sometimes very well organized and politically strong. There are cases in which big industries besides being potential polluters and exploiters, become also collaborators in the development of groundwater projects *e.g.* the Itawa Springs project in Zambia supported by Zambian breweries; and Pepsico in India replenishing and conserving groundwater resources. Some municipalities or local authorities, apart from being sometimes owners of water companies, are at the present also key players in transmitting needs and opinions among stakeholders and decision-makers on practical policies. Several NGOs, in addition to performing their ordinary tasks and monitoring the effectiveness of management, have increased their impact as pressure groups when they are well linked to the media.

The evolution of stakeholder involvement has diversified their role. The complexity inherent to their large number and diversity, as well as the fact that their interests may be partly conflicting, explain why smooth and balanced cooperation in groundwater governance has not automatically been forthcoming, but has to be orchestrated (Global Diagnostic on Groundwater Governance, 2016). Evidently, the specific level of participation and the effectiveness of the roles and responsibilities of the stakeholders depend on their capacity, level of awareness, funding, and the socio-political landscape of the country or region.

2.4.3 Legal and institutional frameworks

The diversity of legal and institutional frameworks around the world reflects the current and historical political ideologies, economic frameworks, and cultural settings of each country and region around the world. The remnants of customary law applied for many centuries are still present in several regions of the world (especially in rural areas of developing countries), but the tendency is to replace it with formal legislation. Formal or 'modern' legislation on groundwater has been adopted now in almost all countries, with occasional redefinitions of the laws in order to adapt to the explosive growth of unregulated groundwater use and to current views on ownership and user rights (Global Diagnostic on Groundwater Governance, 2016). The legal framework of rights and obligations are normally dictated by hydrogeological and socio-economic constraints. Two basic doctrines can be distinguished within this domain. (1) The doctrine of absolute property, where groundwater is the property of the owner of the overlying land. (2) The doctrine of beneficial use, where groundwater is state or public property, but individual organizations can be granted the right of reasonable beneficial use. Absolute ownership normally adheres to groundwater in areas with abundant groundwater resources or in situations of limited use, whereas beneficial use is dominant in areas with limited resources or high exploitation rates. Both doctrines have in the course of time been evolved with a variety of restricting regulations to protect the aquifer against depletion or pollution and/or protect other competing users of the same aquifer (see Box 2.3). These regulation measures are normally executed through licenses and permits and often include the obligation of monitoring the effects of exploitation. Modern legislation however, is not necessarily optimal or sufficient. Many countries of the world lack the capacity, financing, political will or political power to implement and enforce formal legislation and its different instruments; it requires continuous efforts, a great deal of money and strong and capable institutions.

Box 2.3: Evolution of 'modern' groundwater legislation in The Netherlands

The Netherlands did not develop special regulations with respect to groundwater until the mid-20th century. Before that time, every landowner could sink a well and extract water as long as no damage was done to the property of others. Limitations for the exploitation of groundwater were thus set by private law. This posed a problem to water supply- companies who could never be sure of continuity. Therefore, the Groundwater Act Water Supply was enacted in 1954. From then on drinking-water companies needed a license for any abstraction, specifying conditions and damage compensation; groundwater abstractions by others, however, were not authorized under this law. In 1981 an overall Groundwater Act became effective for all abstractions and all activities related to infiltration and groundwater recharge. According to this law, the provinces are the authorities responsible for permission, registration and reporting. Quality aspects are mainly related to protection of recharge areas. Other groundwater-quality issues are dealt with in the Soil Protection Act of 1987, which includes regulations for prevention of subsurface pollution and for remediation of contaminated soils. Within this legislation, the provinces are obliged to set up and maintain a groundwater monitoring and management plan.

Source: Dufour, 2000

At the international level, regional and international legislations such as the European Water Framework Directive (WFD) and the UNECE International Water Convention on Transboundary Watercourses and Lakes are landmarks in promoting improved governance of groundwater and showing how to achieve adequate legislation at an international level.

2.4.4 Policy and planning

Currently, not all countries have dedicated policies on groundwater, but those that do, show a broad diversity in focus, scope, type, degree of detail and other characteristics. The origin or driver of this diversity lies not only in different country-specific physical, cultural, socio-economic and political conditions but also in the differences in the stage of advancement of groundwater management and governance (Chaisemartin, 2017). Compared to laws, policies are easier to modify and/or update in a short period of time in response to changing preferences and goals, and to feedback from the field (learning processes). Set-backs may be observed when changes of parties in power occur, bringing new policies and plans according to the new ideologies of their parties. The policies in countries in a pre-management stage are focused mainly on the supply and allocation of groundwater resources and sanitation. Countries in an initial management stage tend to focus on the control and protection of the resource, acting mainly in a reactive mode and with implementation of principles such as 'polluter pays'. Countries in an advanced management stage have applied more comprehensive and integrated approaches towards groundwater management, taking into consideration other domains and using principles like IWRM and the precautionary principle.

The later states that uncertainty should not be an excuse to postpone addressing emerging serious challenges, and it should prevent decision-makers from taking premature risky decisions. Advanced management is characterized by anticipating on expected developments and challenges in the future (proactive mode).

2.4.5 From single-issue to integrated groundwater management and governance

Certainly, Integrated Water Resource Management (IWRM) has been the principle that brought water/groundwater from being managed within a single sector into the broad context of sustainability sectors and modern approaches. Historically, we can go back centuries, if not millennia, to discover forerunners of the present IWRM, for instance the multi-stakeholder participatory water tribunals of Valencia from approx. the 10th century, or the 'falaj' irrigation systems in Oman from around 2500 BC. For many centuries, these management practices took place around the world without being classified and recognized. It was not until 1977 that these classical approaches themselves received special attention among the international community at the United Nations Conference on Water in Mar del Plata, with the recommendation to incorporate the multiple uses of water resources (Rahaman & Varis, 2005). In many countries, this cross-sectoral approach to water has replaced the traditional, fragmented sectoral approaches that ignored the interconnection between the various water uses and services (Van der Gun, 2012). The degree of implementation of the IWRM approach depends greatly upon the management stage of the country. In countries in a pre- or initial management stage, an initial step (among many others) is the establishment of key IWRM institutions such as river basin or catchment organizations at a national level as well as international level. While in countries in an advanced stage, the focus is on integrating fully the IWRM approach into their already existing groundwater policies. The principle is a good representation of conjunctive management and governance response to a complicated issue such as the adequate use of groundwater resources.

2.4.6 Managing and governing groundwater across jurisdictional borders

Significant progress has been made on activities regarding transboundary aquifers since these were put on the international agenda at the end of the twentieth century. The international aspect of a transboundary aquifer requires a joint management strategy. An informed and sustainable management of shared aquifers asks for adequate knowledge of its characteristics, present state and trends (IGRAC, 2015). Regional inventories and studies of transboundary aquifer systems have been conducted under the umbrella of the United Nations Economic Commission for Europe (UNECE) in Europe (UNECE, 1999) as well as in the Caucasus, central Asia and south-east Europe (UNECE, 2009); and under the umbrella of the ISARM Programme in the Americas (UNESCO, 2008), Africa (UNESCO, 2010a) and Asia (UNESCO, 2010b); (Van der Gun, 2012). To date, 592 transboundary aquifers have been identified throughout the world (IGRAC and UNESCO-IHP, 2015). This number is expected to grow as exploration and assessment of transboundary aquifers continues. New aquifers are mapped and larger ones are investigated in more detail and subdivided into smaller individual aquifers.

The UN Draft Articles of the Law on Transboundary Aquifers are a major effort to appraise the international law in its applicability to transboundary aquifers. The Draft Articles were elaborated by the United Nations International Law Commission (UNILC) and formally acknowledged by the United National General Assembly (UNGA) in 2008. Later (in 2013), the UNGA commended the Draft Articles also formally to States as guidance in the framing of aquifer-specific agreements. The Draft Articles are therefore not binding and individual countries could use them as guidance for their conduct in transboundary aquifer management (Global Diagnostic on Groundwater Governance, 2016).

2.4.7 Making linkages with surface water and non-water policy domains

As stated in paragraph 2.4.2, groundwater management and governance is no longer an exclusive domain of hydrogeologists and engineers, but is also interlinked with economists, sociologists, ecologists, climatologists, lawyers, institutional experts, and communication specialists; among others. Focusing on groundwater certainly does not imply that groundwater systems are self-contained, or that they can be understood and managed on the basis of hydrogeological information only (Van der Gun, 2012). Groundwater is a component of the hydrological cycle that closely interacts with the other components at various temporal and spatial levels. In addition, it is directly and indirectly related to numerous other sectors within a region/country; agriculture being the most pertinent. The increased acknowledgment of groundwater broadens the pressures and impacts on other natural or anthropogenic systems, putting more pressure on adequate responses to achieve a sustainable use of the resource. Management and use of groundwater has significant impact on health issues (in relation to safe drinking water and adequate sanitation), socio-economic development (mainly in relation to irrigation for agriculture and to industry), protection of nature and environment (protecting land and aquatic groundwater-dependent ecosystems), land use (water availability is often a limiting factor in decisions on land use), energy (cooling water for thermal power plants, geothermal energy development), groundwater use for the exploration and production of oil and gas (injection wells for fracking and shale gas development), and use of the subsurface space (mining, and disposal and storage of hazards such as radioactive waste and CO_2). Coordinating groundwater management with other related sectors, taking into account their interactions, is likely to produce better and safe solutions. However, this increases complexity and requires significant legal and organizational capacity of the various lead agencies. Therefore, for most countries, especially countries in a low groundwater management stage, this is an ambition that can only be fulfilled in a very modest way.

2.4.8 Global priorities and programmes

Groundwater management and governance has been brought to the international agenda by different international agencies and global programmes over the last decades. Especially since the 1960s, numerous international projects and initiatives have been carried out in the context of development cooperation and joint projects between 'befriended nations', either in a bilateral or a multilateral setting (Van der Gun, 2007). Global political commitment for action on the world's most important development issues has been mobilized at the highest levels by Earth Summits on Sustainable Development, organized by the United Nations (Van der Gun, 2012). Global targets such as the MDG's and SDG's, apart from placing (ground)water into the international agenda, raise political support and funding, mobilizing all types of programmes, professionals, public and private sectors, NGO's, and local groups for the achievement of these goals. Other initiatives that have proved to be successful in advancing international cooperation have been UNESCO's International Hydrological Decade (1965–1974) and its successor the International Hydrological Programme (IHP), as well as the UN's International Drinking Water Supply and Sanitation Decade (1981–1990). United Nations plans such as the Agenda 21, programs like IHP and intergovernmental bodies as the Intergovernmental Panel on Climate Change (IPCC) have been outstanding contributors to the development of groundwater science, management and governance over the last four or five decades. The combination of all programmes, goals, initiatives, and plans (added to the national initiatives) has significantly contributed to achieving the current level of groundwater knowledge, management and governance worldwide.

2.5 SUMMARY AND CONCLUSIONS

Use of groundwater has been a constant factor in human development. The challenge throughout recent human history has been to regulate this exploitation with adequate groundwater management measures and governance. For many centuries, local management and regulation measures have been developed and effectuated by communities in diverse parts of the world. These communal based controls were stimulated or triggered by a diversity of needs produced by geographical conditions (including natural disasters, notably drought), and ideologies and traditions (location-specific context). In some cases, these early measures were carried out and developed in such an effective way that they are still functioning, as for instance the qanat systems, and the water authorities in The Netherlands. Evidently, in pre-industrial times, the utilization and consumption of groundwater was at a relatively small scale with little impact at aquifer level.

The industrial revolution triggered major developments (both positive and negative) in groundwater exploitation and management. The initial groundwater exploitation and management systems during and after the industrial revolution were often based on the idea of infinite reserves fed by remote sources. These concepts were based on a lack of proper knowledge and on speculation on the origin and movement of groundwater, and thus hampered the development of awareness regarding negative consequences of abstraction, like depletion, environmental degradation, pollution, etc. It was only after these negative aspects were encountered that management provisions gradually developed concurrently with advances in scientific and technical knowledge. This *reactive mode* became a constant in the approach of groundwater management in most countries. Constant adaptation of management practices had to be implemented to cope with the consequences of increasing demand and the associated problems of depletion, pollution, and inequitable and inefficient allocation.

The advances in technology and scientific knowledge of the 19th and 20th century greatly helped to solve quantitative and qualitative groundwater management problems. Concurrently, governance increasingly became based on legislation, which may or may not have acknowledged and incorporated traditional or customary law. Thus, increasing scientific knowledge and level of awareness on groundwater-related issues during the 20th century led to a better understanding of the scope of groundwater and its reach in general. The realization that solutions of identified problems could not be purely technical and that the water issues were linked to other sectors as well, led to the awareness of governance as a "new" concept, incorporating the expanding scope of integrated groundwater management to the economic and socio-political aspects. Evidently, countries with stable economic and socio-political conditions are in a better position to apply their scientifically based knowledge, to strengthen their institutions, to recognize the importance and rights of all stakeholders, to endorse legislation, to adapt to changing conditions and to anticipate to future challenges. This means a transitioning from of a *reactive approach* to an advanced management stage based on a *proactive approach*. In DPSIR terminology: the emergence and enhancement of groundwater governance has resulted in more adequate responses to drivers, pressure, state change and impacts.

The present challenge is to share and implement the advances of the last decades in a global context, and especially in developing countries. The main problem to overcome in this process is a lack of capacity to implement the necessary solutions, because of weak or non-existing institutions to support the developing programs, or simply because of a lack of awareness of groundwater being the important resource that it is.

In broad terms, groundwater management, governance and awareness show around the world a positive tendency over time. But it remains a challenge to simultaneously cope with the increasing exploitation, pollution, inefficient allocation and environmental degradation. There are still many locations in the world which lack coherent governance frameworks, and many others are in need of improvement. Groundwater management and governance are processes that require constant attention and adaptation to an ever-changing world. The paramount aim for the future will be in many countries the transition towards a more advanced stage, coping with the existing and changing economic, socio-political and climatic conditions..

REFERENCES

Al-Zubari, W.K. 2008. Integrated Groundwater Resources Management in the GCC Countries – A Review. *Proceedings of the Water Science and Technology Association Eighth Gulf Water Conference: Water in the GCC.* Towards an Optimal Planning and Economic Perspective, Manama, Bahrain, 2–6 March.

Al-Zubari, W.K. 2013. Groundwater Governance Regional Diagnosis in the Arab Region. *Groundwater Governance: A Global Framework for Action, GEF.*

Anisha, N.F., and Hossain, S. 2014. A case study on water logging problems in an urban area of Bangladesh, and probable analytical solutions. *2nd International Conference on Advances in Civil Engineering 2014 (ICACE-2014)*, At CUET, Bangladesh.

Bannerman, R.R. 2000. Conflict of technologies for water and sanitation in developing countries. In: *Water, Sanitation and Health,* (eds. I. Chorus, U. Ringelband, G. Schlag and O. Schmoll), pp. 167–170, IWA, London.

Barlow, P., and Reichard, E. 2010. Saltwater intrusion in coastal regions of North America. *Hydrogeology Journal*, Vol. 18, No. 1, pp. 247–260.

Chaisemartin, M., Varady, R.G., Megdal, S.B., Conti, K.I., Van der Gun, J., Merla, A., Nijsten, G., and Scheilber, F. 2017. Chapter 11 Addressing the Groundwater Governance Challenge A call from the "Groundwater Governance: A Global Framework for Action" Project. In: *Fresh Water Governance for the 21st Century Global Issues in Water Policy 6,* DOI 10.1007/978-3-319-43350-9_11. Springer International Publishing. ISBN: 978-3-319-43348-6

Chilton, J., and Smidt, E. 2014. Diagnostic Report UNECE Region (2nd Draft). *Groundwater Governance: A Global Framework for Action, GEF.*

EEA, 2007. *Diffuse sources, European Environmental Agency.* Electronic publication note. Available at: http://www.eea.europa.eu/themes/water/water-pollution/diffuse-sources

Foster S. and Loucks D.P. 2006. Non-Renewable Groundwater Resources – A guidebook on socially-sustainable management for water-policy makers. *IHP-VI, Series on Groundwater* No. 10.

GAB Coordinating Committee, 2017. *Great Artesian Basin Sustainability Initiative, Phase 4.* Web page, http://www.gabcc.gov.au/basin-management/sustainability-initiative-gabsi

Glaeser, L.C., Vitt, D.H., and Ebbs, S. 2016. Responses of wetland grass, Backmannia syzigachne, to salinity and soil wetness: Consequences for wetland reclamation in the oil sands area of Alberta, Canada. *Ecological Engineering.* Vol. 86, pg. 24–30.

Global Diagnostic on Groundwater Governance, 2016. *Groundwater Governance – A Global Framework for Action.* March 2016. GEF.

International Groundwater Resources Assessment Centre, 2015. Guidelines for multi-disciplinary assessment of transboundary aquifers. Draft version. *IGRAC Publications*, Delft 2015.

Izzi Dien, M. 2000. The Environmental Dimensions of Islam. *The Lutterworth Press.* ISBN 0 7188 2960 3.

Johnson, T. 2007. Battling sea water intrusion in the Central and West coast basin. *Technical documents on groundwater and groundwater monitoring. WRD Technical bulletin*, Vol. 13 Fall 2007

Kataoka, Y., and Shivakoti, B.R. 2013. Groundwater Governance Regional Diagnosis – Asia and the Pacific Region. *Groundwater Governance: A Global Framework for Action, GEF.*

Kristensen, P. 2004. The DPSIR Framework. Paper Presented at the 27–29 September 2004 Workshop on a Comprehensive/Detailed Assessment of the Vulnerability of Water Resources to Environmental Changes in Africa Using River Basin Approach. UNEP Headquaters, Nairobi, Kenya (2004).

Lee, C.H., Lee, B.Y., Chang, W.K., Hong, S., Song, S.J., Park, J., Kwon, B.O., and Khim, J.S. 2014. Environmental and ecological effects of lake Shihwa reclamation project in South Korea: a review. *Ocean & Coastal Management.* Vol. 102 Part B. pg. 545–558.

MacDonald, A.M., Bonsor, H.C., Ahmed, K.M., Burgess, W.G., Basharat, M., Calow, R.C., Dixit, A., Foster, S.S.D., Gopal, K., Lapworth, D.J., Lark, R.M., Moench, M., Mukherjee, A., Rao, M.S., Shamsudduha, M., Smith, L., Taylor, R.G., Tucker, J., van Steenbergen, F., and Yadav, S.K. 2016. Groundwater quality and depletion in the Indo-Gangetic Basin mapped from in situ observations. *Nature Geoscience 9, 762–766.* DOI 10.1038/ngeo2791

Margat, J., and Van der Gun, J. 2013. *Groundwater around the World: A Geographic Synopsis.* Leiden, The Netherlands, CRC Press/Balkema.

Martin-Alonso, J. 2016. Artificial recharge systems applied in the low Llobregat aquifers (Barcelona, Spain). *Presentation at the 9th International Symposium on Managed Aquifer Recharge.* Mexico City, 2016. Available at: http://linux.iingen.unam.mx/pub/ISMAR9/

Presentations/Artificial%20recharge%20systems%20applied%20in%20the%20Low%20Llobregat%20aquifers%20(Barcelona,%20Spain).pdf

Martínez-Santos, P., and Martínez-Alfaro, P.E. 2012. A Brief Historical Account of Madrid's Qanats. *Ground Water*, Vol. 50, no. 4: 645–653.

Murton, Br., 2000. *"VI.4: Famine", The Cambridge World History of Food, 2, Cambridge.* New York, pp. 1411–27, OCLC 44541840

NHV, 1998. Water in The Netherlands. *Netherlands Hydrological Society.Netherlands National Committee of the International Association of Hydrological Sciences (IAHS).* ISBN 90-803565-2-2

Opie, J. 2000. Ogallala: Water for a Dry Land (2nd ed.).University of Nebraska Press, Lincoln, London.

Queensland Government, 2017. *Great Artesian Basin Sustainability Initiative.* Web page, https://www.dnrm.qld.gov.au/water/catchments-planning/catchments/great-artesian-basin/gabsi

Rahaman, M.R., and Varis, O. 2005. Integrated Water Resources Management: evolution, prospects and future callenges. *Sustainability: Science, Practice, & Policy*, Vol. 1, No. 1, Spring 2005.

Schmoll, O., Howard, G., Chilton, J. and Chorus, I. 2006. Protecting Groundwater for Health: Managing the Quality of Drinking-Water Sources. *WHO Drinking Water Quality Series.* Geneva/London, WHO/IWA Publishing.

Shah, T., Burke, J. and Villholth, K.G. 2007. Groundwater: A global assessment of scale and significance. D. Molden (ed.), *Water for Food, Water for Life: A Comprehensive Assessment of Water Management in Agriculture.* London/Colombo, Earthscan/IWMI, pp. 395–423.

SKM, 2013. *GABSI3 Performance Evaluation and Future Options Report.* Great Artesian Basin Sustainability Initiative Phase 3 Mid-Term Review. SKM, Braddon, Australia.

Petry, B., Van der Gun, J., and Boeriu, P. 1998. Coping with water scarcity: a case study from Oman. In: *Proceedings of the IAHR/UNESCO Conference on Coping with Water Scarcity,* Hurghada, Egypt, 26–28 August 1998, organised by the NWRC/HRI.UNECE (United Nations Economic Commission for Europe). 1999. Inventory of Transboundary Groundwaters, Volume 1. Lelystad, the Netherlands, RIZA.

UNESCO (United Nations Educational, Scientific and Cultural Organization). 2008. *Marco legal e institucional en la gestión de los sistemas acuíferos transfronterizos en las Américas.* Montevideo/Washington, OEA/UNESCO-PHI.

UNESCO (United Nations Educational, Scientific and Cultural Organization). 2010a. *Managing Shared Aquifer Resources in Africa.* Proceedings of the Third International ISARM Conference, Tripoli, 25–27 May 2008. IHP-VII Series on Groundwater No. 1. Paris, UNESCO.

UNESCO (United Nations Educational, Scientific and Cultural Organization). 2010b. *Transboundary Aquifers in Asia: A Preliminary Inventory and Assessment.* IHP-VII Technical Document in Hydrology. Beijing and Jakarta, UNESCO.

Van der Gun, 2007. Sharing groundwater information, knowledge and experience on a worldwide scale. *No H040054, IWMI Books, Reports, International Water Management Institute.*

Van der Gun, 2012. Groundwater and Global Change: Trends, Opportunities and Challenges. United Nations World Water Assessment Programme. *Side Publication Series.*

Van de Ven, G.P. (ed.), 1993. Man-Made Lowlands. *History of Water Management and Land Reclamation in the Netherlands.* Utrecht, Matrijs.

World Bank. 2010. *Deep Wells and Prudence: Towards Pragmatic Action for Addressing Groundwater Exploitation in India.* Washington, DC: World Bank.

Xue, S.F., Sun, T., Zhang, H., and Shao, D. 2016. Suitable habitat mapping in the Yangtze River Estuary influenced by land reclamations. *Ecological Engineering.* Vol. 97, pg. 64–73.

Chapter 3

Understanding groundwater governance through a social ecological system framework – relevance and limits

Marta Rica[1], *Olivier Petit*[2] & *Elena López-Gunn*[1,3]
[1]*ICATALIST, Las Rozas, Madrid, Spain*
[2]*CLERSE, Université d'Artois, Arras, France*
[3]*Cheney Fellow, University of Leeds, UK*

ABSTRACT

This chapter looks at how groundwater governance can be framed and analysed from a social ecological system perspective, which considers the importance of balancing ecosystem flows, health and functions with socioeconomic well-being in an equitable manner, while taking into account issues of power and political economy at different scales. Under this analytical frame, groundwater systems are perceived as having inherent properties like resilience, non-linear feedbacks, redundancy, diversity and modularity composed by human, biophysical and ecological variables and components, which are interdependent on each other. The chapter outlines this approach, its main components as well its main challenges and opportunities to help better understand groundwater governance.

3.1 INTRODUCTION

This chapter deals with the rationale and governance consequences of understanding groundwater governance through a social ecological system framework (SES hereafter). This approach is sometimes neglected in the literature dealing with groundwater governance. The aim of this chapter is to stress the main interests and limitations of such an analytical framework to deal with groundwater governance issues.

One of the main challenges to which groundwater governance is confronted with is how to balance ecosystem health with socioeconomic goals in an equitable manner. Since no panaceas exist to solve these challenges (Ostrom, 2007), an integrative systems approach that tackles the interactions of human and natural variables in each scale and context can be useful to find the most appropriate governance process to achieve sustainable outcomes for context specific situations and scales (Barreteau *et al.*, 2016).

Social ecological systems[1] (SES) are interdependent systems of people and nature, where humans must be seen as a part of, and not apart from, nature (Berkes *et al.*,

[1]Some authors use the terms 'socio ecological systems', 'socio-ecosystems' or 'coupled human-environment systems'.

2003). SES are complex systems with inherent properties like resilience, non-linear feedbacks, redundancy, diversity and modularity (Levin *et al.*, 2012), composed by human, and biophysical variables interdependent on each other. Systems' thinking is a way to picture the complexity of ecosystems and societies, understanding how systems can respond to external disturbances, as well as internal changes – a systems perspective in which all variables are interconnected. Problems do not emerge isolated and there are different scales at which challenges need to be tackled. An advantage of making use of SES conceptual approaches is that they incorporate into their analytical frame polycentric and/or multi-scalar approaches, although with a heavy emphasis on the local scale. It provides the framework to relate the human and the ecological or biophysical variables that each system contains, acknowledging the different scales that governance entails and what key variables and components influence the sustainable or unsustainable outcomes of the governance system. Adopting a social ecological system framework can provide useful tools to analyse groundwater governance, acknowledging the potential vulnerability of this SES and stressing the need for adaptability, robustness and resilience of the institutional arrangements designed to govern groundwater resources.

The chapter outlines the development and main components of a SES framework, as well as its main limitations, especially when considering groundwater dependent systems. Section 3.2 presents a number of related but slightly different approaches closely linked to the SES framework, namely the key elements in groundwater governance, a Common Pool Resource (CPR) collective action approach, the co-evolutionary development model of (informal) groundwater economies, the role of power relationships and finally an ecosystems and resilience approach. The third section then goes into details on the key defining components for a SES analytical framework (namely social, economic and political settings, actors, resource units and systems and governance systems), as well as the key performance criteria for a robust SES. The fourth section then critically analyses how the SES framework could be combined with other approaches that could overcome some of its analytical shortcomings. The final section concludes on the main added value of using a SES framework to analyse groundwater governance, as well as on its limitations.

3.2 GOVERNING GROUNDWATER RESOURCES: EMPIRICAL CHALLENGES AND THEORETICAL APPROACHES

Groundwater is not merely water stored underground. Groundwater bodies or aquifers are biophysical systems with particular flow dynamics, which can be connected to a river basin or to other ecosystems such as wetlands, where use is subjected to social, political and economic drivers. However, while groundwater intensive use is on the rise globally, groundwater governance is often lagging behind. Generally speaking, groundwater governance *comprises the enabling framework and guiding principles for collective management of groundwater for sustainability, equity and efficiency* (Groundwater Governance project, 2016a: 7). According to Ross (2016: 146), *Groundwater governance is defined as the system of formal and informal rules, rule-making systems and actor networks at all levels of society that are set up to*

Table 3.1 Lessons learned from case studies across key governance elements.

Governance Elements	*Lessons Learned*
Institutional Setting	Governing is often a thankless task, yet it requires popularity Legislation does not always translate into implementation Conflict resolution is central to groundwater governance Sufficient funding is of the utmost significance for governance
Availability and Access to Information and Science	Natural systems, social systems, and institutions all have been understudied and would benefit greatly from additional research Trust is a necessary element for all research Urbanized landscapes are critical components of groundwater governance
Robustness of Civil Society	Equity is an essential ingredient of groundwater governance Community-based governance requires deliberate, purposeful intention Leaders can unite stakeholders
Economic and Regulatory Frameworks	Economic incentives can be effective, but may sometimes yield unintended, even opposite results "Indirect" management approaches may be suitable in certain settings, but they should be used cautiously The effectiveness economic incentives as use-control mechanisms depends greatly on the system employed

[Source: Varady *et al.*, 2016: 16]

steer societies towards the control, protection and socially acceptable utilization of groundwater resources and aquifer systems.

In addition to these definitions, a recent analysis of ten case studies of groundwater governance across the globe (Varady *et al.*, 2016) stressed the importance of four cross-cutting governance issues (see Table 3.1). These four elements are: (1) the institutional settings; (2) availability of and access to information and scientific knowledge; (3) civil society and its robustness; and (4) the economic and regulatory frameworks. Despite acknowledging that contextual factors are crucial to determine and shape groundwater governance processes and path dependencies, these authors argue that all groundwater governance approaches will entail the four mentioned elements.

These governance elements put particular emphasis on the regulatory framework and institutional settings for groundwater management, as well as on the collective action processes driven by groundwater users and civil society at large.

The above-mentioned elements are central, but the literature on groundwater governance has also been enriched over the past decades thanks to the work on three distinct approaches (Faysse and Petit, 2012), and a fourth emergent one: i) the study of groundwater governance as collective action initiated by E. Ostrom (1990), ii) the analysis made by T. Shah (2009) on informal groundwater economies and iii) the work of A. Prakash (2005), A. Mukherji (2006) and T. Birkenholtz (2009) drawing on political ecology, and finally iv) an ecosystem services and resilience approach which has recently come to the fore (CGIAR, 2015; Knüppe and Pahl-Wostl, 2011; Knüppe *et al.*, 2015; Custodio *et al.*, 2016).

The first approach considers groundwater resources as common pool resources (CPRs) subject to overexploitation when rules for managing these resources are not

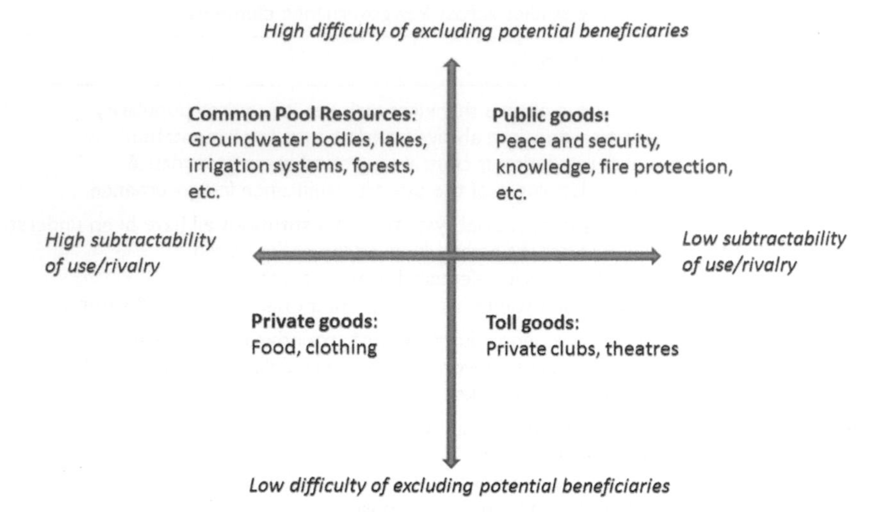

Figure 3.1 Type of goods. Modified from Ostrom (2005).

established and enforced (Ostrom, 1990). Common pool resources are defined by their high subtractability and low excludability. Subtractability by one user limits availability to others[2], and given the nature of CPRs, it remains difficult to exclude potential users from the access to the resource (see Figure 3.1).

Institutional analysts argue that in the case of groundwater governance the emergence of collective action can offer a realistic management model for this particular type of resource, and thus avoid the tragedy of the commons (Blomquist, 1992). The Groundwater Governance project (2016b) also highlights the need to support and recognize groundwater stakeholders' organizations. The nature of groundwater as a CPR therefore poses critical and specific governance challenges. These challenges draw from the ease of access and difficulty in excluding users (or closing the resource to new users), and the fact that the appropriation by one user will affect other users. Good examples are the difficulties that regulators face to quantify groundwater use or enforce regulation compliance (De Stefano and López-Gunn, 2012), and the effect that the intensive use of one pumper can have over the groundwater table of another pumper's well, not realizing about this externality (Shah, 1993). In addition, important inherent resource qualities (low upstart costs, on site availability, resilience to droughts, etc.), combined with increased uncertainty due to climate variability and change, make groundwater an attractive resource.

This attribute of groundwater as a CPR has implications on its use, management, and governance. Decisions triggered by self-interest increase the resource consumption, regardless of the social and environmental consequences unwanted by the group as a whole. When cooperation does not take place, we say we face a 'CPR social

[2]Moreover, one important governance challenge is the slow response of aquifer to external impacts as a result of which users/managers tend to ignore the impact of subtractability that will not become apparent at least in a few years' scale.

dilemma'. However, it has been shown that users can cooperate for the governance of the resources and thus avoid such a social dilemma (Ostrom 1990). Users can self-organize and regulate to share the resource, taking initiative on collective action. This alternative is based on cooperation and self-regulation by users, cooperating as well with regulatory agencies if they exist, as in the Spanish case (López-Gunn, 2003; López-Gunn and Martinez Cortina, 2006; Rica *et al.*, 2014). It is based on different activities, such as negotiating that abstractions are done according to shared priorities and are consistent with groundwater availability, especially when the operation is adversely affecting groundwater levels, river flows, groundwater-dependent wetlands or water quality, as it is increasingly frequently found in coastal areas.

The second approach by contrast, looks at the co-evolution of governance models associated with informal groundwater economies and links to the resource type base (type of aquifer). The analysis of groundwater governance developed by T. Shah (2009) starts from groundwater use at farm level. He then looks at the economic impacts created by the agricultural groundwater use at regional level and studies the co-evolution of groundwater resources levels with the development of groundwater economies. Shah's main focus is on groundwater economies in South Asia, where a variety of trajectories can be identified – the key element explaining these various trajectories is the aquifer types (hard rock aquifers, versus alluvial aquifers for instance).

According to Shah, the informal use of groundwater resources by thousands of individual farmers makes it difficult to control and limit groundwater use (through public or through community-level initiatives). Thus, groundwater governance is mainly limited to pragmatic solutions depending on the type of aquifers. For instance, groundwater governance can focus on the improvement in water availability (*i.e.*, groundwater recharge programs in hard rock aquifers) or on indirect demand management measures like *e.g.* energy pumping costs, to act on groundwater use in alluvial aquifers. In this context, T. Shah is very skeptical about the possibilities to adapt the Integrated Water Resources Management (IWRM) toolbox in South Asia, simply because three important pillars of IWRM (water law, policy and administration) are currently lacking.

The third approach stresses the importance of power relationships between the actors dependent on groundwater resources. This approach draws on the works of political ecology to analyse the intensive use of aquifers for irrigation. The diversity of actors, power relationships and constitution of coalitions between actors is analysed. The design of groundwater governance mechanisms is also analysed, in order to understand how these mechanisms are legitimised, implemented and sometimes contested. The inequalities in access to groundwater and the differentiations between farms are then discussed. The analyses performed by the authors belonging to this third approach (Prakash, 2005; Mukherji, 2006; Birkenholtz, 2009) led to a general critique of top-down policies to regulate groundwater access and use, and to developing a plea for the establishment of bottom-up institutions, which could be better able to take into account the lot of the poorest and marginalized farmers.

The fourth approach is looking at groundwater from an ecosystem services and human well-being perspective. Here the emphasis lies on the functions provided by groundwater systems, the associated ecosystem services and the resilience of groundwater systems to continue to provide these services when faced with *e.g.* intense use or even groundwater mining (Knüppe and Pahl-Wostl, 2011; CGIAR, 2015; Knüppe *et al.*, 2015; Custodio *et al.*, 2016).

Table 3.2 Four approaches to groundwater governance.

Approach and Key authors	*Main features*
CPR/collective action (Ostrom; Bloomington School)	Groundwater as a Common pool resource (subtractability and non-exclusion); self-governing rules to avoid social dilemmas;
Evolution of (Informal) groundwater economies (Shah)	Co-evolution of governance modes and aquifer characteristics; role of informal institutions vis-à-vis formal institutions
The Political ecology of groundwater (Mukherji)	Importance of political ecology- power relationships as determinant factors, equity/inequity of access and use
An Ecosystem services and resilience approach to ground-water systems (Knüppe)	Emphasis on ecosystem functions and services, system approach to groundwater systems resilience, human well-being and groundwater dependent ecosystems

Interestingly, even if the epistemological foundations of the various approaches presented here are rather different, Faysse and Petit (2012: 113) argue, concerning the first three approaches, that *studying the resilience of a groundwater territory, defined as a social ecological system, and assessing the adaptive nature of the governance processes implemented, is one of the issues that would probably benefit from a cross-reading of the authors studied here*. Frameworks to analyze groundwater governance can be enriched integrating different Social Ecological Systems related approaches. Table 3.2 summarises the main features of these four approaches.

3.3 GROUNDWATER GOVERNANCE THROUGH A SOCIAL ECOLOGICAL LENS

Even if the four theoretical approaches presented in the previous section have methodological frameworks globally compatible with Social Ecological Systems (SES), the mostly used – though not perfect – frame of analysis can be found in the adaptation of the Institutional Analysis and Development (IAD) framework developed since the mid-1980's by the Bloomington school (Kiser and Ostrom, 1982; Oakerson, 1992), which has experienced interesting complements when E. Ostrom started to work with several leading figures of the resilience alliance[3] (McGinnis and Ostrom, 2014).

In this perspective, SES can be understood as complex adaptive systems, in which the components, and the structure of interactions between them, adapt over time to internal and external disturbances (Anderies *et al.*, 2004). When analysing SES, certain key components can be identified (see Figure 3.2): actors related to the resource, the governance systems, and resource systems and units. This approach acknowledges the role of the social, economic and political settings, as well as the related ecosystems attributes. All these variables influence the "action situation", where actors interact with each other and jointly influence outcomes that are differentially valued by those actors (McGinnis and Ostrom, 2014).

[3]https://www.resalliance.org/

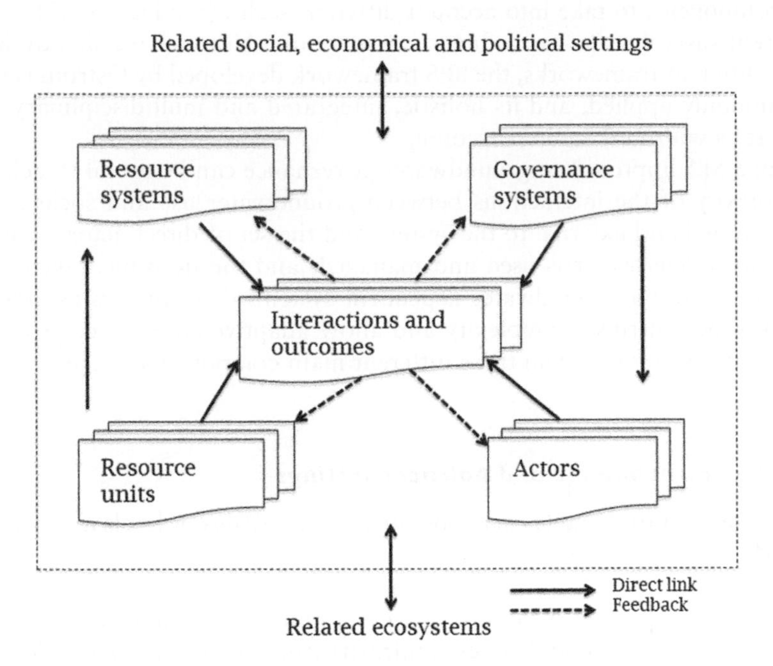

Figure 3.2 SES framework, with multiple components and interactions at different scales. Source: McGinnis and Ostrom (2014).

Though SES approaches are not a panacea for groundwater governance, some authors have claimed that the SES conceptual paradigm can provide a powerful analytical framework for the governance of natural resources (Ostrom, 2007). According to McGinnis and Ostrom (2014), "*A framework provides the basic vocabulary of concepts and terms that may be used to construct the kinds of causal explanations expected of a theory. Frameworks organize diagnostic, descriptive, and prescriptive inquiry. A theory posits specific causal relationships among core variables. In contrast, a model constitutes a more detailed manifestation of a general theoretical explanation in terms of the functional relationships among independent and dependent variables important in a particular setting. Just as different models can be used to represent different aspects of a given theory, different theoretical explanations can be built upon the foundation of a common conceptual framework*". The SES framework has helped facilitate: (1) increased recognition of the dependence of humans on ecosystems; (2) improved collaboration across disciplines, and between science and society; (3) increased methodological pluralism leading to improved systems understanding; and (4) major policy frameworks that now incorporate social-ecological interactions (Fischer *et al.*, 2015).

3.3.1 Main components of the SES framework: settings, actors, governance systems and resource systems and units

The SES framework promoted by the Bloomington school provides a multi-tiered set of variables, with the option to play and combine different subcomponents of each system component, to take into account different scales (See Figure 3.2). Thus there are different variables that can influence an outcome. Although we already stated that there are different frameworks, the SES framework developed by Ostrom is one of the most commonly applied, and its holistic, integrated and multidisciplinary character offers lessons worth taking into account.

Using a SES approach to groundwater governance can be useful to acknowledge the complexity of the interactions between groundwater use and society. Different factors, internal and external to the system and the set of direct actors, are influencing the way groundwater is used and managed, and the decisions taken at various scales. We can frame groundwater dependent systems or territories as SES to better understand their intrinsic complexity and allow adaptive governance processes. The following sub-sections explain these different main components of the SES analytical framework.

3.3.1.1 Social, economic and political settings

This important set of variables may operate at larger scales or levels and involve other actors outside the 'groundwater territory'. Variables include issues related to economic development and economic sectors, demographic dynamics, political trends and stability, governance and governmental frameworks, policies and compliance, such as land use and agriculture trends and policies, infrastructure and technological development, market influence, media interest on social or environmental issues. These variables may apply at the local, regional, national or even at international levels.

3.3.1.2 Actors

Resource users must decide whether it is worthwhile engaging in a collective process to address the problems they are confronted with, given the associated transaction costs (López-Gunn and Martínez-Cortina, 2006) and when incentives of engagement can take a long time. The same users can decide to self-organize in order to share the resource, taking the initiative of the collective action and collectively crafting rules concerning resource use. On the other hand, authorities like *e.g.* central government or development agencies can incentivize the creation of user associations, like the Spanish case where certain Groundwater User Associations were created top down as a measure to avoid intensive groundwater use. However, evidence on the slow and rare emergence of Groundwater User Associations in Spain highlights that these "top-down" measures did not have the expected success, and conflictive issues such as water rights regulation were not solved (Rica *et al.*, 2012). Hence, top-down approaches are not guaranteed to succeed in the long run, as observed in the case of Andhra Pradesh Community Groundwater Management (AP CGM) initiated by World Bank, FAO and other partners. While AP CGM was successful in mobilizing the local community and creating institutional mechanisms to govern groundwater resources in a democratic manner, a recent visit by one of the authors found that external drivers such as drought and increasing water demand had resulted in relatively less active groundwater management communities than anticipated.

Traditionally, in the IAD framework, the main actors of the SES are the resource users (McGinnis and Ostrom, 2014) – for instance farmers irrigating their land with

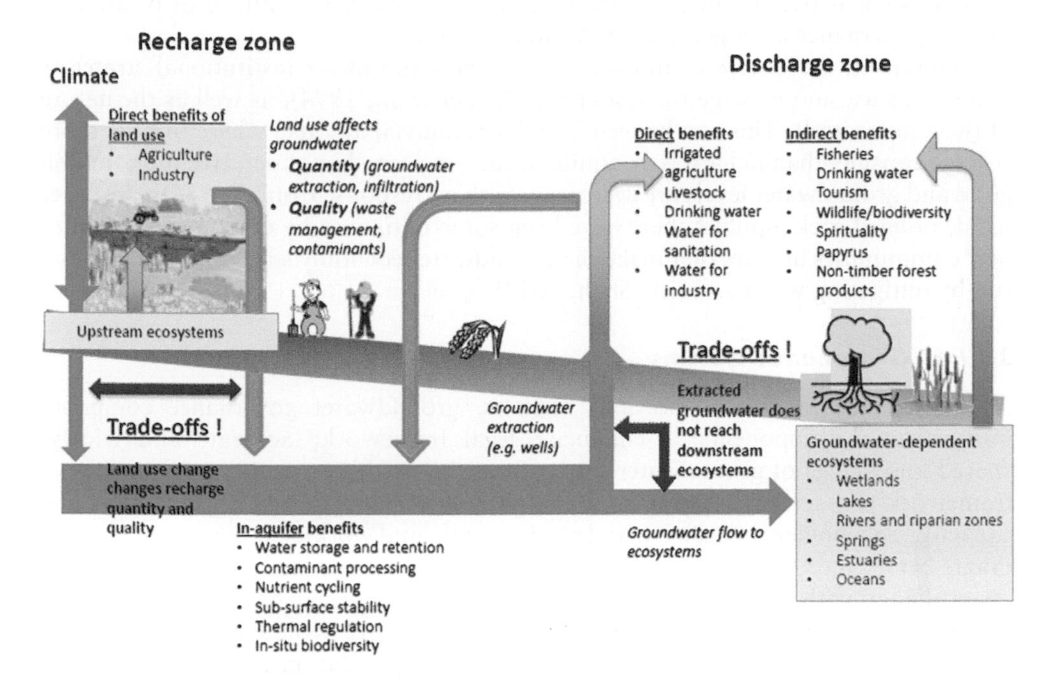

Figure 3.3 Groundwater flows, benefits and tradeoffs (Source: CGIAR, 2015).

groundwater. However, other actors (State representatives from various ministries, local level administrative staff, environmental NGO's, etc.) have to be taken into account. Attention should be paid to the socioeconomic attributes of these actors, their social capital, gender, resource dependency, past experience, leadership patterns and access to technology.

3.3.1.3 Resource systems: their units and related ecosystems

It is common to find these two components of the framework separated. However, due to the need to highlight ecological connectivity, flows and ecological health, we are considering both as a whole. These components comprise variables regarding the biophysical nature of the resource providing ecosystem services, and the dynamics of this process (Figure 3.2). Such variables would include the clarity of the system boundaries, the size of the resource system, the type of human-constructed facilities, system productivity, predictability of system dynamics, storage, type of replacement rate, economic value, spatial and temporal distribution, etc. In the case of groundwater, it is important to track information regarding recharge and discharge flow rates so that impact of an action could be predicted both in space and time.

Hydrogeology and aquifer profiles can determine the governance settings, as observed in the cases of alluvial and hard rock aquifers in India (Shah, 2009). Groundwater-surface water interaction, recharge rates, vulnerability to pollution, groundwater dependent ecosystems, will generate a different governance response for each specific context (Figure 3.3). Aquifers can be embedded in larger social ecological systems, such as river basins, and processes at others scales may affect -or be affected by- the governance arrangements at the aquifer level.

Moreover, the degree of mobility and storage can affect institutional strategies adopted to use and manage the resource (Schlager *et al.*, 1994), as well as the nature of the aquifer itself. This can be seen in India in alluvial aquifers where there is more storage capacity than in hard rock aquifers, and yet users do not perceive the common good and groundwater levels are decreasing with no response from users. On the other hand, in hard rock aquifers there have been some initiatives by certain groundwater user communities in ways that make the groundwater economy sustainable in the long run by mitigating water scarcity (Shah, 2012).

3.3.1.4 Governance systems

As already stressed in the previous sections, groundwater governance comprises four essential components: a conducive legal framework; accurate and widely-shared knowledge of groundwater systems together with awareness; an institutional framework characterized by representation and leadership, sound organizations and capacity, stakeholder engagement and participation, and working mechanisms to coordinate between groundwater and other sectors; and policies, incentive structures and plans aligned with society's goals.

An approach to governance from a SES perspective would need to, at least, determine a) governmental and non-governmental organizations, b) actors' network structure and information sharing, c) property rights systems and bundle of rights, d) different set of rules: operational-choice, collective-choice, constitutional-choice and e) monitoring and sanctioning rules. We will see later how this governance approach can be enriched.

Box 3.1: Polycentric governance and groundwater resources

Groundwater governance can, to a certain extent, be considered polycentric. A political order is polycentric, according to McGinnis (1999:2) "*when there exist many overlapping arenas (or centers) of authority and responsibility. These arenas exist at all scales, from local community groups to national governments to the informal arrangements at the global level*". In the context of groundwater resources, there are different actors at different levels taking decisions on the use of the resource, and where applicable, there are also water authorities with stakes on the resource regulation. Polycentric governance systems for groundwater resources already exist and have been studied in the US in States like California (Blomquist, 1992), Texas and Arizona (Sugg, 2016). In a multi-level governance system, local levels may benefit on the one hand from financial or technical resources only available at supra-local levels, and on the other hand from their own capacity to access and manage the common resource with their local knowledge. In addition, these systems may be more efficient to solve problems related to non-cooperators or local inequities. In this kind of systems it is important that "bridging organizations" mediate between different actors such as users and water authorities.

Thus, to summarise, when using a SES approach, it is central to look at the dependencies and relations between social and biophysical elements. Interactions among system subcomponents will determine what is called "action situations" in the SES framework. Proposed interactions and outcomes, not limited to the SES framework, are the harvesting of water and other metabolic relations, information sharing through different methods by actors, conflicts among actors, investment on the resource, networking and lobbying capacity, rulemaking at different levels, monitoring and control activities, co-management performance, processes of evaluation of the resource situation and of the effects of management initiatives, evolution of access to the resource by different users, but also power dynamics and the effects and responses to market mechanisms.

All these elements are important factors which need to be taken into account when looking at groundwater governance through a SES lens. However, if taken separately, they do not necessarily inform us as much as system dynamics. This is the reason why several key criteria must also be identified to shape groundwater governance through a SES lens.

3.3.2 Selecting criteria to shape groundwater governance outcomes through SES lens

The purpose of using a SES approach is not merely to picture and frame the interactions between groundwater and society, but rather to assess and stimulate the adaptive capacity of the governance process. The diagnosis approach must not be a panacea, but rather based on societal and environmental objectives tailor made for different governance contextual settings, taking into account possible trade-offs between levels, and including linkages out of the water box.

For any diagnosis we need a criterion, to use as a guide for the evaluation of the outcomes of a process. Following the SES framework ontology, this subset of criteria may coincide with (good) groundwater governance principles or indicators (Lautze *et al.*, 2011; Groundwater governance project, 2016b). Delgado-Serrano and Ramos (2015) defined certain outcomes for SES governance: efficiency, socio-economic sustainability, equity, accountability, effects of deliberation processes, empowerment and adaptation strategies. When evaluating the governance of SES, the main desired outcome is resilience, as well as allowing adaptive capacity for the system to be resilient. However, despite the integrative view of the SES approach, and in particular when looking at resilience, it is difficult to the social and the ecosystem or resource subunits to be seen as one undifferentiated system. It may also be important to consider resilience as a neutral outcome, as there can be systems in undesired states, and resilience of these systems would only perpetuate a negative impact (Petit *et al.*, 2017).

The resilience of the social subunit of the system may not go hand in hand with the resilience of the ecosystem subunit, or the system as a whole. It may in fact be contradictory with the definition of SES, but it seems that at certain scales it happens that the equilibrium among subunits is not stable (Rica *et al.*, 2014). There may be a moment when the ecological subunit cannot stand the disturbance but the social subunit draws upon external resources in order to keep resilient. In other words, the "social subunit" breaks the dependency interaction with the groundwater resource, and it would mean a sequential destruction of natural resources (Anderies *et al.*, 2004).

This is another way to explain socioeconomic development or growth at the expenses of groundwater over-abstraction and resource system transformation (López-Gunn *et al.*, 2012).

Box 3.2: Resilience, adaptability, transformability – what are we talking about?

First introduced in the field of ecology by Holling in 1973, nowadays resilience stands as a central interdisciplinary concept which is being used in several research fields besides ecology such as psychology, economics, or sociology. It summarizes the magnitude of disturbance that can be tolerated before a system moves into a different state and a different set of controls (Holling and Meffe, 1996). It has also been called robustness with a similar meaning, "A SES is robust if it prevents the ecological systems upon which it relies from moving into a new domain of attraction that cannot support a human population, or that will induce a transition that causes long-term human suffering" (Anderies *et al.*, 2004). Carpenter *et al.* (2001) highlight three properties for resilience, namely a) the amount of change the system can undergo; b) the degree to which the system is capable of self-organization; c) and the degree to which the system is able to learn and adapt.

"Adaptability" refers to the ability of an SES of learning, combining experience and knowledge to adjust to changing factors, and further develop within a domain of stability (Berkes *et al.*, 2003). Some authors have defined adaptability as "the ability of the actors in the system to influence resilience" (Walker *et al.*, 2006), concept linked to robustness. The "transformability" on the other hand, refers to the ability of the system to transform their internal or external components to create another social ecological system, where ecological, economic or social structures make the existing system non-viable (Folke *et al.*, 2002).

3.4 MAIN CHALLENGES OF A SOCIAL ECOLOGICAL SYSTEMS FRAMEWORK FOR GROUNDWATER GOVERNANCE

The SES framework adapted to groundwater governance comes up against a number of challenges which need to be taken into account in order to better understand the many different aspects of groundwater governance. These challenges and limitations can be complemented by mobilizing the different approaches of groundwater governance presented in the second section.

First, groundwater governance is only rarely a matter of local institutions alone. Various scales and institutional levels are mobilized to understand the relationships between actors, sectors and issues. For instance, focusing only on groundwater resources users can be useful to understand the collective action mechanisms implemented by the end users to share groundwater resources according to the rules they have themselves implemented. However, groundwater access also depends generally on property rights and economic incentives established at the national level and the (often ambiguous) role of the State must be taken into account to study groundwater governance mechanisms (Wester *et al.*, 2011; Fofack *et al.*, 2015; Molle and Closas, 2017). Thus, authorities at different levels can be active participants in collective action processes. Then, we could be talking of co-management, when formal -or informal- responsibilities are shared among different actors (Rica *et al.*, 2012; Molle and Closas 2017).

Groundwater governance is often a question of multi-level governance analysis, which refers to the dispersion of authority away from central government. Decisions are not made at a single level, either only at the top (highest level of government enforcing decisions), or the intermediate (state or provinces enforcing decisions beneficial for their regions), or the individual level (Pahl-Wostl, 2009). The coordination between the various stakeholders in charge of governing groundwater resources is however often difficult. This idea is known as the "problem of interplay" (Young, 2002). According to Theesfeld (2010: 138): *When different authorities need to work together, ambiguity often exists in the definition of their respective central and local responsibilities. Often the central level basically tries to retain control over local decisions while simultaneously reducing expenditures for regional development.*

Second, issues of group heterogeneity may increase governance complexity and therefore need to be well addressed from a social-ecological system perspective. Particularly in larger aquifers, actor's heterogeneity can be high. Eight design principles were identified by Ostrom (1990) as key conditions that lead to optimal resource governance: define clear group boundaries, match rules governing use of common goods to local needs and conditions, ensure that those affected by the rules can participate in modifying the rules, make sure the rule-making rights of community members are respected by outside authorities, develop a system, carried out by community members, for monitoring members' behavior, use graduated sanctions for rule violators, provide accessible, low-cost means for dispute resolution, build responsibility for governing the common resource in nested tiers from the lowest level up to the entire interconnected system. These principles have been tested and replicated in different studies. However, it is unclear whether these would apply to larger-scale environmental governance dilemmas (Fleishmann *et al.*, 2014). These design principles can be useful when analysing or strengthening groundwater governance. However, it is also important to keep in mind that other factors such as system size, or group heterogeneity can increase governance complexity, and in fact can be more determinant than these design principles (Rica *et al.*, 2014). Recent efforts are being made in order to integrate group heterogeneity on the study of larger social ecological systems (Cox, 2014).

Third, power relationships between actors are not properly addressed in most SESs analysis (Fabinyi *et al.*, 2014). It has been shown that power dynamics and the inherent politics of groundwater governance often determine how water is actually accessed and how access to decision making or appropriation of groundwater use is not frequently equal among different actors or social groups (Rica *et al.*, 2014; Kulkarni *et al.*, 2015; Pells, 2015). We need to develop approaches on governance and sustainability that also incorporate power relations in decision making, as considered by political ecology (Mukherji and Shah, 2005; Birkenholtz, 2009).

Fourth, one key issue that is not always addressed in groundwater governance studies is equity (Hoogesteger and Wester, 2015). Access to groundwater and water rights distribution is often conflictive, and less powerful groups or the environment tend to be disadvantaged. Perreault (2014) suggests that equity in water governance must be determined analyzing critically the institutional arrangements of the market, the State and civil society through which water is allocated and accessed. This is where the concepts of social and environmental justice arise, allowing to define a framework to analyse groundwater (in)justice (Hoogesteger and Wester, 2015; Ameur *et al.*, 2017). Some argue that the SES framework does not tackle these questions properly, and it may be better to complement the framework with a political ecology analysis. Political ecology helps to detect problems related to inequities of access embedded at multiple scales and problems related to the exercise of political and economic power (Swyngedouw, 2009). Neal *et al.* (2016) argue why environmental and social justice should be integrated into groundwater governance, including local communities and the environment in the decision-making and allocation process in order to avoid or ameliorate potential social and/or environmental injustices. They provide examples from Northern Australia and Saudi Arabia, highlighting the gap between the meaning of justice and equity for individuals and 'equitable use' in international water law, and the trade-offs between water rights and environmental justice – water rights for some may in effect deny basic human rights to water for others (Mirosa and Harris, 2012). These situations are replicated all around the world, and trigger the discussion on what uses should get priority under different circumstances, and what rationale, from a socio-economic to justice basis, lies behind the decision.

As Fabinyi *et al.* (2015) stress, the contributions of resilience science to societal challenges such as poverty, security or inequity, with intrinsic environmental dimensions, would gain from amplifying the focus towards conflict, contestation, micropower and macrosystems dynamics. Thus, shifting to *political and ethical questions as crucial drivers of social-ecological outcomes rather than 'inconvenient' politics that can be simply sorted out through institutional design* (Côte and Nightingale, 2011:484). This approach would contribute to match the analysis with the criteria of social and environmental justice, and to align governance in this regard.

3.5 CONCLUSION: GROUNDWATER GOVERNANCE FOUNDATIONS THROUGH A SES LENS

This chapter has focused on what a SES lens can bring to the understanding of groundwater governance, adopting a perspective of coupled human and natural systems. It has briefly outlined similar/complementary approaches, as well as the challenges these approaches need to integrate or overcome to frame groundwater governance more effectively.

Under a SES approach a key contribution is that it starts from the acknowledgement that groundwater governance is a complex issue, and thus adopting a complex system approach (like SES) could help to better understand groundwater governance, or complement other approaches. Frameworks that build on social ecological concepts and theories help to structure the complex interactions and feedbacks taking place between different human and biophysical elements, and at different spatial and temporal scales

The chapter though has adopted a critical approach to briefly outline the main lessons learned from previous works dealing with SES theory and groundwater governance, but also to identify the shortcomings of the commonly applied SES approach and frameworks. Thus the chapter has looked at a SES framework to look at groundwater governance from a critical yet constructive perspective that builds on the shortcomings and advantages of the SES framework based on the literature and its evolution and informed by real examples and experience from groundwater in particular.

We reviewed four different approaches, i) the study of groundwater governance as collective action ii) the analysis made by T. Shah (2009) on informal groundwater economies and iii) studies that draw from political ecology, and iv) an ecosystem services and resilience approach. We argue, that far from being alternative approaches, these can complement each other to help us analyze groundwater governance through a Social Ecological lens.

The SES framework became mainstream after the seminal work by Ostrom in the nineties on the role of collective action, followed by many other authors and the Bloomington school. This has enriched our knowledge on the weaknesses and inconsistencies (*i.e.* what we have learnt since the 1990s) which has helped make the SES analytical framework more grounded in reality and thus more usable.

Indeed collective action is not the same as governance. However it is a key element in the process to establish governance rules and structures, particularly for groundwater even if we now know it is always subject to a scale. Through a SES lens we see the key importance of self-organising systems as an emergent property- as a way to understand collective action. In Ostrom's works, collective action is mainly synonymous with self-organising systems. However, economic incentives and command and control mechanisms are also fundamental to understand the dynamics of groundwater governance. Thus, the State (at different levels) can be an active participant in co-management processes. Using frameworks that operationalize the analysis based on SES theories help us tackle issues of scale, which we argue is key to understand groundwater governance processes.

There is an emerging line of argument that defend that political ecology and resilience need to be integrated in the social ecological approach towards groundwater governance. A definition of the criteria used to analyse groundwater governance should be done carefully. This would help to avoid neglecting issues of power imbalances and politics, often inherent to realities of groundwater dependent societies, if the criteria of social and environmental equity are to be met.

REFERENCES

Ameur, F., Kuper, M., Lejars, C. and Dugué, P. (2017). Prosper, survive or exit: Contrasted fortunes of farmers in the groundwater economy in the Saiss plain (Morocco). *Agricultural Water Management* 191: 207–217, DOI: 10.1016/j.agwat.2017.06.014.

Anderies, J.M., Janssen, M.A. and Ostrom, E. (2004). A framework to analyze the robustness of social-ecological systems from an institutional perspective. *Ecology and Society* 9(1): 18. [online] URL: http://www.ecologyandsociety.org/vol9/iss1/art18/

Barreteau O. *et al.* (2016). Disentangling the Complexity of Groundwater Dependent Social-ecological Systems. In: Jakeman, T., Barreteau, O., Hunt, R., Rinaudo, J.D., Ross, A. (Eds.), *Integrated Groundwater Management*, Springer, Dordrecht, The Netherlands: 49–73.

Berkes, F., Colding, J. and Folke, C. (2003). *Navigating social-ecological systems: Building resilience for complexity and change.* Cambridge University Press, Cambridge, UK.

Birkenholtz, T. (2009). Groundwater governmentality: hegemony and technologies of resistance in Rajasthan's (India) groundwater governance. *The Geographical Journal* 175(3): 208–220.

Blomquist, W. (1992). *Dividing the Waters. Governing Groundwater in Southern California*, San Francisco, California: ICS Press.

Carpenter, S., Walker, B., Anderies, J. M., and Abel, N. (2014). From Metaphor to Measurement: Resilience of What to What?. *Ecosystems*, 4(8): 765–781.

CGIAR Research Program on Water, Land and Ecosystems (WLE). (2015). Groundwater and ecosystem services: a framework for managing smallholder groundwater-dependent agrarian socio-ecologies – applying an ecosystem services and resilience approach. Colombo, Sri Lanka: International Water Management Institute (IWMI). CGIAR Research Program on Water, Land and Ecosystems (WLE). 25 p. doi: 10.5337/2015.208

Côte and Nightingale, (2011), Resilience thinking meets social theory: Situating social change in socio-ecological systems (SES) research. *Progress in Human Geography* 36: 475–489.

Cox, M., (2014). Understanding large social-ecological systems: introducing the SESMAD project. *International Journal of the Commons*. 8(2): 265–276.

Custodio, E., Andreu-Rodes, J.M., Aragón, R., Estrela, T., Ferrer, J., García-Aróstegui, J.L., Manzano, M., Rodríguez-Hernández, L. Sahuquillo, A. and del Villar, A. (2016). Groundwater intensive use and mining in south-eastern peninsular Spain: Hydrogeological, economic and social aspects. *Science of the Total Environment* 559: 302–316.

Delgado-Serrano, M. del M., and P. Ramos. (2015) Making Ostrom's framework applicable to characterise social ecological systems at the local level. *International Journal of the Commons* 9:808–830.

De Stefano, L. & López-Gunn, E. (2012) Unauthorized groundwater use: Institutional, social and ethical considerations. *Water Policy*, 14, 147–160

Fabinyi, M., L. Evans, and S. J. Foale. (2014). Social-ecological systems, social diversity, and power: insights from anthropology and political ecology. *Ecology and Society* **19**(4): 28.

Faysse N. and O. Petit, (2012). Convergent Readings of Groundwater Governance? Engaging Exchanges between Different Research Perspectives, *Irrigation and Drainage*, Vol. 61, Suppl. S 1, pp. 106–114.

Fischer, J., T. A. Gardner, E. M. Bennett, P. Balvanera, R. Biggs, S. Carpenter, T. Daw, C. Folke, R. Hill, T. P. Hughes, T. Luthe, M. Maass, M. Meacham, A. V. Norström, G. Peterson, C. Queiroz, R. Seppelt, M. Spierenburg, J. Tenhunen. (2015). Advancing sustainability through mainstreaming a socialecological systems perspective. *Current Opinion in Environmental Sustainability* 14:144–149.

Fleischman, F.D. *et al.* (2014). Governing large-scale social-ecological systems: Lessons from five cases. *International Journal of the Commons*. 8(2).

Fofack, R., Kuper, M., Petit, O. (2015). Hybridation des règles d'accès à l'eau souterraine dans le Saiss (Maroc): entre anarchie et Léviathan? *Études rurales*, n°196, 2015/2: 127–149.

Folke, C., Carpenter, S., Elmqvist, T., Gunderson, L., Holling, C.S, and Walker, B. (2002). Resilience and sustainable development: building adaptive capacity in a world of transformations. *Ambio*, 31, p. 437–440.

Groundwater Governance Project. (2016a). *Global Diagnosis on Groundwater Governance*. GEF, World Bank, UNESCO-IHP, FAO and IAH http://www.groundwatergovernance.org/

Groundwater Governance Project. (2016b). *Global framework for action to achieve the vision on groundwater governance*. Groundwater governance – A global framework for action, GEF, UNESCO-IHP, FAO, World Bank and IAH. http://www.groundwatergovernance.org

Holling, C.S. and Meffe, G. (1996). Command and control and the pathology of natural resource management, *Conservation Biology* 10(2), p. 328–337.

Hoogesteger, J., Wester, P. (2015). Intensive groundwater use and (in) equity: Processes and governance challenges. *Environmental Science & Policy* 51: 117–124.

Kiser, L. L., and Ostrom E. (1982). The three worlds of action: a metatheoretical synthesis of institutional approaches. *in* E. Ostrom, editor. *Strategies of political inquiry.* Sage, Beverly Hills, California, USA, pp. 179–222.

Knüppe, K., Pahl-Wostl C., (2011). A framework for the analysis of governance structures applying to groundwater resources and the requirements for the sustainable management of associated ecosystem services. *Water Resour Manag* 25(13):3387–3411. doi:10.1007/s11269-011-9861-7

Knüppe, K., Pahl-Wostl, C., Vinke-de Kruijf, J. (2015). Sustainable Groundwater Management: A Comparative Study of Local Policy Changes and Ecosystem Services in South Africa and Germany, *Environmental Policy and Governance.* 26(1).

Kulkarni, H., Shah, M., P. S. Vijay Shankar (2015). Shaping the contours of groundwater governance in India, *Journal of Hydrology: Regional Studies,* 4 172–192.

Lautze, J., de Silva, S., Giordano, M. and Sanford, l. (2011). Putting the cart before the horse: Water governance and IWRM. *Natural Resources Forum* 35: 1–8.

Levin, S., T. Xepapadeas, A-S. Crépin, J. Norberg, A. De Zeeuw, C. Folke, T. Hughes, K. Arrow, S. Barret, G. Daily, P. Ehrlich, N. Kautsky, K-G. Mäler, S. Polasky, M. Troell, J. Vincent and B. Walker (2013). Social-ecological systems as complex adaptive systems: modelling and policy implications. *Environment and Development Economics,* 18: 111–132.

López-Gunn, E. (2003). The role of collective action in water governance: a comparative study of groundwater user associations in La Mancha aquifers (Spain) commissioned for Special Issue Water Management in the Iberian Peninsula. *Water International* 28(3) September: 367–378.

López-Gunn, E. and Martinez-Cortina, L. (2006) Is self-regulation a myth? Case study on Spanish groundwater associations and the role of higher level authorities. *Hydrogeology Journal* 14 (3): March, 361–375.

López-Gunn, E., Rica, M., van Cauwenbergh, N. (2012) Taming the groundwater chaos. De Stefano and Llamas (eds). In *Water, Agriculture and the Environment: can we square the circle?* Taylor and Francis.

McGinnis, M. D. (1999). *Polycentric governance and development: readings from the Workshop in Political Theory and Policy Analysis.* Univ of Michigan Press

McGinnis, M. D., and Ostrom, E. (2014). Social-ecological system framework: initial changes and continuing challenges. *Ecology and Society* 19(2): 30. http://dx.doi.org/10.5751/ES-06387-190230

Mirosa, O., and Harris, L. M. (2012). Human right to water: Contemporary challenges and contours of a global debate. *Antipode,* 44, 932–949.

Molle, F., and Closas, A. (2017). Groundwater governance: a synthesis. IWMI project publication – "Groundwater governance in the Arab World – Taking stock and addressing the challenges", Report n°6, IWMI, USAID.

Mukherji, A. and Shah, T. (2005) Groundwater socio-ecology and governance: a review of institutions and policies in selected countries. *Hydrogeology Journal* 13(1): 328–345.

Mukherji, A. (2006). Political ecology of groundwater: the contrasting case of water-abundant West Bengal and water-scarce Gujarat, India. *Hydrogeology Journal* 14: 392–406.

Neal, MJ., Greco, F., Connell, D., Conrad, J. (2016). The social–environmental justice of groundwater governance. in Jakeman, T, Barreteau, O, Hunt R, Rinaudo, JD, Ross, A, (eds), *Integrated Groundwater Management*, Springer, Heidelberg,

Oakerson, R.J. (1992). Analysing the Commons: A Framework, in Bromley D.W. (Ed), *Making the Commons Work*, San Francisco, ICS Press: 41–59.

Ostrom, E. (1990). *Governing the Commons. The evolution of institutions for collective action.* Cambridge, Cambridge University Press.

Ostrom, E. (2005). *Understanding Institutional Diversity.* Princeton University Press.

Ostrom, E. (2007). A diagnostic approach for going beyond panaceas *PNAS* September 25, vol. 104, no. 39, 15181–15187

Pahl-Wostl, C. (2009). A conceptual framework for analyzing adaptive capacity and multi-level learning processes in resource governance regimes. *Global Environmental Change* 19(3): 354–365. DOI: 10.1016/j.gloenvcha.2009.06.001

Pells, C., (2015). Power and the Distribution of Knowledge in a Local Groundwater Association in Guadalupe Valley. In: *Collaborative Governance Regimes*, Publisher: Georgetown Press, Editors: Kirk Emerson and Tina Nabatchi.

Perrault, T. 2014. What kind of governance for what kind of equity? Towards a theorization of justice in water governance. *Water International* Vol. 39 , Iss. 2.

Petit, O., Kuper, M., López-Gunn, E., Rinaudo, J.D., Daoudi, A. and Lejars, C. (2017). Can agricultural groundwater economies collapse? An inquiry into the pathways of four groundwater economies under threat. *Hydrogeology Journal.* DOI 10.1007/s10040-017-1567-3. pp. 1–16.

Prakash, A. (2005). *The dark zone: groundwater irrigation and water scarcity in Gujarat.* PhD thesis submitted to Wageningen University, Wageningen, The Netherlands.

Rica, M., López-Gunn, E. and Llamas, R. (2012). Analysis of the emergence and evolution of collective action: an empirical case of Spanish groundwater user associations. *Irrigation & Drainage*, 61 (Supplement S-1), p. 115–125.

Rica, M., Dumont, A., Villarroya, F., López-Gunn, E., (2014). Whither collective action? Upscalling collective action, politics and groundwater management in the process of regulating an informal water economy. *Water International* 39(4), p. 520–533.

Ross, A. (2016). Groundwater governance in Australia, the European Union and Western USA. In: Jakeman, T., Barreteau, O., Hunt, R., Rinaudo, J.D., Ross, A. (Eds.), *Integrated Groundwater Management*, Springer, Dordrecht, The Netherlands: 145–171.

Shah, T. (1993). Groundwater markets and irrigation development: Political economy and practical policy. Bombay: Oxford University Press.

Shah, T. (2009). *Taming the Anarchy. Groundwater Governance in South Asia.* Resources for the Future Press: Washington, DC.

Shah, T. (2012). Community response to aquifer development: distinct patterns in India's alluvial and hard rock aquifer areas. *Irrigation and Drainage*, 61: 14–25.

Schlager, E., Blomquist, W., and Tang, S. Y. (1994). Mobile Flows, Storage, and Self-Organized Institutions for Governing Common-Pool Resources. *Land Economics*, 70(3): 294–317.

Sugg, Z. (2016). Governing the Unseen: A Comparative Analysis of Arizona and Texas Groundwater Institutions. PhD Diss. University of Arizona.

Swyngedouw, E. (2009). The Political Economy and Political Ecology of the Hydro-Social Cycle. *Journal of Contemporary Water Research & Education*, 142: 56–60.

Theesfeld, I. (2010). Institutional Challenges for National Groundwater Governance: Policies and Issues. *Ground Water* 48(1): 131–142.

Varady R.G., Zuniga-Teran A.A., Gerlak A.K., Megdal S.B. (2016). Modes and Approaches of Groundwater Governance: A Survey of Lessons Learned from Selected Cases across the Globe. Water 8, 417; doi:103390/w8100417

Walker, B. H., Anderies, J.M.; Kinzig, A.P. and Ryan, P. (2006). Exploring resilience in social-ecological systems through comparative studies and theory development: introduction to the special issue. *Ecology and Society* **11**(1): 12.

Wester P., Sandoval Minero R., Hoogesteger J. (2011). Assessment of the development of aquifer management councils (COTAS) for sustainable groundwater management in Guanajuato, Mexico. *Hydrogeology Journal* 19: 889–899.

Young O.R. (2002). *The Institutional Dimensions of Environmental Change: Fit, Interplay, and Scale.* Cambridge, Massachusetts: MIT Press.

Chapter 4

Groundwater management: policy principles & planning practices

Stephen Foster[1,2] *& John Chilton*[3]
[1] *Past President, International Association of Hydrogeologists, UK*
[2] *Visiting Professor of Groundwater Management, University College London, UK*
[3] *Past Executive Manager, International Association of Hydrogeologists, UK*

ABSTRACT

Substantial changes in approach to groundwater governance are widely required (a) to respond to the growing challenges of resource depletion and quality degradation resulting from radical changes in land-use and water-demand associated with global population growth and (b) to incorporate fully consideration of the sustainability of groundwater-dependent ecosystems. Integrated groundwater policy formulation and management planning are absolutely critical facets of sound governance to manage the 'required change process'. They need to be founded on a clear understanding of the key linkages between groundwater systems and surface-water, land-use and other sectors, which are thus discussed in some detail. It is simultaneously recognised that community mobilization and stakeholder organization around a 'shared vision' of resource sustainability are essential prerequisites to formulate and implement groundwater management plans, but detailed discussion of the social dimensions is beyond the present scope.

4.1 GROUNDWATER MANAGEMENT IN A CHANGING WORLD

4.1.1 The evolving paradigm for sound governance

Effective groundwater resource management and protection, and the improved governance arrangements that facilitate them, have become a pressing need worldwide (FAO, 2016a). The term 'governance' when applied to groundwater is generally taken to encompass the promotion of responsible collective action by society to ensure resource sustainability. For each defined resource unit this should include establishing the necessary institutional and participatory arrangements, agreeing policy positions and their translation into specific goals, providing procedures and finance for implementation, assuring compliance and resolving conflicts, and (most importantly) establishing appropriate monitoring and clear accountability for outcomes (Foster *et al.*, 2009).

In recent decades there has been clearer recognition of:

- the groundwater dependence of many aquatic and terrestrial ecosystems
- the increasing socioeconomic importance of groundwater resources for urban and rural water-supply, irrigated agriculture, and industrial and mining enterprise

- the vulnerability of groundwater resources to pollution by contaminant loads generated from urbanisation, agricultural intensification and the manufacturing and extractive industries
- groundwater flow being easily and meaningfully translated into equations applicable at a realistic field scale.

But groundwater is a classic 'common pool resource' – which can be susceptible to its stakeholders acting solely in short-term self-interest, rather than taking long-term communal requirements into account, because of a misperception that personal interests cannot be assured through collective action. Thus ever-increasing pressures on groundwater, for water-supply provision and from polluting activities, have led to serious degradation of the resource, due in essence to inadequate governance arrangements (FAO, 2016b).

Calls for 'groundwater management interventions' usually arise when a decline in water-table and/or waterwell yield and/or groundwater quality seriously affects one group of stakeholders. Where groundwater systems have been subject to only limited anthropogenic stress, traditionally it was deemed sufficient to safeguard the access rights of registered groundwater users against 'third-party derogation'. This was done by regulatory controls on new waterwells and polluting activities and/or providing a legal mechanism for them to recover damages for such derogation. With greater anthropogenic stress, a need to promote 'managed groundwater development' within socially-agreed scenarios is now occurring more widely.

Groundwater resource management has to deal with balancing the exploitation of a complex resource (in terms of quantity, quality and surface water interactions) with the increasing demands of water and land users, who can pose a threat to resource availability and quality. Thus managing groundwater is as much about managing water and land users (the socio-economic dimension), as it is about scientific understanding of resource behaviour under stress (the hydrogeological dimension). Thus it is essential that governance provisions facilitate blending both dimensions, and that structured and informed community engagement be fostered (Steenbergen, 2006). In some situations local stakeholders may wish to 'take complete control of their local resource' (perhaps because of a breakdown of trust in the government acting as the 'resource guardian') – but even where local hydrogeological conditions favour this, success is unlikely to be achieved without some form of private-public partnership (Lopez-Gunn & Cortina, 2006; Garduño *et al.*, 2010).

4.1.2 Context of chapter

There is a fundamental need to move the 'groundwater management target' from individual waterwells or springheads to entire aquifer systems. This 'paradigm shift' involves applying the concept of 'integrated groundwater management' and introducing governance principles that will facilitate such an approach. The EU has been in the vanguard of the 'integrated system approach' with the principles being debated in the 1990s and enshrined in the Water Framework Directive-2000, and then supplemented by the Groundwater Directive-2006 (EC, 2008; Quevauviller, 2008).

Meanwhile in the USA at federal level the advent of CERCLA and Super Fund for contaminated land clean-up placed the focus much more at site scale than system level, except in a few specific states. At the same time other programmes were pioneering a more integrated and participative approach, such as the World Bank GW-MATe Programme of 2001–11 (Foster *et al.*, 2010; Garduño & Foster, 2011; Tuinhof *et al.*, 2011), IWMI projects in Asia (Mukherji & Shah, 2005; Mukherji *et al.*, 2009; Shah *et al.*, 2012) and IUCN initiatives in the Middle East (IUCN, 2016). Much of this experience served as the basis for the GEF Global Groundwater Governance Framework-for-Action (FAO-UN, 2016a).

4.2 POLICY AND PLANNING AS CRITICAL FACETS OF SOUND GOVERNANCE

4.2.1 Shaping the framework

Strengthening groundwater governance must recognise the presence of a highly 'decentralised resource', which is potentially affected by the actions of a large number of waterwell users and potential polluters, and thus needs to be managed at the most local scale compatible with the hydrogeological setting. There can be no simple blueprint for integrated groundwater management (Foster *et al.*, 2013) but rather a framework of principles for policy and planning that foster subsidiarity in the detail of local application and clear *'vertical policy coordination'* between national, provincial and local level. In reality the local hydrogeological setting and socioeconomic circumstances together frame groundwater resource availability and use, and in turn constrain the measures which are likely to be applicable to manage aquifer degradation risks and to resolve potential conflicts (Foster *et al.*, 2009).

There needs to be clear definition of the collective responsibility for the resource, and who is accountable for the outcome of management measures (FAO, 2016a). In turn, this will require specific management objectives defined within a stated time-frame for the local aquifer system or sub-system (groundwater body or management unit) in question. The management objectives would normally include one of the following (in addition to providing a certain level of water-supply security for legitimate existing users):

- conserving the existing groundwater status in terms of maintaining water-levels and water quality within existing fluctuations
- not allowing groundwater levels or quality to decline below a pre-defined condition established on acceptability criteria
- reversing an existing trend of declining groundwater levels and/or quality to achieve a satisfactory status within a given period.

To achieve such outcomes will, of necessity, require as an essential part of 'good groundwater governance' (FAO, 2016a):

- elaboration of an effective *groundwater management plan* for the local aquifer system in question, with agreed targets, desired outcomes, a programme of measures or interventions, financial support, clear time-frame, adequate monitoring and periodic review – with an appropriate level of integration within the overall

hydrologic cycle through co-management with other components of water and land resources
- establishment of *trans-sector policy linkages*, and incorporating groundwater considerations in policies of related sectors – because the principal drivers of change in pressures exerted upon groundwater systems often arise from social development goals enshrined in the national policies of other sectors (such as agriculture, built infrastructure, industrial production, energy, mining, etc.).

4.2.2 Re-orienting financial incentives

In many countries food production and energy use subsidies make up a large proportion of public expenditure. As part of effective groundwater governance, the incentives for waterwell users provided by such subsidies need to be aligned with the needs of sustainable groundwater management, since otherwise they may exert an overriding negative influence on the behavior of groundwater abstractors (FAO, 2016a).

There are numerous examples of guarantee prices for high water-consumption crops and/or subsidies on electrical energy/diesel fuel (or grants for solar-energy panels) for waterwell pumping which create a perverse incentive to abstract even larger volumes of groundwater regardless of the status of the resource. Individual groundwater users often pay only the cost of waterwell drilling and pump installation/maintenance – and do not meet the full cost of pumping energy or contribute to the cost of groundwater management, and thus have no incentive to conserve groundwater.

Existing public finance of subsidies could be better used to help address the problems of groundwater depletion and salinization, and of ecosystem degradation, and/or assisting the neediest members of the local rural community. The concept of farmer payment for 'environmental services' is now gaining ground (Foster & Cherlet, 2014; IUCN, 2016).

4.2.3 Learning from past experience

In the process of strengthening groundwater management policy and planning it will be essential to evaluate why previous policies and plans have failed, through a sound diagnosis of existing governance provisions, social engagement and (sometimes perverse) stakeholder incentives (FAO, 2016b). The absence of political awareness of the issues themselves or of politically-viable solutions and/or the lack of capacity in the responsible government agency to address them will often be more important than deficiencies in the legal framework for resource management and pollution protection.

In some instances, resistance to groundwater governance reform and management strengthening may be encountered because the current status quo is generating major benefits for the vested interests of some well-established groups (Foster *et al.*, 2009). Thus the path from groundwater policy aspirations to delivery of desired management outcomes is often a tortuous one, with many pitfalls that have to be avoided. In this sense achieving context-appropriate timing and sequencing of management interventions can be critical to success or failure.

Moreover, the perception of groundwater resources at the public and political level are frequently inadequate, and need to improve if groundwater governance provisions, policy formulation and management planning are to be strengthened (FAO, 2016b).

Even some water-resource administrators (like many water users) have limited understanding of groundwater and thus both excessive complacency about the sustainability of intensive groundwater use and irrational under-utilization of groundwater resources still occur. At the core of this problem are certain basic misconceptions about the distinctive characteristics of groundwater systems, which are discussed in the following subsection.

4.3 GROUNDWATER CHARACTERISTICS – THE PHYSICAL FRAMEWORK FOR MANAGEMENT

4.3.1 Large aquifer storage compared to recharge

All aquifers have two fundamental characteristics: a capacity for groundwater storage and a capacity for groundwater flow. The vast storage of many groundwater systems (much larger than that of man-made reservoirs) is their most distinctive characteristic, but can result in the false impression that 'groundwater resources are inexhaustible'. Whilst the storage of most aquifers provides a very effective 'natural buffer' against the high variability of rainfall and surface run-off, their contemporary recharge is finite and limited, and it is this which controls the long-term physical sustainability of groundwater supplies.

Different aquifer systems vary widely in storage properties (Foster *et al.*, 2013) because of:

- major differences of their saturated thickness (between 10–1000 m) and spatial extent (from 10–1,000,000 km^2) according to the type of geologic formation
- major variation of aquifer unit storage capacity (storativity), between unconsolidated granular sediments and highly-consolidated fractured rocks.

As regards groundwater flow capacity, different geologic formations also vary widely:

- unconsolidated deposits, such as sand and gravel, permit substantial flow in their porous matrix which comprises up to 30–35% of their volume
- consolidated porous and/or fractured rocks – such as sandstones, that can have porosities of 20–25% but pores so small as to allow only limited drainage, or limestones with water in micro-fractures which rarely occupy more than 1% of rock volume but can enlarge by dissolution in percolating acidic groundwater in so-called 'karst systems' which can transmit groundwater very rapidly.

4.3.2 The dynamics of groundwater flow

As a result of being an 'invisible resource' the flow of groundwater is also still too widely a source of public misconception – with ideas about 'underground rivers' or 'subterranean lakes' persisting. Groundwater normally moves very slowly through the myriad of pores and/or fractures in aquifer systems, from areas of recharge to areas of discharge (determined by the geologic structure). Tens, hundreds or even thousands

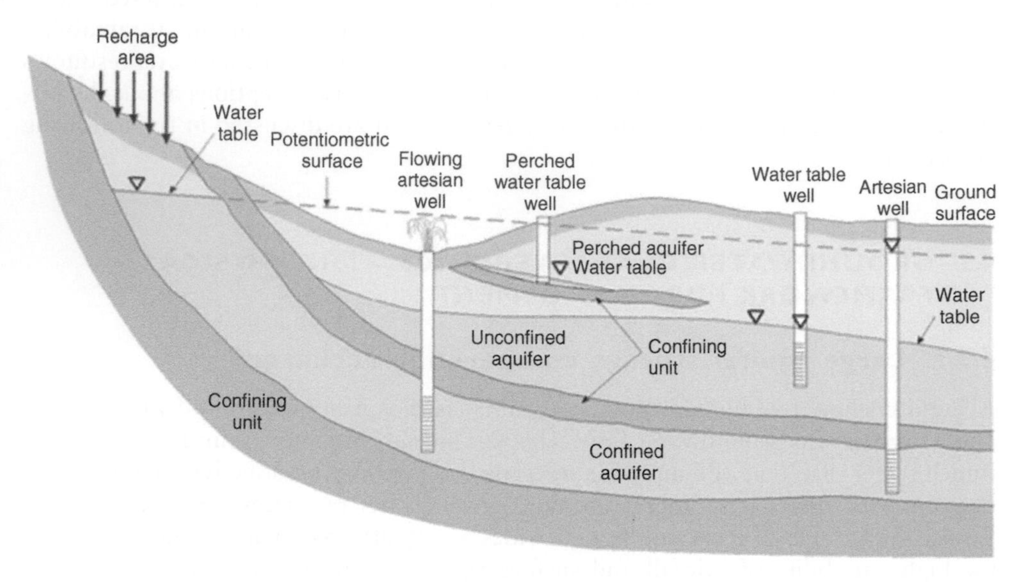

Figure 4.1 Conceptual scheme of groundwater occurrence and waterwell behaviour (developed from World Bank, GW-MATe Strategic Overview Series).

of years can elapse until eventual discharge to a spring, river, aquatic or terrestrial wetland, or directly to the coastal zone.

Slow flow rates and long residence times, consequent upon large aquifer storage, are another distinctive feature of groundwater systems, and they transform highly variable natural recharge regimes into more stable natural discharge regimes. Where aquifers dip beneath much less permeable strata, their groundwater becomes confined (to varying degrees) by overlying layers, and this results in a degree of isolation from the immediately overlying land surface (Figure 4.1), but not from the aquifer system as a whole. In some hydrogeologic settings, shallow unconfined and deep confined aquifer layers are superimposed, with leakage downwards and upwards between layers according to local conditions.

Past episodes of natural climate-change have transformed some large land areas (which formerly had much wetter climates) into deserts, and virtually eliminated all contemporary groundwater recharge whilst some discharge to oases often still occurs. Groundwater reserves which are not being actively recharged are known as *'fossil groundwater'*. These reserves can be, and are being, tapped by waterwells but once pumped out may never be replenished – they are thus termed *'non-renewable groundwater resources'* and as such merit special governance provisions (Foster *et al.*, 2013; Custodio *et al.*, 2016a & 2016b).

Evaluating the relationship of surface water to underlying aquifers is an important component of groundwater system characterisation, distinguishing between:

- streams, rivers and lakes on which an aquifer is dependent as a significant source of its overall recharge

- rivers that in turn depend significantly on aquifer discharge to sustain their dry-weather flow

and having an appreciation of the sensitivity of hydraulic connectivity of the two systems.

4.3.3 Inherent imprecision of evaluating groundwater recharge

Contemporary aquifer recharge rates are fundamental when considering the sustainability of groundwater resource development. The quantification of natural recharge, however, is subject to significant methodological difficulties, data deficiencies and resultant uncertainties (Foster *et al.*, 2013) because of:

- wide spatial and temporal variability of rainfall and runoff events
- widespread lateral variation in soil profiles and hydrogeological conditions

Nevertheless, for most practical purposes it is sufficient to make approximate estimates, and refine these subsequently through monitoring and analysis of aquifer response to abstraction over the medium term.

A number of generic observations can be made about aquifer recharge processes:

- areas of increasing aridity will have a much lower rate and frequency of downward flux to the water-table, with direct rainfall recharge generally becoming progressively less significant than indirect recharge from surface runoff and incidental artificial recharge arising from human activity
- estimates of the direct rainfall recharge component are almost always more reliable than those for the indirect component from runoff recharge.

4.3.4 Difficulty in specifying 'safe yield' and 'resource overexploitation'

All groundwater flow is discharging somewhere, and extraction from waterwells will intercept and reduce these discharges. But the source of pumped groundwater can be complex. Any attempt at defining some form of *'safe yield'* must:

- make value judgements about the importance of maintaining (at least a proportion of) natural 'beneficial' discharge from the aquifer system
- focus mainly on consumptive use and/or catchment export of extracted groundwater, and to some degree discount for non-consumptive uses which generate return water.

In this way, after incorporating an additional allowance for future priority drinking-water abstraction, a *'reasonable cap'* on abstraction can be defined.

'Aquifer overexploitation' is an emotive term not capable of rigorous scientific definition – but one which water resource managers would be wise not to abandon completely, given its clear register at public and political level. Most take it to mean that the 'long-term average rate of groundwater recharge is less than waterwell abstraction'. But

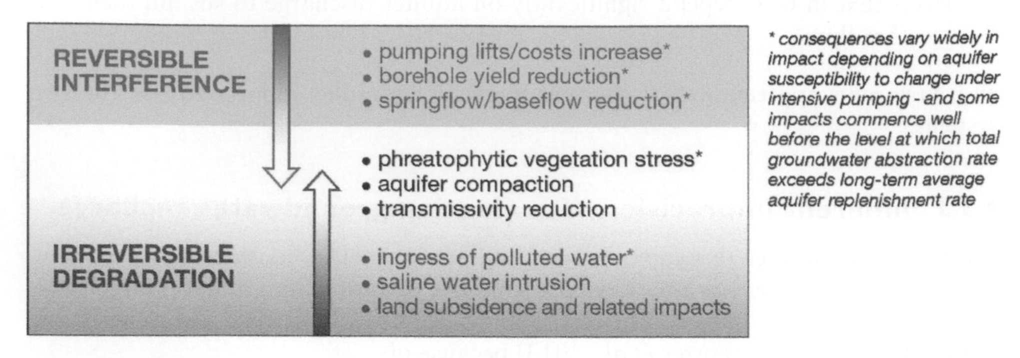

Figure 4.2 Potential impacts of groundwater exploitation on aquifer systems (developed from World Bank GW-MATe Publication SO-4).

problems arise in specifying over what period and which area the groundwater balance should be evaluated – especially in more arid climates where major recharge episodes occur once in decades and pumping effects may also be very unevenly distributed. In practice, when speaking of aquifer overexploitation we are invariably more concerned about the consequences of intensive groundwater abstraction than its absolute level. Thus the most appropriate definition is probably an economic one: that the 'overall cost of the negative impacts of groundwater exploitation exceed the net benefits of groundwater use'. But these impacts can be difficult to predict and cost precisely, and natural susceptibility to (the more important) irreversible side-effects varies widely with aquifer type (Figure 4.2). Amongst the most critical of potential impacts from intensive groundwater use is insidious salinization, which can arise by a number of different mechanisms, only some of which are related to intensive groundwater pumping.

4.3.5 Complexity of groundwater quality controls and pollution vulnerability

Groundwater is for the most part naturally of excellent microbiological and chemical quality. The underlying reasons for this are:

- the capacity of subsoil profiles to filter-out fecal micro-organisms pathogens, and all suspended solids and organic matter, from percolating recharge
- its long sub-surface residence time (decades to millennia) compared to the environmental survival of pathogens (usually <50 days and rarely >300 days)
- the relatively low solubility and non-toxic nature of the matrix of most aquifers.

There are, however, some important exceptions to this since some aquifers exhibit:

- natural groundwater contamination with trace elements that create a health hazard (arsenic and fluoride) or nuisance (dissolved iron and/or manganese)
- elevated vulnerability to pollution from the land surface due to their thin vadose zone and/or the presence of highly-preferential pathways to the water-table (Figure 4.3).

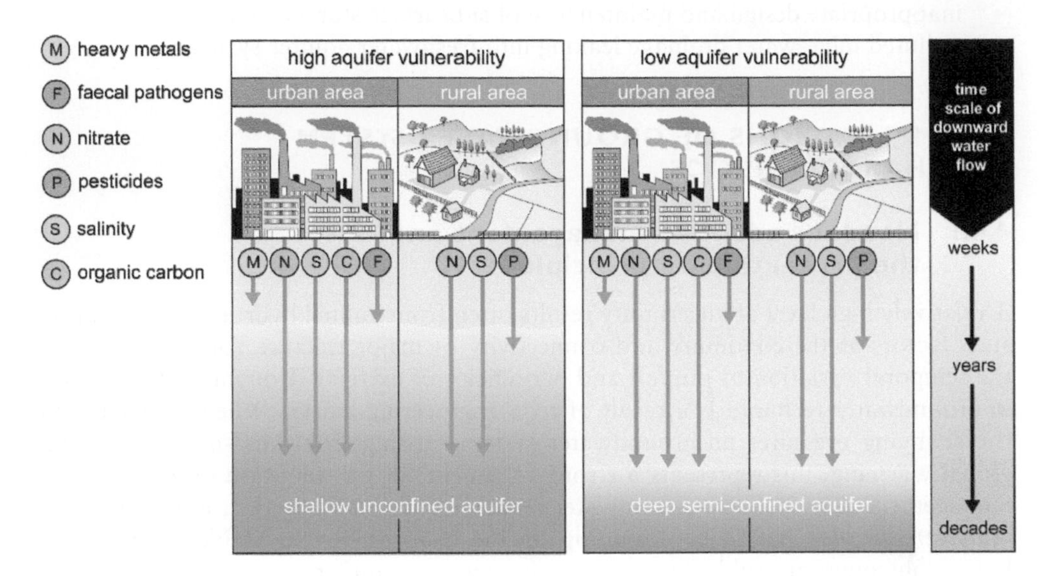

Figure 4.3 Natural variations in groundwater pollution vulnerability (developed from World Bank GW-MATe Groundwater Protection Guide).

Moreover, all aquifers are vulnerable to pollutants that are resistant to subsurface adsorption and/or biodegradation – such as nitrate, salinity and numerous man-made organic chemicals, some of which have serious ecotoxicological impacts in addition to being a serious drinking-water hazard. The massive growth in urbanisation, agricultural and industrial production, together with hydrocarbon development and mining enterprises, over the past 50 years or so, is generating a much greater and more complex contaminant load on the subsurface, in many instances beyond natural self-purification capacity (*e.g.* by excessive application of animal manures and urban wastewater to the ground).

Once polluted, groundwater is extremely difficult to clean-up, given its inaccessibility and very slow rates of movement in finer pores and fractures. In many cases an area of groundwater contamination has to be contained, allowing natural attenuation processes very slowly to take effect. Since groundwater is a very important source of water-supply for public use, sensitive industrial production, terrestrial ecosystems and river baseflow, it is essential that its quality be protected for present and future use. This requires the mapping of zones of high pollution vulnerability and drinking-water source protection (Foster *et al.*, 2013), with application of appropriate controls on hazardous activities in such areas to reduce the risk of major groundwater pollution.

The situation is further complicated in situations where pollutants might gain direct access to below the water-table, for example as a result of:

- poor design, maintenance or misuse of in-situ sanitation units, drainage soakaways or sewerage systems with direct discharge of pollutants to water-table
- inappropriate waterwell design allowing cross-connection of shallow contaminated zones with deeper groundwater

- inappropriate design and maintenance of subsurface storage facilities
- polluted mine-water drainage leaking into freshwater aquifer systems.

4.4 IMPLICATIONS OF GROUNDWATER SYSTEM CHARACTERISTICS

4.4.1 Need for adaptive management and the precautionary principle

A relatively high level of uncertainty results often from limited hydrogeologic data on such factors as the continuity and connectivity of major fracture zones in aquifers, the temporal variation of rainfall and periodicity of extreme drought and reduction in groundwater recharge as a result of river engineering works. When coupled with the changing pressures on groundwater systems arising from land-use change and global warming, this represents a strong argument for the adoption of an 'adaptive management approach' to groundwater resources. To facilitate this, a *groundwater management plan* has to be drawn-up on the basis of best available information, but its outcomes are subject to systematic review of aquifer response after 5 years, with adjustment of the programme of management measures according to the periodic evaluation of resource status and trends (Foster *et al.*, 2015).

One other issue that arises is how to approach certain decisions on waterwell pumping and on conditions for permits for polluting activities that arise at the outset of development or prior to the 5-year periodic review. The elaboration of a 'worst-case scenario' numerical model is recommended to generate data and develop cautious guidance on the related decision.

4.4.2 Supply versus demand side interventions

In groundwater resource management to confront situations of excessive and unstable groundwater resource exploitation, it is helpful to distinguish clearly between:

- *demand-side management interventions* (such as restricting waterwell use at certain times, making savings in consumptive use in irrigation or industry)
- *in-situ supply-side engineering measures* (such as rainwater harvesting, managed aquifer recharge enhancement).

It is important to stress that constraining demand for groundwater abstraction will normally be essential to achieve a groundwater balance, irrespective of what local supply augmentation measures can be economically undertaken. Of course, importing water from outside the groundwater basin will also usually be another option (in some cases even involving desalination of brackish groundwater or sea-water), but unless this is done with full economic cost recovery from local users then it will introduce distortions into the management of local groundwater resources.

The concept of *real water-resource savings* in irrigated agriculture is critical in this regard (Foster & Perry, 2009) – and such savings are only made through reductions in beneficial and/or non-beneficial consumptive use, and in loss to saline water bodies, but not those reductions which would have generated aquifer recharge. In urban areas, real water savings can be made by reducing water-mains leakage and wastewater seepage, but only where they generate discharge to brackish water bodies or create drainage problems.

4.4.3 Potential polluter pays for protection

The economic concept usually prescribed to constrain point-source water pollution is the 'polluter-pays-principle' (Quevauviller, 2008). This incorporates the cost of pollution externalities into the cost of industrial production, rather than leaving them for society to pay. However, in the case of groundwater the burden of proof of pollution is often onerous, because determining who is to blame is made difficult by both the hydraulic complexity and the very large time-lag in pollutant transport (even in some cases just to reach the water-table), which is typical of many (if not all) aquifer systems. Thus the above approach is not readily applicable to groundwater, and would be largely ineffective as regards precautionary protection of aquifers – because of the extreme persistence of some contaminants in the subsurface and the frequent impracticability of clean-up.

For groundwater the 'polluter pays principle' should be interpreted as the 'potential polluter pays the cost of required aquifer protection', which will vary spatially with soil profile, underlying geology and be greatest in important groundwater recharge areas. Moreover, in drinking-water protection zones it will be desirable to exclude hazardous activities, through a combination of regulatory provisions and economic instruments, in preference to controlling their design and operation incrementally. It will be preferable to introduce incentives for potential polluters to improve existing industrial premises and their wastewater handling, treatment, re-use and disposal facilities, and the minimisation and safe disposal of solid wastes, especially in areas where aquifer vulnerability assessments suggest high risk of groundwater pollution. The imposition of strong sanctions for non-compliance, as well as incentives for compliance, will be essential.

4.5 GROUNDWATER MANAGEMENT POLICY – KEY LINKAGES

4.5.1 Principled pragmatism required

The identification of linkages, and assessing how important they are for groundwater management, will involve a diagnosis of current governance provisions, the main drivers of change and their potential impacts in the local hydrogeologic, socio-economic, political and macro-economic setting. The complexity of groundwater governance and management increases as more linkages are considered, and a pragmatic decision will need to be made on which are most relevant (FAO, 2016a).

Groundwater management will need to maintain a reasonable balance between the costs and benefits of interventions, and thus take account of the susceptibility of the system in question to degradation and the legitimate interests of water users, including ecosystems and those dependent on downstream baseflow. Possible interventions need to be put in the context of overall groundwater development, and preventive management approaches are likely to be more cost-effective than reactive ones. Policy development, management options and possible interventions may also be contingent on the legal status and the precise nature of the public or private ownership model of the main water users and the ways in which these determine the interests and influence of various stakeholders.

However, the essential policy linkages that always have to be addressed in groundwater governance and management are those with interrelated surface water features and with land-use in aquifer recharge areas.

4.5.2 Appropriate integration with surface water ecosystems

Groundwater is an inseparable component of the hydrological cycle (interacting directly and indirectly with other components) and thus requires an integrated approach, which takes such linkages fully into account (Foster & Ait-Kadi, 2012), and avoids the potential oversights that can arise from narrower approaches. It must also explicitly consider the needs of groundwater-dependent ecosystems, which are usually characterised by phreatophytic plants deriving a major proportion of their water needs from saturated soils. Long-term groundwater depletion will eliminate these species and their ecosystem function, and their removal (to reduce groundwater evapotranspiration) may cause the water-table to rise, lead to soil waterlogging and other environmental problems.

The level and mode of management integration, however, needs to be appropriate to the hydrogeological setting – with rapidly-connected systems such as karstic limestone formations and major alluvial aquifers requiring a different approach to deep sedimentary aquifers in arid regions. In most circumstances groundwater should be managed conjunctively with surface-water resources, since streams, rivers, reservoirs, lakes and irrigation canals are a major source of groundwater recharge, especially in more arid climatic zones, whilst more widely surface water bodies are fed by natural groundwater discharge (as springs, seepages and baseflow). The policy challenge is to define, for any given setting, the mode of conjunctive use of surface-water and groundwater use that is balanced and complementary (Foster & Steenbergen, 2011).

Moreover, groundwater represents a key resource for climate-change adaptation, providing an opportunity to buffer the increased variability and scarcity of surface-water predicted under many global-warming scenarios (OECD/GWP, 2015). Water resources management policy and planning need to take full advantage of aquifer storage to improve water-supply security, but also recognise natural variability in the resilience of groundwater systems themselves to the pressures arising from global change.

4.5.3 Promoting groundwater-friendly rural land-use

Land-use in groundwater recharge areas exerts a major influence on recharge quality and quantity, and thus needs to be systematically linked with groundwater management (Morris *et al.*, 2003). But this is not straightforward since land-use decisions are usually the domain of local government and strongly influenced by national agricultural policy in rural areas. This linkage is critically important in the capture zones of waterwells, wellfields and springs used as municipal drinking-water sources to:

- prevent or limit certain types of point-source pollution threat, through legal provisions and local regulations
- control the agricultural diffuse pollution threat from intensive cultivation using heavy applications of fertilisers and pesticides, which can often only be influenced indirectly through national agricultural and forestry policy, promulgation of voluntary codes of best practice and/or payment for ecosystem services (Foster & Cherlet, 2014).

Groundwater management requires a consultation mechanism with the planning, investment and management procedures related to land-use in both urban and rural areas. Where groundwater performs a strategic municipal water-supply and/or ecological function, a useful instrument to facilitate such consultation is a regulatory provision to declare special 'groundwater protection zones' (for highly vulnerable recharge areas and/or drinking-water capture zones), which will allow the water-resource agency to exert restrictions on land-use practices and potentially-polluting development in such zones.

4.5.4 Essential integration with irrigation water management

The practice of irrigated agriculture in aquifer recharge areas always results in an intimate linkage with groundwater resources – but the nature of this relation varies considerably with hydrogeological setting (especially water-table depth) and whether groundwater or surface water is the main source of irrigation water-supply (Garduño & Foster, 2010).

Where groundwater is the main source, finding ways to reduce the amount pumped is extremely important because agriculture is widely the largest consumer of groundwater. The replacement of flood irrigation with precision drip or sprinkler technology can reduce the volume of groundwater applied to a specific crop and therefore reduce the energy used for pumping. In addition well-managed precision ferti-irrigation delivers nutrients directly to the root zone, reducing weed growth and increasing crop yields – but it must be stressed that this so-called 'efficient irrigation' is certainly not usually a significant 'water-resource saving measure' and its introduction often has negative consequences for the groundwater system as a whole (Foster & Perry, 2009) through:

- greatly reducing groundwater recharge from irrigation-water returns
- build-up of soil salinity, reducing crop yield or quality and requiring leaching to restore fertility, which in turn causes increasingly saline groundwater recharge.

Moreover, expansion of the irrigated area is often justified by the 'alleged water saving' leading to an overall deterioration of the groundwater balance. If surface water is the water-source the reduction in irrigation-water returns and incidental aquifer recharge may be even greater.

Thus a well-informed and carefully-balanced policy approach is required, and the challenge (particularly in arid areas) is not only to focus on 'efficient water-use' but also to reconcile gross groundwater abstraction with overall average recharge. Irrigation-water management should be founded upon evapotranspiration quotas and include soil management to retain favourable moisture and salt balance (Garduño & Foster, 2010). Governance arrangements are also required that boost crop-water productivity (net income/m^3 evaporated) (Llamas & Martinez-Santos, 2005; Garrido *et al.*, 2006) – but also in the long-run constrain groundwater use to achieve resource sustainability.

Metering of irrigation water-use is a highly-desirable governance requirement and management provision, but one that is often resisted as being too complex and costly. A simpler (and usually adequate) proxy is to meter the energy supply for waterwell pumping, which for example can be facilitated by using electronic smart-card technology for pump activation, with individual card allocations being tradeable, chargeable and annually varied according to aquifer water-level trends. Moreover, rural energy pricing could be used as part of an incentive framework for promoting sustainable groundwater extraction, with joint billing of pump energy consumption and groundwater resource use (with connection depending on payment) (Garduño & Foster, 2010). Rationing of supply from power lines dedicated to irrigation pumping has also been used successfully and facilitated improvements in village domestic electricity supplies (Shah *et al.*, 2012).

In many ways the introduction of solar-panel generated energy to power waterwell pumps is a welcome development, since it will reduce (perhaps greatly in the longer term) dependence on electricity generated from fossil fuels (Shah *et al.*, 2012). However, it will be very important for water resources agencies to work with power companies to introduce 'grid buy-back tariffs' that are sufficiently attractive to avoid solar energy being used for continued over-pumping of groundwater.

4.6 POLICY INTEGRATION BEYOND THE WATER SECTOR

The principle drivers of change and degradation of aquifer systems can also be generated from outside the water-sector. Thus incorporation of groundwater resource and quality considerations in policy formulation of certain related sectors or sub-sectors (so-called *horizontal policy integration*) can be required to avoid national policies with contradictory signals and perverse incentives (Figure 4.4) (FAO, 2016a).

4.6.1 Urban infrastructure including sanitation

Urbanisation has a major impact on the underlying groundwater regime in terms of:

- quantity – with recharge simultaneously being reduced by paving and roofing, and increased by water-mains leakage and seepage from in-situ sanitation units and drainage soakaways
- quality – from large volumes of domestic and/or industrial wastewater and solid waste – with the extent to which this threatens groundwater depending on the adequacy of sanitation, wastewater and waste management arrangements; the groundwater hazards posed by industrial zones usually being reduced by regulations on the use, reuse, treatment and disposal of specific chemical substances.

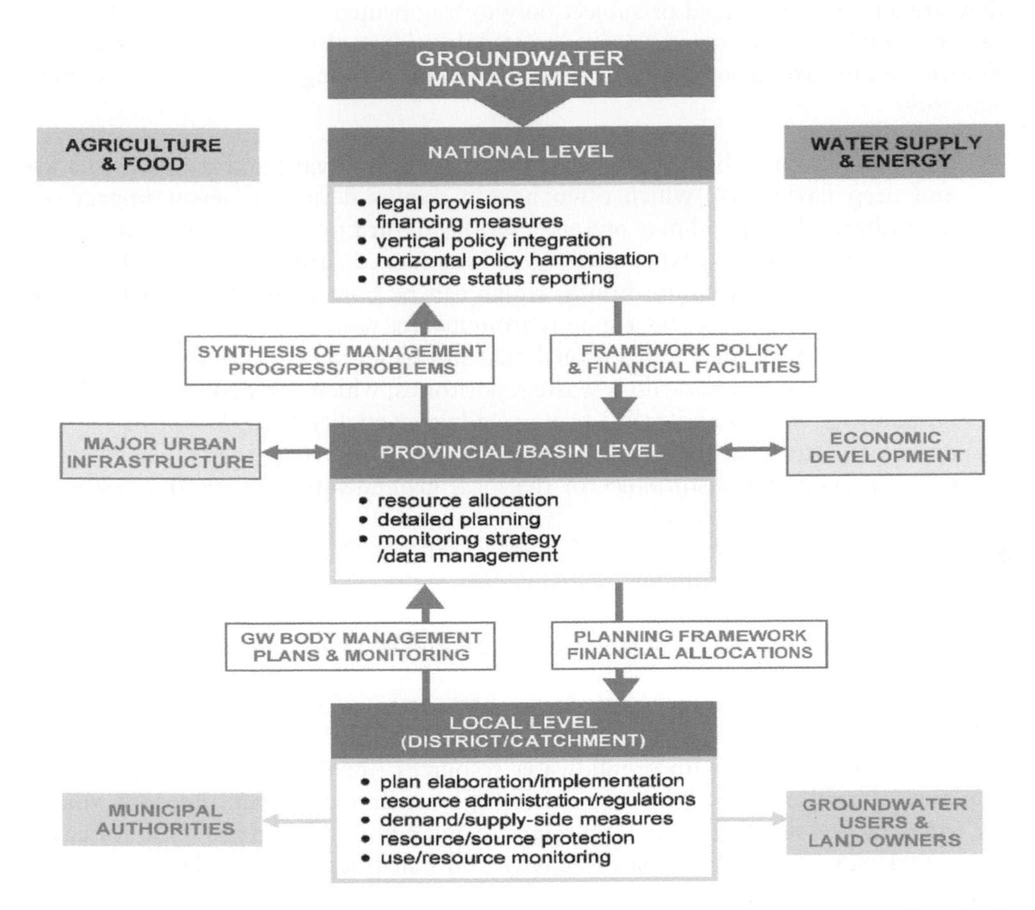

Figure 4.4 Institutional structure for groundwater policy development (developed from World Bank GW-MATe Publication SO-1).

Of particular significance are in-situ sanitation practices and wastewater handling/re-use from mains sewerage systems, which provide a significant component of urban groundwater recharge in more arid climates, but simultaneously pose a serious threat of shallow groundwater pollution (including pathogenic micro-organisms, ammonium or nitrates, toxic community chemicals and pharmaceutical residues). The pollution risk varies widely with the local hydrogeological setting, density of population served, design of in-situ sanitation units or the level of wastewater treatment, and location/mode of wastewater reuse. Thus it is critical that groundwater vulnerability and dependence are taken into consideration in the planning and implementation of sanitation investments – but the governance and operational arrangements for this to occur are still widely lacking.

4.6.2 Subsurface space and extractive industries

The use of subsurface space and the extraction of subsurface resources are both on the increase, and can have strong impacts on groundwater (FAO, 2016a). But widely they are almost unmanaged or subject only to fragmented regulation. Procedures to factor groundwater considerations into related decision-making are thus an important element of effective groundwater governance. The following are the more common activities:

- the construction of buried pipelines, underground railways and roads, car parks and deep basements, which often have a temporal or permanent impact on groundwater levels and may present a serious hazard for groundwater quality
- hydrocarbon (oil and gas) extraction, and the related surface storage and subsurface injection of formation brines, which can be both a significant demand on groundwater resources and a serious groundwater pollution hazard
- hydrocarbon fuel tanks and seasonal heat storage in the subsurface, and subsurface radioactive and hazardous waste repositories, which can perturbate the local groundwater system and/or create a groundwater quality hazard
- open-cast extraction (of sand-and-gravel, coal/lignite, etc.), which usually produces a significant disturbance of the local groundwater regime and can be a groundwater quality hazard
- deeper mining activities (for coal, metals, salt/potash, precious minerals, etc.), which often involve pumping large volumes for drainage modifying the groundwater flow and quality regime, and on abandonment with water-table rebound can lead to the discharge of highly-acidic and polluted groundwater.

All these activities can be very risky for groundwater, unless keen awareness of the risks and great care is taken in technical design and routine operation. Eventually coordinated governance with groundwater resources (involving cross-sector regulation and joint planning) will be required to facilitate harmonisation. Recently voices are emerging that advocate long-term financial and administrative provision for environmental management of all such activity and construction throughout its entire 'life-cycle'.

4.7 GROUNDWATER MANAGEMENT PLANNING PROCESS

4.7.1 General rationale

An effective *groundwater management plan (GW-MaP)* should capture and integrate basic scientific understanding, sustainable management measures and a focused action-plan for a specified priority groundwater management unit in a single document. This is an essential component of groundwater governance (Foster *et al.*, 2015). In some ways groundwater management planning is an art form, and a far from fashionable one – but one which is central to so-called *adaptive management* for groundwater, which is needed to confront the joint challenges of global change and scientific uncertainty. It is important from the outset to emphasise that adaptive management is in no way inconsistent with groundwater planning, since a GW-MaP will have fixed targets, which will be achieved by a programme of measures that will almost always require adjustment following periodic review of their effectiveness.

GWMaPs have another important governance function in that they help to harmonise the groundwater-related activities of all government organisations

(Foster *et al.*, 2015). The sustainability of the groundwater resource base will depend on the technical adequacy, institutional suitability and implementation efficiency of GW-MaPs. Specific management instruments and measures must be tailored to local context as regards:

- the hydrogeologic setting of the groundwater body under consideration
- the social, economic and political position of the country/province concerned.

It will thus vary significantly with position along the developmental cycle.

The groundwater management planning process should be promoted by the responsible national groundwater ministry or agency (through provision of protocols and guidance) and undertaken by the corresponding local groundwater resource agency or office together with all relevant stakeholders. It will require co-mobilisation of financial investment for the demand management and/or pollution control measures required for plan implementation. A GW-MaP should be dynamic in nature and implemented as a structured, step-wise long-term (5–10 year) sequence. Indicators of resource status (for example a predefined groundwater level or quality at a strategic monitoring site) can act as barometers of aquifer condition and facilitate the adaptive management approach.

Groundwater is quintessentially a local resource (with large numbers of actual abstractors and potential polluters). Thus, priority groundwater management units should generally be defined at the lowest rational spatial scale, and managed as close as possible to these local stakeholders—usually differentiating between areas in which the major resource and quality stress is urban development or intensive agriculture. There are, however, some exceptions to this rule—for example, where a larger aquifer system extends across international frontiers and a component of transboundary cooperation will be required for its successful governance, even if many aspects of routine management can be handled locally in groundwater sub-catchments. The same applies to some large aquifers extending across state boundaries in federal countries.

The process proposed conforms in general terms with that adopted by both the EU-Water Framework Directive (EC, 2000) and the GEF Groundwater Governance Programme (FAO-UN, 2016a) and is transparent, consultative and evidence-based, thereby creating a framework for cooperation and accountability. The resulting plans comprise a formal public document with budgeted, time-bound, actions and outcomes that can be evaluated. They also bring all governance provisions together, test them and make it possible to assess their effectiveness.

4.7.2 Fundamental steps of plan elaboration

The **1st Step-Characterisation of Priority Aquifer Systems** (also referred to as 'groundwater management units' or 'groundwater bodies') can be achieved in various ways:

- physical delineation of the system considering groundwater flow regime from natural recharge to discharge zones (thus connecting the landscape with the subsurface system), whilst taking account of major man-made perturbations

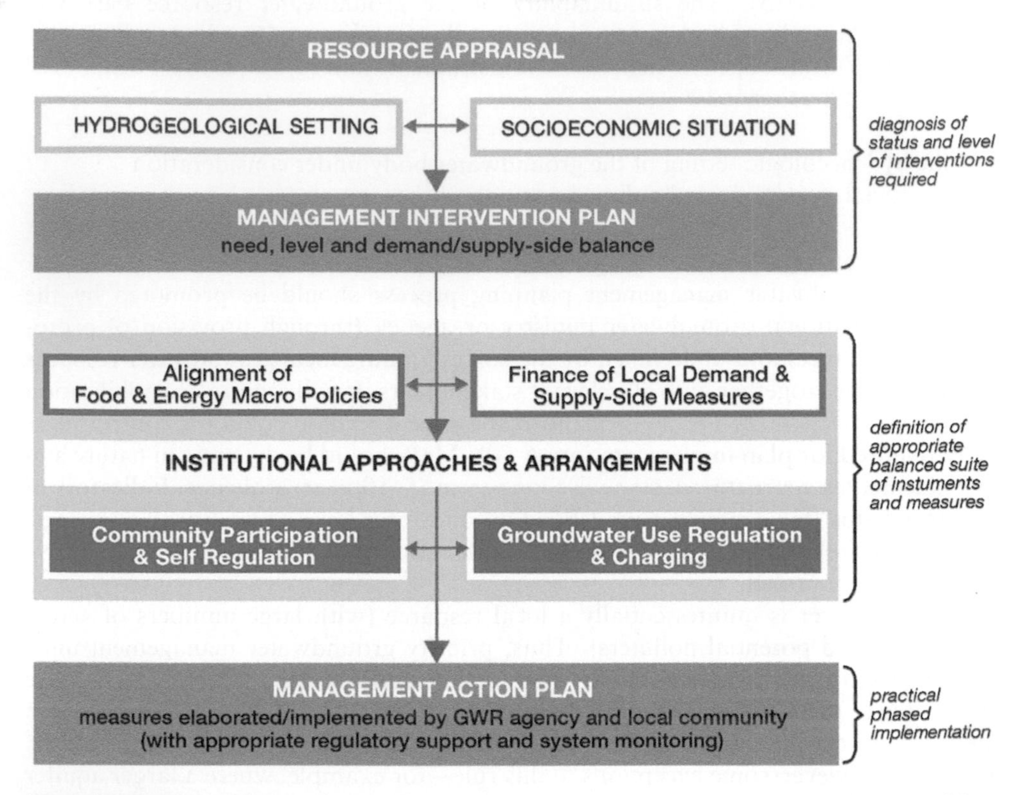

Figure 4.5 Stages and factors in the elaboration of a groundwater management plan (developed from World Bank GW-MATe Publication SO-4).

- evaluating the importance of the system to socioeconomic development and to ecosystem conservation – highlighting where groundwater plays a critical role in public water-supply, irrigated agriculture, industrial production and/or aquatic ecosystem sustainability
- assessing pressures on the system, and in particular its susceptibility and vulnerability to irreversible degradation (through subsidence, salinization and persistent pollution) or tendency to be associated with land water-logging and groundwater flooding.

Priority aquifers systems should be chosen on the basis of sound professional reasoning and broad socio-political support. Whilst they occur in a variety of developmental situations, urban aquifer systems will often be amongst the highest priorities because of their strategic importance in water-supply and the major pressures to which they are subjected. An 'integrated approach' to urban groundwater plan elaboration and implementation is strongly recommended, which will benefit from a powerful champion (*e.g.* municipal mayor or water-utility chief engineer). A plan is an important governance action, even where large-scale water-supply transfers are going to be introduced into urban areas previously heavily-dependent on local groundwater supplies. Groundwater quality is intimately linked with urban land-use, and improved metropolitan/municipal planning based on an integrated vision needs to be aligned with groundwater management planning to avoid persistent and costly problems.

The **2nd Step-Assessment of Groundwater Resource Status** is a basic foundation of all GW-MaPs and requires consideration of:

- the geographical scale of the aquifer system and size of its storage reserve (a function of aquifer type), which will determine how identifiable it will be for local stakeholders and how amenable it might be for user self-regulation
- the degree of connectivity with surface water, determining whether conjunctive management of surface and groundwater is essential to achieve the efficient use and improved conservation of both resources
- the level of contemporary recharge, since if the use of non-renewable groundwater resources is likely to be involved it should be subject to rigorous control given the strategic implications for intergenerational equity
- aquifer susceptibility to irreversible degradation and groundwater vulnerability to pollution, which together will determine the urgency for action and whether comprehensive provisions in a regulatory sense are essential.

This work will normally be undertaken (and in due course owned) by the corresponding local groundwater agency (in collaboration with local specialists and stakeholders), following protocols provided by the responsible national groundwater focal-point.

The **3rd Step-Plan Consultation Process** will be essential to promote dialogue aimed at establishing consensus on the priority services required from the aquifer system under consideration, which could be:

- water-supply security for urban, agricultural or other purposes
- guaranteed access for small private users
- sustaining dependent ecosystems and dry-weather riverflows.

This consultation is, by definition, a participatory process, with final decisions resting with the government agency mandated to manage groundwater. It is very important that the consultations are well informed about current groundwater resource and quality status, any related trends, the potential consequences and costs of 'management inaction', and the options as regard management measures. Such information needs to come from, and be delivered explicitly by, a recognised expert(s) of independent stature.

Some governance provisions (and sets of management instruments/measures) will need to be specifically tailored to certain facets of the socioeconomic situation conditioning groundwater use, dependence, management and protection, such as:

- *density of groundwater abstraction points and/or potential polluters* – since if elevated it will not be realistic for the public administration to promote conventional regulation (unless these users can be brigaded into logical groups)

- *state of institutional evolution* – since regulatory and charging approaches require a public administration with considerable authority, capacity and experience
- *proportion of population directly abstracting groundwater* – since if this is high 'democratic influence' may be exerted for continuing environmentally-perverse subsidies
- *economic significance of groundwater resource use* – since this will influence the ease with which finance can be raised to invest in governance provisions and instruments and the necessary monitoring of their effectiveness

The **4th Step-Planning Document Elaboration** will require stakeholder interaction and/or participation, since the detail of the plan will need to be accepted as a 'balanced way forward' to achieve groundwater resource sustainability. It will thus be important to find the most appropriate form of stakeholder participation and for the public administration to nurture this as a vehicle for plan implementation, including:

- identifying regulatory measures, economic incentives and policy changes to address groundwater management needs within the given legal and institutional framework – and thus achieve a practical balance between top-down administration and bottom-up stakeholder participation
- identifying a technically and economically sound array of demand-side and supply-side measures to re-balance groundwater withdrawals with average recharge, such that the risk of irreversible damage to aquifers and ecosystems is avoided
- definition of stakeholder roles, and specification of how these roles will be factored into planning and management, and be maintained
- recognising any dependence upon essentially non-renewable groundwater resources, requiring additional governance provisions and management strategies
- identifying situations of groundwater over-abundance, in which soil water-logging and land drainage problems need to be avoided through conjunctive management
- identifying monitoring requirements for evaluating the effectiveness of measures or the impact of 'no management action'.

Groundwater management plans will often need to incorporate groundwater pollution control measures, accepting that trade-offs will be necessary to reach balanced agreement. In this context it is important to recognise the difference between:

- attempting to protect all groundwater recharge, or focusing only on protection of the capture zones of major public water-supply sources
- dealing with point-source pollution (which is relatively easy once the problem has been identified), whilst recognising that the control of diffuse-source pollution is likely to take much longer and requires a different approach
- addressing the diffuse-source pollution threat from intensive agricultural land-use through promulgation of 'best farming practices' that guide the application of manures, fertilisers and pesticides based on groundwater protection considerations.

Monitoring requirements for groundwater quality assessment are onerous and data are often insufficient or their interpretation subject to considerable uncertainty.

Substitution of indirect hydrogeologic methods to assess pollution vulnerability are, however, acceptable as a first approximation for initiating the planning process.

The **5th Step-Implementing & Reviewing Plans,** which by definition are consensualised, should be undertaken progressively on a structured basis. The plan must include an operational time-frame and management monitoring network endorsed by the responsible national/local groundwater agency and all relevant stakeholders. Its implementation will often require some strengthening of institutional linkages, raising substantial capital investment, improving groundwater use/protection measures and aquifer response monitoring, promoting more effective public information campaigns and undertaking capacity building. It will also be necessary to pursue inter-ministry cross-sector coordination to avoid agricultural or industrial development plans which are incompatible with groundwater resource constraints and the co-mobilisation of financial investment for the required demand management measures.

The plan must be dynamic providing capacity for adaptation to change in technical knowledge and in external drivers (such as climate-change and land-use). Indicators of groundwater status (such as pre-defined water-table level or quality at a strategic monitoring site) can act as barometers of aquifer condition. Whilst some types of aquifer system are relatively rapid to respond to changes in groundwater pumping and pollution load, and a response can be expected to manifest itself within 2 years, others (especially quality-related responses in thick aquifer systems) can take more than 10 years to become apparent. A carefully-designed monitoring network is highly desirable to avoid falling into a false sense of complacency when considering the initial aquifer response to newly-applied pressures. Feedback from the first cycle of plan implementation should be used to up-grade the GWMaP and, if necessary, to refine the underlying governance provisions.

ACKNOWLEDGEMENTS

The authors acknowledge the contribution made by all colleagues of the GEF Groundwater Governance Programme–International Steering Committee and Core Drafting Team to this topic during 2011–15. Many stimulating discussions were enjoyed with Frank van Steenbergen, Jacob Burke, Mohammed Bazza and Stefano Burchi in particular. The first author also wishes to express his thanks to the World Bank Groundwater Management Advisory Team (GW-MATe), and especially to Hector Garduño & Karin Kemper, for their interest and work on the topic during 2001–11, and the interest of the Global Water Partnership (Mohammed Ait-Kadi & Ania Grobicki) in broadening the related international dialogue during 2011–15. The editors wish to thank the World Bank for permitting four figures from GW-MATe publications to be included in this chapter.

REFERENCES

Custodio E, Andreu-Rodes J M, Aragon R, Estrela T, Ferrer J, Garcia-Arostegui J L, Manzano M, Rodriguez-Hernandez L, Sahuquillo A & Villar A del. (2016) Groundwater intensive use and mining in south-eastern peninsular Spain: hydrogeological, economic and social aspects. *Science of Total Environment* 559: 302–316.

Custodio E, Cabrera M del C, Poncela R, Puga L O, Skupien E & Villar A del (2016) Groundwater intensive exploitation and mining in Gran Canaria & Tenerife, Canary Islands, Spain: hydrogeological, environmental, economic and social aspects. *Science of Total Environment* 557–558: 425–437.

EC-Directorate General for the Environment (2008) *Groundwater protection in Europe – consolidating the regulatory framework.* European Community Office for Publications (Luxembourg).

FAO (2016a) *Global Framework for Action to achieve the vision on Groundwater Governance.* UN-Food & Agricultural Organisation (for GEF, IAH, UNESCO & World Bank) Publication (Rome).

FAO (2016b) *Global Diagnostic on Groundwater Governance.* UN-Food & Agricultural Organisation (for GEF, IAH, UNESCO & World Bank) Publication (Rome).

Foster S, Garduño H, Tuinhof A & Tovey C. (2009) *Groundwater governance – conceptual framework for assessment of provisions and needs.* GW-MATE Strategic Overview Series SO-1 World Bank (Washington DC).

Foster S & Perry C. (2009) Improving groundwater resource accounting in irrigated areas: a prerequisite for promoting sustainable use. *Hydrogeology Journal* 18; 291–294.

Foster S, Hirata R, Misra S & Garduño, H. (2010) Urban groundwater use policy: balancing the benefits and risks in developing countries.GW-MATE Strategic Overview Series 3 World Bank (Washington DC).

Foster S & Steenbergen F van (2011) Conjunctive groundwater use – a 'lost opportunity' for water management in the developing world? *Hydrogeology Journal* 19: 959–962.

Foster S & Ait-Kadi, M. (2012) Integrated Water Resources Management (IWRM) – how does groundwater fit in ? *Hydrogeology Journal* 20: 415–418.

Foster S, Hirata R & Andreo, B. (2013) The aquifer pollution vulnerability concept: aid or impediment in promoting groundwater protection? *Hydrogeology Journal* 21: 1389–1392.

Foster S, Chilton J, Nijsten G-J & Richts, A. (2013) Groundwater–a global focus on the 'local resource'. *Current Opinion on Environmental Sustainability* 5: 685–695.

Foster S & Cherlet, J. (2014) *The links between land use and groundwater: governance provisions and management strategies to secure a 'sustainable harvest'.* Global Water Partnership Perspectives Paper (Stockholm).

Foster S, Evans R & Escolero, O. (2015) The groundwater management plan: in praise of a neglected 'tool of our trade'. *Hydrogeology Journal* 23: 847–850.

Garduño H & Foster, S. (2010) *Sustainable groundwater irrigation–approaches to reconciling demand with resources.* GW-MATE Strategic Overview Series 4. World Bank (Washington DC).

Garduño H, Steenbergen F van & Foster, S. (2010) *Stakeholder participation in groundwater management – enabling and nurturing engagement.* GW-MATE Briefing Note Series 6. World Bank (Washington DC).

Garrido A, Martinez-Santos P & Llamas M R. (2006) Groundwater irrigation and its implications for water-policy in semi-arid countries–the Spanish experience. *Hydrogeology Journal* 14: 340–349.

IUCN (2016) *Spring – managing groundwater sustainably.* International Union for Conservation of Nature Publication (Gland).

Lopez-Gunn E & Martinez-Cortina I . (2006) Is self-regulation a myth – case study on Spanish groundwater users associations and the role of high-level authorities. Hydrogeology Journal 14: 340–349.

Llamas, M R & Martinez-Santos P. (2005) Intensive groundwater use: silent revolution and potential source of social conflicts. ASCE Journal Water Resources Planning & Management 131: 337–341.

Morris B L, Lawrence A R, Chilton P J, Adams B, Calow R C & Klinck B (2003) *Groundwater and its susceptibility to degradation – a global overview of problem and options for management.* UNEP Early Warning Assessment Report Series 03-3 (Nairobi).

Mukherji A, Villholth K G, Sharma B R & Wang J. (2009) *Groundwater governance in the Indo-Gangetic and Yellow River Basins–realities and challenges.* IAH Selected Papers in Hydrogeology 15.

OECD/GWP (2015) *Securing Water & Sustaining Growth. Report of the Task Force on Water Security & Sustainable Growth.* University of Oxford Publication (Oxford).

Quevauviller P (2008) *European Union Groundwater Policy. Groundwater Science & Policy – an International Overview.* Royal Society of Chemistry (RSC) Publishing (London) 85–106.

Shah T, Giordano M. & Mukherji A 2012 Political economy of energy-groundwater nexus in India: exploring issues and assessing policy options. Hydrogeology Journal 20: 995–1006.

Steenbergen F van (2006) Promoting participatory groundwater management. *Hydrogeology Journal* 14: 380–391.

Tuinhof A, Foster S, Steenbergen F van, Talbi A & Wishart M. (2011) Appropriate groundwater management policy for Sub-Saharan Africa – in face of demographic pressure and climatic variability. GW-MATE Strategic Overview Series 5. World Bank (Washington DC).

Part 2

Key elements of groundwater governance

Chapter 5

Leadership and political will for groundwater governance: indispensable for meeting the new Sustainable Development Goals (SDGs)

Alfred M. Duda

Retired, World Bank Group and Global Environment Facility Secretariat, Washington, DC, USA

ABSTRACT

Lack of attention to groundwater governance and to water resources management is irreversibly depleting and degrading the strategic resource, not only threatening national economies but also contributing to social unrest, civil war and millions of refugees. This paper identifies needed governance reforms and obstacles that undermine those reforms. Leadership has been insufficient and is needed at all levels to make progress. A number of cases are presented to illustrate ways to foster such leadership, ranging from NGOs and academia to communities, local governments, national sector ministries as well as transboundary and global institutions. Case studies funded by the Global Environment Facility (GEF) are described illustrating five GEF processes and tools that can be used to foster partnerships and leadership. Integrated approaches to land and water resources management (integrating surface water, aquifers, and recharge areas) represent keys to balancing competing water uses in basins and aquifers. Other reforms such as pricing water use, land tenure reform, water allocation systems based on consumptive use have been piloted, and now need scaling-up. Professionals need to be ready with reforms when disasters like drought strike, which provide political driving forces for leaders to finally exercise political will for improving groundwater governance.

5.1 INTRODUCTION

Inter-linked crises of land degradation, food security, ecosystem decline, water quality, and surface/groundwater depletion still stand in the way of poverty reduction and achieving the new Sustainable Development Goals (SDGs) adopted by Heads of State (Box 5.1). New concerns about climatic extremes just make matters worse. Surface and groundwater management has been abysmal in many rich and poor countries with serious governance failures as irrigation, hydropower, mining, or drinking water projects compete for what little water is left. For 40 years, since the green revolution, sector by sector grabs for available water continue with little serious consideration of downstream or aquifer impacts and little priority for water ecosystems.

With business as usual, catastrophic shortages of food and water coupled with seriously degraded ecosystems lie on the horizon. Estimates are that 60% more food is needed by 2050 to meet growing needs (World Resources Institute, 2013); this challenge will be difficult given increased water use for industry, energy, people, and ecosystems. Many solutions integrating land and water management are known and were demonstrated in pilot situations and published in reports. Integrated management will prove critical to free up excessive water used in irrigation and reallocate it to other damaged uses in basins and aquifers to meet multiple SDGs. Otherwise, SDGs may serve as ministry silos, each trying to grab water to achieve their related SDG.

Meanwhile, estimates of people living in basins under water stress have risen from 1.7 billion (Johnson *et al.*, 2001) to at least 2.6 billion for the period 1996–2005 based on more accurate assessments related to consumptive use on a monthly basis (Hoekstra *et al.*, 2012). Recently, Mekonnen and Hoekstra (2016) utilized updated techniques and estimated that 4 billion people, two-thirds of the planet, experience severe water scarcity at least one month a year. With so many people being inconvenienced, the SDGs adopted by the UN in 2015 provide a new, potentially significant driving force for improving groundwater governance as noted in Box 5.1.

Box 5.1: Sustainable Development Goals (SDGs) to Catalyze Transformational Change

In 2015, Heads of States met at the United Nations and adopted the SDGs, known officially as: *Transforming our World: the 2030 Agenda for Sustainable Development.* It is a set of 17 aspirational goals with 169 targets that replaces the older Millennium Development Goals and is intended for all nations on the planet as a guide for coordinated investment. The goals range from "end poverty" and "end hunger" to the water-related Goal 6 "ensuring the availability and sustainable management of water and sanitation for all". Goal 6 has 6 targets including: *6.1 By 2030, achieve universal and equitable access to safe and affordable drinking water for all; 6.5 By 2030, implement integrated water resources management at all levels, including through transboundary cooperation as appropriate;* and *6.6 By 2020, protect and restore water-related ecosystems, including mountains, forests, wetlands, rivers, aquifers and lakes.* Through the elements of protecting and sustaining aquifers for drinking water and through integrated management and protection of aquifers, there is new impetus for improving both water and groundwater governance. The SDGs are intended to align all governments, sectors, civil society and international organizations to cooperate toward transforming our world. More information on the SDGs is available from the UNGA (2015) and the UN website for SDG 6 (UN Department of Economic & Social Affairs, 2016).

5.2 SCALES OF GOVERNANCE IMPROVEMENTS

Improved groundwater governance is needed at all scales, from the very local farmer's field or city catchment to sub-national and local government, hydrologic units such as river basins or aquifers to national water resource management and sector management policies and ministries, transboundary water systems, and global institutions. Different actors may work at the different scales to provide leadership and foster needed political will. For example, much capability exists in university systems to undertake investigations and analyses that may catalyze action at every scale. The work of Hoekstra *et al.* (2012) on a global scale for analyzing water scarcity from a consumptive use estimate and monthly time scale represents a great case. This new approach identified more people living with water scarcity compared to historic ones that continually used water withdrawals as the basis for global estimates that under-predicted the severity of the global crisis. Similarly with NGOs, campaigns based on science and advocacy can make a difference in pressure for political will from local groups to global NGOs helping build constituencies for action at the national and global scales.

The Global Environment Facility (GEF) worked to encourage the World Bank and FAO as GEF agencies to collaborate with UNESCO on a GEF International Waters (IW) project to assess the state of groundwater governance regionally and develop a global vision and action framework for improved governance (FAO, 2016b and 2016c). The project's diagnostic (FAO, 2016a) notes that most every country has some groundwater-related legal framework but most have fragmented, poorly funded, capacity-poor systems not linking well horizontally with other sectors (like irrigation) or vertically to local institutions. Even rich countries could use improvement to address gaps, overlaps, and resource constraints if SDGs are to be met. Each of the types of governance institutions requires immediate attention as described in following sections.

5.2.1 National sector interventions supporting local action

National sector policies, legal frameworks, programs of support, and resources are all critical for supporting not only sector decisions but also local actions. First and foremost, groundwater is a local issue. It is often related to land use, and some countries do not have national laws to directly manage groundwater because it is the purview of lower levels of government – for example in the US. However, sub-national governments many times do not have capacity or resources to improve management; national ministries must take leadership and secure resources to build capacity if action is to be expected. This support begins with policies and legal frameworks, which have been reviewed to be mostly inadequate by the Global Groundwater Governance project (FAO, 2016a). Clearly leadership starts here in the national ministries to ensure sustainability for precious aquifers even if sub-national institutions are the regulatory actors.

Two types of national legal frameworks seem appropriate – one aimed at specific sector activities that may impact groundwater and a second type aimed at the groundwater resource itself. Examples of the first would cover mining, irrigation projects, oil/gas drilling, waste disposal, underground injection, remediation of already chemically polluted aquifers to minimize extent of damage, etc. National environmental impact assessment legislation that most countries have can also assist. The US Superfund represents an example of the remedial type of national law aimed at old waste sites while the US Resource Conservation and Recovery Act is another example in requiring sub-national governments to regulate waste disposal and authorize national funding to sub-national governments. National, legal frameworks under EU legislation represent another example, with policies and directives aimed at national actions.

The second type, focused on the water resource, is also critical for national policy to protect groundwater in conjunction with the sector laws. Once again, the EU represents a best case with its "nitrates directive" and the newer Groundwater Directive that was spurred by existing gaps. Another example is US national water quality legislation, which sets standards for water quality, and the related Safe Drinking Water Act. Ultimately, legislation creates ministries and agencies with program missions to set rules and undertake valuable monitoring of surface and groundwater quality and quantity. Annually agencies and ministries compete for funding to support their work and face perennial cuts in budget as political figures desire to place resources elsewhere.

The EU approach toward groundwater resource protection has been termed state of the art in the recent global review (FAO, 2016a); the United Nations Economic Commission for Europe Transboundary Water convention also provides opportunities for improved governance for waters within cross border water systems. Other regional economic bodies like Southern Africa Development Community or Association of South East Asian Nations need to contribute to improved groundwater governance by fostering political will for national action because economic competition creates unfair advantages for countries refusing to protect water. The EU approach provides political will for members to act nationally to overcome special interests.

5.2.2 Local land use planning and aquifer protection

Well-head protection programs, sub-national and local land use planning and zoning, permitting programs, water user organizations, and implementation of programs authorized by national law are at the front-line of improving groundwater governance. As noted earlier, NGOs, community groups, and academic interests can all help raise awareness and demonstrate leadership in creating the political will to act. Communication programs and the media are key tools for the campaign. As noted in section 5.5, donors and international organizations have helped in this regard in developing countries to foster local leadership leading to some political will for local action.

A key decision is whether or not to protect an aquifer's recharge area. Small Island States have virtually no choice but to protect what they have left with much of their shallow groundwater polluted. Government must work with local officials and declaration of various areas as parks can help. One example is from St. Kitts and Nevis in the Caribbean. Funded by the GEF through a UNDP/UNEP project, the Basseterre aquifer provides 60% of the drinking water for the capital. Development pressures were changing the valley with urbanization. The GEF helped the communities protect the recharge area for the drinking water aquifer in reducing agriculture, urban and sewage pollution, and successfully implemented a GEF requirement for a national cross-department committee along with the communities and NGOs. In addition to zoning and best practices, the national Cabinet has approved the recommendation to formally declare the lower part of the Valley and unconfined aquifer a protected area in advance of the processes of establishing a National Park to protect the aquifer recharge zones (Duda *et al.*, 2009).

5.2.3 Local aquifer/recharge area-specific management needed

Ultimately, aquifer-specific interventions will be needed to sustain the resource, involving both resource management as well as sector reforms. In terms of developed countries, support from legislation or creation of ad hoc programs still is needed. In Spain, there is a long history of water and aquifer management. With its dry climate, more and more groundwater withdrawals for irrigation still pose problems. National law and programs promote aquifer management with aquifer management bodies and water user associations, but the largest use for irrigation still is not adequately addressed (Stefano *et al.*, 2015). In the US, the California legislature finally found the political will to enact a recent groundwater law as a result of the latest drought but implementation still has a decade to go.

Morocco has reformed its water law in a lengthy process to address governance of groundwater after it found that water depletion was the greatest threat to economic development. It has implemented improvements in a national irrigation water savings program (African Development Bank, 2009) as part of its Green Morocco Plan to help balance water uses such as in the Sousse Valley where a strategy has been legislatively adopted through aquifer contracts to reduce over-pumping. Measures ranged from increased fees for water use and new technology like drip irrigation to closing illegal wells and limiting irrigation expansion. This approach illustrated leadership and political will of the King and ministers to do what is best for the country as noted by the World Bank (2009a). A recent review of actual implementation of the excellent government reforms has shown that elites have still dragged their feet and frustrated the restoration of over-exploited aquifers so that they can continue to over-pump to maintain their agricultural exports (Closas and Villholth, 2016). Political will must extend to enforcement of legislative reforms.

5.2.4 Community to cabinet aquifer protection: top-down meets bottom-up

GEF IW interventions in the Pacific Islands have resulted in coining a phrase referring to combining top-down interventions with bottom-up: from community to cabinet as outlined across the GEF waters focal area by Duda *et al.* (2012). The saying describes the GEF strategy involving top-down interventions where inter-ministry committees for the project (required by GEF) work across sectors influencing surface water, groundwater, and recharge areas along with bottom-up action with local communities. These committees are one key to GEF's promotion of IWRM as described by Duda and El Ashry (2000) and actually are a key tool for integrating land management with surface and groundwater management.

This strategy for generating political will illustrates benefits of new practices and measures to both villagers and ministers to conserve water resources and protect its quality while supporting local income. This approach results in ministers from cabinet visiting local demo sites and seeing how local people benefit. The ministers then go back to cabinet and push for the budget envelope to scale up interventions or press donors for funding, or enact the needed reforms.

Two other examples come to mind of top-down meets bottom-up in the Pacific. Majuro Atoll is the capital of the Marshall Islands with its primary drinking water coming from rain on the airport runway and groundwater as a reserve source. Majuro is one of the most densely populated places in the world and is no wider than 400 meters in places. Part of the island is known as Laura's Village and was being urbanized. The GEF/UNDP/UNEP project aimed to utilize zoning, best practices, and community involvement to reduce density and protect the recharge area so that it can supply safe drinking water. A Pacific Drinking Water Safety Guide makes the protection of water sources understandable for communities, and the Laura Water Lens Protection Coordinating Committee leads the inter-ministerial coordination committee (Duda *et al.*, 2009).

In the Pacific country of Tuvalu, septic tanks and latrines contaminate groundwater. The capital island of Funafuti is regularly affected by long and difficult droughts. Rainwater is the only cheap and reliable source of potable water in recharging groundwater. As a direct result of the Tuvalu GEF IWRM demonstration project there has been a remarkable increase in community demand for the waterless, composting toilet to reduce the use of water and adverse impacts. After the first 40 demonstration composting toilets were installed, a survey of 530 households found 100% of respondents answered "yes" to composting toilets. Community engagement work, publicizing advances of composting toilets with their savings of 30% in water use, a community road show, less cost, and continuing droughts convinced the residents, cabinet, and Prime Minister that the risks of typhoid and other diseases a were just too much. The Tuvalu government approached New Zealand for help in scaling up as reported Duda *et al.* (2009).

5.2.5 Transboundary aquifer and basin systems

Transboundary aquifers underlie up to 40 percent of the world with at least 367 systems identified. Governance for transboundary aquifers has been analyzed in the GEF supported Global Groundwater Governance project (FAO, 2016a) and in the recently completed GEF Transboundary Waters Assessment Program (UNESCO, 2016). The analysis reported that most large aquifers are away from intense irrigation and are still in good shape for future use. However, the aquifers will need joint management institutions to keep sustaining benefits.

One shared aquifer between France and Switzerland has such an institution while fledgling cooperative frameworks have been fostered by the GEF for the Northwest Sahara, Iullemeden, Nubian, and Guarani aquifers the last decade. Integrated approaches to conjunctively manage surface and groundwater represent an important avenue for improved governance. GEF also tried to catalyze surface and groundwater management in its Nile Basin projects but the countries focused mostly on surface water. GEF funded one project in Egypt on groundwater as an alternative to surface water and then a small Nile Basin-wide groundwater project with UNDP (2008). Countries still focus most attention on surface waters and the Nile treaty that is open for signature and ratification. A new initiative has been funded by the GEF for the combined Niger River Basin through their treaty/basin organization linking with the Iullemeden and Taoudeni-Tanezrouft Aquifer System that underlies it (UNDP and UNEP, 2014).

5.3 GEF CASE STUDIES OF PROCESSES AND TOOLS TO FOSTER LEADERSHIP/POLITICAL WILL

The GEF is best known as the financial mechanism for a number of global environment conventions such as the climate change and biological diversity. The GEF's mandate is to provide incremental cost finance to address these global environment issues that include international waters (IW) – which covers transboundary surface/groundwater systems, marine waters and basins draining to those coastal and marine waters. The only new funding to emerge from the 1992 Earth Summit, GEF has allocated in its IW focal area over $US 100 million the last 15 years in helping 49 countries that requested work toward improved groundwater governance. Processes recommended by GEF that build trust and confidence and fill information gaps have been tested in GEF initiatives for aquifers like the North-west Sahara, Iullemeden, Guarani, Dinaric Karst, and Nubian. They help build sector and ministerial leadership for making the case in capitals that better governance is essential. The GEF strategy involves work at various levels of government – from multi-country to national sector as well as sub-national and local levels. Box 5.2 outlines these processes that can catalyze leadership to improve groundwater governance.

Box 5.2: GEF International Waters (IW) Project Processes that Help Catalyze Leadership

The GEF IW focal area provides grants to countries to address different water-related conflicts in basins and aquifers toward improved governance. GEF's practical experiences have validated three recommended processes and two GEF requirements fostering IWRM. The processes are: (1) formation of national inter-ministry committees; (2) production of an analysis for the basin of concern on status of the river or aquifer basin, different sector water uses, conflicts, and future projections (known as a Transboundary Diagnostic for multi-country water systems or a diagnostic analysis for single country systems), and (3) development of a strategic action program of policy, legal and institutional reforms and investments that address the priorities in the analysis through multi-stakeholder participative processes across sectors to balance the competing uses, make trade-offs and form partnerships for action. Additionally two requirements also help: required stakeholder participation and funding for local demonstration projects in hot spots to engage local communities and spur action among officials in the capitals.

If ministers agree to implement the action program, GEF may follow up with multiple projects to help implement solutions to the priorities identified by the analysis and included in the action program. Establishment of the functioning national inter-ministry committee is an indicator to be reported in project monitoring and evaluation. The analysis and draft program of action provide tools for engaging stakeholders in the process to foster leadership. Both are intended to simplify complex situations in order to better understand them. For ease of implementation, actions are divided into individual sector pieces to be included in budgets of local or national institutions, and local leaders can facilitate pushing for needed political will in the capitals. The GEF processes coupled with stakeholder participation and locally relevant demonstrations projects have been shown to catalyze leadership toward the MDGs and now SDGs. Further explanation of the GEF International Waters Strategy can be found in Duda and La Roche (1997).

5.3.1 National integrated river basin/aquifer management

Integrated land and water resources management (IL&WRM or IWRM) and conjunctive management has been fostered by the GEF with the recommended processes outlined in Box 5.2 and stakeholder participation anchored not only in the other processes but also in local demo projects. There is one notable case with a GEF project that illustrates success in a basin with intense irrigation. The government of China requested a GEF IW project through the World Bank to test innovative means of IWRM for surface water and underlying aquifers for a degraded basin, the Hai River basin from Beijing to Tianjin in the irrigated and intensively cropped North China Plain. Box 5.3 illustrates features of the $110 million project ($17 million GEF) that represents a best practice case for issuing water allocations based on consumptive use, water rights for farmer user associations, use of satellite technology for allocations and enforcing compliance with rights and allocations under the water law. Additional features of the GEF Hai Basin project include use of pricing to reduce irrigation for real waters savings, conjunctive management of surface and groundwater—both quantity and quality, and increased farmer income while using much less water that left the stakeholders satisfied. China is now upscaling the approach, and with the experiences in Australia for the Murray-Darling basin, the Hai Basin example illustrates paths forward on conjunctive management for basins with intensive irrigation to meet the SGDs.

Box 5.3: GEF Hai River Basin and Aquifer Project Illustrates Key Governance Reforms

The GEF/World Bank Hai River Basin IWRM project involved many facets of water and land reforms for improved quality in the river and aquifer as well as reduced water use in irrigation so more will be available for environmental flows to the river and reduced overdraft of the aquifers. The 7-year project, with $74 million from the Government of China, $41 million from the World

Bank, and $17 million in grants from the GEF water focal area, pioneered water and land management reforms in an integrated manner. The Hai Basin project demonstrated the utility of increased charges for irrigation water, a new water rights/allocation system under Chinese Law based on consumptive use (estimates of evapotranspiration or ET) and not standard withdrawal amounts, satellite technology for integrated land-water planning to support issuing and then enforcing the water allocations under law, and other water saving irrigation technologies to begin the process of rebalancing food security and water and environmental uses in the basin and its depleted aquifer system.

A truly innovative approach to IL&WRM, the project also included water quality improvement measures, capacity building for the basin water resources commission under Chinese law, and use of pre-paid cards for individual farmers with only enough allocations for their pumping to meet the reduced allocations needed for the real water savings. Satellite data on estimated evapotranspiration at a 30 meter by 30 meter scale was utilized with simulation models to provide reduced allocations to farmer-led water user associations, which in turn distributed the quotas to over 100,000 farmer households through pre-paid cards for pumps that they pay for. Once the allocation was gone, no more water could be pumped. Extension services assisted with practices for green water savings, best management practices (mulching, plastic, cropping patterns, drip technology), and alternative crops for increased farmer income. The result of the project was per capita income increases of 193%, water productivity increases of 82%, and a 27% decrease in consumptive use–with the real water savings available to stabilize the aquifer draw-down and leaving more water in the river for ecosystem use (World Bank, 2011; and Duda *et al.*, 2012).

5.3.2 Fostering the Guarani transboundary aquifer cooperative framework

With very few legally binding cooperative frameworks for transboundary aquifers, the time is right for leadership in negotiating those frameworks to avoid future degradation and depletion. One good example funded by GEF water area is the Guarani Aquifer. Shared by Argentina, Brazil, Paraguay, and Uruguay, the Guarani is the largest aquifer in South America. Little attention was paid to the groundwater in the water-rich area, and no regional cooperative framework was in place when the countries approached the GEF for assistance to better understand and manage the system which provides more than 15 million people with drinking water. The GEF encouraged the World Bank as the appropriate GEF agency for the IW project, as described in Box 5.4 and by the World Bank (2009b). The project utilized all five processes outlined in Box 5.2 in the GEF International Waters Strategy, and fostered leadership at all scales resulting in the Heads of States signing the Guarani Agreement. It is the first joint aquifer management agreement that is consistent with the 2010 United Nations Resolution on the Law of Transboundary Aquifers and provides an important model to improve governance of shared groundwater.

Box 5.4: Guarani Aquifer GEF IW Project Creates Opportunities for Leadership in Many Ways

The Guarani supplies drinking water to some 15 million people in the four countries including about 500 cities and towns in Brazil. In the 1990s pumping on border areas between Argentina and Uruguay had led to increased tensions while agriculture was threatening potential contamination of sensitive recharge areas in all countries. The four countries approached the GEF for support to develop an integrated management framework with the objective to "implement a shared institutional, legal, and technical framework to preserve and manage the Guarani Aquifer System for current and future generations". Consistent with GEF practice, all countries established national inter-ministry committees to promote cross-sector action to protect the Guarani. In Brazil similar committees were also established by sub-national state governments to ensure a more integrated management approach by the many cities and towns depending on the aquifer. The transboundary analysis was produced to engage the technical community and enhance understanding of the aquifer system, providing a basis for different stakeholders and policymakers to discuss the draft analysis that describes complexity in simple terms.

The formulation of a Strategic Action Program (SAP) also helped to increase stakeholder dialogue and utilize existing political processes to help determine a shared vision for the future transboundary management of the resource. Given the sheer size of the Guarani, countries targeted management actions on important recharge zones and sensitive border areas. The SAP included measures aimed at national reforms in addition to the legally binding cooperative framework that was negotiated then signed by all 4 Heads of State. All countries have now taken practical measures to protect the Guarani at the provincial and national levels as noted by Duda *et al.* (2012): Brazil integrating groundwater considerations into its National Water Resources Plan with funding allocated to support the implementation of its Surface and Groundwater Integrated Management Program; all six Guarani aquifer provinces in Argentina are now represented on the Argentina Federal Water Resources Council; Paraguay's new Water Resources Law now includes groundwater; and Uruguay has established a national Guarani Management Unit. Participation and awareness were advanced among the wider public and indigenous communities through a dedicated "Citizen's Fund" established to encourage community- based NGOs in participating.

The Guarani Aquifer is one of the largest aquifers on the planet. Little attention was paid to the groundwater in the water-rich area and no regional cooperative framework was in place when the countries approached the GEF for assistance to better understand and manage the system which provides more than 15 million people with drinking water. There was a hint of concerns with agricultural chemicals being applied in suspected recharge areas and excessive pumping on the border of Argentina and Uruguay. The World Bank and GEF helped the countries obtain a GEF IW project to support he countries to follow the GEF recommended procedures to build trust and confidence and provide analysis products for stakeholder dialogue to catalyze leadership opportunities. Each country established a national inter-ministry committee and Brazil went down to the next level and established state inter-ministry committees to ensure vertical and horizontal collaboration, especially with cities and towns and business stakeholders.

The results of the project are covered in Duda *et al.* (2012) and World Bank (2005).

5.4 REFORMS TO IMPROVE GROUNDWATER GOVERNANCE AND IL&WRM

Past experiences point the way toward strategies for meeting multiple SDGs by scaling up conjunctive management of surface and subsurface waters and their catchments through IL&WRM as a first step toward balancing competing uses of water. Only place-based, integrated approaches on the landscape combined with sector reforms will work. Key reforms include land tenure reform (and women's rights to land) and water pricing (charges from irrigation, mining, and urban water users). As noted many years ago by Duda and El-Ashry (2000), sufficient water use charges are a critical tool to encourage conservation and support management. The importance of pricing has also been shown in the Hai Basin example and other places such as the Murray Darling Basin for irrigation, urban water tariff increases with the recent California drought that saved 25% of water (California Water Board, 2016), and the urban example of Windhoek, Namibia with its comprehensive tariff approach (Lahnsteiner & Glempert, 2007).

Public grants and development assistance should include provisions for adopting: water rights/allocation systems; water charges for larger irrigation, mining, and urban uses (with the funding staying in the aquifer or basin); national inter-ministry committees (with memoranda of understanding among ministries) for projects and programs with integrated approaches to ensure horizontal and vertical government cooperation; basin and aquifer-specific drought management planning to mitigate climatic variability; and significant increases in development assistance support to enact the needed reforms and operationalize IL&WRM to meet SDGs.

Other driving forces for local and national reforms for better subsurface governance include subsidence of infrastructure/buildings and aquifer compaction with over-pumping that can damage aquifer structure. Serious economic costs result globally from subsidence as subsurface space is compacted; more information about the issue is available from van der Gun *et al.* (2012). Vast damage can result with remedial actions being very expensive. For example, the Norfolk-Virginia Beach coastal area of Virginia has over-pumping its aquifer for many decades producing serious subsidence, infrastructure damage, and sea level incursions into the four-city area as noted by the US Geological Survey (Eagleston and Pope, 2013), making the urban area the second most susceptible city in the US to storm inundation from the sea next to low-lying New Orleans. The solution under development is large-scale treatment of wastewater with injection into the aquifer to prop it up (Fears, 2016) which will run into the hundreds of millions of dollars. Limiting fracking and injection of fluids for energy extraction to appropriate areas and banning hazardous disposal into the subsurface are also needed and require national and local reforms.

Integrating groundwater management with surface water represents another reform necessary to meet SDGs. Ministries and agencies often have separate authorities for quality, quantity, surface and subsurface waters, which erect barriers to integrated water and land resources management. The Australian government presents good examples of conjunctive management of surface and groundwater and the background for application (National Water Commission, 2014). The GEF global groundwater governance project supported a paper covering features of this important integrated approach (Evans *et al.*, 2013). Likewise, aquifers serve as important tools in reducing droughts with water harvesting adopted as part of climate smart agriculture and managed aquifer recharge becoming common, with guides available from UNESCO (Gale, 2005).

Unfortunately, conjunctive management legislation has not been popular. Approaches such as described by Duda (1989) for integrating groundwater into surface water legislation in the US have gone nowhere with special interests being successful in keeping the status quo in the US Congress. Leadership must be exercised in most countries to authorize integrated management of the much larger groundwater resource with the much smaller surface water resource that gets all the attention. With adoption of the SDGs by Heads of States in 2015 and climate adaptation funding becoming available for drought management planning as a result of increased droughts, opportunities are becoming available to catalyze needed action for integrating surface and groundwater governance with agricultural land management in catchments and recharge areas.

At the multi-country, transboundary scale, GEF experiences detailed herein and by Duda *et al.* (2009) illustrate that processes have been developed to build trust and confidence among countries to enter into cooperative management frameworks ranging from exchanges of information to management arrangements signed by Heads of States. The GEF pilot now underway for the Niger Basin and the Iullemeden and Taoudeni-Tanezrouft Aquifer System that underlies it will provide an indication of practicality on a transboundary scale. At a global scale, a whole host of issues related to trade policy, global markets, land grabs by foreign nations, prices for commodities, and distortions caused by agricultural and energy subsidies in many countries have accompanied globalization. Agricultural subsidies in the North reduce prices for food produced in developing countries so that farmers find it hard to compete with those in rich countries. The World Bank (2008) estimated global agriculture subsidies to exceed $US 245 billion annually. Other damaging subsidies include subsidized electricity to run irrigation pumps (half of India's irrigated land comes from pumping groundwater) that discourage spending on water savings or restoration of overly-exploited aquifers (World Bank, 2008). Fertilizer subsidies that divert money for conservation or water savings to benefit elite farmers also need elimination. While improved IL&WRM measures and polices need up-scaling, there is a set of interventions and campaigns needing to be waged on a global scale for these politically sensitive policy issues.

5.5 HIGH LEVEL POLITICAL WILL, LEADERSHIP AND PARTNERSHIPS

At a global scale and within international institutions, obstacles exist to coherence on action among national government sectors, international organizations, donors, civil society, and the private sector for catalyzing sustainable surface and groundwater management. Heads of States and national legislatures sometimes need encouragement to generate political will. Agenda 21 from the 1992 Rio Conference and the Millennium Development Goals (MDGs) in 2000 represented a "light touch" on the part of Heads of States. Given these experiences, the need for more comprehensive, integrated approaches was recognized by Heads of States in their adopting the SDGs. This may prove to be the driving force necessary for political will.

In support Agenda 21 and the MDGs, the GEF was tasked with testing and demonstrating on-the-ground measures and accompanying policy reforms to achieve sustainable development. The GEF water area has provided leadership on groundwater governance and integrated approaches since the 1990s with several dozen groundwater projects. Even today, current funding priorities for GEF through 2017 include transboundary aquifer systems and conjunctive management of surface and groundwater (GEF, 2014). More importantly, the GEF-supported Global Groundwater Governance Project that was successfully concluded in early 2016 and described in Box 5.5 has generated valuable governance analyses regionally as well as a global vision with a consensus global framework for action that provides a pragmatic roadmap for countries, businesses, and organizations to improve groundwater governance (FAO, 2016b and 2016c).

Box 5.5: Global Vision and Framework for Action – GEF GLOBAL GROUNDWATER GOVERNANCE INTERNATIONAL WATERS (IW) PROJECT

Following the late 1990s development of a vision on water, which was mostly surface water based, the GEF realized the gap in such processes for groundwater and worked with the World Bank, FAO, UNESCO, and the International Association of Hydrogeologists to develop a similar global dialogue process for improving groundwater governance. The GEF Council approved the proposed global GEF IW project to undertake the work, and after five years, it successfully concluded in early 2016. The diagnostic phase of the project undertook development of thematic papers from experts, regional diagnostic analyses of the state of groundwater governance in countries, five regional consultations with hundreds of interested professionals, and a global synthesis of shortfalls, gaps, good governance examples, and options for improving governance with policy responses and financial needs as reported by FAO (2016a). A global visioning process was also undertaken and a consensus ***Shared Groundwater Governance Vision*** detailed critical aspects of good governance that can help support sustainable development as reported by FAO (2016b). More importantly, the ***Global Framework for Action*** represents a call for action by all segments of society to take urgent steps to conserve and sustain invaluable groundwater resources. The Framework

(FAO, 2016c) covers elements for good governance, building of effective institutions at all levels, groundwater governance in the context of IWRM, developing linkages with other sectors that complement well the new 17 SDGs adopted by Heads of States, redirecting finance, and pragmatic steps to be taken toward improved groundwater governance. The reform process doesn't start from zero. Background analyses, good practice cases, key elements, and consensus strategies already exist for immediate application.

As GEF has demonstrated, grants can clearly catalyze some leadership, but much more than GEF pilot funding needs to be allocated. Specialized UN agencies, regional development banks and the World Bank, country development assistance agencies, global NGOs, and universities as well as foundations need to work together coherently in partnerships with national governments (not in competition) to achieve shared, multiple goals. Adoption of the SDGs now makes this a possible driving force for improved groundwater governance.

In the key area of water resources that underpin achieving most SDGs, The UN Secretary General and the President of the World Bank recently took a valuable step with leadership in operationalizing the new political will of the SDGs by creating a High Level Panel on Water. As described in Box 5.6, the Panel may help support countries that wish to make the transformational governance reforms with capacity building, technical assistance, and funding for investments.

Still, countries of the North that may be influenced by special interests may not enthusiastically work toward the SDGs. In these cases, civil society organizations, NGOs, local governments, and the university community need to stand up and provide leadership in calling attention to the impediments to sustainable surface and groundwater governance.

Box 5.6: UN High Level Panel on Water (HLPW) Created to Support Water SDGs

Realizing that water resources underpin achieving most all of the SGDs through inter-sectoral linkages, the Secretary-General of the UN established the HLPW with the President of the World Bank Group. Consisting of 11 Heads of States and Governments and an Advisor, the Panel is designed to advocate a comprehensive and coordinated approach to water resources as well as increase attention and investment in water-related services. As Heads of States and Governments, the Panel has committed to lead the way forward with a new policy priority and funding resources from the UN and development banks to support countries.

Achieving the water dependent SDGs requires a comprehensive and transformative approach across sectors (both in capitals and vertically at the surface/groundwater basin scale) and with development assistance organizations. Leadership in terms of political will for this comprehensive approach is required for progress. The HLPW is aimed at motivating partnerships, governance reforms, and local action across governments, civil society, and the private sector. For more information on the HLPW refer to High Level Panel on Water (2016).

Perhaps the most important catalytic tool for leadership lies with partnerships of global NGOS and multi-national corporations. Since the turn of the century, a steady drum beat has come from the private sector about degraded and depleted water resources becoming a serious risk to their profits and sustainability. Coca Cola, Heineken, and other companies established partnerships with NGOs the last decade and set corporate goals for water use sustainability. For several years, the World Economic Forum has been discussing water constraints to global business. The formation in 2008 of the 2030 Water Resources Group of private sector leaders supported by the International Finance Corporation of the World Bank Group led to its landmark call to action for business entitled "Charting our Water Future" (McKinsey & Company, 2009). Now, as a result of recent risk analyses, new warnings from the World Economic Forum (2015) have been issued that water problems represent the number one risk to the global economy. Corporate water stewardship has become an important element (Rozza, 2013). Many companies are adopting measures to reduce water use at facilities and work through their agricultural supply chains (irrigation) to save water by using climate smart agriculture. This new interest in protecting profits can be harnessed to provide leadership on improving groundwater governance.

NGOs and governments of the North can catalyze the partnerships starting with multi-national corporations and their supply chains in both the North and South toward supporting reforms related to integrated surface and groundwater management to meet multiple SDGs. If multi-national corporations walk the talk with their operations and agricultural supply chains to influence high-level government officials with whom they interact, leadership and political will for reforms, programs, and investments can be catalyzed by these business contacts.

The potential for political will to exert leadership is at an all-time high with many elements in place. Added to this new impetus for coherent action is the roadmap associated with five years of global dialogue under the GEF Global Groundwater Governance Project. The resulting diagnostic analyses, assessments, shared global vision, and consensus framework for pragmatic actions now need to be utilized to improve groundwater governance. The specter of a changing climate with extreme droughts will energize civil society, academia, local officials, NGOs, and international organizations to make sure this revolution toward integrated land, surface and groundwater management takes place within governments and the business community.

5.6 CONCLUSIONS

Lack of attention to governance of aquifers, their recharge areas and the larger issue of integrated water resources management is irreversibly depleting and degrading precious groundwater in many countries with devastating social and economic results that will constrain future prosperity. Leadership has just been insufficient and governance has been weak. Two thirds of the people on our planet now suffer water scarcity at least one month per year. There is no way that multiple goals of the new Sustainable Development Goals adopted by Heads of States in 2015 can be achieved without improved surface and groundwater governance. That means that strong leadership and political will is necessary not only for effective groundwater governance but also to achieve multiple sustainable development goals.

Standing in the way of governance reforms have been policy failures not only in water management but also in sectors that compete for water. This paper has identified obstacles undermining efforts to improve groundwater governance and presented case studies of approaches in developing and developed countries that can foster leadership to begin overcoming those obstacles. The obstacles fall into a number of categories ranging from physical to economic and social. From a physical standpoint, surface water, groundwater, aquifers, and catchments pose enormous complexities and physical unknowns. Attempts at managing these open access and common property systems are impeded by variable geology, lack of data/information on unseen groundwater, a changing climate, and complex cause-effect inter-linkages that support meeting multiple SDGs. Economic obstacles run the gamut from poverty and low government priority for assessment and management to generous sector subsidies stressing groundwater and low funding priorities by foundations, NGOs, and international aid agencies.

By far, social/institutional obstacles are most important for standing in the way of progress. Culture, political interference, missing political will, lack of capacity and awareness, and the challenge of mobilizing civil society participation are only a start. A whole array of institutional failures conspires against action, including: lack of policies, legislation gaps, lack of enforcement, corruption, few incentives and disincentives, competition among ministries for water use, difficulty in cooperation among ministries (horizontal), challenges in national and subnational (vertical) collaboration, and the nagging lack of transparency and accountability in government.

Despite the obstacles, progress has been made in many countries but much remains to be accomplished. Improved groundwater governance is needed at all scales, from the very local farmer's field or city catchment to sub-national government like provinces or states, hydrologic units such as river basins or aquifers and their recharge areas to national water resource management and sector management policies and ministries, transboundary water systems, and global institutions. Different actors may work at the different scales to provide leadership and foster needed political will. A changing climate provides impetus in developed countries for new leadership fostered by university and scientific capacity, NGOs, community groups, foundations, and local governments. Even in developing countries these groups must be ready with campaigns and coherent partnerships aimed at creating political will when extreme events hit.

In developing countries, programs of rich country development assistance agencies and international organizations, including NGOs, must exercise leadership, including grant funding devoted to needed reforms for improving governance. Case studies illustrate that grant funding from the GEF International Waters focal area has been successful in facilitating progress toward improved groundwater governance. Three recommended GEF processes seem critical: establishing national inter-ministry committees, producing a diagnostic analysis, and formulating an action program based on visioning and participation to set new priority for programs, reforms and investments. Two additional GEF requirements (civil society participation and on-the-ground local demonstration pilots) have proven to be key elements to foster local leadership and national commitment to action. This GEF concept termed "from community to cabinet" is simply a combination of top-down and bottom-up processes that have proven to catalyze integrated surface and groundwater reforms for basins and aquifers. Other cases of local aquifer action, sector regulatory action, best practices, and piloted governance reforms also contribute to integrating approaches to land, surface and groundwater management Additionally, pricing water use, land tenure reform, water rights and allocation systems, water quality protection institutions, utilization of climate smart agriculture, and removal of agricultural subsidies that distort food prices and trade will be necessary to improve groundwater governance. Grant resources from national governments to subnational and local governments have proven invaluable to spur progress and are needed for targeting resources to important priority aquifers.

For the first time in 30–40 years since the irrigation and input-dependent agricultural "Green Revolution" began devastating surface and groundwater resources and their associated biological diversity, sufficient driving forces have come together to stimulate the opportunity for real action toward improved and integrated surface and groundwater governance. Lessons have been learned from the partially successful attempts at sustainable development through Agenda 21 and the Millennium Development Goals, resulting in a more transformative and integrated set of 17 Sustainable Development Goals (SDGs) adopted in 2015 by Heads of State at the UN. On-the-ground measures, tactics, and policy reforms for sustainability have been implemented in some developed countries and pilot tested globally. The GEF has funded these integrated approaches for two decades, and now leadership is needed from individuals in every organization globally and locally to scale-up what has worked.

In response to the SDGs, the development assistance community has come together more coherently to sustain water resources through leadership of the UN and the World Bank in the High Level Panel on Water and through their organizational programs. The private sector has now awoken to its potential loss of profits and sustainability with corporate water stewardship becoming a necessity. Climate change adaptation funding provides another driving force for action to convince ministers and Heads of States to exert political will for reforms and funding to support integrated basin and aquifer/recharge area participatory management. The GEF Global Groundwater Governance Project has successfully concluded dialogues with thousands of people to produce materials, strategies, a shared vision and key elements of a Framework for Action in which we all can play leadership roles to improve groundwater governance. It is a new day, and it will take everyone working coherently together to catalyze action through partnerships that build political will for improving groundwater governance and attaining the new SDGs.

REFERENCES

African Development Bank. (2009) *National irrigation water saving programme support project (PAPNEEI)*. Kingdom of Morocco. Project Appraisal Document.

California Water Board. 2016. *Drought Update* Thursday, July 21, 2016

Closas, A. & Villholth, G. (2016) *Aquifer contracts: A means to solving groundwater overexploitation in Morocco?* International Water Management Institute, GRIPP Case Study Report Number: 001.

Duda, A. M. (1989) Unified management of surface and ground-water quality through Clean Water Act authorities. *Ground Water*, 27(3), 351–362.

Duda, A. M. (2003) Integrated management of land and water resources based on a collective approach to fragmented international conventions. *Philosophical Transactions of the Royal Society of London B*, (358), 2051–2062.

Duda, A. M. & La Roche, D. (1997) Sustainable development of international waters and their basins: Implementing the GEF Operational Strategy. *Water Resources Development,* 13(3), 383–401.

Duda, A. M. & El-Ashry, M. T. (2000) Addressing the global water and environment crises through integrated approaches to the management of land, water, and ecological resources. *Water International*, 25(1), 115–126.

Duda, A. M., Severin, C., Bjornsen, P., Zavadsky, I. & Menzies, S. (2009) *From ridge to reef: water environment and community security: GEF action on transboundary water resources.* Global Environment Facility.

Duda, A. M., Zavadsky, I., Severin, C., Hume, A., Menzies, S., & Donaldson, R. (2012) *From community to cabinet: Two decades of GEF action to secure transboundary river basins and aquifers*. GEF IW:LEARN and United Nations University.

Eagleston, J. & Pope, J. (2013) *Land Subsidence and Relative Sea-Level Rise in the Southern Chesapeake Bay Region.* US Geological Survey Circular Number: 1392.

Evans, W. R. & Evans, R. (2014) *Conjunctive use and management of groundwater and surface water*. FAO-GEF Groundwater Governance Project Thematic Paper Number: 2.

Fears, D. (2016) Hopes that wastewater can conserve land in coastal Virginia: Purified wastewater will be used to help restore aquifer. *The Washington Post.* October 21, 2016.

Food & Agriculture Organization. (2016a) *Global diagnostic on groundwater governance.* FAO-GEF Groundwater Governance Project.

Food & Agriculture Organization. (2016b) *Shared global vision for groundwater governance 2030 and a call-for-action.* FAO-GEF Groundwater Governance Project.

Food & Agriculture Organization. (2016c) *Global framework for action to achieve the vision on groundwater governance.* FAO-GEF Groundwater Governance Project.

Foster, S. & Ait-Kadi, M. (2012) Integrated water resources management (IWRM): How does groundwater fit in? *Hydrogeology Journal*, 20, 415–418.

Gale, I. (ed) (2005) *Strategies for managed aquifer recharge (MAR) in semi-arid areas.* United Nations Education Science & Cultural Organization and United Nations Environmental Programme.

Global Environment Facility. (2014) *GEF-6 programming directions*. GEF Document Number: GEF/A.5/07/Rev.01.

High Level Panel on Water. (2016) *Mobilizing action towards a water secure world for all.* https://sustainabledevelopment.un.org/HLPWater

Hoekstra, A. Y., Mekonnen, M.M., Chapagain, A. K., Mathews, R. E., & Richter, B. D. (2012) Global monthly water scarcity: Blue water footprints vs blue water availability. *Plos One* [online] (7), 1–9. doi: 10.1371/journal.pone.0032688.

Johnson, N., Revenga, C., Escheverria, J. (2001) Managing water for people and nature. *Science,* (292), 1071–1072.

Lahnsteiner, J. & Glempert, L. (2007) Water management in Windhoek, Namibia. *Water Science & Technology* (55) 441–448.

Mekonnen, M. M. & Hoekstra, A. Y. (2016) Four billion people facing severe water scarcity. *Sci. Adv.*, (2), 1–6.

McKinsey & Company. (2009) *Charting Our Water Future: Economic frameworks to inform decision-making.* 2030 Water Resources Group. International Finance Corporation.

National Water Commission. (2014) *Integrating groundwater and surface water management in Australia.* Government of Australia.

Rozza, J. (2013) Corporate water stewardship: Achieving a sustainable balance. *Journal of Management and Sustainability,* 3 (4), 41–52.

Stefano, L de, Fornés, J.M., López-Geta, J.A. & Villarroya, F. (2015) Groundwater use in Spain: an overview in light of the EU Water Framework Directive. *International Journal of Water Resources Development.* 31(4) 640–656.

United Nations Department of Economic and Social Affairs. (2016) *Sustainable development knowledge platform: Sustainable development goal 6.* [Online] Available from: http:sustainabledeveleopment.oun.org/sdg6. [Accessed 20 October 2016].

United Nations Development Program. (2008) *Mainstreaming groundwater considerations into the integrated management of the Nile river basin.* Global Environment Facility Project Number: 3321.

United Nations Development Program and United Nations Environment Program. (2014) *Improving IWRM, knowledge-based management and governance of the Niger basin and the Iullemeden-Taoudeni/Tanezrouft aquifer system (ITTAS).* Global Environment Facility Project Number: 5335.

United Nations Education Science & Cultural Organization and United Nations Environmental Programme. (2016). *Transboundary aquifers and groundwater systems of small island developing states: Status and trends.* United Nations Environmental Programme.

UN General Assembly (UNGA). 2015. *Transforming our world: the 2030 Agenda for Sustainable Development.* Resolution adopted by the UNGA on 25 September 2015. Resolution Number: A/RES/70/1.

van der Gun, J., Merla, A., Jones, M. & Burke, J. (2012) *Governance of the subsurface: Space and groundwater frontiers.* FAO-GEF Groundwater Governance Project Thematic Paper Number: 10.

World Bank. (2008) *Chapter 4 Reforming trade, price, and subsidies.* World Development Report 2008.

World Bank. (2009a) *Making the most of scarcity: Accountability for better water management in the Middle East and North Africa.* Water Sector Board Water P Notes Issue Number: 40.

World Bank. (2009b) *Implementation completion and results report: Environmental protection and sustainable development of the Guarani Aquifer system project.* World Bank Number: ICR00001198.

World Bank. (2011) Implementation completion and results report (TF 53183) for Hai River Basin integrated water and environmental management project. World Bank Report Number: ICR1962.

World Economic Forum. (2015) *The Global Risks Report,* 10th Edition.

World Resources Institute. (2013) *Creating a sustainable food future: A menu of solutions to sustainably feed more than 9 billion people by 2050.* World Resources Report 2013–2014. Interim Findings.

Chapter 6

Legal principles and legal frameworks related to groundwater

Stefano Burchi
International Association for Water Law – AIDA, Rome, Italy

ABSTRACT

A mix of established and emerging trends in domestic groundwater-related legislation is illustrated from a comparative law perspective. The analysis is arranged around a number of questions or issues the reader may wish to ask or explore, from who owns groundwater to what is regulated and how, from what role for groundwater users in resource governance to concern for the ecosystem-support function of groundwater, from the land/subsurface space/groundwater nexus to conjunctive use and managed aquifer recharge, to where groundwater regulation intersects the customary water practices of traditional communities. Also, the body of international law that governs relations between States as regards groundwater and aquifers that straddle international or interstate boundary lines is illustrated. Despite the scanty evidence on record, the implementation and enforcement of domestic groundwater legislation is briefly addressed, and the importance of such legislation to meet States' obligations regarding transboundary groundwater and aquifers is emphasized. Advances in discrete segments of the groundwater regulatory spectrum, and strides in reaching out to non-groundwater regulation in view of its relevance to groundwater, are highlighted in conclusion, alongside advances in the codification of norms of States' behaviour, and in the negotiation of treaties and agreements between States, regarding transboundary aquifers.

6.1 INTRODUCTION

In this chapter, the legal practices – established and emerging – which underpin, and are an integral part of, the governance of groundwater domestically and in a transboundary inter-State context are illustrated, and discernible advances highlighted. Well-established legal practices are illustrated first, and advances highlighted, in relation to groundwater ownership, regulation of groundwater extraction, control of groundwater pollution from "point" sources, and the increasingly prominent role of groundwater users in groundwater governance. Novel and emerging legal practices and regulatory trends are illustrated next, signalling advances from mainstream groundwater governance and regulation towards other facets of the groundwater governance spectrum engaging, notably, the environment-support function of groundwater, the land/groundwater interface and interaction, the surface water/groundwater interface and interaction, and the interface and interaction of customary groundwater rights and practices with formal groundwater rights. The transboundary dimension of groundwater governance is addressed separately, with attention directed at advances in the legal frameworks and in the practice of States as regards the governance of aquifers common to two or more sovereign States, or also to two or more states or provinces of federal countries. The chapter casts the net further, by also exploring the much neglected and largely un-charted governance territory of implementation and law enforcement. Finally, the main features and advances in the legal practices and frameworks for groundwater governance are summarized in a concluding section of the chapter.

6.2 ESTABLISHED TRENDS AND PRACTICES

6.2.1 Whose groundwater is it?

One of the cornerstones of contemporary groundwater governance is severing the link that has long existed in law between ownership of the land and ownership – and control – of groundwater lying below. This is a powerful link, also – and perhaps foremost – in the minds of landowners, who tend to think of groundwater as something intensely private, and to behave accordingly, regardless of what the law says (Mechlem, 2016). In view of the common-pool nature of groundwater, all landowners can stake equally valid land property-based claims, fuelling conflict and litigation compounded by the limited knowledge and poorly understood dynamics of groundwater, and risking over-abstraction and eventual depletion of the resource[1]. This circular, highly conflictive and potentially destructive situation has effectively been tackled in a vast majority of the countries the world over by removing groundwater from the exclusive ownership and control of landowners, and by placing it under the stewardship of the State. This way, opportunities for groundwater extraction and development have opened up for non-landowners as well.

The shift from private ownership to State stewardship of groundwater is nowadays firmly entrenched in contemporary water legislation. There persist, however, a few isolated pockets where groundwater is regarded by law as the property of the overlying landowner, notably India, Pakistan, and the state of Texas (USA). There, the so-called rule of capture by the overlying landowners prevails[2], negating effective governance,

[1]In a famous 1904 USA case (*Houston & Texas Central Railway Company v. East*, 81 SW 279) the Texas Supreme Court ruled that a landowner was not entitled to damages from a railroad company whose drilling of wells for its steam locomotives had caused his well to go dry (reported in Nowlan, 2005). To-date, Texas is the only USA state to maintain the so-called "rule of capture" (see footnote 2 below). In two Indian Union cases dating back to 1923 [*Babaji Ramlin Gurav v. Appa Vithavja Sutar*, AIR 1924 Bom 154 (High Court of Bombay, 23 February 1923)] and 1930 [*Malyam Patel Basavana Gowd (dead) v. Lakka Narayana Reddi*, AIR 1931 Mad 284 [High Court of Madras, 23 October 1930], the courts spent considerable time and effort in attempts to unravel certain physical features of the groundwater in dispute between adjoining landowners (reported in Cullet, 2012)

[2]The Texas Supreme Court has upheld the rule of capture in a contemporary landmark case adjudicated in 2012 (*Edwards Aquifer Authority v. Day*, 369 SW 3rd 814, discussed in https://scholar.google.com/scholar_case?case=17654424129403106972&hl=en&as_sdt=6&as_vis=1&oi=scholarr%5C) [accessed 18 April 2017]. That same ruling was subsequently

as the sorry state of the resource in those jurisdictions amply demonstrates. The shift from private ownership to State stewardship has not always been immune to legal challenge before the courts of law, on grounds of a takings of constitutionally protected private property rights, triggering the issue of compensation. The available record shows however that claims of compensation have been consistently rejected by the courts in the cases on record, notably, the USA (Arizona, New Mexico), Italy and Spain, on mixed grounds of law and policy (Burchi, 1999; Burchi & Nanni, 2003). As a result, and also on the strength of the experience of countries like South Africa, where apprehensions about the constitutionality of dispossessing landowners of their private property rights in water resources, surface and underground, turned out to be ill-founded, any serious advance in groundwater governance is predicated on the State asserting its role of steward or trustee, let alone owner, of the resource on behalf of the public (GEF, FAO, UNESCO-IHP, World Bank, IAH, 2015 & 2016).

6.2.2 What is regulated, and how? From well-drilling and groundwater extraction to groundwater pollution from point-sources

It is well-established groundwater governance practice to routinely regulate the extraction of groundwater and the instrument to do so, *i.e.*, the drilling of wells, by means of administrative permits or licenses or other equivalent legal instruments. It is also standard practice that such legal instruments, and the rights they carry, be time-bound, and qualified as to, notably, volumes and rates of extraction. Groundwater extraction rights are also adjustable through their lifetime to reflect changing circumstances, however the issue of compensation looms large in this regard[3]. Well-drilling and groundwater extraction regulation is also selectively targeted at aquifers under stress from over-extraction (the phenomenon is commonly known as groundwater "mining"[4]). In the state of Texas (USA), for instance, permitting, well spacing and setting extraction limits, are available inside the perimeter of Groundwater Conservation Districts. Restrictions, however, are not mandatory as most of the districts which have been established have worked to get landowners to implement conservation measures voluntarily through educational programmes and by providing data on available supply, annual withdrawals, recharge, soil conditions, and waste. In Spain, groundwater extractions may be curtailed in areas which are declared groundwater mining areas, until a plan for the recovery of the aquifer is made and adopted (Burchi & Nanni, 2003). A similar approach is in effect in Algeria, however the government's authority to curtail extractions in aquifers under stress is not qualified as in Spain. In the states of

---

interpreted and applied by the Texas Court of Appeals in *Edwards Aquifer Authority v. Bragg*, 421 SW 3rd 821 (2013).

[3]Under Spain's Water Act (2001), for instance, water abstraction licences can be curtailed to accommodate the exigencies of the environment. Compensation has been ruled out by the action of the courts (Brufao Curiel, 2008). A similar provision exists in the Water Resources Management Act of Namibia (2013) with specific regard also to well drilling licences, however the issue of compensation is not addressed in the Act, nor has case law emerged so far clarifying it.

[4]The term groundwater "mining" is also used in connection with the extraction of fossil groundwater, *i.e.*, groundwater stored in non-recharging aquifers.

Punjab and Haryana (India), it is the practice of paddy rice cultivation which is the target of regulatory attention aimed at slowing down groundwater mining for irrigation. Well-established governance practices often include the regulation of the well-drilling trade and profession, like in most Western states of the United States, where moreover it is a legal requirement to contract with duly licensed well drillers only.

Building upon baseline regulatory practices, mature groundwater governance systems tie the initial grant and/or management by government of extraction rights to water planning instruments and determinations (*e.g.*, France, Spain, the state of California (USA)). In the state of Arizona (USA), extractions are tied by the state water law to the maximum water duty or allotment on each farm, which is based upon the crops historically grown and assuming increasingly stringent measures for the efficient application of irrigation water, such as lining of irrigation canals and the use of laser leveling fields. Other mature water governance systems, such as that in effect in the Australian state of New South Wales, have turned the quantum of extractions from a volumetric allocation to a variable share in the available groundwater from a given aquifer. Relevant licences are thus made up of two parts: a "share component", which entitles the licence holder to a share in the available groundwater from the aquifer; and an "extraction component", which entitles the licence holder to take groundwater at specified times, rates and at specified locations from the given aquifer. The share component of a licence is the kingpin to this sophisticated governance regime, and it is determined on the basis of water sharing "rules" negotiated in cyclical ten-year aquifer management plans. The sharing rules may undergo change during the life of the relevant plan, however access/extraction rights holders may claim compensation if their rights suffer diminution as a result (Burchi, 2001; Burchi & Nanni, 2003).

The more advanced governance practices allow for a delicate balancing act where the legitimate interests of groundwater developers are weighed at regular intervals against the sustainability of extraction rates and against the survival of groundwater-dependent habitats, thus taking on board also the interests of the environment and those of future generations. Moreover, mature groundwater governance systems take on board the economic value of groundwater by implementing the "user pays" principle, and charging for the extraction of groundwater a price which generally reflects the higher scarcity value of high-quality groundwater compared to surface water. In the state of Arizona (USA), for example, a tax is levied on all users of groundwater according to the volume which is consumed, and the proceeds from the collection of the tax are directed to purchasing groundwater rights and retiring them from use.

As regards man-made pollution of groundwater, it is well-established governance practice to regulate "point" sources of pollution, *i.e.*, industrial outfalls and municipal sewers discharging underground, notably through injection wells, by means of administrative permits, licences or other equivalent instruments. It is also standard practice that such legal instruments, and the rights they carry, be time-bound, and qualified as to, notably, the quality of the wastewater discharged (also termed "effluent") and the required treatment, and the timing and rate of discharge. The more sophisticated governance systems attune the quality standards of discharges to the achievement of pre-determined "ambient" water quality objectives for the recipient water body, and complement discharge permits with payment of charges aimed at penalizing polluting discharges ("polluter pays" principle). In view of the vulnerability of groundwater to irreversible pollution, however, permit-based and charge-based approaches to point-source pollution are qualified by paramount prohibitions as regards, in particular, the discharge of hazardous or toxic waste underground. In the European Union, all direct discharges of pollutants to groundwater have been outlawed since 2012 (Mechlem, 2012).

6.2.3 What role for groundwater users in groundwater governance?

Groundwater users are increasingly attracted by law in the making and implementation of decisions which affect them, and have become as a result a steady feature of contemporary groundwater governance systems. Their active participation in the governance of groundwater is widely seen and practised as an effective vehicle to build support for, and eventual compliance with, unpopular decisions. The participation of groundwater users is a well-entrenched feature of the traditional governance systems in place in a number of countries. In Yemen, for instance, local communities manage water supply systems, and a few have implemented schemes to protect groundwater used for drinking purposes from intensive agricultural use. In the state of Gujarat (India), there is a large farmer movement based on Hindu tradition to recharge dug wells in hard rock areas (Burke & Moench, 2000).

In the formal groundwater governance systems, regulated groundwater planning processes routinely provide opportunities for groundwater users' participation in the formation and adoption of plans, directly and through their elected representatives on the committees tasked with the formation of the plans. In the French sophisticated water planning system, detailed water master plans covering specific basins, sub-basins or aquifers (SAGE) are formed and adopted by an ad hoc Local Water Commission, one-fourth of whose members consist of representatives of water users. Water users participate also in the adoption of general water resources plans (SDAGE) through their one-third share in the membership structure of the Basin Committees (Burchi & Nanni, 2003).

Users' participation is further fostered by legislation governing the direct involvement of water users in the management of groundwater resources under stress from accelerated depletion (also known as groundwater mining) and/or from pollution. In Spain and Chile, for example, Water Users' Groups must be formed from among the users of overexploited aquifers. These groups are to share in the groundwater management responsibilities of the government and, in particular, in the management and policing of groundwater extraction rights. In the state of New South Wales (Australia), the Minister has the power to declare aquifers that are classified as being under environmental stress as groundwater management areas, and to establish groundwater management committees to advise him or her on the necessary measures. Government, local councils, groundwater users and interest groups present in the groundwater management areas must be represented in the groundwater management committees. Amongst other things, the committees are responsible for developing draft aquifer management plans in consultation with the community. Approved plans are binding on government and on groundwater users alike. In the state of Guanajuato (Mexico), where groundwater overdraft is particularly severe, "Groundwater Technical Committees" (*Comités Técnicos de Aguas Subterráneas* – COTAS) have been actively promoted and have become part of the state groundwater governance system, canvassing all groundwater users and stakeholders within an aquifer. COTAS are civil society organizations functioning in a consultative, consensus-building capacity, and in support of Basin Councils and of the government groundwater rights administration (CNA, 2006). In yet another twist on the same theme, France's innovative use of contractual instruments for the management of aquifers under stress is noteworthy. The contract between government and groundwater users (*contrat de nappe*) is seen and used as an instrument binding groundwater users to remedy the vulnerability of an aquifer to over-exploitation or pollution, by adopting such aquifer management measures as are agreed among them and with government. Contracts cannot curtail existing groundwater users' rights, and in actual practice the government insists that contracts align with approved water plans (Burchi & Nanni, 2003; Burchi, 2012; supplemented by personal communications to the author). The contractual approach to engaging with groundwater users in the management of aquifers under stress is practised also in Morocco, with mixed success (Closas & Villholth, 2016).

Groundwater users are a feature of contemporary groundwater governance structures also through their represention in the organs which make up the internal structure of river basin authorities and agencies. For instance, Spain's River Basin Authorities (*Confederaciones Hidrográficas*) include water users' representatives in their decision-making and advisory organs. Similarly, users' representatives make up at least two-thirds of the total membership of the board of directors of France's Water Agencies (*Agences de l'eau*). They are also represented on the Agencies' advisory Basin Committees. Irrigators hold a minority of seats on the board of directors of Morocco's Basin Authorities. In South Africa and Panama, water users and environmental interest groups are represented in the decision-making structure of the Catchment Management Agencies. In Brazil, water users are represented in the basin committees, alongside the representatives of civil society and of the federal, state and municipal governments. In Mexico, water users form no less than one-half the total membership of Basin Councils. In the two countries, committees and councils have a consensus-building remit (Burchi, 2012). In all the examples above-mentioned, however, the representation of groundwater users is implied, as no seats are reserved for them explicitly.

6.3 NOVEL TRENDS AND EMERGING PRACTICES

6.3.1 Where does "environmental" groundwater stand? The "greening" of groundwater laws

The environment-support function of water resources in general, and of groundwater in particular at the point where it discharges feeding wetlands, or also as it provides the baseflow of watercourses at times of low flow, has grown nowadays to the point where it stands on a par with water's entrenched utilitarian function and associated water-related needs and wants. The resulting process of re-direction of water governance practices as reflected in the relevant laws has been described as a "greening" of water laws (Burchi, 2010; Eckstein, 2010). Whereas a majority of the regulatory mechanisms bearing out a "greening" trend are aimed at water resources in general, including groundwater by implication – notably, the systematic recourse to an environmental impact assessment (EIA) of water (and groundwater) development projects, the reservation of water volumes or stocks for an environmental purpose, the priority ranking of environmental water allocations, and the independent standing of environmental water allocations at the hands of an Environmental Water Holder – a few have a distinct groundwater connotation. For instance, in the state of Texas (USA) a cap on extractions from the Edwards Aquifer has been imposed, also for environment-related purposes, under the authority of a dedicated law passed in 1993. A more articulate approach is in effect in the state of New South Wales (Australia), where an "aquifer interference activity" approval by the government is required of activities that interfere with groundwater, and a management plan is required for the relevant area where such controlled activities occur. The plan must identify the nature of the aquifer interference having any effect, including cumulative impacts, on water sources or their dependent ecosystems, and the extent of those impacts. Plans for such controlled activity also deal with undertaking work with a view to rehabilitating the water source or its dependent ecosystems and habitats. Aquifer "safe yield" determinations protecting the health and functions of aquifers are also relevant in this context. In Tanzania and Namibia, for example, the "safe yield" of an aquifer is the amount and rate of extraction which does not damage the aquifer, the quality of groundwater or the environment, and safe yield determinations are to guide the grant of groundwater extraction permits. "Greening" is also at work where, like in Namibia, the government has authority to scale down duly authorized borehole drilling operations in progress in response to any adverse effect such operations display on a water-dependent aquatic ecosystem, or on an environmental water reservation or allocation (Burchi, 2012). Finally, environmental considerations may be required to enter the formation of groundwater plans, as in the state of California (USA), where "groundwater sustainability plans", which are to guide the management and recovery of the state's groundwater resources, must include impacts on groundwater-dependent ecosystems.

6.3.2 How about the land/subsurface space/groundwater nexus?

Land use regulation, and regulation of uses of the sub-surface space, are critical to ensuring the proper functioning of aquifers, notably through the protection of the relevant natural recharge and discharge processes and areas from interferences originating from uses of the land above or from uses of the sub-surface space. Land use regulation, and regulation of uses of the sub-surface space, are equally critical to ensuring that the quality of groundwater is not polluted by "diffuse" (or "non-point") sources originating from rural and urban land uses. Equally so when pollution sources can be pinpointed with accuracy, however are not in the nature of intentional discharges – notably, landfills and waste dumps, and the underground storage of dangerous substances, where the threat of pollution originates from un-intentional leakage. Zoning of the recharge areas of aquifers, and attendant restrictions on land uses, hold much promise in this regard, and are beginning to be part of a systemic groundwater governance response to these issues. Restrictions on cultivation practices and on the use and storage of pesticides, fertilizers and animal manure with a view to preventing and abating groundwater pollution from "diffuse" sources are a steady feature of mature groundwater governance systems, as in the European Union.

As regulation of the uses of the land and of the sub-surface space is the province of town & country planning, construction & building, mining, waste disposal, and handling & storage of dangerous substances legislation, reconciling and harmonizing or simply coordinating and interlinking mainstream groundwater resources regulation with other legislation, and the relevant respective administration on the ground, remains a challenge. The Flemish Region of Belgium has responded to the challenge by linking water governance and regulation to town & country planning governance and regulation. In the specifics, a "water assessment" of construction projects is prescribed by the water law to prevent, avoid or minimize the harmful effects of town and country plans and of relevant permit decisions on the water systems, including by implication groundwater. The prescription is mirrored in the town and country planning law, which directs that a water assessment must be taken into account before a permit is granted under that law (Herman, 2010).

The land/groundwater link is also borne out of water abstraction charging schemes which include an environmental services payment component, generally targeted at the preservation of the groundwater-replenishment function of forested properties in the upper watersheds. A notable example is Costa Rica, where part of the proceeds from the collection of kind-of-water-use and volume-based water abstraction and (higher) groundwater extraction charges is returned to municipalities to fund the purchase of private property for the specific purpose of protecting the recharge areas of groundwater, feeding water supply systems in particular.

6.3.3 Conjunctive use and managed aquifer recharge (MAR): does the law hinder or help?

The term "conjunctive use" of surface and groundwater has several different meanings but basically stands for maximizing the beneficial use and economic benefits of both surface water and groundwater through coordinated use. Methods include augmentation of supplies, allocation of costs, managed aquifer recharge and storage of surface water underground, and the coordination of rights reflecting the interconnection between the two kinds of sources. Mature groundwater governance systems have adapted available regulatory instruments to enable conjunctive use practices, and to reap the relevant benefits. In the Western states of the United States, for instance, the rule of prior appropriation whereby s/he who is first in taking water from a source has priority to keep taking it as a matter of right, is applied to interconnected surface and groundwater. As a result, priorities of rights to the use of interconnected waters are correlated and subject to a single set of priorities that encompasses the whole common water supply. In practice, new abstraction permits can be refused in the area, permissible total withdrawals can be apportioned among appropriators, junior appropriators can be restricted or curtailed in their withdrawals, the extraction and use of groundwater can be subjected to a rotation system and well spacing requirements can be introduced for new wells. By no means is conjunctive use predicated on the rule of prior appropriation. For conjunctive use is practised elsewhere in the United States where prior appropriation does not control. In the state of Texas (USA), for instance, irrigators using groundwater can move return flows to natural surface streams and divert and use such flows further downstream, without fear of losing their water as a result of seemingly abandoning it. In the states of California and Arizona (USA) water users may store excess water underground when there is surplus flow available. The water is recharged underground subject to call or trade when needed. In addition, Arizona law allows any person to carry out groundwater recharge projects in return for groundwater recharge credits – something resembling a groundwater "banking" mechanism. These credits may either be used by the recharger or sold to other water users. Arizona law further allows a person to deliver water directly to a farmer to be used by that farmer in lieu of water he would have pumped from under the ground (known as "in lieu recharge"). This effectively leaves underground the groundwater the farmer would have pumped. The "in lieu" recharger receives groundwater credits which again can be used by the recharger or traded (Burchi, 1999).

The managed recharge of aquifers (MAR) with, in particular, treated wastewater for eventual recovery and use or for environmental benefit is at the cutting edge of the conjunctive use discourse and debate. The governance response of the few countries showing significant levels of MAR activity with treated wastewater – *i.e.*, Israel, South Africa, Spain, and the states of Arizona (USA) and of Western and South Australia – indicates that the regulation of the MAR cycle and activities is achieved in piece-meal fashion, through the application of available strands of regulation to discrete components/phases of the MAR cycle. These include wastewater discharge regulation for pollution control purposes, and treated wastewater reuse/recycling regulation from a public health perspective, at the recharge end of the cycle; and mainstream groundwater extraction regulation, coupled with regulation of the kind of permissible end-uses of recovered groundwater from a public health perspective, at the recovery end of the MAR cycle. Land use planning, building and environmental impact assessment legislation add to the complexity of the regulatory frameworks currently in use for MAR with treated wastewater. Exceptionally, mature groundwater governance systems have sought to consolidate the regulation of discrete segments of the MAR cycle, and have achieved varying levels of regulatory integration, like the state of Arizona (USA), and the state of Western Australia (Burchi, 2014). Further strides in the direction of the integrated governance of the MAR cycle from a regulatory perspective are being contemplated in Palestine, under the guise of a dedicated regulation of MAR with treated wastewater which builds upon various strands of regulatory legislation and aligns them to the requirements of MAR, in addition to supplementing the MAR-specific elements which are missing from the available legislation (Burchi, 2015).

6.3.4 Where groundwater regulation intersects the customary water practices of traditional communities: conflict or co-existence?

Customary law in many countries plays an important role in water governance, particularly at the community level in the rural areas. Despite the social and economic significance of customary water-related systems and practices, water laws have tended to ignore them, or have pushed them to the margin of the regulatory spectrum and relieved them of formal administrative requirements as to their uses of water. The net result is that such uses and users are left without legal protection before formal water rights holders (Hodgson, 2016). Also the blanket statutory recognition of customary rights and practices on the ground has the same effect, as such rights are separated out of the mainstream "modern" water rights regulated by statute, and a separate legal space is created for customary rights which, for want of statutory particulars, comes close to a legal limbo. As a result, the limbo in which customary rights float leaves the potential for conflict with formal rights intact (Burchi, 2005). The more mature water and groundwater governance systems seek to reckon with customary users on the ground, including groundwater users, in a more articulate, less conflict-prone fashion. In Chile for instance, the government is under a statutory duty not to grant a water abstraction concession if the water rights of designated traditional peoples are affected, and no alternative source of supplying their properties with water has been provided first. In Mozambique, the traditional and customary rights practised in the rural areas are accorded priority of allocation of available water resources. Moreover, the government is under an affirmative duty to facilitate the enjoyment of the rights by creating the necessary easements of access to the relevant water sources. The customary rights of traditional communities have by law priority call on available water resources also in Peru and in Paraguay, and statutory grants of water abstraction rights are subject to such customary rights. In Mali, the appropriation of water for a private purpose is subject to recognized customary rights. In Namibia, the government is required by law to factor the impact of a proposed borehole drilling/groundwater extraction licence on existing customary groundwater practices and rights on the ground, prior to granting such licence. A similar approach is reflected, if in relation to water abstractions in general, in the water laws of Bhutan and Zambia.

Opportunities for customary water rights to be reckoned with by the formal water governance systems are routinely afforded, at least on paper, in the process of scrutinizing applications for a water abstraction/development licence or concession, or for a wastewater disposal permit, and of litigating relevant administrative decisions through the administrative or judicial review processes. As the rich experience in this specific matter of Canadian Provinces, and in particular of British Columbia, proves, settlement of customary water rights via conflict and litigation with formal water rights is painful and costly, and the outcome un-predictable (Nowlan, 2004).

6.4 AQUIFERS WHICH STRADDLE THE BOUNDARIES OF STATES: WHAT LAW FOR THEM?

Aquifers may extend across the international borders of two or more countries. They may also extend across the borders of two or more states or provinces of federal countries like Argentina, Australia, Canada, Germany, India, Malaysia and the USA. The governance of transboundary aquifers extending across international as well as inter-state (or inter-provincial) boundary lines is underpinned by rules aimed at equitably apportioning the beneficial use of groundwater stocks, at preventing harm, or at remedying the consequences of harm. The fundamental rule of behaviour among States is that States are entitled to a reasonable and equitable share in the uses of groundwater they have in common. This rule is complemented by the other, equally fundamental, rule that no State has the right to inflict "significant" harm across the international border, through its own actions or through those of its subjects. In fact, States must take measures with a view to preventing "significant" cross-border harm. If such harm occurs nonetheless, the responsible State must take measures to eliminate or to mitigate the cross-border harm. These substantive rules are complemented by the procedural duties of States to exchange information and data, and to provide prior notification of planned measures likely to have a cross-border impact on neighbouring aquifer States. These four cardinal rules of inter-State groundwater-related behaviour are at the heart of what is commonly referred to as customary international law, and form the backbone of transboundary groundwater governance. They are crystallized in the UN Convention on the Non-navigational Uses of International Watercourses, which covers transboundary surface water bodies and linked groundwater, and which is binding since August 2014 on the States which have ratified it. They are also crystallized in the Convention on the Protection and Use of Transboundary Watercourses and International Lakes, which was brokered in 1992 by the UN Economic Commission for Europe (UNECE). Compared to the UN Convention, the UNECE Convention has a distinct pollution prevention & control connotation, and a broader scope in that it covers all kinds of groundwater, whether they are linked to a surface water system, or are de-linked from it. In practice, whereas non-recharging aquifers are excluded from the ambit of the UN Convention, they are included in the ambit of the UNECE Convention. Moreover, since 2013 the UNECE Convention has the ambition to reach out to a global membership from the original regional membership on which the Convention is currently binding. As for all other States, the four basic rules of customary international water law are binding regardless, and have inspired the few treaties and agreements on record, made by States having a transboundary aquifer in common (see below). These same four cardinal rules have been relied upon, and gained additional authoritativeness as a result, in the few cases adjudicated by the International Court of Justice, and by international arbitral panels, between disputing States. All such cases, however, have regarded transboundary rivers[5]. UN Resolution 63/124 of 11 December 2008, carrying "Draft articles on the law of transboundary aquifers", has added to the complexity and articulation of the governance of transboundary aquifers, and has pointed in the direction of significant advances. So do the "Model Provisions on Transboundary Groundwaters" adopted in 2012 by the Parties to the UNECE Convention mentioned earlier, and which closely mirror the UN Draft Articles. In addition to crystallizing the four cardinal rules of inter-State behaviour illustrated above, the UN Draft Articles cast the net much wider by adding to their scope non-recharging (also known as "fossil") aquifers and norms about their management, norms about the recharge and discharge areas of aquifers, uses of aquifers other than the extraction of groundwater, the ecosystem-support function of aquifers, and joint institutional arrangements. The UNECE Model Provisions follow a very similar path. It must be borne in mind however that, with the exception of the four cardinal rules illustrated earlier, which are binding regardless, the balance of the rules cast in the UN Draft Articles and in the UNECE Model Provisions are not binding and, as a result, have an aspirational value and a moral weight only. Such weight is not to be under-estimated, however, in view of the authoritativeness of the source the two instruments originate from, respectively.

A handful treaties and agreements between or among sovereign States are so far on record, addressing specifically a transboundary aquifer. These are the 2007 agreement

[5]Notably, the Danube (the Gabcikovo-Nagymaros project case between Hungary and Slovakia, adjudicated in 1997 by the International Court of Justice (ICJ)), the Uruguay (the Pulp Mills pollution case between Argentina and Uruguay, adjudicated in 2010 by the ICJ), and the Indus (the Baglihar hydropower dam case between India and Pakistan, adjudicated in 2007 by an appointed neutral expert; and the Kishenganga hydropower dam case also between India and Pakistan, adjudicated in 2013 by an arbitral court).

on the Geneva Aquifer, shared by France and Switzerland; three agreements on the Nubian Sandstone Aquifer System (1992 and 2000), shared by Chad, Egypt, Libya and Sudan, providing for a Joint Authority, and for monitoring and data collection and exchange; the agreement for the establishment of a tri-lateral consultative arrangement for the North-Western Sahara Aquifer System, shared by Algeria, Libya, and Tunisia (2002–2008); the 2010 agreement on the Guarani Aquifer, shared by Argentina, Brazil, Paraguay and Uruguay, and the 2015 agreement on the Al-Sag/Al Disi Aquifer, shared by Jordan and Saudi Arabia. An agreement made in 2014 by Algeria, Benin, Burkina Faso, Mali, Mauritania, Niger and Nigeria for the establishment of a Consultation Mechanism for the management of the groundwater resources of the Iullemeden and Taoudeni/Tanezrouft Aquifer Systems is awaiting the signature of three Party States to become effective. The Geneva Aquifer agreement is a complex instrument covering controlled groundwater extractions, controlled artificial aquifer recharge operations, pollution control, the apportionment of all relevant costs, and a permanent bi-lateral institution for the administration and implementation of the obligations undertaken by the Parties. The Al-Sag/Al-Disi Aquifer agreement concerns a fossil aquifer, and places severe restraints on groundwater withdrawals and on the kind of uses of extracted groundwater, to be monitored by a joint body. By contrast, the other three are framework-type agreements, whose centerpiece is an inter-State institution which is to administer aquifer monitoring, data collection and exchange and, limited to the Iullemeden and Taoudeni/Tanezrouft Aquifer Systems agreement, joint aquifer management, obligations.

Local-level arrangements between local authorities are also on record, regarding international aquifers. The French Party to the 2007 Geneva Aquifer Agreement cited earlier, for instance, are two communities and a unit of local-level government. Elsewhere in Europe, most small international aquifers are managed by local authorities under local transboundary arrangements. These are facilitated by the 1980 European Outline Convention on the Transfrontier Cooperation between Territorial Communities or Authorities (Sohnle, 2006). In the USA, a Memorandum of Understanding is on record between the water utilities of the adjoining cities of El Paso, Texas and Ciudad Juarez, Mexico, regarding withdrawals of groundwater from the Hueco Bolson aquifer straddling the USA-Mexico border. However, the USA federal government has provided little if any guidance on such local initiatives, let alone shown any interest in them. The resulting legal limbo casts a shadow of doubt as to the legal effects of local-level arrangements made by US local authorities with their homologues across the international boundary lines with Mexico and Canada (Eckstein & Hardberger, 2007).

More often, States have included groundwater in agreements on transboundary surface waters or river/lake basins, and have extended to linked transboundary groundwater the commitments made in regard to surface waters or river/lake basins. Examples include the River Danube Convention (1994), the Rhine Protection Convention (1999), the Sava River Basin Framework Agreement (2002), the Lake Tanganyika Convention (2003), the Lake Victoria Convention (2003), the Peace Treaty between Israel and Jordan (1994), and the Great Lakes Water Quality Protocol between Canada and the United States of America (1983), amending the Great Lakes Water Quality Agreement (1978) (Burchi & Mechlem 2005).

The basic governance norms of aquifers extending across the borders of two or more states or provinces of federal countries are no different to those of aquifers extending across the international borders of two or more sovereign States. Such basic norms, and additional, negotiated inter-state or inter-provincial aquifer-specific governance norms are crystallized in inter-state or inter-provincial agreements or compacts. Seldom, however, are such agreements or compacts aquifer-specific, as more often than not groundwater governance rules are included in agreements or compacts regarding a surface water body, and connected groundwater. Known examples of aquifer-specific agreements are the 1992 Pullman-Moscow Aquifer agreement between the states of Idaho and Washington (USA), and the 1985 Border Groundwater agreement between the states of South Australia and Victoria (Australia) (updated in 2005). In Australia, a Consultative Council of representatives of the states of Queensland, New South Wales, South Australia and the Northern Territory, all of which overlie the Great Artesian Basin (GAB), and of the federal government, has been established to monitor the implementation of an agreed GAB Strategic Management Plan, and to facilitate the exchange of information among the GAB states (Caponera, 2007). Groundwater is included in the 2001 Lake Eyre Basin agreement between the Australian federal government and the states of Queensland and South Australia, in the 2001 Murray-Darling Basin agreement between the Australian federal government and the states of New South Wales, Victoria and South Australia, in the 2003 Paroo River agreement between the Australian states of New South Wales and Queensland (Burchi & Mechlem, 2005), in the 2005 Great Lakes-St. Lawrence River Basin Water Resources Compact among the eight Great Lakes Basin states on the US side of the border, and in the 2007 Water Charter for Sustainable and Equitable Management of the Hadejia-Jama'are-Komadugu-Yobe Basin made by the federal government of Nigeria and the six basin states. As they make explicit provision for the governance of surface water and connected groundwater, these ostensibly surface water agreements reflect an integrated resource pool perspective where, however, groundwater is "junior" to surface water. The basic norms of governance of inter-state and inter-provincial rivers, lakes and aquifers in a federal context have been relied upon and gained as a result added authoritativeness as a result of inter-state litigation before the concerned countries' highest judicial bodies. The United States Supreme Court, for instance, has adjudicated several water disputes between or among states of the USA. Whereas the available case law concerns exclusively rivers or lakes, the first ever inter-state groundwater dispute is now pending before the US Supreme Court. The case regards an aquifer underlying large tracts of Arkansas, Mississippi and Tennessee, and a small part of Kentucky, with Mississippi disputing Tennessee's alleged groundwater "theft" from the aquifer.

6.5 HOW DOES DOMESTIC REGULATION GET IMPLEMENTED AND ENFORCED? AND WHAT ABOUT INTER-STATE OBLIGATIONS REGARDING AQUIFERS STRADDLING STATE BORDERS?

Implementation, administration and, in the event, enforcement of domestic groundwater regulation are critical, if often neglected, aspects of groundwater governance.

What scanty evidence is documented on this score lends credit to the widespread belief that there is much room for improvement in all the three areas of water and groundwater law and governance. In an effort to advance the knowledge and practice of these neglected areas of water law and governance, the FAO of the United Nations has provided a blueprint for addressing the implementation and administration of formal administrative water abstraction and groundwater extraction licences. This hinges on a structured and costed multi-year advance planning approach, and preparation of government capacity ahead of new governance arrangements coming on stream (Garduño, 2001). Regarding the enforcement of water and groundwater regulation, the patchy and anecdotical evidence available suggests that the chances of success increase with the involvement of water/groundwater users. Effective policing by groundwater users of negotiated groundwater extraction limits has halted the depletion of Spain's largest aquifer, feeding the Daimiel Tablas wetland in Central Spain, which is a Ramsar site. At the government end of the governance spectrum, equally anecdotical evidence indicates that the availability of environmental skills in the police corps, in the public prosecutor offices, and in the judiciary; the coordination or consolidation of environmental police functions, coordinated law enforcement action by the police corps and the judiciary as regards criminal offences in particular, and the innovative use of law enforcement mechanisms, all help water and groundwater regulation enforcement. For instance, in 2013 intensive and coordinated law enforcement action by the three public prosecutor offices having jurisdiction on the Calanque National Park in Southern France reportedly led to the detection and successful criminal prosecution of uncontrolled waste discharge and uncontrolled pesticides use in the local rice fields, threatening irreversible pollution of the underlying aquifer.

Domestic groundwater regulation is critical to the implementation of transboundary aquifer obligations, in an international but also in a federal context. As obligations between and among sovereign States always take precedence over domestic regulation, the latter must always be aligned with the former. The availability of groundwater regulation domestically, therefore, and its implementation and eventual enforcement are the prime instrument for compliance by States with their obligations regarding transboundary aquifers, and for the implementation of the same by the concerned States. For example, if an agreement provides for the sharing or apportionment of river flows or of groundwater stocks between or among States, the domestic water governance of such States must include regulation of water and groundwater abstractions in general within each State, reverberating on the States' ability to honor transboundary river- or aquifer-specific sharing or apportionment obligations. And, if an agreement provides for pollution control of a river or lake or aquifer, the domestic water governance of the States Party to the agreement must include regulation of waste discharges to rivers, lakes and aquifers in general, reverberating on the States' ability to honor transboundary river-, lake- or aquifer-specific pollution control obligations. Unless such domestic regulation is (a) available on the statute books and (b) operational in actual practice, *i.e.*, implemented and in the event enforced, transboundary obligations risk remaining dead letter, and risk engaging the concerned States' responsibility for non-compliance eventually (UNESCO-IHP, 2016). Despite its importance in the general economy of transboundary groundwater governance, the alignment of the domestic groundwater governance and regulation, including implementation and enforcement, of the States which have stipulated a transboundary aquifer agreement, with obligations stemming from such agreements, and State compliance with such obligations, is un-charted groundwater governance territory.

6.6 CONCLUSIONS

At the legal end of the governance spectrum, mature trends have been solidifying, notably in regard to the State's stewardship of groundwater, to the State-administered allocation of available groundwater for extraction and use, and for the disposal of wastes underground, through administrative grants, and to the role of groundwater users, who have become a steady feature of mature groundwater governance systems. At the same time, new trends are emerging which point in the direction of significant advances in other areas of the groundwater governance and regulatory continuum. The solidification process of well-established regulatory trends and practices, however, is all but immune to innovation and advancement. The switch from volume-based to share-based administrative allocations and extraction rights, and from fixed to adjustable allocations and extraction rights during the latter's life, signal significant advances in the well-established practice of regulating groundwater allocation and extraction with the instrument of administrative licences or concessions. So does the use of groundwater extraction regulation for the purpose of preserving the environment-support function of aquifers from competition from productive uses of groundwater. Advances are also reported in the direction of ensuring a meaningful representation of water users in the governing bodies of river basin organizations, including groundwater users by implication. The novel and emerging trends point in the direction of growing attention to facets of groundwater governance which have traditionally escaped the radar screen of mainstream groundwater regulation, and of linking non-groundwater areas of governance and regulation with mainstream groundwater governance and regulation. Groundwater regulation-mandated recourse to zoning the recharge areas of aquifers for controlled land uses, and regulation of discrete land uses – notably, cultivation and livestock rearing – impacting on groundwater quality, signal significant advances in the direction of recouping the connection with land use regulation. So do instances of linking land use determinations to the relevant groundwater impact, from the perspective of land use regulation and of groundwater regulation combined. Water regulation-mandated EIA of groundwater development projects signals advances in the direction of linking with environmental regulation, in the larger context of the environment-support function of aquifers gaining centre stage on the radar screen of groundwater governance and regulation.

Opportunities for the conjunctive use of surface water and groundwater, and for managed aquifer recharge, are pursued in mature groundwater governance systems through the creative adaptation of available regulatory mechanisms, coupled with water credit instruments echoing the commercial banking system. Advances are reported, admittedly in isolated instances, in the direction of the integrated governance of the managed recharge of aquifers with treated wastewater. The intersection of formal groundwater rights with the customary groundwater rights and practices of traditional communities is another area of groundwater governance showing significant advances on paper, as modalities of peaceful coexistence have been sketched out, which are premised on the legitimacy but also on the primacy of traditional rights and practices over formal rights and the relevant administrative grants. The effectiveness on the ground of primacy-based coexistence, however, remains to be tested and documented.

As regards transboundary aquifers, significant strides have been made towards the codification of norms of inter-State aquifer governance reaching past the core norms of equitable and reasonable use of groundwater, of not causing significant cross-border harm to an aquifer, of exchanging aquifer-relevant information and data, and of providing prior notification of planned measures regarding or impacting an aquifer. Whereas these core norms are binding on all nations, the more articulate and advanced norms, crystallized in particular in a dedicated 2008 UN resolution, and canvassing other facets of transboundary aquifer governance, are not. Advances have also been made in the negotiation of aquifer-specific agreements and governance norms by sovereign states, with local-level arrangements between local authorities also beginning to emerge where the legal circumstances permit, notably in Europe. Aquifer-specific governance norms, and the agreements recording them, however, are in the minority compared to surface water governance norms and agreements. It is noteworthy that a number of these canvass groundwater which is hydraulically linked to a surface water system. As surface water dominates in such agreements, however, groundwater inevitably trails behind in the implementation agenda of the agreements.

Implementation and eventual enforcement of domestic groundwater regulation remain to-date largely un-mapped groundwater governance territory. Authoritative international wisdom emphasizes the importance of building up the government's administrative response capability in advance of new water and groundwater governance arrangements coming on stream. Evidence of actual advances on this score is scanty and anecdotical, offering nonetheless some useful pointers as regards, in particular, law enforcement.

As a final note, experience suggests that advances in the regulatory systems of groundwater governance, and the sophistication the more advanced such systems imply, are inextricably linked to, and premised on, the availability of comprehensive and reliable sets of data, enabling a good understanding of the aquifer and reliable assessment of the groundwater in it, and providing a credible science-based bedrock for informed, quality decision-making.

REFERENCES

Brufao Curiel, P. (2008) *La revisión ambiental de las concesiones y autorizaciones de aguas.* Zaragoza, Fundación Nueva Cultura del Agua.

Burchi, S. (1999) National regulations for groundwater: options, issues and best practices, in: S. Salman (ed.), *Groundwater: Legal and Policy Perspectives*, World Bank Technical Paper No. 456. pp. 55–67.

Burchi, S. (2005) *The interface between cusotmary and statutory water rights – a statutory perspective.* FAO Legal Papers Online No.45. Available from http://www.fao.org/Legal/prs-ol/paper-e.htm [Accessed 9 September 2016].

Burchi, S. (2001) Year-end Review. *Journal of Water Law*, 12(6), 330–337.

Burchi, S. (2003) Year-end Review. *Journal of Water Law*, 14(1), 7–15, and 14(6), 281–290.

Burchi, S. (2005b) Year-end Review. *Journal of Water Law*, 16(1), 5–13.

Burchi, S. (2005c) Year-end Review. *Journal of Water Law*, 16(5), 155–166.
Burchi, S. (2006) Year-end Review. *Journal of Water Law*, 17(6), 223–230.
Burchi, S. (2007) Year-end Review. *Journal of Water Law*, 18(5), 149–158.
Burchi, S. (2008) Year-end Review. *Journal of Water Law*, 19(6), 223–228.
Burchi, S. (2009) Inter-state and inter-provincial water resources agreements – A comparative overview. In Embid, A. (ed.) *Gestión del agua y descentralización política: Proceedings of the International Conference on the Management of Water Resources in Federal and Quasi-federal Countries, Zaragoza, Spain, 9–11 July 2008*. Thomson Reuters. pp. 518–542.
Burchi, S. (2010) Balancing development and environmental conservation and protection of the water resources base: the "greening" of water laws. In Cullet, P., Gowlland-Gualteri, A., Madhav, R., Ramanathan, U. (eds.) *Water Governance in Motion – Towards Socially and Environmentally Sustainable Water Laws*. New Delhi, Cambridge University Press India. pp. 333–358. Prior to publication, the paper had been posted online as FAO Legal Papers Online No. 66 (2007), available at http://www.fao.org/legal/prs-ol/lpo66.pdf [Accessed 9 September 2016].
Burchi, S. (2011) Year-end Review. *Journal of Water Law*, 22(6), 248–265.
Burchi, S. (2012) A comparative review of contemporary water resources legislation: trends, developments and an agenda for reform. *Water International*, 37(6), 613–627.
Burchi, S. (2014) *Study on the Legislative Framework Regulating the Recharge of Aquifers with Adequately Treated Wastewater*. GWP-Med, Sustainable Water Integrated Management – Support Mechanism (SWIM-SM) – available from http://www.swim-sm.eu/index.php/en/resources/assessments. [Accessed 9 September 2016].
Burchi, S. (2015) *Guidelines for developing national regulation for managed aquifer recharge in Palestine*. GWP-Med, Sustainable Water Integrated Management – Support Mechanism (SWIM-SM) – available from http://www.swim-sm.eu/index.php/en/resources/assessments [Accessed 9 September 2016].
Burchi, S. & Mechlem, K. (2005) *Groundwater in international law – Compilation of treaties and other legal instruments*. Rome, Food and Agriculture Organization of the United Nations (FAO) Legislative Study No.86. Available from http://www.fao.org/docrep/008/y5739e/y5739e00.htm [Accessed 9 September 2016].
Burchi, S. & Nanni, M. (2003) How groundwater ownership and rights influence groundwater intensive use management. In: Llamas, R. & Custodio, E. (eds.) *Intensive Use of Groundwater – Challenges and Opportunities*. Lisse, Balkema Publishers. pp. 227–240.
Burke, J.J. & Moench, M.H. (2000) *Groundwater and Society: Resources, Tensions and Opportunities*, Themes in groundwater management for the twenty-first century, New York, United Nations Publication ST/ESA/265.
Caponera, D. (2007) *Principles of Water Law and Administration, National and International*. 2nd edition, revised and updated by Nanni, M. London, Taylor & Francis.
Closas, A. & Villholth, K. (2016) *Aquifer contracts – A means to solving groundwater over-exploitation in Morocco?*. Colombo, Sri Lanka: International Water Management Institute (IWMI).
CNA (Mexico, Comisión Nacional de Aguas) (2006). *Los comités técnicos de aguas subterráneas en México*. [Presentation] EXPO Agua, Guanajuato (Mexico) 18 August.
Cullet, P. (2012) The Groundwater Model Bill – Rethinking Regulation for the Primary Source of Water. *Economic and Political Weekly*, 47/45, 40–47.
Eckstein, G. & Hardberger, A. (2007) State Practice in the Management and Allocation of Transboundary Groundwater Resources in North America. *Yearbook of International Environmental Law*, 18, 99–125.
Eckstein, G. (2010) *The greening of water law: managing freshwater resources for people and for the environment*. Nairobi, UNEP. Available from http://www.unep.org/delc/Portals/119/UNEP_Greening_water_law.pdf [Accessed 9 September 2016].

Garduño Velasco, H. (2001) *Water rights administration. Experience, issues and guidelines.* Rome, FAO (Legislative Study No.70). Available from http://www.fao.org/docrep/003/x9419e/x9419e00.htm [Accessed 9 September 2016].

GEF, FAO, UNESCO-IHP, World Bank, IAH (2015) *Shared Global Vision for Groundwater Governance 2030.* Rome, FAO. Available at http://www.groundwatergovernance.org/ [accessed 15 April 2017].

GEF, FAO, UNESCO-IHP, World Bank, IAH (2016) *Global Framework for Action to achieve the vision on Groundwater Governance.* Rome, FAO. Available at http://www.groundwatergovernance.org/ [accessed 15 April 2017].

Herman, C. (2010) Will the Floods Directive keep our feet dry? Policies and regulations in the Flemish region and Scotland. *Journal of Water Law* 21(4), 156–166.

Hodgson, S. (2016) *Exploring the concept of water tenure.* Rome, FAO (Land and Water Discussion Paper No.10). Available from http://www.fao.org/3/I5435E [Accessed 26 September 2016].

Mechlem, K. (2012) *Thematic Paper 6: Legal and Institutional Frameworks.* Global Environment Facility (GEF) – Groundwater Governance: A Global Framework for Country Action. Available from http://www.groundwatergovernance.org/fileadmin/user_upload/groundwatergovernance/docs/Thematic_papers/GWG_Thematic_Paper_6.pdf [accessed 30 September 2016].

Mechlem, K. (2016) Groundwater Governance: The Role of Legal Frameworks at the Local and National Level—Established Practice and Emerging Trends. *Water* 2016, 8, 347.

Nowlan, L. (2004) *Customary Water Laws and Practices in Canada* (un-published report).

Nowlan, L. (2005) *Buried Treasure – Groundwater Permitting and Pricing in Canada.* The Walter & Duncan Gordon Foundation.

O'Donnell, E. (2011) Institutional reform in environmental water management: the new Victorian Environmental Water Holder. *Journal of Water Law*, 22(2/3), 73–84.

Sohnle, J. (2006), *Transboundary Aquifers and Local Transfrontier Cooperation in Europe* (unpublished report).

UNESCO-IHP (2016) *Hydrodiplomacy, Legal and Institutional Aspects of Water Resources Governance – From the International to the Domestic Perspective, Training Manual.* Paris. Available at http://unesdoc.unesco.org/images/0024/002452/245262e.pdf (accessed 23 March 2017).

Chapter 7

Participation of stakeholders and citizens in groundwater management: the role of collective action

Zachary Sugg[1,2] *& Edella Schlager*[3]
[1]*Martin Daniel Gould Center for Conflict Resolution, Stanford Law School, Stanford, CA, USA*
[2]*Water in the West Program, Stanford University, Stanford, CA, USA*
[3]*School of Government and Public Policy, University of Arizona, Tucson, USA*

ABSTRACT

Groundwater basins provide diverse benefits, from consumptive water uses, to water storage, to riparian and aquatic habitat. Realizing these multiple benefits while managing the negative spillover effects requires the engagement and participation of diverse stakeholders and citizens. In this chapter we examine how institutional arrangements encourage different forms of participation and the realization of different values. We illustrate the effects of institutions on participation and governance through two case studies from the western U.S.: the Edwards Aquifer in south central Texas, and the Arkansas River Basin in southeastern Colorado. We conclude with lessons about how to encourage and sustain stakeholder participation in groundwater governance.

7.1 INTRODUCTION: GROUNDWATER BASINS AND COLLECTIVE ACTION DILEMMAS

Participation of stakeholders and citizens in groundwater governance is challenging to organize because of the multiple, and sometimes competing uses made of groundwater. Groundwater basins, as common pool resources, are subject to the tragedy of the commons. The boundaries of many groundwater basins are difficult and costly to identify making exclusion prior to the emergence of intensive use very unlikely. Furthermore, groundwater is rival – the water pumped by one resource user is not available to be pumped by other resource users (Ostrom and Ostrom, 1977). Given costly exclusion and rivalness of units, users of groundwater basins may engage in a race to harvest, especially during times when demand for water may be especially high, and surface water scarce, such as droughts.

Avoiding the mining of groundwater basins is not the only collective action dilemma resource users face. Groundwater aquifers are also sources for water storage. Surface water, whether it is flood waters, reclaimed waters, or unallocated river flows, may be stored underground to be used at some future point in time. The capacity of groundwater aquifers to store water may be threatened by a race to pump as declining water tables may be followed by subsidence and compaction which undermine storage capacity.

Groundwater aquifers may also be hydrologically connected to surface water sources, such as rivers, streams, and springs, contributing to surface water flows and supporting riparian and aquatic habitat and species. Riparian and aquatic habitat and species represent ecosystem services that require careful governance if they are to remain intact and productive. Connected surface and groundwater sources pose coordination challenges as too much groundwater pumping may dry up surface water sources and the diversion of surface water sources may deprive groundwater basins of important sources of recharge.

Avoiding the depletion of groundwater basins by governing them sustainably and in ways that tap into their full potential, such as a buffer against extreme disturbances, and as key contributors to ecosystem services, requires the participation of resource users and other stakeholders in their governance. Not only do resource users have a stake in sustaining groundwater basins to support their livelihoods, but they also have considerable time and place information about the response of groundwater levels to different types of uses, as well as the effects of drought and flooding. Most importantly, however, if groundwater aquifers are to be governed sustainably, resources users must be committed to changing their behavior, if necessary, following the rules of governance, and participating in devising, implementing, and monitoring policies. Involving stakeholders in the development of governing arrangements opens the possibility of better matching institutional arrangements with context and encouraging rule following behavior, both of which are likely to reduce conflicts (Ostrom, 1990).

How citizens and stakeholders participate in groundwater governance is structured by institutional arrangements. Existing arrangements affect the willingness, ability, and capacity of stakeholders to participate in groundwater governance. In this chapter we focus on different forms of participation in groundwater governance in the western U.S., how those forms have changed over time, and the effect participation has on the values and goals realized. The next section considers the different forms of institutional arrangements that structure and condition participation in groundwater governance, from water rights to planning processes. Section three consists of two case studies that explore in depth different forms of participation and its importance. We conclude with lessons learned.

7.2 INSTITUTIONAL ARRANGEMENTS AND PARTICIPATION

The spillover effects of many actors sharing and enjoying the benefits of groundwater aquifers are affected by and addressed through water and environmental laws and administration as well as water planning processes. Each of these approaches frames and encourages participation in distinct ways as we discuss below.

7.2.1 Water laws

U.S. states possess and exercise extensive control over the water resources within their boundaries (Sax *et al.*, 2006; Getches, 2009). Most constitutions of western states contain clauses identifying the state as the owner of water resources and the water law doctrines that define how water will be allocated and consequently administered

(Hobbs, 1997; Sax *et al.*, 2006). Individuals who are allocated water rights have rights of use, but states retain ownership and, thus, regulatory authority.

In general, all western states have adopted some form of the prior appropriation doctrine to govern surface water flows (Getches, 2009). The prior appropriation system governs water based on when appropriation first occurs, with those earlier in time exercising the more secure right. Most western U.S. states, at least on paper, if not in practice, apply the prior appropriation system to both surface and groundwater (Schlager, 2006).

A few states use a dual system of surface and groundwater laws, with surface water governed by the prior appropriation system and groundwater governed by beneficial use or correlative use (Joshi, 2005). Beneficial use allows the landowner to capture and use the water found under his or her lands as long as the water is devoted to beneficial uses as defined in law and not wasted (Sax *et al.*, 2006). Correlative rights, a doctrine used by two western states, requires landowners to share groundwater, limiting use to reasonable amounts and sharing water reductions in times of scarcity (Blomquist, 1992).

7.2.1.1 States' water laws and participation

Allocating property rights in water to users has had the effect of privileging rights holders over other stakeholders in governing water. Water rights holders develop, protect, and contest water rights through administrative proceedings and court cases. For instance, the State of Colorado uses specialized water courts to develop, change, and enforce water rights (Vranesh, 1987). Besides the state's water engineers, only people seeking to develop or acquire water rights or who hold water rights are allowed to participate in water court proceedings (Hobbs, 1997). States that use administrative proceedings to govern water rights likewise limit who is allowed to participate. New Mexico, for example, only allows individuals or organizations who hold water rights that they believe will be substantially impaired to protest the granting of a water right (New Mexico Office of the State Engineer, no date).

Privileging water rights holders has created challenges to governing groundwater aquifers sustainably. Groundwater rights are initially allocated with little attention given to sustainability. Rather, states require landowners to obtain well permits, but permits do not place limits on pumping. Efforts to limit groundwater pumping typically occur after spillover effects build up and conflicts emerge, often between surface water rights holders and groundwater pumpers. Conflicts center around limiting and more strictly regulating groundwater pumping to buffer its effects on surface water flows (Blomquist *et al.*, 2004; Heikkila and Schlager, 2012; Sugg *et al.*, 2016).

Addressing and managing such conflicts entails considerable time and effort on the part of state officials and water rights holders to develop regulations that most rights holders find acceptable. The structure and design of the regulations vary from state to state, but they typically involve well moratoria, well spacing requirements, and limits on amounts of water that may be pumped (Schlager and Blomquist, 2008). Such regulations also provide opportunities for water rights holders to actively search out and utilize opportunities to store surplus surface water underground (Blomquist *et al.*, 2004; Sugg *et al.*, 2016). Engaging in conjunctive water management is made possible as surface water rights holders are assured that if they store water underground others will not be allowed to pump it out, as would have been the case under unregulated rule of capture or beneficial use doctrines.

Given the privileged position that water rights holders have in governing ground and surface water, environmental organizations and interest groups have advocated, with some success, for recognition of environmental uses of water as a beneficial use by state water laws (Sax *et al.*, 2006; Garrick, 2015). This has allowed environmental groups to purchase or develop water rights and devote the associated water to environmental purposes (Garrick, 2015). Extending what is considered a beneficial use has expanded the values that are recognized in governing water; however, realizing those values requires the acquisition of water rights.

In sum, water rights privilege the participation of rights holders and exclude the participation and the values of other actors. Western states have recognized the protection of environmental values as a beneficial use, which allows water rights to be devoted to environmental purposes, but water rights still have to be acquired in order to realize such values.

7.2.2 Water planning processes in western U.S. states

Western U.S. states have turned to water planning processes as a means of addressing a host of issues and concerns while also promoting more extensive forms of participation in water decision making. Some states such as California have engaged in formal water planning for decades. Other states have recently adopted the practice, such as Colorado, which approved its first state water plan in 2015.

Planning processes are notable for the diversity of participants. From groundwater users to environmentalists, from water managers to recreationalists, many types of users and the values they represent are involved. Although consumptive uses, such municipal water supplies, are often front and center, some attention is given to non-consumptive uses.

Most plans assess demands for and supplies of surface and groundwater, how demand and supply is likely to change under different future scenarios, identify sensitive watersheds that require additional protections, and suggest legislative and administrative actions to address problems, issues, and opportunities identified in the plans. The extent to which plans are actively embraced by water users and decision makers varies. Some states, such as Colorado, provide funding for pilot projects identified in the plan, with the possibility of having successful pilots made available statewide (Arkansas Roundtable Basin Implementation Plan, 2015). Texas provides incentives for participation by requiring organizations (water utilities, irrigation districts, etc.) to participate in planning processes in order to be eligible for state funding for water projects.

7.2.3 U.S. environmental laws and participation

Environmental laws, from the U.S. Clean Water Act to the U.S. Endangered Species Act, are used as avenues to broaden participation beyond water rights holders and, consequently, the values that are considered in water governance decisions. U.S. environmental laws do not undermine or overturn states' water laws or administration;

however, they do provide opportunities for participation by groups who value environmental protection (Kundis Craig, 2014). The opportunities range from "notice and comment" proceedings, to petitioning for the listing of endangered species, to the creation of collaboratives and partnerships that actively work to realize the requirements of environmental laws and regulations while protecting or enhancing local livelihoods (Sabatier *et al.*, 2006).

U.S. laws require federal agencies to engage in a notice and comment period when taking regulatory actions (Houck, 1993). Anyone may submit comments, allowing a wide range of issues, evidence, and values to be represented. In addition, the federal agency must respond to all substantive comments (Brandon, 2015:321). As Brandon (2015) notes, notice and comment is especially important for the Endangered Species Act as it is required for all decisions relating to species listing and designating critical habitat.

In addition to notice and comment as a form of participation, citizens may petition the two agencies charged with implementing the Endangered Species Act – the U.S. Fish and Wildlife Service and the National Marine Fisheries Service – to list species as endangered, another important form of participation, especially for environmental groups who do not hold water rights in a watershed (Brandon, 2015). In a study comparing citizen petitions for listing endangered species to federal agency listing of species, Brosi and Biber (2012:803) find that species for which citizens sought protection are more biologically threatened than those proposed for listing by the U.S. Fish and Wildlife Service, and that such species are more likely to be in conflict with development projects and activities. Thus, citizen participation in listing of species provides more species protection than if federal agencies were the sole decision makers.

Finally, in order to realize the requirements and goals of federal environmental laws, federal and state agencies as well as citizens and interest groups have formed watershed partnerships and collaboratives (Leach *et al.*, 2002; Koontz *et al.*, 2004; Sabatier *et al.*, 2006). Watershed partnerships and collaboratives attempt to encourage widespread participation among many different actors and interests in developing biological, hydrological, and socio-economic data, monitoring of selected environmental indicators, and development and implementation of plans and projects targeted at maintaining and enhancing stream flow and recovering species. Watershed partnerships are common across the western U.S. For instance, the New Mexico Environment Department listed over 30 watershed partnerships in the state as of 2014. The Oregon Network of Watershed Councils lists over 80 councils or partnerships as of 2016 (http://www.oregonwatersheds.org/).

Participation in groundwater management is guided and constrained by institutional arrangements. Historically, states and water rights holders were and continue to be the central participants. However, state water planning processes expanded who is allowed to participate in decision making to include recreational and environmental interests. Furthermore, U.S. federal environmental laws also provide opportunities for citizens and interest groups, who value environmental quality and protection, to participate in decision making processes of federal agencies implementing the laws. Citizens and stakeholders have formed partnerships to address the goals and requirements of environmental laws. Thus, over the last several decades groundwater governance has been transformed from the domain of water users to the domain of different interests cooperating, competing, and conflicting over realizing their preferred water values.

7.3 PARTICIPATION IN PRACTICE

To illustrate how the above institutional arrangements permit and limit participation in groundwater governance, and water governance more generally, two case studies are presented in this section. These cases were selected to show how participation is organized and how it has changed over time as disturbances and crises required water officials and water users to reconsider who should be involved in groundwater governance. The first case involves a concerted effort to protect endangered species reliant on aquatic habitat fed by groundwater. The second case illustrates how participation evolves over time as the uses made of water in a river basin move from irrigated agriculture to municipal, industrial, and environmental protection.

7.3.1 Collaborative groundwater planning for the Texas Edwards Aquifer

The case of the regulation and management of the Edwards Aquifer in south-central Texas is an example of groundwater governance developing over many years out of claims brought under the Endangered Species Act. Because of a litigious history during the 1990s (Votteler, 1998; Donahue, 1998) and into the 2000s, the Edwards Aquifer is perhaps most widely known for conflict rather than cooperation. Less well known are the important efforts in recent years to develop a program of management measures to meet social, economic, and ecological needs on the aquifer. More specifically, a protracted formal stakeholder participation process took place to develop a suite of springflow and habitat protection measures. As such, it adds a new chapter to a long-running saga and presents an interesting example of addressing a collective action challenge.

7.3.1.1 Physical and historical context for the development of participatory groundwater planning in the Edwards Aquifer

The Edwards (Balcones Fault Zone) Aquifer is a unique hydrologic system in which groundwater and surface water are linked in very visible ways in the form of artesian springs (Figure 7.1). The aquifer itself is a complex, faulted limestone system with highly variable transmissivities in the confined portion of the aquifer (Maclay, 1995). The major springs at Comal and San Marcos form streams that are sustained by springflows during times of little rainfall.

The unique hydrologic conditions of the aquifer have over time fostered unique ecosystems which are host to several endemic species that have been designated as threatened or endangered by the U.S. Environmental Protection Agency (USEPA). The aquifer remains the primary water supply for more than 2 million people in the San Antonio metroplex and other towns in the region. It also supports an irrigated agriculture economy to the west of San Antonio. Increasing demands, historically few legal restrictions on pumping, and periodic drought eventually led to the jeopardization of

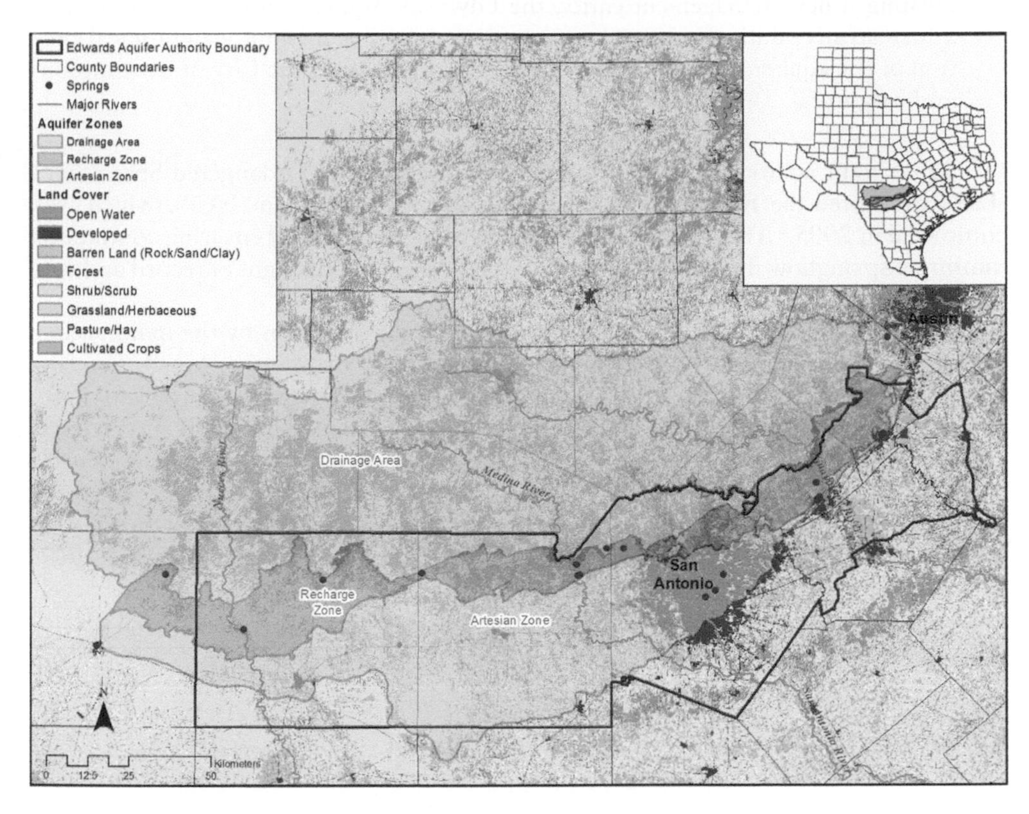

Figure 7.1 Edwards (Balcones Fault Zone) Aquifer region of Texas.

several endangered and threatened endemic species. The crux of the issue lay in the interception of groundwater by the cities and farmers before it could reach the springs, thus reducing springflows and degrading riparian habitat, especially during drought conditions; for example, Comal springs ceased to flow for several months in 1956 during the drought of record.

During drought conditions in 1991, the Sierra Club[1] filed suit against the U.S. Fish and Wildlife Service (USFWS) to force the imposition of pumping limits. In 1993 a federal judge ruled in favor of the Sierra Club, requiring USFWS to determine minimum flow requirements for the aquifer while leaving it up to the state to devise a plan for meeting the requirements. The legislative response was Senate Bill 1477, also known as the Edwards Aquifer Act of 1993. SB 1477 contained several notable provisions, including:

- a statutory cap on total annual withdrawals from the aquifer
- replacing the open access situation created by the law on the books (rule of capture) with a managed tradable groundwater permit system

[1] *Sierra Club v. Lujan*, No. MO-91-CA-069, 1993 WL 151353 (W.D. Tex. 1993).

- creating a new management entity, the Edwards Aquifer Authority (EAA), with administrative duties and regulatory powers over a significant geographical portion of the aquifer that included the major pumpers (*e.g.*, the City of San Antonio and irrigators)

Subsequent additional threats of legal action under the Endangered Species Act (ESA) prompted the EAA to draft a Habitat Conservation Plan (HCP), which was completed in 2005.[2] That HCP was determined to be incapable of ensuring continuous minimum springflow during a recurrence of the 1947–1957 drought of record and was thus rejected by the USFWS.

In 2006, however, USFWS began spearheading the adoption by the major water users and state regulators of a type of formal stakeholder planning process for endangered species protection dilemmas known as a recovery implementation program (RIP). RIPs have been used to address similar challenges with balancing human and ecological water needs in heavily used rivers in the Western U.S. such as the San Juan River[3] and the Platte River[4].

Initially a voluntary effort, the RIP for the Edwards Aquifer (known as the EARIP) became mandatory following the Texas Legislature's passage of Senate Bill 3 in 2007. This was followed by 5 years of participatory stakeholder deliberations towards developing a plan to adequately protect springflows and threatened and endangered species that would meet USFWS approval.

7.3.1.2 *Participants involved in Edwards Aquifer groundwater planning*

Because of the unique configuration of the aquifer and the importance of the springs, something of an upstream-downstream alignment of interests had developed that was more typical of a river basin-scale dispute. The "upstream" interests were generally those who wanted to withdraw groundwater before it reached the springs, *e.g.*, farming communities and some municipalities, most importantly the City of San Antonio. The "downstream" interests were generally those groups concerned primarily with maintaining the springflows, *e.g.*, conservation and environmental groups, municipalities in the spring areas, and the Guadalupe-Blanco River Basin Authority. Many of these groups had years of acrimonious history behind them and so simply getting them all together in a room was no mean feat.

To start, a 26-member Steering Committee was formed, representing a wide range of regional interests that included "environmental, water authority and purveyor, industrial, municipal, public utility, state agency, and agricultural interests" (Gulley and Cantwell, 2013:11). An even larger set of interest groups (40 in total) signed a memorandum of understanding in 2008 specifying how the EARIP would proceed.

---

[2]See Gulley and Cantwell (2013) for a discussion of the key provisions of the ESA related to the use of the Edwards Aquifer.
[3]https://www.fws.gov/southwest/sjrip/index.cfm
[4]https://www.platteriverprogram.org/Pages/Default.aspx

7.3.1.3 The groundwater planning process for the Edwards Aquifer

The best available account of the EARIP has been written by its program manager, Robert Gulley.[5] According to Gulley, the chief problems confronting the group were "the issues related to the mandated continuous minimum [spring]flows – what minimum flows were required, how those flows could be achieved, and how the cost of achieving those flows would be funded" (Gulley, 2015:100). Any solutions had to be effective during a repeat of the 1950s drought; otherwise, USFWS would not approve of the plan.

Meetings began with collaborative learning sessions and sessions focused on respectful communication styles. According to Gulley, the latter were helpful for establishing conditions of openness and transparency in discussions among the different stakeholders (Gulley, 2015). The Steering Committee established a two-tier system of decision-making criteria. Tier 1 decisions, including all decisions made by the Steering Committee, were most important and required a consensus, defined as an absence of opposition[6]. Tier 2 decisions were of a more routine nature and required a majority vote for approval (Gulley, 2015).

With the ground rules in place and Gulley acting as mediator and facilitator, the EARIP proceeded in the form of monthly or more-than-monthly meetings attended by between 50 and 80 people representing the 40 different stakeholder groups (RECON Environmental Inc. *et al.*, 2012). Subcommittees were created on: Science; Recharge Feasibility; Public Outreach; and Ecosystem Restoration (RECON Environmental Inc. *et al.*, 2012). The Science committee was notable for being tasked with determining the necessary flow regimes for maintaining the springs (Gulley, 2015). Additionally, 16 work groups were formed as needed to tackle specific issues such as funding; aquifer storage and recovery; and implementation of HCP provisions. According to Gulley and Cantwell (2013:11), these work groups "proved very effective in facilitating resolution of complex or contentious issues in the decision-making process."

That is not to say that there were no disagreements; in fact, there were points at which the entire process nearly ground to a halt. One nearly fatal impasse arose late in the process over how the pumpers and non-pumpers would share the cost of implementing the package of measures in the draft HCP, which was anticipated to be about $27 million on average per year for the first seven years (Gulley, 2015:145). In Gulley's estimation, this was "the most contentious obstacle the EARIP faced" (Gulley, 2015:137). When federal funding and a regional sales tax were no longer viable options, the issue of how to distribute financial burden raised difficult questions of fairness among the stakeholders themselves: who would derive the greater benefit from the positive effects of HCP measures on water availability, the pumpers or non-pumpers? And based on that, what level of financial obligation would be fair for the different types of stakeholders? (Gulley, 2015). A work group with representation from both pumpers and non-pumpers met to resolve these questions but the problem

---

[5]See Gulley (2015) for a detailed account of the issues surrounding the Edwards Aquifer and the story of the EARIP itself.

[6]The Habitat Conservation Plan states that "... in practice, decisions generally were made without opposition and without the need for a vote by Steering Committee members" (RECON Environmental Inc. *et al.*, 2012:1-19–1-20).

proved intractable. The entire EARIP process nearly came to a halt when San Antonio Water System proposed postponing completion of the HCP over the issue, a critical delay which Gulley believes the EARIP could not have survived (Gulley, 2015). Yet representatives of the principal interests continued to return to the negotiating table and finally hammered out an agreement that both pumpers and non-pumpers could live with. Gulley and Cantwell (2013) suggest that the series of decision making deadlines spelled out in Senate Bill 3 was important to the success of the EARIP by keeping the process moving forward when it might have otherwise run out of steam.

7.3.1.4 What resulted from the participation process?

After several years of meetings, the HCP was finally completed in 2012[7] and approved by USFWS in 2013. These were landmark events, especially in light of the highly fraught historical backdrop. Although the implementation of the HCP is perhaps too early to allow a fuller assessment of this example of participatory groundwater decision-making, the contents of the HCP itself and the existence of oversight are reasons for a positive appraisal.

There are two main categories of measures contained in the HCP: flow protection and habitat protection. Long-term biological objectives and measures designed to achieve them were established for particular species in particular ecosystems. They are too numerous to list here, but an example will illustrate. For instance, the biological goal for the endangered fountain darter fish in the Comal Springs/River ecosystem was defined in terms of densities of particular types of vegetation in certain areas in addition to densities of the fish itself. Management practices to achieve desired densities of fountain darters were determined to involve the protection and restoration of several native vegetation types in certain areas, as well as water quality criteria such as temperature and dissolved oxygen concentrations. Flow objectives were also set in the form of minimum and long-term average discharge rates from Comal Springs (RECON Environmental Inc. *et al.*, 2012).

Broader categories of measures to achieve various objectives of the HCP include: a Voluntary Irrigation Suspension Program for farmers; a Regional Water Conservation Program; Critical Management Period provisions whereby pumping is reduced during severe drought conditions; and an aquifer storage and recovery system (RECON Environmental Inc. *et al.*, 2012). The HCP also provided for the creation of an adaptive management program based upon continual monitoring, research, and modeling. The adaptive management process is intended to evaluate the success of the various activities conducted under the auspices of the HCP in relation to stated goals.

There is also oversight of the HCP in the form of an implementation committee and a formal review by a U.S. National Research Council "blue ribbon" expert committee appointed by the National Academy of Sciences. At the time of this writing the review is ongoing, but the committee's first report – an appraisal of the scientific aspects of the HCP – was largely positive (National Research Council, 2015).

Whether all of the programs and activities will be enough to achieve the stated objectives in the long term remains to be seen. Although modeling exercises have

---

[7]The Habitat Conservation Plan for the Edwards Aquifer is available at http://www.eahcp.org/files/uploads/Final%20HCP%20November%202012.pdf

predicted that in a repeat of the 1950s drought of record the key springs would not cease to flow, a drought more severe and/or prolonged would not be out of line with future projections or past events that are known from paleoclimate proxies (Cook *et al.*, 2015; Cleaveland *et al.*, 2011; Woodhouse *et al.*, 2010; Cook *et al.*, 2004; Loáiciga *et al.*, 2000). However, from the narrower perspective of participation and collective action, the EARIP can be judged a remarkable success. The creation of an effective participatory process was essential to success where numerous previous attempts had failed.

7.3.2 Groundwater participation in the Arkansas River Basin, Colorado

Participation often evolves over time, and may eventually coalesce at the river basin level as groundwater driven conflicts spark adaptations to changing contexts. The Arkansas River Basin is a prime example.

The Arkansas River rises in the Rocky Mountains south of Denver, Colorado and flows south before turning to the east, passing through the City of Pueblo, Colorado, and then through miles of farm and ranch land before crossing the border into the state of Kansas (Figure 7.2). The river is hydrologically connected to an alluvial groundwater aquifer and over the past century as demand for water has exceeded supplies, conjunctive management of the two sources has become increasingly important. Conjunctive management is a means for addressing spillover effects between surface and groundwater uses, as well as allowing new and additional uses of water (Blomquist *et al.*, 2004). As water uses have diversified, so too have participation patterns. Water rights holders of both surface and ground water were the principal participants in water governance. This changed over the past three decades as demands for new, non-agricultural uses of water emerge, often from users and uses with no affiliated water rights. Responding to these new demands has opened up different opportunities for participation in water governance.

7.3.2.1 Early issues and forms of participation

The Arkansas River Valley from Pueblo, Colorado east to the state boundary is laced with hundreds of miles of canals, more than a dozen reservoirs large and small, and many hundreds of groundwater wells, most of which are dedicated to irrigating thousands of acres of land (Arkansas River Basin Implementation Plan, 2015). This infrastructure, and the water diversions it represents, was created beginning in the latter part of the 19th century and into the early decades of the 20th century. The predictability and security that encouraged such agriculture development and investment was the system of water rights and administration (Vranesh, 1987).

Early on surface water rights, governed by the prior appropriation doctrine, were developed and perfected through water courts and administered by the State Water Engineer's Office and water commissioners (Blomquist *et al.*, 2004). Water commissioners administered water rights in order of priority, and individual commissioners were often from local farming families (Vranesh, 1987). Only water rights holders, or those attempting to obtain water rights, were allowed in water court, and water rights holders were actively engaged in administering and monitoring water diversions and

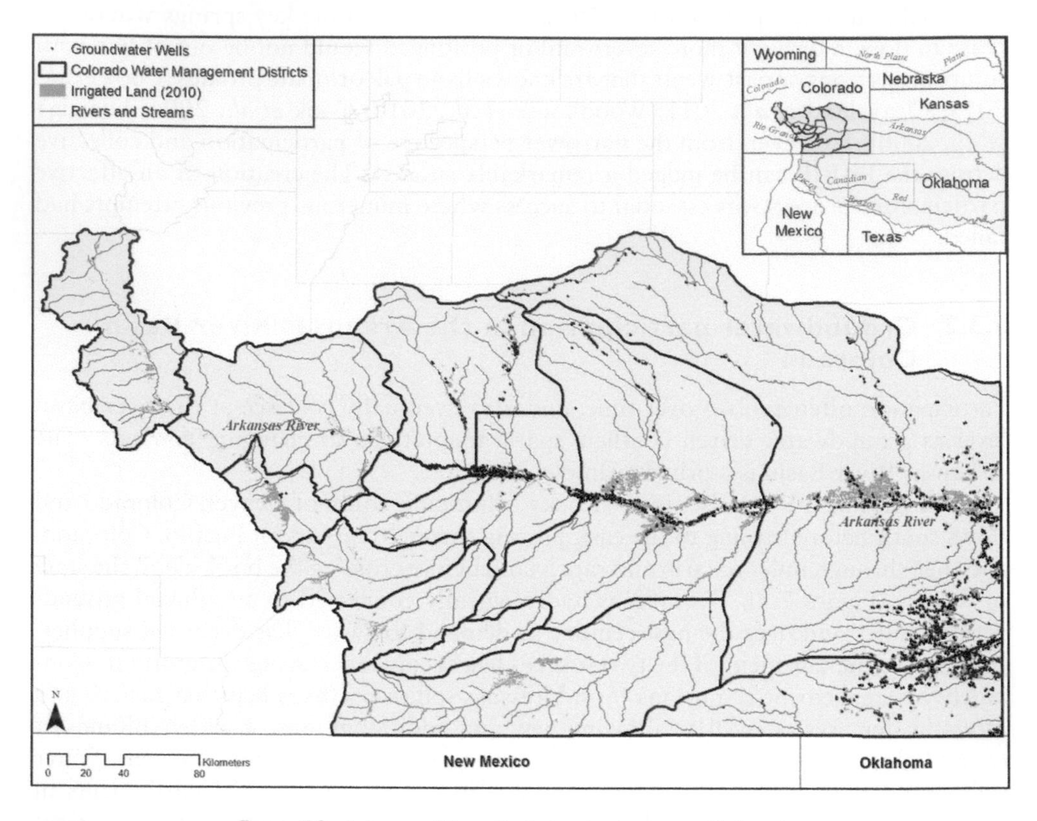

Figure 7.2 Arkansas River Basin in southeastern Colorado.

uses. Participation centered on actors who held water rights, and those actors were largely Colorado surface water irrigators.

Serious water conflicts emerged in the decade following a historic drought. In the 1950s Arkansas River flows were significantly reduced and irrigators turned to groundwater pumping to supplement meager surface water supplies. While the drought ended by 1957, surface water rights holders began to suspect that pumping was affecting river flows. Senior (in time) surface water rights holders began demanding that the State Water Engineer and water commissioners shut down junior (in time) groundwater pumpers in order to protect their water (Blomquist *et al.*, 2004:100).

It took two decades, several legislative acts, and multiple court cases to develop workable solutions for coordinating the use of groundwater and surface water. The length of time it took to find workable solutions reflects the seriousness of the issues involved. If the prior appropriation doctrine were to be carefully followed, most groundwater pumping would have to cease as most groundwater rights are junior to surface water rights. However, severely restricting groundwater pumping would mean that the estimated two million acre feet of water stored in the alluvial aquifers in the Arkansas River Basin could not be tapped (Blomquist *et al.*, 2004:94). How to leave the prior appropriation system in place while still allowing groundwater pumping was the puzzle.

The fierce resistance of well owners had to first be overcome. The threat of groundwater regulations spurred the creation of groundwater users associations. Made up of well owners, the associations actively resisted groundwater regulations that would place any meaningful limitations on pumping and they found a sympathetic Colorado Supreme Court that twice found that the State Water Engineer inappropriately exercised his rule making authority and struck down groundwater regulations (Blomquist *et al.*, 2004:107).

This was the status quo until the mid-1980s when an external threat to groundwater pumping emerged. The Arkansas Interstate River Compact was entered into by the States of Colorado and Kansas in 1948. Under the compact the two states allocated the water of the Arkansas River and devised a set of operating rules for the John Martin Reservoir, whose water was owned by Colorado and Kansas irrigators (Arkansas Basin Roundtable, 2015:119). Kansas water officials had long complained that groundwater pumping in Colorado was intercepting river water that belonged to Kansas irrigators. In 1985, Kansas filed suit with the U.S. Supreme Court, which ruled in 1995 that Colorado was in violation of the compact because of groundwater pumping and that Colorado would have to pay millions of dollars in damages to Kansas. This time the well owners associations collaborated with the State Water Engineer and State Attorney General and, utilizing the authority of the water court, developed a set of rules for addressing the effects of groundwater pumping on surface water flows (Blomquist *et al.*, 2004:109).

The regulatory solution was labeled “replacement plans”. The State Engineer, using a hydrologic model that accounts for ground and surface water interactions, estimates the effects of groundwater pumping on surface water flows as a function of a well’s distance from the river and the amount of water pumped. That information is used to determine how much “replacement” water well owners have to provide to cover the effects of their pumping on river flows. The well owner associations are key to the implementation of replacement plans. They estimate how much replacement water is necessary to cover their members’ pumping. They then lease surplus surface water or devote the water associated with surface water rights that they hold and provide it to the State Engineer to be released at appropriate times and places (Arkansas Basin Roundtable, 2015:91). Currently, in an average water year, the members of the associations pump just over 100,000 acre-feet annually and provide almost 20,000 acre-feet of replacement water (Arkansas River Roundtable, 2015:91–92).

7.3.2.2 Contemporary issues and forms of participation

Over the decades it took to address conflicts between groundwater and surface water users, the water context changed. Non-agricultural water uses driven by growing municipalities now posed the greatest threat to agricultural water users. Not only did cities require additional supplies of water, but their growing populations were seeking out and demanding water-based recreational activities. Growing demands for water to be used for non-agricultural purposes have sparked additional and ongoing water governance adaptations that have changed the patterns of participation.

“Buy and dry” is the name of the practice whereby cities purchase agricultural land and the water rights used to irrigate that land. Cities fallow the land and transfer the use of the water rights to municipal purposes (Howe and Goemans, 2003). The practice has multiple negative spillover effects, such as loss of tax revenue for local jurisdictions, the weakening of rural economies as demand drops for agricultural inputs, and critically, it becomes much more costly to find and acquire surface water to replace the effects of groundwater pumping (Howe and Goemans, 2003).

Two large "buy and dry" events occurred in the mid-1980s that placed water rights holders in the basin on notice that agricultural uses of Arkansas River basin water were increasingly viewed as a major source of water for cities outside of the basin. The City of Aurora, Colorado located in the South Platte River Basin, very near the northwestern edge of the Arkansas River Basin, purchased the rights to just under 112,000 acre-feet of water from two irrigation canals (Howe and Goemans, 2003:1060). Although Aurora agreed to a number of programs to ameliorate the effects of fallowing thousands of acres of farmland, such as payments in lieu of taxes, buying, drying and transferring the water out of the basin has significant economic impacts on the basin of origin, especially if the regional economy is not robust (Howe and Goemans, 2003).

The response by residents of the basin was to find means of limiting "buy and dry" projects, which required more broad-based participation by a variety of interests. Under Colorado state law, local residents may create water conservancy districts that have the authority to invest in the development of water sources and that have taxing and bonding powers to pay for water projects. The Lower Arkansas Water Conservancy District, which was overwhelmingly approved by voters in the proposed district, was created to limit out of basin water transfers, both by purchasing and/or leasing water rights, but also by creating programs that support the viability of agriculture (LAWCD no date). The district's board of directors consists of people from different backgrounds, not just agriculture but urban interests, too (LAWCD no date).

Since its creation in 2002, the LAWCD has purchased and leased water rights, created a land conservation easement program, and is working with several major ditch companies to create a 'Super Ditch' (Campbell, 2015). The conservation easement program promotes the use of land and its water rights to preserve habitat, open space, and irrigated agriculture. The Super Ditch program recognizes the tremendous market pressure to move water rights from agriculture to municipalities. To avoid the practice of "buy and dry", the participating ditch companies are developing a plan by which cities may lease water from the "Super Ditch" and the participating ditch companies commit to delivering that water by rotating the fallowing of irrigated lands. Municipalities gain access to reliable sources of water while farmers and farm communities maintain their rural economies (Campbell, 2015).

In 2002, a drought of record spurred the Colorado Water Conservation Board (CWCB) to conduct a detailed analysis of current and future statewide water demands and water supplies, which revealed the need for statewide water planning (CWCB, 2011). In 2005 Colorado passed legislation that initiated a statewide collaborative water planning process that involved each of the state's nine basins. Called basin roundtables, participants include representatives from local governments, agriculture, industry, recreation, and the environment. In other words, many actors who historically would not have been allowed or invited to participate in water governance were now included.

The Basin Implementation Plan developed by the Arkansas Basin Roundtable addresses the water demands and supplies of the basin as a whole and the interactions of the basin with those adjoining it. It addresses a wide range of issues, such as watershed health, municipal water conservation, and multi-use water projects.

Over the last several decades who participates in water governance in the Arkansas River Basin has dramatically changed. Initially, consumptive uses took precedence. The past fifteen years, however, have witnessed dramatic changes in participation. With the recognition of the close hydrologic connections between surface and groundwater sources and the emergence of many different uses and values, the Arkansas River Basin is governed in a more integrated fashion. The Arkansas Basin Roundtable and the Basin Implementation Plan assess the full range of water demands and supplies in the basin, funding a host of experimental projects, such as the development of the Super Ditch. The Roundtable also provides the venue through which many different water uses and values are recognized and coordinated. No longer is participation in water governance, including groundwater, restricted to water rights holders.

7.4 CONCLUSION

Since participatory groundwater governance is still in its formative stages, it is worthwhile to consider noteworthy experiments. In this chapter we have used two exemplary case studies to illustrate the types of complex socioecological problems that are connected to the use of groundwater resources and how diverse groups of interests have tried to address them. Taken together, these cases from the Western U.S. illustrate the importance of participation in water governance in general and groundwater governance specifically. Below, we draw lessons about participation in groundwater governance.

7.4.1 How and why participation happens in complex situations

These examples show that participation happens in a variety of forms, indicating that there is no single best approach. Participatory planning processes may happen for a variety of reasons. In some cases, a collaborative stakeholder deliberation process may be seen by the interested parties as more appealing than resorting to formal conflict resolution through the court system. Participation may also be motivated or imposed by a higher political authority such as a federal judge or state legislature, as was the case for the Edwards Aquifer and the Arkansas River Basin. In the Edwards Aquifer case, the participants' ultimate impetus for collaborating was to develop a Habitat Conservation Plan that would achieve protection of the parties from further litigation under the Endangered Species Act while also avoiding the imposition of a specific solution from "outside" by the federal government or court. In the case of the Arkansas River Basin, direct and indirect water users sought common ground in order to keep water in the Basin to support local economies.

7.4.2 Importance of common understanding

In both cases, groundwater use and management were central issues but never the only issues. Each case clearly shows the importance of groundwater's hydrologic connection to surface water sources and by implication its critical role in ecological conditions.

A common central dilemma is how to balance consumptive and non-consumptive uses of groundwater in order to avoid unacceptable ecological degradation. In such cases, there may be a large number of interest groups involved with competing visions, needs, and values. Yet, it is possible for them to be reconciled to the relative satisfaction of those involved.

There are aspects of participation which seem necessary or at least conducive, if not sufficient, for such a positive outcome. Because the physical dynamics of groundwater aquifers and the nature of their connections to surface water bodies are often quite complex and in many cases poorly or at least inadequately understood when a participatory process begins, potentially significant investments in scientific studies of various kinds may be of fundamental importance. In the case of the Arkansas River Basin, the U.S. Supreme Court ordered the development of a hydrologic model before it would move forward with a decision. That hydrologic model forms the foundation for water governance and is used to monitor the impacts of human activities on the river system. The EARIP found it necessary to create technical committees to compile and evaluate existing studies and identify knowledge gaps for additional research needed to support informed decision making. Unless the parties can arrive at some consensus view about how an aquifer system is likely to behave under certain conditions of use and natural climatic variability, negotiations over policies and practices are unlikely to succeed. The development of a hydrologic model and/or refinement of an existing one is critically important in this regard.

7.4.3 Opening up participation to a broader set of interests

As we noted in the introduction, working within and around water laws alone tends to narrow the number of parties who may participate. The meaningful inclusion of environmental and ecological water needs requires opening decision-making and planning to interests historically excluded from the table.

In the case of the Edwards Aquifer, historically excluded interests resorted to suing the federal government in order to exert some influence. Before 2007, a great deal of legal conflict occurred between relatively narrow interests. It was not until a new form of participation was tried that a solution could finally be achieved. With the EARIP, conservation and recreation interests had a formal seat at the table along with the more politically powerful water management and provisioning entities. In the Arkansas Basin, expanding participation in water governance has created opportunities for innovative problem solving, such as the creation of the Super Ditch, which hold promise for meeting the values and demands of different users.

However, it should be noted that simply increasing the number of participants involved and the range of interests represented is not a guarantee of success. In the case of the Edwards Aquifer, for example, earlier collaborations had failed and it seems likely that the EARIP would have as well without a legislatively mandated decision-making timetable.

Finally, achieving some kind of consensus can be very difficult, and we caution against viewing experiments in participatory groundwater planning in overly binary terms of success or failure. For example, it took more than two decades for water users in the Arkansas Basin to develop a series of institutional arrangements for addressing the most pressing water problems. Surprisingly, the case with the most fraught and bitter history, the Edwards Aquifer, was able to eventually come to consensus agreements about many different issues and was ultimately able to achieve the goals the participants set out to accomplish. However, we are unlikely to know whether the provisions of the HCP are truly adequate until a repeat of the drought of record puts them to the test. Relatedly, we emphasize the importance of having an adaptive decision-making process capable of revising mitigation and management measures based on new scientific findings and hydrologic and ecological monitoring and assessments.

REFERENCES

Arkansas River Roundtable. (2015). *Basin Implementation Plan.* [Online] Available from: http://www.arkansasbasin.com/uploads/2/7/1/8/27188421/arkansas_bip_executive_summary.pdf [Accessed 7th December 2016].

Blanton & Associates, Inc. (2016) *Edwards Aquifer Habitat Conservation Plan 2015 Annual Report.* [Online] Available from: http://eahcp.org/documents/EAHCP%202015%20Annual%20Report%20Final.pdf [Accessed 10th December 2016].

Blomquist, W. (1992) *Dividing the waters: governing groundwater in Southern California.* San Francisco, CA, ICS Press Institute for Contemporary Studies.

Blomquist, W., Schlager, E. and Heikkila, T. (2004). *Common waters, diverging streams: linking institutions to water management in Arizona, California, and Colorado.* Washington, DC: Resources for the Future.

Brandon, T. (2015) Fearful asymmetry: how the absence of public participation in section 7 of the ESA can make the 'best available science' unavailable for judicial review. *UCLA Law Review,* 39, 311–369.

Brosi, B.J. & Biber, E.G.N. (2012). Citizen involvement in the U.S. Endangered Species Act. *Science* 802, 803.

Campbell, S. 2015. The Super Ditch: can water become a cash crop in the west? *Landlines* 10–17.

Cleaveland, M. K., Votteler, T. H., Stahle, D. K., Casteel, R. C., & Banner, J. L. (2011) Extended chronology of drought in South Central, Southeastern and West Texas. *Texas Water Journal,* 2(1), 54–96.

Colorado Foundation for Water Conservation. (2004) *Citizens guide to Colorado water law.* 2nd edition. [Online] Available from: www.cfwc.org [Accessed December 12, 2016]

Colorado Water Conservation Board. (2011) *Statewide water initiative 2010: Colorado's water supply future.* [Online] Available from: http://cwcb.state.co.us/water-management/water-supply-planning/Pages/SWSI2010.aspx [Accessed 12th December 2016]

Cook, B. I., Ault, T. R., & Smerdon, J. E. (2015) Unprecedented 21st century drought risk in the American Southwest and Central Plains. *Scientific Advances,* 1(1), 7. Available from: doi.org/10.1126/sciadv.1400082 [Accessed 8th December 2016]

Cook, E. R., Woodhouse, C. A., Eakin, C. M., Meko, D. M., & Stahle, D. W. (2004) Long-term aridity changes in the western United States. *Science,* 306(5698), 1015–1018.

Donahue, J. M. (1998) Water wars in south Texas: managing the Edwards Aquifer. In: Donahue, J.M. & B. R. Johnston (Eds.), *Water, culture, and power: local struggles in a global context.* Island Pr.

Duane, T. & Opperman, J. (2010) Comparing the conservation effectiveness of private water transactions and public policy reforms in the conserving California landscapes initiative. *Water Policy,* 12, 913–931.

Freethey, G. W. (1982. *Hydrologic analysis of the Upper San Pedro Basin from the Mexico-United States boundary to Fairbank, Arizona* (U.S. Geological Survey Open-File Report No. 82–752) (p. 64). U.S. Geological Survey.

Garrick, D. E. (2015). Water Allocation in Rivers under Pressure: Water Trading, Transaction Costs and Transboundary Governance in the Western US and Australia. Edward Elgar Publishing.

Getches, D.H. (2009) *Water Law in a Nutshell*. 4th edn. St. Paul, MN, Thompson West.

Gulley, R. L. (2015) *Heads above Water*. College Station, TX, Texas A&M University Press.

Gulley, R., & Cantwell, J. B. (2013) The Edwards aquifer water wars: the final chapter? *Texas Water Journal*, 4(1), 1–21.

Hobbs, G.J. (1997). "Colorado Water Law: A Historical Overview" University of Denver Water Law Review 1(1):1–138.

Houck, O.A. (1993). The Endangered Species Act and its implementation by the U.S. Department of Interior and Commerce. *University of Colorado Law Review*, 64, 277.

Howe, C.W. & Goemans, C. (2003) Water transfers and their impacts: lessons from three Colorado water markets. *Journal of the American Water Resources Association*, 1055–1065.

Kundis Craig, R. (2014) Does the Endangered Species Act preempt state water law? U. Kan. L. Rev., 62, 851.

Loáiciga, H. A., Maidment, D. R., & Valdes, J. B. (2000) Climate-change impacts in a regional karst aquifer, Texas, USA. *Journal of Hydrology*, 227(1), 173–194.

Lower Arkansas Water Conservancy District (no date) "Mission" [Online] Available at: http://www.lavwcd.com/ [Accessed December 12, 2016].

Maclay, R. W. (1995) *Geology and hydrology of the Edwards Aquifer in the San Antonio area, Texas* (USGS Numbered Series No. 95–4186). Austin, TX: U.S. Geological Survey. [Online] Available at: http://pubs.er.usgs.gov/publication/wri954186 [Accessed November 23, 2016].

National Research Council. (2015) *Review of the Edwards Aquifer habitat conservation plan: report 1*. Washington, D.C.: National Academies Press. [Online] Available at: https://www.nap.edu/catalog/21699/review-of-the-edwards-aquifer-habitat-conservation-plan-report-1 [Accessed November 23, 2016].

New Mexico Office of the State Engineer. (2013) *New Mexico water plan handbook*. [Online] Available at: http://www.ose.state.nm.us/Planning/RWP/PDF/Revised%20RWP%20Handbook%20ISC_Dec_2013_Final.pdf [Accessed 7th December 2016].

New Mexico Office of the State Engineer. no date. *New Mexico regional water plans*. [Online] Available at: http://www.ose.state.nm.us/Planning/regional_planning.php [Accessed 7th December 2016].

Ostrom, E. (1990). *Governing the commons*. Cambridge: Cambridge University Press.

Ostrom, V. & Ostrom, E. (1977) Public Goods and Public Choices. In: *Alternatives for delivering public services: toward improved performance*, ed. E. S. Savas, 7–49. Boulder, CO: Westview Press.

Pielke, R.A. Sr, N. Doesken, O. Bliss, T. Green, C. Chaffin, J.D. Salas, C.A. Woodhouse, J.J. Lukas, & K. Wolter. (2005) Drought 2002 in Colorado: an unprecedented drought or a routine drought? *Pure and Applied Geophysics*, 16(8–9), 1455–1479.

Radosevich, G.E., K.C. Nobe, D. Allardice, & C. Kirkwood. (1976) *Evolution and administration of Colorado water law: 1876–1976*. Fort Collins, CO: Water Resources.

RECON Environmental Inc., Hicks & Company, Zara Environmental LLC, & BIO-WEST. (2012) *Edwards aquifer recovery implementation program habitat conservation plan*. [Online] Available at: http://www.eahcp.org/files/uploads/Final%20HCP%20November%202012.pdf [Accessed 28th November 2016].

Sax, J.L., B.H. Thompson, Jr, & J.D. Leshy. (2006) *Legal control of water resources*. 4th edn. St. Paul MN: West Group.

Schlager, E. (2006) Challenges of governing groundwater in U.S. western states. *Hydrogeology*, 14(3), 350–360.

Schlager, E. & W. Blomquist. (2008) *Embracing Watershed Politics*. Boulder, CO: University Press of Colorado.

Votteler, T. (1998) The little fish that roared: the Endangered Species Act, state groundwater law, and private property rights collide over the Texas Edwards aquifer. *Environmental Law*, 28(4), 845–846.

Vranesh, G. (1987) *Colorado water law.* 3 vols. Boulder, CO: Natural Resources Law Center.

Woodhouse, C. A., Meko, D. M., MacDonald, G. M., Stahle, D. W., & Cook, E. R. (2010) A 1,200-year perspective of 21st century drought in southwestern North America. *Proceedings of the National Academy of Sciences*, 107(50), 21283–21288. Available from doi.org/10.1073/pnas.0911197107 [Accessed 28th November 2016].

Chapter 8

Economic instruments, behaviour and incentives in groundwater management

Phoebe Koundouri[1,2,3], Ebun Akinsete[3,4], Nikolaos Englezos[3,5], Xanti I. Kartala[3,5], Ioannis Souliotis[3,6] & Josef Adler[3]
[1]*School of Economics, Athens University of Economics and Business, Greece*
[2]*Grantham Research Institute, London School of Economics and Political Science, UK*
[3]*International Center for Research on the Environment and the Economy (ICRE8), Athens, Greece*
[4]*School of Applied Social Studies, Robert Gordon University, Aberdeen, UK*
[5]*Department of Banking and Financial Management, University of Piraeus, Greece*
[6]*Centre for Environmental Policy, Imperial College London, UK*

ABSTRACT

This chapter will provide an overview of the contemporary groundwater literature and will show what is currently being done to achieve sustainable groundwater management. First, models for groundwater management will be presented focusing on resource modelling under uncertainty, in particular uncertainty surrounding the effect of climate change on groundwater resources and trans-boundary frameworks. Then, the ecosystem services approach and the concept of the TEV of water will be presented in more detail. Furthermore, we will show where and how these new concepts are integrated in policy frameworks and will present applied examples of sustainable water governance. Last but not least, we will venture out to the future of sustainable groundwater management and have a look at upcoming challenges, opportunities, and cutting-edge research.

8.1 AN INTRODUCTION

Groundwater quantity and quality are exposed to a multitude stressors (Navarro-Ortega *et al.*, 2015). Due to heavy usage as potable water and input in economic sectors – households, industry, tourism, and agriculture – groundwater has been over-exploited, polluted, and degraded. Since groundwater is a pivotal input for all the above mentioned, there have been calls to manage it more efficiently. Gisser and Sánchez (1980) question, however, whether managing groundwater resources will increase social welfare. They show that there is no quantitative difference between temporal optimal control of groundwater and competitive, myopic usage. This apparent paradox, the so-called *Gisser-Sánchez-Effect*, vanishes, however, if one considers water quality issues and their externalities (Kundzewicz & Döll, 2009), allows non-linearity in water demand and supply in the model (Koundouri, 2004b), and considers uncertainties surrounding future availability (Maqsood, Huang, & Yeomans, 2005), due to climate change (Taylor *et al.*, 2013), shortcomings in data on interaction between surface and groundwater, the hydrological cycle (Li, Huang, & Nie, 2006), and unknown recharge rates of aquifers (Brouyere, Carabin, & Dassargues, 2004). Once one adds trans-boundary aquifers to the model, groundwater management issue are exacerbated due to institutional and legal concerns (Koundouri & Groom, 2002). All these issues should be considered when trying to determine the value of groundwater.

The ecosystem services approach (ESA) tries to provide a holistic methodology that identifies benefits and costs ecosystems create, illustrates problems concerning services and trade-offs between them, and finally assigns a monetary value to them which adds up to the total economic value (TEV) of water (Benayas, Newton, Diaz, & Bullock, 2009; Koundouri *et al.*, 2015). An ESA has already been included in policy frameworks such as the European Water Framework Directive (European Commission, 2000; P. 12) or the Marine Strategy Framework Directive (Koundouri & Dávila, 2015). The goal being to design and assess measures, *i.e.* economic instruments providing incentives (*i.e.* taxes, permits, subsidies, pollution fees etc.), to recover the full cost of groundwater services and to choose the most economically efficient one of them by applying a cost-benefit analysis (Birol, Koundouri, & Kountouris, 2010). Since groundwater may exhibit non-market characteristics, it is crucial to consider all different aspects that contribute to water value – its use value, such as irrigation, and its non-use value, such as a subjective value a person may attribute to improvements in, *e.g.* a wildlife habitat (Bateman, Brouwer, *et al.*, 2006). However, due to unobservable or unavailable prices for water, in general other means have to be found to assign a monetary value to groundwater services. Energy, water, or fuel subsidies to the agricultural sector to spur rural development promote groundwater usage further and complicate this estimation (Shah, 2007). Apart from leading to groundwater overexploitation, those subsidies, taxes, and other policy instruments exacerbate the non-market characteristics of groundwater by distorting its price (Groom, Koundouri, & Swanson, 2005; Koundouri *et al.*, 2015). Consequently, in the past years a range of non-market valuation methods have been used to estimate the TEV of groundwater resources (Birol, Koundouri, & Kountouris, 2006; Brouwer, 2008). The results are supposed to help policy makers in deciding on how to allocate water in the future and to design economic incentives to induce more efficient use (Koundouri & Dávila, 2015), in order to ensure sustainable resource management (Kløve *et al.*, 2011).

8.2 GROUNDWATER MANAGEMENT MODELING

Integrated hydro-economic models are formal mathematical models which aim to quantify the complex structure of groundwater management along the lines of the fundamental economic principles of demand and supply. In these models, optimal groundwater management is treated as an optimization problem of an objective function which considers TEV, subject to specific constraints rising from predetermined control criteria on the groundwater resource evolution. A critical literature overview of the available economic models of groundwater use and their potential benefits from optimal groundwater management was provided by Koundouri (2004a). This study analyzed the Gisser-Sanchez model which is the basic representation of economic, hydrologic and agronomic facts that occur due to the irrigator's choice of water pumping. The environmental constraint of the problem derives from the change in the height of the water table which is given by the following differential equation

$$\dot{H} = \frac{1}{AS}[R + (a - 1)w], \quad H(0) = H_0, \tag{8.1}$$

where R is the constant recharge measured in acre feet per year, α is the constant return flow coefficient which is a pure number, H_0 is the initial level of the water table measured in feet above sea level, A is the surface area of the aquifer (uniform at all depths) measured in acres per year, S is the specific yield of the aquifer which is a pure number and w is the water extraction measured in acre-feet per unit of time. In order to model the case of a non-constant river recharge due to stochastic rainfall or a possible exogenous and reversible shock to the groundwater resource, one could consider that R is a random variable (cf. Laukkanen & Koundouri, 2006 and De Frutos Cachorro *et al.*, 2014) or a stochastic process (cf. Zeitouni, 2004). Hence, in this section, we shall present recent advances in such hydro-economic according to different aspects of groundwater management, such as coastal aquifer water management, conjunctive use of surface and subsurface water resources, and game theoretical approaches, including stochastic frameworks imposed by climate change conditions, both in a boundary and a transboundary scale.

In the literature (Tsur & Zemel, 2014) the first type of *uncertainty* that enters into the resource management problems corresponds to the limited knowledge of certain parameters of the resource (for instance abrupt system behavior when the stock process crosses some unknown threshold) and the second one is the exogenous uncertainty that takes into account random environmental elements (for example weather variability). According to these types of uncertainty, many studies dealt with the relationship between precautionary behavior and an increase in uncertainty (see Brozovic & Schlenker, 2011 and Zemel, 2012). Assuming a stochastic recharge rate, Zeitouni (2004) argued that it is optimal to keep the water stock at a certain positive threshold in the case of a limited aquifer capacity. Considering a known decrease in the recharge rate as an exogenous shock, De Frutos Cachorro *et al.* (2014) showed that the optimal adapted extraction of a groundwater aquifer decreases in the short–run for a deterministic occurrence date of the shock and vice versa for a stochastic one.

Groundwater management in coastal regions has been widely studied due to the rapid demand for fresh water and the groundwater quality deterioration from *seawater intrusion*. Karterakis *et al.* (2007) compared the classical linear programming (LP) optimization algorithm of the SM and the Differential Evolution (DE) algorithm, used to compute the optimal hydraulic control of the saltwater intrusion in an unconfined coastal karstic aquifer, concerning the computation time and the values of the water volume flow rates. Katsifarakis & Petala (2006) and Kentel & Aral (2007) studied simulation-optimization coastal aquifer problems subject to a penalty term regulated by the seawater intrusion due to the applied pumping scheme and by the limited groundwater resources in the region, respectively. In order to reduce computation burden and capture the uncertainty in the physical system, Sreekanth & Datta (2014) substituted the numerical simulation model with a genetic programming (GP) stochastic surrogate model to characterize coastal aquifer water quality regarding to pumping, under parameter uncertainty, and obtain a stochastic and robust optimization of groundwater management. Additionally, Koundouri and Christou (2006) analyzed the optimal management of groundwater resources with stock-dependent extraction cost and a backstop substitute. The developed model considers heterogeneous sectors and use multistage dynamic optimal control.

Proper conjunctive use of water, namely *the integrated use of surface and groundwater resources,* is an essential issue due to the increasing water demands of the agricultural sector. An integrated dynamic approach was employed by Chang *et al.* (2011) to simulate the interaction between surface and subsurface water as a system, where the natural groundwater recharge is considered as a water source to the system and its volume is estimated using geographic information system (GIS) tools, a groundwater modular-dimensional groundwater flow (MODFLOW) model, and a parameter identification model. On another strand, Yang *et al.* (2009), Peralta *et al.* (2011) and Rezapour & Soltani (2013) applied genetic algorithms (GAs) and constrained differential dynamic programming (CDDP) techniques to study multi-objective problems associated with the performance of a conjunctive use surface and subsurface water system, considering issues of maximizing the minimum reliability of the system as well as minimizing both the fixed and the time varying operating costs due to water supply. In a different study Peralta *et al.* (2011) quantified limits and acceptable impacts on selected water resources indicators, and developed a new simulation–optimization algorithm with limits to compute optimal safe yield groundwater extraction policies.

Several studies developed an analytical game-theoretic formulation to calculate sustainable groundwater extraction rates in both cooperative and non-cooperative conflict-resolution approaches (Loaiciga, 2004), to find an optimal balance between positive economic benefits and negative environmental impacts among alternative groundwater extraction scenarios (Salazar *et al.*, 2007), to compute cooperative optimal allocation policies in a multi-objective finite difference aquifer subject to water provision costs (Siegfried & Kinzelbach, 2006), and to address the problem of optimal groundwater extraction by multiple spatially distributed users from an aquifer (Brozovic *et al.*, 2006). Bazargan-Lari *et al.* (2009) proposed a new GA methodology for the conflict-resolution conjunctive water use with different users, Saleh *et al.* (2011) investigated both cooperative and myopic groundwater inventory management schemes with multiple users via a dynamic game-theoretic formulation, and Wang & Segarra (2011) studied the game-theoretic common-pool resource dilemma in extracting nonrenewable groundwater resources when water demand is perfectly inelastic and water productivity is heterogeneous. The game-theoretical framework was also employed to conflict-resolution groundwater management in irrigated agriculture (Latinopoulos & Sartzetakis, 2011), in assessing the value of cooperation under the presence of environmental externalities (Esteban & Dinar, 2012), in common pool resources by cooperative (Madani & Dinar, 2012a) and non-cooperative (Madani & Dinar, 2012b) institutions.

In a river basin scale, several hydro-economic models were used to integrate riparian zones and wetlands (Hattermann *et al.*, 2006) and optimize the conjunctive management of surface and groundwater systems (Pulido-Velazquez *et al.*, 2007, 2008, Safavi *et al.*, 2010, Wu *et al.*, 2015, and Nasim & Helfand, 2015), as well as under uncertainty analysis (Wu *et al.*, 2014). The conflict-resolution issues on water scarcity and infrastructure operations concerning river basin management in transboundary water resources allocation, *i.e.* the river is a common water resource to multiple countries, is addressed by the game theoretic approach. Wu and Whittington (2006) investigated the incentive structure of both cooperative and no cooperative policies for different riparian countries that share an international river basin. Eleftheriadou & Mylopoulos (2008) quantified the consequences caused by water flow decrease for different scenarios to estimate compromising solutions acceptable by two countries. Under the effects of climate change, Bhaduri *et al.* (2011) presented a stochastic non-cooperative differential game to obtain sustainable transboundary water allocation by linking transboundary flows to hydropower exports, whereas Girard *et al.* (2016) compared cooperative game theory and social justice approaches with respect to cost allocation of adaptation measures at the river basin scale.

8.3 CALCULATING THE TOTAL ECONOMIC VALUE OF GROUNDWATER

The total economic value (TEV) comprises different types of use and non-use values. The first relates to actual or potential use values (option value) which derive from the direct or indirect use of an environmental resource (*e.g.* water irrigated from a groundwater aquifer that is used in agriculture refers to direct benefits, whereas the increase in jobs this yields in the agricultural sector refers to indirect use). Option value relates to the value that might accrue in the future from the existence of the resource *i.e.* willingness to pay for maintaining a resource although it is possible that it will not be used in the future. For example, the discovery of new species of plants might lead to the development of drugs that fight diseases. Non-use values are grouped into three main categories; bequest value relates to the value individuals place on the fact the future generations will have access to the same benefits. Existence value, refers to individuals' willingness to pay to preserve the characteristics of the resource as it stands. Finally, altruistic value corresponds to the utility that individuals obtain, by knowing that others users in the community obtain benefits from a specific resource.

Koundouri, Palma, and Englezos (2017) examine various valuation methods in detail, extensively reviewing existing for determining the TEV of groundwater. Revealed preference techniques base their results on data drawn from existing markets or actions (*e.g.* driving to visit a natural site) that encapsulate the value of environmental benefits These techniques however, can only estimate the use values of environmental resources. Such techniques are the *hedonic pricing method*, the *travel cost method* and *cost of replacement*. The first aims at tracing the footprint of the value of an environmental good, by observing the prices in markets. In many applications this has been done by observing the real estate markets in two areas with similar characteristics and varied levels of environmental amenities (*e.g.* The second, considers several parameters that relate to traveling to a destination (*e.g.* a park). Such parameters are travel expenses (fuel, overnight stay etc.), time spent traveling, frequency of traveling, distance from the destination, substitutes in the vicinity and characteristics of the destination. Considering these factors, the method can estimate the value that individuals place on the recreational benefits provided. The second family of methods is the stated preferences techniques, which include *contingent valuation method* and *choice modelling*. These techniques can elicit both use and non-use values through structured surveys that ask respondents to state their WTP. A difference between the two approaches is that contingent valuation can elicit the value of whole goods, whereas choice modelling can estimate the value of both whole goods and their specific characteristics. Similar to the above method, Choice Modelling also uses surveys to obtain information from respondents. This method is heavily based on the theory of Lancaster (1996), which ascribes that goods are a bundle of different characteristics.

Besides the above, benefit transfer methods use results from earlier primary studies in areas similar to that under investigation. By first adjusting the value for the differences in the socioeconomic characteristics (income, prices, currency, etc.) between areas, the value is transferred to express the preferences of the users of the study area. Koundouri *et al.* (2016) used this method to assess the value of four ecosystem services of the Anglian river basin in the UK. In order to estimate these values, several other studies had been considered, such as choice experiments and hedonic pricing. Another study by Koundouri *et al.* (2014) used this approach to estimate the benefits of mitigating industrial pollution. They valued the change in water quality from "bad" to "very good" as set by the Directive 2000/60/EC. This was found to vary between 88.28 and 116.94 euros. In relation to this approach, several studies have combined its methodology with GIS (Geographical Information System) data to assess the economic value of conservation and restoration projects (*e.g.* Jenkins *et al.*, 2010), to estimate value of ecosystem services (Plummer, 2009) and to aggregate benefits from non-market environmental goods (Bateman *et al.*, 2006) among others. Finally, other experimental and market techniques exist, such as laboratory experiments. These are techniques that are implemented in a controlled environment (laboratory) and ask respondents to make choices following a well-structured scenario. For example, Drichoutis *et al.* (2014) implemented this technique by engaging respondents in a 6 auction rounds (three of them were hypothetical and three real). Respondents had to choose if they would exchange their endowment with an amount of a good from a river basin with good ecological status and a river basin with bad ecological status that could potential raise health concerns. The results indicated that people would bid higher for the goods that were produced in the region that had water of good ecological status, showing aversion to potential health issues stemming from heavily polluted water. Another study by Carson *et al.* (2011) assessed the economic consequences of the effects of arsenic contamination. The study was concerned about the effect on labor supply in Bangladesh. For this reason, a labor supply model was estimated that used labor data from local households, which was matched with data on arsenic contamination. The results indicated that labor hours are lost, due to the fact that individuals try to hedge against contamination dangers. Also, meta-analysis is a method that is widely used. Such studies include statistical analysis of combined results of previous studies. For example, Van Houtven *et al.* (2006) identified 300 studies that relate to water quality improvements, most of which were stated preference studies. Table 8.1 depicts studies which focus on estimating the value of several services provided by groundwater.

Through the years several ecosystem services classifications have been suggested, such as the Millennium Ecosystem Assessment (MEA 2005) that recognizes four broad types of ecosystem services: *provisioning*, *regulating*, *supporting*, and *cultural* services. While the MEA provides a straightforward connection between the natural environment and the processes that take place within it and welfare, a major disadvantage is

Table 8.1 Summary table of economic valuation studies.

Paper	*Resource*	*Method*	*Values*
Hedonic price analysis and selectivity bias: water salinity and demand for land. (Koundouri & Pashardes, 2002)	Groundwater	Hedonic pricing	£11.5 per hectare
Arsenic mitigation in Bangladesh: A household labour market approach. (Carson, Koundouri, & Nauges, 2011)	Groundwater	Labor Market Approach	$18–38 household/year
Environmental cost of groundwater: A contingent valuation approach. (Martínez-Paz, & Perni, 2011)	Groundwater	Contingent Valuation	€23.52 person/year
The value of scientific information on climate change: a choice experiment on Rokua esker, Finland. (Koundouri, 2012)	Groundwater	Choice Experiment	€9.71–36.92 per household/year
A Value Transfer Approach for the Economic Estimation of Industrial Pollution: Policy Recommendations. *Water Resources Management Sustaining Socio-Economic Welfare*, 7, 113–128. (Koundouri, 2013)	River, Groundwater	Benefit Transfer	€88.28–116.94 household/year

that the framework does not distinguish between intermediate and final services which might lead to double-counting of ecosystem services (Kontogianni *et al.*, 2010; Boyd & Krupnick, 2009).

8.4 INSTITUTIONAL FRAMEWORKS AND POLICY

The governance and management of water as a resource has been at the fore of global environmental and political efforts for decades. The idea of Global Water Governance emerged as a result of a growing consensus that water management was reaching a crisis point and needed to be made a priority (Rogers and Hall, 2003; Cooley *et al.*, 2013). In 2003 the United Nation issued its first Water Development Report, within which water management is identified as a "social, economic and political" challenge (United Nations, 2003). In the wake of the acute impact of climate change felt across the globe today, water management remains a global priority and features prominently in the United Nations Agenda 2030. In addition the issue of water management is embedded within the Sustainable Development Goals (SDGs), addressed both as Goal (#6: Clean Water and Sanitation) in its own right, as well as a cross cutting theme (Sustainable Development Solutions Network, 2015; United Nations, 2016). 2016 saw the convention of the United Nations High Level Panel on Water (HLPW) which has a remit to "ensure availability and sustainable management of water and sanitation for all, as well as to contribute to the achievement of the other SDGs that rely on the development and management of water resources". The panel is expected to provide global leadership in the collaborative effort for inclusive and sustainable water resource management at all scales (HLPW, 2016).

At European level, a number of policies have been introduced in order to regulate the quality of groundwater across the continent. In 1979, the Commission issued

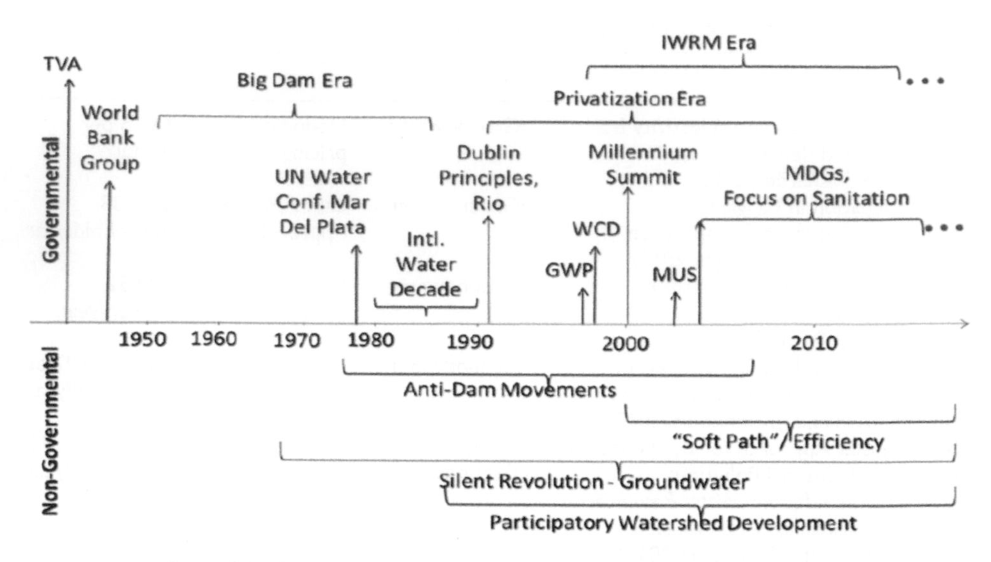

Figure 8.1 Global Water Governance Timeline (Source: Cooley *et al.*)

'Directive 80/68EEC' which aimed at preventing the pollution of groundwater by toxic, persistent and bioaccumulable substances including metalloids and their compounds (European Commission, 1979). Since then several other Directives which consider the preservation of groundwater quality in one way or another have been developed and come into force; these include the Drinking Water Directive (1980), the Urban Wastewater Treatment Directive (1991), the New Drinking Water Quality Directive (1991), the Nitrates Directive (1991), the Plant Protection Products Directive (1991), the Directive for Integrated Pollution and Prevention Control (1996), the Biocides Directive (1998) the Groundwater Directive (2006) and the Directive on Industrial Emissions (2010) (European Commission, 2017).

In 2000, the Water Framework Directive (European Commission, 2000) introduced an integrated legal framework for the protection of European freshwater ecosystems, as well as the means to achieve that which are crystallized within its objectives. The ultimate objective of the Directive is to achieve Good Ecological Status (GES) in all freshwater ecosystems (rivers, lakes, transitional waters, groundwater, etc.) across Europe. In order to achieve that member states must adopt the Directive, define River Basin District and set out a plan of action that will lead to the achievement of GES. The WFD not only assesses the chemical, biological and morphological status of surface water, but it stresses the importance of the social and economic status of each river basin district. It considers economic aspects of the basins in articles 5, 9, 11 and Annex III (Koundouri & Davila 2013). According to these, member states must define the water uses in each river basin district, estimate the total economic cost of water services and design measures that assist in achieving full recovery of this cost.

Within Saleth and Dinar's (2004) framework (see Figure 8.2), endogenous and exogenous factors of change are identified and assessed. These factors are important for the design and implementation of coordinating mechanisms among ministries and

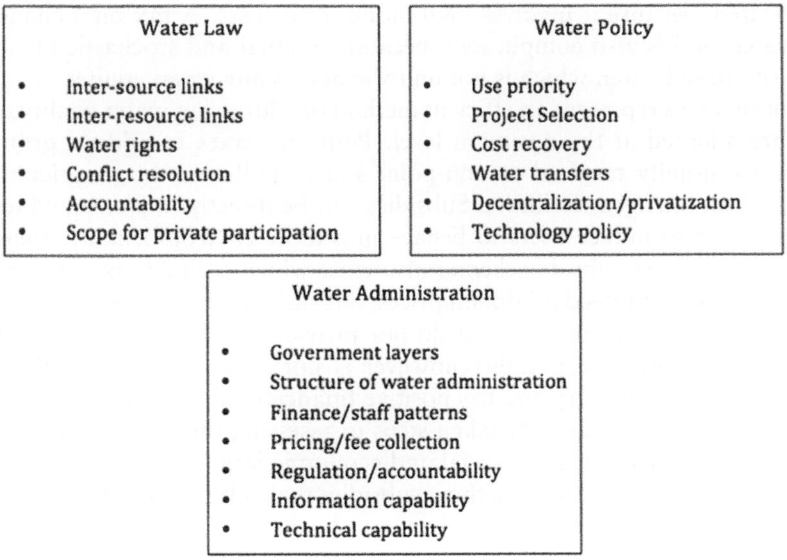

Water Law

- Inter-source links
- Inter-resource links
- Water rights
- Conflict resolution
- Accountability
- Scope for private participation

Water Policy

- Use priority
- Project Selection
- Cost recovery
- Water transfers
- Decentralization/privatization
- Technology policy

Water Administration

- Government layers
- Structure of water administration
- Finance/staff patterns
- Pricing/fee collection
- Regulation/accountability
- Information capability
- Technical capability

Figure 8.2 Saleth and Dinar's Analytical Framework (Source: Saleth and Dinar, 2004).

may allow or hinder cross-sectoral collaboration between diverse bureaus in the water and green growth fields. Saleth and Dinar's institutional framework is re-categorized into state, market, and community to take into account the arguments about the drivers and instruments of economic and social development and environmental conservation based on the state, the market, and community.

Water Abstraction taxes are taxes that can be used to restrain water users from lowering the water level below a certain standard. Area pricing is the most common form of water pricing whereby users are charged for water used per irrigated area. Output pricing methods involve charging a fee for each unit of output produced per user whereas, input pricing involves charging users for water consumption through a tax on inputs. The efficiency of water abstraction taxes is relative and depends on technical and institutional factors. Volumetric pricing is the optimal water tariff[1] where price is equal to marginal cost of supplying the last unit. The effectiveness of a tax depends on the correct estimation of the marginal tax level and on how risk-averse farmers are with respect to damage from reduced water availability (both in quality and quantity terms). A differentiated tax level has to be created, because of local differences in both the monetary value of reserves and the vulnerability of the environment to changes in the groundwater level. An advantage of a tax is that it improves both economic and technical efficiency. Administrative costs are high, since

[1] A water tariff is a price assigned to water supplied by a public utility through a piped network to its customers. Prices paid for water itself are different from water tariffs. They exist in a few countries and are called water abstraction charges or fees. Water tariffs vary widely in their structure and level between countries, cities and sometimes between user categories (residential, commercial, industrial or public buildings). The mechanisms to adjust tariffs also vary widely.

a differentiated tax is not easy to control and monitor. A volumetric tax on extraction is complicated, because it involves high monitoring costs. A tax on a change in the groundwater level is also complicated, because external and stochastic factors affect the level of groundwater, which is not uniform across any given aquifer.

Pollution taxes represent an efficient method of addressing water quality problems if these are adopted at the optimum level. Pollution taxes to address groundwater pollution are usually targeted at non-point source pollution from agriculture, and are imposed on nitrogen fertilizers. Subsidies can be directly implemented for water-saving measures to induce users to behave in a more environmentally friendly way. Alternatively, indirect subsidy schemes also exist which include tax concessions and allowances, and guaranteed minimum prices. Subsidies however are not economically efficient, they create distortions and do not provide incentives for the adoption of modern technologies. Acceptability however is not an issue, since participation in subsidy schemes is voluntary and has positive financial implications.

Some countries have already taken steps in assessing their subsidies programmes in terms of their environmental, social and economic impacts and in reforming their harmful policies, towards reducing those subsidies that enhance fossil-fuel use and thus act as a hurdle to combating climate change and achieving more sustainable development paths. As discussed in Zilberman *et al.* (2008), rising energy prices however, will alter water allocation and distribution. Water extraction will become more costly and demand for hydroelectric power will grow. The higher cost of energy will substantially increase the cost of groundwater, whereas increasing demand for hydroelectric power may reduce the price and increase supply of surface water. Thus, rising energy prices will alter the allocation of water, increase the price of food and may have negative distributional effects.

Groundwater tradable permits assume the introduction of water markets (Howitt, 1997) in which water rights, or permits, can be traded to address different aspects of the water resource problem (Kraemer and Banholzer, 1999); *e.g.* water abstraction rights, discharge permits and tradable permits for use of water-borne resources such as fish or potential energy. Generally, the government will determine the optimal level of water resource use over a specified time period and will allocate an appropriate number of permits. The financial impact on affected parties and related acceptability of tradable permits depends on the initial allocation of rights. These can either be distributed for free (for example depending on historical use or other criteria), or auctioned off to the highest bidders. While there are some examples of its implementation, the use of tradable rights for groundwater seems to be complicated in practice, since the impact of changes in the groundwater level on agricultural production and nature depends on location-specific circumstances. To avoid transferring rights among areas with heterogeneous characteristics, trading has to be restricted. Tradable water permit systems have been implemented in a number of countries including Chile, Mexico, Peru, Brazil, Spain, several states in Australia and the Northern Colorado Water Conservancy District in the USA (Marino and Kemper, 1999).

Voluntary agreements try to convince farmers (through education) of the advantages of fine-tuned groundwater control. Voluntary agreements on controlling groundwater use are in principle efficient, since they rely on specialized knowledge of participants about local conditions. The principle of allowing the individual members of agricultural organizations and water boards to make decisions on issues that affect them rather than leaving those decisions to be made by the whole group, the so called 'principle of subsidiary', is widely accepted. Environmental liability systems intend to internalize and recover the costs of environmental damage through legal action and to make polluters pay for the damage their pollution causes. If the penalties are sufficiently high, and enforcement is effective, liability for damage can provide incentives for taking preventative measures. For liability to be effective there need to be one or more identifiable actors (polluters); the damage needs to be concrete and quantifiable and a causal link needs to be established between the damage and the identified polluter.

8.5 APPLICATIONS TO GROUNDWATER MANAGEMENT

All around the world a variety of projects have focused on applying hydro-economic, game theoretical, and optimization models to groundwater management issues. On top of that a number of strategies to achieve sustainable groundwater management have focused on an ecosystem services approach to calculate the TEV of surface and groundwater. So too, has some of our research been applying these new models and concepts to projects and case studies. In the following, we will present some of the work on groundwater management done at the International Center for Research on the Environment and the Economy (ICRE8: www.icre8.eu) and the Research Team on Socio-Economic and Environmental Sustainability (ReSEES: http://www.icre8.eu/resees) of the Athens University of Economics and Business, which is part of ICRE8's research cluster structure. Since September 2016, the International Center for Research on the Environment and the Economy (ICRE8: www.icre8.eu) is part of a Horizon 2020 (European Commission) project that will establish a Decision Analytic Framework to explore the water-energy-food Nexus (DAFNE) in complex trans-boundary water resources of fast developing countries. ICRE8 is responsible for developing socio-economic models in a complex trans-boundary framework – the two case studies are river basins that link eight African countries, two respectively – considering uncertainties due to climate change (DAFNE Project, 2017). Game theoretical models will be applied to construct interactions and competing interests in the river basin. Further, the concept of TEV of water will be applied to estimate the value of the resources.

GLOBAQUA is an ongoing project which ATHENA Research and Innovation Center (https://www.athena-innovation.gr/en.html) currently participates in. The project is funded by the European Commission 7th Framework Program. In six case study regions the project aims at identifying multiple stressors, including water scarcity, which affect biodiversity and the services which the ecosystem provides. In a latter step, the project wants to establish socio-economically and environmentally sustainable management strategies in each of the case study regions, consistent with the goals of the Water Framework Directive. A range of models, including the River Water Quality Model and InVest, is consulted to estimate the value of ecosystem services (Navarro-Ortega *et al.*, 2015).

Apart from these two ongoing projects that take into consideration the interlinkage between surface and groundwater, there are a number of ReSEES (Laboratory on Research on Socioeconomic and Environmental Sustainability at the Athens University of Economics and Business: http://www.icre8.eu/resees) projects on groundwater management that have been completed already. These include, among others, GENESIS funded by the 7th Framework Program of the European Commission, which developed concepts, methods, and tools to improve groundwater management (Bioforsk (a)). In the project ReSEES preformed game theoretical and economic-mathematical modeling of surface and groundwater interaction under uncertainty and risk, used non-market valuation methods to assess the TEV of groundwater, and performed cost-benefit analyses on the proposed management strategies (Bioforsk (b)). THESEUS (Innovative technologies for safer European coasts in a changing climate) project that is funded by the European Commission, 7th Framework Program, which examines the application of innovative coastal mitigation and adaptation technologies aiming at delivering safe coasts for human use and development. The primary objective is to provide an integrated methodology for planning sustainable defense strategies for the management of coastal erosion and flooding which addresses technical, social, economic and environmental aspects. Other projects include project funded by the European Commission, 6th Framework Program, such as EUROLIMPACS (Evaluate Impacts of Global Change on Freshwater Ecosystems), AQUASTRESS (Solving Water Stress Problems by Integrating New Management Economic and Institutional Instruments). Also projects funded by the 5th Framework Program of the European Commission, such as ARID CLUSTER (Strengthening complementarity and exploitation of results of related RTD projects dealing with water resources use and management in arid and semi-arid regions) and Sustainable Use of Water on Mediterranean Islands: Conditions, Obstacles and Perspectives; and the CYPRUS (Integrated Water Management in Cyprus: Economic and Institutional Foundations) project, funded by the 4th Framework Program.

In addition to the aforementioned projects, ICRE8 and ReSEES participated in a number of projects funded by non-European sources, such as the World Bank: The significance of subsidized electric energy tariffs on the behavior of groundwater users for agriculture in India in general and in Rajasthan in particular (2003), Bangladesh Arsenic Mitigation Water Supply Project: Water Tariffication Re-structuring in Rural Bangladesh (2003), Water Pricing and Management in Urban China: Welfare Implications (2003–2004), World Bank Desk Work: A Report on the Economics of Arsenic Mitigation: Valuing Cost and Benefits Under Uncertainty and Health Risk (2003-2004); Governments: The Implementation of the Economic Aspects of Article 11 of the Water Framework Directive in Cyprus (Government of Cyprus, 2009–2010), The Implementation of the Economic Aspects of Article 5 of the Water Framework Directive in Greece (the Greek Government, 2007–2008); Sustainable Management of the South East Kalahari Aquifer System (Government of the Republic of Namibia, 2002); The Economic Value of Groundwater (the United Kingdom Environment Agency, 2012); Integrated Management for the ASOPOS River Basin (Greece): Economic Efficiency, Social Equity and Environmental Sustainability (Andreas Papandreou Foundation and National Bank of Greece, 2010); A Methodology for Integrated Watershed Management (International Institute for Environment and Development, 2008); Economic Valuation of Groundwater Review (Environment Agency–Aby Dhabi, 2014); Economic Instruments to Protect Freshwater Resources in the Republic of Buryatia, Lake Baikal Basin (Organization for Economic Co-Operation and Development, 2013–2014); Water and Green Growth Program – Phase 2 (World Water Council, 2014).

All of the above projects combine the aspects of groundwater resources management that we have considered in this paper. Specifically, they integrate stochastic hydro-economic models of groundwater use under different institutional and policy frameworks, and estimate the parameter values of these model using market and non-market estimation methods. Dynamic comparison of status quo values with respective optimal values, defines the level of needed interventions in terms of economic, legal and policy instruments, over time and space. The challenge of achieving environmental-economic-social sustainability in groundwater allocation over time, space and people is huge, multi-dimensional and should be treated in an interdisciplinary dynamics systems approach that can accommodate efficiently the involved complications.

8.6 CONCLUDING REMARKS

The application of economic instruments in the context of groundwater management requires that at least two strong limitations be considered. The first one refers to the set of non-market benefits and dimensions related to groundwater resources. Theoretical models, from which prescriptions to define instruments' design are drawn, should consider the Total Economic Value of the resources. Secondly, economic instruments – tariffs, tradeable permits or some other incentives – are deployed over a space of institutional aspects: customs, laws, decision making procedures, distribution and quality of information, distribution of rights and permits, some of which are far more important than the economic impact of the instruments themselves. These two dimensions jointly configure the set of possible elements for defining and implementing economic instruments. Numerous examples of feasible cooperation mechanisms, which provide tangible benefits for sustainable groundwater management, have been reported in the literature. This chapter concludes by highlighting the role of inter-disciplinary research projects and initiatives in offering useful information about the interrelated – social, economic, environmental – aspects surrounding groundwater management. Modeling these interrelations can help identify the likely impacts of alternative economic instruments, and avoid omitting unexpected effects or consequences. We thus conclude raising the importance of considering any economic instrument, or any combination of some instruments, within the larger sphere of dimensions and interrelations in which they operate.

ACKNOWLEDGEMENTS

This work has been supported by the European Communities 7th Framework Programme Funding under Grant agreement no. 603629-ENV-2013-6.2.1-Globaqua and by the EU Framework Programme for Research and Innovation, Horizon 2020, Funding under Grant agreement no. 690268-2-WATER-5c-2015-DAFNE.

REFERENCES

Bateman, I. J., Day, B. H., Georgiou, S., & Lake, I. (2006). The aggregation of environmental benefit values: welfare measures, distance decay and total WTP. *Ecological Economics*, 60(2), 450–460.

Bazargan-Lari, M., Kerachian, R. & Mansoori, A. (2009). A conflict-resolution model for the conjunctive use of surface and groundwater resources that considers water-quality issues: a case study. *Environmental Management*, 43(3), 470–482.

Benayas, R. J. M., Newton, A. C., Diaz, A., & Bullock, J. M. (2009). Enhancement of biodiversity and ecosystem services by ecological restoration: a meta-analysis. *Science, 325*(5944), 1121–1124.

Bhaduri, A., Manna, U., Barbier, E. & Liebe, J. (2011). Climate change and cooperation in transboundary water-sharing: an application of stochastic Stackelberg differential games in Volta River Basin. *Natural Resource Modeling*, 24 (4), 409–444.

Bioforsk (a): "Background". [Online] Available: http://www.bioforsk.no/ikbViewer/page/prosjekt/tema?p_dimension_id=16858&p_menu_id=16904&p_sub_id=16859&p_dim2=18920 (Accessed: 31/1/17)

Bioforsk (b): "Partners, AUEB-RC, Greece". [Online] Available: http://www.bioforsk.no/ikbViewer/page/prosjekt/tema/artikkel?p_dimension_id=16858&p_menu_id=16904&p_sub_id=16859&p_document_id=46333&p_dim2=18912 (Accessed: 31/1/17)

Birol, E., & Koundouri, P. (Eds.) (2008). Choice experiments informing environmental policy: a European perspective. Edward Elgar Publishing: Cheltenham, UK.

Birol, E., Koundouri, P., & Kountouris, Y. (2010). Assessing the economic viability of alternative water resources in water-scarce regions: Combining economic valuation, cost-benefit analysis and discounting. *Ecological Economics*, 69(4), 839–847.

Boyd, J., & Krupnick, A. (2009). The definition and choice of environmental commodities for nonmarket valuation. Resources for the Future: Washington, DC.

Brouwer, R. (2008). The potential role of stated preference methods in the Water Framework Directive to assess disproportionate costs. *Journal of Environmental Planning and Management*, 51(5), 597–614.

Brouyere, S., Carabin, G., & Dassargues, A. (2004). Climate change impacts on groundwater resources: modelled deficits in a chalky aquifer, Geer basin, Belgium. *Hydrogeology Journal*, 12(2), 123–134.

Brozović, N. & Schlenker, W. (2011). Optimal management of an ecosystem with an unknown threshold. *Ecological economics*, 70, 627–640.

Carson, R. T., Koundouri, P., & Nauges, C. (2011). Arsenic mitigation in Bangladesh: A household labor market approach. *American Journal of Agricultural Economics*, 93(2), 407–414.

Chang, L.-C., Ho, C.-C., Yeh, M.-S. & Yang, C.-C. (2011). An integrated approach for conjunctive use planning of surface and subsurface water system. *Water Resource Management*, 25(1), 59–78.

Colby, B. & Wishart, S. (2002). *Riparian areas generate property value premium for landowners. Agricultural & Resource Economics*, University of Arizona, College of Agriculture and Life Sciences. [Online] Available: https://pdfs.semanticscholar.org/8731/49701615f0d770202e2a4e5212b542d35192.pdf (Accessed: 22/3/17)

Cooley, H., Ajami, N., Ha, M., Srinivasan, V., Morrison, J., Donnelly, K. & Christian-Smith, J. (2013). *Global Water Governance in the 21st Century.* [Online] Available: http://pacinst.org/app/uploads/2013/07/pacinst-global-water-governance-in-the-21st-century.pdf (Accessed: 22/3/17)

DAFNE Project (2017). *Case Studies.* [Online] Available: https://dafne.ethz.ch/casestudies/ (Accessed: 31/01/17)

Defra (2007). *An introductory guide to valuing ecosystem services* [Online] Available: http://ec.europa.eu/environment/nature/biodiversity/economics/pdf/valuing_ecosystems.pdf (Accessed: 29/03/17)

De Frutos Cachorro, J., Erdlenbruch, K. & Tidball, M. (2014). Optimal adaptation strategies to face shocks on groundwater resources. *Journal of Economic Dynamics and Control*, 40, 134–153.

De Groot, R. S., Wilson, M. A., & Boumans, R. M. (2002). A typology for the classification, description and valuation of ecosystem functions, goods and services. *Ecological economics*, 41(3), 393–408.

Drichoutis, A., Koundouri, P., & Remoundou, K. (2014). A Laboratory experiment for the estimation of health risks: Policy recommendations. In: Koundouri, P., & Papandreou, N. A. (Eds) *Water Resources Management Sustaining Socio-Economic Welfare* (pp. 129–137). Springer: Netherlands.

Eleftheriadou, E. & Mylopoulos Y. (2008). Game theoretical approach to conflict resolution in transboundary water resources management. *Journal of Water Resources Planning and Management*, 134(5), 466–473.

Esteban, E. & Dinar A. (2012). Cooperative management of groundwater resources in the presence of environmental externalities. *Environmental and Resource Economics*, 54, 443–469.

European Commission (1979). "Directive 80/68/EEC", *Official Journal of European Communities*. [Online] Available: http://eur-lex.europa.eu/legal-content/EN/TXT/PDF/?uri=CELEX:31980L0068&from=EN (Accessed: 28/01/17)

European Commission (2000). "Directive 2000/60/EC of the European Parliament and of the Council", *Official Journal of European Communities*. [Online] Available: http://eur-lex.europa.eu/resource.html?uri = cellar:5c835afb-2ec6-4577-bdf8-756d3d694eeb.0004.02/DOC_1&format=PDF (Accessed: 28/01/17)

European Commission (2017). "*Groundwater: Current Legislative Framework*". [Online] Available: http://ec.europa.eu/environment/water/water-framework/groundwater/framework.htm (Accessed: 28/01/17)

Girard, C., Rinaudo, J.-D. & Pulido-Velazquez, M. (2016). Sharing the cost of river basin adaptation portfolios to climate change: Insights from social justice and cooperative game theory. *Water Resource Research*, 52(10), 7945–7962.

Gisser, M., & Sánchez, D. (1980). Competition versus optimal control in groundwater pumping. *Water Resources Research*, 16(4), 638–642.

Groom, B., Koundouri, P., & Swanson, T. (2005). Cost – benefit analysis and eifficient water allocation in Cyprus. In Brouwer, R. & Pearce, D. (Eds.) *Cost-Benefit Analysis and Water Resource Management*. Edward Elgar Publishing: Cheltenham, UK.

Hatterman, F. F., Kyrsanova, V., Habeck, A. & Bronstert, A. (2006). Integrating wetlands and riparian zones in river basin modeling. *Ecological Modelling*, 199(4), 379–392.

Howitt, R. E. (1998). Spot prices, option prices, and water markets: An analysis of emerging markets in California. In *Markets for water* (pp. 119–140). Springer, US: New York.

Jenkins, W. A., Murray, B. C., Kramer, R. A., & Faulkner, S. P. (2010). Valuing ecosystem services from wetlands restoration in the Mississippi Alluvial Valley. *Ecological Economics*, 69(5), 1051–1061.

Karterakis, S. M., Karatzas, G. P., Nikolos, I. K., & Papadopoulou, M. P. (2007). Application of linear programming and differential evolutionary optimization methodologies for the solution of coastal subsurface water management problems subject to environmental criteria. *Journal of Hydrology*, 342(3), 270–282.

Katsifarakis, K. L. & Petala, Z. (2006). Combining genetic algorithms and boundary elements to optimize coastal aquifers' management. *Journal of Hydrology*, 327(1–2), 200–207.

Kentel, E. & Aral, M. (2007). Fuzzy multiobjective decision-making approach for groundwater resources management. *Journal of Hydrologic Engineering*, 12(2), 206–217.

Kløve, B., Allan, A., Bertrand, G., Druzynska, E., Ertürk, A., Goldscheider, N., Henry, S., Karakaya, N., Karjalainen, T. P., Koundouri, P., Kupfersberger, H., Kvœrner, J., Lundberg, A., Muotka, T., Preda, E., Pulido Velázquez, M., & Schipper, P. (2011). Groundwater dependent ecosystems. Part II. Ecosystem services and management in Europe under risk of climate change and land use intensification. *Environmental Science and Policy*, 14(7), 782–793.

Kontogianni, A., Luck, G. W., & Skourtos, M. (2010). Valuing ecosystem services on the basis of service-providing units: A potential approach to address the 'endpoint problem' and improve stated preference methods. *Ecological Economics*, 69(7), 1479–1487.

Koundouri, P. (2004a). Current issues in the economics of groundwater resource management. *Journal of Economic Surveys*, 18, 1–38.

Koundouri, P. (2004b). Potential for groundwater management: Gisser-Sanchez effect reconsidered. *Water Resources Research*, 40(6), 1–13.

Koundouri, P., & Christou, C. (2006). Dynamic adaptation to resource scarcity and backstop availability: theory and application to groundwater. Australian Journal of Agricultural and Resource Economics, 50(2), 227–245.

Koundouri, P. & Dávila, O. (2015). The use of the ecosystem services approach in guiding water valuation and management: Inland and coastal waters. *Handbook of Water Economics*, 126–149.

Koundouri, P., & Groom, B. (2002). Groundwater Management: An Overview of Hydrogeology, Economic Values, and Principles of. In: Silveira, L., & Usunoff, E. J. *Groundwater – Vol. III*, pp. 101–134, EOLSS Publishers: Oxford.

Koundouri, P. & Groom, B. (2009). Groundwater Management: An Overview of Hydrogeology. In Silveira, L. & Ursunoff, J. Groundwater Vol. III. Eolss Publishers Co.: Oxford, UK.

Koundouri, P., & Pashardes, P. (2002). Hedonic price analysis and selectivity bias: water salinity and demand for land. In Pashardes, P., Swanson, T. & Xepapadeas, A. (Eds.) Current Issues in the Economics of Water Resource Management (pp. 69–80). Springer Netherlands, doi 10.1007/978-94-015-9984-9_4

Koundouri, P., & Skianis, V. (2015). Socio-Economics and Water Management: Revisiting the Contribution of Economics in the Implementation of the Water Framework Directive in Greece and Cyprus. No 1506, DEOS Working Papers from Athens University of Economics and Business, 1–34.

Koundouri, P., Ker Rault, P., Pergamalis, V., Skianis, V., & Souliotis, I. (2016). Development of an integrated methodology for the sustainable environmental and socio-economic management of river ecosystems. *Science of the Total Environment*, 540, 90–100.

Koundouri, P., Palma, C. R., and Englezos, N. (Forthcoming 2017) 'Out of sight, not out of mind: developments in economic models of groundwater management', *International Review of Environmental and Resource Economics.*

Koundouri, P., Papandreou, N., Stithou, M., & Dávila, O. G. (2014). A Value Transfer Approach for the Economic Estimation of Industrial Pollution: Policy Recommendations. *Water Resources Management Sustaining Socio-Economic Welfare,* 7, 113–128.

Koundouri, P., Scarpa, R., & Stithou, M. (2014). A choice experiment for the estimation of the economic value of the river ecosystem: Management policies for sustaining NATURA (2000) species and the coastal environment. In *Water Resources Management Sustaining Socio-Economic Welfare* (pp. 101–112). Springer: Netherlands.

Kraemer, R. A., & Banholzer, K. M. (1999). *Tradable permits in water resource management and water pollution control.* In: OECD (1999) Implementing Domestic Tradable Permits for Environmental Protection (75–107). OECD Publishing: Paris.

Kundzewicz, Z. W., & Döll, P. (2009). Will groundwater ease freshwater stress under climate change? *Hydrological Sciences Journal*, 54(4), 665–675.

Lancaster, K. J. (1966). A new approach to consumer theory. *Journal of political economy*, 74(2), 132–157.

Latinopoulos, D. & Sartzetakis, E. (2011). Optimal exploitation of groundwater and the potential for a tradable permit system in irrigated agriculture. Milan: Fondazione Eni Enrico Mattei Research Paper Series. FEEM Working Paper No. 26.2010.

Laukkanen, M., & Koundouri, P. (2006). Competition versus Cooperation in Groundwater Extraction: A Stochastic Framework with Heterogeneous Agents. In Koundouri, P., Karousakis, K., Assimacopoulos, D., Jeffrey, P., & Lange, M. A. (Eds.) Water Management in Arid and Semi-Arid Regions: Inter-disciplinary Perspectives. Edward Elgar Publishing: Cheltenham, UK.

Li, Y.-P., Huang, G.-H., & Nie, S.-L. (2006) An interval-parameter multi-stage stochastic programming model for water resources management under uncertainty, *Advances in Water Resources*, 29, 776–789.

Loaiciga, H. A. (2004) Analytic game-theoretic approach to groundwater extraction. *Journal of Hydrology*, 297, 22–33.

Madani, K. & Dinar, A. (2012a). Cooperative institutions for sustainable common pool resource management: Application to groundwater. *Water Resources Research*, 48, 2012a.

Madani, K. & Dinar, A. (2012b). Non-cooperative institutions for sustainable common pool resource management: Application to groundwater. *Ecological Economics*, 74:34–45.

Marino, M., & Kemper, K. E. (1999). *Institutional Frameworks in Successful Water Markets. Brazil, Spain, and Colorado, USA* (WTP 427) The World Bank: Washington DC

Martínez-Paz, J. M., & Perni, A. (2011). Environmental cost of groundwater: A contingent valuation approach. *International Journal of Environmental Research*, 5(3), 603–612.

Maqsood, I., Huang, G. H., & Yeomans, J. S. (2005). An interval-parameter fuzzy two-stage stochastic program for water resources management under uncertainty. *European Journal of Operational Research*, 167(1), 208–225. https://doi.org/10.1016/j.ejor.2003.08.068

Millennium Ecosystem Assessment (2005) *Ecosystems and Human Well-Being: Biodiversity Synthesis*. [Online] Available: http://www.millenniumassessment.org/documents/document.354.aspx.pdf (Accessed: 03/04/2017)

Nasim, S. & Helfand S. (2015). Pakistan Strategy Support Program, Optimal Groundwater Management in Pakistan's Indus Water Basin. International Food Policy Research Institute. [Online] Available: http://economics.ucr.edu/people/faculty/helfand/PSSP%20and%20IFPRI%20WP%2034%202016.pdf (Accessed: 23/3/17).

Navarro-Ortega, A., Acuña, V., Bellin, A., Burek, P., Cassiani, G., Choukr-Allah, R., Dolédec, S., Elosegi, A., Ferrari, F., Ginebreda, A., Grathwohl, P., Jones, C., Ker Rault, P., Kokm, K., Koundouri, P., Ludwig, R. P., Merz, R., Milacic, R., Muñoz, I., Nikulin, G., Paniconi, C., Paunoviæ, M., Petrovic, M., Sabater, L., Sabater, S., Skoulikidis, N. Th., Slob, A., Teutsch, G., Voulvoulis, N., & Barceló, D. (2015). Managing the effects of multiple stressors on aquatic ecosystems under water scarcity. The GLOBAQUA project. *Science of the Total Environment*, 503–504, 3–9.

Peralta, R., Timani, B. & Das, R. (2011). Optimizing safe yield policy implementation. *Water Res. Manage.*, 25 (2), 483–508.

Pearce, D., Atkinson, G. & Mourato, S. (2006). *Cost-benefit analysis and the environment: recent developments*. [Online] Available: http://www.oecd.org/environment/tools-evaluation/36190261.pdf (Accessed: 03/04/2017)

Peralta, R., Timani, B., & Das, R. (2011). Optimizing safe yield policy implementation. *Water resources management*, 25(2), 483–508.

Pulido-Velazquez, M., Andreu, J. & Sahuquillo, A. (2007). Economic optimization of conjunctive use of surface and groundwater at the basin scale. *Journal of Water Resources Planning and Management*, 132 (6), 454–467.

Pulido-Velazquez, M., Andreu, J., Sahuquillo, A. & Pulido-Velazquez, D. (2008). Hydro-economic river basin modelling: the application of a holistic surface-groundwater model to assess opportunity costs of water use in Spain. *Ecological Economics*, 66(1), 51–65.

Rezapour Tabari, M. M. & Soltani, J. (2013). Multi-objective optimal model for conjunctive use management using SGAs and NSGA-II models. *Water Res. Management.*, 27, 37–53.

Rogers, P. & Hall, A.W. (2003). Effective Water Governance, *TEC Background Papers* No.7, Global Water Partnership, Stockholm. [Online] Available: https://www.researchgate.net/publication/42765754_Effective_Water_Governance (Accessed: 28/04/2017)

Safavi, H. R., Darzi, F. & Mariño, M. A. (2010). Simulation-optimization modeling of conjunctive use of surface water and groundwater. *Water Resourource Management,* 24(10), 1965–1988.

Salazar, R., Szidarovszky, F., Coppola, E. Jr. & Rojana, A. (2007). Application of game theory for a groundwater conflict in Mexico. *Journal of Environmental Management*, 84(4): 560–571.

Saleth, R. M. & Dinar, A. (2004). The Institutional Economics of Water: A Cross-Country Analysis of Institutions and Performance. The World Bank, Washington, D. C., USA & Edward Elgar Publishing, Cheltenham, UK (Co-Publication). Available: http://documents.worldbank.org/curated/en/782011468780549996/pdf/ 302620PAPER0In1l0eco nomics0of0water.pdf (Accessed: 23/3/17).

Saleh, Y., Gurler, U. & Berk, E. (2011) Centralized and decentralized management of groundwater with multiple users. *European Journal of Operational Research*, 215(1), 244–256.

Shah, T. (2007). The Groundwater Economy of South Asia: An Assessment of Size, Significance and Socio-ecological Impacts. In Giordano, M. & Villholth, K. G. (Eds.) The agricultural groundwater revolution: opportunities and threats to development. Wallingford, UK: CABI. pp. 7–36 (Comprehensive Assessment of Water Management in Agriculture Series 3).

Siegfried, T., & Kinzelbach, W. (2006). A multi-objective discrete stochastic optimization approach to shared aquifer management: Methodology and application. *Water Resources Research*, 42(2).

Sreekanth, J. & Datta, B. (2014). Stochastic and robust multi-objective optimal management of pumping from coastal aquifers under parameter uncertainty. *Water Resourcees Management*, 28(7), 2005–2019.

Sustainable Development Solutions Network (2015). *Indicators and a Monitoring Framework for the Sustainable Development Goals.* United Nations: New York, USA.

Taylor, R. G., Scanlon, B., Doell, P., Rodell, M., van Beek, R., Wada, Y., Longuevergne, L., Leblanc, M., Famiglietti, J. S., Edmunds, M., Konikow, L., Green, T. R., Chen, J., Taniguchi, M., Bierkens, M. F. P., MacDonald, A., Fan, Y., Maxwell, R. M., Yechieli, Y., Gurdak, J. J., Allen, D. M., Shamsudduha, M., Hiscock, K., Yeh, P. J.-F., Holman, I., & Treidel, H. (2013). Ground water and climate change. *Nature Climate Change*, 3(4), 322–329.

Tsur, Y. & Zemel, A. (2014). Dynamic and stochastic analysis of environmental and natural resources. In Fischer, M. M., Nijkamp, P. (Eds.) Handbook of Regional Science. Springer: Berlin/Heidelberg, DE.

United Nations (2003). "*Water for People, Water for Life. World Water Development Report 1*" New York: UNESCO Publishing and Berghahn Books.

United Nations (2016) *Sustainable Development Goals.* [Online] Available: https://sustainable development.un.org/sdgs (Accessed: 26/9/16)

United Nations High Level Panel on Water (2016). *High Level Panel on Water – Background Note.* [Online] Available: https://sustainabledevelopment.un.org/content/documents/10004 High%20Level%20Panel%20on%20Waterbackground%20note.pdf (Accessed: 26/9/16)

Van Houtven, G., Powers, J., & Pattanayak, S. K. (2006). Valuing water quality improvements using meta-analysis: Is the glass half-full or half-empty for national policy analysis. *Resource and Energy Economics,* 29, 206–228.

Wang, C. & Segarra, E. (2011). The economics of commonly owned groundwater when user demand is perfectly inelastic. *Journal of Agricultural and Resource Economics.* 36 (1), 95–120.

Wu, X. & Whittington, D. (2006). Incentive compatibility and conflict resolution in international river basins: a case study of the Nile basin. *Water Resources Research,* 42, 1–15.

Wu, B., Zheng, Y., Tian, Y., Wu, X., Yao, Y., Han, F., Liu, J. & Zheng, C. (2014). Systematic assessment of the uncertainty in integrated surface water-groundwater modeling based on the probabilistic collocation method. *Water Resources Research,* 50, 5848–5865.

Wu, B., Zheng, Y., Wu, X., Tian, Y., Yao, Y., Han, F., Liu, J. & Zheng C. (2015). Optimizing water resources management in large river basins with integrated surface water–groundwater modeling: a surrogate-based approach. *Water Resources Research*, 51(4), 2153–2173.

Yang, C.C., Chang, L.C., Chen, C.S. & Yeh, M.S. (2009). Multi-objective planning for conjunctive use of surface and subsurface water using genetic algorithm and dynamics programming. *Water Resources Management*, 23(3), 417–437.

Zeitouni, N. (2004). Optimal extraction from a renewable groundwater aquifer with stochastic recharge. *Water Resourources Research,* 40(6).

Zemel, A. (2012). Precaution under mixed uncertainty: implications for environmental management. *Resource and Energy Economics*, 34, 188–197.

Zilberman, D., Sproul, T., Rajagopal, D., Sexton, S., & Hellegers, P. (2008). Rising energy prices and the economics of water in agriculture. *Water Policy*, 10(1), 11–21.

Chapter 9

Cooperation and conflict resolution in groundwater and aquifer management

W. Todd Jarvis
Institute for Water & Watersheds and College of Earth, Ocean and Atmospheric Sciences, Oregon State University, Corvallis, OR, USA

ABSTRACT

Conflicts related to groundwater and aquifers are peculiar. Groundwater is inconvenient to water law and water diplomacy because it is hidden and referenced differently than surface water. The groundwater container, the aquifer storing groundwater, is part and parcel of the groundwater debate, but rarely is discussed at the same table of surface water and groundwater. Part of the problem focuses on the value of groundwater by scientists and engineers trained in the tradition of multiple working hypotheses. Tension exists between technical training and traditional knowledge of groundwater through local stories and myths. Boundaries used to develop and manage groundwater and aquifers are conflictive because of the lack of standard methodologies, as well as the social and technical differences in developing hydrogeologic conceptual models. As the subsurface is increasingly relied upon for changing conditions for energy, waste, and water management, pracademics in water negotiations need to integrate multiple frameworks and transdisciplinary skills-building situations to enhance cooperation over groundwater and aquifers.

9.1 INTRODUCTION

Conflicts over water depend on the characteristics of the resource. Conflicts over groundwater and aquifers are very different from those posed by surface water resources. Surface water negotiations typically focus on allocations and flows; negotiations over groundwater typically focus on storage and water quality. Klein (2017) dubs the conflict as "groundwater exceptionalism", because law often addresses groundwater differently than surface water. Whereas surface watersheds, the common boundary for integrated water resource management, are static, groundwater boundaries are value laden and constantly change during development. The resources are oftentimes managed separately even though both resources are hydraulically connected.

Formal groundwater hydrology practiced by scientists and engineers differs dramatically from popular groundwater hydrology practiced by water users and water diviners. Conflicting conceptual models of the subsurface are commonplace for both permeability architecture and groundwater circulation. Dueling experts can easily overtake conflicts focusing on identity, interests, and the investments and risks connected to groundwater and aquifers.

Science remains at the core of groundwater and aquifer disputes. Disagreements over groundwater science and engineering are not easily defined without the assistance of trained experts that also exercise skills in process. Cooperation on groundwater and aquifer governance takes many forms by first dealing with the dueling experts through scientific mediation. Learning and experiencing different water negotiation frameworks through serious gaming enhances participatory approaches to adapting existing subsurface governance.

9.2 TRIGGERS OF CONFLICT

A contentious groundwater situation can be classified as a wicked problem, employing features of complexity – it is unpredictable, uncontrollable, and it has several, often contradictory interpretations (Kurki, 2016). The perception that conflicts or negotiations over groundwater and aquifers are all about allocation and ownership is misinformed. Some of the most contentious battles over groundwater focus on the perceived threats to the quality of groundwater. There also exist many myths, paradoxes and misunderstandings of the tenets within hydrogeology that ultimately lead to conflicts between groundwater professionals as well as a lack of trust by decision makers (Jarvis, 2014).

Looking beyond the internal conflicts within the field of hydrogeology, Delli Priscoli & Wolf (2009) indicate the interpersonal causes of conflict that may conflate the hydrogeologic confusion include:

- Relationships (poor communication, negative behavior)
- Data (interpretation, misinformation, procedures)
- Interests (perceived competition, procedural interests)
- Structural (unequal power in terms of bargaining, material and ideational power, time, destructive behavior, geography)
- Values (ideology, spirituality)

Jarvis (2014) added identity as a basis for conflict that is especially unique to groundwater and aquifers. Identity in this context includes history and control – the dueling experts steeped in the formal training of hydrologists, and the folk beliefs (*e.g.*, dowsing or water divination) common to practitioners of popular hydrology.

Groundwater professionals have a strong personal affinity and identity to their work given that imagination and creativity are key parts of developing their working hypotheses. The ownership of the creativity associated with imagining what is going on in the subsurface can lead to a dueling experts situation. Conflicting conceptual hydrogeologic models are also part of the formal training of hydrogeologists focusing on the intellectual method of multiple working hypotheses introduced in the late 1890s by the first hydrogeologist in the US, Thomas Chamberlain. The structure of the method of multiple working hypotheses revolves around the development of several hypotheses to explain the phenomena under study. The antithesis of multiple ways of knowing is considered a ruling theory, often espoused by individuals who consider the geology and hydrology of where they live and work as so complex and unique that only a local professional would understand how their hydrogeology works (Jarvis, 2014). As a consequence, conventional groundwater management approaches, drawing from expert-based instrumental rationality, often are insufficient for successful project planning and implementation (Kurki, 2016).

Resolving groundwater conflicts can be particularly tricky due to many other factors including a lack of aquifer performance data, spotty water quality data, traditional and preferred easy access to water, the extensive variety of draws on a single aquifer, historical water rights in conflict with the needs of new population and economic growth, exemptions for domestic wells, and the list goes on (Vinett & Jarvis, 2012). Superimposed on all of these drivers of conflict is the administrative separation of laws governing groundwater, surface water, and seawater, all of which imply the physical, biological, and political boundaries between the groundwater, surface water, and seawater are easily delineated.

9.3 GROUNDWATER BOUNDARY CONUNDRUM

Defining boundaries around groundwater resource domains is conflictive because boundaries represent different interpretations of key issues, such as water quality, water quantity, nature, economics, politics, and history. Boundaries define who is in, who is out; what is permissible, what is not; what needs to be protected and what is already protected (Jarvis, 2014).

Some might argue that defining boundaries around a hidden resource is fuzzy and perhaps impossible to do with any degree of certainty. However, the literature is replete with boundaries for groundwater domains. Careful examination of the literature reveals three groundwater domains: (1) traditional approaches to defining groundwater domains that focus on predevelopment conditions; (2) groundwater development creates new boundaries, that meshes hydrology, hydraulics, property rights and economics; and (3) the social and cultural values of the groundwater and aquifer resources (Jarvis, 2014). For example, Aladjem (2015) identified one of the unexpected conflictual issues that needs to be addressed while implementing the California Sustainable Groundwater Management Act of 2014 is the question of defining the boundaries of the groundwater basins. By legislation, Groundwater Sustainability Agencies (GSA) and related plans are part of complying with the efforts to sustainably develop groundwater. The costs associated with organizing a GSA and preparation of the plans may create a situation where developing a new groundwater basin boundary may occur to save money.

9.4 GEOGRAPHY OF CONVENTIONAL GROUNDWATER CONFLICTS

While a comprehensive geographic inventory of groundwater conflicts is beyond the scope of this chapter, the following vignettes are used to illustrate groundwater conflicts that receive frequent attention in the news media.

9.4.1 Groundwater for agriculture

It is well known that groundwater represents a large share of water used for agriculture irrigation (OECD, 2015). Ease of point of use and water on demand are key drivers to global groundwater use.

California serves as a good example of the tension associated with agricultural use of groundwater given the importance of the state for providing a significant share of the US food supply. Groundwater provides about 40 to 60 percent of all water used in California. The Earth Security Group (2016) identified the situation in California as a global aquifer hotspot with conflicts between large groundwater users and domestic well owners over pumping water levels leaving some domestic well owners without usable wells. Long-term groundwater pumping has also caused land subsidence, damaging infrastructure such as canals. And the detailed lithologic logs of drilled wells are considered confidential, thus leading to incomplete and conflicting conceptual hydrogeologic models and computer-generated predictions.

Groundwater in California went unregulated until passage of the Sustainable Groundwater Management Act (SGMA) in 2014 (Aladjem, 2015; Moran *et al.*, 2016). Prior to the SGMA, water rights were acquired through an adjudication process that was largely driven by the goal of attaining safe yield – a concept with multiple policy definitions, sometimes defined as the quantity of groundwater that can be pumped from an aquifer without exceeding the recharge to an aquifer, or as the quantity of water that can be pumped without getting into trouble. Despite the concept being discredited in the academic literature as discussed by Jarvis (2014), safe yield of a basin as defined by existing California case law was not the same as the sustainability yield of basin outlined in the SGMA (Aladjem, 2015).

Beyond the boundary issues and confusion over safe yield versus sustainability yield, other factors contributing to conflict over groundwater in California identified by Moran & Cravens (2015) that probably sound familiar to practitioners in water conflicts across the globe include: fragmented groundwater management, voluntary groundwater management, legal uncertainty in the SGMA, property rights and existing legal rights to water, data, information, models, and dissemination of data, and funding and support.

9.4.2 Groundwater for growth

The fragmented nature of water and land use laws at the level of individual counties, states, provinces, and countries is leading to a new paradigm in water planning and management that focuses on a "bottom-up" approach instead of the traditional top-down approach. Concurrency laws for proposed land use have evolved over the past 15 years to address groundwater recoverability and aquifer mechanics. Jurisdictions across the United States are crafting policies that specifically require "proving" water availability for housing developments (California, Colorado, Texas, Utah) and new agricultural uses (California). Some counties are also weighing interference between proposed developments and senior surface water rights through uncontrolled pumping of groundwater through domestic wells (Washington). Elsewhere, counties are asked by state governments to develop groundwater management plans to ascertain the availability for other high value uses such as permitting short term sales of groundwater appropriated for agricultural uses to the drilling industry for hydrofracking (Wyoming).

Implementation of concurrency ordinances, as well as groundwater sustainability initiatives such as California's SGMA, require making decisions in the face of uncertain data. Funding shortfalls, the uncertainty associated with the quantitative characteristics of groundwater systems, increased use of numeric groundwater models as necessary components for informed groundwater management decisions, yield a growing frustration with the dueling expert situation (Jarvis, 2014).

9.4.3 Groundwater for ecosystems

Conflicting hydrogeologic conceptual models connected to how groundwater is valued are best exemplified by the situation where the Santa Cruz Aquifer is shared between the US and Mexico. The Santa Cruz Aquifer is not the subject of any treaties (Delgado, 2013). Nevertheless, the communities in the Mexican state of Sonora and the US state of Arizona are heavily reliant on groundwater for agricultural, municipal, and industrial uses. For the state of Arizona, the goal is to maintain safe yield, prevent a decrease in the water table, and preserve the riparian areas used by endangered species. In contrast, the goals of the state of Sonora focus more on the general wellbeing of the population, including the development of basic water services, as well as extension and improvement of the existing groundwater-based supplies.

Water agencies in both countries operate independently with little coordination regarding the data collection and conceptual models of the aquifer. The obvious results of such fragmented coordination are different interpretations of water availability, impacts of groundwater use, recharge and protection activities.

The aquifer use by each country yields conflicting hydrological conceptual models that have led to disagreements of the physical conditions and availability of groundwater. While open dialogue has yielded a modicum of cooperation on some scientific information, there is still no agreement on a collaborative assessment or management of the Santa Cruz Aquifer for either development for use by residents of Mexico or for preservation of ecosystems and related endangered species by residents of the United States (Delgado, 2013).

9.5 GEOGRAPHY OF EMERGING GROUNDWATER CONFLICTS

9.5.1 Nitrate wars

Excess nitrate concentrations in aquatic systems, in combination with other nutrients such as phosphorus, lead to algae blooms in ponds, lakes, streams and rivers. Large algae blooms also contribute to hypoxia, or low dissolved oxygen, in lakes and rivers that negatively impacts many fish species. The World Health Organization drinking water standard for nitrate is 10 milligrams per liter to prevent nitrate toxicity.

Agricultural fertilizer use, onsite wastewater systems (septic tanks) and animal wastes are typically associated with rural residents. Urban dwellers many times rely on drinking water supplies that are transmitted long distances to the point of use. Rural dwellings are sometimes located upstream from urban areas where rivers and lakes are valued for water amenities and fisheries; sometimes rural residential developments are located in aquifer recharge areas thus creating tension between urban and rural communities. A good case study of a nitrate war across the urban-rural divide in Wyoming continues after over 20 years of dueling experts and a general lack of appreciation for the role of a neutral third party with skills in both process and substance is summarized by Jarvis (2014).

Community cohesion and civility becomes fragmented and deepens the urban-rural divide when it comes to the issue of delineating protection areas for wellheads, springs, and recharge areas for public drinking water supplies. There is also the concern that onsite wastewater systems contaminate groundwater and surface water, thus impacting the water quality of rivers and streams, as well as the drinking water supplies, utilized by urban areas. The antagonism between urban residents who feel an affinity to the greater good versus rural residents who value independence and wide open spaces and who just want to be left alone is real (Jarvis, 2014). These types of conflicts are becoming more commonplace with exurban development. Conflicts over nitrate and the urban-rural divide can last decades. Kurki (2016) provides a case study of a comparable urban-rural divide situation where tensions continue to flare in Finland as described in a later section.

9.5.2 Groundwater flooding

Neighbor wars come in many shapes and sizes. Border disputes range from barking dogs, noisy neighbors, nosy neighbors, fencing or lack thereof, fugitive trees and vegetation, neighborhood blight, attractive nuisances such as pools, private lakes, wildlife, episodic stormwater runoff, and increasingly, groundwater flooding. Groundwater flooding is an emerging problem globally with changes in land use (deforestation, impervious surfaces) and changes in precipitation patterns (more rain, less snow). Groundwater flooding is a frequent problem in areas where the depth to groundwater is shallow. The problem is a supercharging of shallow aquifers, resulting in full ditches and small ponds. Groundwater flooding is common in rainy climates and urban areas such as the United Kingdom. Yet the situation is increasingly described in the news media in both urban and rural areas that receive moderate precipitation such as Rocky Mountain and Midwest states of the US, arid regions in the Middle East, and rainy, deforested regions in the Pacific Northwest. As such, groundwater flooding issues and related conflicts are not easily resolved through the lens of land use planning because the conflicts emerge as a function of changing hydrologic conditions that may, or may not be, associated with climate change.

The conflicts resulting from the perceived solutions to fugitive water drainage often leads to long-term conflict over the episodic efforts to drain supersaturated land. Stormwater flooding oftentimes is controlled through engineered structures such as culverts, gabions, and ditches that direct flow to creeks and rivers. Stormwater situations become unfriendly once the engineered features direct flow to a neighbor resulting in damaged property. Groundwater flooding is a stealth variety of stormwater flooding. Groundwater flooding is perceived as stormwater that is controllable by collection, diversion, and discharge. Yet the control of groundwater flooding through traditional approaches is a mirage. Digging ditches deeper to increase drainage only permits more groundwater to flow into the excavations. Efforts to drain one property owner's lands through drainage ditches only exacerbate the collection of stormwater on their neighbor's land (Kemper, 2016).

9.5.3 Managed recharge

Managed aquifer recharge (MAR) is increasingly used to combat water scarcity. Large MAR projects exist in the Middle East, Australia, Europe, Jamaica, and across the US. Given that MAR is considered a solution to a water scarcity problem, it is surprising to learn of how conflictual the practice is in some locations in the world. Given the large investments associated with MAR, government entities typically initiate the projects. Tension can escalate due to many factors associated with MAR beyond debates over the financial costs, including bulk water transfer and treatment, underground storage, overlying land use to protect the stored water, chemical and microbiological mixing of treated waters and native groundwaters, and the discharge of MAR water to surface water sources.

Finland has abundant water resources, yet community water supplies are increasingly reliant on mixtures of natural and artificially recharged water to improve water quality for municipal and industrial uses. However, potential areas for groundwater development and supplemental MAR are sparsely situated (Kurki, 2016). Therefore, urban areas often convey developed groundwater long distances from rural areas. Like the Nitrate Wars, tensions across the urban-rural divide create long-term conflicts, oftentimes leading to extended litigation. Kurki (2016) indicates that local history matters when it comes to assessing conflicts over groundwater and alternative uses of aquifers, as well as anticipating conflicts, through stakeholder analyses.

9.5.4 Subsea aquifers

Perhaps the most relevant to the challenge of groundwater governance is the recognition of fresh and brackish groundwater reserves stored below the sea floor. The potential volume of fresh and brackish water stored in offshore aquifers may be two orders of magnitude greater than the approximately 4,500 km^3 estimated to have been extracted globally from continental aquifers since 1900 (Post *et al.*, 2013).

This is an important discovery that begs the question as to how to deal with the anticipated conflicts associated with future development of subsea aquifers. The ongoing debate between legal and groundwater governance scholars focuses on the role of the global commons, through the Law of the Sea, or perhaps the development of a Law of the Hidden Sea, or through some form of contract, or operating agreement. Martin-Nagle (2015) argues that even if the challenges regarding accessibility and financial return can be negotiated, jurisdictional issues and ownership of the water needs to explore how domestic law, international water law, or the Law of the Sea fit into the puzzle. "Aquifers lying under the territorial sea of one nation would doubtless be governed by its domestic laws, but questions would arise for transboundary aquifers. If international water law principles were to guide ownership and use, a further determination would have to be made about which guidelines to follow. Rather than ownership of water following national boundaries and territorial seas, a new regime might be constructed whereby the reserves would be viewed as a common asset belonging to all peoples." (Martin-Nagle, 2015).

9.5.5 Hydrofracking

The news media hysteria regarding the threat of hydrofracking to local, regional, and national water supplies has brought the concerns of the general public to the fore regarding conflicts over groundwater quantity and quality. Clearly, the unconventional exploration and development of fossil fuels by hydrofracking is perceived as a unique threat to groundwater users beyond other industrial pressures. For example, documentary film is increasingly serving as a medium for the hydrofracking discourse. Consider, for example, the documentary *Gasland* (2010) that portrays the global efforts of hydraulic fracturing for natural gas in unconventional shale reservoirs as contaminating groundwater and impacting private wells. *FrackNation* (2013) counters many of the assertions in *Gasland* (2010). *Gasland, Part II* (2013) was filmed to counter *FrackNation* (2013).

The hydrofracking debate is a scalable conflict ranging from *micro* with some communities voting to ban fracking, to *meso* with counties, states, and provinces passing legislation to prohibit fracking or even related industries (*e.g.*, sand mining) from operating within their boundaries. A few nations, such as France and Bulgaria, represent the *macro* scale for banning hydrofracking.

The conflicts over hydrofracking are so wicked that the situation is best viewed through the lens of systems thinking as depicted on Figure 9.1. Systems thinking serves as the best method of analyzing the hydrofracking controversy because it (1) promotes a holistic understanding that is both accessible and pluralistic, (2) transforms a single issue focus into a multi-issue view, (3) clearly illustrates that complex situations cannot be fully managed/controlled, (4) corresponds well to natural resource management, (5) encourages agencies to think beyond their default formulation of the situation paradigms that have emerged in the past 25 years (Daniels & Walker, 2012).

Fracking bans are evolving out of fear over direct versus indirect impacts to air quality, land quality, surface water quality, groundwater, and earthquakes associated with both the fracking process and injection of the produced waters. The interface between the different natural media and humans best classifies these conflicts as an interest-based Coupled Human-Nature Complex (Figure 9.1).

The investments and associated risks with fracking create a form of Regulatory-Industrial Complex. Tension between industry and regulatory agencies many times lead to lags in regulatory frameworks. Conflicts within the industrial domain are manifold, ranging from multiple working hypotheses associated with conceptual geologic models that dictate some of the fracking technology. The regulatory domain juggles conflicts with water use, familiarity with conventional fracking technologies common in vertical wells to the new unconventional fracking approaches associated with horizontal wells.

Well drillers and drilling engineers take great pride in their work and take umbrage when wells of all varieties are targeted as part of the hysteria associated with hydrofracking. A form of a Socio-Technical Complex creates an identity-based conflict because the drilling industry values a shared emphasis on their achievement of both excellence in technical performance and quality in their work.

Clearly, no single approach to conflict resolution or water negotiations framework can be applied to the wicked local, regional, national, or international hydrofracking situation. The value of transdisciplinarity continues to be acknowledged as

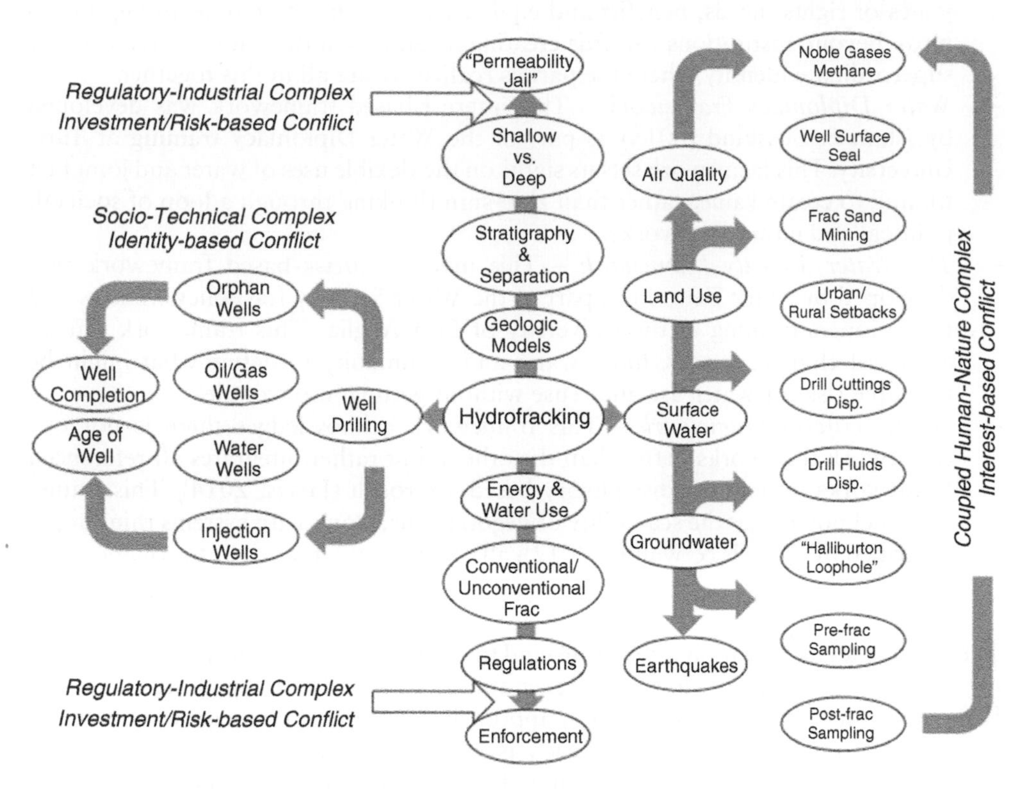

Figure 9.1 Hydrofracking situation map.

key to groundwater conflict resolution, groundwater negotiations, and groundwater governance; however, limited guidance is available on achieving it in practice.

9.6 PATHS FORWARD FOR COOPERATION

9.6.1 Water negotiation frameworks

What are the best approaches to negotiations over water, especially groundwater and related aquifers? The answer to this question mimics the problem of defining the vague concepts of safe yield and sustainability discussed in earlier sections – the best approach depends on whom you ask and when you ask. Water negotiation frameworks come in many names and forms. The following is a brief summary of a few water negotiation frameworks described in the literature.

- *Four Worlds Framework* – This identity-based framework was developed by Aaron Wolf as part of the Program in Water Conflict Management and Transformation at Oregon State University and is described more fully by Jarvis & Wolf (2010). This water conflict transformation approach points disputants towards topics of issues of rights, needs, benefits and equity, while at the same time attempting to move beyond institutions towards creating incentives in the quest to create a new superordinate identity where the parties realize we are all in this together.

- *Water Diplomacy Framework* – This interest-based framework was developed by Islam & Susskind (2013) as part of the Water Diplomacy training at Tufts University. This framework sets its sights on the flexible uses of water and joint fact finding to create value, rather than zero-sum thinking through a loop of societal, political and natural networks.
- *The Water Security Framework* – This investment/risk-based framework was developed by Mark Zeitoun as part of the Water Security for Policy Makers and Practitioners training at the University of East Anglia. This framework utilizes a web of climate, energy, food, water and community to define what might be tolerable risk for water use and reuse without getting into trouble.
- *Hydro-Trifecta Framework* – This framework acknowledges there is not one framework that works better than the others, but rather integrates all referenced frameworks into a transdisciplinary-based approach (Jarvis, 2014). This framework acknowledges the scalability of negotiations, along with systems thinking as described by Daniels & Walker (2012), all of which are important to collaborative learning, building competencies or acquiring new skills, to invent new science.

The hydrofracking situation is fertile ground for future research on how systems thinking and integrative negotiation approaches are key to just about any wicked problem. Likewise, other integrative negotiation approaches are used to manage related issues in subsurface reservoirs (or aquifers) such as oil, gas, geothermal, and carbon sequestration through unitization. Unitization is the well-known collective action approach of managing oil or gas reservoirs by all the owners of rights in the separate tracts overlying the reservoirs that has been in practice for over 100 years (Jarvis, 2014). "Pooling" is sometimes referred to as unitization. Unitization as employed in the oil industry is designed to be collectively beneficial, and is practiced in 38 states and 13 countries.

More recently, unitization is the favored approach for sharing transboundary hydrocarbon resources in the Gulf of Mexico as outlined in the US-Mexico Transboundary Hydrocarbons Agreement of 2012. Clearly, boundaries of the groundwater bodies including legal, political, hydrological, geological, biological, financial, and technical, will be an important facet of developing this new subsea resource. However, unitization could serve as the ideal approach for managing conflicts over subsea aquifers, as well as both developed and undeveloped terrestrial aquifers, given that unitization was initially designed for dispute prevention as opposed to conflict resolution.

9.6.2 Scientific mediation

Scientific mediation is used by scientists as part of outreach to the general public in matters where technical jargon and high levels of uncertainty lead to a stalemate on decision making, or to resolve disputes between scientists. As used herein, "scientific mediators attempt to tread the path between Merchants of Doom and

Merchants of Doubt as Merchants of Discourse using multiple working hypotheses and multiple ways of knowing as their moral compass" (Moore *et al.*, 2015). While at first glance it appears silly that water professionals cannot get along, but first-hand experience by Moore *et al.* (2015) revealed that water scientists and engineers are like other people with personal and political opinions that can affect their work. The danger of not addressing a dueling expert situation in an effective manner leads to distrust in groundwater science and engineering by the public and policy makers.

The scientific mediation process depicted in Figure 9.2 attempts to reach agreement on the merits of the disagreement as opposed to having personal and political biases cloud the scientific process. Jarvis (2014) describes a situation where groundwater scientists have debated the role of geologic faults on groundwater circulation in the recharge area of a municipal water supply where domestic wells and onsite wastewater systems are also located over a 20-year period.

However, the problems associated with dueling experts is not limited to the policy making process. Large multi-year, multidisciplinary projects undertaken in the academies can also become similarly entrenched leading to a schism among different factions within the research enterprise. While scientific mediation is a process that sounds rather utopian, it is garnering much interest by conflict resolution pracademics because it moves beyond the tired and overused cliché of agreeing to disagree used by entrenched expert egos.

9.6.3 Serious gaming for groundwater cooperation

Serious games are useful because they provide a structured environment in which learning, research, and joint fact finding can occur – and they are fun. When it comes to training students and professionals in water negotiations, everybody likes to play a game (Workman, 2016). In his blog, *The Consensus Building Approach*, Susskind (2012) writes "There are various ways games can be used to inform, and even alter, high-stakes policy negotiations ... but this only works when the actual negotiators take part in the game in advance of undertaking their own "real life" interactions."

Serious games introduce the different types of negotiation styles even in situations where language or cultural barriers exist. Many countries are just beginning the organization of alternative dispute resolution systems; computer-based and online games enhance their online competency in water negotiations. Collaborative modeling is a form of serious game playing with participants developing various groundwater management scenarios.

One of the tried and true approaches to negotiation training is a role play. Nearly every academic or professional training program in water negotiations uses role plays. Most focus on surface water allocations, water rights, benefit sharing, how to move water, and the benefits associated with water, across political boundaries.

However, the topics of the groundwater-related simulations are becoming increasingly diverse as groundwater professionals become more involved with both the technical substance of hydrogeology and the process of conflict prevention and resolution. Table 9.1 lists many different role playing games and their applications to

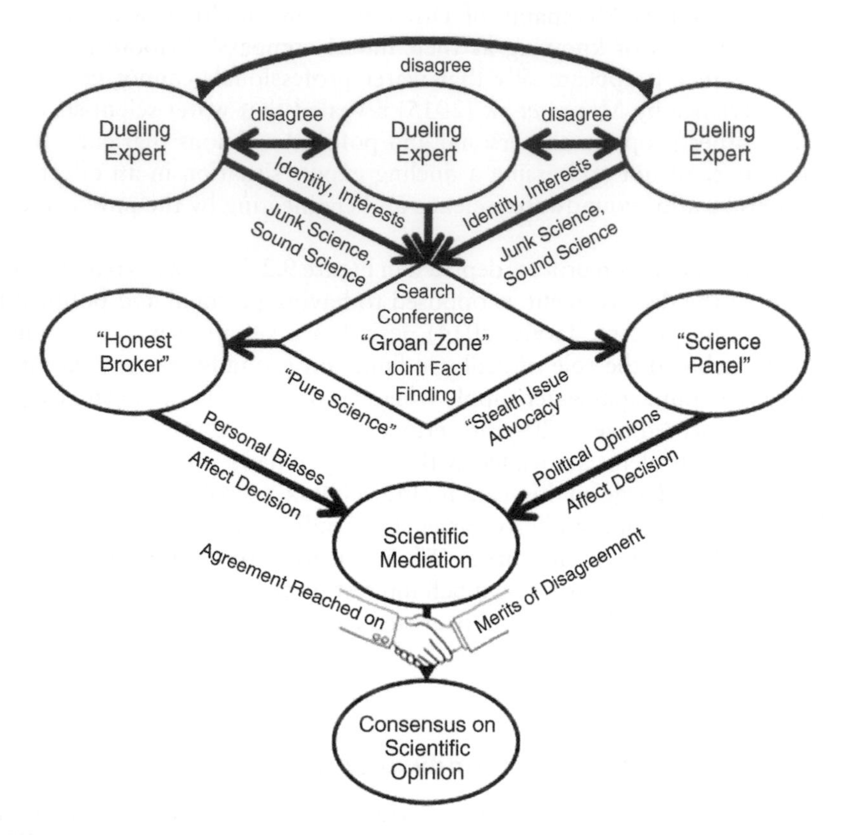

Figure 9.2 Scientific mediation framework (Moore *et al.*, 2015).

groundwater disputes. The groundwater protection dueling expert role play was developed by Jarvis (2014) who works as a groundwater hydrologist that teaches and practices conflict resolution in groundwater and water well construction. It provides an applied situation of the conflicts associated with multiple working hypotheses and the emerging field of scientific mediation. Likewise, the Edwards Aquifer Case was developed by a government scientist and academic collaborative governance practitioner for the complex situation of groundwater as a private property right grounded by the Endangered Species Act.

Board games with monies permit negotiations around a table where multiple languages are spoken. *Santiago* is a water allocation board game with farms, fleeting fidelities, and that fiddles with bribery. The groundwater counterpart to *Santiago* is the *California Water Crisis Game* – a groundwater board game where the winner is the player with the best reputation.

A pioneer in computer games is the *Tragedy of the Groundwater Commons Groundwater Game* developed by the U.N. International Groundwater Resources Assessment Centre (IGRAC). This game is part of IGRAC's GroFutures – Groundwater futures in Sub-Saharan Africa project. The game uses of a spreadsheet model to

Table 9.1 List of serious groundwater games.

Game	*Situation*	*Reference*
Water on the West Bank Role Play	Water well siting, aquifer depletion	Harvard Program on Negotiation (1988)
Managing Groundwater Beneath the Pablo-Burford Border Role Play	Agriculture water quantity and quality	Harvard Program on Negotiation (1996)
Santiago Board Game	Diversion of spring water to canals for plantations	AMIGO Spiel http://www.amigo-spiele.de/
International Groundwater Negotiation Role Play	Hydraulic connection to a transboundary water resources	Paisley (2007)
Tragedy of the Groundwater Commons Computer Assisted Role Play	Hydrologic capture analysis and economics of pumping wells	IGRAC http://www.un-igrac.org/downloads
Groundwater Protection Dueling Expert Role Play	Wellhead protection and aquifer protection boundaries	Jarvis (2014)
California Water Crisis Board Game	Groundwater use and depletion for agriculture, ecosystems, and urban growth	Firstcultural Games https://www.thegamecrafter.com/games/california-water-crisis
The Edwards Aquifer Case Role Play	Groundwater, common law rule of capture, Endangered Species Act, and role of science	Zerrenner & Gulley (2016)

analyze well development impacts and economics to neighboring water users. Isaak (2012) provides an excellent review of the game played by a transdisciplinary group of university students in the US.

9.7 SUMMARY

This chapter provides an overview of the challenges associated with conflict and cooperation over groundwater and aquifers. Scholars in water law and pracademics alike recognize that conflicts over groundwater and aquifers are markedly different than conflicts over surface water resources, in part because of the hidden nature of aquifers, the conceptual models developed by practitioners of popular and formal hydrology, uncertainty and fragmentation of data, data collection, data interpretation, and application of data to address the schizophrenic approaches to managing surface water, groundwater, and seawater at different scales. Threats to groundwater quality are more conflictive than disputes over allocations that typify surface water resources. While disputes over interference between surface water rights and groundwater development and groundwater depletion will continue with increases in population and climate change causing a redistribution of precipitation, emerging conflictive situations connected to groundwater and aquifers include continued influx of nitrate, groundwater flooding, development of subsea aquifers, and hydrofracking. Cooperation can be enhanced through a transdisciplinary approach to water negotiations, refusing to accept tired clichés such as agreeing to disagree uttered by dueling experts through scientific mediation, and embracing the challenge of having fun and learning from each other through serious gaming.

REFERENCES

Aladjem, D. (2015) Into the trenches: an early assessment of California's new groundwater legislation. *The Water Report*, 135, 20–26.

Daniels, S.E. & Walker, G.B. (2012) Lessons from the trenches: Twenty years of using systems thinking in natural resource conflict situations, systems research and behavioral science. *Systems Research and Behavioral Science*, 29, 104–115.

Delgado, P.A.T. (2013) *Collaborative Assessment and Management of Transboundary Aquifers In North America*. Master of Science in Environmental Science Thesis [Online], Corvallis, Oregon State University. http://hdl.handle.net/1957/45046 [Accessed 10th June 2017].

Delli Priscoli, J. & Wolf, A. (2009) *Managing and Transforming Water Conflicts*. New York, Cambridge University Press.

Earth Security Group (2016) *Earth Security Index: Business Diplomacy for Sustainable Development*, [Online] earthsecuritygroup.com/esi2016 [Accessed 10th June 2017].

FrackNation (2013) A film by Magdalena Segieda, Phelim McAleer, Ann McElhinney, [Online] Available from: http://fracknation.com/ [Accessed 10th June 2017].

Gasland (2010) A film by Josh Fox, [Online] Available from: http://gaslandthemovie.com/ [Accessed 10th June 2017].

Gasland II (2013) A film by Josh Fox, [Online] Available from: http://gaslandthemovie.com/ [Accessed 10th June 2017].

Isaak, M.T. (2012) Tragedy of the Commons Game. *Arizona Water Resource Newsletter*. Winter 2012. [Online] Available from: https://wrrc.arizona.edu/awr/w12/commonsgame [Accessed 10th June 2017].

Islam, S. & Susskind, L.E. (2013) *Water Diplomacy: A Negotiated Approach to Managing Complex Water Networks*. New York, NY, RFF Press, Routledge.

Jarvis, T. & Wolf, A. (2010) Managing Water Negotiations: Theory and approaches to water resources conflict and cooperation. In: Earle, A., Jägerskog, A. & Öjendal, J. (eds.) *Transboundary Water Management: From Principles to Practice*. London, Earthscan. pp. 125–154.

Jarvis, W.T. (2014) *Contesting Hidden Waters: Conflict Resolution for Groundwater and Aquifers*. London, Earthscan.

Kemper, J.B. (2016) *Groundwater Flooding and Guerilla Trenches: a Participatory Approach for Flood Control*. Master of Sciences in Water Resources Engineering Thesis [Online] Corvallis, Oregon State University. http://ir.library.oregonstate.edu/xmlui/handle/1957/59209 [Accessed 10th June 2017].

Klein, C.A. (2017) Owning Groundwater: The Example of Mississippi v. Tennessee (March 30, 2017). *Virginia Environmental Law Journal*, 35, (Forthcoming). [Online] Available from: https://ssrn.com/abstract=2943321 [Accessed 10th June 2017].

Kurki, V.O. (2016) *Negotiating Groundwater Governance: Lessons from Contentious Aquifer Recharge Projects*. Ph.D. Dissertation [Online] Tampere, Finland. Tampere University of Technology. 1387. https://tutcris.tut.fi/portal/en/publications/negotiating-groundwater-governance(a3bde410-fc67-49b2-9d31-c77c3c0d9a13).html [Accessed 10th June 2017].

Martin-Nagle, R. (2015) Offshore Aquifers. *The Environmental FORUM*, 32 (3) 26–31.

Moran, T. & Cravens, A. (2015) *California's Sustainable Groundwater Management Act of 2014: Recommendations for Preventing and Resolving Groundwater Conflicts*. [Online]

Water in the West, Stanford University. http://waterinthewest.stanford.edu/sites/default/files/SGMA_RecommendationsforGWConflicts_2.pdf [Accessed 10th June 2017].

Moran, T., Cravens, A., Martinez J. & Szeptycki, L. (2016) *From the Ground Down: Understanding Local Groundwater Data Collection and Sharing Practices in California*. [Online] Water in the West, Stanford University. http://waterinthewest.stanford.edu/sites/default/files/GW-DataSurveyReport.pdf [Accessed 10th June 2017].

Moore, C., Jarvis, T. & Wentworth, A. (2015) Scientific Mediation, *mediate.com*. [Online] Available from: http://www.mediate.com/articles/JarvisT1.cfm [Accessed 10th June 2017].

OECD (2015) *Drying Wells, Rising Stakes: Towards Sustainable Agricultural Groundwater Use, OECD Studies on Water*. [Online] Paris OECD Publishing http://www.oecd-ilibrary.org/agriculture-and-food/drying-wells-rising-stakes_9789264238701-en [Accessed 10th June 2017].

Paisley, R.K. (2007) *FAO training manual for international watercourses/river basins including law, negotiation, conflict resolution and simulation training exercises*. [Online] Food Agriculture Organization of the United Nations (FAO). Available from: http://iwlearn.net/resources/documents/5134 [Accessed 10th June 2017].

Post, V.E.A., Groen, J., Kooi, H., Person, M., Ge, S. & Edmunds, W.M. (2013) Offshore fresh groundwater reserves as a global phenomenon. *Nature*. [Online] 504, 71–78. Available from: http://www.nature.com/nature/journal/v504/n7478/full/nature12858.html [Accessed 10th June 2017].

Susskind, L. (2012) Learning from Games: The Debate over Role-Play Simulations. Available from: http://theconsensusbuildingapproach.blogspot.com/ [Accessed 10th June 2017].

Vinett, M. & Jarvis, T. (2012) Conflicts Associated with Exempt Wells: A Spaghetti Western Water War. *Journal of Contemporary Water Research & Education*. 148, 10-16. Available from: http://ucowr.org/files/Achieved_Journal_Issues/148/148_3.pdf [Accessed 10th June 2017].

Workman, J. (2016) Games about frontiers: Building skills to negotiate the politics of water. *The Source Magazine*. June 10, 2016 [Online] Available from: http://www.thesourcemagazine.org/games-about-frontiers/ [Accessed 10th June 2017].

Zerrenner, A. & Gulley, R. (2016) The Edwards Aquifer. [Online] The Program for the Advancement of Research on Conflict and Collaboration (PARCC), Maxwell School of Citizenship and Public Affairs, Syracuse University, https://www.maxwell.syr.edu/parcc/eparcc/cases/The_Edwards_Aquifer/ [Accessed 10th June 2017].

Chapter 10

Data, information, knowledge and diagnostics on groundwater

Jac van der Gun
Van der Gun Hydro-Consulting, Schalkhaar, The Netherlands

ABSTRACT

Data, information and knowledge on the local groundwater systems and their context are indispensable for effective groundwater resources management, thus they form basic components of good groundwater governance. Therefore, this chapter briefly addresses the following subjects: (i) the role of data, information and knowledge, and how it varies according to actor and activity; (ii) the categories of data and information that are most relevant; (iii) mechanisms and provisions for generating data, information and knowledge; (iv) presenting, sharing and disseminating data and information; and (v) diagnostics as an important step from understanding groundwater systems towards decision-making on groundwater resources management. In spite of unprecedented progress made since the middle of the 20th century, the current availability of data, information and knowledge on the local groundwater conditions is in many parts of the world still insufficient for reliable diagnostics and optimal decisions on groundwater resources management policy and measures. Here lies a challenge for the entire groundwater community to mobilize support from politicians and other decision-makers for investing structurally and permanently in essential data, information and knowledge on their valuable groundwater systems.

10.1 DATA, INFORMATION AND KNOWLEDGE IN GROUNDWATER DEVELOPMENT AND MANAGEMENT

10.1.1 Role of information and knowledge

Our decisions and behaviour in daily life are governed by needs, preferences, goals and ambitions, but guided by our *perceptions* about the world around us. These perceptions, in turn, are based on observations (using our 'sensory systems') and on how the sensory data or information is interpreted and used for better understanding. It is obvious that human decisions and behaviour tend to become haphazard, ineffective or even counterproductive if perceptions are ambiguous or significantly diverging from reality. Knowledge – fed by reliable information – is essential for developing meaningful perceptions and thus is crucial for adequate decision-making.

Information and knowledge make a real difference for the quality of the many decisions taken on groundwater-related activities. Among others, adequate information and knowledge enable aquifers containing good-quality groundwater to be located, wells or well fields to be sited successfully, optimal parameters to be defined for well construction and pump selection, or for the design of drainage systems if groundwater levels need to be lowered temporarily or permanently. Information and knowledge are also essential inputs to complex studies and planning projects, such as on improving the beneficial use of groundwater and the related allocation of the benefits, on protecting groundwater systems against overexploitation, salinity, pollution and other threats to sustainability, on the impacts of pursued changes in behaviour or practices (*e.g.* related to groundwater use, land use practices, wastewater disposal), and on much more. It is not exaggerated to conclude that no good groundwater governance can exist without adequate information and knowledge on the local groundwater systems and their context.

10.1.2 The DIKW hierarchy

The terms data, information and knowledge are used in numerous reports and publications, sometimes more or less indiscriminately and usually without defining them. Unambiguous definitions of these terms are hard to find and even among information science specialists there is no full consensus on the concepts. Nevertheless, some typical features of these concepts as well as the linkages and differences between them will be outlined below, as selectively interpreted from various papers, including Ackoff (1999), Bellinger *et al.* (2004), Clark (2004), Hey (2004) and Liew (2007).

Data consists of captured and stored records of past or current situations or events, in the form of symbols (words, numbers, diagrams, images) or signals (sensory readings of light, sound, smell, taste or touch). Data as such is commonly assumed to have no particular meaning of itself. Information, on the other hand, contains a message. It is generated by processing and relating relevant data, and putting it into a context; which introduces a certain degree of subjectivity, but enhances its usefulness for practical purposes. Information answers ‘who’, ‘what’, ‘where’ and ‘when’ questions. Knowledge resides within the human mind or brain, thus it is internalized by individuals. It is produced by combining information with personal perceptions and previously acquired experience. Knowledge answers in particular ‘how’ and ‘why’ questions. The hierarchical relation between data, information and knowledge, together with ‘wisdom’ at the top-end of the hierarchy (DIKW hierarchy), is shown in Figure 10.1. Complexity and understanding are increasing when moving upwards to the higher levels of the chain. Wisdom represents the ultimate level of understanding and like knowledge it operates within the human mind. It enables people to judge on and to synthesize patterns in their knowledge base, to extrapolate them and use them in innovative ways. Whereas data and information refer to the past, and knowledge to the past or present, wisdom may produce projections and action-oriented solutions for the future.

Data forms building-blocks for generating information, information can be absorbed by the human mind for developing knowledge, while knowledge is one of the pillars on which wisdom is built. But there are also relationships acting in opposite direction. Information may be captured and stored as data, and part of the knowledge (explicit knowledge) may be externalized and transferred to other persons in the form of information – as opposed to so-called tacit knowledge that is hard to formalize and communicate. The close interrelationships between the components of the DIKW chain imply that activities intended to augment data, information or knowledge are

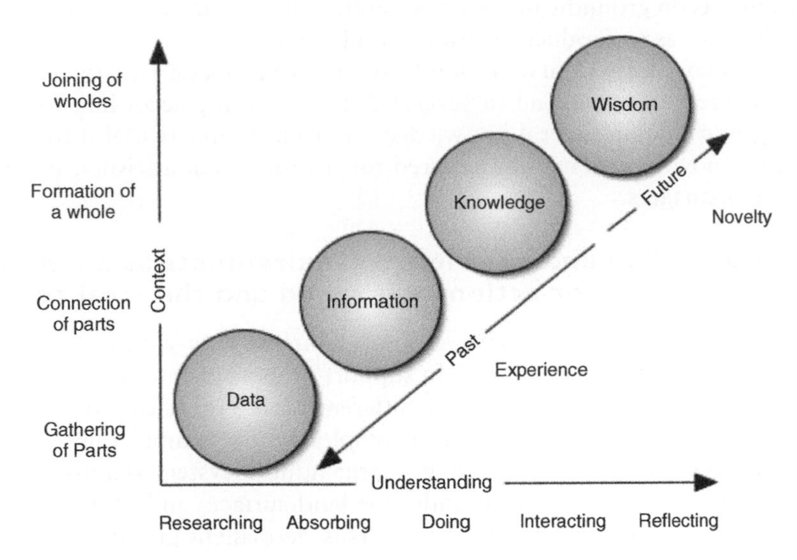

Figure 10.1 The DIKW hierarchy depicted as a linear chain (Clark, 2004).

seldom carried out in isolation from each other, but usually are combined to a certain degree for two or three of these components.

10.1.3 Information and knowledge requirements of different categories of actors

Many different actors are involved in groundwater development, management and governance: hydrogeologists and other groundwater specialists or investigators, social scientists, lawyers, planners, groundwater managers, politicians and other decision-makers, water suppliers, private groundwater well owners/users, groundwater irrigators, water-using industries, households, environmentalists, the general public, etc. Each category of actors has its own specific information demands, in terms of subjects or content, degree of detail and format or mode of presentation.

For instance, actors in charge of investigating, advising, planning and management will often be interested in the full extent and detail of available relevant information within the scope of their professional disciplines. They also will be keen on having access to the raw data on which the information is based, which will allow them to assess the reliability of the information and the degree of uncertainty. This group of actors will need the background and skills to deal with data and information present in formats that are too complex for outsiders, and the knowledge to interpret and use them adequately. In contrast, politicians and other decision-makers should not be bothered with large quantities of information. Rather, they should get tailor-made information that briefly summarises what is at stake and that focuses on pros and cons of alternative options for decision-making. This information has to be presented in an easily digestible way that is compatible with their usually limited knowledge and background on groundwater matters. Groundwater users and those who potentially pollute groundwater need mainly information to make them aware of their current or potential impacts on groundwater and they preferably should be educated on changes in their behaviour as to produce lowest harmful impacts.

It is clear that information on groundwater has to be made available or presented in varying degrees of detail and in several different forms, according to the different target groups. Some general knowledge on groundwater is useful for all actors, but in-depth knowledge is mainly required for investigators, advisors, planners and groundwater managers.

10.1.4 Information and knowledge requirements as a function of the activities or actions envisaged and the local context

Groundwater governance deals with all human interactions with groundwater and its uses, as well as with the provisions in support of groundwater resources management. These encompass activities of very different nature (physical interventions and non-physical action), of different size and complexity (for example: use of a shallow well by a single family versus protecting a large aquifer system against all potential pollution sources located above and under the land surface) and of widely diverging duration (temporary building-pit drainage versus permanent groundwater resources management). Information and knowledge is needed to facilitate all these activities and actions, but each of the potential activities or actions will have its specific requirements in terms of type of data and information, their spatial and temporal resolution, and in terms of knowledge.

The local context puts also its mark on the information and knowledge requirements, and in particular it is relevant for defining which data should get priority if scarcity of available resources does not allow more than only modest data acquisition efforts. This local context does not only include the physical environment (groundwater systems and relevant interlinked systems), but also aspects such as the demographic and socio-economic conditions, land use and land use practices, the environment, exploitable subsurface resources, legal and political setting, etc. Priorities in the development of information and knowledge follow societal goals as defined by the government, identified opportunities and risks, and the current stage of groundwater development and management.

It may be concluded that the diversity of actors, activities, actions and local contexts, as outlined in this section and the previous one, implies that a 'one-size-fits-all' approach to information and knowledge management will not be satisfactory for good groundwater governance. This diversity calls for differentiated information provisions and for a broad palette of knowledge.

10.2 WHAT CATEGORIES OF DATA AND INFORMATION ARE MOST RELEVANT?

10.2.1 Conceptual models and a framework of analysis as tools to guide the acquisition of data and the development of information

Table 10.1 presents a generic list of data and information categories that are in principle relevant for underpinning groundwater resources management. For any specific

Table 10.1 Key data and information categories for underpinning groundwater resources management.

		Is information on variations over space and time crucial?			
		Spatial variation		*Variation over time*	
	Type of information	*Yes*	*No*	*Yes*	*No*
1	**Groundwater systems**				
1.1	Horizontal and vertical boundaries of spatial hydrogeological units (aquifers, aquitards, aquicludes and aquifuges)	X			X
1.2	Hydraulic properties of the main aquifers and aquitards	X			X
1.3	Groundwater piezometric levels (by aquifer)	X		X	
1.4	Groundwater quality parameters (by aquifer)	X		(X)	
1.5	Natural groundwater discharge (type, locations and fluxes)	X		(X)	
1.6	Groundwater abstraction (type, locations and fluxes)	X		X	
1.7	Natural groundwater recharge (type, locations, fluxes)	X		(X)	
1.8	Artificial groundwater recharge (type, locations, fluxes)	X		(X)	
1.9	Groundwater-related water-logged zones (location)	X		X	
2	**Use, in-situ functions and benefits of groundwater**				
2.1	Total population in the area		X	X	
2.2	Groundwater abstracted for domestic purposes		X	X	
2.3	Groundwater abstracted for agricultural purposes		X	X	
2.4	Groundwater abstracted for industrial & other purposes		X	X	
2.5	Use of springs and baseflows (differentiated by use sector)		X	X	
2.6	Total surface water abstracted (differentiated by use sector)		X	X	
2.7	Groundwater-fed wetlands (location, total area, importance)	X			X
2.8	Water-table fed agricultural lands (location, total area)	X			X
2.9	Economic and other benefits of groundwater in the area (preferably differentiated by sector)		X	(X)	
3	**Current or potential interactions and threats**				
3.1	Zones of groundwater-surface water exchange of fluxes (influent and effluent; strong or weak)	X			X
3.2	Land use pattern and land use practices	X			X
3.3	Location of main subsurface resource exploitation and main use of subsurface space (type, location, depth, characterisation)	X			X
3.4	Current and potential sources of pollution above land surface	X			X
3.5	Current and potential sources of pollution due to subsurface resources exploitation and the use of subsurface space	X			X
3.6	Zones where fresh groundwater is threatened by salinization (sea-water intrusion, saline/brackish groundwater upconing or horizontal migration, irrigation return flows)	X			X
3.7	Zones prone to land subsidence	X			X
3.8	Infrastructural projects having a major impact on the local or regional groundwater regime	X		X	
4	**Groundwater governance aspects and provisions**				
4.1	Past or current groundwater assessment activities	X			X
4.2	Monitoring programmes (type, variables, length of records)	X		X	
4.3	Legal and institutional frameworks on groundwater	N/A			X
4.4	Current policies and planning on groundwater	N/A			X
4.5	Status of groundwater management in the area (implementation of policy, regulations and planning; stakeholder involvement)	X			X

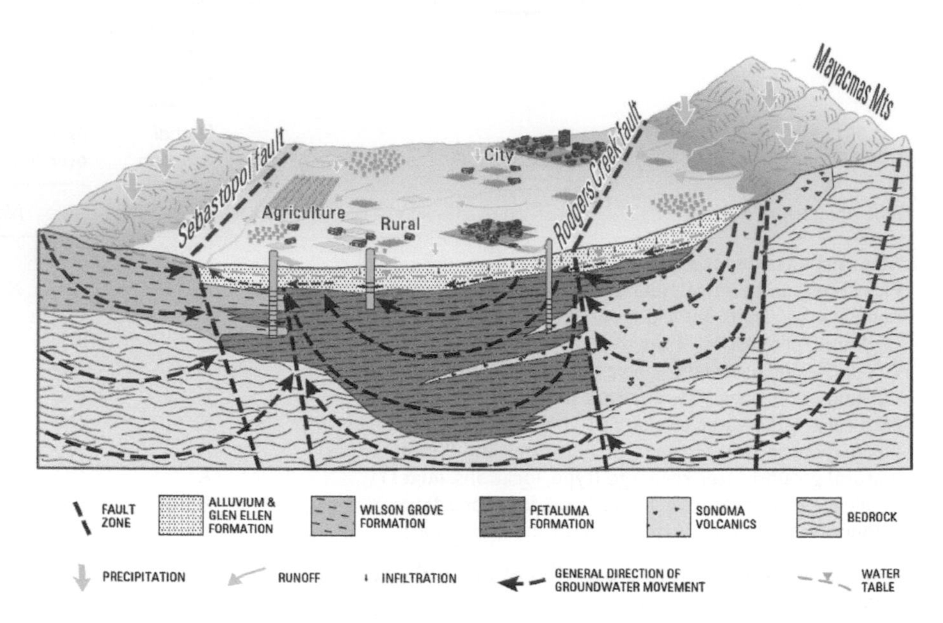

Figure 10.2 Conceptual model of the Santa Rosa Plain groundwater system, California (Source: USGS, 2014).

area at a given moment of time, the list can be improved by a making a break-down of the items shown, deleting some items and adding new ones, as required. Conceptual models can be helpful to tailor such a list to the specific characteristics of the area considered. A conceptual model of a groundwater system is here understood as a qualitative representation of a groundwater system – often in the form of a picture or a diagram – that conforms to hydrogeological principles and summarises the current understanding of the system. Initially, limited information may allow only a very simple conceptual model to be developed, for instance in the form of a geological or hydrogeological cross-section or block diagram. When more data and information become available, the conceptual model may be periodically revised; thus in an upgraded version it may also show components and features of the groundwater flow and/or transport processes. Figure 10.2 is an example of a simple conceptual model.

A conceptual model of a groundwater system is more suitable for depicting the overall setting and dynamics of the physical components than for highlighting interactions with humans. To compensate for this, one may resort in addition to a tool such as the DPSIR framework (Figure 10.3), adopted by the European Environmental Agency for causal-chain analysis (EEA, 1995). This framework can be applied to groundwater resources management and shows the dynamic links between drivers of change (D), pressures (P), groundwater state (S), impacts (I) and responses (R).

Data acquisition and generating information on groundwater are in most parts of the world steadily progressing, usually at a slow pace due to limitations in professional capacity and allocated resources. Priorities among the types of data and information depend on perceived critical information gaps and on emerging groundwater development and management issues. Change over time needs to be addressed by monitoring activities.

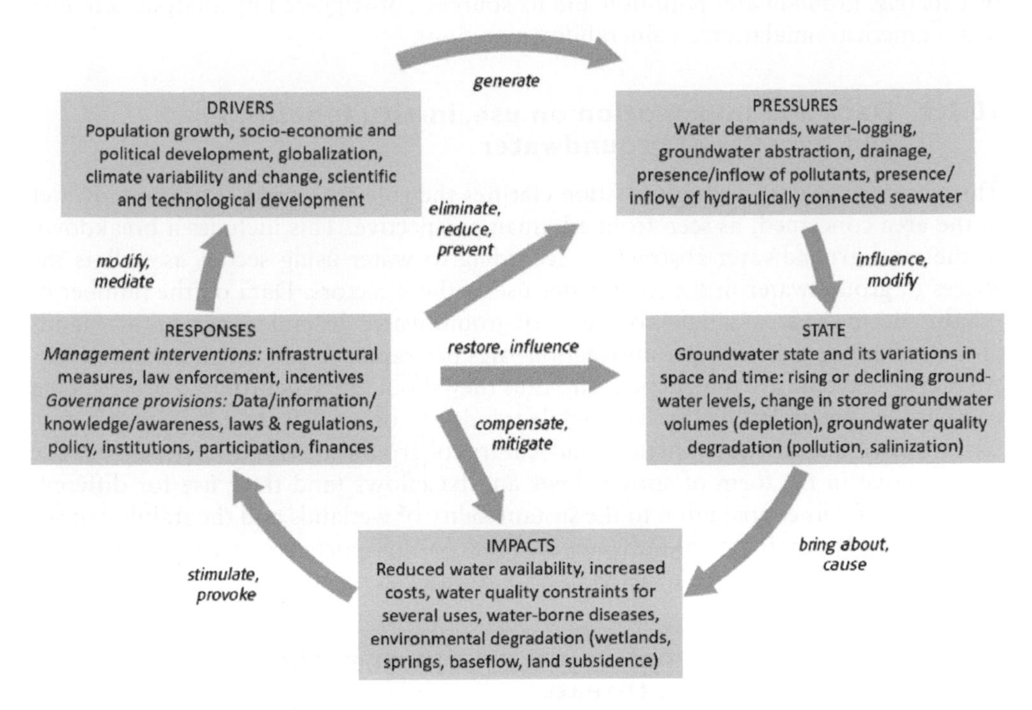

Figure 10.3 The DPSIR framework of analysis, tailored to groundwater resources management.

10.2.2 Data and information on groundwater systems

Data and information on groundwater systems should in the first place provide answers to practical questions such as: Where in the area and at what depths can groundwater of suitable quality be abstracted? What is the expected yield of wells drilled in each prospective zone and at approximately what depth below the ground surface will the static water level be encountered? Next, in order to gain an overall understanding of the state and dynamics of the local or regional groundwater systems – which is a precondition for adequate groundwater resources management – a more comprehensive, systems-oriented perspective has to be adopted. This requires in particular the systematic assessment of the following: (a) the geometry, lithology and hydraulic characteristics of aquifers and aquitards in the area; (b) the piezometric patterns in the different aquifers; (c) the patterns of background groundwater quality in each aquifer separately; (d) the components of groundwater recharge – both natural and anthropogenic – with their locations and estimates of their mean annual rates; (d) the components of groundwater discharge – both natural and anthropogenic – with their locations and estimates of their mean annual rates. For critical time-dependent variables (piezometric levels, groundwater abstraction, groundwater quality in shallow aquifers, etc.) monitoring is essential to enable proper control of groundwater quantity or quality. Additional data and information requirements on the groundwater systems will for each specific area emerge in response to identified interactions and threats (*e.g.* groundwater pollution and its sources), or triggered by analysis activities (*e.g.* numerical simulations, vulnerability mapping).

10.2.3 Data and information on use, in-situ functions and benefits of groundwater

This category of data and information clarifies the role and impacts of groundwater in the area concerned, as seen from a human perspective. This includes a breakdown of the total groundwater abstraction according to water using sector, as well as the shares of groundwater in the total water use of these sectors. Data on the number of inhabitants in the area and the total area of groundwater-dependent agricultural lands allow evaluating groundwater abstraction rates per capita or per hectare, respectively. All mentioned data are time-dependent, thus they should be synchronised before being compared or interrelated otherwise; while trends can be identified and assessed if time series are available. In-situ functions and features of groundwater to be assessed include its discharge in the form of spring flows and baseflows (and their use for different purposes), and its contribution to the sustainability of wetlands and the stability of the land surface. Profits from groundwater use for economic purposes and other indicators may be used to evaluate the benefits of groundwater in the area.

10.2.4 Data and information on current or potential interactions and threats

Interactions inside the water cycle are implicitly taken into account already by collecting data and information on the groundwater systems (the groundwater recharge and discharge components form the links). Intensive groundwater abstraction modifies the groundwater quantity regime, usually with negative impacts on the services and functions of the exploited groundwater system. Special attention is also needed for identifying zones prone to land subsidence due to groundwater abstraction and for exploring zones of contact with the sea or with rivers, where the quality of fresh groundwater may be threatened. In addition, groundwater is interacting also with systems and activities extending beyond the water system. Most important are land use, land use practices, extraction of subsurface resources, use of the subsurface space and infrastructural construction projects (see the chapters on these subjects elsewhere in this book). These external systems and activities need to be assessed since they may affect groundwater quantity and quality, and thus harbour potential threats to the services and functions of the groundwater systems.

10.2.5 Data and information on groundwater governance aspects and provisions

Finally, any entity or person playing a role in groundwater management and governance may benefit from information on the current groundwater governance setting and governance provisions in the area concerned. This should include meta-data and meta-information on groundwater-related data, information and knowledge in the area; information on the current roles of governmental and non-governmental agencies, local stakeholders and any other actors involved in groundwater governance;

information on the legal and regulatory frameworks; on groundwater policies and planning; and on the current state-of-affairs of groundwater management in the area, including law enforcement and plan implementation.

10.3 METHODS, MECHANISMS AND PROVISIONS FOR DATA ACQUISITION, GENERATING INFORMATION AND DEVELOPING KNOWLEDGE

10.3.1 Commonly used methods and techniques

10.3.1.1 Overview

Table 10.2 presents an overview of typical activities intended for data acquisition and for generating information and knowledge on groundwater systems and their context. A distinction is made between activities that are carried out at the office and those that take place primarily in the field (and are followed by processing and interpretation at the office). The latter are subdivided in two groups: fieldwork for conducting enquiries among the local population and fieldwork for direct data acquisition by specialized professional observations and measurements. Most of the activities mentioned in Table 10.2 cannot be ranked exclusively under either data acquisition, generation of information or development of knowledge, but contain elements of two of these categories or all or three. Below, brief clarifications and comments follow on some of the listed activities. More information can be found in specialized handbooks, such as Walton (1970), Griffiths & King (1981), Kruseman & De Ridder (1990), Struckmeier & Margat (1996), Brassington (2007) and Moore (2012).

10.3.1.2 Borehole and well records

Keeping a record of observations during borehole drilling and well construction activities is the most direct method of collecting data on local groundwater and aquifer conditions. Some drilling companies do not keep such records, many others do. Of particular importance are the driller's log (lithological column, sometimes with stratigraphic interpretation), data on water struck level and static water level, water quality, geophysical well-logs and observations made during well development and well testing.

Geological surveys or other groundwater agencies often have duplicates of the borehole and well records regarding their territory.

10.3.1.3 Geological and hydrogeological mapping

Since the beginning of the 19th century, when in the United Kingdom William Smith produced the first geological map (Winchester, 2001), geological cross-sections and geological maps have proven to be an excellent tool to portray the geology of an area. Such maps, encapsulating both field observations and remotely sensed information, form the point of departure for developing hydrogeological maps, that likewise are powerful in summarizing the hydrogeological features of an area in an easily accessible format. Hydrogeological maps do not only provide an interpretation of geological units in terms of their capacity to store and transmit water, but usually they also contain information on many other relevant groundwater-related features and properties.

Table 10.2 Typical activities for data acquisition and for generating information and knowledge on groundwater systems and their context.

		Operational characteristic of the activity		
Category	*Sub-category*	*Office activity*	*Field work tapping data or information from local population*	*Field work focusing on technical measurements and/or drilling*
Production of field records by drillers, well operators/users, etc.	Logs/tests of boreholes and wells			X
	Groundwater level data			X
	Groundwater quality data			X
	Water abstraction/use/injection			X
Inventory and interpretation of existing data and information	Previous studies in relevant fields	X		
	Records of drillers/operators/users	X		
	Demographic and socio-economic data and information	X		
	Other contextual information	X		
	Preliminary mapping	X		
Field surveys	Groundwater reconnaissance			X
	Detailed hydrogeological mapping	X		X
	Exploratory drilling			X
	Geophysical surveys			X
	Aquifer testing surveys			X
	Water use surveys		X	(X)
	Socio-economic surveys		X	
	Hydrological surveys			X
	Hydro-ecological surveys			X
	Water quality surveys			X
Enquiries (on-line and otherwise)	Various themes	X		
Remote sensing	Aerial photographs	X		(X)
	Satellite imagery	X		(X)
Water resources assessment	Combination of above-mentioned activities	X	X	X
Monitoring	Groundwater levels			X
	Groundwater quality			X
	Groundwater abstraction and use		X	
	Relevant flows & quality of connected streams			X
	Environmental impacts			X
Desk studies	Groundwater balance studies	X		
	Groundwater modelling	X		
	Groundwater quality and pollution studies	X		
	Groundwater vulnerability studies	X		
	Environmental impact studies	X		
	Documenting institutional and legal frameworks on groundwater	X		
	Inventory of groundwater policy, planning and management	X		

Nowadays, most hydrogeological maps are prepared on the basis of a standardised methodology and legend promoted by the International Association of Hydrogeologists and UNESCO (Struckmeier & Margat, 1995). Low-resolution hydrogeological maps are available at the global scale, for each of the continents, and in the form of national hydrogeological maps for most countries of the world (WHYMAP, 2016; Margat & Van der Gun, 2013). Hydrogeological maps showing more detail (say, at scales 1:20 000 to 1:200 000) are much more sparsely available and often have to be elaborated from scratch if they are required for water resources management planning or other projects. Apart from elaborating more or less detailed hydrogeological maps as a consolidated picture of local hydrogeological information, it may be very useful to prepare simple preliminary maps during early stages of groundwater investigations. They are a convenient tool for integrating collected or inventoried data and thus contribute to the advancement of local knowledge.

10.3.1.4 Surveys

Surveys tend to focus rather narrowly on specific subjects or techniques used. The majority of all groundwater-related surveys is conducted in the field. Part of these field surveys deals with physical components of the groundwater systems and for this purpose makes use of technological equipment for measurements, sampling and analysis. Examples are geophysical surveys, aquifer testing campaigns and hydro-chemical surveys. Other field surveys either consist of interviewing people on water-related matters (for instance on land/water use practices, customary water rights, income from water or perceived water-related problems and opportunities) or are a mix of interviewing and measuring (such as inventories of wells, springs or potential sources of pollution). Some types of surveys can be conducted at the office, such as enquiries by questionnaires via mail or the internet, literature searches, or compiling data and information collected by third parties.

10.3.1.5 Reconnaissance and assessments

A groundwater reconnaissance and a groundwater resources assessment have in common that they both intend to give an overall picture of the groundwater resources conditions in the area considered. The difference is that a reconnaissance is only a preliminary step, resulting in a provisional conceptual model, while an assessment is expected to produce a detailed and more or less consolidated picture, preferably including quantitative information on essential parameters and variables. A groundwater resources assessment includes usually the inventory and analysis of existing information, various surveys (most of which conducted in the field), often an exploratory drilling programme, usually also hydrogeological mapping, and in all cases the processing and interpretation of all obtained data and information. Many intensely exploited aquifer systems in the world have been assessed already.

10.3.1.6 Monitoring

Groundwater resources management is about controlling groundwater quantity and quality, and it pursues also a wise use and optimal allocation of abstracted groundwater, while protecting nature and the environment. It is essential to monitor the variation in time of the corresponding key variables (groundwater levels, water quality variables, groundwater abstraction and use for different purposes, natural groundwater discharge, etc.). This is done by means of monitoring networks. Installation and continued operation of such networks demands considerable resources and their benefits often are not recognized by those who decide on funding and financing. As a result, in spite of positive exceptions, the state of groundwater monitoring around the world is generally poor.

10.3.1.7 Desk studies

Desk studies are based on analysis of the data and information collected in the field, or obtained by remote sensing techniques. There is a huge variety of such studies and their purposes. Typical examples are quantifying the hydrogeological budgets of specified hydrogeological units, predicting the evolution of groundwater levels and storage, explaining the spatial hydro-chemical variations in groundwater, mapping the vulnerability of groundwater systems against pollution, establishing values of groundwater indicators to support management decisions and evaluating the impact of implemented groundwater resources management measures. All these studies go beyond reporting on data and generating information; they contribute to the development of knowledge on the local groundwater systems. Numerical models are a powerful tool that forms the core of many of these desk studies; these models simulate groundwater flow and/or solute transport for observed boundary conditions and for hypothetical boundary conditions in the future (Anderson *et al.*, 2015). Noteworthy among the more recent developments in remote sensing is the Gravity Recovery and Climate Experiment (GRACE) satellite mission; its observations can – under certain restrictions – be used for estimating changes in groundwater storage, albeit with low spatial resolution (Richey *et al.*, 2015; Alley & Konikow, 2015).

10.3.2 The role of agencies or other actors, programmes and projects

Data acquisition, generating information and developing knowledge on groundwater and its context requires very significant inputs of human capacity, time and financial resources. Except for relatively small local projects, these inputs are normally beyond what a private person or party can afford. Consequently, dominant actors in this domain are governmental agencies and non-governmental agencies entrusted with a special responsibility, mandate or mission related to groundwater, companies carrying out groundwater-related activities and international agencies active in the field of water.

Governments play in principle a main role. In most countries, they entrust a geological survey, hydrological service or a similar agency with the responsibility for groundwater data acquisition, for establishing a centralised groundwater data repository (database or information centre) and for generating the information and knowledge as required for adequate use and management of the country's groundwater resources. Governments have also the option to issue regulations that oblige entities and individuals dealing with groundwater (drilling, well construction, groundwater abstraction, applying for a license, etc.) to share relevant data with the mandated groundwater agency, *e.g.* borehole and well-construction data and monitoring data on groundwater abstraction, groundwater quality and groundwater levels. The corresponding entities and individuals are numerous and represent together an enormous potential source of groundwater data. Rather than operating in an ad-hoc way, the agencies responsible for groundwater data, information and knowledge should adopt pro-active approaches embedded in a strategy endorsed by the government, usually the main financier. Optimal conditions for repetitive components such as monitoring activities are created if these are incorporated in the regular working programmes of adequately functioning agencies. Most activities of a non-permanent nature (*e.g.* mapping, water resources assessments, studies on themes that currently have political priority) are likely to be most successful if carried out in the framework of well-designed and properly funded projects. Some of these projects may be carried out by governmental groundwater agencies, other ones by non-governmental agencies, or by a combination of domestic agencies with international partners in bilateral or multilateral co-operation projects and programmes.

10.3.3 Constraints and options for improvement

In spite of significant progress made in many countries during recent decades regarding the acquisition of data, generation of information and development of knowledge on groundwater systems, there is still much to be desired. First, relatively few countries so far have invested in information and knowledge beyond a very general and spatially aggregated level, or beyond scattered local cases. Second, lack of monitoring data is in most countries a major obstacle to effective groundwater management. The state of groundwater monitoring activities is globally on the decline and the operation of many monitoring systems often is ended already after a limited number of years (FAO, 2016a).

These flaws and gaps can be attributed to a number of constraints. The more fundamental ones among these constraints are limited government budgets, low political priority for groundwater and a generally poor understanding of the role and value of groundwater, and in particular of the need for managing it carefully. As a result, key groundwater agencies in many countries are weak and poorly performing, due to insufficient budgets for adequate staffing and operations, lack of a vision regarding groundwater, and limited capacity and expertise of their professional staff. Obviously, it is crucial that such agencies are strengthened, which requires in the first place strong lobbies at the government level in order to raise their budgets, to be followed by upgrading professional staff (both in number and in expertise), by capacity building and by developing a new operational strategy and a corresponding plan of activities. This should allow such agencies to develop true leadership on the subject, to design and implement relevant assessment and monitoring programmes, to promote synergy between the different entities and individuals involved in groundwater, to define and implement mechanisms for data exchange, to embark upon international co-operation programmes and projects, to provide valuable advisory services, and to become a centre of knowledge on groundwater in their territory. Bilateral and multilateral international co-operation has for many decades proven to be a valuable catalyst, and it still continues to be so at present. Examples of international cooperation efforts that have very significantly boosted groundwater monitoring activities are the Indian Hydrology Project and the EU Water Framework Directory.

10.4 PRESENTING, SHARING AND DISSEMINATING DATA AND INFORMATION

10.4.1 The many ways of presenting data and information

As illustrated in Figure 10.4, there is a diversity of forms commonly used for presenting data and information related to groundwater, making use of different media for storage and dissemination (paper, electronic files, radio, TV, internet, mobile phones, meetings). In practice, these are complementary, in the sense that they reflect the differences in information requirements between the main target groups and also inside each of these groups. How relevant the individual presentation forms are for each of the three main target groups is schematically indicated by their position in Figure 10.4. The more technical and scientific products are meant in particular for use by groundwater professionals (enabling studies and the development of knowledge) and planners (providing a solid basis for groundwater development and management plans). Stakeholders and the general public should be made aware of what is at stake regarding groundwater and why certain measures are needed, but they need information about these aspects in an easily accessible and digestible form, preferably with some options (mass media, social media, meetings) to communicate about it with groundwater professionals, planners and decision-makers. Decision-makers usually have no time to read and analyse detailed information, therefore the information presented to them should be very short and focus on awareness raising and briefing on groundwater policy issues and responses. In practice, groundwater professionals will define the content of almost all presented information. The styling and final presentation of information intended for decision-makers, local stakeholders and the general public, however, will often benefit greatly from the intervention of a communication specialist or a journalist.

10.4.2 Sharing and disseminating data and information

Sharing data and information on groundwater obviously has enormous advantages for the parties involved. Nevertheless, half a century ago it was still a rarely observed practice, for a number of reasons. Data often was less well-organised than nowadays and dispersed over many different offices, all data and information were on paper, copying had to be done by hand, and interinstitutional rivalry made most agencies and companies reluctant to share data or motivated them to charge a significant fee for supplying data. The advent and development of information and communication technology and the proliferation of the internet have produced major changes. Reports, papers, datasets and maps nowadays can be easily supplied at virtually no cost in the form of digital files, or can be consulted at or downloaded from the internet. Geographic Information Systems allow to generate – without much effort – all kinds of maps on the basis of the latest available data, sometimes even with user-defined legend.

This huge technological progress has caused the dissemination of hard-copies of reports and data files to become dwarfed by the massive exchange of their electronic equivalents, many of which are downloadable from the internet. It has also enabled and catalysed the development of internet-based groundwater information systems, focusing at different spatial levels. Interesting examples of national groundwater

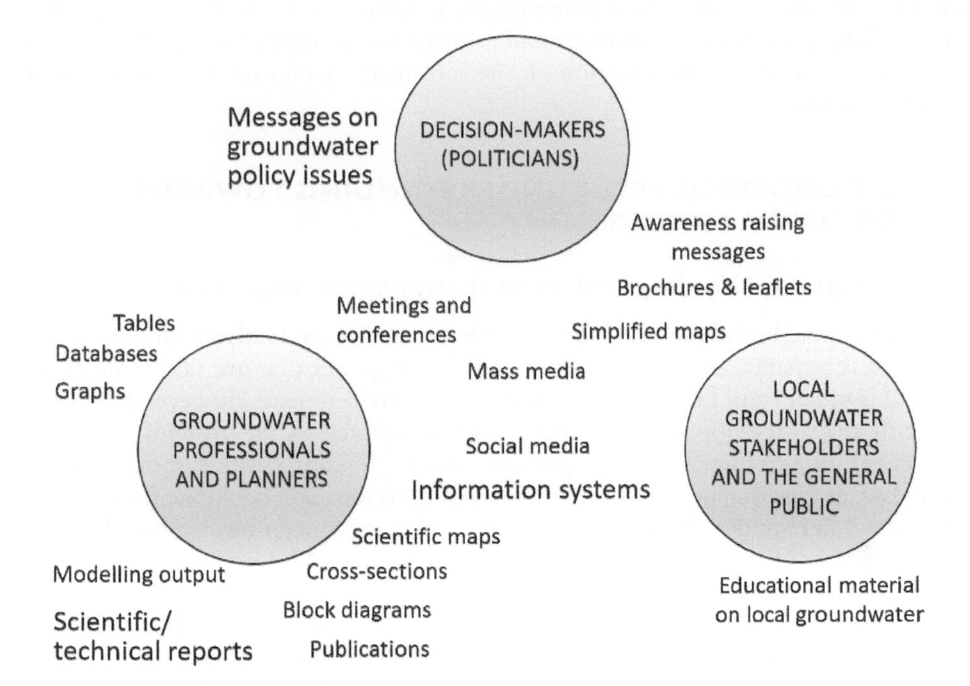

Figure 10.4 Selected forms of presenting groundwater data and information, in relation to envisaged users.

information systems are those of the United States (USGS, 2016a, 2016b), Australia (Bureau of Meteorology, 2016), India (CGWB, 2016), the United Kingdom (BGS, 2016) and The Netherlands (TNO, 2016); some of these focus on technical data and information mainly for groundwater professionals, other ones are also addressing the information needs of the layman and thus contain easily digestible information. At the global level, the GGIS, WHYMAP and AQUASTAT information systems can be mentioned (IGRAC, 2016; WHYMAP, 2016; FAO, 2016b), largely complementary in terms of content. In addition, several groundwater projects around the world, including projects on transboundary aquifers, have their own internet portals, providing access to groundwater data and information at the local or aquifer level.

In most of the less wealthy countries there is still limited use of these modern technologies for sharing and disseminating groundwater data and information. Lack of funds, staffing and expertise are the main reasons, often in combination with available data and information still being scattered over many offices. Reluctance of agencies and institutes to share information has not yet disappeared completely and is still strong in some countries. Government regulations on sharing data and information may be helpful to change this. Similarly, smart regulations (combined with smart technology) may also improve the data flow from individual well owners/users (*e.g.* on abstraction, water quality and water levels) towards centralized public groundwater databases, which is in most countries still insignificant or non-existing. Negotiating data and information sharing with oil companies and other private sector entities involved in subsurface resources other than groundwater is another very interesting opportunity for enriching groundwater databases and information systems. These companies own impressive quantities of information on the subsurface, so far mostly inaccessible for external parties.

10.5 DIAGNOSTICS: FROM UNDERSTANDING TOWARDS DECISION-MAKING

10.5.1 Role of the diagnostics and suggested approach

Information and knowledge are indispensable for getting a reliable picture of a specific groundwater system and for understanding the processes that are taking place both inside this system and between the system and its environment. However, they do not explicitly reveal the opportunities and challenges offered by the groundwater system considered, nor do they present suggestions on how to address these in the context of groundwater governance. These latter activities form the main components of what we understand here under the term 'diagnostics'. Diagnostics can be associated with the fourth member of the earlier presented DIKW hierarchy ('Wisdom', see section 10.1.2) and is looking into the future. At the one hand, it requires analytical capacity and judgement – to interpret the implications of available information and knowledge for achieving societal goals; at the other hand, it calls for creativity – to define measures and behaviour that contribute to meeting these goals as closely as possible. The outcomes thus provide a rational basis for developing a groundwater resources management strategy. In other words: diagnostics is a crucial step towards informed decision-making on groundwater systems and their resources.

Diagnostics forms an initial step towards developing a groundwater management plan and should unambiguously assess: (i) the value and functions of the groundwater systems to be governed; (ii) opportunities and challenges offered by the groundwater systems concerned (including potential responses); (iii) uncertainty and related risks; (iv) groundwater governance provisions and constraints. Each of these aspects will be briefly described below.

The diagnostic process is no longer the exclusive domain of groundwater professionals who assist in defining content for water resources planning. The voice of local stakeholders is steadily becoming louder. In several countries stakeholder involvement is changing from zero or from the rather passive role of 'being heard' (*e.g.* in the form of enquiries or occasional meetings) to active participation in the discussion on critical or controversial issues in local groundwater management. In general, the focus of this public discussion tends to be less on the identification of groundwater-related challenges than on how to respond to these. This process of stakeholder involvement is facilitated by the mass media and by the rapidly increasing role of social media, including the proliferation of mobile devices.

10.5.2 Value and functions of the groundwater system to be governed

Making a judgement on the relevance and value of the groundwater system to be governed is a logical first step in the diagnostic. Does groundwater locally represent an essential source of water, without which life and socio-economic development in the area would be seriously constrained, or is the groundwater system only of marginal importance and can its services be easily substituted by available surface water? And even if groundwater is locally the main source of water: is it abundant, or only very scarce? How much groundwater is currently abstracted on an annual basis, for which purposes and for which categories of beneficiaries? How is the demand for groundwater likely to develop during the coming tens of years? What types of in-situ functions does the groundwater system support? Answers to these questions give insight in the groundwater system's relevance and value, hence will give guidance to defining the priority assigned to its governance and management.

10.5.3 Opportunities and challenges offered by the groundwater system considered

By interfacing information and knowledge on the groundwater system concerned, on current groundwater abstraction and other services, on projected groundwater demands and on changing autonomous boundary conditions (such as demography, economic development, climate change, production of pollutants, etc.) it is in principle possible to identify, map and even quantify in a systematic way the opportunities and challenges offered by the groundwater system considered.

In the category 'opportunities', the most interesting questions are whether there are options and scope for augmenting the groundwater resources (by artificial or induced recharge), for increasing groundwater abstraction without unacceptable side-effects, or for increasing overall benefits from groundwater by modifying groundwater abstraction allocations.

In the category 'challenges' a systematic assessment has to be made of the current and expected future threats to the sustainable use and quality of the groundwater resources, as well as of undesired socio-economic and environmental impacts and trends. Typical examples of such challenges are issues such as progressive groundwater level declines resulting from intensive pumping (in non-renewable aquifers even at low pumping rates), climate change, seawater intrusion and saline water upconing in coastal zones, groundwater pollution (differentiated by source of pollution) and waterlogging; as well as impacts such as water shortages, increasing cost of groundwater development, decreasing water quality and environmental impacts of groundwater abstraction (degradation of wetlands, land subsidence). What in the diagnostic phase should be done is identifying and mapping these challenges, providing an indication of their severity, and suggesting how each of these challenges can be addressed successfully. Numerical simulation models are a strong tool to support the analysis.

10.5.4 Uncertainty and risk

How reliable and accurate the outcomes of the diagnostics are depends both on the analyst's capacities and on the reliability and accuracy of the information used. The commonly large margins of error of groundwater data cause information often to be inaccurate, while lack of data or data scarcity translates into limited reliability of information intending to offer a spatially and temporally continuous picture of the groundwater system. Therefore, one has to take into account that the outcomes of a diagnostic analysis on groundwater are characterised by uncertainty, and that this results in some risks.

One of the risks is that opportunities for increasing abstraction or for augmenting the groundwater resources may be under- or overestimated. A second one is that critical issues might be overlooked and only be noticed when some problems have progressed already. Another type of risk is that predictions by model simulations may diverge significantly from reality if insufficient data of good quality is available, but that modellers, planners and decision-makers nevertheless have blind confidence in the numerical outcomes.

Uncertainty and risk may be reduced by collecting more data, but in the case of groundwater systems this is usually expensive and time-consuming, thus on the short run only feasible to a limited extent. Groundwater management planners have to accept that decisions have to be made under uncertainty, hence they have to examine the robustness of the conclusions on the identified groundwater management issues and the proposed measures to address them, irrespective of data deficiencies. Sensitivity analysis and stochastic approaches may be helpful to conduct this examination properly.

10.5.5 Groundwater governance provisions and constraints

In addition to the aspects mentioned above, most of them related to physical conditions and processes at field level, it is worthwhile to include in the diagnostics also the local groundwater governance conditions and provisions. Briefly, governance encompasses the following components (FAO, 2016a): actors (government agencies, NGOs, private sector, local stakeholders); legal, regulatory and institutional frameworks; policies and plans; and information, knowledge and science. With the outcomes of a diagnostic analysis on groundwater governance it can be judged much better which ones of the potential groundwater management interventions or measures are likely to be feasible in the area considered. Typical constraints identified by such an analysis may include – among others – weak leadership, poor performance of public agencies, limited stakeholder involvement, shortage of government funds, inadequacy and poor implementation of legal frameworks, fragmented or even conflicting institutional mandates, incoherent policies, lack of awareness on groundwater and insufficient data (FAO, 2016a). Assessing the current groundwater governance setting and provisions, and clearly defining the corresponding constraints, will undoubtedly contribute to making groundwater management strategies and planning as realistic as possible.

10.6 CONCLUSIONS

Adequate decision-making on groundwater resources requires proper understanding of the local groundwater systems and their context. This understanding should not be limited to the local hydrogeology, but also include the socio-economic and ecological functions of the local groundwater systems, as well as their interactions with other natural systems (surface water, environment) and human activities (land use, exploitation and use of subsurface resources). In addition, for decisions of a complex nature one should be acquainted with the local groundwater governance conditions and provisions.

According to the DIKW paradigm, understanding is based on four hierarchically linked components: data, information, knowledge and wisdom. Each of these four components has its own role and merits. Data represent the lowest level in this hierarchy and is indispensable for understanding being rooted firmly in reality. Information, obtained by processing and organizing data, adds meaning and reveals patterns, often supported by visualisation in the form of maps and graphs. Access to information is crucial for professionals and other individuals who want to acquire the local knowledge that enables them to analyse and understand relevant processes and relationships in the local setting. Wisdom, finally, links this knowledge with societal demands and aspirations, looks into the future and produces recommendations for informed decisions on action. In other words: it facilitates diagnostic analysis, which forms the interface between knowledge and planning.

The current state of affairs related to groundwater-related data, information, knowledge and diagnostics is highly variable across the globe. Nevertheless, a few rather general characteristics and trends can be observed. In the first place, very significant groundwater resources field assessment and mapping efforts have been made during the last half a century and have produced coherent sets of time-independent data and information on the majority of the world's most intensively exploited aquifers. Monitoring, on the other hand, is a widely neglected branch of data acquisition, with the result that the dynamics of groundwater quantity and quality is documented for very few aquifers only. Lack of monitoring data forms a critical obstacle to informed decision-making on groundwater in most parts of the world. Information on groundwater has usually kept pace with groundwater data acquisition activities, while the access to information has been improved in a revolutionary way by modern information and communication technology. The awareness that information requirements are user-specific gradually leads to tailoring content and presentation of the information according to the envisaged target groups. The development of knowledge and wisdom requires institutions that create favourable conditions and that give priority to capacity development related to groundwater activities. It is no surprise that the wealthier countries in general perform much better in this respect than poor countries, since the latter usually cannot afford building and sustaining strong groundwater agencies, and therefore often remain dependent on support by foreign experts.

Uncertainty related to decisions on groundwater cannot be banned completely, but governments and other key players in groundwater governance should be aware that the benefits from investing sufficiently in groundwater assessment and monitoring, information systems and capacity building (for knowledge and diagnostics) tend to be many times larger than the costs. Here lies a challenge for professional members of the groundwater community. As a first step, they should actively raise the awareness of politicians, other decision-makers and the general public on the role and importance of groundwater and the multiple benefits it produces. Furthermore, they should be advocates for governing and managing the groundwater resources wisely and convince decision-makers that this is only possible if sufficient local data, information and knowledge are available. This implies that arrangements have to be made for structural and permanent investments in the acquisition of groundwater-related field data (in particular by operating monitoring networks) and in strong and active professional institutions in charge of the groundwater-related data, information and knowledge required to underpin groundwater resources management.

ACKNOWLEDGEMENTS

The author would like to thank Tibor Stigter and Kirstin I. Conti for reviewing the first draft of this chapter and providing valuable comments.

REFERENCES

Acoff, R. (1999). From Data to Wisdom. In: *Acoff's Best* (1999). New York, John Wiley & Sons, pp. 170–172.

Alley, W. M. & Konikow, L.F. (2015). Bringing GRACE Down to Earth. *Groundwater*, Vol. 53, No. 6, pp. 826–829.

Anderson, M.P., Woessner, W.W. & Hunt, R.J. (2015). *Applied Groundwater Modeling: Simulation of Flow and Advective Transport*. Second Edition. London, Academic Press, Elsevier.

BGS (2016). *British Geological Survey portal. Groundwater Data and Information.* http://www.bgs.ac.uk/research/groundwater/datainfo/dataInformation.html (Accessed: September 2016).

Bellinger, G., Castro, D. & Mills, A. (2004). *Data, Information, Knowledge and Wisdom.* http://www.systems-thinking.org/dikw/dikw.htm (Accessed: September 2016).

Brassington, R. (2007). *Field Hydrogeology*, Third Edition. John Wiley and Sons.

Bureau of Meteorology (2016). *Groundwater information portal, Australian Government.* http://www.bom.gov.au/water/groundwater/index.shtml (Accessed: September 2016).

Clark, D. (2004). *Understanding and Performance.* http://www.nwlink.com/~donclark/performance/understanding.html (Accessed: September 2016).

EEA (1995). *A general strategy for integrated environmental assessment at the EEA*. Unpublished report for the EEA by RIVM.

FAO (2016a). *Global Diagnostic on Groundwater Governance*. Groundwater Governance – a Global Framework for Action, http://www.groundwatergovernance.org.

FAO (2016b). *AQUASTAT, Global Water Information System.* http://www.fao.org/nr/water/aquastat/main/index.stm (Accessed: September 2016).

Griffiths, D.H. & King, R.F. (1981). *Applied Geophysics for Geologists and Engineers*, Second Edition. Oxford, Pergamon Press.

GWB (2016). *Central Groundwater Board Portal, India. Groundwater Data Access.* http://cgwb.gov.in/GW-data-access.html (Accessed: September 2016).

Hey, J. (2004). *The Data, Information, Knowledge, Wisdom Chain: The Metaphorical Link.* http://www.dataschemata.com/uploads/7/4/8/7/7487334/dikwchain.pdf (Accessed: Sept. 2016).

IGRAC (2016). *IGRAC portal, Global Groundwater Information System (GGIS).* https://www.un-igrac.org/global-groundwater-information-system-ggis (Accessed: September 2016).

Indian Hydrology Project (2014). *Groundwater Handbook: Groundwater level.* Report HP II, http://www.slideshare.net/hydrologyproject0/final-gw-handbook-190514 (Accessed: September 2016).

Kruseman, G.P. & De Ridder, N.A. (1990). *Analysis and Evaluation of Pumping Test Data.* Second Edition. Wageningen, The Netherlands: ILRI Publication 47, 377 p.

Liew, A. (2007). Understanding Data, Information, Knowledge and Their Interrelationships. *Journal of Knowledge Management Practice*, Vol. 8, No. 2, June 2007. http://www.tlainc.com/articl134.htm.

Margat, J. & Van der Gun, J. (2013). *Groundwater around the World: A Geographic Synopsis*. Leiden, The Netherlands, CRC Press/Balkema, 348 p.

Moore, J. (2012). *Field Hydrogeology: A Guide for Site investigations and Report Preparation*. Second Edition., Boca Raton (USA), CRC Press, 2012.

Riche, A.S., Thomas, B.F., Lo, M.-H., Reager, J.T., Famiglietti, J.S., Vosss, K., Wensson, S. & Rodell, M. (2015). Quantifying renewable groundwater stress with GRACE. Water Resources Research, DOI: 10.1002/2015WR017349. http://onlinelibrary.wiley.com/doi/10.1002/2015WR017349/full.

Struckmeier, W. & Margat, J. (1995). Hydrogeological Maps: A Guide and Standard Legend. *IAH International Contributions to Hydrogeology*, Volume 17, Heise.

TNO (2016). *DINOLoket. Data and Information on the Dutch Subsurface*. https://www.dinoloket.nl/en (Accessed: September 2016).

USGS (2014). *Simulation of groundwater and surface-water resources of the Santa Rosa Plain Watershed, Sonoma County, California*. Scientific Investigations Report 2014-5052, US Dep of the Interior. https://pubs.usgs.gov/sir/2014/5052/pdf/sir2014-5052.pdf.

USGS (2016a). *USGS Groundwater Information Pages*. http://water.usgs.gov/ogw/ (Accessed: September 2016).

USGS (2016b). *Groundwater Atlas of the United States*. http://pubs.usgs.gov/ha/ha730/ (Accessed: September 2016).

Walton, W.C. (1970). *Groundwater Resource Evaluation*. McGraw-Hill Series in Water Resources and Environmental Engineering, New York, McGraw-Hill Books, 664 p.

WHYMAP (2016). *Portal, World-wide Hydrogeological Mapping and Assessment Programme*. http://www.whymap.org/whymap/EN/Home/whymap_node.html (Accessed: September 2016).

Winchester, S. (2001). *The Map that Changed the World*. London, Penguin Books.

WWAP (2006). *The United Nations World Water Development Report 2: Water, a shared responsibility*. Paris, UNESCO, Paris & New York, Berghahn Books, 584 p.

Chapter 11

Education and capacity development for groundwater resources management

Viviana Re[1] & Bruce Misstear[2]
[1]Department of Earth and Environmental Sciences, University of Pavia, Italy
[2]School of Engineering, Trinity College Dublin, Ireland

ABSTRACT

Education and capacity development are crucial for successful groundwater management. There are different actors and educational needs to be considered. The education of groundwater professionals should cover the basics of groundwater science as well as the many other technical and socio-economic issues at the interface between groundwater and other disciplines. To achieve these aims, there are complementary educational roles for universities, individual practitioners, NGOs, consultancies, scientific associations and international development agencies. In addition, there is a need for the education of policy makers and the general public, including well owners, who are key stakeholders in groundwater governance. There is also a need for properly-functioning institutions for the management of groundwater resources, yet the reality is that many of the institutions are under-resourced, both in expertise about groundwater and in financial resources. Capacity development is therefore crucial for improved groundwater management. Different communication strategies are required for different stakeholders, and there are opportunities for groundwater scientists to make more use of social media and visual art in their outreach activities. Capacity development can be enhanced by knowledge-sharing through the establishment of participatory fora, and through schemes involving the assignment of local counterpart staff to work alongside international experts.

11.1 INTRODUCTION

Education and capacity development[1] are fundamental to good governance and management of groundwater resources. Whilst education is obviously concerned with teaching hydrogeologists the basic principles of their science, it also involves increasing the knowledge and awareness of groundwater issues amongst all the stakeholders in groundwater governance, including policy makers, engineers and the general public.

[1]In this chapter "capacity building" and "capacity development" are used synonymously to indicate "the process through which individuals, organizations and societies obtain, strengthen and maintain the capabilities to set and achieve their own development objectives over time (UNDP, 2009).

Capacity development deals with building institutions that are capable of performing their designated roles: good management of groundwater resources relies on the effectiveness of institutions as well as individuals. The GEF project (Groundwater Governance – A Global Framework for Action [2011–2015]) identified the problem that Government organizations responsible for the management of groundwater resources are generally understaffed and have inadequate budgets to cope with the problems they face (FAO, 2016a, 2016b). Thus, there is a need to grow the capacity of such organizations for meeting the many challenges pertaining to groundwater management.

This chapter 11 explores some of the main issues surrounding education and capacity development, addressing the needs of the groundwater scientist, the policy maker and the general public, including the well owner. Section 11.2 considers the education of hydrogeologists and the roles of individual hydrogeologists and organizations in professional career development and in increasing the awareness of policy makers and the public about groundwater issues. Section 11.3 considers the institutional dimension of capacity development, whilst in Section 11.4 new educational and outreach tools are analyzed, including their effectiveness for capacity building in an era dominated by social media and visual arts. Some of the main lessons gained by groundwater professionals in education and capacity development are reviewed, with examples, in Section 11.5, and the chapter conclusions are presented in Section 11.6.

11.2 GROUNDWATER EDUCATION

In the first part of this section we consider the education and training of groundwater professionals, including the role of universities in education. We then discuss the ways in which these groundwater professionals can contribute to further education and training of future generations of hydrogeologists, for example through mentoring, and more widely by acting as advocates for science-based groundwater management. The final part of this section on groundwater education looks briefly at the roles of organizations such as scientific associations and international agencies in education.

11.2.1 The education and training of groundwater scientists and professionals

It is during their university studies that aspiring hydrogeologists learn the fundamentals of groundwater science. Whereas undergraduate degree programmes in geology, earth science and civil engineering sometimes include one or more modules on hydrogeology, the education of hydrogeologists in many countries is achieved mainly through graduate degree programmes, either taught-course masters or doctoral degrees.

An excellent review of hydrogeology teaching was carried out by Gleeson *et al.* (2012). This review included a survey amongst academic hydrogeologists which identified 15 topics that are considered essential for an introductory, undergraduate level hydrogeology course (Table 11.1). Interestingly, all of the topics listed in the table are concerned with the basic principles of groundwater occurrence and flow, which are clearly fundamental building blocks in the education of a hydrogeologist. Gleeson *et al.* (2012) also point out that learning outcomes from the classroom are enhanced

Table 11.1 The top 15 most important hydrogeology topics for an undergraduate hydrogeology course, as identified in a survey of 68 academic hydrogeologists (based on information in Gleeson et al., 2012).

Topic	
1. Hydraulic conductivity/intrinsic permeability	9. Wells and piezometers
2. Darcy's law and its applicability	10. Transmissivity
3. Aquifers and confining units	11. Specific discharge and average linear velocity
4. Water table and mapping	12. Primary and secondary porosity
5. Gradient and head	13. Homogeneity and isotropy
6. Water table	14. Recharge and discharge areas
7. Hydraulic head	15. Steady flow in aquifers
8. Specific yield and storativity	

through field and laboratory exercises. At graduate level, the basic topics are usually supplemented by modules covering subjects such as groundwater contamination, groundwater protection, and numerical modelling of aquifer systems.

Based on his experiences both in the university class room and outside academia – where data limitations often constrain the extent of analysis possible – Siegel (2008) identified a top ten list of what students and practicing hydrogeologists '*fundamentally* need to know':

1. Don't push the data farther than they can be pushed and be honest with respect to what can be done
2. Darcy's law needs to be understood at the 'gut' level
3. Potentiometric surfaces are different from the water table
4. Surface water is an 'outcrop' of the water table
5. Groundwater occurs in nested flow systems, separated by hydraulic boundaries
6. Groundwater chemistry is predictable from first principles
7. Chemical oxidation and reduction control many important groundwater and contaminant chemical compositions
8. As a working approximation, contaminant plumes should be considered narrow and no wider than a few times the width of the source at their heads
9. Contour using your head, and not your computer
10. Explore simple bivariate plots as an analysis tool.

This list contains a lot of good advice for the practicing hydrogeologist. Importantly, a full appreciation of the fundamentals should reduce the risk of the hydrogeologist following a "recipe-book approach" in their work, whereby standard techniques are sometimes applied in an unthinking manner to the particular problem at hand, resulting in poor project outcomes (Possin, 2002; Nyer *et al.*, 2002; Misstear, 2016). It may also help avoid the misapplication of complex numerical groundwater models in situations where there are few data and/or a poor conceptual understanding of the groundwater system.

If the groundwater professional does not learn the basic principles of groundwater science at university, then it is unlikely that he/she will be able to pick up this knowledge later in their careers through mentoring, short courses or other training opportunities in the workplace. However, some universities and funding agencies prioritise research over teaching, and perhaps undervalue the importance of delivering graduates in subjects like hydrogeology. In the United Kingdom, for example, and partly because of the priority given to research, specialist taught masters programmes in hydrogeology have been in a state of flux over the past 20 years, with several courses coming and going (Misstear, 2013) – and even the internationally-renowned Birmingham University masters in hydrogeology programme was placed under threat recently (Misstear, 2016). It is imperative that professionals engage with universities about the importance of continuing to educate hydrogeologists and other water scientists, highlighting the key role such graduates play in addressing the global challenges surrounding water and environmental sustainability.

As well as learning the fundamentals of hydrogeological science, groundwater professionals also need to be educated in subjects that lie at the boundary between groundwater science and other disciplines that are crucial to good groundwater management. Many of these subjects are covered in other chapters of this book, including groundwater and ecosystems, groundwater and the water-food-energy nexus, and the linkages between groundwater and land use. It is unrealistic to expect a university course to be able to cover all of these topics. Indeed, university degree programmes may not be the most appropriate fora for such teaching, given the large range of stakeholders involved and the close linkages with professional practice. Nevertheless, it is desirable that scientific programmes in hydrogeology do introduce students to some of the socio-economic issues that are so important in the practice of hydrogeology. For example, a groundwater professional working on rural water schemes in Africa must be aware of gender issues and local water governance and maintenance arrangements when siting and constructing new well schemes, otherwise such schemes are unlikely to be successful. Again, a study of private wells in Ireland highlighted the lack of awareness among well owners about their wells and the risks to water quality and hence potentially to their health (Hynds *et al.*, 2013); this study advocated the need for educational strategies targeted at well owners, strategies that the hydrogeologist will have a key role in formulating.

11.2.2 The role of individual groundwater professionals in education and training

11.2.2.1 Life-long learning

One of the key considerations for any groundwater professional is that his/her education does not cease after graduating with their university degree. Life-long learning through continuing professional development programmes is essential to ensure that the hydrogeologist keeps up to date with new knowledge in his/her field. To illustrate the importance of this, whilst a groundwater professional graduating in the 1970s would have learnt about anthropogenic contaminants such as nitrates from agricultural fertilisers, he/she would not have been aware of the threat of arsenic contamination in certain alluvial aquifers in Asia and South America, since this knowledge did not become available until much later in the 20th century (BGS/DPHE, 2001; Ravenscroft *et al.*, 2009). The groundwater professional working in such environments must keep up to date with respect to new information on the causes, extent and management strategies for tackling the arsenic problem, issues about which our understanding continues to evolve. Again, there are many anthropogenic contaminants where the risks to human health and the environment are not well understood (Sutherson *et al.*, 2016). These so-called "contaminants of emerging concern" (CECs) include various pharmaceutical compounds, water treatment disinfection by-products and nanoparticles. A continuing professional development programme for a hydrogeologist working on industrial contamination projects must therefore include keeping abreast of the latest scientific literature and attending conferences, workshops and short training courses on the subject of these CECs.

11.2.2.2 Mentoring

Mentoring of early career professionals can make an important contribution to lifelong learning. Many private companies have effective staff mentoring schemes, but these are perhaps less common in public sector organisations, including universities. Scientific associations and professional accreditation institutions may also operate mentoring schemes. The International Association of Hydrogeologists (IAH), for example, has recently introduced a mentoring scheme (see https://iah.org/knowledge/mentoring) which aims to help its members in three areas: a) providing advice on scientific topics; b) giving guidance on career options and pathways; and c) offering the benefits of practical experience from case studies and knowledge of local hydrogeological issues.

11.2.2.3 Hydrogeologists as advocates for science-based groundwater management

Hydrogeologists have an important role to play in increasing the awareness of groundwater science amongst policy makers, water managers and members of the public. Whatever the target audience, one of the key issues for the groundwater professional is communication. In order to get our message across, we must use appropriate language (Harris, 2016). We should avoid unnecessary scientific jargon, which can cloud the key messages. We should also remember that senior policy makers are busy people, so the advice needs to be delivered in concise documents or presentations, and at the right time to influence emerging policies.

The general public is a key stakeholder in the environment, including our groundwater resources (see the chapter on stakeholder participation). The recent GEF Groundwater Governance project found that "The effective involvement of stakeholders in groundwater resources management is, in most countries, non-existent or still in its infancy" (FAO, 2016a). This report identifies the local water user associations (WUAs) which have been established in some countries as an exception to this non-involvement. However, these WUAs often face significant challenges in order to be able to ensure effective community participation. A recent research project in Uganda, for example, identified some of the problems facing WUAs which include lack of financial resources, inadequate access to well and pump maintenance facilities, and insufficient leadership roles for women (Fagan *et al.*, 2015).

One of the key threats facing groundwater resources in many countries is pollution from agriculture. It is increasingly recognised that, in order to be able to manage this problem, farmers are key stakeholders. A top-down regulatory strategy is unlikely to be effective. Farmers and other stakeholders need to be involved in a bottom-up decision making process, and this should be undertaken at the catchment scale as part of an integrated catchment management strategy (Boyden, 2015). Groundwater professionals therefore need to engage with farmers at the local level, explaining the links between groundwater quality and agricultural activities like arable farming and milk and beef production, and how, for example, the correct timing and application rate of fertilisers will not only reduce the pollution risk but also potentially lead to more efficient food production and save the farmer money. Indeed, although many groundwater professionals have discussions with farmers and well operators while they perform fieldwork activities, this precious contact time could be used more profitably if framed within a structured socio-hydrogeological approach (Re, 2015). For example, a recent project in Tunisia, involving the integration of specific social analysis (namely stakeholders' network analysis and semi-structured interviews administration) to hydrogeological assessments, proved to be effective not only in identifying and presenting explanations for otherwise unexplained social and political dynamics, but also for fostering the dialogue among the concerned stakeholders, which is important for the effective implementation of new groundwater management strategies (Tringali *et al.*, 2017).

11.2.3 Role of scientific associations and international agencies in education

The previous section focused on the role of the individual groundwater professional in raising awareness about groundwater issues. However, organizations ranging from companies and NGOs, to government agencies and international scientific organisations and associations also play a key part in education.

In terms of international bodies, UNESCO has been involved for many decades in educational and other activities that promote a greater awareness of groundwater issues. These include the UNESCO-IHE Institute for Water Education in Delft, which provides postgraduate water education, the UNESCO IHP programme with its GRAPHIC (Groundwater Resources Assessment under the Pressures of Humanity and Climate Change) component, and IGRAC (International Groundwater Assessment Centre, which operates under the auspices of UNESCO IHP).

There are many scientific associations concerned with groundwater, including the IAH, the International Association of Hydrological Sciences (IAHS), and the International Water Association (IWA). These associations may work independently on certain activities, but they also collaborate on specific initiatives such as conferences and cooperate with international agencies on issues surrounding water resources. The mission of the IAH, for example, is to "further the understanding, wise use and protection of groundwater resources throughout the world" – which is entirely compatible with the good groundwater governance theme of this book. To achieve this, IAH engages with policy makers and international organizations. Examples of project-related outputs include support for the production of the WHYMAP (World Hydrogeological Map), which involved several other organizations including the German Geological Survey (BGR) and UNESCO, and inputs to the GEF Groundwater Governance project led by FAO and also involving UNESCO and the World Bank (FAO, 2016a, 2016b).

A recent educational initiative of IAH is the preparation of a series of strategic overview papers. These take a look at the linkages between groundwater and issues such as global change, food security, energy, ecosystem conservation, resilient cities and human health (IAH, 2015; 2016), and thus deal with several of the topics covered in the chapters of this book that deal with integration and policy linkages beyond the local groundwater system. These short papers aim to inform professionals in other sectors about these key interactions with groundwater resources and hydrogeological science. They are available for download at https://iah.org/knowledge/strategic_overview_series. The most recent thematic paper "The UN-SDGs for 2030: Essential Indicators for Groundwater" explains the role of groundwater in meeting the Sustainable Development Goals (especially SDG 6 on water and sanitation), and provides advice to stakeholders on the complexities surrounding groundwater monitoring (IAH, 2017).

11.3 THE INSTITUTIONAL DIMENSION OF CAPACITY BUILDING

In addition to education and training for both scientists and practitioners, specific attention should be paid to enhancing the capacity of government agencies, non-governmental organizations and local communities to manage groundwater resources effectively. This implies addressing strategic and long-term capacity building targeted at: (i) promoting science-based decision making; (ii) enhancing stakeholder engagement and fostering networking among all the different concerned institutions and actors; (iii) supporting knowledge transfer to strengthen both 'horizontal' (*e.g.*, between state and non-state actors, or across sectors such as agriculture or energy) and 'vertical' (*e.g.*, between various levels within any given sector) cooperation.

Groundwater experts can play an important role when capacity building is at stake. In particular, they can collaborate with national and local authorities in providing training for specialists on technical issues like well construction and piezometer management, groundwater mapping, design and implementation of water quality monitoring programmes, contaminant transport modelling, etc. Moreover, by adequately sharing the outcomes of their research (as discussed in the following section), they can contribute to (or even establish new) platforms for discussing the implementation of institutional arrangements targeted at groundwater management. Indeed, in this context, it is also necessary to consider the role of private companies, NGOs, trade unions and water users' associations in helping to promote a greater awareness of groundwater resources, and improving the collaboration among institutional stakeholders and water end-users.

It must be stressed that the weaknesses of groundwater management often lie in budget constraints, weak mandates, fragmentation of responsibilities and, more generally, structural governance problems that require reforms beyond the reach of capacity building approaches, as highlighted by BGR *et al.* (2007) in relation to different case studies in Africa. However, as the same authors point out, capacity development can be the driver for different processes that may lead to the necessary changes in the groundwater sector. Another topic of concern, one that can be addressed by enhancing knowledge transfer among all the relevant stakeholders, is groundwater legislation and regulation. In many cases, groundwater legislation (if it exists) might need to be updated as current knowledge progresses, and this clearly requires a more effective interaction between scientists and governmental agencies, especially when multiple and complex interests are at stake (*e.g.* groundwater ownership, transboundary aquifers issues).

One approach to capacity development (with which one of the authors of this chapter has some first-hand experience) is the assignment of counterpart staff within a local water agency to the international staff engaged on a groundwater development project. The various counterpart and international personnel may have roles such as project manager, hydrogeologist, irrigation engineer, soil scientist, agronomist, etc. The main aim is for the counterpart staff to learn from working alongside the experienced international personnel. These programmes may include on-the-job training activities such as lectures and production of educational materials. They may also include more informal hands-on training in specific tasks, for example drilling, geophysical logging and pumping tests. Of course, the educational exchanges take place in both directions – the international groundwater expert will learn a considerable amount about the local hydrogeology and practices, as well as about the important socio-hydrogeological factors that apply in the particular country (see Section 11.2.2).

11.4 ENHANCING COMMUNICATION STRATEGIES TO BRIDGE THE GAP BETWEEN SCIENCE AND SOCIETY

Recent years have witnessed an expanding debate on sustainability, with a frequent call to civil society to both understand and take action towards contributing to solving the new environmental protection challenges. Indeed, in relation to groundwater resources, the "hidden component of the water cycle", the media attention is rapidly growing, although the entertainment component often dominates over the scientific one, and thus fails to fully explain the causes of the main hydrogeology-related issues (*e.g.* fracking, groundwater and climate change, aquifer overexploitation and pollution). Linking scientific knowledge to action for sustainability clearly requires a new approach for both research and education, with the first generating information that goes beyond the purely scientific domain and the second enabling students to be visionary and capable of tackling future challenges (Wiek *et al.*, 2011).

11.4.1 How can groundwater scientists and practitioners make the best use of their findings?

Among the main challenges facing groundwater scientists are how to improve our knowledge of the state of global groundwater resources, and our ability to model and predict aquifer behaviour in diverse hydrogeological environments. For scientists, peer to peer information exchange is a fundamental component of the scientific process, permitting more robust results to be achieved through confrontation with other experts in the same field. However, dissemination at thematic conferences and/or through scientific journals are merely two examples of communication. No less important is communicating with the general public and, in a broad sense, all possible (ground)water stakeholders and end-users. Reliable information is an essential requirement for any decision-making and management process (Baldwin *et al.*, 2012), but only when knowledge is adequately transferred to final users and relevant actors. Only a wide-ranging communication strategy, adequately targeted to reach all potential stakeholders, can ensure that groundwater professionals can make the best use of their findings, so that these are used effectively to support sound science-based management practices. In addition, a multi-faceted communication approach helps bridge the gap between science and society, hence contributing to the creation of a network of mutual trust and collaboration with water users (Re, 2015), and to the avoidance of common misunderstandings among water managers and researchers (Borowski and Hare, 2007). Effective communication can therefore be a step towards public engagement activities, as a basis for the implementation of new governance structures that take into account both scientific outcomes and the real needs of concerned stakeholders.

For successful communication with the general public, different tools and language registers than those used among groundwater experts may be required, while at the same time the pitfall of trivializing the scientific message must be avoided. Thus groundwater professionals must engage with new tools for education and capacity building, permitting more effective information sharing with any specific target audience (*e.g.* based on different age and education level, but also cultural, regional background and needs).

11.4.2 The role of social media in supporting education and capacity building: a useful tool or a waste of time?

Groundwater investigations are generally targeted at solving real world problems, although results are often poorly disseminated outside of the specialists' arena. Translating science to the general public and policy makers is therefore a challenge which groundwater experts must tackle to ensure effective management of groundwater resources. Besides public talks and direct engagement with stakeholders, social media are increasingly becoming a key channel for bridging this communication gap. With around 2.3 billion active social media users (of a total of 3.4 billion internet users), growing at an annual rate of 10% (brandwatch.com, 2016), it is clear that social networks have enormous potential to help in disseminating scientific outcomes to a wider audience. Also in the groundwater sector, social media are increasingly being used for sharing information, research updates and new scientific outcomes (especially via Facebook and Twitter), as well as for seeking advice on solving groundwater problems and on career development (for example, through Facebook, LinkedIn or ResearchGate). To some extent, social media can be seen as a useful tool for fostering international and cross-sectoral discussion about groundwater issues worldwide, although some people, especially senior groundwater experts, might not have sufficient time or the necessary IT skills for contributing to web-based discussions. This may result in the use of social media for disseminating results only, rather than taking advantage of their full potential as platforms for capacity development.

Furthermore, considering the ever-increasing amount of information available, it might be difficult, especially for non-experts, to disentangle all this information (and particularly to identify the reliable sources). Therefore, when starting a groundwater-related discussion on social media, we may first have to understand what criteria people use when looking at posts. It is possible that social network users will only be attracted by discussions from people they know, by those from users recognized as the most authoritative (as perhaps indicated by the number of followers or ascertained popularity), or by posts dealing with the trend-topic of the moment. Therefore, the

Table 11.2 Pros and cons of the use of social media for scientific dissemination and knowledge sharing.

Pros	*Cons*
• Improve access to information • Favour participation in conversation among scientists, general public and policy makers • Create potential for collaborations • Facilitate promotion of research outcomes • Promote dissemination of knowledge • Encourage independent thinking • Engage young people	• Reduce time available for in-depth analysis of information available • Can lead to a superficial or even biased analysis of a given issue • Difficulty in identifying which information to trust • Absence of content quality check prior to information sharing • Potential for publishing biased information • Experts might be more interested in sharing information to promote their own work rather than contributing to objective and independent discussion

question can be posed as to whether social media are really helpful for groundwater experts to spread science and knowledge to the general public, or if this is just another tool for peer-to-peer confrontation. Given these uncertainties – and considering both pros and cons of social media use (Table 11.2) – some resistance from the hydrogeological community towards this additional outreach effort might be expected (and notwithstanding the possible attractiveness of these platforms to future generations of (ground)water scientists). In practice, the Twitter (and more generally the social media) conversation is happening already whether or not we participate, and "ensuring credible scientific evidence […] must be considered a responsibility of the research community to keep that dialog balanced" (Kapp *et al.*, 2015).

11.4.3 The role of e-learning in groundwater education: crossing the borders of learning

E-learning and distance learning are becoming increasingly popular worldwide, as they offer a more flexible alternative for those willing to be trained in groundwater sciences. With distance learning students are offered the opportunity to attend training and participate in discussions with teachers and experts in their home and, most importantly, in their home countries, hence representing a more affordable option, especially in low income countries.

For e-learning to be effective, proper interaction with teacher and trainers must be guaranteed. Both in the case of synchronous (*i.e.* real-time training with all participants interacting at the same time) or asynchronous learning (*i.e.* self-paced) it is fundamental to allow all participants to engage in the knowledge exchange, either with direct discussion in virtual classrooms or with the support of technologies such as emails, blogs, forums, webinars and discussion boards. In both cases, the use of Web 2.0 technologies (*e.g.* social networks, video sharing and web applications) that favour the interaction and collaboration among users, is an important asset. Among the numerous tools available online, The Water Channel (www.thewaterchannel.tv) is perhaps worth mentioning here, as it contains a whole section dedicated to groundwater, with around 150 videos that can be used for education and training.

Whilst e-learning has many advantages, we would like to end this brief discussion by stating that e-learning should not replace traditional learning in the classroom or workplace. For groundwater professionals, there is no proper substitute for on-the-job training, including the taking of field measurements.

11.4.4 The role of visual art in groundwater education: when science meets art

Living in a visually-intensive society poses additional challenges for effective groundwater outreach. The easy and widespread access to art, graphics, videos and other visual media in this digital age means that people are more and more used to visually-striking images and so it becomes increasingly difficult to attract (and hold) their attention, even when it comes to conveying important messages relevant to their lives. For this reason, it is necessary to strengthen the interaction between science and art to produce outreach materials that are both scientifically authoritative and visually attractive (some simple examples of effective cartoons are shown in Figure 11.1). By doing this, not only will it be possible to reach a broader audience and increase awareness of groundwater issues, but also working with visual artists can help raise different viewpoints and reveal hidden assumptions within scientific communication.

Good examples of the integration of sound scientific information and high quality visual content are provided by the United Nations World Water Assessment

No, not enough
VRRRRR
STOP
Groundwater: the hidden resource
Who owns the groundwater?

Figure 11.1 Examples of some of the most well-known and widely used groundwater cartoons available on the Dutch portal to International Hydrology (Hydrology.nl, 2017; Courtesy of The Netherlands Chapter of IAH).

Programme (UNESCO-WWAP) "The Water Rooms". These are short animated movies which trigger interest and encourage learning about freshwater resources and their responsible management in the context of sustainable development (http://thewaterooms.org/). The "Water dialogue posters" by the International Water Management Institute (IWMI) are another interesting example of outreach materials (https://waterandfood.org/water-posters/) targeted at educators and trainers at all levels, promoting a better understanding of the link between water and food production. Concerning groundwater specifically, the American Groundwater Trust (https://agwt.org/) is actively engaged in promoting efficient and effective groundwater management, also producing interesting educational materials for children (*e.g.* the illustrated book "Well ... What's all that drilling about", 2007). Another valuable example is the book "Groundwater – Our Hidden Asset" produced by the UK Groundwater Forum (Downing, 1998), aimed at providing a wide audience with basic information about groundwater in an interesting and informal manner, supported by numerous attractive illustrations. Nevertheless, examples of effective integration of groundwater sciences and visual arts are still few and far between.

11.5 LESSONS LEARNED BY GROUNDWATER SCIENTISTS AND PROFESSIONALS

In the previous paragraphs the reciprocal benefits of education and capacity development for demystifying science and creating a network of mutual trust among experts, institutions and groundwater end-users have been presented. Indeed, education and capacity building are not only effective in raising awareness about groundwater issues and management needs, but also represent a powerful tool for the scientific community to gain a better understanding of the real needs of local people. Together they can result in a more conscious use of water resources, and eventually foster the implementation of new, participatory and bottom-up actions for long-term groundwater protection.

One such example of a participatory approach is the successful Andhra Pradesh collaborative project (Andhra Pradesh Farmer Managed Groundwater Systems, APFMGS, 2016) developed in Southern India with the support of the Food and Agriculture Organization of the United Nations (FAO). In an area subject to droughts and uneven distribution of water resources, farmers were actively involved in groundwater monitoring activities after receiving basic training on hydrogeology and water management practices (also through traditional theatre and poetry). The result of knowledge sharing and direct engagement was an increase in water efficiency that gradually led to an improvement in both crop production and farmers' incomes.

A similar approach is being used in Burkina Faso by Water Aid UK to secure for drought-prone local communities access to water during the whole year (Water Aid UK, 2016). In achieving this, local people were trained to become water experts themselves (able to measure water levels, monitor rainfall and spot emerging data patterns), while the government is also involved to guarantee the sustainability of the project in the long-term. By "putting the community at the heart of managing their own water supply" local householders and farmers are now better equipped to make informed, collective decisions about their water supply, and achieve an overall improvement to their wellbeing. In addition, data retrieved by local groundwater experts are collected into the national monitoring schemes, to help build a more cohesive national assessment of climate patterns across the country, thus avoiding a potential loss of information resulting from decentralized monitoring actions.

Community-based groundwater monitoring using a citizen-science approach has also proven to be effective in complementing existing government-run monitoring programmes in other regional and hydrogeological contexts. Little *et al.* (2016), for example, highlighted that these kinds of approaches can be implemented in large municipal districts, like Rocky View County (3900 km^2) in Alberta, Canada. In this pilot case study, community volunteers were trained to directly measure the water level in their wells, and the data collected were made available for consultation by the general public through an online database. The close collaboration among scientists, county staff members and volunteers, made possible the successful implementation and operation of the network for the 5-year pilot period, although some volunteers abandoned the programme for different reasons (*e.g.* loss of interest, moving house, change of lifestyle), and showed the potential for replication in other municipalities and watersheds. Compared to the approach adopted in the previously mentioned Burkina Faso project, this activity involved the use of different items of equipment that were given to volunteers, hence implying a significant budget for the project. This may clearly be an obstacle for its implementation in lower-income regions worldwide, and illustrates the need for the development of different participatory frameworks for different environments (see the chapter in this book on stakeholder participation). Other key elements of this monitoring project were the focus on education and outreach activities, and the need for frequent discussions of the project's results to keep the volunteers motivated, highlighting the importance of science dissemination for the effective implementation of new participatory management practices.

Concerning capacity development, a successful example at international level is the Global Environmental Facility (GEF) International Waters Learning Exchange and Resource Management (GEF IW:LEARN, 2016). The GEF proposes to "strengthen transboundary water management by collecting and sharing best practices, lessons learned and innovative solutions to common problems" through the development of a participatory forum. Since 2000 the IW:LEARN forum has given 300 participants from 80 countries the opportunity to share their studies, experiences and knowledge on international water management. In this way, this project promotes dialogue, knowledge sharing, and replication between projects, allowing appropriate solutions to be found to common problems at a global level.

In the previous paragraphs we cited some of the many examples of successful case studies and investigations where different capacity building strategies resulted in improved groundwater management plans. However, this is not always the case and, as previously highlighted, weak governance still remains a significant issue worldwide, one that requires innovative and collaborative approaches to education and capacity development for a better engagement of the future generations of groundwater managers and scientists.

11.6 CONCLUSIONS

This chapter has reviewed the roles of the different stakeholders in education and capacity development, together with the various communication tools and participatory approaches that are available. The main conclusions are:

a) Universities continue to play a vital role in teaching the fundamental principles of groundwater science, since it is difficult for practitioners to catch up on this core knowledge later during their careers. University courses should also introduce students to subjects at the interface between groundwater science and other disciplines that are necessary for good groundwater management, including the social aspects of water development.
b) Individual groundwater scientists, scientific associations, professional institutes and international agencies all perform important roles in the career development of groundwater professionals ("life-long learning"), through short courses, seminars, mentoring schemes, etc. These individual scientists and organizations also need to engage with other stakeholders in groundwater development such as water managers, planners, policy makers and members of the public, in order to raise awareness of groundwater issues.
c) For effective groundwater governance all the dimensions of education and capacity building should be considered (as summarized in Table 11.3):

    - Scientific and technical level, aimed at training groundwater specialists to become the future groundwater leaders capable of implementing integrated approaches for solving groundwater issues;

Table 11.3 Goals and dimensions of capacity development according to the different target groups involved in groundwater management and protection.

Target group	*Dimension*	*Goal*	*Improved education and capacity development challenge*
Groundwater experts	Scientific	Ensure that hydrogeological conditions are fully considered in any new water management plan	Promote multidisciplinary investigations and act as advocates for integrated groundwater management and protection
National and local authorities	Institutional	Ensure that groundwater is fully included in the political agenda	Foster the implementation of new science-based groundwater management practices, also taking into account the real needs and issues of all the concerned stakeholders
General public	Water end-users/ well owners	Contribute to science-demystification and bridge the gap between science and society	Raise awareness about the role of groundwater end-users and increase their ability to take action
NGOs, water companies	Private	Provide practical support for protecting local and regional groundwater resources	Identify adequate strategies for capacity development for the private sector

- Institutional level, specifically involving decision makers and all the relevant stakeholders;
- Groundwater end-user level, to raise awareness about the importance of shared actions for long-term groundwater management;
- Private sector and NGO dimension.

d) It is essential to address all four dimensions in order to support new policies for sound groundwater management in the long run. In addition, it is crucial to go beyond a purely sectoral approach, by promoting multidisciplinary actions to simultaneously address all the dimensions of groundwater governance.

e) All possible communication tools (including social media and visual art) must be wisely and effectively used in order to enhance capacity development among all the relevant groundwater users and managers.

f) Different education and capacity development approaches are already being implemented successfully, but more specific actions, such as increasing concerned stakeholder engagement and developing innovative, multidisciplinary approaches, are required to ensure sound science-based management actions, fully taking into account the real needs of groundwater users.

REFERENCES

APFMGS (2016) *Andhra Pradesh Farmer Managed Groundwater Systems – Demystifying Science for Sustainable Development.* www.fao.org/nr/water/apfarms/index.htm (accessed online September 2016).

American Groundwater Trust (2007) *Well ... What's all that drilling about.* Illustrated by Rachel Pender. 35p.

Baldwin C, Tan PL, White I, Hoverman S and Burry K (2012) How scientific knowledge informs community understanding of groundwater. *Journal of Hydrology*, 474: 74–83.

BGR, Cap-Net, WA-Net and WateNET (2007) *Capacity Building for Groundwater Management in West and Southern Africa.* http://www.geozentrum-hannover.de/DE/Themen/Zusammenarbeit / TechnZusammenarbeit / Politikberatung _GW / Downloads / groundwater_capacity building.pdf?__blob=publicationFile&v=2 (accessed online September, 2016).

BGS and DPHE (2001) *Arsenic contamination of groundwater in Bangladesh.* Kinniburgh DG and Smedley PL (eds), British Geological Survey Report WC/00/19, Keyworth, UK.

Borowski I and Hare M (2007) Exploring the Gap Between Water Managers and Researchers: Difficulties of Model-Based Tools to Support Practical Water Management. *Water Resources Management* 21: 1049–1074.

Boyden M (2015) *Public Engagement in Integrated Catchment Management: Streamscapes Recommendations.* Environmental Protection Agency, Wexford, Ireland.

Brandwatch.org (2016) https://www.brandwatch.com/2016/03/96-amazing-social-media-statistics-and-facts-for-2016/ (accessed online July 2016).

Downing RA (1998) *Groundwater – Our Hidden Asset.* British Geological Survey, Keyworth, UK, 61p.

Fagan H, Linnane S, McGuigan K and Rugumayo A (eds) (2015) *Water is Life: Progress to secure safe water provision in rural Uganda.* Practical Action publishing, Rugby, UK.

FAO (2016a). *Groundwater Governance Synthesis Report – Draft.* FAO, Rome.

FAO (2016b). *Global Framework for Action to achieve the vision on Groundwater Governance.* FAO, Rome.

GEF IW:LEARN.NET (2016) *International Waters Learning Exchange and Resource Management*. http://iwlearn.net/ (accessed online September 2016).

Gleeson T, Allen DM and Ferguson G (2012) Teaching hydrogeology: a review of current practice. *Hydrology and Earth System Science* 16: 2159–2168.

Harris R (2016) Getting the buggers to listen – some observations of the trials, tribulations and successes in delivering (and using) relevant scientific research. In *Proc. 36th annual groundwater seminar on 'Sustaining Ireland's Water Future: The Role of Groundwater'*, International Association of Hydrogeologists (Irish Group), Tullamore, 12–13 April 2016, VII-1–VII-2.

Hydrology.nl. 2017. *Dutch portal to International Hydrology*. http://www.hydrology.nl/iahpublications/201-groundwater-cartoons.html.

Hynds P, Misstear BDR and Gill LW (2013) Unregulated private wells in the Republic of Ireland: Consumer awareness, source susceptibility and protective actions. *Journal of Environmental Management*, 127C: 278–288.

IAH (2015, 2016 & 2017) *Strategic Overview Papers*. https://iah.org/knowledge/strategic_overview_series.

Kapp JM, Hensel B and Schnoring KT (2015) Is Twitter a forum for disseminating research to health policy makers? *Annals of Epidemiology* http://dx.doi.org/10.1016/j.annepidem.2015.09.002

Little KE, Hayashi M and Liang S (2016) Community-Based Groundwater Monitoring Network Using a Citizen-Science Approach. *Groundwater*, 54:317–324.

Misstear BDR (2013) Water, but not on the brain. *Geoscientist*, 23(1): 9.

Misstear BDR (2016) Looking to the future: Opportunities and challenges in hydrogeology education and training. In *Proc. 36th annual groundwater seminar on 'Sustaining Ireland's Water Future: The Role of Groundwater'*, International Association of Hydrogeologists (Irish Group), Tullamore, 12–13 April 2016, II-11–II-18.

Nyer EK, Fierro P and Guillette B (2002) A lost generation. *Ground Water Monitoring and Remediation*, 22: 34–38.

Possin BN (2002). Editorial: The lost tribe of hydrogeologists? *Ground Water* 40: 329–330.

Ravenscroft P, Brammer H and Richards K (2009) *Arsenic pollution: a global synthesis*. Oxford, UK, John Wiley & Sons.

Re V (2015). Incorporating the social dimension into hydrogeochemical investigations for rural development: the Bir Al-Nas approach for socio-hydrogeology. *Hydrogeology Journal* 23: 1293–1304.

Siegel D (2008) Reductionist hydrogeology: ten fundamental principles. *Hydrological Processes* 22: 4967–4970.

Sutherson S, Quinnan J, Horst J, Ross I, Kalve E, Bell C and Pancras T (2016) Making Strides in the Management of "Emerging Contaminants". *Groundwater Monitoring & Remediation* 36: 15–25.

Tringali C, Re V, Siciliano G., Zouari K, Chkir N, Tuci C (2017) Insights and participatory actions driven by a socio-hydrogeological approach for groundwater management: The Grombalia Basin case study (Tunisia). Hydrogeology Journal, doi:10.1007/s10040-017-1542-z.

UNDP (2009) *Capacity Development: A UNDP Primer*. http://www.undp.org/content/dam/aplaws/publication/en/publications/capacity-development/capacity-development-a-undp-primer/CDG_PrimerReport_final_web.pdf

Water Aid UK (2016) *Clean water today — and every day*. https://medium.com/@WaterAidUK/clean-water-today-and-every-day-53808efbbc9b#.tw6aywuzu (Accessed online September 2016).

Wiek A, Farioli F, Fukushi K and Yarime M (2011). Sustainability science: bridging the gap between science and society. *Sustainability Sciences* 7 (Supplement 1): 1–4.

Chapter 12

Groundwater governance – impact of awareness-raising and citizen pressure on groundwater management authority in the United States

Andrew Stone
Executive Director, American Ground Water Trust, Concord, New Hampshire, USA

ABSTRACT

Many state and local jurisdictions have authority that impact how groundwater is managed. Awareness, information and education are essential for science based decisions about groundwater allocation, protection, and sustainability. Examples are provided that show the role of citizen organizations and interest groups in framing issues and providing a voice for stakeholders in groundwater governance in the United States.

12.1 INTRODUCTION

For this chapter, groundwater governance is considered to represent a dynamic and evolutionary process involving collective influences on policy and water management decisions with many different interests involved. Groundwater governance does not start from a blank page. Today's governance is built on the palimpsest of the past. This chapter shows some of the ways in which citizens, non-governmental organizations (NGOs), professional associations, and industry interest groups influence the way in which groundwater resources are managed. A key challenge for success in groundwater governance is to raise the level of scientific understanding of groundwater to help reconcile divergent values, interests and preconceptions. Information (data) that are the basis for decisions must be seen as credible with clear definition of hydrologic facts.

In addition to state and local agencies with responsibility and authority over water issues there are typically other agencies, for example forestry, transport, agriculture and health that have overlapping responsibilities that could also influence policy decisions about groundwater allocation and management. Stakeholders in groundwater governance outcomes include those directly involved as end-users (water utilities, irrigators) or as indirect economic beneficiaries in the community such as agricultural suppliers, engineering companies and developers. Stakeholders in groundwater management issues also include individuals and organizations with environmental, ecological, health related and socioeconomic priorities that could be affected by decisions about groundwater use and source protection priorities.

Decisions about the use of groundwater today are rooted in complex connections of political structure, historical precedent, hydrogeological conditions, legal rights, vested interests and perceptions of future need. The United States has diverse types of aquifer systems, regional differences in climate and topography, great variation in historical water use and regional differences in the evolution of water policy over the last 200 years.

The fifty states of the US each have independent political oversight of most natural resource issues. Water management strategies and the associated governance which provides management authority have typically developed in reaction to supply and demand concerns of major water users with resource competition and drought as major drivers.

It is only in the last few decades, that citizens have become active in water and environmental issues related to groundwater. The public, not just end-users, now have the power to influence policy. Public awareness of cause and effect in deteriorating water quality, the growth of strong citizen-based environmental organizations and the impact of demographic pressures have helped prioritize the need for political response to find solutions that will maximize the benefits of sustainable groundwater use among all citizens.

The devolution of authority from federal to state to local control over the use and protection of aquifers can be controversial if it generates "turf wars" over who has jurisdictional authority. Competing vested interests may clash over characterizing problems of overuse or deteriorating water quality and in proposing solutions. Promoting a science based understanding of groundwater has become a major education challenge for independent NGOs, agencies and academics striving for groundwater resources protection and sustainable use.

The chapter is organized under the headings of control of water, evolution of governance, informing the public, local authority, and concludes with examples of awareness raising and citizen pressure on groundwater decisions and policy.

12.2 CONTROL OF WATER

Control and authority over water resources is an evolutionary process, with groundwater management issues a relative late-comer as an essential part of overall water management. The political power that results from authority (given or taken) over the control of water is a well-documented phenomenon. For example Karl Wittfogel's thesis about the rise of "hydraulic societies" throughout history, (Wittfogel, 1957) and examples in Donald Worster's book, Rivers of Empire, subtitled Water Aridity & The Growth of the American West (Worster, 1985).

Perhaps even more important as an influence of how groundwater is actually managed locally are the many elected and appointed boards, commissions, agencies and associations with direct or indirect authority over planning and environmental issues. In addition to direct citizen involvement with these officially recognized groups, there are many NGOs such as watershed associations with a focus on local or regional environmental issues. There are thousands of environmental organizations in the US. Most can be accessed via a web-portal provided by the US Environmental Protection Agency that has a state by state listing and provides links to environmental organizations for America's 43,000 postal zip code areas (www.cfpub.epa.gov/surf). Some major environmental organizations such as the Sierra Club the Nature Conservancy, and the Environmental Defense Fund operate nationwide and have programs related to resource protection and have members and professional staff who exert powerful influence on shaping policy. Groundwater may not be a primary stated concern, but the broad environmental interest focus of these groups will often have an indirect influence of policies of groundwater resources use and aquifer protection.

There are professional membership groups such as the National Ground Water Association and the American Water Resources Association that do specifically exert influence on groundwater policy, and there are specialist non-profit groups not organized to represent any particular profession, such as the Groundwater Foundation and the American Ground Water Trust that have education about groundwater as the principal means of fulfilling their mission to protect resources and achieve sustainable use of groundwater.

Indirect impacts of groundwater governance are locally and nationally influenced by public pressure on decision makers arising from issues that have mobilized citizens to become actively involved. Other impacts come from groups that have economic interests tied to policy outcomes. Some specific examples are outlined in more detail later in this chapter. They show how groundwater management policy is directly or indirectly influenced by the involvement of NGOs in raising awareness, framing issues and in facilitating dialogue among stakeholders, groundwater professionals, agencies, regulators and political decision-makers. For example:

- Citizen and agricultural industry concern about drought impacts in California which have caused increased over pumping of aquifers for irrigation and dried domestic water wells.
- The loss of irrigated agricultural production in Colorado with urban water users buying agricultural water rights in so-called "buy and dry" deals leading to social impacts on rural communities because of reduced agricultural activity.
- Water quality concerns in states where there is oil and gas development involving hydraulic fracturing, pipeline construction and produce-water disposal via injection wells.
- Concerns that bottled water companies are depleting aquifers and making "profit" from local groundwater at the expense of local communities.
- Concerns about alterations to groundwater quality resulting from water recharged in Aquifer Storage Recovery projects.
- Litigation by water suppliers and end-users to assign responsibility and take action over legacy contaminants from past industrial activities.

12.3 EVOLUTION OF GROUNDWATER GOVERNANCE IN THE US

In any one place and in any one snapshot in time, management decisions such as groundwater pumping, source protection and water allocations are the result of past complex interactions among landowners, citizens and the legal authority vested in federal, state and local units of government.

The US Federal system distributes authority geographically among the fifty states. Each state claims jurisdiction over groundwater because they administer water law and water rights. The independence of state authority is complicated by laws related to federally funded reclamation projects, the interstate complications of transboundary aquifers and rivers, federal oversight of public lands, military bases and the commitment to treaty rights of Indian tribes. There is also federal legislation such as the Clean Water Act (1972), the Endangered Species Act (1973) and the Comprehensive Environmental Response, Compensation and Liability Act (1980) that may reduce the independence of state authority.

The historical record of the development of America's water resources in the 19th century and in the first half of the 20th century, particularly in states west of the Mississippi, is one of federal government investment in structures to move water from where it occurs to where it is needed. The dams, diversions and canal structures were virtually all based on surface water. The implementation of large scale groundwater pumping for agriculture really began in the 1950s based on the completion of electrification of rural areas and the development of high capacity turbine pumps.

Citizen influence on groundwater governance has for most places in the US evolved over the centuries since early settlement. These influences include factors such as land ownership, water rights, perceptions of the economic value of groundwater, changing economic demands, environmental concerns, innovations in water technology, litigation and case-law and long and short-term changes in weather patterns. In recent years where groundwater scarcity has created a challenge for decision-making, the US has continued to vacillate over the conundrum of the sanctity of "water rights" and the need to make water allocation and management decisions in the public interest. An additional challenge, if economic value and costs are used as an element of governance, is that there are complex distortions in price because of the legacy of federal and state interference through subsidies, direct financing, interest forgiveness or by direct government construction of water storage and distribution systems.

Towards the end of the 20th century groundwater use was clearly established as a vital ingredient of the US economy with approximately half the nation's drinking water and a third of irrigated agriculture from wells (Gollehon, 2000). The increasing importance of groundwater was recognized by the first Environmental Protection Agency Director, William Reilly, "*Ground Water resources are of vital importance to this country – to the health of our citizens, the integrity of many of our ecosystems, and the vigor of the economy*." (US EPA, 1991).

Objectives of groundwater governance strategies are principally to achieve sustainability while protecting a diverse range of vested interests. When there is plenty of water to go around then controls are not an issue. When demands exceed supply (in reality or perception) then balancing economic, environmental and social issues within institutional political frameworks raises the issue of governance. Who has the authority at federal, state or local level to make policy decisions and develop regulations? What local units of government at county, parish, city or township level have authority that can influence resource management? Who has reliable groundwater data sets? What role is played by NGOs with environmental and social interests and groups with vested financial interests?

Public awareness can have an impact on the political process. While groundwater may not be the principal focus of citizen attention, there are contemporary issues where citizen pressure is directly or indirectly creating attention and political will for stronger and more consistent control of aquifers and groundwater pumping. There is a need to maintain citizen pressure on policy-makers. As Sonnenfeld remarked in the context of environmental governance "*Persistent efforts by interested parties are required to retain salience, maintain momentum, and extend effectiveness.*" (Sonnenfeld, 2002).

12.4 INFORMING THE PUBLIC

Water allocation and management policies work best with the support and cooperation of individuals and communities. A traditional top-down approach is that authorities know what is best and only need to inform the public what they are doing. A bottom-up approach requires that the public knows what it wants its elected representatives and state officials to achieve.

The key to achieving science based groundwater management policy is to make the sub-surface hydrologic system understandable to five principal groups:

- Policy-makers (elected representatives)
- Current and potential groundwater-user stakeholders
- The public in general and the myriad of citizen organizations and interest groups that have opinions and perspectives
- Journalists, publishers and broadcasters involved with radio, TV and print media
- Organizations and individuals active with social media via blogs and other postings.

Awareness, information and education are important elements that can assist all groups (politicians, groundwater end-users the media and citizens) better understand the important, allocation, protection, and sustainability issues that are involved with groundwater governance. Where the current system of water management and governance is not working, collaborative governance is critical for better decision-making.

Sustainability has moved from being a scientific exercise to becoming a political reality that recognizes the need to reduce stress between human population and natural resources. An informed public is better able to understand issues and recognize potential conflicts and can therefore voice support or opposition to decisions affecting the local use of resources. An informed public will have a basic level of hydrologic literacy and understand basic groundwater terminology such as aquifer, drawdown and recharge.

The first learning objective about groundwater should be the recognition that groundwater is that part of the hydrologic system that occurs in a geologic environment. Citizens informed about hydrology basics are empowered to demand and ensure that groundwater users, legislators, regulators and the news/communication media do not believe, receive, or dispense incorrect concepts or information. Education of all the constituent groups is essential to reduce the effects of incorrect information, "spin" from lobbyists and misinformed interest groups seeking to influence policy. To be effective, groundwater governance needs objective information and considered opinions from verifiable data sets, testable hypotheses and predictive models of water scientists. However, groundwater policy is not developed in academic isolation and there are many peripheral influences that result in groundwater management policy being a hybrid.

12.5 LOCAL AUTHORITY IMPACTING GROUNDWATER DEVELOPMENT, USE AND PROTECTION

In the US there is considerable local authority and control delegated at local town and county level regarding development and land use. There is a plethora of rules and ordinances that are in place to protect water resources, the application of which can have a major impact on whether or not groundwater resources may be developed. Groundwater regulation issues involving who can pump, where wells are sited, how much pumping, and resource protection needs etc. are much wider than the simple concept of managing aquifers for sustainability. Much of the regulatory oversight that is related to local planning, environmental concerns, building codes and land-use is administered by a complex web of state, county and local agencies, boards and government entities. These entities may not have aquifer management as a principal focus but never the less they may have regulatory oversight that on the local level affects groundwater management decisions. For example:

12.5.1 Boards that develop and oversee well construction standards

The standards for well depth, diameter, casing materials and grouting requirements for different groundwater uses are typically regulated as a means of protecting aquifers. Design specifications are intended to prevent aquifer interconnection and to ensure that there is no possible conduit between surface water and groundwater. Siting requirements for new wells and protocols for well abandonment have cost implications that affect the economics of potential groundwater use and source protection.

12.5.2 Training and licensing standards for well contractors

Mandating that water wells can only be drilled by licensed contractors provides a means of maintaining professional standards. Continuing education requirements can enhance the contractors' knowledge of geology and local aquifers and the best-management practices that are effective in preserving groundwater integrity. Professional integrity and ethical standards are essential for well construction because of the costs and logistical challenges that would be required for regulatory supervision during the drilling process at every installation.

12.5.3 Zoning boards planning boards

Virtually all US towns have boards with the responsibility of controlling development. These boards provide an important element of groundwater governance by protecting aquifers and known groundwater recharge areas from industrial development and preventing land use changes that could impact hydrology. Local zoning boards generally have subdivision and site-plan review regulations. Determining housing density and setting the minimum area for property size can be important to ensure the sustainability of on-site wells. For municipal drinking water supply using groundwater there are requirements for protection zones and set-back distances from wells.

12.5.4 Building codes for onsite wastewater treatment and disposal systems

In the US, decentralized waste water systems (often called septic systems) collect, treat, and release about fifteen million cubic meters of effluent per day from an estimated 26 million facilities nationwide, (NEIWPCC, 2017). Although not designed specifically to do so, all of them, by design, are efficient aquifer recharge devices! Having siting criteria and codes for the correct design and construction of septic systems is important for reducing the risk of aquifer contamination. Virtually all in-house water use for homes with on-site disposal is recharged to groundwater.

12.5.5 Building code design criteria for onsite water wells

There are an estimated 40 million people in the US with on-site private wells for drinking water supply (Stone, 2013). The well is part of the equity value of the home and developers and home builders are often (although not always) required to prove a minimum well yield before an occupancy permit is issued. Typical minimum required yields are in the order of 10 to 20 liters per minute. In many instances where wells are obtaining water from bedrock fractures, actual yields may be accepted at a lower rate (two liters a minute gives close to three cubic meters a day). Of significance for overall groundwater protection is the permitted minimum property size allowed for each home. In places with low yielding aquifers, housing density of more than one home per hectare could result in over exploitation of groundwater.

12.5.6 Health requirement for water quality

Most community health authorities have jurisdiction over drinking water quality. Health authorities have the power to shut down municipal wells that supply public drinking water and issue advisories for private well owners. As for example in recent instances of aquifer contamination from perfluorochemicals, outlined in section 12.6.6 later in this chapter. In many jurisdictions, health authorities require full water quality testing for wells at the time of property transfer where there is an on-site source of water supply. To help achieve awareness of the importance of a safe and properly functioning water system the NGOs, Water Systems Council and American Ground Water Trust have developed education training programs about groundwater, wells and water treatment equipment specifically for real estate professionals, public health officials and home inspectors.

Some examples of other elements of local government and regulatory authority that impact decisions governing groundwater development and aquifer protection include conservation commissions that advocate for preservation of open-space, transport engineers that design detention ponds for storm water disposal and requirements from water utilities in northern states about restricting winter road-salt application in groundwater recharge areas.

As the public becomes more aware of groundwater and as groups such as health officials, real estate professionals and home inspectors are more involved with groundwater as a supply source, then the base of groundwater governance responsibility becomes wider. A wider base can provide more governance stability while retaining flexibility for changing local conditions.

12.6 AWARENESS-RAISING AND CITIZEN PRESSURE ON GROUNDWATER MANAGEMENT AUTHORITY

In this section, six examples are provided to show the ways in which citizen and NGO concerns are involved with the resolution of groundwater issues.

12.6.1 Citizen concern about drought impacts in California

California is in the process of trying to establish comprehensive groundwater governance that will apply to all 515 of its alluvial basins. The state has over one thousand regional and local water agencies operating complex storage and delivery systems. For decades the state did not have the political will to establish groundwater pumping controls despite evidence of continued aquifer depletions. However, from 2010 onward, a sustained drought stimulated political awareness from diverse pressure points. There has been ongoing involvement from environmental activists, anti-growth forces, and the farm lobby (with essential irrigation needs for tree and vine crops). There has been constant media coverage of aquifer declines, frenetic drilling, threats of farm bankruptcies water restrictions in cities, drying residential wells, damage from land subsidence caused by aquifer dewatering and concerns for potential declines in food production from the state's multi-billion dollar agribusiness.

In addition to greatly decreased stream flows, surface water transfers to Central Valley farms from northern California via the state aqueduct system were reduced in order to maintain environmental flows to Chinook salmon and delta smelt, two fish species listed as "Endangered" in 1994. Because of the effects of drought and the associated reduction in spring and summer meltwater from mountain snowpack that traditionally provided surface water, between 2010 and 2014 irrigators increasingly supplemented their needs by pumping groundwater.

The California Sustainable Groundwater Management Act (SGMA), signed in 2014 has the intent of developing a state-wide uniform standard of sustainable management that will be applied at the local level. Unlike the Central Valley, in Southern California many of the groundwater basin and sub-basins have had some controls on pumping resulting from court adjudications, most of which occurred between 1960 and 1990. The limits on groundwater pumping stabilized groundwater levels. However over the same time period in the Central Valley, groundwater levels continued to decline as water was "mined" from the aquifers. The graphs in Figure 12.1 illustrate the differences in cumulative aquifer drawdown.

The basis for future California groundwater governance as proposed by SGMA is via a regulatory structure that gives local control of management, allocation and restrictions to achieve sustainability in identified groundwater basins. Stakeholders (a word which can have a wide inclusive interpretation) must get together and form a local groundwater agency with the responsibility to form a Sustainable Groundwater Management Plan for their basin by 2020. SGMA has statutory requirements for stakeholder engagement to encourage active involvement of social and cultural elements of the population. Most water agencies and local units of government have produced informational fact sheets and have held thousands of citizen meetings discussing the need for SGMA's groundwater governance initiative. One of the major challenges is

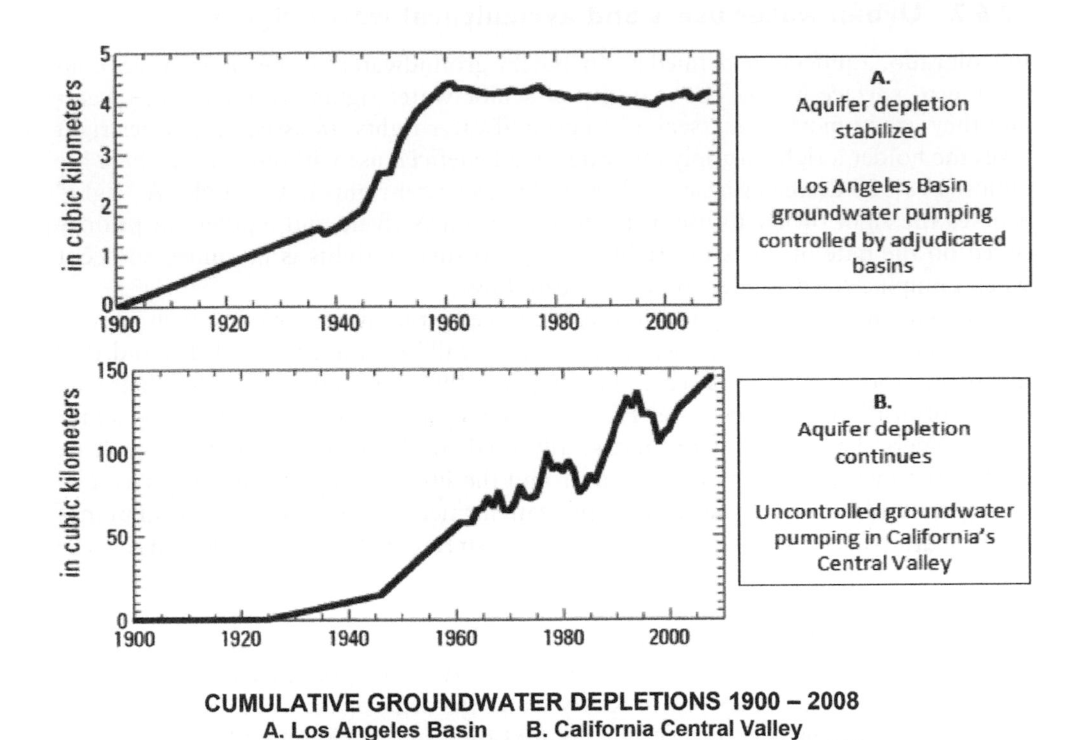

Figure 12.1 Comparison of cumulative aquifer drawdown.

that the management strategies have to be harmonized with common law and this has the potential for conflict with water rights.

A 2014 initiative from the California Water Foundation, developed through stakeholder dialogue, provided recommendations for sustainable groundwater management to allow California's diverse groundwater users and managers to balance supply and demand, protect private property rights, and meet the future needs of farms, cities, and the environment, (California Water Foundation, 2014). The Foundation also developed an online Information Bank available to the public to promote transparency and understanding about groundwater management in California.

With the opportunity to develop local control over California groundwater resources there are several NGOs that are actively involved in awareness raising and assistance by providing hydrologic and economic expertise, providing insight about data handling and management software, promoting open source code for groundwater models, helping coordinate stakeholders, and facilitating the interface of groundwater end-users with legal experts to ensure compliance with SGMA requirements while protecting established rights. Some of the NGOs actively involved with assisting California's new groundwater governance process are: Consensus Building Institute, Groundwater Resources Association, American Ground Water Trust, Clean Water Action, Public Policy Institute of California, and the Pacific Institute.

12.6.2 Urban water users and agricultural water rights

In Colorado, aquifers are defined as "tributary groundwater" if they have direct connection to surface water. The majority of senior water rights are for surface water and they are in most cases "senior" to groundwater rights. In essence, a water right gives the holder a right to apply the water to a beneficial use without waste. There are almost 180,000 decreed surface and groundwater water rights in Colorado. A "right" enables the right owner to use water from the state's rivers and aquifers in priority based on the date of the water right. "Injury" to surface rights is presumed to occur when pumping from aquifers reduces stream flow.

As part of the complex decision-making over groundwater management, downstream senior water right holders can make a "call" on the river and demand that upstream groundwater users cease pumping in order to be compensated for their past out of priority groundwater use. However, the augmentation water that the senior right holders then receive is not necessarily used by them for irrigation but may be sold to the metropolitan areas of Denver and the Front Range. Selling water can be more profitable than growing crops. Upstream farmers may be prevented from pumping (for drainage or for irrigation) while down-stream water right holders derive the benefit. This occurrence is "legal" under the existing water law but the augmentation requirements may be based on groundwater models which overestimate the required augmentation volumes (Gates *et al.*, 2012).

Colorado is a state with complex water rights, some going back to the 1860s. Groundwater governance, while officially under the aegis of the Colorado State Engineer, is in fact strongly influenced by decades of water litigation and court decrees. Colorado has one eighth of the population of California but has more law firms (16 compared with 13) specializing in water law (U.S. News, 2017). Denver is said to have the world's greatest concentration of water rights lawyers!

Meanwhile, rising groundwater levels where pumping is prohibited are causing flooded fields and crop failure (D'Elgin, 2016). The calculations made for restricting pumping and the time period required for augmentation are based on the volumes of past out of priority pumping by the junior right holder. Sticking rigidly to the replacement volume and timing calculations does not make hydrologic sense because the aquifers are demonstrably over-full but the augmentation law requirements in essence say "keep filling them up."

This abuse of common sense about this aspect of groundwater governance is having social and economic consequences in affected areas (Fryar, 2012). Organizations such as the Lower Arkansas Valley Water Conservancy District, the Family Farm Alliance, and Weld County Farm Bureau, are working to overcome these results from a governance system that honours water rights while ignoring the basics of groundwater science. NGOs such as the American Ground Water Trust and the Colorado Water Education Foundation, have provided regular workshops and conferences that help frame the issues and highlight the challenges of effectively managing groundwater against the rigidity of the state's legal code.

12.6.3 Groundwater and oil and gas development

There are four aspects of fossil fuel extraction via drilled bores that have potential groundwater impact. Firstly, finding water to use in the process (many fossil fuel deposits are in arid areas where groundwater is the only possible source). Secondly, there are potential contamination risks during drilling, well stimulation, and the operation processes which extract oil or gas. Thirdly, there are challenges for treatment and disposal of contaminated water (often accomplished by deep injection) and fourthly there are possible risks from accidents during transport of fuels away from the drill site by pipeline, road or rail. Hydraulic fracturing to stimulate the flow of oil and gas wells was first used in the United States in 1947. The process is now carried out on the horizontal portion of directionally drilled wells and typically requires thousands of cubic meters of water and results in the return of contaminated process water.

Over the last decade, public protests about oil & gas development using directional drilling and hydraulic fracturing have often focused on the risks (or perceived risks) to groundwater quality, although there may also be other concerns with climate change or globalization of energy companies. Whatever the protest reasons, the effect of consistent activism has forced authorities in affected areas to be more vigilant about groundwater protection rules and had resulted in increased regulatory oversight and reporting on drilling, well construction and the various chemicals that may be used. Examples of some of the local and national organizations that have been effective in exposing contamination, increasing awareness of possible risks to groundwater, making legal challenges and holding the industry and regulatory authorities accountable include: The Groundwater Protection Council, Living Rivers, Center for Biological Diversity, Marcellus Shale Coalition, Environmental Defense Center, and extensive independent reporting by ProPublica and recognized newspapers such as the New York Times and Wall Street Journal.

In practice, the deep zones where the gas or oil bearing rocks are actually fractured do not represent risks to groundwater because thick overlying rock units are a barrier to vertical propagation of fractures. Ensuring the integrity of the well casings and seals used in the upper portions of vertical bores has been shown to be important in reducing risks of methane migration into aquifers. However, the risks to freshwater aquifers and virtually all of the reported contamination issues result from problems on the surface from accidents and spills in the storage, handling, and treatment of fuels, chemicals and process water, or in the transport of the extracted product by road, rail or pipeline. The vast majority of gas wells do not have any reportable environmental violations, (Soeder, 2017). Additional disruptions in affected areas are disturbances associated with a noisy industrial process, road building, and truck traffic etc.

Benefits to groundwater science from well-orchestrated citizen pressure have been a plethora of research reports providing detailed knowledge of the potential aquifers for source water and detailed studies of the boundaries and properties of drinking water aquifers that require particular regulatory oversight vigilance. The Groundwater Protection Council has developed a publicly accessible database for reporting of hydraulic fracturing in gas development. The data base includes information on water quality, the volumes of water used and the fracking additives. This independent data source can inform groundwater management decisions and provide a basis for science-based regulatory guidelines and will hopefully help the fossil fuel industry water utility managers and environmental organizations to cooperatively coexist.

Citizen activism has put fossil fuel companies in the spotlight. Citizen watchdog groups have forced a high degree of operational transparency and environmental responsibility. It seems likely that it will be many decades before the use of fossil fuels can be phased out. Going forward, the citizen pressure has strengthened the hand of regulators responsible for groundwater protection and has resulted in increased research and technology investment in using saline water for fracking, developing non-water based hydraulic fracture options, and in the development of equipment for comprehensive on-site treatment of return flow (produce) water.

Improving groundwater governance by promoting objective science as the basis for safe oil & gas operations and development of groundwater protection policy has been assisted by the work of professional associations such as the American Institute of Professional Geologists, the National Ground Water Association, and the American Ground Water Trust. These organizations, and others, have helped frame issues at the water and energy interface and have facilitated meetings and conferences among land owners, environmental groups, regulators, the oil & gas industry and groundwater experts.

12.6.4 Concerns that bottled water companies are depleting aquifers

Public opposition to bottled water enterprises can result in demands for tighter controls over pumping permits. Such pressures can strengthen the hand of local regulatory authorities. The US ranks 6th in the world in per-capita annual consumption of bottled water (138 liters). Mexico is first with an annual average consumption of 247 liters, (Rodwan, 2016). According to the Beverage Marketing Corp., quoted in the Wall Street Journal, March 9, 2017, the US now consumes more bottled water per capita (39.3 gallons) than carbonated soft drinks (38.5 gallons). MANTA, a business directory firm, lists 628 companies the US with mineral and spring water bottling as their major enterprise.

The rise of bottled water consumption has produced great environmental controversy, much of which is related to concerns about disposal of containers and alleged misleading marketing. Opposition to the industry (for whatever reason) has given attention to the water sources used in bottling operations. In the US, 70% of all bottled water is from groundwater sources which may be a flowing spring, a well tapping the aquifer supplying the spring or a borehole that may have natural artesian flow or be pumped.

Local concerns, expressed in protest meetings, yard-signs and litigation, typically claim that bottling plants are drying up aquifers and disrupting aquatic ecology. In some instances, wider issues such as protecting the sanctity of water against any form of privatization are a driving force for opposition. The comments (below) from citizens in Michigan, (Cited in Business Insider, November 2016) show the passion that can be aroused by a bottled water facility, "The rape of our Michigan inland fresh water sources is a cause for concern, especially when it is done by a private company for profit." "Trying to privatize water is NOT acceptable. You're an evil corporation and just want you to know there isn't and will never again be a product of yours in our house."

Regulating agencies may run the risk of litigation unless they strictly follow laws and regulations. The focus of many of the opposition protests is to influence local units of government that have planning, zoning and permit oversight over development or water withdrawals and/or jurisdictional authority over environmental protection. A widely publicized case, Michigan Citizens for Water Conservation v. Nestlé Waters

North America Inc., (Michigan, 2007), prompted the state legislature to make reforms in groundwater law. An example of how citizen pressure and legal action can bring legislative changes that impact groundwater governance.

Companies involved with or investing in a bottled water enterprise would presumably have source protection and sustainability as an important business criteria. The larger bottled water enterprises employ groundwater experts who present studies to demonstrate that their pumping will not deplete the resource or negatively impact springs or surface water. The fact that many of the plants have remained in operation for many years, apparently without any serious negative aquifer impacts, shows the importance of thorough initial hydrogeological investigations. It seems that for some people, it is the use to which the water is put that is more cause for concern that the use of the water. Although in the Michigan case the court did not treat the water bottler any differently than other commercial water users.

12.6.5 Concerns over impacts of aquifer storage recovery projects

Aquifer storage and recovery (ASR) systems are a sub-set of the many aquifer recharge technologies that are used to enhance groundwater storage. Engineered river diversions spreading basins and surface detention ponds are also used to increase the infiltration of surface water to groundwater storage. Water for recharge can be storm water, treated waste water or any other source where there is a surplus. Storing water underground has many economic and environmental benefits over the construction of surface impoundments. ASR is a particular technology where water (when available at time of surplus) is introduced via a well into a target storage zone in a receiving aquifer and then pumped for use via the same well when there is a demand. Books by Peter Dillon (Dillon, 2002) and David Pyne (Pyne, 2005) are frequently used references on aquifer recharge.

The first ASR well in the US began operation at Wildwood, New Jersey in 1969 and there are now many states in the US with operating systems. The US EPA reports project sites with a total of 307 ASR wells, (US EPA, 2016a). The technology has had to overcome considerable public resistance, regulatory roadblocks and skepticism from some traditional water engineers. Much of the public resistance has centered on potential aquifer contamination from microbiota such as bacteria virus and protozoa; possible impacts of disinfection by-products if chlorinated water is used for recharge; potential leaching of metals such as arsenic mercury and uranium, and ownership and liability issues if the recharged water moves off site from the point of recharge. Politicians, sensitive to public concerns have been slow to accept the technology and management protocols. Regulatory guidelines for ASR have evolved project by project and state by state, often involving the need to resolve problems of overlapping jurisdictional authority among federal and state agencies.

Non-profits and professional organizations have helped alleviate environmental and health based concerns, interfaced scientists, end-users, legislators and regulators and facilitated information exchange about the progress of ASR projects among the US states. While several US professional groundwater organizations have been involved, in promoting aquifer recharge in the US there have also been international exchanges of technology via programs of the International Symposium on Managed Aquifer

Recharge (ISMAR) and the International Association of Hydrogeologists. The NGO, American Ground Water Trust (AGWT) has been particularly active in promoting ASR solutions in the US.

Since 2001 the AGWT has regularly convened conferences in Florida, California, Texas and Colorado that have focused on aquifer recharge and ASR issues. A combined total of over 500 technical, scientific, engineering and policy presentations have been made at these AGWT events to audiences comprised of political and regulatory agency decision-makers, water district and utility managers and their scientific, engineering and legal advisors. These programs have become the de facto information exchange venue on aquifer recharge for water professionals, environmental groups and elected representatives. The issues presented have stimulated research, the results of which have incrementally led to a regulatory response at federal and state level that is now less restrictive about issuing project permits. Key research findings prompted by conference discussions that have led to regulatory changes include: natural attenuation with aquifer residence time of microorganisms in recharge water, lowering oxygen levels in recharge water by degasification to inhibit metals mobilization, and cycle testing of recharge and recovery to demonstrate progressive reduction of metal leaching to levels that comply with health standards.

12.6.6 Perfluorochemicals – legacy contamination from industrial activities

Perfluorochemicals (called PFCs) are a group of synthetic chemical compounds, not found naturally in the environment but which are of growing concern as a groundwater contaminant. PFC compounds have been used in the manufacture of products such as stain-resistant carpets and clothing, food packaging, non-stick cookware, cosmetics and cleaning products. In commercial use, PFCs have been used for photo imaging, semiconductor coatings, firefighting foam, plastics and hydraulic fluids. So ubiquitous has been their use over the last fifty years that virtually the whole US population carries in their blood some very small but measurable amount of a PFC compound (CDC, 2016).

Citizen action, pressure from environmental groups and litigation are influencing regulatory and management responses to PFC contamination. Recent reports show impacts to private wells and public water supply wells. In 2015 the Environmental Working Group reported contamination in 94 public water systems in 27 states (EWG, 2015). Although private wells are not directly regulated by state agencies, the health of all citizens is a state and local government responsibility. With increased public awareness and increased testing of groundwater quality over PFCs, governance for private wells has now become an issue for local and state government. The Federal Government has established 70 parts per trillion as a guideline threshold for PFCs (EPA, 2016b). Local jurisdictions may have lower thresholds, for example the Vermont Department of Health's drinking water health advisory level is 20 parts per trillion, but many citizen groups are demanding even lower thresholds.

There are treatment technologies available that will remove PFOA compounds from drinking water, such as granular activated carbon; anion exchange, reverse osmosis and specialized membrane filtration. A policy dilemma is to choose between treating water pumped from affected wells or closing the wells and establishing a pipeline connection to a utility supply. Closing wells, sealing abandoned wells, de-designating aquifers as drinking water sources are decisions that impact groundwater management options.

Agency response in advising citizens with wells affected by PFCs has been swift and effective, and to date has been helped with tax dollars and in some cases with voluntary financial payments from former PFC industrial users. For example in New Hampshire, where the Department of Environmental Services has been proactive (NH DES, 2017), a former manufacturer that used PFC products has agreed to pay for a permanent treatment plant for town wells in Merrimack New Hampshire, (MVDWW, 2017). However, when it comes to long term financial redress for costs, litigants may have a vested interest in promoting the most expensive option. Independent objective information from NGOs about the extent and severity of the contamination may help moderate reaction.

12.7 CONCLUSION

Over forty years ago, in a paper on planning theory, Rittel and Webber outlined two types of problems, "Benign problems" which have a clear and logical definition and "Wicked problems" with multiple and conflicting criteria for defining solutions (Rittel & Webber, 1973). There is little doubt that groundwater governance falls in the "wicked" category because of the complexities of the mix of historical precedent, vested interests, social and economic pressures, water allocation disputes, legal and political opinion and competing jurisdictional authority. Impacting every part of the "wicked" mix is ignorance and misunderstanding of the basics of groundwater science. NGOs and professional organizations can play an important role in framing issues, providing objective information, facilitating information exchange and helping technical, academic and engineering professionals integrate into the political process where "governance" is generated. Many different agencies and local units of government play a complicated role in creating, implementing and policing groundwater regulations which are the basic building blocks of groundwater governance. Citizen pressure and the interventions of associations and NGOs can have a major influence on regulations and policy. Acceptance by the regulated that there is a rational need for regulations is an important prerequisite to making rules workable and education needs to be a key element of regulation in order to achieve cooperation and compliance.

REFERENCES

California Water Foundation (2014) Recommendations for Sustainable Groundwater Management: Developed Through a Stakeholder Dialogue – Groundwater Report-5-2014.

CDC (2016), https://www.cdc.gov/biomonitoring/PFCs_FactSheet.html, Centers for Disease Control Fact Sheet.

D'Elgin, T. (2016) *The Man who thought he owned water.* Fort Collins: University Press of Colorado, ISBN 978-1-60732-495-9.

Dillon, P.J. (Ed.) (2002) *Management of Aquifer Recharge for Sustainability.* Lisse: A.A. Balkema. ISBN 90 5809 527 4.

EWG. (2015) Environmental Working Group. [Online] Available from: http://www.ewg.org/research/teflon-chemical-harmful-smallest-doses/pfoa-found-94-public-water-systems-27-states.

Fryar, J. (2012) Longmont News. [Online] Available from:http://www.timescall.com/news/longmont-local-news/ci_20809436/newstip.

Gates, T. K., Garcia, L. A., Hemphill, R. A., Morway, E. D., & Elhaddad, A. (2012) *Irrigation Practices, Water Consumption, & Return Flows in Colorado's Lower Arkansas River Valley: Field and Model Investigations*. Fort Collins, CO: Colorado Water Institute.

Gollehon, N. l. & Quinby W. (2000) Irrigation in the American West: Area, Water and Economic Activity. *Water Resources Development*, Vol. 16, No. 2, 187–195.

Konikow, L. (2013) Groundwater depletion in the United States (1900–2008): U.S. Geological Survey Scientific Investigations Report 2013–5079. [Online] Available from: http://pubs.usgs.gov/sir/2013/5079.

Michigan (2007) *Michigan Citizens for Water Conservation v. Nestlé Waters North America* Inc. 737 N.W.2d 447 (Mich.2007).

MVDWW. (2017) PFOA and Wells 4 and 5 update, Merrimack Village District Water Works, Public Notice, Merrimack, NH. **12-13-16.** [Online] Available from: http://www.mvdwater.org/.

NH DES (2017) Investigation into Perfluorooctanoic Acid (PFOA) Found in Southern New Hampshire Drinking Water, Department of Environmental Services. [Online] Available from: https://www.des.nh.gov/organization/commissioner/pfoa.htm.

NEIWPCC (2017) Protecting Drinking Water Sources in Your Community: Tools for Municipal Officials Chapter 6, New England Interstate Water Pollution Control Commission, Massachusetts.

Pyne, R. D. G. (2005) Aquifer Storage Recovery, Gainesville: ASR Systems LLC. ISBN 0 9774337 090000.

Rittel, H. & Webber, M. (1973) Dilemmas in a General Theory of Planning, *Policy Sciences*, 4: 155–169.

Rodwan, J. (2016). *Bottled Water Statistics Market Report*, New York: Beverage Marketing Corporation,

Soeder, D. J. (2017) Unconventional: The Development of Natural Gas from the Marcellus Shale *Geological Society of America Special Paper* 527 SPE527, 143 p.

Sonnenfeld, D. A. & Mol, A.P.J. (2002) Globalization and the Transformation of Environmental Governance. *American Behavioral Scientist* 45(9): 1318–1339.

Stone, A. W. (2013) Water resource issues in the United States and the changing focus on groundwater, Chapter 1 in, *Assessing and Managing Groundwater in Different Environments*, (Eds Cobbing, J. et al.) Volume 19, Selected papers on hydrogeology, Boca Raton: CRC Press. ISBN 978-1-138-00100.

US EPA, (1991) Protecting the Nation's Ground-Water: EPA's Strategy for the 1990s, Final Report of The EPA Ground-Water Task Force, (21Z-1020), Office of the Administrator, Washington DC.

US EPA, (2016a) Aquifer Recharge and Aquifer Storage and Recovery, Available from: https://www.epa.gov/uic/aquifer-recharge-and-aquifer-storage-and-recovery, United States Environmental Protection Agency.

US EPA, (2016b) Drinking Water Health Advisories. Available from: https://www.epa.gov/sites/production/files/2016-06/documents/drinkingwaterhealthadvisories_pfoa_pfos_updated_5.31.16.pdf PFOA & PFOS, United States Environmental Protection Agency.

U.S. News (2017) Best Law Firms. Available from: http://bestlawfirms.usnews.com/methodology.aspx, US News.

Chapter 13

Assessing and monitoring groundwater governance

Aziza Akhmouch & Delphine Clavreul
OECD Water Governance Programme, Paris, France

ABSTRACT

Despite years of documenting groundwater management experiences, the practice of monitoring and evaluating groundwater governance is still in its infancy. More work is needed to reach agreement on a common monitoring framework and a set of evaluation criteria, and to evaluate the role of contextual variables in shaping and influencing groundwater governance. This chapter explores what could be the defining features of a monitoring and evaluation framework for groundwater governance. It conceptualises twelve mutually reinforcing and complementary components of such a reference framework that could be tailored to local contexts. The chapter builds on the OECD Principles on Water Governance to provide a prism for looking at the main characteristics of groundwater governance that owe to be monitored and evaluated. In doing so, the chapter rounds up, in a systemic way, several of the key governance topics discussed in-depth in previous chapters.

13.1 PREVAILING TRENDS CALL FOR "ASSESSING WHAT NEEDS TO BE IMPROVED" IN GROUNDWATER GOVERNANCE

Groundwater[1] is by far the largest freshwater resource on Earth (not counting water stored as ice). It represents over 90% of the world's readily available freshwater resource (UNEP, 2008; Boswinkel, 2000). Groundwater is an important source of water supply for drinking, irrigation and industry in many parts of the world and also contributes to sustaining groundwater-dependent ecosystems.

However, pressures on the quantity and quality of the resource have increased significantly. Globally, groundwater withdrawals have risen almost tenfold in the past 50 years (Shah *et al.*, 2007), while the resource is becoming increasingly degraded due to pollution and saline intrusion[2]. The total global withdrawal of groundwater

[1] Groundwater is water present in the earth's crust in a saturated or non-saturated soil, weathered mantel or consolidated rock formation. Groundwater is moving in and out of these relatively "static" geological layers – sometimes making clear cut distinctions between surface and groundwater impossible (GEF *et al.*, 2015).

[2] Sea level rise due to climate change, and unsustainable withdrawals (often related to tourism and irrigation activities) contribute to saline intrusion in coastal and land aquifers.

was estimated at 8% of the mean global groundwater renewal in 2010, but this may reach up to 50% in some countries (GEF *et al.*, 2015). The role of groundwater as a water source is becoming increasingly prominent as modern extraction technologies become commonplace and more accessible surface water resources are gradually over-exploited. In the European Union, the fraction of groundwater supply for domestic water use is approximately 70%. The volume of groundwater used by irrigators is substantially above recharge rates in some regions of Australia, Mexico and the United States, undermining the economic viability of farming (OECD, 2015a). Over-exploited aquifers, especially in semi-arid and arid regions, lead to environmental problems (poor water quality, reduced stream flows, drying up of wetlands), higher pumping costs and the loss of a resource for future generations (Shah *et al.*, 2007).

The reality is that, in many regions of the world, groundwater is being exploited faster than it can be replenished and may become the greatest shortcoming to sustainable agriculture and urban water supplies in several regions in the coming decades (OECD, 2012). However, groundwater governance has generally not kept pace with these increasing pressures. In most parts of the world, groundwater governance[3] is generally poor or absent (GEF *et al.*, 2015a). Government leadership, knowledge of the resource and awareness of long-term risks were found to be lacking in certain regions (OECD, 2017 forthcoming). As intensive development of groundwater is relatively recent, institutional responses tend to lag behind the tasks created by new legislation. Groundwater and surface water systems are closely interlinked in most places on Earth and human actions, such as water abstraction, irrigation and artificial drainage, have intensified these interactions (GEF *et al.*, 2015a). In confined aquifers, a substantial portion of groundwater flow often emerges to join surface water, supporting the base flow of surface water bodies (Margat and van der Gun, 2013), and groundwater withdrawals may be used as a substitute for surface water withdrawals, and vice versa. Thus, groundwater and surface water policies need to be designed and implemented conjunctively, where possible. In addition, in most countries, groundwater management policies are strongly conditioned by historical preferences and usage patterns, locking groundwater resources to uses that are no longer as valuable today as they were years ago (OECD, 2015). However, one cannot improve what is not measured. Whilst not an end per se, monitoring and evaluating groundwater governance can help assess whether existing frameworks effectively support groundwater policy outcomes. A core question that precedes groundwater governance analysis is the justification for policy intervention. Building on public economic theory that supports actions in the presence of market imperfections and market failures (OECD, 2015a), this chapter considers that policy intervention in groundwater resources can help addressing the latter to transform groundwater to a long-term, climate insulated, sustainable reservoir, wherever possible.

[3]We define (ground)water governance as encompassing political, institutional and administrative rules, practices, and processes (formal and informal) through which decisions are taken and implemented, stakeholders can articulate their interests and have their concerns considered, and decision-makers are held accountable (OECD, 2015c).

13.2 WHY MONITOR AND EVALUATE GROUNDWATER GOVERNANCE

A compelling argument for monitoring and evaluating groundwater governance is one of accountability (to ensure the proper use of institutional resources). There are other reasons: evaluation provides the opportunity to determine whether the governance system works or to learn from past experiences for the purposes of making future improvements either in the policy itself or in the way that it is designed and implemented.

M&E can provide a tangible, consensual and objective base that can trigger collective action and improvements of the (ground)water governance cycle (Figure 13.1). Measuring whether or not certain conditions are in place is the first crucial step to identify what can hinder effective groundwater policy design and implementation (*e.g.* roles and responsibilities are unclear or overlapping), what is missing (*e.g.* lack or insufficient coordination with other policy fields such as agriculture or land use) and what can be improved (*e.g.* tools for collective groundwater management). As such, monitoring and evaluation (M&E) produce information that can improve decision making, enhance resource allocation, and increase accountability.

A distinction should be made between monitoring and evaluation. *Monitoring* is typically an ongoing process of collecting and assessing qualitative and quantitative information on the inputs, processes, and outputs of groundwater governance arrangements, and the outcomes they aim to address. It may involve assessment against established targets, benchmarks or relevant comparable phenomena and the

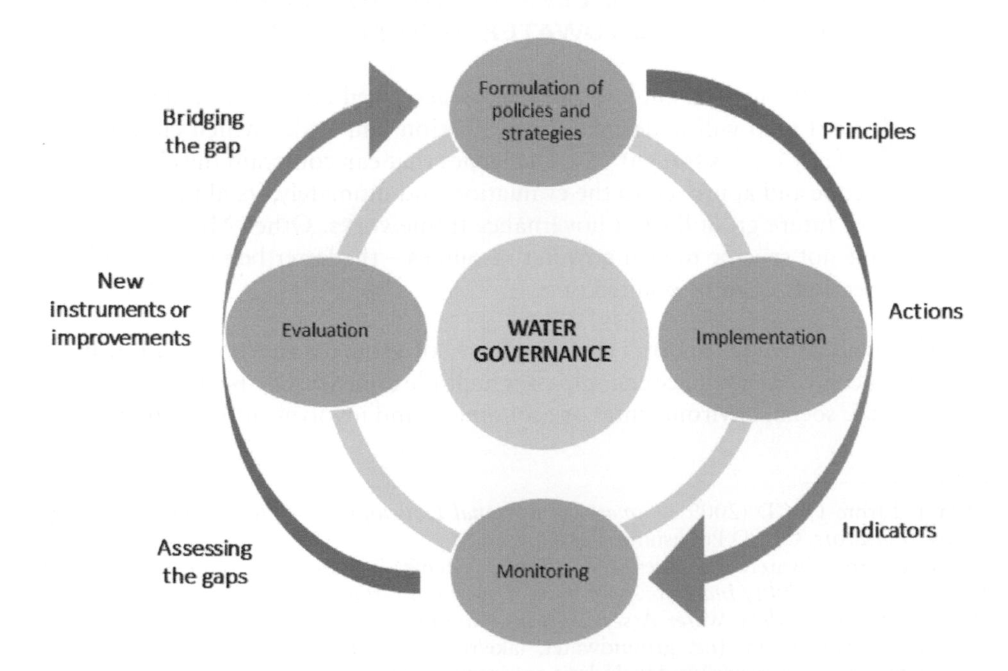

Figure 13.1 The water governance cycle (OECD, 2015c).

integration of incentives for actors to achieve targets. *Evaluation* aims to assess if particular objectives of groundwater governance have been achieved. Evaluation frequently makes a specific attempt to link cause and effect and to attribute changes in outcomes to governance choices. Thus, assessing the impact of groundwater governance approaches on environmental protection outcomes, on reduction of social disparities, and economic growth generally falls under the domain of evaluation[4]. On the one hand, because the purposes of M&E differ, the two activities tend to rely on different methodologies. On the other hand, M&E are often discussed together because they are complementary, and a combination of both activities provides a comprehensive approach to enhancing policy performance.

Groundwater *resources* are progressively the subject of monitoring systems[5] (*e.g.* NASA scientists combine GRACE data with ground-based measurements to map ground water) to keep track of groundwater levels, quality, lithology, and well construction. In addition, a growing number of environmental policies include monitoring requirements for what concerns chemical and/or biological parameters of groundwater resources, as the Groundwater Directive (2006/118/EC) of the European Union and the National Framework for Compliance and Enforcement Systems for Water Resource Management of Australia illustrate. However, apart from few exceptions[6], the same attention has not been extended to groundwater *governance*. Experts who have studied groundwater use around the world tend to agree that too little is known about the institutional and policy frameworks that govern groundwater use (Mukherji and Shah, 2005), and let alone the capacity of these frameworks to perform optimally.

13.3 UNPACKING THE CHALLENGES OF MONITORING AND EVALUATING GROUNDWATER GOVERNANCE

The arguments for undertaking evaluation, as described above, are tightly linked to questions about who will undertake the evaluation and under which circumstances. Evaluation of any kind is fraught with challenges that can constrain the choice of evaluator, the scope and approach to the evaluation and ultimately, its ability to influence the design of future groundwater governance frameworks. Other M&E-related challenges, while not specific to groundwater resources – the latter being discussed in the following section –, can be pointed out:

- The first challenge concerns the *complexity* of groundwater governance. Indeed, the concept of governance encompasses multiple dimensions – be they institutional, political, social, environmental or economic – and involves different management

[4]Adapted from OECD (2009), *Governing Regional Development Policy: The Use of Performance Indicators*, OECD Publishing, Paris

[5]Existing groundwater resources monitoring systems include UNESCO's *Groundwater Resources Sustainability Indicators* and IGRAC's *Global Groundwater Information System.*

[6]UNEP's *Transboundary Water Assessment programme* is divided into five interlinked transboundary water systems (*i.e.* groundwater, lake/reservoir basins, river basins, large marine ecosystems, and open ocean) and includes a governance indicator that looks at the assessment of three levels of governance: legal frameworks, hydro-political tensions and enabling environment.

models (*i.e.* groundwater resources being an individual or shared resources depending on the region)[7] and a multitude of actors at different levels of government, in the public and in the private sector.

- The completeness of the information and the ability of gathering data thus represent a great challenge[8].
- The second challenge is the *uncertainty of the context*: policy makers have limited control on factors that might affect the effectiveness of governance (*e.g.* fiscal crisis, climate change conditions, etc.). The uncertainty of the context might require a certain degree of adaptability, affecting choices and capacity of policy makers and planners to implement proper policies and strategies for efficient groundwater governance at different scales[9].
- The third challenge is about the *scarce availability and comparability of data* on groundwater governance. In turn, the scarce availability of data can hinder monitoring on a regular basis. Last but not least, *causality* between instruments and results can be difficult to establish and should not be underestimated as an M&E exercise might not be able to assess whether or not benefits are the results of certain actions implemented to achieve effective groundwater governance. As for other policies, understanding the causal linkages between policies and results is critical in the water sector. However, an established evaluation system might not be able to assess whether or not benefits are the results of certain actions implemented to achieve "effective groundwater governance". This is specially the case when evaluation is not only used as a *tick boxes exercise*, but as a tool through which evaluating linkages between inputs and outputs.

As a contribution to bridge the M&E "gap" in groundwater governance, this chapter explores how the *OECD Principles on Water Governance* can serve as a baseline for monitoring and evaluating whether needed institutional, regulatory and legal frameworks are in place to allow technical solutions to be efficiently implemented.

13.4 THE OECD PRINCIPLES ON WATER GOVERNANCE

The *OECD Principles on Water Governance* were approved by all (35) member countries, backed at Ministerial level, and endorsed by 130+ major stakeholder groups since 2015. Their development relied on a two-year bottom-up and multi-stakeholder process within the Water Governance Initiative, a network of 100+ delegates from public, private and non-profit sectors gathering twice a year in a Policy Forum.

[7]Groundwater is often considered a common pool resource (Foster *et al.*, 2009; Lopez-Gunn *et al.*, 2012a), *i.e.* defined by the presence of costly exclusion and subtractability of units. Each unit that is extracted by a user is not available for others (Schlager, 2007). However, strict common pool and private property resources are end-members along a spectrum, and most aquifers will fall somewhere along the continuum. As such, attributing a common pool resource status to groundwater is often not applicable as it depends on the nature of the aquifer (Brozovic *et al.*, 2006).

[8]Romano, O., Akhmouch, A. (2016).

[9]Romano, O., Akhmouch, A. (2016).

The Principles aim to enhance water governance systems that help manage "too much", "too little" and "too polluted" water in a sustainable, integrated and inclusive way, at an acceptable cost, and in a reasonable time-frame. They consider that governance is good if it can help to solve key water challenges, using a combination of bottom-up and top-down processes while fostering constructive state-society relations. It is bad if it generates undue transaction costs and does not respond to place-based needs.

The Principles were developed on the premise that there is no one-size-fits-all solution to water challenges worldwide, but a menu of options building on the diversity of legal, administrative and organisational systems within and across countries. The OECD Principles on Water Governance recognise that governance is highly contextual, that water policies need to be tailored to different water resources and places, and that governance responses have to adapt to changing circumstances. The Principles are relevant for all levels of government and for OECD countries, emerging economies and developing countries alike. They acknowledge the diversity of situations within and across countries in terms of legal and institutional frameworks, cultural practices, as well as climatic, geographic and economic conditions at the origin of diverse water challenges and policy responses. Therefore, the Principles can be used by countries to design and implement their national policies in light of country-specific circumstances. The Principles also acknowledge that water governance is a shared responsibility between levels of government, public, private and non-profit stakeholders (Figure 13.2).

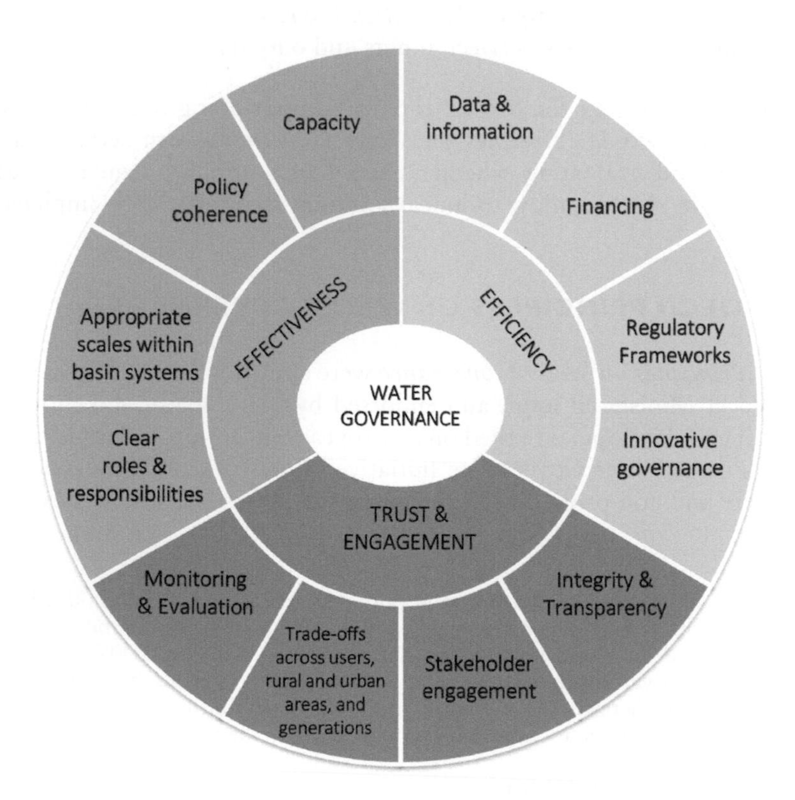

Figure 13.2 Overview of the 12 OECD Principles on Water Governance (OECD, 2015c).

13.5 TOWARDS A HOLISTIC M&E FRAMEWORK FOR GROUNDWATER GOVERNANCE

Groundwater governance faces many of the same challenges as surface water governance, but at times to a greater degree. Inherently, groundwater governance can be considered as more complicated than that for surface water. Groundwater governance is fraught with three specific challenges. First, resources are highly differentiated in terms of aquifer geology (with different degrees of permeability) and water quality, which both affect recharge[10], abstraction and end uses. In addition, some resources are non-renewable. Second, there are many gaps in data on groundwater resources and their connections with surface water, which leads to uncertainty about sustainable water use. Finally, many of the impacts of groundwater overuse only become apparent in the mid to long-term; and when corrective actions are taken, resources take a long time to recover. This section discusses what should make up an M&E framework for groundwater governance, by looking at the main characteristics of groundwater governance through the lens of the twelve OECD Principles on Water Governance. Thus, it suggests twelve mutually reinforcing and complementary components that can provide a framework of reference and be tailored to local contexts.

13.5.1 Principle 1 – clearly allocate and distinguish roles and responsibilities for water policymaking, policy implementation, operational management and regulation, and foster co-ordination across these responsible authorities

Monitoring and evaluating groundwater governance should begin by considering whether roles and responsibilities for groundwater policymaking, policy implementation, operational management and regulation are clearly allocated. Indeed, similar to surface water management, the management of groundwater resources is decentralised in a majority of countries, and requires leadership at both central and local level to set-up and maintain sound groundwater governance frameworks.

Groundwater legislation is usually comprised of rules on ownership, abstraction and use based on entitlements, protection from pollution, and assignment of roles and responsibilities to competent authorities. Laws and enforcement responsibilities related to quality are often distinct from other aspects of groundwater management (GEF *et al.*, 2015a). The main tasks of an authority dealing with groundwater typically comprise allocating and pricing groundwater resources (*e.g.* issuing and administrating groundwater rights, including the maintenance of registers); planning and modelling future demands and impacts on groundwater resources; monitoring groundwater quality and quantity; and implementing and enforcing the groundwater-related law and groundwater rights regimes.

The tendency to decentralise groundwater functions has resulted in a dynamic relationship between multiple public authorities, but also at times in redundancies and grey areas. An additional level of institutional complexity comes from the fact that

[10] In some permeable materials, groundwater may move several metres in a day; in other places, it moves only a few centimetres in a century.

groundwater resources are often managed collectively by local users. Groundwater, unlike surface water, is subject to the individual decisions of hundreds or even perhaps thousands of independent users with direct access to the resource. Because of this, user communities are often advocated as the most plausible solution to ensure adequate groundwater resources management. As a result, policymaking and policy implementation for groundwater resources often require the combination of top-down control and bottom-up approaches. These dual approaches are referred to as a "tripod" in OECD (2015a) whereby groundwater policies should combine regulatory, economic and collective management instruments. An M&E framework should strive to diagnose who is responsible for groundwater governance and at which level, whether groundwater governance is impeded by institutional fragmentation, and if so what are some of the coordination mechanisms in place to help solve gaps and overlaps regarding roles and responsibilities.

13.5.2 Principle 2 – manage water at the appropriate scale(s) within integrated basin governance systems to reflect local conditions, and foster co-ordination between the different scales

Assessing whether groundwater is managed at the appropriate scale(s) should be the second pillar of an M&E framework. Challenges for both surface water and groundwater governance arise from the mismatch between administrative boundaries and the relevant scale for governance, typically river basins. In the case of groundwater, particular attention should also be paid to aquifer boundaries, which do not always correspond to river basins. While the river basin is the fundamental spatial unit for application of integrated water resources management, this has to be reconciled with the fact that coherent groundwater units defined by hydrogeological criteria are the appropriate spatial framework within which to manage and protect groundwater (Foster, 2013).

Groundwater is a widely-distributed but essentially local resource. Thus, to assess whether effective governance arrangements are in place, one has to get down to subnational level – where most groundwater bodies[11] exist – and which can be related as necessary to the overall basin in which they occur. Furthermore, it is also important to consider that while some aquifers are very large, water using communities understand groundwater as a local resource. Individual groundwater users organise themselves to manage issues with an immediate impact on their production such as new wells, well depth and seasonal timing. They are less likely to take action to manage remote and long-term impacts on streamflow, water tables and wetlands (Schlager, 2007). Therefore, an M&E framework should investigate whether groundwater policies are fitted to places and take account of multi-level dynamics across the national, federal/state, basin and local levels, and whether there are co-ordination challenges between the different scales.

[11]Groundwater bodies are defined as resource management units with clearly-defined and scientifically-sound boundaries (usually parts of aquifer systems) (Foster *et al.*, 2010).

13.5.3 Principle 3 – encourage policy coherence through effective cross-sectoral co-ordination, especially between policies for water and the environment, health, energy, agriculture, industry, spatial planning and land use

Groundwater management has both significant externalities on other activities and is affected by other policy fields. As a result, M&E should gauge groundwater governance by looking at its close relationship with various sectoral policies:

- Groundwater use is very closely related to agricultural practices. In fact, the rise of intensive use of groundwater by millions of small-scale farmers has been so striking that it has been dubbed a "silent revolution" (Llamas and Martinez-Santos, 2005). In agricultural areas, external drivers often include factors such as the unreliability or absence of alternative surface water-supplies; highly subsidised or flat-rate electrical energy tariffs for water well pumping; general subsidies on water-well construction, irrigation technology, fertilisers and pesticides, etc.; and guaranteed prices for certain crop types (OECD, 2015a).
- Groundwater resources are highly dependent upon land-use (and changes in land-use) in the main aquifer recharge area, which exert a direct influence on the rates and quality of recharge and pollution. Groundwater depletion can cause land subsidence, which permanently reduces aquifer storage capacity and increases susceptibility to flood damage. In urban environs, land-use classification and control are generally the domain of municipal or local government, and the absence of mechanisms whereby water resource agencies can participate in the process is a frequent governance weakness. Moreover, in many countries, legislation to cope with undesirable land-use practices is often weakly enforced or even non-existent – and progress with implementing controls in the interest of groundwater are highly dependent upon stakeholder awareness and participation. As such groundwater governance cannot be addressed in isolation from consideration of the processes determining or controlling land-use.
- Groundwater resources are affected by agricultural pollution (nitrates and pesticides), mining and industrial pollution, as well as energy production (geothermal energy, carbon dioxide capture and storage, and hydro-fracturing ("fracking")).
- Other common external drivers comprise the process of urbanisation, especially given the potential "coupling" between in-situ sanitation and groundwater, and the frequent inadequacy of utility water supplies (Foster *et al.*, 2010); political boundaries impeding the rational development of protected urban well fields for major water-supply and influencing the discharge points of potentially-polluting effluents; the planning and development of industrial and mining enterprises; and the development of tourism facilities where this is a major source of income.

The assessment of policy coherence within groundwater governance systems should also look into the complementarities between groundwater and surface water policies when these are devised separately. Human action, such as water abstraction, irrigation and artificial drainage, has intensified these interactions. The depletion of even a small portion of the total volume of groundwater (in some cases only a few percent) has a substantial effect on water resources. Where groundwater discharges to streams and lakes, even a small amount of groundwater depletion reduces stream flow and lowers lake water levels, reducing the amount of surface water available for use by humans or riparian and aquatic ecosystems. These external effects can in turn become limiting factors to the further development of groundwater resources and put stress on groundwater-dependent ecosystems such as wetlands. Thus, groundwater and surface water resources need to be managed conjunctively, not in isolation, where relevant.

An M&E framework for groundwater governance should investigate the barriers to integration between groundwater policies, surface water policies and other closely related policy fields (*e.g.* agriculture, energy, territorial development, etc.), and foster the adoption of cross-sectoral coordination mechanisms to mitigate conflicts and encourage coordinated management.

13.5.4 Principle 4 – adapt the level of capacity of responsible authorities to the complexity of water challenges to be met, and to the set of competencies required to carry out their duties

Capacity – in terms of planning, rule-making, project development, finance and budgeting, cross-sector co-operation, stakeholder engagement, implementation and evaluation – conditions effective (ground)water governance. Assessing whether the needed capacities – both technical and non-technical – are secured to shoulder responsibilities for groundwater management should be the fourth pillar of an M&E framework. Public agencies managing groundwater resources tend to be underfinanced and lack the capacity to do an adequate job (Wijnen, 2012). Governments often fail to provide the capacity and budgets needed for implementing the groundwater parts of water management plans. Pressing capacity needs concern particularly the implementation, administration and enforcement of groundwater legislation and regulations. An M&E framework should enquire into such capacity gaps as well as into challenges to capacity development and how capacity needs can be filled (*e.g.* specialised training courses, etc.).

13.5.5 Principle 5 – produce, update, and share timely, consistent, comparable and policy-relevant water and water-related data and information, and use it to guide, assess and improve water policy

Data and information are the main "currency" of groundwater governance. An M&E framework for groundwater governance should therefore determine whether data and information are produced, updated and shared, in a timely and policy-relevant manner. However, data on and observability of groundwater resources, as well as knowledge of the institutional diversity and management practices in groundwater governance are scarce and incomplete, which can be explained by several factors:

- First, there is greater scientific uncertainty about the state (quality and quantity) of groundwater resources as compared to surface water, due in no small part to the fact that it is an "invisible" resource stored underground. Full mapping and assessment of larger, deeper aquifer systems have generally only been carried out in developed countries.

- Second, the intrinsic complexity and heterogeneity of groundwater, and the relative inaccessibility for monitoring results in cases where many of the groundwater system and aquifer characteristics like its dimensions, storage and conductivity properties and rates of recharge and discharge, are poorly known (Varady *et al.*, 2013).
- Third, relevant socio-economic information related to groundwater is especially missing and where the administrative capacity to register and monitor the groundwater development by new users is missing, the accumulated groundwater demand is hard to estimate.

Information is needed not only on aquifer characteristics but on uses and users in order to understand behaviour and trends. Good information should be collected on local contexts and the dominant drivers (and their projected impact) on groundwater resources in the future. Such information needs to be converted into knowledge in order to enable public authorities and stakeholders to take informed management decisions; develop effective rights and allocation regimes; prevent conflicts; and protect and groundwater quality in the long term.

An M&E framework should attempt to identify the information gaps regarding groundwater, looking not only for fundamental technical data on resource status, trends and vulnerabilities, but also for information on the complex network of public agencies, groundwater users and other stakeholders involved. It should also look at other information-related challenges, beyond gaps, such as dispersion (*i.e.* when data is scattered across multiple sources/government agencies and difficult to access) and overload (*i.e.* when policymakers receive too much information of poor quality that does not help to understand an issue or make decisions) and how these affect groundwater governance. Ultimately, an M&E framework should foster improvement in data acquisition, analysis, information sharing, and dissemination of useful knowledge as the key steps to guide groundwater governance.

13.5.6 Principle 6 – ensure that governance arrangements help mobilise water finance and allocate financial resources in an efficient, transparent and timely manner

The next building block of an M&E framework for groundwater governance would look into the financial resources mobilised to support the effective implementation and sustainability of groundwater policies and the ability of responsible authorities to carry out their tasks effectively. It is crucial to keep operation costs as low as possible for the effectiveness of (ground)water governance models, guaranteeing good value for money, and to manage trade-offs between economic efficiency and social objectives. Groundwater resources management implies significant investments from the "hard" end of construction, operation and maintenance of infrastructure (*e.g.* networks, pumps) to the "soft" end of the design, implementation, monitoring and enforcement of groundwater policy. The role of governance in this context is to provide an institutional framework underpinning incentives for cost-efficiency as well as public counter-guarantees that strive to ensure the lowest cost for society in conjunction with affordability and equity issues.

The M&E framework would investigate whether the performance of groundwater management agencies has been impaired by shortages of finance, and identify where investment needs are most pressing in groundwater governance to ensure that fundamental functions of regulation, planning and monitoring are not under-resourced. It would also help identify where better governance can contribute to efficiency gains across the groundwater chain, for example through improved collaboration between authorities and pooling of financial resources and capacities at the relevant scale for optimising resources.

13.5.7 Principle 7 – ensure that sound water management regulatory frameworks are effectively implemented and enforced in pursuit of the public interest

Legal and regulatory frameworks play a crucial role for effective groundwater governance. They provide the basis and starting point for policy development and they turn policy decisions into rights and obligations. However, historically, groundwater legislation has lagged behind legislation for surface water, remaining fragmented, incoherent or simply ignored in many countries. In many jurisdictions, groundwater is still regulated inadequately: the water legislation does not address groundwater in a technically and comprehensive manner, or is outdated, or contains gaps and inconsistencies. Among other reasons, this is because the state of groundwater is unseen, the resource is ubiquitous and aquifer systems respond over time creating less immediate regulatory pressures (Mechlem, 2012).

The problems are also weak regulatory capacity, widespread lack of adherence to the objectives and practices of regulation. Water users in Chile often exceed their assigned amount of water, mostly because of the lack of enforcement and monitoring of water abstractions from groundwater and surface sources (OECD, 2017). In China, according to a survey carried out across the country in 2004, only 10% of Chinese sample drillers hold an extraction permit, despite this being nearly a universal requirement across the country (Wang *et al.*, 2009). The same survey demonstrated that no abstraction charges or quantity limitations were imposed on well owners in any of the sample villages. Only 5% of sample community leaders believed that well drilling decisions required considerations of well spacing requirements[12]. What is more, the issue or unregulated wells is prevalent in several regions, such as Southern Europe (OECD, 2010). In European Mediterranean countries, as many as half of the wells could be unregistered or illegal (EASAC, 2010).

An M&E framework for groundwater governance should promote formal legally binding documents applicable to groundwater that include the principles necessary to achieve sustainable groundwater governance. The M&E framework should diagnose whether such regulatory texts include i) a common terminology that is rooted

[12]More information on groundwater management in China can be found in OECD (2017, forthcoming).

either in the state-of-the art hydrogeology or legal norms, as appropriate; ii) definitions and scope that recognise the duality of groundwater being both part of and apart from the contemporary hydrologic cycle, thus including aquifer of all types whether non-recharging, layered, or linked to surface water[13]; and iii) norms presently under-represented in legally binding texts. In addition, regulatory frameworks should provide for mechanisms to reconcile tensions between groundwater uses and to cope with the effects of climate change (Conti and Gupta, 2015). Groundwater laws should also be coherent with land laws that have important implications for access to groundwater and its protection.

13.5.8 Principle 8 – promote the adoption and implementation of innovative water governance practices across responsible authorities, levels of government and relevant stakeholders

In light of future trends (*i.e.* climate change projections, increase in water demand, etc.), current policy instruments and governance frameworks for groundwater will have to be increasingly adaptive and innovative. Incomplete groundwater policies that are badly enforced, and or relatively rigid in their implementation are likely to prevent the sustainable exploitation of groundwater for agriculture in the future (OECD, 2015a).

One avenue for action concerns surface and groundwater bodies, which remain subject to separate management regimes when an increased emphasis on conjunctive use is needed. Countries should also exploit the potential promises of innovative local collective mechanisms, which combined with national or regional regulatory and economic policies seem to be among the most successful in tackling critical groundwater scarcity challenges (OECD, 2015a). An example of collective groundwater management (*i.e.* information sharing, financing, regulation) is California's Sustainable Groundwater Management Act in the United States, which requires – since the adoption of the law in 2014 – the formation of regional groups of users to set up a monitoring and management system to reach sustainable use of groundwater resources (Gruère, 2015). Another example concerns innovative information sharing strategies, which can trigger cost-effective voluntary conservation measures. Some experiences in the United States for example showcase how better mapping and monitoring groundwater have triggered new conservation measures such as a voluntary 20% reduction of withdrawals by farmers in Kansas, or incentives for landowners and ranchers to jointly manage water resources in Utah (Struzik, 2013).

An M&E framework for groundwater governance should assess whether the enabling environment is in place for innovative approaches (*e.g.* identifying barriers to and mechanisms for innovation). The overall objective should be to transform groundwater governance so it is fit for future challenges.

[13]An example can be found in the Júcar river basin (Spain) where the river basin management plan include registrations of both groundwater and surface water users (see: http://www.chj.es).

13.5.9 Principle 9 – mainstream integrity and transparency practices across water policies, water institutions and water governance frameworks for greater accountability and trust in decision-making

Both politics and power can have a major influence on whether and how (ground)water governance reforms actually happen. Indeed, the status quo tends (by definition) to benefit the vested interest of some "well-established" constituencies. Vested interest can often lead to repeated failure of soundly-based reform of groundwater governance to facilitate sustainable resource use and effective resource protection, and usually can be classified under the heading of either 'rational policy distortion' or 'biased resource management' (Foster *et al.*, 2010).

Biased or corrupt behaviour can also impede community-based groundwater management when organised minority groups favour their friends and family by, for example, not denouncing illegal drilling of new water wells in groundwater conservation areas or the ground disposal of pollutants in groundwater protection zones. This is why an M&E framework for groundwater governance should be a guardian of integrity and transparency. It should diagnose what makes the breeding ground for corruption in groundwater governance, and promote enhanced understanding of groundwater constraints and vulnerabilities amongst all stakeholders, and information transparency through open access to data on groundwater resource and quality status, and water well abstraction licenses. It should also foster accountability of decision makers and stakeholders as regards the way they manage, access and use groundwater resources (*e.g.* procurement processes for irrigation systems) by calling for codes of conduct and charters.

13.5.10 Principle 10 – promote stakeholder engagement for informed and outcome-oriented contributions to water policy design and implementation

There is a diversity of actors in groundwater governance, coming from the public, private and non-profit sectors that ought to be involved in groundwater-related policy making and implementation. However, despite this plethora of stakeholders, there is overall little effective co-operation between government agencies, the private sector and other stakeholders regarding groundwater, in part because their objectives are typically at variance (GEF *et al.*, 2015a).

A factor, among others, which can explain the lack of stakeholder engagement in groundwater governance is an awareness gap regarding the nature and challenges of groundwater (in large part due to the fact that groundwater resources are "invisible" and often perceived as private regardless of what the law says), which fails to motivate stakeholders to align their behaviour with the objectives of good groundwater management. User participation is complicated by the physical invisibility of groundwater systems, which make it harder to agree on the problems and the responses and make monitoring more difficult. In fact, unless people agree there is a problem, stakeholders may not see the point of cooperating. Furthermore, legal and institutional provisions that empower collective management institutions are insufficient. For example, water user associations may be consulted over basin plans, but they rarely have any power over final decisions. The process of enabling and nurturing stakeholder engagement is a critical groundwater governance instrument because:

- Management decisions taken unilaterally by a regulatory agency without social consensus are often impossible to implement;
- Essential management activities (*e.g.* including monitoring, inspection, fee collection, etc.) can be carried out more effectively and economically through cooperative efforts and shared burdens; and
- The integration and coordination of decisions relating to groundwater resources, land use and waste management is facilitated.

An M&E framework for groundwater governance should help identify the wide range of stakeholders concerned with groundwater policymaking and implementation (including an inventory of groundwater users and uses), as well as check whether the appropriate mechanisms are in place for stakeholder engagement.

Engaging stakeholders in processes leading to new legislation or planning or in day-to-day aquifer management, contributes to build acceptance of the inevitable trade-offs of effective groundwater governance, and to align individual behaviour with common goals. France introduced collective management bodies as an attempt to reduce over-exploitation of groundwater for agriculture: *les organismes uniques de gestion collective* (OUGCs, or collective management bodies). The OUGCs are composed of different groups or institutions, including agricultural chambers, groups of local irrigators, owners of land used for irrigation, local legal groups or territorial associations; They are in charge of collecting water withdrawal requests from irrigating farmers in a defined water apportionment zone (*e.g.* a basin), and, based on these requests, propose annual plans for the allocation of the total abstractable volume of water among the irrigators. While in practice, the implementation of the OUGCs has sparked some controversy due to the conflictual relations between those exercising the tasks and those that are meant to benefit from them, as well as decision-making procedures which seem to limit the influence of some stakeholders, it illustrates substantial efforts from a government to strengthen the conservation and allocation of groundwater resources through collective management[14].

Despite the fact that engaging in inclusive approaches is costly and requires long-term commitment from public services and communities, communication with stakeholders is the key to developing (ground)water governance systems with which stakeholders feel invested. The more bottom-up the approach, the stronger the engagement and empowerment of local stakeholders. At the same time, there is a need for an "institutionalised" approach to stakeholder engagement, such as an aquifer management organisation, in which all groundwater users and other main categories of stakeholder are represented.

[14]More information on France's OUGCs can be found in OECD (2017, forthcoming).

13.5.11 Principle 11 – encourage water governance frameworks that help manage trade-offs across water users, rural and urban areas, and generations

When diverging claims of different groundwater users cannot be satisfied simultaneously, when the use of groundwater for agricultural purposes may reduce the amount of groundwater available for domestic use, or when the increase in provisioning services to meet short-term societal needs is detrimental to long-term needs of groundwater ecosystems, groundwater trade-offs become apparent and need to be managed. Often, public governance tries to balance these trade-offs by devising formal and informal institutions and governance modes which influence societal behaviour in such a way that the normative goal of human well-being and sustainability may be achieved (Beckh, 2013).

An M&E framework for groundwater governance should look into the main trade-offs relating to groundwater management and measure their distributional consequences on users and places. It should also assess whether mechanisms are in place to address these trade-offs.

13.5.12 Principle 12 – promote regular monitoring and evaluation of water policy and governance where appropriate, share the results with the public and make adjustments when needed

There has been increasing attention to the need for monitoring and evaluating groundwater resources. The main piece of water legislation usually contains obligations to monitor groundwater use and status (Foster *et al.*, 2010). However, the same cannot be said of monitoring and evaluating groundwater policies and practices. In only a few countries has it been sustained over many years.

It is the role of an M&E framework for groundwater governance to instil a culture of assessment and accountability when it comes to groundwater policy choices and their implementation. It should review whether dedicated institutions are charged with monitoring and evaluation responsibilities, and whether they make the results of their assessments transparent.

13.6 CONCLUSION

The chapter discusses the twelve critical components of an M&E framework for groundwater governance, building on the *OECD Principles on Water Governance*, with the intention to assist decision makers and stakeholders in developing a systemic process for reviewing and following-up groundwater policy choices and policy implementation.

The chapter stresses that monitoring and evaluating groundwater governance is important to render groundwater governance processes more responsive and resilient to future challenges posed by bio-physical, socio-economic and institutional changes. Harmful practices often originate from the failure of policies, laws and decision-making processes to reflect back on how they perform and to determine whether a change is due. Monitoring and evaluation are a *means* to an *end*: they help assess governance gaps as well as new instruments needed to bridge these gaps. As such, monitoring and evaluation primarily seek to encourage a *process* that can trigger improvements of the (ground)water governance cycle (see Figure 13.1 above).

Adaptive management of groundwater resources is on the rise in many countries, in part because of a desire to move from crisis management to risk management. In addition, in a climate of citizens' suspicion and low level of trust in institutions, governments are increasingly interested in accountability. They want to know how well a governance system supports the achievement of established groundwater management goals and objectives. They also want to see how the results generated compare with the effort expended and the resources committed. Therefore, monitoring and evaluation can help ensure accountability, signal that the time is ripe to bring about changes in groundwater governance approaches, and point towards more appropriate ones.

The chapter does not intend to provide a rigid method. Rather, it proposes a framework of reference that can be tailored to specific situations. There are vast differences between regions in terms of groundwater ecosystems, cultures, political and social settings, history, access to information and available resources. The proposed framework can offer insights for all these environments, and even enable sharing of lessons with others using this same framework. The search for a single "best" groundwater governance approach that can be applied to any situation is unlikely to bear fruit but through rigorous evaluation, it is possible to identify *better* methods than others, methods that are better suited to different situations and perhaps even a "best" method for different but definable contexts.

Looking ahead, one can rejoice at the opportunity the Sustainable Development Goals offer to advance the M&E agenda for water governance at large. The water-related goal features governance chiefly in two targets on integrated water resources management (target 6.5) and the participation of local communities (6.a). Ongoing reflections on the SDG monitoring framework provide a window to contribute to the broader agenda and maximise synergies with current efforts in terms of water-related data collection and analyses. Furthermore, the OECD is developing water governance indicators to support the implementation of the *OECD Principles on Water Governance* that can contribute to monitoring and evaluating groundwater governance.

ACKNOWLEDGEMENTS

The authors would like to thank Dr Guillaume Gruère, Senior Natural Resources Policy Analyst at OECD; Dr Andrew Ross, Visiting Research Fellow and Consultant at the Fenner School of Environment and Society of the Australian National University; and Dr Stefano Burchi, Executive Chairman of the International Association for Water Law, for their valuable comments on an earlier draft of the chapter.

REFERENCES

Australian Government (2012), *National Framework for Compliance and Enforcement Systems for Water Resource Management*, Canberra http://www.environment.gov.au/resource/national-framework-compliance-and-enforcement-systems-water-resource-management.

Beckh, C. (2013), *Governance of groundwater ecosystem service trade-offs in Gauteng, South Africa – An institutional analysis,* Institute for Advanced Sustainability Studies Potsdam.

Brosovic, N., Sunding, D.L. and Zilberman, D. (2006), "Optimal management of groundwater over space and time", in R. U. Goetz and D. Berga (eds.), Frontiers in water resource economics, Springer, New York.

Clifton, C. (2010), "Water and Climate Change: Impacts on Groundwater Resources and Adaptation Options", Water Working Notes, Note No. 25, June 2010, World Bank Group.

Conti, K.I. and Gupta, J. (2015), *Global governance principles for the sustainable development of groundwater resources.*

Conti, K.I., Velis, M., Antoniou, A. and Nijsten, G.-J. (2016), *Groundwater in the Context of the Sustainable Development Goals: Fundamental Policy Considerations*, Brief to the GSDR (2016 update).

EASAC (European Academies Science Advisory Council) (2010), "Groundwater in the Southern Member States of the European Union", EASAC Policy Report 12, EASAC, Halle.

European Communities (2007), *Common Implementation Strategy for the Water Framework Directive (2000/60/CC): Guidance Document No 15*, Guidance on Groundwater Monitoring. Technical Report-002-2007, Brussels. https://circabc.europa.eu/sd/a/e409710d-f1c1-4672-9480-e2b9e93f30ad/Groundwater%20Monitoring%20Guidance%20Nov-2006_FINAL-2.pdf.

Foster, S. *et al.* (2010), *Groundwater Governance: conceptual framework for assessment of provisions and needs*, GW-MATE Strategic Overview Series Number 1.

Foster, S. *et al.* (2013), "Groundwater- a global focus on the 'local resource'", *Current Opinion in Environmental Sustainability*, (5) 685–695.

Foster, S. S. D. and Chilton, P. J. (2003), "Groundwater: The processes and global significance of aquifer degradation", *Philosophical Transactions of the Royal Society of London*, Series B-Biological Sciences 358, no. 1440: 1957–1972.

GEF *et al.* (2015a), "Global Diagnostic on Groundwater Governance".

GEF *et al.* (2015b), "Global Framework for Action: to achieve the vision on Groundwater Governance".

Gruère, G. (2015), "A Californian enigma: Record-high agricultural revenues during the most severe drought in history", in OECD *Insights* Series, 9 December 2015, http://oecdinsights.org/2015/12/09/a-californian-enigma-record-high-agricultural-revenues-during-the-most-severe-drought-in-history/.

IGRAC (2015), *Groundwater in the Sustainable Development Goals – emphasizing Groundwater in the Negotiation of the Final Goals*, Position Paper 2, https://www.un-igrac.org/resource/groundwater-sustainable-development-goals-emphasizing-groundwater-negotiation-final-goals.

IGRAC (2015), *Groundwater in the Sustainable Development Goals – Including Groundwater in the Draft Goals*, Position Paper 1, https://www.un-igrac.org/resource/groundwater-sustainable-development-goals-including-groundwater-draft-goals.

Llamas, M. R. and Martínez-Santos, P. (2005), "Intensive Groundwater Use: silent Revolution and Potential Source of Social Conflicts", Journal of Water Resources Planning and Management, September–October 2005, pp. 337–341.

Margat, J. and van der Gun, J. (2013), *Groundwater around the World: A Geographic Synopsis*, CRC Press/Balkema, Taylor and Francis, London.

Mechlem, K. (2012), "Legal and Institutional Frameworks", Thematic Paper 6, Groundwater Governance: A Global Framework for Action, GEF, FAO, UNESCO-IHP, IAH, and the World Bank.

Morris, B.L. *et al.* (2003), *Groundwater and Its Susceptibility to Degradation: A Global Assessment of the Problem of Options for Management*, Early Warning and Assessment Report Series, RS. 03-3, UNEP, Nairobi, Kenya.

OECD (2009), *Governing Regional Development Policy: The Use of Performance Indicators*, OECD Publishing, Paris. DOI: http://dx.doi.org/10.1787/9789264056299-en.
OECD (2010), *Sustainable Management of Water Resources in Agriculture*, OECD, Paris, France, www.oecd-ilibrary.org/environment/oecd-studies-on-water_22245081.
OECD (2012), *OECD Environmental Outlook to 2050: The Consequences of Inaction*, OECD Publishing, Paris. DOI: http://dx.doi.org/10.1787/9789264122246-en.
OECD (2015a), *Drying Wells, Rising Stakes: Towards Sustainable Agricultural Groundwater Use*, OECD, Studies on Water, OECD Publishing, Paris. DOI: http://dx.doi.org/10.1787/9789264238701-en.
OECD (2015b), *OECD Inventory: Water Governance Indicators and Measurement Framework*, http://www.oecd.org/cfe/regional-policy/Inventory_Indicators.pdf (visited on 19 June 2017).
OECD (2015c), *OECD Principles on Water Governance*, available online at http://www.oecd.org/governance/oecd-principles-on-water-governance.htm.
OECD (2017), *Infrastructure Governance Review: Chile – Gaps and Governance Standards of Public Infrastructure*, OECD Publishing, Paris.
OECD (forthcoming), *Groundwater Allocation*, OECD Publishing, Paris.
Romano, O. and Akhmouch, A. (2016), *Scoping Note: OECD Water Governance Indicators.*
Schlager, E. (2007a), *Community Management of Groundwater, M. Giordano and K.Villholth (ed.), The Agricultural Groundwater Revolution: Opportunities and Threats to Development*, Wallingford, CABI Publishing.
Shah, T. *et al.* (2007b), *Groundwater: A Global Assessment of Scale and Significance*, International Water Management Institute (IWMI), 2007.
Struzik, E. (2013), *Underground Intelligence: The need to map, monitor, and manage Canada's groundwater resources in an era of drought and climate change*, Munk School of Global Affairs, University of Toronto, Toronto.
Varady *et al.* (2013), *Groundwater Policy and Governance*, Thematic Paper No. 5.
Wang *et al.* (2014), "Assessment of the Development of Groundwater Market in Rural China", in K.W. Easter and Q. Huang (eds.), *Water Markets for the 21st Century: What Have We Learned?*, Global Issues in Water Policy 11, DOI 10.1007/978-94-017-9081-9__14, Springer Science+Business Media Dordrecht 2014.
Wijnen, M. (2012), *Managing the invisible – Understanding and improving groundwater governance*, Thematic paper n° 11.

Part 3

Integration and policy linkages beyond the local groundwater system

Chapter 14

Groundwater governance for poverty eradication, social equity and health

Sean G. Furey
Rural Water Supply Network (RWSN), Skat Consulting Ltd., St. Gallen, Switzerland

ABSTRACT

Groundwater use and its governance should serve a purpose that is well defined and has a broadly accepted mandate, without it, there is a risk that benefits will accrue to existing elites only for their own benefits. Access to safe, affordable water is a recognised Human Right and a Sustainable Development Goal because it is critical for the health and wellbeing of every person in the world. Groundwater represents 96% of all liquid freshwater in the world and so any discussion about groundwater is also a discussion about human rights, development, health and social equity. Groundwater is used in many different ways, many uncontrolled and unmonitored and this can cause substantial problems – even causing cities to sink below sea level. Recent recommendations on improving groundwater governance may not be adequately aligned with the Human Right to Water or giving sufficient priority to poverty alleviation. However, groundwater use unlocks the potential of human ingenuity, cooperation and enterprise that can build the foundations for health, resilient livelihoods in the face of growing global uncertainties. The three areas identified for further focus are: increase understanding of the links between groundwater use and poverty; improve understanding and management of private 'self supply' groundwater sources; improve the training and professionalisation around groundwater technology innovation and scaling up.

14.1 INTRODUCTION

"Water is our greatest challenge, but water is life." Abdul Hakim is the Secretary to the Jahudanga Union Parishad in the coastal belt of southwest Bangladesh. Looking around, there is lush tropical vegetation; rice paddies knee high in water; there are ponds in-between houses; and handpumps scattered throughout the village, each with red paint flaking off their cast-iron spouts. It seems hard to believe that water scarcity could be a problem here. Mr Hakim explains that ponds and the rainwater tanks don't last through the dry season, the river water and deep groundwater are saline and the shallow groundwater is rich in arsenic – and hence painted red by the Department for Public Health Engineering. Water, water everywhere but not a drop to drink.

Until the 1970s the microbial quality of surface water, on which most people depended, was generally poor and the cause of morbidity and mortality, particularly in children. This prompted UNICEF and Bangladesh's Department for Public Health

Engineering (DPHE) to undertake an ambitious drilling programme across the country to get people using 'safe' groundwater. By the mid-1980's, the private sector had taken over and 'self-supply' tubewells with cheap handpumps became commonplace. Then in 1993, when the presence of arsenic in tubewells was confirmed, it was realised that what had been seen an immensely successful programme for using groundwater to tackle poverty and health was actually "the largest poisoning of a population in history" (Smith, Lingas & Rahman, 2000). Of 4.7 million tubewells tested in Bangladesh, 1.4 million were found to have arsenic levels higher than the government standard of 50 parts per billion (UNICEF, 2008).

This story is well known and huge efforts have been taken by UNICEF, DPHE and many other agencies to address the problem. However, these efforts are increasingly confounded by other trends: declining fresh groundwater levels, particularly in the north; increased salinity and saline intrusion as a result of over-abstraction; sea-level rise; expansion of salt-water shrimp farming; rapid population growth and economic growth that have increased pollution to surface and groundwater, despite (or perhaps even because of) dramatic reductions in open defecation and widespread use of pit latrines. Groundwater governance and good science is needed more than ever.

In this Chapter we take a practitioner's perspective on the practical challenges of ensuring groundwater resources are used wisely and pragmatically so that all citizens in a given country can achieve higher levels of public health and wellbeing.

14.2 GROUNDWATER GOALS & GOVERNANCE

The Sustainable Development Goals have the overarching aim to eliminate all poverty. Many of the goals depend on water of the right quality to meet demands for ecological needs and ecosystem services so that there is safe drinking water, agriculture and economic development so that poverty is eliminated, and that risks from hazards, such as floods, drought, pollution and land subsidence (Box 14.1), are minimised and mitigated. However groundwater attracts curiously little attention in the wording of the goals and targets, despite it being 96% of all liquid freshwater (IAH, 2017).

Box 14.1: A future Atlantis? How groundwater abstraction is threatening to submerge Indonesia's capital

In Indonesia, the capital city, Jakarta, has a population of over 10 million, centred on a deltaic plain by the north coast of the Island of Java. Studies since at least 1993 have identified land subsidence across much of the city, at a rate of 3–10 cm year^{-1} and the recorded impacts have included cracking and damage to buildings and infrastructure, increased flood risk and reduced effectiveness of drainage systems and channels. The four main causes of land subsidence have been identified, in order of importance, as (Abidin *et al.*, 2015):

1 Excessive groundwater abstraction
2 Load of buildings and construction

3 Natural consolidation of alluvium soil
4 Tectonic activity

The municipal piped water system is fed by inland surface water sources, but this only able to meet 30% of demand, so the remaining demand is met by private boreholes (Delinom, 2008). During a fieldtrip in May 2015, as part of Indonesia International Water Week, staff from the Ministry of Public Works explained that because the of difficulties providing a reliable water supply to the whole city from the municipal water system continued to drive groundwater abstraction across the city. Furthermore, the poorer areas of the city, near the sea, depend on shallower wells, which are suffering increasingly from saline intrusion, and these areas are at most risk from flooding.

Figure 14.1 Tengah Pumping Station, Jakarta, Indonesia, refit completed in 2014, lifts water up from the Waduk Pluit lagoon.

Figure 14.2 Waduk Pluit lagoon, into which city storm water drains, up into the sea (Photos: Author, 2015).

Reaching these ambitious goals requires an interconnected approach that brings together technical, legal, institutional, cultural, economic, ecological and human resource dimensions. A widespread approach to this is Integrated Water Resource

Management (IWRM). However, the low visibility of groundwater in IWRM implementation has been criticised:

> *"Although ideally basin organizations should take full responsibility for groundwater management within their IWRM mandate, in practice they largely manage surface resources, so that cooperative mechanisms with groundwater agencies are needed."* (FAO, 2016A)

At a practical level, in the Author's experience of rural water supply in Sub Saharan Africa and Asia, there is often limited interaction between water supply practitioners and water resource planners because they operate at different scales and often fall under the mandates of different government departments. However, coordination mechanisms such as Joint Sector Reviews can help (Danert *et al.*, 2015).

The Groundwater Governance project (2011–2016)[1] concluded that the key groundwater governance weaknesses included:

- Inadequate leadership from government agencies
- Lack of awareness of long-term risks
- Lack of knowledge of the resource and its status
- Non-performing legal systems
- Insufficient stakeholder engagement
- Poor integration with related national policies

The project's extensive global consultation concluded with a framework for groundwater governance (FAO, 2016A) that includes the following principles:

- Understanding the context
- Creating a basis for governance
- Building effective institutions
- Making essential linkages: in particular with IWRM, land use planning, the use of sub-surface resources, energy sector and mainstreaming groundwater within the policies of other sectors and sub-sectors
- Redirecting finance and aligning incentive systems
- Establishing a process of planning and management

These recommendations would fit within an IWRM process, but the challenge remains to ensure that groundwater is given sufficient priority alongside surface water resource management issues.

There is also no guarantee that the implementation of such a framework would make a contribution to addressing poverty alleviation, health or social equity objectives. Therefore important objectives around these are set and routinely monitored.

14.3 DEFINING AND MEASURING SUCCESS

Groundwater folk are, on the whole, a practical bunch who like to get their hands and boots dirty. However, action can easily cause unintended consequences and creates

[1] www.groundwatergovernance.org

winners and losers therefore it is incumbent on the groundwater professional, like a medical professional, to endeavour to do no harm, whether it is drilling a borehole or drafting a national policy. Therefore the science and evidence-based approach needs to be done within a framework of commonly understood terms and norms, which ideally can be measured. These include:

1 The Human Rights to Water and Sanitation
2 Equity, Equality and Equitable
3 Water Rights
4 Safe Water

14.3.1 Human rights to water and sanitation

The Human Right to Water "entitles everyone to sufficient, safe, acceptable, physically accessible and affordable water for personal and domestic use" (Albuquerque, 2014:p.6) The human rights to water and sanitation (the two are separate, but linked) were ratified by the UN General Assembly in 2010 (UN, 2010). In 2016, the UN General Assembly Resolution A/RES/70/169 (UN, 2016B) reaffirmed the importance of the Human Rights to Water and Sanitation within the context of the SDGs. This included the following urgings:

> *9. Reaffirms that States have the primary responsibility to ensure the full realization of all human rights and to endeavour to take steps ... to the maximum of their available resources, with a view to progressively achieving the full realization of the rights to safe drinking water and sanitation by all appropriate means ...;*
>
> *10. Stresses the important role of the international cooperation and technical assistance ... and urges development partners to adopt a human rights-based approach when designing and implementing development programmes in support of national initiatives and plans of action related to the rights to safe drinking water and sanitation;* (UN, 2016B, pp. 5–6).

Making this human right real is organised around five principles:

1 Non-discrimination and equality
2 Access to information and transparency
3 Right to participation
4 Accountability
5 Sustainability and non-retrogression

14.3.2 Equality, equity and equitable

The human right to water, and its principles, has implications for how groundwater is governed and used, and can create tensions with groundwater resource sustainability and exploitation of groundwater for economic growth.

From the outset it is essential to differentiate between "equality" and "equity". In everyday communication, they are often used synonymously, but for our purposes they are distinct. Equality is a legal term that is enshrined in the Universal Declaration Human Rights, article 1 of which states "All humans are born free and equal in dignity and rights". The first UN Special Rapporteur on the Human Right to Water, Catarina de Albuquerque defined the two thus:

> "***Equality*** *entails a legally binding obligation to ensure that everyone enjoys equal enjoyment of her or his rights. Substantive equality requires a focus on all groups in society experiencing direct or indirect discrimination, and the adoption of targeted measures to support these groups when barriers persist, including affirmative action or temporary special measures."*
>
> "***Equity*** *– is the moral imperative to dismantle unjust differences. It is based on principles of fairness and justice. In the context of water, sanitation and hygiene, equity, like equality, requires a focus on the most disadvantaged and the poorest. Many organisations in the sector have made equity a central part of their agenda; however, from a human rights perspective, relying on equity carries certain risks because it is a malleable concept that is not legally binding."* (Albuquerque, 2014c: p. 7)

However, equity is also used in international law: *"[E]quity as a legal concept is a direct emanation of the idea of justice. The Court whose task is by definition to administer justice is bound to apply it"* MacIntyre (2009). The use of the term 'equity' is controversial particularly when related to shared natural resources, according to MacIntyre (2009) who notes that despite its use in settling international disputes, *"there exists no universally accepted meaning of equity in international law."* He relates that the Statute of the International Court of Justice (ICJ) can apply equity in two different ways: the first is 'use principles' and the body of law found in 'civilised nations'; the second is to disregard existing legal rules so that a Court may decide what is meant by equity, but this can only be done with the permission of both parties in the dispute in question.

The UN Special Rapporteur on the Human Rights to Water and to Sanitation plays a key, and daunting, role in monitoring the decisions and actions of governments, and other powerful interests, and calling them out when they fall short of their obligations to the poor and vulnerable.

In the context of groundwater governance 'equity' can be used for two distinct purposes.

The first is the fundamental principle that everyone has right to safe water, regardless of their gender, age, disability, ethnicity or socio-economic status, but bearing in mind that UN Special Rapporteurs, first Catarina de Albuquerque and more recently, Leo Heller, have repeatedly emphasised that this right does not mean that everyone should have free access to water services:

> *"The right to water requires water services to be affordable for all and nobody to be deprived of access because of an inability to pay. As such, the human rights framework does not provide for a right to free water. However, in certain circumstances, access to safe drinking water and sanitation might have to be provided free of charge if the person or household is unable to pay for it. It is a State's core obligation to ensure the satisfaction of, at the very least, minimum essential levels of the right, which includes access to a minimum essential quantity of water."* (UNHCR, 2010)

Secondly, equity is used in terms of equitable rights to the use of groundwater resources. The case studies presented show that making both types of fairness a reality can be challenging, but not impossible.

Finally, there is the word 'equitable' which is used in the wording of SDG Target 6.1. The JMP interpretation of this "[i]mplies progressive reduction and the elimination and inequalities between population sub-groups." (WHO/UNICEF, 2017).

14.3.3 What do we mean by 'safe' water?

A big shift between the Millennium Development Goals (MDGs) and the Sustainable Development Goals (SDGs) has been the elevation of water and water, sanitation & hygiene (WASH) from a target within a Goal to a Goal in its own right – SDG6. In the MDGs, water for human health and domestic use was a target that was defined in terms of access to an 'improved' source, such as a protected well, borehole or spring, as opposed to an 'unimproved' water source such as an open well or a surface water body.

SDG6, target 1 states *"By 2030, achieve universal and equitable access to safe and affordable drinking water for all"*. This emphasis on "safe" water aligns with the Human Right to Water, which requires that *"Water must be safe for human consumption and for personal and domestic hygiene. It must be free from microorganisms, chemical substances and radiological hazards that constitute a threat to a person's health"* (Albuquerque, 2014A: p.35). The UNICEF/WHO Joint Monitoring Programme normative interpretation is that: *"Safe drinking water is free from pathogens and elevated levels of toxic chemicals at all times."* (WHO/UNICEF, 2017) This is an important change because under the MDG's, water from an 'improved' source was not necessarily safe to drink, and quite often further treatment is necessary.

This has important implications for water service providers and regulators because it raises the bar for what is an acceptable minimum level of service. For example, an 'improved' groundwater source may protect a borehole from surface water contamination but will not protect the water user if the groundwater quality itself is unsafe, due to geogenic or anthropogenic contaminants. In Malawi, although 89% of the population is reported as having access to an 'improved water source, this does not reflect the functionality of the water point, good water quality or a positive impact of the quality of life of the water user (Rieger *et al.*, 2016). At a practical level this means that the metrics used for the MDGs do not have much relevance in the new SDG era.

14.3.4 Links with groundwater governance

Returning the discussion in section 14.2 about the current published thinking on groundwater governance, it is notable that the Groundwater Governance Framework (FAO, 2016A) report talks about groundwater use rights, but makes no explicit reference to the Human Right to Water, or gender and lists poverty alleviation as only one of several possible policy objectives. The work pre-dates the UN ratification of the 2030 Agenda and the SDGs, however, at a policy level it would appear that there is still some work to do to marry groundwater governance to Human Rights and the SDG agenda.

14.4 PRACTICAL CHALLENGES OF ACHIEVING THE HUMAN RIGHT TO WATER AND SDG6

So can groundwater development protect human rights and lead to higher wellbeing across a country or region? There are a number of challenges and tensions that need to be overcome, through dialogue, trust-building, capacity development and clear policy.

14.4.1 Human rights versus water rights

Water rights are legal agreements, such as a permit or licence to abstract water, from the government that allows a private individual or entity to take or use water from a specified point, water resource type or land area. From the perspective of the UN Special Rapporteur on the Human Right to Water:

> "*Someone availing themselves of their water rights may be violating another person's human rights to water and sanitation, for example, in cases of overextraction or pollution. ... Priority must always be given to water required for the realisation of the human right to water, and water resources must be protected from over-use or pollution to this end.*" (Albuquerque, 2014A: p. 20)

There is a potential conflict between water right owners and the wider society if water abstraction or use-rights benefit from tighter and better enforced national laws and regulations than human rights to safe and affordable water. A further conflict, in some contexts, is land access rights and groundwater use rights where a land owner (FAO, 2016a).

14.4.2 Potable water versus agriculture

The human right to water focuses on water for personal use in the home, workplace, public spaces and institutions, such as schools, hospitals and prisons. However, in the broader context of groundwater-use in development it is important to consider that in absolute terms the volumes of water involved are generally very small compared to that used – or impacted – by agriculture, industry, commerce, transport, energy, mining and forestry. Furthermore, these facets of the human water cycle take place within the much broader environmental fluxes of water, energy and nutrients on which our societies depend.

Severe stresses around groundwater, health, social equity and health are apparent in low and middle-income countries where infrastructure is inadequate and institutional capacity for monitoring and regulating groundwater use is weak. However, problems can occur in high-income, industrialised economies as well: the most well-publicised in the recent years has been the prolonged drought in California[2] which has led to widespread water restrictions and intense debate regarding what equality means when the needs of domestic water users is put in conflict with an agricultural sector that employs relatively few people but is an important pillar of the state economy.

[2] California translated the Human Right to Water into State law in 2012: http://srwatersanitation.ohchr.org/en/pressrelease_california.html [accessed 10.05.2017]

The tension between agriculture and public water supply manifests in different ways in many parts of the world, however, what is often overlooked is that water is often put to multiple uses, particularly in rural areas. A farmer with a well in rural Zambia doesn't necessarily differentiate between using that well water for her family, or her livestock or irrigating a plot of land. Conversely, many households in rural and urban areas use multiple water sources, for different purposes or at different times of year depending on cost, convenience, quality and availability – and some well owners in urban Kenya sell their water to neighbours (Kumamaru *et al.*, 2011; Okotto *et al.*, 2015). Such complex patterns of behaviour can challenge the conventional ways of governing and balancing the interests of different water users. In terms of the human right to water in relation to other human rights, notably:

> *"Access to water is essential for agriculture in order to realise the right to adequate food. While the recognition of the human rights to water and sanitation has brought attention to the requirement to prioritise access to water for personal and domestic use for marginalised individuals and groups, there is also a requirement to ensure access to sufficient water for marginalised and poor farmers for subsistence and small-scale farming."*
>
> *"Human rights law includes environmental obligations. Finite resources must be protected from overexploitation and pollution, and facilities and services dealing with excreta and wastewater should ensure a clean and healthy living environment."* (Albuquerque, 2014: p.38)

14.4.3 Rights versus economics

Water economics is essential for understanding the behaviour and incentives around groundwater use – from how it is used to why, sometimes, it is not used. As a topic, it deserves its own chapter, but for our purposes it is important to point out some opportunities – and risks.

Used wisely, groundwater can be used to generate value – either directly in agricultural, industrial or energy outputs that can be sold, on indirectly through multiple benefits to all parts of society and the economy. However, export earnings from highly water-dependent products is coming under increasing scrutiny. For example, WWF Switzerland have identified that although the alpine country itself enjoys high water security, it imports large amounts of water-intensive products from water-scarce countries (WWF, 2016). Some of that 'virtual' water footprint is physically transferred (for example the water in tomatoes from Spain, or grapes from Greece) or indirectly through products, such as beef, that result in high rates of evapo-transpirative loss of water from catchments. The Swiss Agency for Development & Cooperation (SDC) is increasingly looking at mechanisms, such as payment for watershed services (SDC, n.d.), that de-couple economic growth from water use. This is possible: the US economy has continued to grow but *per capita* water use peaked in the 1980s and has largely plateaued since – the link has been broken (Donnelly & Cooley, 2015; Gleick, 2014). However, this may hide the wider virtual water footprint of the US economy, such as the overexploitation of groundwater in Mexico (Scott, 2013) for crops that subsequently imported by the US.

Therefore the challenge for development, whether in an emerging economy in Sub-Saharan Africa, a transition economy in South East Asia or a high income economy in Europe, is how to maximise the economic value created through groundwater use, whilst maintaining sufficient buffer to meet the needs of all groundwater users through dry periods and when facing increasing uncertainties from climate change.

A further challenge, in relation to equity and human rights, is that economics, by its nature, focuses on metrics and monetisation. Water (and groundwater) regulation also tends to focus on what can be measured and what generates the most income (in the form of abstraction licences and tariffs) and poses the greatest risk to others (large abstractors and dischargers). For a cash-strapped regulator, it is easier to deal with a few bigger stakeholders, than thousands of small ones. The net result of both biases is that the poor become invisible to the formal financial economy and water-cycle economy. This can be dangerous because it pushes the allocation of water resources towards more powerful interests (even discounting corruption, patronage and political interference). The view of the poor – particularly the rural poor – based on participation in the cash economy is also skewed because a worker on a plantation, with little livelihood security, will be viewed as being richer than a subsistence farmer who is largely self-sufficient by his own means or through a local barter economy.

This where the recommendations on groundwater governance (FAO, 2016a) make the critical point about fully understanding the context before looking to improve the basis for governance or improving institutions, and needs to be linked to the five principles of realising the Human Right to Water so that when citizens exercise their rights they are treated fairly by the governing authorities.

Enterprises (state or private-owned) that generate employment are attractive – both politically and where water economics is a driving factor behind water right allocations. However, without sound employment and welfare laws and support mechanisms, such water uses may not improve quality life, happiness or wellbeing – it may do the opposite. Although, not groundwater, an extreme case of economically and politically driven water allocations can be found in Central Asia, particularly in the cotton-growing areas of Uzbekistan. Here, the government – through the Ministry of Irrigation – focuses water allocation almost exclusively on cotton, which accounts for 25% of the country's GDP and is linked to poor worker conditions. Public water supply has very limited voice within government and access to improved water sources has been in decline since the break-up of the Soviet Union (Mukhamedova & Wegerich, 2014).

It can be attractive for governments to think that economic development from groundwater use can be done in a straightforward manner where the resource is owned, used and controlled by a few powerful actors. Therefore a strong human rights watchdog is vital to protect the rights of the poor.

An alternative is decentralisation – and IWRM guidance emphasises the importance of subsidiarity principle: take decisions at the lowest appropriate level. Whilst decentralisation can create a greater sense of local ownership and legitimacy over decision-making, it can also create complicated bureaucratic structures, unclear responsibilities and mandates and inconsistent decision-making. For groundwater, this may mean that decisions over water allocations and access rights are decided in a different way at different timescales (to differing levels of competence) based on administrative boundaries, which rarely – if ever – coincide with hydrological or hydrogeological boundaries. Once again, the Groundwater Governance principle of ensuring that context informs the governance structures and processes is essential.

14.4.4 Universal access: Is the job ever done?

Universal access is an attractive goal but it is not a static end point that can be achieved and then ticked off the list before move onto the next challenge. This way of thinking is partly encouraged by specific challenges that have definitive, measurable end points, such as the eradication of small pox, polio or *schistosomisias*. But even in the disease-eradication goals there can be sudden and unexpected set-backs, perhaps best illustrated by the Ebola outbreak in West Africa in 2013–15 and Cholera in Haiti in 2010 after a major earthquake. So this seductive end-point way of thinking is problematic for several reasons.

Firstly, the closer we get to the target, at any given scale from district to continental to global, the harder and more challenging – and expensive – it becomes to reach the last few. Therefore the approaches that improve the situation for 95–99% of any given population is unlikely to work the last 1–5%. In the case of Polio eradication the challenge has been in the last places where it occurs (Nigeria, Pakistan, Afghanistan) has come from beneficiaries being unwilling to participate (due social norms and beliefs that include extreme ideologies, vaccination fatigue or misunderstandings about the vaccine's safety and effectiveness), rather than finance or logistical constraints (Toole, 2016).

These 'last mile' difficulties are likely to incentivise governments, donors and NGOs to focus on the 'low-hanging fruit' first and deal with the hard-to-reach last, where the cost per beneficiary is higher and thus harder to justify to funders. In reality a balanced approach is needed, because targeting the hard-to-reach will take time and patient investment, but this needs to be balanced against some tangible, visible improvements to livelihoods and quality of life across many groups so that morale and confidence is sustained. Groundwater is well placed because of its distributed nature and because it often already being used, via springs or hand-dug wells, by remote households and communities, or low-income households in urban, slum and peri-urban areas. However, the groundwater governance techniques applied for reaching the last 1–5% may need to different to that needed for rest of the population. For example, for physically remote populations at a low population density it may not be appropriate – or feasible – to conduct the same intensity of monitoring, planning, regulation and enforcement that is required in a densely populated area where there are many competing interests that are influencing a limited groundwater resource.

Secondly, improving a water supply (as with any change) creates new opportunities and fresh problems, some of which can be anticipated and others that will be unexpected. In groundwater use we see this where energy is a constraining factor – water poor people are generally energy poor as well. In many areas of India, for example, energy for pumping has received subsidies or is charged at a flat rate – this encouraged groundwater use by small farmers, and the additional water for agriculture helped improve incomes without the farmer incurring high costs or risks (Casey, 2013). This is fine if the groundwater resource and recharge are plentiful, but what has happened in many areas is long-terms decline in groundwater levels, and it is generally the poorer farmers who can't afford to drill deeper, who lose out first and hardest and have least voice in how the resource is used or regulated (CGIAR, 2015). A similar, story may happen with a rapid uptake of solar-pumping – with energy removed as the constraining factor, the critical bottleneck moves elsewhere, which maybe the groundwater resource availability, or something less expected. The groundwater-energy nexus is important to consider, particularly when considering policy prescriptions for groundwater management and agricultural economic support (Scott, 2013).

14.4.5 Groundwater poverty and quality of life – do we really understand the links?

Throughout this, it is important to keep an eye on the overarching goal which is to improve the quality of life of all citizens, particularly those facing most severe hardship. It is generally assumed that increasing access to groundwater will reduce poverty improve health and this assumption drives many WASH implementation programmes.

To begin to test this assumption, a short study on Groundwater and Poverty in Sub-Saharan Africa (Carter *et al.*, 2017) was undertaken as part of the Unlocking the Potential of Groundwater for the Poor (UPGro) programme. Evidence was sought on four pillars: health, education, livelihoods and education. The evidence from a meta-analysis of published material and more detailed integration of data from eight African countries and six African cities indicated that groundwater access can bring a variety of benefits for households in urban and rural areas, but that the poorest households don't necessarily capture an equitable share of the gains, and in some cases existing inequalities may be exacerbated. For example, in urban areas, it is generally richer households that can invest in private boreholes, thus depriving municipal water utilities of income that would allow them to cross-subsidise services to poorer households.

14.5 BUILDING A VISION OF SUCCESS

So where are we going? Groundwater is a flexible resource, but as users and managers, we need to be flexible too. The amount of water available to use will vary from year-to-year, decade-to-decade. Aquifers are ultimately self-regulating – when the water is gone, or is too polluted or saline, then it can no-longer be used and usage will reduce, or stop altogether. However, this *laissez-faire* approach is pessimistic and the collapse of a groundwater resource will disproportionately impact the poorest in society, which could bring about conflict or humanitarian crises – either *in situ*, or through migration.

So what is needed is for all users of a given aquifer, to have some understanding of what groundwater – and water resources more generally – is available, how it is allocated, how much is available to them, and what are the consequences of over-abstracting. How such a management regime will look, will vary from context to context – in some an egalitarian approach based on traditional governance and conflict resolution may work, in others a more authoritarian, centralised approach may be the only way to get things done. Whatever the set-up, there will some common elements that are referred to throughout this book:

Physical science: Hydrogeology is as complex and diverse as the geology of our planet. The science of how water flows, and is stored, in all the different rock and formation types is being understood better all the time, but there is still scope for surprise – particularly in regions of the world that have not been studied in great detail. For example, recharge processes in tropical zones are not as well understood as they are in more temperate latitudes and therefore there is still space for fundamental research.

Groundwater and poverty: The UPGro study (Carter *et al.*, 2017) highlighted the following knowledge gaps from Sub-Saharan Africa, which may also be relevant for other regions:

- A paucity of high quality research on unravelling the causal linkages and assessing the longer term impacts on poverty trajectories.
- Few studies of risks and threats to the quality and quantity of groundwater resources, which are likely to disproportionately impact the poor.
- Weaknesses in the publicly available data: there are limitations in the categorisation of water sources in household surveys and these surveys fail to fully address peoples use of multiple sources of water for multiple uses.
- Limited understanding of country-specific causative factors, in particular the differing political economies.
- There is very limited data available on what proportion of urban piped water supplies is derived from groundwater.
- The implications of self-supply, both for the financial viability of utilities, and for the health of the poor (and to a lesser extent the well-off), are probably highly context-specific, and little known.

Technology transfer: water lifting and monitoring technologies are crucial to groundwater use and monitoring. Many pump technologies are necessarily simple and robust but through the introduction of newer technologies, such as motion-sensors, they can unlock new insights for groundwater monitoring and new business models for reliable water service delivery (Colchester *et al.*, 2017; Katuva *et al.*, 2016). However, technology development and transfer have to be sensitive to context, and that is particularly the case with groundwater technologies where there are many complex environmental, socio-cultural, legal and economic factors and perspectives (Olschewski & Casey, 2016).

Fieldwork, Data & Mapping: Groundwater is hidden, secret. It is rarely seen directly and difficult to measure. Therefore investment in monitoring networks is both necessary and needs to be consistent. Too often, data collection is project or study based and so even if the data is publicly released it only often represents a narrow window of 2–3 years of behaviour – which can be fine for addressing a specific question or problem and can be the fruitful basis of collaboration between academic researchers and practitioners (see Box 14.2).

Long term records: In most areas of the world, multi-decadal groundwater level series are rare and this has implications for our understanding of how climate change and long term changes in groundwater use and land use affect groundwater resources. The Chronicles Consortium is an initiative trying to address this by assembling groundwater records from across Africa (IGRAC, 2016). Although remote sensing data, such that from the Gravity Recovery And Climate Experiment (GRACE) has some potential to fill in the gaps, ground-truthing such data will always be essential to avoid over or underestimation of groundwater resources (Long *et al.*, 2016). Cost effective ways of doing this are being explored through remote monitoring that takes advantage of increasing mobile phone connectivity and coverage. Another promising approach, being tried in Ethiopia, is community-based hydrometeorological monitoring (Walker *et al.*, 2016).

Box 14.2: Aid Agencies & Researchers collaborating – Northwest Cambodia: groundwater for drinking and agriculture (Vouillamoz *et al.*, 2016)

An important role for hydrogeology is to answer pressing, practical questions. An example of this can be found in northwest Cambodia, where displaced people have been returning since the end of the Cambodian Civil War, in 1998. This rapid population growth has put pressure on water infrastructure to deliver what is needed for drinking and irrigation water. A key question, therefore, for aid agencies and researchers was: can the aquifers in the area support the water demanded by both priorities?

The difficulty was that groundwater in that area of Oddar Meanchey Province had never been studied before. To a tackle this, a partnership was established between an aid agency (French Red Cross), and researchers (IRD and University of Tokyo) to carry out a four-year study (January 2011–December 2014). The team concluded that to achieve useful results for both parties, the following conditions had to be met:

1 The scientific activities and expected results need to be clearly defined and integrated into the development project.
2 There needs to be dedicated staff and budget allocation to implement the scientific fieldwork.
3 They study period needs to be sufficiently long (4–5 years) to study the hydrological cycle.

The last point may cause some tension between the partners because, in general, the implementing aid agency partner will need to make decisions and take action quicker and earlier than the scientific findings allow. Therefore, such a collaboration is better suited to development cooperation context with a planning horizon of 10+ years and less suitable where the threat to human life is immediate or imminent.

They study undertook the following work to create an understanding of aquifer properties and behaviour. From the data collected, the team was able to build up an understanding of the shallow sandstone aquifer and concluded that it could easily support 100 litres per person per day the projected population in 2030, but that it would not be able to – in general – support large-scale supplementary irrigation for rice cultivation. Nevertheless, localised, small-scale irrigation may be possible due to the heterogeneous properties found.

Allocation rules: Who can take groundwater, from where, how much, over what period of time? Such rules may be localised and informal, or codified through hierarchy of national statutes, laws, regulations, policies and advisory notices – there may be special event rules that are triggered by a pollution incident or during times of drought or flood. There maybe blanket bans on groundwater use in particular areas, or from particular aquifers at certain depths.

Conflict resolution: If one person's use of groundwater derogates another, then a way is needed for cases to be presented and heard and judgements passed and enforced that conflict is avoided and damages are compensated. A difficulty in groundwater arbitration is often the lack of certainty between cause and effect; and the mismatch of resources and power between different groundwater users. Both these can mean that even where laws and legal processes are in place, justice for the poor is not served, or in contexts where political power favours small groundwater users, the legal risk from spurious cases deters private sector investment that could bring employment and resources for effective groundwater monitoring and management.

Human resources: Any groundwater management system is only as good as the people running it. Not everyone can be a hydrogeologist, but everyone should be able to have an understanding of aquifers and how they work to a level that allows them to perform their role in the activities above. Water is life, so everyone needs some level of understanding – and an ability to anticipate and avert risk, to be able to adapt to changing conditions, and be able to solve problems for themselves and others. This is not achieved through a few workshops and online courses, it needs to be in national education systems from school geography lessons to specialist post-graduate courses in the science, law, economics, politics and cultural uses so that there is a pipeline of youth and adult education that produces a steady flow of talent to meet the future challenges. That is the start, and some progress is reported in Malawi where the output of graduates from Mzuzu University is being claimed to improve water sector efficiency across the country (Holm *et al.*, 2016). Professionalism in groundwater management needs professions that young people can enter and career paths that they can follow. Education, skills and experience are not enough – motivation is critical.

14.6 CONCLUSIONS

Groundwater use has the potential to improve the lives of tens of millions of people across the world. In launching the SDGs there have been high-level calls to do things differently and that 'business as usual' for organisation like the UN just will not do (Møller, 2016). Such calls will go unheeded unless there is a space for debate and experimentation on what these new ways of working could look like at a practical level. For groundwater to fulfil its potential there needs to be greater dialogue and cooperation – and a common language – between the worlds of water resources, water services and health, and between the worlds of water users, academia, practitioners and political leaders. Such an idea is not new but we are now in a time of unparalleled and rapidly increasing connectivity. New frontiers have opened up for knowledge sharing and collaboration which should be explored.

It is encouraging that there is growing recognition of the importance of groundwater as a strategic resource for achieving the goals around sustainable development and human rights, however, this brings both opportunities and challenges. While improved access to groundwater can improve quality of life of people, this is not guaranteed. There are three broad areas where attention is needed:

> Firstly, more work is needed to properly understand the link between groundwater use and the quality of life of poor people otherwise there is a risk that well-meaning interventions will create unforeseen and harmful consequences.

Secondly, further work, in relation to groundwater, health, equity and development is around the existing patterns of self-supply and private groundwater development and use in both rural and urban areas. The behaviour patterns around water use are complex large schemes, whether in rural Ethiopia or urban Indonesia have met with considerable set-backs and collateral water security problems. Uncontrolled, decentralised groundwater use also creates equity concerns because without a regulating authority, benefits of groundwater use – even at a village or city ward level – are likely to accrue to the richest first and most, leaving the most vulnerable behind. However, rather than ignoring self-supply groundwater use as an unknowable problem, we should instead see it as an opportunity to harness the creativity and investment of people themselves to solve their own problems and create benefits and opportunities for themselves and their communities.

Thirdly, technology and market approaches can play important roles in unlocking previously difficult or impossible opportunities to provide health and income benefits for the poor. However, such interventions often bring new risks and rather than solve problems completely, they often just move the problem elsewhere. For example, solar pumps that make water lifting easier and cheaper, but increase the risk over-abstraction of the aquifer. Technology development, introduction and scaling up through enterprise needs to be sensitive to context and needs to be implemented professionally. Therefore investment is needed in education and training of current and future generations of hydrogeologists, drillers, water engineers and water service providers.

For communities, such as Mr Hakim's in coastal Bangladesh there may be no easy answers. For his community, the water resources were so constrained that engineers from HYSAWA, a national WASH fund, recommended piloting a reverse osmosis plant. Today, many of the people in Jahudanga Union Parishad, including a school, get their drinking water for this plant, which uses a mix of rainwater, pond water and deep saline groundwater, depending on which is the best quality and most plentiful at a particular time of year. It is an expensive option and not without financial and environmental impact, but sometimes we have to acknowledge that real life is messy, unpredictable and needs creative problem solving.

For Mr Hakim, and for millions of other people, it is essential national and local governments have some form of groundwater governance system in place – and the skilled people to make it work – so that groundwater can continue to be part of the vision of ensuring that everyone has access to a safe, affordable water supply.

ACKNOWLEDGEMENTS AND DECLARATION OF INTERESTS

This chapter was written with thanks to financial support from the Skat Consulting Ltd. Project Fund, and UPGro: Unlocking the Potential of Groundwater for the Poor, a 7-year research programme funded by DFID, NERC and ESRC, for which the author leads the Knowledge Broker team. The author is co-lead of the Rural Water Supply Network (RWSN) theme, Sustainable Groundwater Development. The experiences of witnessing groundwater use in Bangladesh came whilst undertaking a consultancy assignment for the Swiss Agency for Development and Cooperation (SDC).

REFERENCES

Abidin, H. Z., Andreas, H., Gumilar, I. and Brinkman, J. J. (2015) Study on the risk and impacts of land subsidence in Jakarta, *Proc. IAHS*, 372, 115–120, 2015, proc-iahs.net/372/115/2015/, doi:10.5194/piahs-372-115-2015 [Accessed 10.05.2017].

Albuquerque, C. (2014A) *Handbook on the Human Right to Water & Sanitation, Volume: Introduction*, [Online] Available: http://www.righttowater.info/handbook/ [Accessed 21.12.2016].

Albuquerque, C. (2014B) *Handbook on the Human Right to Water & Sanitation, Volume: Principles*, [Online] Available: http://www.righttowater.info/handbook/ [Accessed 21.12.2016].

Albuquerque, C. (2014C) *Handbook on the Human Right to Water & Sanitation, Volume: Sources*, [Online] Available: http://www.righttowater.info/handbook/ [Accessed 21.12.2016].

Casey, A. (2013) *Reforming Energy Subsidies Could Curb India's Water Stress*, World Water Institute, [Online] http://www.worldwatch.org/reforming-energy-subsidies-could-curb-india%E2%80%99s-water-stress-0 [Accessed 21.12.2016].

Carter, R.C., Foster, S., Foster, T., Cavill, S., Simons, A., Bizoza, A. R., Baguma, A., Jobbins, G., Shepherd, A., Katuva, J., Koehler, J., Hope, R., Theis, S. and Furey S. G. (2017) *Groundwater and poverty in sub-Saharan Africa: a short investigation highlighting outstanding knowledge maps*, Unlocking the Potential of Groundwater for the Poor (UPGro), Technical Report Skat Foundation, St. Gallen, Switzerland. [Online] https://upgro.files.wordpress.com/2015/09/groundwater-and-poverty-report_0003.pdf [accessed 20.06.2017].

CGIAR (2015) *Groundwater and ecosystem services: a framework for managing smallholder groundwater-dependent agrarian socio-ecologies – applying an ecosystem services and resilience approach*. Colombo, Sri Lanka: International Water Management Institute (IWMI). CGIAR Research Program on Water, Land and Ecosystems (WLE). 25p. doi: 10.5337/2015.208 [Accessed 10.05.2017].

Colchester, F. E., Marais H. G., Thomson P., Hope, R. and Clifton D. A. (2017) Accidental infrastructure for groundwater monitoring in Africa, *Environmental Modelling & Software* 91 (2017) 241–250 [Accessed 10.05.2017].

Danert, K., Furey, S., Mechta, M. and Gupta, S. (2016) Effective Joint Sector Reviews for Water, Sanitation and Hygiene (WASH). A Study and Guidance – 2016, The World Bank Global Water Practice, http://www.rural-water-supply.net/en/resources/details/757 [Accessed 10.05.2017].

Delinom, R. M. (2009) Groundwater management issues in the Greater Jakarta area, Indonesia, *Proceedings of International Workshop on Integrated Watershed Management for Sustainable Water Use in a Humid Tropical, Region, JSPS-DGHE Joint Research Project, Tsukuba, October 2007. Bull. TERC, Univ. Tsukuba, No.8 Supplement, no. 2*, 2008 [Online] Available from: http://citeseerx.ist.psu.edu/viewdoc/download?doi=10.1.1.655.23&rep=rep1&type=pdf [Accessed 21.12.2016].

Donnelly, K. and Cooley H. (2015) *Water Use Trends in the United States*, Pacific Institute [Online] Available from: http://pacinst.org/app/uploads/2015/04/Water-Use-Trends-Report.pdf [Accessed 21.12.2016].

FAO (2016A) *Global Framework for Action to achieve the Vision on Groundwater Governance*, FAO, Rome, March 2016. ISBN 978-92-5-109258-3 Available from: http://www.groundwatergovernance.org [Accessed 10.05.17].

FAO (2016B) *Global Diagnostic on Groundwater Governance*, FAO, Rome, March 2016. ISBN 978-92-5-109259-0 Available from: http://www.groundwatergovernance.org [Accessed 10.05.17].

Gleick, P. (2014) National Geographic Science Blogs: Peak Water: United States Water Use Drops to Lowest Level in 40 Years, *Pacific Institute Insights*, November 5, 2014

[Online] Available from: http://pacinst.org/national-geographic-scienceblogs-peak-water-united-states-water-use-drops-to-lowest-level-in-40-years/ [Accessed 21.12.2016].

Grönwall, J. (2016) Self-supply and accountability: to govern or not to govern groundwater for the (peri-) urban poor in Accra, Ghana, *Environ. Earth Sci* (2016)75:1163, DOI 10.1007/s12665-016-5978-6 [Accessed 10.05.2017].

Holm, R., Singini, W. and Gwayi, S. (2016) Comparative evaluation of the cost of water in northern Malawi: from rural water wells to science education, *Applied Economics*, Volume 48, 2016 – Issue 47 http://dx.doi.org/10.1080/00036846.2016.1161719 [Accessed 10.05.2017].

IAH (2017) *The UN-SDGs for 2030 Essential Indicators For Groundwater*, International Association of Hydrogeologists, Briefing Note Available from: https://iah.org/news/groundwater-un-sdgs-essential-indicators-groundwater-strategic-overview-series [Accessed 10.05.2017].

IGRAC (2016) The Chronicles Consortium, [Online] Available from: https://www.un-igrac.org/special-project/chronicles-consortium [Accessed 23.12.2016].

Katuva, J., Goodall, S., Harvey, P. Hope, R. and Trevett, A. (2016) FundiFix: exploring a new model for maintenance of rural water supplies, *7th RWSN Forum, Abidjan, Côte d'Ivoire 29th Nov–2 Dec 2016*. Available from: https://rwsnforum7.files.wordpress.com/2016/11/full_paper_0224_submitter_0276_goodall_susanna1.pdf [Accessed 10.05.2017].

Kumamaru, K., Odhiambo, F. and Smout, I. (2011) Self Supply Dynamic Mapping, 6th *Rural Water Supply Network Forum, Kampala, Uganda, 2011* [Online] Available at: https://rwsnforum.files.wordpress.com/2011/11/153-self-supply-dynamic-mapping.pdf [Accessed 23 December 2016].

Long, D., Chen, X., Scanlon, B. R., Wada, Y., Hong, Y., Singh, V.P., Chen, Y., Wang, C., Han, Z. and Yang, W. (2016) Have GRACE satellites overestimated groundwater depletion in the Northwest India Aquifer? *Sci. Rep.* **6**, 24398; doi: 10.1038/srep24398 (2016).

McIntyre, O. (2009) 'Utilisation and Environmental Protection of Shared International Freshwater Resources: The Role of Equity' In: *J. Benidickson, A. Benjamin, B. Boer and K. Morrow (eds.), Proceedings of the Fifth IUCN Academy of Environmental Law Colloquium. Brazil:* Cambridge University Press.

Møller, M. (2016) 'Business as usual' is not an option anymore, *World Economic Forum* [Online] Available from: https://www.weforum.org/agenda/2016/01/why-do-we-need-a-multi-stakeholder-approach-to-sustainable-development [Accessed 21 December 2016].

Mukhamedova, Nozilakhon and Wegerich, Kai. 2014. Integration of villages into WUAs-the rising challenge for local water management in Uzbekistan. International Journal of Water Governance, 2:153-170. doi: http://dx.doi.org/10.7564/13-IJWG19 [Accessed 10.05.2017].

Okotto, L., Okotto-Okotto, J., Price, H., Pedley, S. and Wright J. (2015) Socio-economic aspects of domestic groundwater consumption, vending and use in Kisumu, Kenya, *Applied Geography*, 58, 189–197.

Olschewski, A. and Casey, V. (2016) Supporting service delivery and business innovation through TAF application, *7th RWSN Forum, Abidjan, Côte d'Ivoire 29th Nov–2 Dec 2016* https://rwsnforum7.files.wordpress.com/2016/11/full_paper_0043_submitter_0121_olschewski_andre.pdf [Accessed 14.03.2017].

Rieger, K., Holm, R. H. and Sheridan, H. (2016) Access to groundwater and link to the impact on quality of life: A look at the past, present and future public health needs in Mzimba District, Malawi, *Groundwater for Sustainable Development* 2–3 (2016) 117–129.

Scott, C. A. (2013) Electricity for groundwater use: constraints and opportunities for adaptive response to climate change, *Environ. Res. Lett.* 8 (2013) 035005 (8pp).

SDC (undated) *Payments for Ecosystem Services: Scaling Up Payments and Investments in Watershed Services*, Factsheet [Online] https://www.shareweb.ch/site/Water/Documents/Factsheet_PWS_final_05.14.pdf [Accessed 21.12.2016].

Smith, A. H., Lingas, E. O. and Rahman, M. (2000) Contamination of drinking-water by arsenic in Bangladesh: a public health emergency, *Bulletin of the World Health Organization*, 2000, 78 (9).

Toole, M. J. (2016) So close: remaining challenges to eradicating polio, *BMC Med*. 2016; 14: 43. Published online 2016 Mar 14. doi: 10.1186/s12916-016-0594-6 [Accessed 21.12.2016].

UN (2010) Resolution adopted by the General Assembly on 28 July 2010: 64/292. *The human right to water and sanitation*, United Nations General Assembly [Online] Available: http://www.un.org/es/comun/docs/?symbol=A/RES/64/292&lang=E [Accessed 21.12.2016].

UN (2016A) Sustainable Development Goals [Online] http://www.un.org/sustainable development/sustainable-development-goals/ [Accessed 21.12.2016].

UN (2016B) Resolution adopted by the General Assembly on 17 December 2015, 70/169. The human rights to safe drinking water and sanitation, United Nations General Assembly, 22 February 2016 Available: http://www.un.org/en/ga/search/view_doc.asp?symbol=A/RES/70/169 [Accessed 21.12.2016].

UNHCR (2010) *Human Rights Fact Sheet 35: The right to water*, United Nations, Geneva, ISSN 1014-556. Available http://www.ohchr.org/Documents/Publications/FactSheet35en.pdf [Accessed 10.05.2017].

UNICEF (2008) Arsenic Mitigation in Bangladesh [Online] https://www.unicef.org/bangladesh/Arsenic.pdf [Accessed 20.02.2017].

UNICEF/WHO (2015) Joint Monitoring Programme (JMP) for Water Supply & Sanitation [Online]: www.wssinfo.org [Accessed 21.12.2016].

UNICEF/WHO (2016) *WASH Post-2015: Proposed indicators for drinking water, sanitation and hygiene*, JMP briefing note [Online] Available at: http://www.who.int/water_sanitation_health/monitoring/coverage/wash-post-2015-rev.pdf [Accessed 23.12.2016].

UNICEF/WHO (2017) *Safely managed drinking water- thematic report on drinking water 2017*, World Health Organisation, Geneva, Available at: https://data.unicef.org/resources/safely-managed-drinking-water/ [Accessed 14.03.2016].

Vouillamoz, J.M., Valois, R., Lun, S., Caron, D. and Arnout, L. (2016) Can groundwater secure drinking water supply and supplementary irrigation in new settlements of North-West Cambodia? *Hydrogeol. J.*, 24:195-209, DOI 10.1007/s10040-015-1322-6.

Walker, D., Forsythe, N., Parkin, G. and J. Gowing (2016) "Filling the observational void: Scientific value and quantitative validation of hydrometeorological data from a community-based monitoring programme" *Hydrology Journal*, Volume 538, July 2016, Pages 713–725.

WWF (2016) *The Imported Risk Switzerland's Water Risk in Times of Globalisation*, [Online] Available from: https://assets.wwf.ch/downloads/study_imported_risk.pdf [Accessed 21.12.2016].

Chapter 15

Managing energy-irrigation nexus: insights from Karnataka and Punjab states in India

Aditi Mukherji
Theme Leader, Water and Air, International Centre for Integrated Mountain Development, Kathmandu, Nepal

ABSTRACT

India is the world's largest groundwater user and current regime of power subsidy without metering is often thought to be the main reason for such rapid growth in groundwater use. Agriculture, groundwater and electricity sectors in much of India are now bound in nexus of mutual dependence where the growth in agriculture sector is being supported by unsustainable trends in groundwater and electricity sectors. All three components of the nexus – groundwater, electricity and agriculture – are state (provincial) subjects and as such only state governments can legislate on these issues. Different states in India have adopted different ways of managing this nexus. This paper describes the divergent experiences of groundwater management through electricity policies in two Indian states, Karnataka and Punjab. Both states face severe problems of groundwater over-exploitation and their electricity utilities are saddled with huge losses. Yet, they have managed this conundrum differently, showing that political will is needed to tackle the wicked problem of water-energy-food nexus in India.

15.1 INTRODUCTION

Many countries like the United States, China, Mexico, Spain, Pakistan, Bangladesh, Iran and India have long history of groundwater irrigation. Before the onset of the Green Revolution, groundwater irrigated area in India was quite small as compared to many other countries. But since then, India's groundwater use has grown at a much faster rate, making India the world's largest user of groundwater for irrigation (Shah *et al.*, 2007). Various factors such as initial stimulus provided by India's massive public tube well program; low cost of pumps and drilling equipment; rural electrification; ease of institutional finance; small land holding and need to intensify cropping to eke out a living; and lack of canal irrigation in most places explain such rapid surge in groundwater irrigation. However, what best explains this runaway growth is the regime of flat rate (unmetered) and zero electricity tariff and power subsidies prevalent in India since the 1980s. At present, India's agriculture is overwhelmingly dependent on groundwater (Figure 15.1).

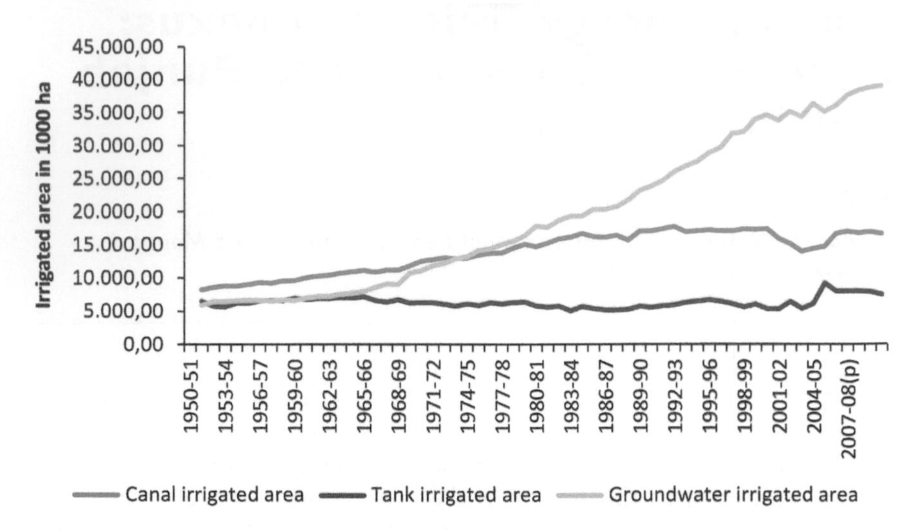

Figure 15.1 Area irrigated by different sources in India, 1950–51 to 2009–10.
Source: Drawn on the basis of data made publicly accessible by the Government of India, various years.

15.2 UNIQUENESS OF THE ENERGY-IRRIGATION NEXUS IN INDIA AND ITS IMPACT ON AGRICULTURE, GROUNDWATER AND ELECTRICITY SECTORS

The policy decision in many states to supply unmetered and subsidized power to the agricultural sector in the early 1980s ushers the beginning of a unique energy-irrigation nexus in India. This policy, in tandem with others like guaranteed food procurement policies and lack of investment in canal infrastructure, ensured that farmers in India have high dependence on groundwater. They actively lobby for continuation of subsidized electricity, and farmers being a large vote bank in India, politicians find it impossible to ignore their demands. But rather than electricity subsidy per se, what makes India's energy-irrigation nexus really complex is the practice of providing unmetered power supply to farmers. This makes it difficult to do proper energy accounting leading to poor service on the part of the electricity utility on the one hand, and gives perverse incentives to farmers to intensify groundwater use on the other hand. In this chapter, we contend that it is the lack of energy accounting due to unmetered supply that is at the heart of this unique energy-irrigation nexus in India and the only way to tackle this nexus is to ensure transparent energy accounting. So, while other countries like Mexico also subsidize electricity for farmers, what makes India's case unique is providing this subsidy through unmetered connections. By doing so, electricity consumption in agriculture sectors becomes difficult to calculate and all kinds of malpractices on the side of the consumers (for example, unauthorized connection, illegal tapping of electricity, increasing pump horse power without informing the electricity utility) and the electricity utility (over-reporting energy provided to farmers and claiming higher subsidies from the government) hiding their inefficiencies and electricity thefts happening in other sectors in the garb of becomes the norm. Table 15.1 (adapted from Shah, Giordano and Mukherji, 2012) shows the evolution of this nexus in India.

India's unique energy-irrigation nexus has had far reaching impacts on agriculture, groundwater and power sectors. For instance, while droughts still plague Indian agriculture, food production is no longer at the mercy of weather due to widespread subsidized groundwater irrigation in major food producing states in India. For example, during the drought year 2009 when rainfall deficit was 33% and 66% in Punjab and Haryana, the reduction in irrigated paddy area was only 0.7% and 10% respectively, because of intensive groundwater use supported by current power policies (Ministry of Agriculture, Crops Weather Watch Group, August 28, 2009). However,

Table 15.1 Evolution of energy-irrigation nexus in India.

Phases in energy-irrigation nexus	*Policy imperative*	*Outcome and challenges*
Phase I: Drive for rural electrification (1950s to mid-1970s)	To promote electricity use in all sectors of rural economy, including agriculture and incentivizing farmers to install electric tubewells	Slow progress in rates of tubewell electrification because costs of tubewell were still prohibitively high compared to crop income. This was the pre-Green revolution period.
Phase II: Introduction of flat electricity tariff and free power (late 1970s onwards)	As the number of electricity connections increased, transaction costs of meter reading and billing increased and most utilities introduced flat tariff whereby amount paid, if any, got delinked with quantity supplied. This brought in lack of accountability on the part of both farmers and utilities. This also coincided with the Green Revolution	The current invidious energy-irrigation nexus problems owe its origin to supplying unmetered power to agriculture. This gave perverse incentives to farmers to over-exploit groundwater and to utilities to hide their inefficiencies in the garb of agricultural power.
Phase III: Attempts at containing Power Subsidies and Groundwater Depletion (since early 2000s)	In 9 critical states where groundwater is over-abstracted, power subsidies and groundwater abstraction got caught in a vicious downward spiral necessitating renewed efforts at managing this nexus	Free or subsidized and unmetered power led to rapid increase in groundwater demand and unmetered supply led to lack of accountability on the part of the farmers and utilities. Farmers' dependence on electricity for pumping groundwater increased, partially as water table lowered due to over exploitation. They organized themselves into powerful lobbies for maintaining power subsidies and electricity for pumping groundwater emerged as an important political issue.

Source: Adapted from Shah, Giordano and Mukherji, 2012.

such drought resilience has come at the cost of long term groundwater sustainability – which means that long term resilience is being seriously undermined. All the nine critical groundwater states (Punjab, Andhra Pradesh, Karnataka, Haryana, Gujarat, Rajasthan, Madhya Pradesh, Maharashtra and Tamilnadu) have an alarming groundwater situation which is getting worse every year. It is these very states which also get unmetered and highly subsidized electricity supply for irrigation. The electricity sector has also become unviable in these states due to such high for subsidies. In early 2000s, the World Bank estimated farm power subsidies to be around "US$6 billion a year—equivalent to about 25 percent of India's fiscal deficit, twice the annual public spending on health or rural development, and two and a half times the yearly expenditure on irrigation." (Monari, 2002:1). Removal of meters on tube wells has undermined energy accounting systems in power utilities and impaired their internal accountability ethics. According to the Economic Survey 2006-07, the major portion of utility losses to the tune of INR 200 billion (1 USD~INR 50) is due to theft and pilferage. More than 75%–80% of the total technical loss and almost the entire commercial loss occur at the distribution stage. It is now common knowledge that utilities hide their technical losses and pilferages under the garb of agricultural power supply since it is unmetered (World Bank, 2001). Agriculture, groundwater and electricity sectors in much of India are now bound in a nexus of mutual dependence where the growth in agriculture is being supported by unsustainable trends groundwater and electricity sectors.

15.3 MANAGING THE ENERGY-IRRIGATION NEXUS FROM THE GROUNDWATER AND AGRICULTURE PERSPECTIVES

The problems facing the electricity sector due to unmetered supply to agriculture and consequent lack of incentives among farmers to make efficient use of electricity and among the utilities to do robust energy accounting is now widely acknowledged. Since this nexus involves three sectors of the economy, efforts are being made on all these three fronts. On the groundwater front, many states, including Andhra Pradesh have promulgated groundwater laws. Punjab introduced a law banning paddy transplantation before 14th of June and this reportedly had the effect of reducing groundwater withdrawals by up to 9% (Singh, 2009), but at the same time exacerbated the energy problem by increasing peak demand during that limited window of transplanting. Other initiatives include community management of groundwater (van Steenbergen, 2006), introduction of efficient irrigation technologies and government or community led initiatives of managed aquifer recharge (Sakthivadivel, 2007). On the agricultural front, initiatives including attempts to lure farmers away from water intensive paddy crops through diversification and better on-farm water management practices such as mulching or laser leveling (Humphreys *et al.*, 2010) are being tried. However, in absence of suitable energy policies that give incentives to farmers to use groundwater efficiently, none of the above measures are likely to be successful. The ultimate aim of all these interventions is to bring back groundwater levels to sustainable levels, without seriously jeopardizing either the agricultural or the energy sectors in the long to medium term, though some short term impacts are inevitable. In this chapter, we look into energy sector related policies which have implications for groundwater management in greater detail.

15.4 MANAGING THE ENERGY-IRRIGATION NEXUS FROM THE ELECTRICITY PERSPECTIVE

All three components of the nexus – groundwater, electricity and agriculture – are state subjects. Different states in India have adopted different ways of managing this nexus and some states have implemented these solutions better than the others. As mentioned earlier, the main problem facing the agricultural electricity sector is unmetered supply to agriculture and all the problems it creates, such as lack of energy accounting, lumping of other losses and inefficiencies to the agricultural sector and the perverse incentives it gives to the farmers to over-extract groundwater. In this chapter, we look at ways in which two states with similar groundwater over-exploitation issues, but with different types of aquifers, have undertaken electricity reforms to tackle the problem of groundwater over-exploitation. These states are Karnataka (hard rock aquifer) and Punjab (alluvial aquifer).

15.4.1 Karnataka: A hard rock aquifer state with free unmetered agricultural electricity supply with poor quality and ineffective rationing

Karnataka is one of the major states in India with a geographical area of 191,976 square kilometers and a population of 61.13 million accounting for 5.13 per cent of the country's total population (Census of India, 2011). Out of 27 districts, 18 districts are drought prone with annual average rainfall of less than 750 mm. Because rainfall is highly variable over space and time and irrigation is limited, agricultural production is correspondingly variable. The main crops grown here are rice, millets (*ragi* and *jowar*), maize and pulses. Rainfed farming is still the dominant mode of cultivation in Karnataka and only 29% of the net sown area in the state is irrigated. In irrigated areas, crops like cotton, sugarcane, tobacco, vegetables and plantation crops (coconut, arecanut, cashewnut, cardamom etc.) are grown. Of the total irrigated area, 46% of the area is irrigated using groundwater (tubewells and dugwells) and the rest from tanks and canals. Tanks have been the dominant source of irrigation in the state in the past, but its coverage has declined in the last two decades as a direct result of the increase in groundwater irrigation (Government of Karnataka, several years).

The state has an annual replenishable groundwater resource of 15.30 billion cubic meters (BCM), of which 10.71 BCM are extracted annually. However, this state level average hides inter district variations. Of the 175 *talukas* (lowest administrative units) in the state for which groundwater assessment was undertaken, 37% of the *talukas* are in the over-exploited category, 10% are in critical and semi-critical category and the remaining 53% are in safe groundwater category according to government estimates (CGWB 2010). With 47% of the blocks above the semi critical level in the state (CGWB 2006), well failure is a rampant phenomenon in Karnataka.

The state of Karnataka has been a pioneer in agricultural electricity use. In 1918, the first ever electric pump in the country was run in this state using electricity from a hydropower plant. By 1947, there were 1,453 electric irrigation pump sets in the state. From 1947 to 1971, the state made rapid investments in electricity generation capacity so that by 1965, there was surplus power in the state. To utilize this excess power, rural electrification and electrification of pump sets was given priority and a separate

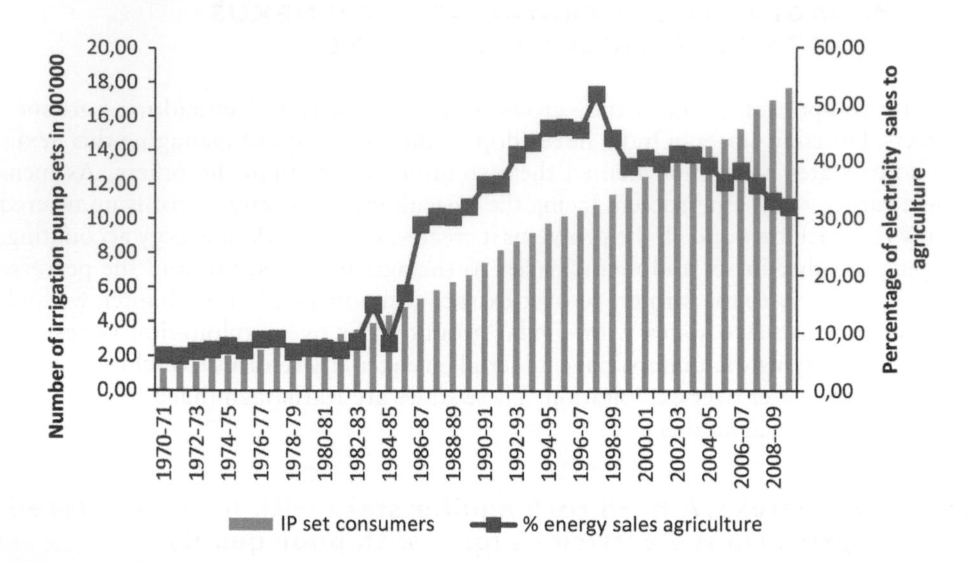

Figure 15.2 Number of irrigation pump sets and percentage of electricity sales to agriculture, 1970 to 2009–10.
Source: Karnataka Electricity Board (KEB) and Electricity Regulatory Commission (ERC) filings of Karnataka Power Transmission Corporation Limited (KPTCL) & Electricity Supply Companies (ESCOMS), several years.

department called the *rural electrification and irrigation pump set sub-division* was formed within Karnataka Electricity Board (KEB). Till 1981, irrigation pumps were metered and farmers paid electricity bills fairly regularly. In 1981, what later proved to be a disastrous precedent, all irrigation sets were de-metered. It was decided to charge farmers a flat electricity tariff on the grounds that costs of reading meters, preparing bills and collection were higher than the revenue so generated. What followed was a steep increase in the number of electric irrigation pump sets (Figure 15.2) and electricity consumption in agriculture.

Having said that, one needs to keep in mind that accounting for agricultural electricity consumption is fraught with methodological inaccuracies and deliberate misreporting. This deviation is as high as 27% to 50% when compared with similar numbers from other sources like the Minor Irrigation Census, Agricultural Census and Input Surveys (Table 15.2). This is because over-reporting of the number of electric pump sets works in favor of electricity utilities since they can inflate their subsidy claims from the government. In absence of metering of agricultural connections, it is near impossible to settle these claims one way or the other.

In the meanwhile, rapid expansion in the number of irrigation pump sets, without a commensurate increase in power production, led to deterioration in quality of the electricity supply. Till 1996, rural areas got 3-phase power supply for about 18 hours a day. This came down to 6 hours of 3 phase supply followed by 10–12 hours of single phase supply (when pumps cannot be run in theory, but in practice, as we will see later, farmers have found a way of running pumps with single-phase electricity), and 6-8 hours of load shedding. In addition, unscheduled load shedding is common

Table 15.2 Discrepancy in number of electric IP sets as reported by Karnataka electricity utilities and agriculture census, minor irrigation census and input use surveys.

	Number of electric irrigation pump sets in thousands from various sources			
Year	*KEB and utility estimates*	*Agriculture Census (includes all electric pumps)*	*Minor Irrigation Census (includes electric groundwater structures and surface lift schemes)*	*Input survey (includes all electric pumps)*
1993–94	917	–	255	–
1995–96	973	708	–	–
1996–97	1049	–	–	676
2000–01	1247	796	860	–
2001–02	1300	–	–	1014
2005–06	1465	735	–	–
2006–07	1525	–	1012	1108

Source: ERC filings of KPTCL & ESCOMs, several years; Agricultural Census of 1995–96, 2000–01 and 2005–06; 2nd, 3rd and 4th Minor Irrigation Census of 1993–94, 2000–01 and 2006–07 and All India Input Surveys of 1996–97, 2001–02 and 2006–07.

almost throughout the year, particularly in the months from January to May when power supply falls short of demand. Farmers report high rates of pump burnout and electricity supply companies report high rates of transformer burn out. On an average, it takes a week to 15 days to repair burnt transformers and this delay is often enough to severely damage standing crops.

15.4.1.1 Electricity reforms and its implications for groundwater in Karnataka

Given the enormity of the issues, the electricity utilities have embarked on a series of reforms. This section takes a critical look at these reforms and whether these reforms had their intended consequences in managing the irrigation-energy nexus better.

15.4.1.1.1 Failed attempts at metering agricultural electricity

During the year 2000, it was decided to provide energy meters to all categories of unmetered connections including agricultural consumers. Wide publicity was given throughout the state clearly stating that energy meters would be fixed only for the purpose of measuring consumption to facilitate subsidy calculation and not for billing purposes, and that farmers would continue to get free power. In spite of such assurances, farmers removed the energy meters and delivered them to local police stations or local utility offices, as per the directions of powerful farmer's associations. With this ended any further attempts to install meters on agricultural pump sets.

15.4.1.1.2 Attempts to ration 3-phase power supply to limit hours available to farmers to pump groundwater

A pilot scheme called *Grama Jyothi* scheme was launched in the year 2004–05. Under this scheme, special switches were provided to disconnect 2 phases at substations and only provide one phase electricity. Since most pumps require 3 phase electricity, this was an attempt meant to prevent farmers from running their pumps. However, this scheme was not successful as it involved high capital cost investment for replacing existing 3 phase distribution transformers by small capacity single phase distribution transformers. This also negatively affected rural industrial and drinking water installations which could not be operated without 3 phase electricity supply. But most interestingly, farmers got ingenious and managed to run their pumps using single phase electricity through specially designed capacitors, thereby defeating the very purpose of the scheme. Rural Load Management Scheme (RLMS) was launched in late 2000s. The salient feature of the scheme was that agricultural pump loads from each distribution transformer were virtually (as opposed to physically) bifurcated into two sets and programmed to provide power supply alternatingly for 6 to 8 hours in a day. This reduced feeder load by 50% and consequently helped reduce technical losses. This scheme also meant that there was no further load shedding in rural feeders. While technically and financially feasible, this scheme failed because it was alleged that farmers tampered with switches on the secondary side of the distribution transformers with the intention of getting continuous power supply.

The latest attempt to segregate agricultural consumers from other rural consumers and provide them with rationed electricity comes in the form of *Niranthara Jyoti* scheme. It is a physical feeder segregation scheme designed on lines of the *Jyotirgram Yojana* of Gujarat (Shah and Verma, 2008). The main purpose of this scheme is to provide 24*7, 3 phase reliable and quality power supply to non-agricultural loads in villages at par with urban areas, and the secondary purpose is to provide limited 3 phase, high quality power supply to agricultural pump sets. However, as a concession to rural habitations and farm houses located within agricultural fields, single phase power supply is also supplied for 12–14 hours in 3 phase agricultural feeders using a specially designed transformer.

As already mentioned earlier, it is possible to run 3 phase motors using single phase supply through special converters. These converters are locally fabricated and available easily. Farmers are already using single phase electricity in those 3 phase agricultural feeders which have specially designed transformers to supply single phase electricity for certain number of hours in a day. While the provision of single phase supply in agricultural feeders is supposed to be limited only for remote habitation and farm houses, in effect, all agricultural feeders are being provided with single phase electricity due to widespread protests by farmers. In other words, provision of 3-phase power as well as single phase power in agricultural feeders is undermining the very objective of this scheme, which was to limit agricultural use, but not at the cost of rural and domestic use. Since feeders will be physically segregated, all agricultural pumps will be connected to agricultural feeder, which will get 6–8 hours of 3 phase power supply. While the rest of rural load will have 24 hours of 3 phase supply from a separate feeder. However, the undoing of this scheme has been that in the agricultural feeder too, the utility is being forced to give 10–12 hours of single phase supply and it is well known that farmers can convert single phase to three phase. Farmers continue to use single phase electricity for running their motors and this leads to increased transformer burn out and overall poor quality of electricity supply to tail end farmers.

Table 15.3 Agricultural electricity consumption estimates by electricity utilities and as approved by KERC, 2000 to 2005.

Year	*Agricultural electricity consumption claimed by electricity companies in Karnataka in Million Units (MU)*	*Agricultural electricity consumption approved by KERC in MU*
2000–01	7343	7343
2001–02	7519	7442
2002–03	8732	8450
2003–04	8039	8007
2004–05	8118	8395

Source: Annual Revenue Requirement filings by electricity companies in Karnataka, several years.

15.4.1.1.3 Failed attempts by the State Electricity Regulatory Commission (SERC) to streamline agricultural power subsidies

In 2010–11, the government of Karnataka paid the various electricity utilities a sum of INR 40.5 billion as reimbursement for free electricity given to the farmers. However, calculation of this subsidy amount is fraught with inaccuracies given that there is high degree of uncertainty about total number of electric pump sets and the number of hours they operate. These are the two critical data that is used to calculate the total amount of electricity provided to farmers and hence the total subsidy. Since 2000, all electricity distribution companies have been estimating their agricultural power consumption based on sample metering of mixed distribution transformers (and not individual pump sets – as attempts to meter them even for monitoring purposes failed, as noted earlier) and then multiplying it by the total number of irrigation pump set consumers. This is problematic because of two reasons – first, as we already mentioned, the number of irrigation pumps has been overestimated by 27–50% and second, the distribution transformers cater to agricultural as well as non-agricultural loads and if the sample distribution transformers have predominantly agricultural load, then agricultural electricity consumption is inflated. Both these errors inflate the agricultural power subsidies that the utilities receive, but these subsidies do not correctly reflect the quality and quantity of electricity that farmers actually receive at their end. This, in turn, makes farmers even more distrustful of the electricity companies. Even among farmers a disproportionate amount of the subsidies goes to large scale farmers because they tend to own and operate a larger number of electric pumps than others (Howes and Murgai, 2003).

The SERCs are mandated to see that subsidy claims by the utilities are reasonable and is backed by verifiable data. The Karnataka Electricity Regulatory Commission (the SERC of Karnataka), has asked the electricity supply companies in the state to improve their method of computation, but progress has been negligible. Table 15.3 shows that agricultural electricity consumption estimates submitted by electricity utilities are approved by KERC with very minor adjustments, showing that it has been unable to influence them to provide better quality data.

An analysis of long term trends in groundwater development and agricultural growth in Karnataka shows two things. First, the growth in irrigation was largely triggered by the expansion of groundwater irrigation. Second, along with the groundwater development, the state experienced a quantum jump in electricity supply to agriculture. In 1981, electric tube wells were de-metered and a flat tariff regime was introduced. This in turn brought its own share of problems, including the difficulty in accounting for agricultural electricity consumption mounting losses of electricity utilities and deterioration of the quality of supply to farmers over the years. The KERC was formed in 2000 and since then it has urged the electricity companies to improve their agricultural energy consumption estimates, but their methodologies remain as fraught with problems as before. As a result, agricultural electricity consumption is grossly over-estimated as are the number of electric pump sets in the state. The current subsidy delivery mechanism is not effective in ensuring that farmers get the subsidy claimed in their name, and even then the net amount of subsidized electricity falls short of farmers' current irrigation demands thereby forcing them to keep large portions of their land un-irrigated or fallow. The farmers in turn resort to illegal connections and under-reporting of their pump capacity making it even more difficult to arrive at independent estimates of agricultural power consumption. Farmers are also reluctant to invest in more expensive energy efficient pumps because of frequent voltage fluctuation and pump burnouts. Overall, there is anarchy below the feeder level – an anarchy that leaves farmers, utilities and the state government much worse off than they need be. A collateral damage of this anarchy is the deep level of mistrust between the farmers and the utility staff, which makes it even more difficult to find mutually acceptable and robust solutions.

15.4.2 Punjab: A state with free agricultural electricity supply but with good quality and effective rationing

Punjab is one of the major states in India and its food basket, with a geographical area of 50,362 square kilometers and a population of 27.7 million accounting for 2.03 per cent of the country's total population (Census of India, 2011). The average annual rainfall is around 500 mm, making Punjab one of the semi-arid states of India. However, the state is endowed with one of the best alluvial aquifers and a good canal network, which ensures that 99% of the state's net sown area is irrigated. This is in sharp contrast with Karnataka, where only 29% of the net sown area is irrigated. Here, tube wells (groundwater) and canals are the main sources of irrigation, but over the years, groundwater has replaced canal water as a source of irrigation and now canals mostly serve to recharge groundwater aquifers. Punjab is the heart of India's green revolution and a majority of farmers here adopted high yielding varieties of crops way back in the 1960s and the 1970s. Punjab is also India's food basket – accounting for 45% of wheat and 25% of paddy production in India. This has been made possible through assured irrigation from groundwater (Sarkar, 2011).

There have been two major shifts in cropping patterns in the state after the green revolution. First, from a diversified cropping system, the state moved to a cereal-based crop system. Second, even within cereals, there emerged specialization in wheat and rice production at the cost of coarse cereals and pulses. Punjab, which was never a rice growing state, is now one of the largest producers of rice in India. Rice cultivation in semi-arid Punjab is completely dependent on groundwater irrigation and hence water availability becomes the major determinant of rice cultivation (Singh, 2004).

Punjab's agricultural growth and shifts in cropping pattern to a rice-wheat system have been made possible by rapid expansion in irrigated area. Punjab has achieved irrigation coverage of 98% of net sown area, which is the highest in the country. There are only two major irrigation sources in Punjab, the tube wells and the canals, which together contribute 99.75% of irrigation in the state. Canal irrigated area was reduced from 45% in 1970 to 22% in 2012 while tube-well irrigated area increased from 54% to 78% during the same period (Government of Punjab, several years). This trend became particularly pronounced after 1997, when electricity was made free for agriculture in the state (Kaul *et al.*, 1991).

A collateral damage to fast growing agriculture supported by subsidized electricity has been groundwater. As per CGWB website (http://cgwb.gov.in/gw_profiles/st_Punjab.htm downloaded on 26th December 2011), the state has a net annual groundwater availability of 21.44 BCM, while the annual groundwater draft is 31.16 BCM – that is a level of groundwater development of 145%. Of the 138 blocks in the state, 103 blocks are in over-exploited category and 9 blocks are in critical and semi-critical category. In almost all the districts in the state, water levels are declining by 1 m to 4 m annually. Since the increase in groundwater irrigation was made possible through the electrification of tube wells, Punjab's agriculture got highly dependent on electricity supply.

Agricultural power consumption in Punjab has increased from 4,253 million units in 1994–95 to 10,843 million units in 2010–11, an increase of 155% during this period of 17 years. Consumption of electricity per ha of gross cropped area has similarly increased from 552 kWh/ha in 1994–95 to 1370 kWh/ha in 2010–11, an increase of 150% again. This increase is due to a number of reasons such as an increase in the area under paddy crop, a shift in pumping technology from mono-block to submersible pumps and a decline in water levels. At the same time, the number of agricultural power consumers has gone up from 0.7 million to 1.15 million during the same period, an increase of 63%. But more importantly, connected load has gone up from 2,412 MW to 7,728 MW during the same period, an increase of 220%. Such rapid increase in agricultural consumption within a span of 17 years can be explained by two factors. First, towards the end of 1990s, a technological shift took place in response to rapid decline in groundwater levels. Farmers started replacing their mono-block pumps with submersible pumps. Since these submersible pumps drew water from deeper levels, electricity consumption went up. In 1997, the government of Punjab announced free electricity and since then demand for new connections has gone up substantially. Corresponding to an increase in agricultural power consumption, the quantum of subsidy has gone up by 130%, from INR 11.37 billion in 2002–03 to Rs. 26.16 billion in 2010–11 (both at constant 2004–05 prices).

15.4.2.1 Electricity reforms and its implications for groundwater in Punjab

Punjab faces similar problems of groundwater over-exploitation as Karnataka. But unlike Karnataka, in Punjab, irrigation is much more intensive, and farmers' incomes are considerably higher. Yet, agricultural electricity is just as politically charged as in Karnataka. Punjab has adopted somewhat similar measures to curb use of electricity for agriculture. But their implementation approach has been different leading to better results.

15.4.2.1.1 Successful attempts by the State Electricity Regulatory Commission (SERC) to streamline agricultural power subsidies

The biggest bone of contention vis-à-vis power subsidy to agriculture is the uncertainty in calculations of agricultural power use in the absence of metering of agricultural tube wells. Past studies have shown that utilities hide their inefficiencies in the garb of agricultural power and that agricultural consumption is much inflated – sometimes to an extent of 20–30% over actual figures (Sant & Dixit, 1996; World Bank, 2001). However, since the setting up of Punjab State Electricity Regulatory Commission (PSERC) in 2000, there have been changes in the way agricultural power consumption is estimated. Up to the year 1999–2000, agricultural consumption was assessed by deducting the metered energy and predetermined transmission and distribution losses from the energy input into the system; the balance was shown as agricultural consumption. This was fraught with errors given that all inefficiencies in electricity supply (including theft by other consumers) were lumped under the head of agricultural electricity and an inflated amount of subsidy was being claimed by the electricity company from the state government. From 2001 onwards, PSERC insisted that the state electricity utility furnish estimates of agricultural electricity use by farmers based on a small sample of metered tube wells monitored by the utility. This is somewhat similar to the approach followed by Karnataka at present. However, PSERC found that estimates by the electricity company were unreliable and recommended that the company adopt a larger sample of metered tube wells for electricity consumption calculations. The Commission also recommended that these sample tube wells be distributed across the state in such a way that all agro-ecological zones are correctly represented. They also enlisted services of an independent third party to do random checks on these meter readings furnished by the state electricity utility (PSERC, several years). All these efforts by the PSERC have ensured that the electricity utility cannot inflate the amount of electricity supplied to the farmers and therefore, cannot claim a higher subsidy. As a result, farmers here are less distrustful of the electricity company than they are in Karnataka.

15.4.2.1.2 Attempts to provide better quality power supply to farmers

Like Karnataka, Punjab also limits agricultural power supply to 6–8 hours in a day. However, unlike Karnataka, where power supply is also of poor quality leading to motor burn-outs, in Punjab the electricity company undertook various schemes to improve the quality of power supply to farmers. First, through a program called Urban Pattern Supply (UPS), all mixed feeders (feeders that supply electricity to both rural houses and for agriculture) were physically segregated into feeders that supplied electricity only to agricultural pumps and feeders that catered to the other electricity needs of the villages. This is akin to Karnataka's *Niranthara Jyoti* scheme. But unlike Karnataka, there was no provision kept for single phase supply in 3 phase agricultural feeders, thereby making it impossible for farmers to run their pumps outside of the hours of 3 phase electricity supplied to these agricultural feeders. In the feeders supplying electricity to rural houses, 24*7 electricity was provided, while electricity was rationed to 6–8 hours in feeders that connected irrigation pumps. Farmers were willing to receive less hours of electricity provided this electricity was of good quality – that is, it was uninterrupted and without voltage fluctuations. To ensure this, the electricity utility converted feeders supplying electricity to agricultural pumps from a low voltage distribution system (LVDS) to a high voltage distribution system (HVDS). Anything from 1 to 5 pumps are connected to one HVDS transformer. HVDS has several advantages: – it provides better quality power supply, while reducing line losses and because only a few pumps are connected to a HVDS transformer, any malpractices by farmers such as an unauthorized increase in pump horse power are easier to detect and penalize. Given that quality of power supply to agricultural pump sets has improved drastically after the introduction of HVDS transformers, farmers are willing to pay a flat rate for electricity unlike their counterparts in Karnataka who receive poor quality electricity (IWMI, 2012). In both states, quantity of electricity is rationed, but Punjab, through its various initiatives, has managed to improve the quality of that rationed supply. However, while farmers, especially in HVDS areas, are happy with quality of power supply, the hours of supply of electricity in *kharif* season (rainy season) when paddy is grown fall short of their irrigation requirements. This forces them to use diesel generators and tractors for extracting groundwater and this is quite expensive. So, even though electricity is free, it is not sufficient and farmers need to spend on diesel, which incentivizes them to use energy efficient pumps and adopt laser levelling for better water management. The current model of subsidy delivery is also relatively effective in that the quantum of subsidy claimed in name of farmers does indeed reach them and margin of error in agricultural power consumption is not more than 3–5% given that PSERC closely scrutinizes all claims by the electricity company.

So, overall, the kind of anarchy below the feeder level that we find in Karnataka is conspicuous by its absence in Punjab. However, groundwater over-exploitation has not been halted because farmers continue to grow paddy due to the very lucrative food procurement system which fetches them good returns (Sarkar & Das, 2014). That lucrative minimum support price for paddy is the main reason behind groundwater over-exploitation was recognized way back in the mid-1980s (Johl, 1986), yet no tangible steps have been taken to address this issue due to both political sensitivity and farmers' livelihoods imperatives. It is unlikely that farmers will stop growing paddy unless procurement policies are changed. It is also unlikely that given India's dependence on Punjab state to grow as much as 1/4th of its paddy, food procurement policies will be changed any time soon. So, in effect, groundwater over-exploitation problems can be dealt only on the margins – so to say, and a plethora of electricity related regulations mentioned earlier has tried to do just that. These electricity policies and regulations have incentivized farmers to increase energy use efficiency on the one hand (through adoption of energy efficient pumps and laser levelling), and on the other hand has forced the electricity utility to do a better job of energy accounting and reducing their own systemic losses which were being passed on to the farmers. This, we contend, even though a marginal step in the overall groundwater sustainability challenge of Punjab, is still a step in the right direction.

15.5 CONCLUSION: MANAGING ENERGY-IRRIGATION NEXUS IN INDIA IS A GOVERNANCE ISSUE

In this chapter, we argue that at the heart of a unique irrigation-energy nexus in India is the de-metering of agricultural pump sets which makes it difficult to do proper energy accounting and provides perverse incentives to both farmers and the electricity utilities, who in turn, undertake malpractices that further worsens the viability of the electricity utility and also leads to groundwater over-exploitation. While subsidized or free electricity is also a part of the problem, it was the decision to de-meter all electric the tube wells around the early 1980s that makes India's energy-irrigation nexus more intractable than anywhere else. Over the years, a poor energy accounting system led to poor financial health of most state owned electricity utilities leading to overall poor infrastructure and electricity supply conditions. The worse affected were the farmers, who were seen as the major cause for financial non-viability of the electricity sector. By early 2000s, most of the state owned electricity utilities were unbundled and separated into generation, transmissions and distribution companies to promote efficiency and productivity. Independent state electricity regulatory commissions were also set up around this time to ensure transparency in operations of electricity utilities. While the basic rules of the game remained the same for all Indian states, the very fact that water, electricity and agriculture are all state subjects, meant that individual states implemented these policies and regulations differently.

We have discussed the case of two states – Karnataka and Punjab – which faces very similar groundwater and electricity issues, and yet, the state of Punjab implemented electricity related policies in such a way that it had a considerable impact on farmers' behavior. First, they successfully limited hours of electricity available to farmers through segregating agricultural and rural feeders. Second, to compensate for the reduction in the number of hours of electricity, the electricity company also installed HVDS systems that substantially improved the quality of power supply. These two measures incentivized farmers to adopt water and energy efficient technologies like energy efficient pumps and laser levelling of fields. The electricity company in turn was incentivized to adopt these measures because the state electricity regulatory commission constantly scrutinized their subsidy claims and pushed them to provide reliable estimates of energy supplied to farmers. In Karnataka, however, even though the electricity utility also adopted measures like feeder segregation, they designed it in a way that was bound to fail and undermined the very purpose of segregation. While agricultural feeders have been physically segregated from non-agricultural feeders and given 3 phase electricity for 6–8 hours in a day, these segregated agricultural feeders have also been provided with 10–12 hours of additional electricity though single phase supply. It is well known that farmers are able to convert single phase electricity through locally fabricated capacitors and in the process can undermine the entire system. At the same time, we also saw that the KERC has not been able to exert enough influence on the electricity utilities in the state to improve their agricultural electricity accounting system and hence, overall, the utilities in Karnataka do not have as much incentive as utilities in Punjab to do reduce their transmission, distribution and commercial losses.

The difference in outcomes between the two states is related to good governance and political will to enforce unpopular decisions. In Punjab, the PSERC has been able to use its constitutional powers and ensured compliance by the electricity utility. PSERC did so by consistently rejecting all claims for subsidy that they thought were not justified. The KERC on the other hand, has been historically approving inflated claims for subsidy by the state utilities, while expressing their dismay and reservation over the same. We conclude by noting that while de-metering of electric tube wells makes India's energy irrigation nexus particularly difficult to manage, yet, some positive changes are possible if only the states have political willingness to take difficult decisions and follow it thoroughly in a way that allays farmers' fears and meets their needs as much as possible, while ensuring transparency in functioning of electricity utilities.

ACKNOWLEDGEMENTS

This paper is based on primary work conducted under ESMAP funded project to IWMI which looked at alternate ways of delivering power subsidies to agricultural consumers in India with a special focus on states of Karnataka and Punjab. The author is now employed by ICIMOD and acknowledges the core funds provided by the governments of Afghanistan, Australia, Austria, Bangladesh, Bhutan, China, India, Myanmar, Nepal, Norway, Pakistan, Switzerland, and the United Kingdom to ICIMOD. The views and interpretations in this publication are those of the authors and are not necessarily attributable to ICIMOD.

REFERENCES

Agricultural Census (1995–96), *Report of the Agricultural Census*, 1995–96, available at http://agcensus.dacnet.nic.in/statesummarytype.aspx, Ministry of Agriculture, Government of India, New Delhi.

Agricultural Census (2000-01), *Report of the Agricultural Census*, 2000–01, available at http://agcensus.dacnet.nic.in/statesummarytype.aspx, Ministry of Agriculture, Government of India, New Delhi.

Agricultural Census (2005–06), *Report of the Agricultural Census*, 2005–06, available at http://agcensus.dacnet.nic.in/statesummarytype.aspx, Ministry of Agriculture, Government of India, New Delhi.

Census of India (2011), Population Census Report, 2011, data downloaded from http://www.censusindia.gov.in/2011census/population_enumeration. html on 10th August, 2011.

Central Groundwater Board, (2006), *Dynamic ground water resources of India (as on March 2004)*, Central Groundwater Board, Ministry of Water Resources, Government of India, New Delhi, India.

Central Groundwater Board, (2007), *Dynamic Groundwater Resources of Punjab State (as on March 2007)*, Central Groundwater Board, North Western Region, Ministry of Water Resources, Government of India Chandigarh, India.

Central Groundwater Board, (2010), *Ground Water Scenario of India 2009–10*, available at http://cgwb.gov.in/documents/Ground%20Water%20Year%20Book%202009-10.pdf on 10th October 2012.

Government of India, (several years), *Agricultural Statistics of India*, Ministry of Agriculture, Government of India.

Government of India. (2005), *Report on Third Census of Minor Irrigation Schemes* (2000–01). New Delhi: Ministry of Water Resources, Minor Irrigation Division.

Government of Karnataka, (several years), Statistical Abstract of Karnataka, Directorate of Economics and Statistics, Bangalore.

Government of Punjab. (several years), *Statistical Abstract of Punjab*, Department of Agriculture, Government of Punjab, Chandigarh

Howes S. and Murgai R. (2003), Karnataka: Incidence of Agricultural Power Subsidies, *Economic and Political Weekly,* (38): 16, April 19–April 25.

Humphreys, E., S. S. Kukal, E. W. Christen, G. S. Hira, B. Singh, S. Yadav, and R. K. Sharma. (2010), Halting the Groundwater Decline in North–West India—Which Crop Technologies will be Winners? *Advances in Agronomy*, 109(5): 155–217.

Input Survey (2006–07), Report of the Input Survey, 2006–07, available at http://inputsurvey.dacnet.nic.in/statetables.aspx.

Input Survey (1996–97), Report of the Input Survey, 1996–97, available at http://inputsurvey.dacnet.nic.in/statetables.aspx.

Input Survey (2001–02), Report of the Input Survey, 2001–02, available at http://inputsurvey.dacnet.nic.in/statetables.aspx.

International Water Management Institute (2012). Alternate ways of delivering electricity subsidy to agricultural consumers in Punjab and Karnataka, Final Project Report submitted to Energy Sector Management Assistance Program (ESMAP), May 10th, 2012. IWMI, New Delhi.

Johl, S.S. *et al.* (1986). *Report of the Expert Committee on Diversification of Agriculture in Punjab*, Government of Punjab.

Karnataka Electricity Board (KEB) and Electricity Regulatory Commission (ERC) filings of Karnataka Power Transmission Corporation Limited (KPTCL) & Electricity Supply Companies (ESCOMS), several years.

Ministry of Agriculture, Crops Weather Watch Group, August 28, (2009). Minutes of the meeting of the crop weather watch group held on 28.08.2009. http://agricoop.nic.in/ncfcweather/ncfcasAug-28-2009.pdf.

Monari, L. (2002), *Power Subsidies: A Reality Check on Subsidizing Power for Irrigation in India*, The World Bank Group, April 2002, Note # 244, Washington DC.

Punjab State Electricity Regulatory Commission, (several years). *Annual revenue requirement filed by the Punjab State Power Corporation Limited.* Dowloaded from http://www.pserc.nic.in/pages on 10th August 2011.

Sakthivadivel, R. (2007). The groundwater recharge movement in India. In Giordano, M. & K.G. Villholth (Eds). *The Agricultural Groundwater Revolution: Opportunities and Threats to Development*, CABI and International Water Management Institute, Colombo.

Sant, G. and S. Dixit (1996), Beneficiaries of IPS subsidy and impact of tariff hike, *Economic and Political Weekly*, 31 (51): 3315–3321.

Sarkar, A. (2011). Socio-economic Implications of Depleting Groundwater Resource in Punjab: A Comparative Analysis of Different Irrigation Systems. *Economic and Political Weekly,* 46 (7): 59–66.

Shah, Tushaar; Burke, J.; Villholth, K.; Angelica, M.; Custodio, E.; Daibes, F.; Hoogesteger, J.; Giordano, Mark; Girman, J.; van der Gun, J.; Kendy, E.; Kijne, J.; Llamas, R.; Masiyandima, Mutsa; Margat, J.; Marin, L.; Peck, J.; Rozelle, S.; Sharma, Bharat R.; Vincent, L.; Wang, J. (2007). Groundwater: a global assessment of scale and significance. In Molden, David (Ed.). Water for food, water for life: a Comprehensive Assessment of Water Management in Agriculture. London, UK: Earthscan; Colombo, Sri Lanka: International Water Management Institute (IWMI). pp. 395–423.

Shah, T. and Verma, S. (2008). Co-management of Electricity and Groundwater: An Assessment of Gujarat's Jyotirgram Scheme. *Economic and Political Weekly,* 43(7): 59–66.

Shah, T., M. Giordano and A. Mukherji. (2012) Political economy of the energy-groundwater nexus in India: exploring issues and assessing policy options. *Hydrogeology Journal,* 20(5):933–941.

Singh, K. (2009). Act to save groundwater in Punjab: Its impact on water table, electricity subsidy and environment, *Agricultural Economics Research Review,* 22 (Conference Number): 365–386.

Singh, S. (2004). Crisis and Diversification in Punjab Agriculture: Role of State and Agribusiness. *Economic and Political Weekly*. 39 (52): 5583–5589.

Sinha, S., Sharma, B.R. and Scott, C.A. (2006). Understanding and managing the water-energy nexus: Moving beyond the energy debate, In: Sharma, B.R., Villholth, K.G. and Sharma, K.D. (eds.). *Groundwater Research and Management: Integrating Science into Management Decisions*, International Water Management Institute, Colombo, Sri Lanka, (2006), pp. 242–257.

van Steenbergen, F. (2006). Promoting local management in groundwater, *Hydrogeology Journal, 14(3)*3, 380–391.

Chapter 16

Steps towards groundwater-sensitive land use governance and management practices

Daniel A. Wiegant & Frank van Steenbergen
MetaMeta Research, 's-Hertogenbosch, The Netherlands

ABSTRACT

Land use patterns and management practices have a great influence on the quantity and quality of groundwater. Conversely, groundwater is instrumental in the development of land use activities by providing a source of water and secure moisture. This facilitates productive and consumptive water uses, but goes further by affecting natural soil fertility and the microclimate. As certain land uses and management practices have long-lasting and sometimes difficult to reverse impacts on groundwater, these are crucial factors in groundwater management. Proper land use management ensures that services like groundwater recharge and purification are safeguarded.

This chapter focuses on the impact of specific land use types and management practices on groundwater resources. Firstly, the main trends are highlighted that have come to define the relation between land use and groundwater, followed by the impact of land use planning on groundwater quantity and quality. Focus is then placed on practices that can bring groundwater tables and quality rates to an appropriate level. In doing so, 3R techniques to promote the retention, recharge and reuse of water are highlighted to address the water holding capacity of areas and expand the diversity of opportunities that communities have at hand. The chapter concludes by outlining a variety of policy instruments that exist to influence drivers that cause groundwater depletion and quality degradation.

16.1 LAND USE AND GROUNDWATER: THE MAIN TRENDS

Land use, land management and groundwater are inseparable. A place where this has become very clear is California's San Joaquin Valley, where several years of large-scale dry conditions – starting in 2012 – have altered the landscape for decades to come (BBC News, 2017). During the long drought, farmers dug more and deeper wells to reach the continuously declining groundwater and sustain their water-dependent land uses. In 2015 alone, 2,500 new wells were drilled, resulting in great pressure on the groundwater table and subsequent land subsidence of up to 60 cm a year. This has caused adverse effects on the future capacity of aquifers, making the Valley less prepared for future droughts.

Precipitation water that enters the soil's unsaturated zone is stored around soil particles where some of it is consumed for plant production and evaporation, while the rest further infiltrates to recharge groundwater. Groundwater accounts for 99 percent of the world's freshwater when snow and ice are excluded (Lerner & Harris, 2009). This massive water reservoir in the subsurface provides the base flows needed to keep most rivers flowing and ensure the integrity of groundwater-dependent wetlands (Scanlon *et al.*, 2012). Besides safeguarding ecological functions, groundwater fulfils a variety of human needs, of which agriculture, industry and urban water supply are the most prominent (EA, 2013). When land use at the surface, groundwater abstraction and climatic conditions remain unaltered, groundwater inflow and outflow are in a long-term equilibrium. However, today's reality is that groundwater quantity and quality are heavily influenced by land use, as the San Joaquin Valley example shows. The exact impact of land uses on groundwater depends on a variety of factors, which include the original land cover, the new land use, and accompanying land management practices. Land use change pathways are complex and shaped by numerous economic, political and technological forces.

16.1.1 The main land use drivers of groundwater quantity change

The conversion of natural areas – forests and grasslands – into human-dominated land use types – cropland, pastureland and urban areas – has transformed a large part of Earth's surface. Over the past 300 years, rain-fed cropland has expanded by 460 percent and pastureland by 560 percent at the global level, largely at the expense of natural forests and grasslands. In this conversion process, groundwater is mainly affected by 1) a conversion to land uses that require considerable groundwater abstraction and 2) changes in land cover that alter recharge rates. Scanlon *et al.* (2007) claim that the effects of these land use changes on water resources may rival or exceed the threats posed by climate change on surface and groundwater resources.

When natural vegetation is cleared, runoff and peak discharge into streams generally increases, as the soil's water holding capacity, and evapotranspiration by the canopy decrease. There are considerable differences in evapotranspiration demand of trees and shrubs on the one hand, and cropland and pastures on the other. Deforestation has reduced evapotranspiration by 4 percent at the global level, increasing water availability for other hydrological processes. In Niger, the conversion of natural savanna to rain-fed cropland increased recharge by one order of magnitude from 1–5 mm a year to 10–47 mm a year, resulting in large groundwater level rises between 1963–1999 of 0.01–0.45 m a year (Scanlon *et al.*, 2007).

During the twentieth century, irrigated agriculture expanded by 480 percent, from 47.3 Mha (1900) to 276.3 Mha (2000) and is projected to further increase with 20 percent by 2030 in the global South. Scanlon *et al.* (2007) claim that irrigated agriculture consumes about 90 percent of the annual flux of global water resources. There are large differences between irrigation that is fed by surface water and irrigation based on groundwater. In case of the first, surface water based irrigation has made groundwater levels to rise as a result of leakage and higher recharge in many systems. In canal-irrigated systems in China's Shaanxi Province for example, groundwater levels rose from 15–30 m (1930s) to 1–2 m (1950s), while in the Xinjiang Province water rose from over 7 m (1950s) to 1 m (1989), after diversion of river water for irrigation.

The reality is quite different in irrigation systems that are fed by groundwater. Particularly in semi-arid and arid areas with high population growth and socio-economic development, interest has turned to groundwater use due to its relative reliability. The western United States are an example of how land use change and management practices have drastically influenced groundwater levels. The groundwater-irrigated parts of the High Plains and the Central Valley in California have accounted for about 50 percent of all groundwater depletion in the United States since 1900 (Scanlon *et al.*, 2012). This resulted in water table declines of over 120 m in parts of the Central Valley, and of 43 m in parts of the Southern High Plains.

Groundwater has brought many economic gains but when groundwater abstraction goes beyond long-term recharge rates it results in groundwater storage depletion. Aquifers around the world are increasingly under pressure due to almost universally unsustainable extraction rates. Wada *et al.* (2010) estimate that over two billion people face the consequences of severe water stress, particularly in parts of the Indian sub-continent, Arabian Peninsula, Sahel, North-East China and western United States. Groundwater depletion in these semi-arid regions is likely to jeopardise future water security as declining groundwater levels result in the drying of base flows and groundwater-dependent wetlands, and the subsequent die-off of sand fixing and wind breaking vegetation, with desertification as an outcome (Scanlon *et al.*, 2007). This trend shows the need to focus on total system performance, in which both increasing groundwater recharge and reducing water demand can reduce groundwater table depletion.

16.1.2 The main land use drivers of groundwater quality change

There is a range of land uses and management practices with growing pollution impacts on groundwater, despite the fact that groundwater is relatively well-protected against pollution compared to surface water (EA, 2013). Evidence is pointing to groundwater quality degradation as a result of salinification, fertilisers and urban and industrial effluents (FAO, 2015b). Some pollutants occur naturally, but become harmful when they are concentrated as a result of human activities. Others are man-made and normally do not occur in nature. Groundwater pollution is a major concern that limits future land use options in a landscape, particularly when the pollution is persistent and too costly or technically difficult to mitigate. Land use can create two main types of pollution that impact on groundwater resources:

- *Point source pollution* is a localised form of pollution that is relatively easy to trace. It is caused by well-defined events like spills, leaks and discharges in a small area (EA, 2013). Examples are nitrate leakage from intensive livestock husbandry, leaking sewers and septic tanks, or contaminated land and landfills in urban areas. Point source pollution mainly affects water supplies services when the two are relatively close to each other. A variation to point source pollution is line source pollution along canals, highways or pipelines;
- *Diffuse pollution* is spread across much larger areas and time periods, and tends to be the cumulative effect of many individual events. Separately, these events may be small and hard to detect. Altogether however, these events have a notable impact on groundwater quality (EA, 2013). Agricultural land management practices primarily cause diffuse pollution through the use of nutrients and pesticides, particularly so in areas where intensification is taking place that is based on high-yielding

monocultures. Although there are differences in their chemical and physical composition, pesticides can easily pollute groundwater. The risk lies in their level of persistence and although some pesticides may degrade, their breakdown products can still be toxic. Once it has occurred it is difficult to treat diffuse pollution, and therefore prevention is the only realistic option to safeguard the groundwater's services (Lerner & Harris, 2009).

Large volumes of pollutants can be stored in the unsaturated zone of the soil before such pollution starts having noticeable effects on the ecosystem. The process of percolating into the groundwater can take years, decades or even centuries, depending on the characteristics of the aquifer and the distance the contaminants must travel (Lerner & Harris, 2009). Even when pollutant leaching would stop within the foreseeable future through improved agricultural practices, it may still take decades and even centuries for concentrations to drop, as a large load of pollutants is slowly making its way through the unsaturated zone into groundwater aquifers.

16.1.2.1 Salinification

Salt is a natural groundwater pollutant with a high solubility. It has accumulated in the subsurface in many semi-arid regions, and primarily impacts groundwater quality when it is mobilised in the conversion process from natural to agricultural land uses, and the associated increased recharge. This results in more discharge of saline groundwater and higher salt loads in streams and rivers (Scanlon *et al.*, 2007). In the southern High Plains of the United States, increases in groundwater salinity – 43 percent, from 150 to 214 mg/L – were linked to land use change, and the related flushing of salts that had accumulated in the soil in the past 15,000 years.

Salinification is a particular concern in waterlogged areas, where water tables have risen as a result of intensive irrigation. In parts of China's Shaanxi, Xinjiang and Henan Provinces, water tables have risen to less than a meter of the surface, with salinification occurring as a result of the upward movement of groundwater in combination with evapotranspiration. Also in parts of Australia, land use change has greatly increased water quantities, while negatively impacting on water quality. Scanlon *et al.* (2007) describe an assessment in Australia's dryland areas where 5.7 Mha of cropland and pastureland were affected by salinification from shallow water tables. However, salinification is an issue in all areas where irrigation takes place, as the water produces an increase of salts in the subsurface of which a part is flushed towards the groundwater.

16.1.2.2 Nitrate loading

Another widespread and problematic pollutant is nitrate. As a soluble compound of nitrogen and oxygen, nitrate is a natural substance that occurs in the soil (EA, 2013). It is the main form of nitrogen, being an essential nutrient for plant production. Like salt, nitrate deposits have naturally built up for millennia as a result of evapotranspiration concentrations from precipitation. However, recent land use change has increased recharge and nutrient flushing into underlying aquifers. More importantly, the global consumption of nitrogen fertiliser – in the form of chemical fertilisers or livestock manure – increased by 600 percent between 1961–2000 (Scanlon *et al.*, 2007). Nutrient inputs from applied fertilisers have now exceeded natural sources of nitrate and cause large-scale effects on groundwater quality (Foley *et al.*, 2005). Any nitrate not used by crops or vegetation is susceptible to leach into the groundwater, which may restrict the suitability of groundwater for several uses. Such leaching may contaminate drinking water and result in health risks as well as the eutrophication of surface water.

Scanlon *et al.* (2007) describe how land use change in the Mississippi Basin has impacted nutrient loading into groundwater, streams and coastal waters. Between 1950–1970 and 1980–1996, a 500 percent increase in nitrogen fertiliser application was observed, as well as a 200 percent increase in nitrate export from the Basin into the Gulf of Mexico, that led to seasonal oxygen deficiencies in parts of the Gulf. The Abbotsford-Sumas aquifer on the border between Canada and the United States is another area where unchecked irrigation practices and dairy industry intensification have caused high nitrate concentrations. This has resulted in the groundwater to be of unacceptable quality for human consumption (Norman & Melious, 2004). A 1999 USGS study concluded that 21 percent of surveyed wells exceeded the maximum contamination levels set by public agencies from both countries.

16.2 DESIGNING GROUNDWATER-SENSITIVE LAND USES

16.2.1 Groundwater-related ecosystem services

Groundwater degradation is just one symptom of a broader trend of land use change with the aim to fulfil immediate human needs (Foley *et al.*, 2005). Such land use conversion trends have adversely affected the resilience of entire ecosystems, and the provision of ecosystem services like clean air, fresh water and healthy soils (SER, 2008). Ecosystem services are subdivided into (i) *supporting* – nutrient cycling, soil formation, primary production –, (ii) *regulating* – erosion control, microclimate balancing, water purification, water retention, recharge, flow regulation – , (iii) *provisioning* – food, water, wood, fibre, fuel – and (iv) *cultural* – well-being, spiritual.

Indications are mounting that vital services like groundwater recharge are compromised in highly intensified agricultural landscapes (Landis, 2016). Such intensification has driven these areas on a path in which ecosystem provisioning services like food and timber production have come at the cost of regulating and supporting services (Bouma & Van Beukering, 2015). This trend results in land degradation, evidenced through soil compaction, erosion and loss of soil organic matter. This negatively affects the soil's hydrologic properties such as its permeability and water holding capacity (Öborn *et al.*, 2015). While, economic gains of certain land uses may benefit specific communities or individual farmers, the resulting depletion in ecosystem services is felt across multiple scales and communities.

Different land use and management types show varying degrees of ecosystem service coverage. Figure 16.1 shows clearly how natural ecosystems support a large variety of ecosystem services on a high level, except for crop production, being a provisioning service. Intensive cropland on the other hand produces food in abundance, but this happens at the detriment of other ecosystem services. This shows clearly that the conversion of natural land to human-dominated land uses has resulted in the degradation of the same ecosystem services upon which humanity depends, while solely focusing on

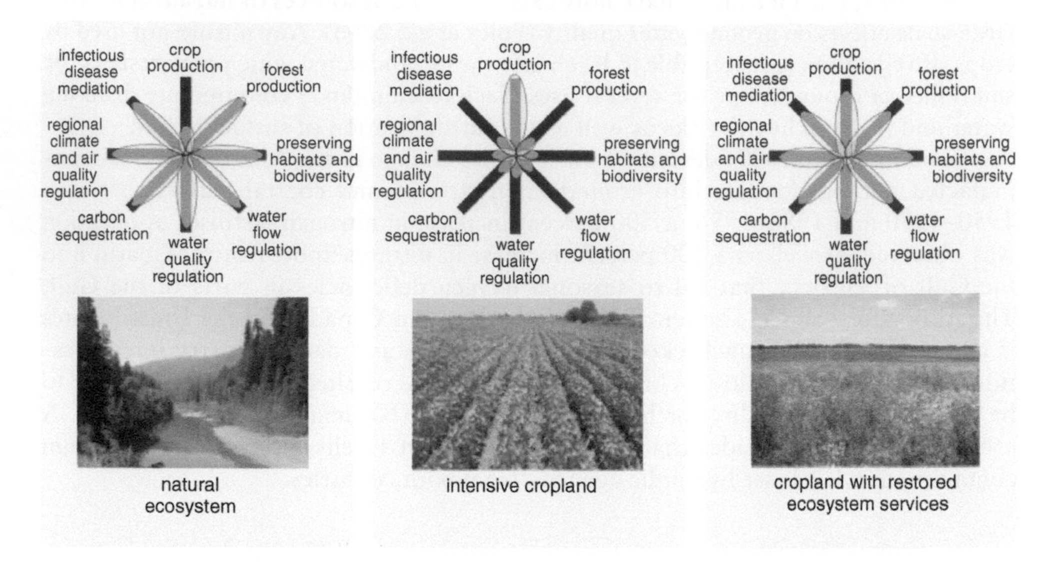

Figure 16.1 The provision of ecosystem services under different land uses. (Source: Foley *et al.*, *2005*).

the fulfilment of human needs. In the third example, cropland is purposefully managed to maintain ecosystem services, while also providing food, albeit to a lesser degree. Most important from a groundwater point of view, cropland with restored ecosystem services is better able to regulate water quality and water flow, greatly benefiting the water holding capacity and balanced base flow discharge of an area.

The availability of high-quality and sufficient water has for long been taken for granted. However, the consequences of groundwater depletion and quality deterioration on the livelihood options that communities have at hand are currently far-reaching. There is a need to assess and manage trade-offs between meeting human needs for food, fodder, fibre, and fuel, and to maintain the capacity of ecosystems to provide water buffering and water purification (Foley *et al.*, 2005). Both approaches and measures that improve water infiltration while maintaining groundwater quality are important now. The restoration of ecosystem services will prove to be more difficult and take more time as a system further degrades. To safeguard groundwater-related ecosystem services, it will be crucial to adopt a holistic planning perspective that transcends specific land uses, while at the same time efforts are made to improve management practices within such land use types.

16.2.2 Adopting a landscape perspective to safeguard groundwater flows

The land use composition and land management practices in a landscape greatly determine its natural recharge and water availability across spatial and temporal scales (Öborn *et al.*, 2015). A key result of agricultural intensification is the simplification of once heterogeneous landscapes, and the related reduction in ecosystem services

(Landis, 2016) including recharge, buffering and water purification. The intensification of agriculture has been characterised by the enlargement of agricultural fields and has greatly reduced the amount of natural areas that once surrounded farmland to provide such services. Forests have been cut, grasslands tilled and wetlands drained, creating simplified, monotonous land uses. In the Upper Mississippi Basin, 14.1 Mha of wetlands were lost, mostly as a result of drainage for agriculture between 1780–1980 (Scanlon *et al.*, 2007). With denitrification being a key function of wetlands, their disappearance critically reduces the ecosystem's capacity to purify nitrate-contaminated groundwater. In reaction to this, Foley *et al.* (2005) suggest that a strategic placement of managed and natural land uses can ensure that ecosystem services like denitrification and runoff infiltration are available across the landscape mosaic.

This means that land use design needs to focus on increasing the overall multi-functionality of landscapes to provide a balanced set of services. As seen in the example of the cropland with restored ecosystem services, yields are typically lower in more heterogeneous areas. However, they are sustainable on the long term, as water purification and the soil's water holding capacity have the opportunity to expand. Wetlands, grasslands and forests can bring back functions that were lost in intensification processes, like base flow regulation and water purification. Restored wetland and riparian buffers between agricultural land and adjacent streams can reduce nitrate loading into the water (Scanlon *et al.*, 2007). Giving room to streams instead of constraining them in narrow embankments can reduce their velocity and enhance groundwater recharge through floodplain expansion (Steenbergen & Tuinhof, 2009). Trees and wetlands that surround fields can also modify the micro-climate, which is beneficial for crop production (Öborn *et al.*, 2015). By providing the example of a typical agriculture landscape in Michigan (Figure 16.2) Landis (2016) shows that the landscape composition can change substantially within short distances and can hence be moulded according to planning priorities that consider multi-functionality.

16.3 IMPROVING SOIL'S WATER HOLDING CAPACITY AND WATER QUALITY

16.3.1 Water buffer management

Besides a focus on land use composition, a variety of land management practices is instrumental to build a water buffer for bridging the temporal disconnections that exist between water supply and demand. Water buffer management helps an area by increasing natural recharge into the subsurface when water is abundant and flows need to be managed to prevent erosion, while making water available during periods when it is scarce. The magnitude of dry season flows is an important indicator for the strength of the water buffer and the ability of an area to deal with variability and uncertain circumstances. With this aim to promote water buffering, the 3R concept was developed, focused on the retention, recharge and reuse water (Steenbergen *et al.*, 2011):

- *Retention* aims to reduce the discharge of groundwater. It helps raise the groundwater table and create large wet buffers. Slowing down the discharge or run-off of water positively affects soil moisture, which can have a large impact on the health of the vegetation cover and on agricultural productivity;

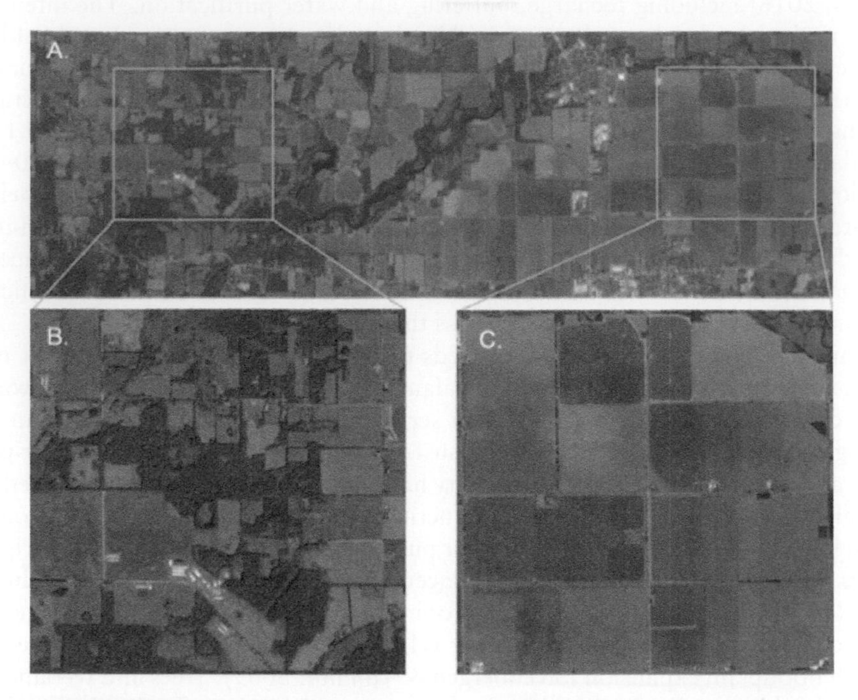

Figure 16.2 A) Differences in land use. B) Complex land use pattern that includes land uses that favour groundwater infiltration in between agricultural production areas. C) Intensive agricultural land use with fewer possibilities for groundwater infiltration and less water holding capacity. (Source: Landis, 2016).

- *Recharge* adds water to the buffer. This can be natural through the infiltration of rainwater and run-off, but it can also be artificial, through the construction of water harvesting structures or considerate planning of roads. Storing excess surface water underground reduces evaporation impacts;
- *Reuse* is the action of cycling the water in the same area as much as possible prior to discharging it. It entails an extension of the chain of water uses for humans and the environment, in a way that minimises the requirement to tap into new water resources.

3R measures focus on groundwater and soil moisture storage, and range from low cost structures to sophisticated techniques (Steenbergen & Tuinhof, 2009). The landscape and livelihood settings in which 3R can be applied is vast and almost universal, ranging from arid and humid climates, as well as from hilly areas to flat plains. There is a wide variety of practices that can be chosen from. Some are well-known and others innovative, ranging from on-site water harvesting structures to improved water infiltration such as terraces, contour bunds, tree lines, recharge wells, basins and infiltration trenches to interventions at the landscape level, like reforestation, riparian buffers, infiltration trenches, water spreading structures and spate irrigation (Scanlon *et al.*, 2012).

Subsurface water storage has the advantage of low evaporation loss and relative protection against water pollution as the water percolates through the various soil

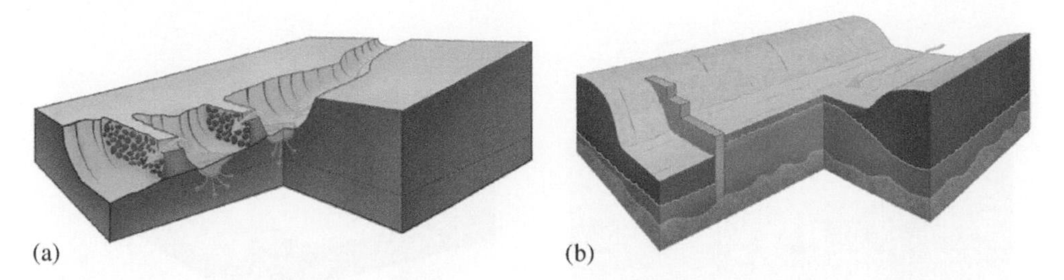

Figure 16.3 Groundwater effects of a) gully plugs and b) sand dams. (Source: *MetaMeta*).

layers. Suspended solids are absorbed by the soil, and with sufficient retention time in warm aquifers many pathogenic bacteria and viruses are eliminated (Steenbergen & Tuinhof, 2009). In addition, soils can reduce acidity, and chemical and biological processes can also change and neutralise hazardous components, thereby purifying the captured river water, surface water or storm water run-off. In the following, a number of common measures are elaborated that promote riverbed and land surface infiltration, and reduce run-off and evaporation.

16.3.1.1 Riverbed infiltration

Riverbeds are the primary drains through which water leaves an area. By constructing the following structures water can be better retained and shallow aquifers recharged, particularly those aquifers that are not covered by an impermeable layer of clay (Tuinhof *et al.*, 2012):

- *Gully plugs and check dams* are constructed in natural drains of an area, where they replenish the groundwater by increasing the infiltration and at the same time reduce erosion and stabilise the gully bed. As gullies are the lowest drains in an area, this measure raises the groundwater table;
- *Sand dams and subsurface dams* are relatively small and constructed in ephemeral riverbeds. Sediments that are transported by the river in the rainy season accumulate behind the dam. The sandy layer that is formed acts as a very small and shallow aquifer that is recharged by water running through the river. This subsurface reservoir can be used during the dry season;

16.3.1.2 Interception of overland flow

Before it enters riverbeds, runoff moves over the land surface (overland flow) during and shortly after heavy precipitation events. In this stage, it can be diverted for infiltration through various structures:

- *Floodwater spreading* is practiced in places where runoff is concentrated in the valley, where heavy floodwater brings the risk of washing away fertile soil. Flood-spreading structures span the entire width of a valley and spread peak flows to decrease the force of the water, while increasing the land on which infiltration takes place. Such weirs are usually built in series;

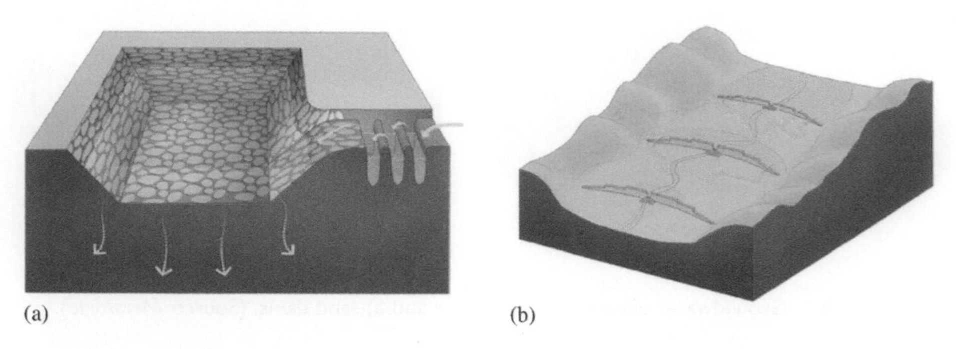

Figure 16.4 Groundwater effects of (a) infiltration ponds and (b) floodwater spreading. (Source: *MetaMeta*).

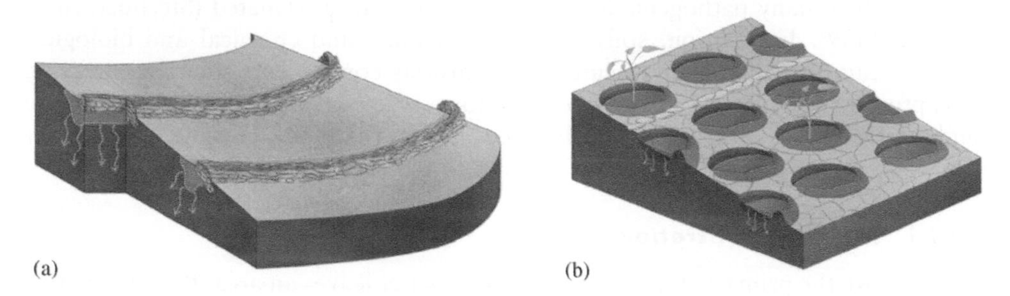

Figure 16.5 Groundwater effects of (a) stone bunds and (b) planting pits. (Source: *MetaMeta*).

- *Infiltration ponds and trenches* are dug along the slope in a way that they capture the overland flow. By holding water in one spot, they facilitate groundwater recharge in rangelands and for tree stands;

Other techniques that have the dual purpose of enhancing the storage of soil water and groundwater, and reducing the destructive force of flooding by overland flow runoff, are:

- *Terraces* locally reduce the hill slope and increase the water retention time for improved infiltration. A less costly option is construction of bunds and grass strips that provide obstacles to water that runs downhill. Such structures are important, as water falling on steep-sloped land runs of more quickly and infiltrates less than water falling on flat land;
- *Planting pits* exist in various sizes and are most suitable on soil with low permeability. They can be combined adjacent to bunds to create a micro-environment for annual and perennial plants, as well as specific trees (Tuinhof *et al.*, 2012);

16.3.1.3 Evaporation reduction

Water that evaporates can no longer circulate in the subsurface system. Therefore, besides improving water retention it is crucial to manage non-beneficial evaporation to facilitate water reuse. There is a fine balance between maintaining sufficient soil moisture and avoiding evaporation losses from the soil. Composting and mulching are important land management practices associated with conservation agriculture that enhance water availability for crops through minimised soil disturbance and improved infiltration capacity of the soil (Öborn *et al.*, 2015). Adding organic matter to the soil by composting and mulching increases the soil's water holding capacity, structure and the vigour of soil microorganism populations. Conservation agriculture decreases water loss as the soil is not overturned and has resulted in 50 percent higher crop yields in Kenya (Scanlon *et al.*, 2007). In addition, organic inputs greatly decrease the use of chemical fertilisers that bring the risk of nitrate loading in the soil.

16.3.1.4 Regreening

Besides the structures and techniques discussed above, the regeneration of natural vegetation is an important approach to increase the water buffer. Particularly in the mountains, slope land deforestation and structurally damaged soils have contributed considerably to high run-off and limited infiltration, and resulted in skewed spatial and temporal distributions of water (Tuinhof *et al.*, 2012). Tree roots and enhanced levels of soil organic matter from litter enhance the soil structure and micro-organic life, which again determine the porosity and water holding capacity of soils (Ismangil *et al.*, 2016). A dense vegetation cover particularly protects the soil from raindrop impact that increases the sealing of the soil and decreases infiltration. Litter on the soil in forests serves as soil protection from such splash erosion.

Reforestation is often highlighted as causing substantial losses in recharge, while deforestation results in increased recharge, suggesting a trade-off between carbon sequestration on the one hand, and groundwater recharge on the other. According to Ellison *et al.* (2017) this trade-off is biased as too much emphasis is placed on the total annual base flow, and too little attention is paid to dry season flows and groundwater recharge dynamics. In addition, long-term and large-scale water cycle relationships, like cloud formation and precipitation from vegetative evapotranspiration are neglected, while at least 40 percent of rainfall over land is said to originate from evapotranspiration. This suggests that a gain in vegetation cover can increase the reliability of rainfall.

Box 16.1: Managing the microclimate

When changes are made in a landscape, changes are made to the microclimate. When farmers plant trees in or around their field, and when communities dig bunds to improve water retention, they change the local climate around them. Microclimates are the localised, dynamic interplays between different processes in the surface layer, like energy and matter exchange, radiation processes and effects of the underlying surface (Ismangil *et al.*, 2016). These again are determined by the specific morphology, soil conditions, vegetation, land use and water retention. The microclimate determines the moisture available in the soil and air to the different ecosystems, the presence of dew and frost, the actual temperatures for plant growth and germination, the vigour of soil biotic life, capacity to fixate nitrogen by soil biota and the occurrence of pests and diseases.

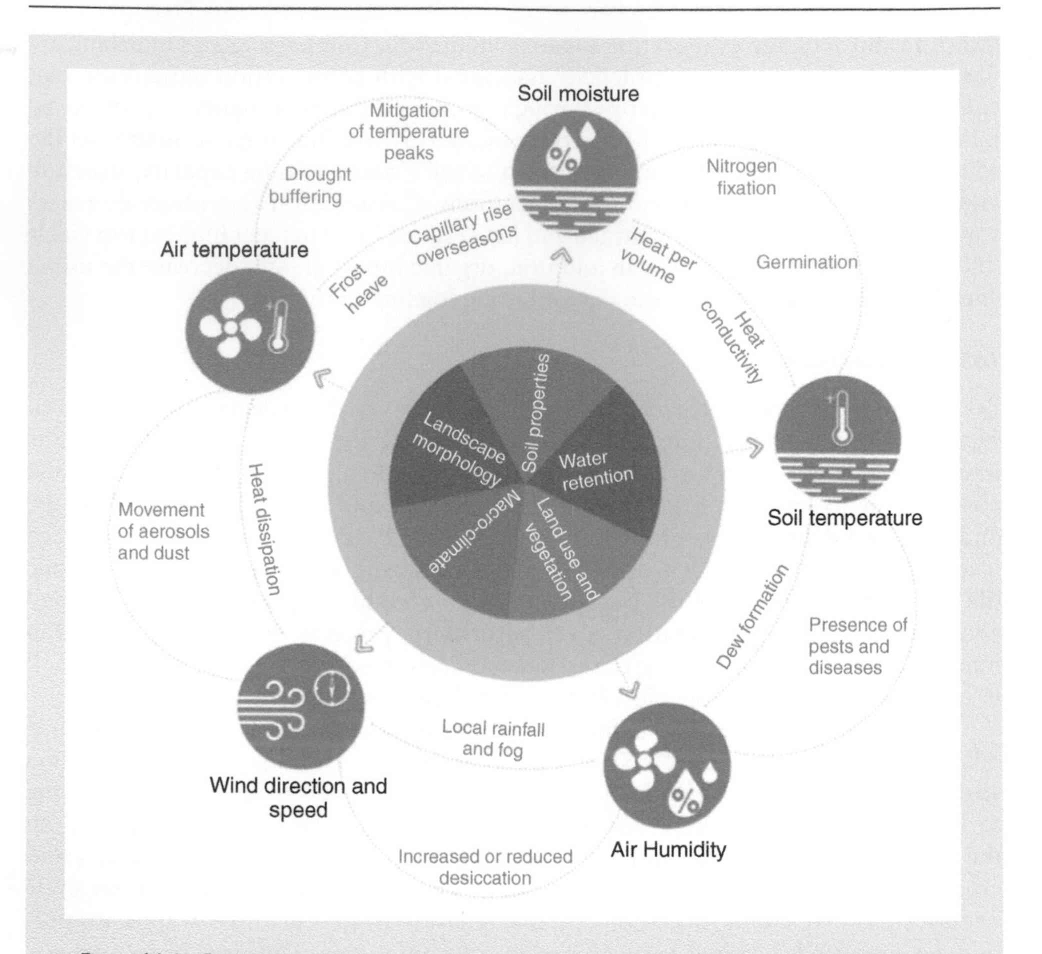

Figure 16.6 Components and interlinkages of microclimate. (Source: Ismangil *et al.*, *2016*).

The microclimate is composed out of a myriad of climatic conditions that come together in localised areas. These conditions include soil moisture and temperature, air temperature, air humidity and wind direction and speed (Figure 16.6). In view of the climate change that exists today, microclimates are a huge opportunity as they can buffer against droughts, temperature peaks, or more irregular or delayed rainfall, all with adverse effects on water availability. It is important to know the dynamics between microclimate components and across scales to adopt microclimate management as a pro-active approach to improve a landscape.

Besides water quantity, regreening also has positive effect on water quality. In parts of Australia, reforestation is considered as an option to reverse salinification challenges by lowering water tables. It was found that reforesting 70–80 percent of a catchment would be needed to achieve significant reductions in the groundwater table and salinity control, an evaluation of 80 sites in West Australia suggests (Scanlon *et al.*, 2007). Similarly, it was found that reforesting 45 percent of a catchment in south-eastern Australia would reduce salt loads by 30 percent. Infiltration through multiple soil layers causes forested catchments to deliver purified ground and surface water. Regreening also reduces the potential of nitrate leaching and deterioration of downstream water quality, as the growing grass and other vegetation takes up larger quantities of nutrients.

16.3.2 Cascading effects to achieve local transformation through improved recharge

Large-scale efforts are required to capture water through soil and water conservation measures (GWP, 2014). The purposeful design and planning of groundwater-sensitive land use and land management practices can create cascading or multiplier effects to support a landscape's hydrological functions and to re-establish connectivity between ecosystem services in places where they were once weakened or lost. Through a high density of measures at the landscape scale, it is possible to reach tipping points that change soil and water processes and make the difference between areas that are vulnerable to drought, wash-outs and erosion, or those that are highly productive landscapes (Tuinhof *et al.*, 2012). Higher water buffering in an area reduces the costs of a drought or an unusually wet period.

Covering a large part of an area with 3R-inspired interventions is important to generate on-stream benefits (secure base flows and groundwater table; reduce erosion sedimentation and flooding; control nutrient discharge; protect fisheries) and off-stream benefits (carbon sequestration through increased biomass; ecosystem connectivity; peace and stability). 3R measures can be scaled up in different ways. Replication of a single measure can be beneficial in areas where a specific technique is particularly useful, such as a cascade of sand dams or massive application of planting pits. Other areas might be more suitable to receive a variety of carefully planned and designed 3R applications as a package, like in areas where landscape morphology and land use composition varies greatly. To get measures precisely right for a local context greatly depends on local innovators, as experience from Niger and Burkina Faso with local experimentation and experience sharing related to zaï pits[1] shows. Investments in 3R methods do not have to differ from investments in roads, harbours or irrigation systems. Although the opportunities for better water buffering vary between landscapes, many examples show that they pay off economically – to society as a whole – and financially – to individual investors (Tuinhof *et al.*, 2012). Working in a way in which a multitude of 3R-inspired structures is implemented in the same area reduces costs, as knowledge and skills are effectively disseminated, supply chains for materials established and general changes to the livelihood system made. Large-scale recharge and buffer management projects can be found in parts of China, India, Ethiopia and Kenya, amongst others. In such areas, the water holding capacity was comprehensively transformed, allowing crop yields and animal stocking rates to increase dramatically and facilitating the introduction of new crops that are more sensitive to stress (Steenbergen *et al.*, 2011).

---

[1] A farming technique to dig pits in the soil of 20–30 cm long and deep and 90 cm apart to harvest water and to concentrate (organic) fertilizer.

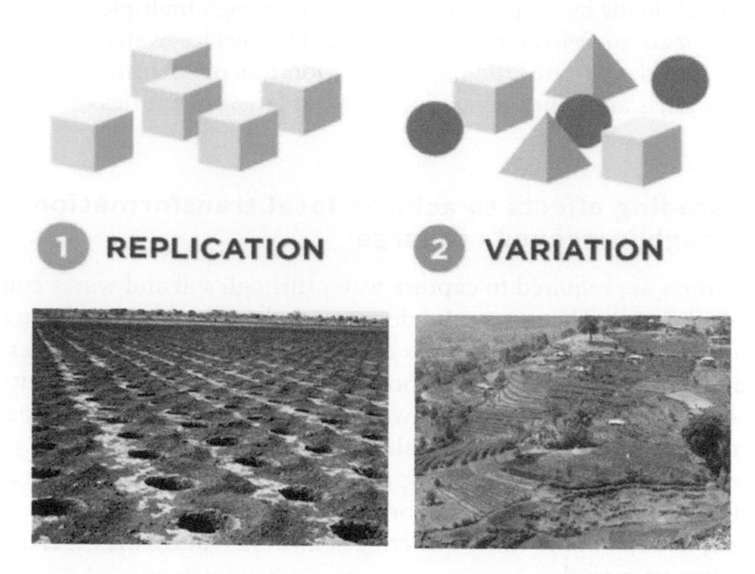

Figure 16.7 Different pathways to bring 3R to scale (Source: adapted from Tuinhof *et al.* (2012)).

16.4 IMPLICATIONS FOR GROUNDWATER MANAGEMENT AND GOVERNANCE APPROACHES

16.4.1 Adopting an integrated approach to link groundwater and land use planning

The role of land use planning in groundwater management has often been ignored, and external costs of certain land use practices on groundwater not considered. However, not giving due attention to the qualitative and quantitative recharge has detrimental effects on the ecological infrastructure, like base flows, groundwater quality and dependent ecosystems in the long run (GWP, 2014). Due to the common good characteristics of regulatory and supporting ecosystem services, individual farmers are reluctant to make investments when the individual benefits are not made clear. The implication of this is that 3R measures and groundwater management often need to be catalysed and largely taken up by government agencies, who on behalf of the public have to enforce sustainable use (Bouma & Van Beukering, 2015). The loss of ecosystem services as a result of human-dominated land use and land management practices can only be addressed through concerted efforts to fundamentally redesign land use composition (Landis, 2016). Given the interconnectedness of landscape processes, of which groundwater resources are a crucial part, a shift is needed from resource-focused policies to ecosystem-focused approaches that work towards creating heterogeneous, groundwater-sensitive land use and land management practices.

The Groundwater Governance Shared Global Vision for 2030 (FAO, 2015a) sets out a number of principles that guide how groundwater governance can be translated into practice. The first principle emphasizes that groundwater should not be managed in isolation but together with other sources of water. It is the interplay between surface and groundwater availability that can cover seasonal variations in human and ecological water needs. The second principle is to co-manage groundwater quantity and quality, and to do this groundwater management should be harmonised with land use planning and management practices. By integrating groundwater as a co-benefit in land use planning, ecosystem protection, infrastructure design and regional development, land and groundwater users are encouraged to adopt techniques and approaches holistically (Steenbergen & Tuinhof, 2009).

To develop landscape designs that recognise short and long-term needs in terms of groundwater quantity and quality issues, balance the spectrum of ecosystem services, and increase landscape resilience will require new models of research and extension, and close collaboration between researchers and practitioners (Landis, 2016). Groundwater-sensitive land use planning and land management practices need to be informed by hydrologists, geologists, ecologists, farmers and other land managers to develop context specific solutions. It requires scientists to work with practitioners to determine the desired mix of ecosystem services, evaluate the current landscape composition in light of that, and implement changes in land use and land management practices to achieve that. A range of skills is needed to better balance the continued flow of ecosystem services with human needs. For this, land managers, extension workers, environmental agencies and farmers need to understand advantages and trade-offs of landscape design.

16.4.1.1 Groundwater-sensitive land use planning process

Future land use planning needs to consider potential groundwater impacts, and consider trade-offs between water, salt and nutrient levels when balancing human and ecosystem needs (Scanlon *et al.*, 2007). A starting point is to understand natural recharge, storage and discharge processes: knowing what the porosity and water holding capacity is of local soils; how shallow groundwater travels through the soil; how it feeds rivers and base flows; and how it links to soil moisture to create conditions that are adequate for certain land uses in an area. Whether specific land uses and management practices work for groundwater in one context depends on landscape characteristics like climate, geology, soil, hydrology, current land use, availability of materials and the local capacity to get organised (Steenbergen *et al.*, 2011). The connection between groundwater processes and land use and land management practices is often poorly understood by policy makers, managers and farmers with no geological training, and it is crucial to safeguard and integrate such knowledge in policy and management processes.

Landscape development is inherently tailor-made as it builds on an area's unique characteristics. Buffer management needs to be based on an understanding of local livelihoods, priorities, needs and potential of people living in an area (Steenbergen & Tuinhof, 2009). According to Öborn *et al.* (2015) a first step in the direction to restore ecosystem services is to recognize the ways in which landscapes were traditionally utilized. An active learning loop is needed that builds on local experience. Many interactive tools and methods are available to set the basis for detailed discussions with local and regional stakeholders on how to improve an area's water quality and water holding capacity (Tuinhof *et al.*, 2012). In this way, the analysis and planning of land use integrates a 'theory of place' (understanding the land use and groundwater situation in a certain context) and a 'theory of change' (understanding how a system can be influenced through stakeholder involvement and management arrangements).

As a result of groundwater dynamics, long-time lags need to be considered between land use change and the impact on groundwater (Scanlon *et al.*, 2007). Impacts of land use change can be sudden, but are often gradual and delayed. For example, when an area is deforested, the repercussions of land use change and groundwater resources are immediate, and are felt in days to months. With reforestation however, it takes longer – years to decades – as trees require time to grow and reach maximum water use, and build the soil's water holding capacity. Hence, the full impact of historic land use changes might not always have been realised yet, and since impact mitigation also takes long, the incentive to focus on groundwater-sensitive land use management should come as early as possible. This implies the need for monitoring.

16.4.2 Different approaches for groundwater-sensitive land use and land management

Broadly, three categories of instruments can be identified to influence and control land use and land management practices, and assist the integration of groundwater and land use: policy and planning instruments; regulatory instruments; and participatory instruments. How adequate these instruments are, often depends on the strength of political and legal frameworks in a specific context, and on the existing threats to groundwater quality and quantity concerns (GWP, 2014).

16.4.2.1 Policy and planning instruments

Groundwater conservation and protection are an issue that is frequently referred to in agriculture, water and environmental policy guidelines at various levels. Informed by such guidelines, land use planning can be applied to support the ecosystem services provided by groundwater (water supply, base flow support and aquatic ecosystem health) (Lerner & Harris, 2009). A determining factor for sustainable groundwater management often lies in the quality in which groundwater concerns and opportunities are integrated into wider policies (FAO, 2015a). Particularly in areas that show serious competing claims and impact on groundwater resources to fulfil human needs, specific land use and management policy frameworks and guidelines need to be put in place to create an enabling environment that safeguards quality and quantity of groundwater recharge. National policy can for example influence the retention of natural forest areas or the promotion of extensive agriculture, with positive outcomes on groundwater conservation and protection. However, when forest protection in the upstream is enforced for downstream benefits, this can potentially restrict the livelihoods of upstream communities, giving rise to the need to fairly distribute benefits and burdens of land use throughout an area (Ellison *et al.*, 2017).

Groundwater protection zones can be included in local land use planning, when guidelines ensure acceptable land use. Groundwater-sensitive land use planning can demand that activities are placed in a preferred land use pattern that creates win-win situations for groundwater (Lerner & Harris, 2009). Here, a lack of ownership and a fragmentation of responsibility to manage groundwater can often be observed

(Tuinhof *et al.*, 2012). Although local government, farmers, water supply companies and environmental agencies all have part of the responsibility, it often happens that none has the overview or takes the lead. As groundwater is a local resource, effective local governance arrangements need to be established that have the agency to identify objectives, assign dedicated personnel and budget, promote stakeholder participation, and take the responsibility to materialise outcomes that consider issues in the wider geographical context (FAO, 2015a).

A key challenge in aligning land use and water vulnerability is the conflict between agricultural diffuse pollution and groundwater protection, as radical land use changes will be required in areas where farming is widespread, intensive and groundwater vulnerable. The issue is that external costs of agricultural practices on groundwater conditions and the ecosystem are often not properly accounted for. In nitrate-vulnerable zones, land use and land management planning could limit the amount of fertiliser and pesticides that farmers can apply to reduce pollutant leaching in the groundwater (EA, 2013). However, a more effective way to protect groundwater lies in policy processes that prevent pollution, *e.g.* those policies that regulate the agricultural products that may be used and which not.

16.4.2.2 Regulatory instruments

Regulations and statutory requirements are crucial to follow-up on policy and planning instruments. In places where groundwater protection zones are declared, it is needed that designated agencies can apply tight regulatory control on any land use and management restrictions imposed. Regulation could relate to the specific agricultural inputs like fertilisers and pesticides that are allowed, and the conditions under which these can be applied to ensure groundwater quality. Lerner & Harris (2009) see as a main challenge the general lack of control over most forms of diffuse pollution, including agriculture. Land users may not comply with the legal constraints to their land use and land management choices, either because of a lack of understanding of groundwater protection rules, or because of a lack of interest to meet the requirements.

Groundwater protection zones are useful to set out land use and land management priorities, agricultural pollution control and groundwater quality monitoring. Associated regulations often require changing the nature of most land management practices, with certain types of production being banned,

Box 16.2: The role of politics in groundwater management and governance

The political organisation of a society greatly influences groundwater management, as the cases of Ethiopia and Yemen show. In Ethiopia, the government has the ultimate claim on land and water resources, while in Yemen land and water ownership is privatised and anchored in local community claims (Steenbergen *et al.*, 2015). Both countries are in different stages of groundwater development. Groundwater use in Ethiopia is still in its infancy. In Yemen, the use of groundwater has been intense since the 1970s, with groundwater being the source for 80 percent of irrigated agriculture. This has led to a high depletion in the groundwater table and competition between different groundwater users.

Cooperation and conflict in groundwater is not only determined by resource competition, but more so by the interest or the indifference of political elites in groundwater development and management. The more directive role of the Ethiopian government and its strong presence at the local level to deal with resource management and conflict resolution, provide the needed agency to undertake extensive soil and water conservation measures that bring strong benefits to groundwater. In Yemen, such cooperation has not been sparked, despite the fact that water resource scarcity has reaching very high levels.

To reach transformation at scale, it is useful to understand what the political will is, and where it comes from in a given context. Understanding the political domain helps to grasp the 'logic' of decision-making mechanisms, regulatory frameworks, and the influence of interest groups in groundwater management. It helps to understand seemingly irrational phenomena, like the planning for seemingly unattainable targets in Ethiopia, or subsidies to large groundwater users in Yemen (Steenbergen *et al.*, 2015). Strong links exist between the political system, access to power, the role and effectiveness of different layers of government, the propensity to plan and form a vision, the scope for private initiative, and interest for groundwater management in general.

modified or displaced to other localities. It is suggested that it should be possible for public water supply utilities to compulsorily purchase significant parts of their groundwater source capture zones and only permit certain land uses that are subject to specific controls (GWP, 2014). Environmental impact assessments should then have a particular groundwater focus, to properly assess major land use change impacts. It could also be possible to establish statutory procedures to make it a legal requirement for local land use planning departments to consult with groundwater management agencies on significant land use changes.

In certain cases, groundwater could be brought under public guardianship, *e.g.* to license the development of water wells and to control localised point pollution. This can facilitate a better balancing of competing and conflicting interests among stakeholders, and enhance coordination between different land uses (FAO, 2015a). Limitations can also be imposed when aiming to ensure a certain quantity of water (hands off flow) for the environment (Lerner & Harris, 2009). When environmental flows fall below the hands-off level, abstractions need to be reduced or stopped. If regulations are very stringent and relatively widespread, they can have considerable repercussions on the local economy however.

16.4.2.3 Participatory instruments

Besides planning and enforcement, there are various instruments that provide incentives to land users to alter their land use and management practices towards ways that are more sustainable from a groundwater perspective. Agro-environmental management and stewardship schemes are examples of ways to stimulate land users to adopt different measures that enhance water quantity and quality, through land use planning, ecosystem restoration, removal of alien tree species and cropping practices, including mulching and composting. Land users are however likely to raise objections to practices that may reduce land productivity (GWP, 2014). Such situations can exist as long as the externalities to groundwater-insensitive land use are not properly accounted for. In addition, it is key to make farmers feel that they are the main beneficiaries of actions taken to conserve and protect groundwater. For example, farmers are interested and willing to plant trees when they see the co-benefits for their livelihood, microclimate, and water quality improvement (Ellison *et al.*, 2017). In such a case, buffer management measures can spread spontaneously, as is the case with the farmer managed natural regeneration of tree cover in Niger and Mali, each involving an area greater than 1 Mha (Tuinhof *et al.*, 2012).

Participatory instruments that mobilise individual farmers and communities need to consider the fact that some buffer investments are directly done at the landscape level, with costs that cannot immediately be attributed to a single user group (road water harvesting, retention measures in larger streams, gully plugging and hill top reforestation) (Steenbergen *et al.*, 2011). In that sense, it is crucial to develop support mechanisms and create synergies with public investment programmes and subsidies that are focused on watershed management. This needs to mobilise action and overcome long payback periods for individual farmers and local communities to carry out 3R practices on their farm and in their landscape (Steenbergen & Tuinhof, 2009).

Sharply focused land use management practices can produce great groundwater benefits with a relatively modest cost (GWP, 2014), ranging from nearly zero to over USD 2500 per ha. This range is quite low compared to irrigated agriculture, where investments range from USD 300 to 8000 per ha, with the lower costs being larger systems and the top end encompassing small-scale systems that require more tailor-making (Steenbergen *et al.*, 2011). Investment in natural resource management makes business sense as it helps to maintain or restore ecosystem services, optimize returns from sustainable agriculture and improve the livelihoods of those dependent on the landscape.

16.5 CONCLUSION

Land use changes with considerable impacts on groundwater recharge and quality are the clearing of natural vegetation, conversion to arable land, extension of irrigated agriculture, and intensification of agriculture. To give an answer to concerns that relate to land-use driven groundwater quantity depletion and quality degradation, large-scale and 3R-inspired water buffering, regreening, and purposeful land use planning are proposed. 3R measures will have a positive effect on groundwater recharge, as they hold soil moisture in the ground for longer and strengthen the water buffer. To duplicate this requires both governance and business models based on the benefits that integrated landscape management bring.

As a result of continued changes in drivers and pressures however, traditional land management measures alone are no longer sufficient, making it needed to adopt an ecosystem perspective in which a variety of land uses are mixed and purposefully designed and composed to ensure the provision of ecosystem services across a landscape. Land use influences groundwater infiltration at different scale levels, and to improve natural recharge and safeguard groundwater quality it is needed to understand both the characteristics of specific land uses and practices, and land use processes at the landscape level. Portfolios need to be developed to invest in building knowledge, monitoring, landscape improvement, groundwater substitution and the protection of recharge zones.

ACKNOWLEDGEMENTS

The authors would like to thank Kirstin I. Conti and Jac van der Gun for their valuable comments. This paper was prepared with the support of research done as part of the 'Africa to Asia: Testing Adaptation in Flood Resource Management' project, supported by a grant from IFAD and the European Commission.

LITERATURE

British Broadcasting Service [BBC] News (2017) *California's drought is over. Now what?* Available on the internet: http://www.bbc.com/news/world-us-canada-39459592. Cited on April 4th 2017.

Bouma, J.A. & P.J.H. van Beukering (2015) *Ecosystem services: from concept to practice*. United Kingdom: Cambridge University Press;

EA [Environment Agency] (2013) Groundwater Protection; Principles and practice (GP3). August 2013 Version 1.1;

Ellison, D., C.E. Morris, B. Locatelli, D. Sheil, J. Cohen, D. Murdiyarso, V. Gutierrez, M. van Noordwijk, I.F. Creed, J. Pokorny, D. Gaveau, D.V. Spracklen, A. Bargués Tobella, U. Ilstedt, A.J. Teuling, S.G. Gebrehiwot, D.C. Sands, B. Muys, B. Verbist, E. Springgay, Y. Sugandi & C.A. Sullivan (2017) *Trees, forests and water: Cool insights for a hot world*. Global Environmental Change 43, 51–61;

GWP [Global Water Partnership] (2014) *The links between land use and groundwater – Governance provisions and management strategies to secure a 'sustainable harvest'*. Perspectives Paper;

FAO [Food and Agriculture Organisation of the United Nations] (2015a) *Groundwater Governance, a call for action: A Shared Global Vision for 2030*. Special Edition for World Water Forum 7. FAO, World Bank, GEF, UNESCO, IAH;

FAO [Food and Agriculture Organisation of the United Nations] (2015b) *Global Framework for Action to achieve the vision on Groundwater Governance*. Special Edition for World Water Forum 7. FAO, World Bank, GEF, UNESCO, IAH;

Foley, J.A., R. DeFries, G.P. Asner, C. Barford, G. Bonan, S.R. Carpenter, F.S. Chapin, M.T. Coe, G.C. Daily, H.K. Gibbs, J.H. Helkowski, T. Holloway, E.A. Howard, C.J. Kucharik, C. Monfreda, J.A. Patz, I.C. Prentice, N. Ramankutty & P.K. Snyder (2005) *Global Consequences of Land Use*. Science, Vol. 309: 570–574;

Ismangil, D., D. Wiegant, Eyasu Hagos, F. van Steenbergen, M. Kool, F. Sambalino, G. Castelli & E. Bresci (2016) *Managing the Microclimate*. Practical Note 27. Spate Irrigation Network Foundation;

Landis, D.A. (2016) *Designing agricultural landscapes for biodiversity-based ecosystem services*. Basic and Applied Ecology (2016). http://doi.org/10.1016/j.baae.2016.07.005;

Lerner, D.N. & B. Harris (2009) *The relationship between land use and groundwater resources and quality*. Land Use Policy 26S: S265–S273;

Norman, E.S. & J.O. Melious (2004) Transboundary Environmental Management: A Study of the Abbotsford-Sumas Aquifer in British Columbia and Western Washington. Journal of Borderlands Studies, Vol.19, No.2, Fall 2004;

Öborn, I., S. Kuyah, M. Jonsson, A. Sigrun Dahlin, H. Mwangi & J. de Leeuw (2015) Landscape-level constraints and opportunities for sustainable intensification in smallholder systems in the tropics. Chapter 12 in: Minang, P.A., M. van Noordwijk, O.E. Freeman, C. Mbow, J. de Leeuw & D. Catacutan (eds) *Climate-Smart Landscapes: Multifunctionality in Practice.* World Agroforestry Cente (ICRAF);

Scanlon, B.R., I. Jolly, M. Sophocleous & L. Zhang (2007) *Global impacts of conversions from natural to agricultural ecosystems on water resources: Quantity versus quality.* Water Resources Research, Vol.43, W03437;

Scanlon, B.R., C.C. Faunt, L. Longuevergne, R.C. Reedy, W.M. Alley, V.L. McGuire & P.B. McMahon (2012) *Groundwater depletion and sustainability of irrigation in the US High Plains and Central Valley.* PNAS, June 12, 2012, vol. 109, no. 24;

SER [Society for Ecological Restoration International] (2008) *Ecological Restoration as a Tool for Reversing Ecosystem Fragmentation.* Policy Position Statement;

Steenbergen, F. van, & A. Tuinhof (2009) *Managing the Water Buffer for Development and Climate Change Adaptation.* Groundwater recharge, retention, reuse and rainwater storage. Wageningen, the Netherlands: 3R Water Secretariat;

Steenbergen, F. van, A. Tuinhof & L. Knoop (2011) *Transforming Landscapes Transforming Lives; The Business of Sustainable Water Buffer Management.* Wageningen, the Netherlands: 3R Water Secretariat;

Steenbergen, F. van, Assefa Kumsa & N. Al-Awlaki (2015) *Understanding political will in groundwater management: Comparing Yemen and Ethiopia.* Water Alternatives Vol. 8, Issue 1: 774–799;

Tuinhof, A., F. van Steenbergen, P. Vos & L. Tolk (2012) *Profit from Storage; the Costs and Benefits of Water Buffering.* Wageningen, the Netherlands: 3R Water Secretariat;

Wada, Y., L.P.H. van Beek, C.M. van Kempen, J.W.T.M. Reckman, S. Vasak & M.F.P. Bierkens (2010) *Global depletion of groundwater resources.* Geophysical Research Letters, Vol. 37, L20402;

Chapter 17

Linking groundwater and surface water: conjunctive water management

Richard S. Evans[1] *& Peter Dillon*[2,3]
[1]*Infrastructure and Environment, Jacobs, Melbourne, Australia*
[2]*CSIRO Land and Water, Adelaide, Australia*
[3]*NCGRT, Flinders University, Adelaide, Australia*

ABSTRACT

In most cases throughout the world groundwater and surface water are intimately linked through the hydrologic cycle. More importantly, there are strong institutional and economic imperatives to plan, develop and manage groundwater and surface water in an integrated manner. Significantly increased water use efficiency can be achieved. Surprisingly, this has seldom occurred. Hence this chapter presents the compelling reasons why this must occur for the future use of the world's total water resources. The significant difference between unplanned and planned conjunctive use (the latter as a component of conjunctive water management), and the approach governance must take to maximize the potential benefits from such use, is explored. The differences in management between highly connected and poorly connected systems are explained. In the highly connected case the potential for double counting of groundwater and surface water is evident. New institutional structures are usually required to achieve effective conjunctive water management. This includes policy, legislation and planning reforms, including the use of a broad range of market based and financial instruments. A suggested set of principles to achieve good conjunctive water management governance is proposed. A powerful technique that embodies the advantages of conjunctive water management is Managed Aquifer Recharge. This is used to address aquifer overdraft, meet seasonal supplies and to increase drought resilience of groundwater through water banking. The world cannot continue to ignore the considerable economic and social advantages of planned conjunctive water management.

17.1 INTRODUCTION

Conjunctive use of groundwater and surface water in an irrigation or urban setting is the process of using water from the two different sources for consumptive purposes. It can refer to the practice at the local level of spontaneously sourcing water from both a well and an irrigation delivery canal, or can refer to a strategic approach at the irrigation command, aquifer, river basin or city level where surface water and groundwater inputs are centrally managed as an input to irrigation or water supply systems. The latter approach ranks under the paradigm of Conjunctive Water Management (CWM). Accordingly, CWM can be characterized as being planned (where it is practiced as a direct result of management intention – generally with a top down approach) compared with spontaneous use (where it occurs at a grass roots level – generally with a bottom up approach). The significant difference between unplanned and planned conjunctive use, and the approach governance must take to maximize the potential benefits from such use, is explored within this chapter. Where both surface and groundwater sources are directly available to the end user, spontaneous conjunctive use generally proliferates, with individuals opportunistically able to make decisions about water sources at the local scale.

The planned conjunctive management of groundwater and surface water has the potential to offer benefits in terms of economic and social outcomes through significantly increased water use efficiency. It supports greater food and fibre yield per unit of water use, an important consideration within the international policy arena given the critical concerns for food security that prevail in many parts of the world. At the resource level, groundwater pumping for irrigation used in conjunction with surface water provides benefits that increase the water supply or mitigate undesirable fluctuations in the supply (Tsur, 1990), control shallow water-table levels and consequent soil salinity, and improve fairness of access to water across a catchment or basin. Various requirements, such as improved fairness, can be a defined objective if this is stipulated.

The absence of planners and of a strategic agenda within governments, to capitalize on the potential for planned CWM to support these needs, is generally a significant impediment to meeting national and international objectives as they pertain to food and fibre security. There is an urgent need to maximize production within the context of the sustainable management of groundwater and surface water.

Many existing irrigation commands source their water supply from both the capture of catchment runoff and aquifer systems. Typically, water has been sourced from either surface or groundwater supplies, with the primary supply supplemented by the alternative source over time. Accordingly, governance settings, infrastructure provisions and water management arrangements have emphasized the requirements of the primary source of supply, inevitably requiring the "retrofitting" of management approaches onto existing irrigation commands or urban supply systems to incorporate supplementary water sources over time. Optimizing the management and use of such resources, which have been developed separately will in some situations require substantial investment in capital infrastructure and reform of institutional structures. Put simply, planned CWM is relatively simple with greenfield (or new development sites), but harder to achieve within existing hydro-physical and institutional/social systems.

Whilst these challenges and the associated benefits of a strategically planned approach are well understood and the subject of numerous reports on CWM, the current status of water management and planning around the world suggests that little has been achieved in its widespread implementation. This chapter explores the reasons underpinning the apparent poor approach to full integration in the management and use of both water sources, and the absence of more coordinated planning. There remain significant gaps in water managers' understanding as to what aspects of the contemporary management regime require overhaul to achieve integrated management and the improved outcomes that could be expected, as compared with separate management arrangements. Such lack of understanding is an important impediment to the governance, institutional and physical infrastructure reforms whereby planned CWM could improve existing management and regulatory arrangements. Reforms may also be impeded by different 'ownership' models of groundwater and surface water delivery infrastructure and the associated entitlement regime (*i.e.* private and/or public); a situation that has implications for social and institutional behaviour and ultimately the adoption of a CWM approach.

This chapter is intended to provide insight into these barriers to adoption and hence provide a new focus on an old paradigm; a focus intended to make progress with the objective of improved water management and water use efficiency and so support longer term outcomes in the form of improved food and water supply security in critical parts of the world.

Even though the primary focus of this chapter is on conjunctive water use and managed aquifer recharge there are important aspect of conjunctive management of groundwater and surface water which involve other aspects of the hydrologic cycle. Conjunctive water level management relates to large scale areas of the world where artificial drainage intentionally lowers the groundwater level to combat water logging and salinization. The artificial drainage of generally flat land results in the conversion of groundwater into surface water. In this case groundwater is usually (but not always) an environmental factor rather than a usable resource. These situations of successful groundwater level management can only exist in conjunction with surface water management. A special subset of conjunctive water level management are "polders" where artificial drainage keeps groundwater below a certain level, but also maintains groundwater above a certain prescribed level. In this case water is drained from the land during wet periods and supplied to the same land during dry periods. Where groundwater levels are in decline, intermittent surface water sources may be used to augment recharge to sustain supplies and prevent saline intrusion.

CWM also covers joint water quality management. Many different mechanisms may operate, but a common mechanism is where surface water groundwater interaction directly influences water quality, primarily in terms of salt levels, but also, especially in urban environments, in terms of pollution management. Groundwater quality may be improved by recharge with fresh water thereby restoring otherwise unproductive aquifers. Consequently CWM also includes recycling of used waters and wastewater, although this aspect is not discussed in this chapter. Another aspect of CWM is in relation to surface water groundwater interaction where the need to avoid double counting is fundamental to total water management. This issue is discussed later in this chapter. CWM also strongly influences environmental water requirements for ecosystem maintenance and protection.

17.2 CONCEPTS AND PRINCIPLES OF CONJUNCTIVE WATER MANAGEMENT

17.2.1 Systems that occur spontaneously and systems that are planned

Conjunctive use', is the subject of a range of definitions. It is defined by Foster *et al.* (2010) as a situation where "both groundwater and surface water are developed (or co-exist and can be developed) to supply a given … irrigation canal-command – although not necessarily using both sources continuously over time nor providing each individual water user from both sources". Alternatively FAO (1995) describes it as follows: "Conjunctive use of surface water and groundwater consists of harmoniously combining the use of both sources of water in order to minimize the undesirable physical, environmental and economic effects of each solution and to optimise the water demand/supply balance". Considering both of these definitions, the aim of CWM is to maximize the benefits arising from the innate characteristics of surface and groundwater water use; characteristics that, through planned integration of both water sources, provide complementary and optimal productivity and water use efficiency outcomes. It is recognised that the 'optimal' situation frequently depends on the stakeholder perspective and this is why good stakeholder engagement, where the different interests are balanced, is necessary.

The introductory section of this chapter highlights the two fundamentally different approaches to conjunctive use, however, there is a continuum in the way that CWM evolves from spontaneous (or incidental/unplanned) conjunctive use at one end, to planned conjunctive management and use at the other. For example, improved catchment management to reduce erosion and flooding and increase soil moisture also has the effect of enhancing recharge, is at one end of the CWM planning spectrum. At the other is intentional recharge enhancement called managed aquifer recharge, designed to increase groundwater supplies, improve their quality, or sustain groundwater-dependent ecosystems (Dillon *et al.,* 2009a).

Planned conjunctive management of surface water and groundwater are usually practiced at the State or regional level and can optimize water allocation with respect to surface water availability and distribution, thus reducing evaporative losses in surface water storages and minimizing energy costs of irrigation in terms of kWhr/ha (Foster *et al.*, 2010). Planned CWM is best implemented at the commencement of a development although experience has shown optimal outcomes may be difficult to achieve when attempts are made to redesign and retro- fit the approach, once water resource development is well advanced.

Where groundwater and surface water are used conjunctively in various parts of the world, spontaneous use prevails. Foster and Van Steenbergen (2011) emphasize that spontaneous conjunctive use of shallow aquifers in irrigation-canal-commands is driven by the capacity for groundwater to buffer the variability of surface water availability enabling:

- greater water supply security;
- securing existing crops and permitting new crop types to be established;
- better timing for irrigation, including extension of the cropping season;
- larger water yield than would generally be possible using only one source;
- reduced environmental impact; and
- avoidance of excessive surface water or groundwater depletion.

Foster *et al.* (2010) report that the most common situation in which spontaneous conjunctive use of surface water and groundwater resources occurs is where canal-based irrigation commands are:

- inadequately maintained and unable to sustain design flows throughout the system;
- poorly administered, allowing unauthorized or excessive off takes;

- over stretched with respect to surface water availability for dry season diversion; and
- tied to rigid canal water delivery schedules and unable to respond to crop needs.

Additionally, spontaneous conjunctive use is also driven to a large degree by poor reliability of water quality in surface water supply cannels. Wells become an insurance against this unreliability. Poor water quality is a common factor at the tail of most irrigation canal systems and usually reflects poor infrastructure maintenance. These factors lead to inadequate irrigation services. As a consequence, the drilling of private waterwells usually proliferates, and a high reliance on groundwater often follows (Foster *et al.*, 2010).

Foster and Van Steenbergen (2011) report spontaneous conjunctive groundwater and surface water use in Indian, Pakistani, Moroccan and Argentinean irrigation-canal commands which have largely arisen due to inadequate surface-water supply to meet irrigation demand. Many other examples from developed countries also show that it is not simply a developing country problem – it is an inherent problem wherever canal-based irrigation is practiced and where there are challenges in terms of reliability of water supply and quality.

In summary, the spontaneous approach to the conjunctive use of surface and groundwater sources reflects a 'legacy of history'. The focus for green-field irrigation developments is primarily access to water, rather than the efficient and optimal use of that water; a consideration that does not gain attention until competition for water resources intensifies. Advancing beyond the farm-scale spontaneous access to each water source to a planned CWM approach entails significant technical, economic, institutional and social challenges that can only be overcome with an effective governance model.

17.2.2 Highly versus poorly connected systems

When groundwater and surface water are hydrologically connected, the interchange of the resource between the systems requires consideration during the management process. Accordingly, it is an aspect that must be considered within a CWM framework, as it can shape the available options and hence define the optimal approach to achieve CWM.

Connectivity comprises two important components: the degree of connection between the two resources (see Figure 17.1) and the time lag for extraction from one resource to impact upon the other. A highly connected resource would be one where the degree of connection is high and the time lag for transmission of impacts is very fast. A fundamental tenet of connectivity understanding is that, essentially, all surface water and groundwater systems are connected and that it is just a matter of time for impacts to be felt across the connection. Important exceptions to these truisms are that of canal-dominated irrigation commands, where the water table is below the water level in the canal system or where the water table is shallow and groundwater extraction is capturing losses to evapotranspiration. In such areas recharge may also be dominated by irrigation-induced root zone drainage, and hence vertical unsaturated zone processes may control the interaction/connectivity process. In these latter areas, the canal distribution systems may provide a significantly reduced contribution to groundwater

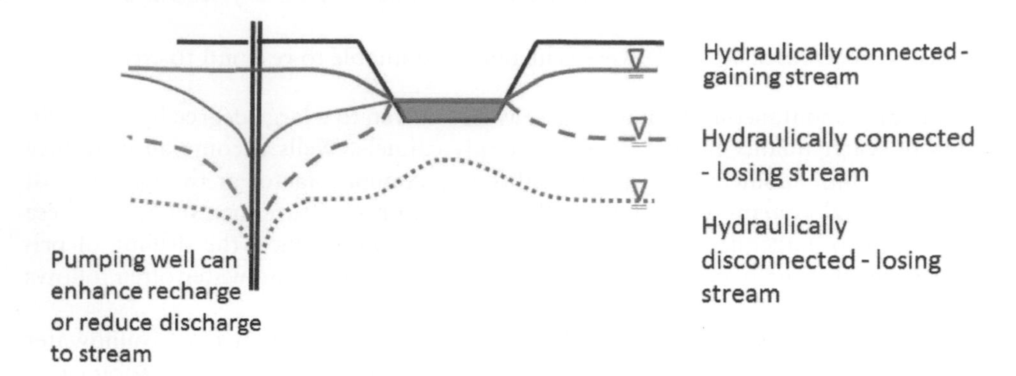

Figure 17.1 Stream-aquifer interaction connections.

extraction. The salient issue for CWM, especially in a planned environment, is to understand the nature of connectivity as a factor in resource use optimization and to ensure that connectivity is understood when considering water resource accounting in a conjunctively managed water system.

The timing of the impacts is also very important. Bredehoeft (2011) has shown that timing is important to water resources managers whether the impacts from groundwater pumping on a stream occur within an irrigation season, or over a longer period. Connectivity will control the timing for groundwater recharge and the timing of changes in discharge from groundwater to the streams due to groundwater abstraction.

In connected systems a serious issue often arises which is called double accounting. Double accounting is where the same volume of water is potentially attributed to both the surface and groundwater resource. It is a common occurrence throughout the world due to the evolution of water resource development and associated regulatory arrangements, and it may reflect an absence of a proper water resource assessment, poor understanding of the water balance, or the undertaking of independent resource assessments for surface water and groundwater. Two common situations occur.

Firstly, when surface-water-based irrigation canals cause recharge to the groundwater system. The groundwater recharge is seen as a 'loss' from a surface-water point of view. A typical water resource management response may be to invest in improved sealing canals or constructing pipelines, however this may not be the most efficient response. In situations where groundwater recovery is financially viable, a more efficient approach may be to utilize aquifer storage capacity and the diffuse distribution of the resource provided by the groundwater system. If in such situations, canal leakage has already been allocated to surface water users, then it should not also be allocated to groundwater users. Instead, mechanisms such as trade should be used to transfer entitlements from one user to another, and hence maintain the integrity of the water accounting framework. Furthermore, any decision to reduce leakage through canal lining, and hence reduce recharge, would require revision to the water resource assessment and may require appropriate adjustments to entitlements, particularly if such recharge had been allocated to groundwater users.

Secondly, the classical surface water/groundwater interaction situation is where groundwater discharges to become base flow. Considered in isolation, this may be deemed as a "loss" from a groundwater management perspective and a basis for allowing groundwater pumping to substantially reduce stream flow. Similarly from a surface-water management perspective, the significance of groundwater discharge in maintaining stream flow during the dry season may be poorly recognized. There are many examples where the implications of not recognizing such interaction have contributed to the depletion of rivers around the world. The assessment of the interaction requires an integrated resource assessment, with the water balance taking into account all extraction regimes and the consequential impacts on both groundwater and surface water resources.

Eliminating double accounting requires integrating water entitlements with a water balance that reflects the full hydrological cycle, and hence fully appreciating the amount and timing of the interaction between groundwater and surface water. It is also critical to appreciate the temporal variability of the process. In this case, it is important that the conjunctive planning time frame be usually long term, for example 50 years, although in some cases a variable planning horizon is appropriate. Short-term planning to meet political or social objectives will not achieve effective CWM. There are some relatively rare situations where there is effectively no interaction between groundwater and surface water. In such situations, CWM is relatively less complicated, but nonetheless important in terms of achieving optimal water management outcomes.

In cases where the two water resources are highly connected with short time lags, CWM may be supported by a transparent water-accounting framework that can be reported on for both surface and groundwater on an annual basis. It may provide flexibility in the way in which surface and groundwater is allocated on an annual basis, and could facilitate the development of a robust two-way water-trading regime between the groundwater and surface water system, providing third-party impacts are understood and effectively managed.

CWM within an environment where surface and groundwater systems are poorly connected is unlikely to provide such a degree of integration. Whilst there are opportunities for integration (such as the application of MAR discussed later) and for taking advantage of the unique attributes of groundwater and surface water, such as storage, distribution and reliability in dry periods, the opportunities and benefits that have the potential to arise from CWM will be different, reflecting differences in the hydrological environment. In other words, within poorly connected systems, CWM will be framed around the task of complementary and integrated management of water use, without the need for such integration to consider natural hydrological linkages of the water sources. This is modified, however, where engineered solutions enable better (anthropogenic) connection between the two parts of the water system.

17.2.3 Technical and management differences between surface water and groundwater

The characteristics of the two primary water sources associated with CWM (*i.e.* groundwater compared with surface water) are inherently different; differences that must be appreciated when optimizing their use. A summary of typical characteristics associated with groundwater and surface water resources is provided in Table 17.1.

Given the extent and diversity of irrigation and urban supply systems covering a vast range of physical environments throughout the world, there are many situations where the characteristics of the surface/groundwater components of local

Table 17.1 Typical characteristics of groundwater and surface water.

Characteristic	*Groundwater*	*Surface Water*
Response time	Slow	Quick
Time lag	Long	Short
Size of storage	Large	Small
Security of supply	High	Low
Spatial management scale	Diffuse	Generally linear
Flexibility of supply	Very flexible	Not flexible

water resources are not represented by the 'typical' characteristics presented above. Nonetheless, physical differences and differences in the history of development of the two resource types provide both challenges and benefits to CWM.

They also reveal how enhancing groundwater with occasionally available surface water can secure the resource and improve its quality. To make progress on CWM, the specific characteristics of groundwater and surface water in the target region must be assessed. Such an assessment includes the social (and cultural), economic and environmental aspects (the so-called 'triple bottom line') so as to evaluate how the particular characteristics of the hydrological environment can be integrated to achieve optimum outcomes. It is almost mandatory in current times to ensure that water resources management is undertaken not only in an integrated manner, but also cognizant of triple-bottom-line issues.

17.2.4 The role of managed aquifer recharge

An increasingly important tool used in CWM is managed aquifer recharge (MAR). MAR is the intentional recharge of water to aquifers for subsequent human use or environmental benefit. In many cases the primary water source is excess wet season surface waters which can be stored in aquifers to secure or supplement dry season supplies and improve groundwater quality. It can be used by individual farmers to refresh their wells when there is flow in a nearby ephemeral stream, but is increasingly used in planned approaches to aquifer replenishment. MAR has emerged as an important linking technique which can often be used to encourage conjunctive management, as depicted in Figure 17.2 particularly where aquifers and surface water systems are poorly connected. MAR is a supply side measure which can increase the available total water resource, but it is not a substitute for effective demand side measures. Thus governance issues often surround MAR schemes. The use of MAR to restore over developed aquifers is also discussed in this chapter.

17.3 SCOPE FOR SECURING SOCIAL, ENVIRONMENTAL AND ECONOMIC BENEFITS THROUGH CONJUNCTIVE WATER MANAGEMENT SCHEMES

Planned CWM should be a clear objective wherever both surface and groundwater resources are available. There are, however, few examples that demonstrate effective implementation of planned CWM; spontaneous conjunctive use is common. The

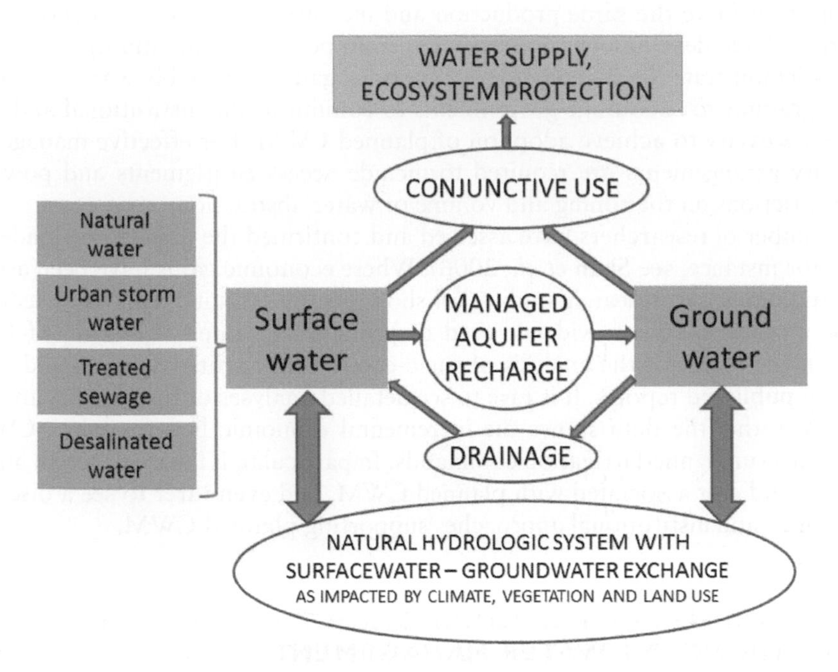

Figure 17.2 The role of Managed Aquifer Recharge and Drainage in Conjunctive Water Management (adapted from Dillon & Arshad 2016). (Note, there are other aspects to Conjunctive Water Management, other than MAR and Conjunctive Use.).

widespread adoption of spontaneous conjunctive use suggests that substantial financial benefits are being realized, otherwise the practice would not prevail.

Evans *et al.* (2012) describes several case examples of the social, environmental and economic successes and failures of conjunctive water management. They point to institutional structures as being key to the successful implementation of CWM schemes. Bredehoeft (2011) emphasises: "Effective conjunctive management can probably only be accomplished by an approach that integrates the groundwater and surface water into a single institutional framework; they must be managed together to be efficient." Successful institutional structures vary from the local to the sovereign level. In most cases the local level controls the management arrangements. However, the optimum approach may prove to be purely theoretical, if implementation is inhibited by existing institutional or policy structures. This specifically applies to the legal 'ownership' of water rights, the ability of local bodies or water user associations to make day-to-day decisions and the ability to undertake effective planning for CWM.

It is clear that there needs to be economic incentives to justify the adoption of CWM at both (or either of) the sovereign or individual level, independent of whether there are strong market drivers operating. The examples in Evans *et al.* (2012) indicate that economic gain is made – where it has been assessed and reported – as a result of the adoption of CWM. This has usually been at the farm-gate level in the form of reduced costs and increased income, however economic returns may also be achieved at the sovereign level through more efficient use of the available water resource, lower subsidies to achieve the same production and increased levels of production leading to more regional development opportunities from post-farm gate multipliers. Further work to demonstrate the sovereign-level economic gains is probably warranted as part of a programme to encourage governments to commit to the institutional and policy reforms necessary to achieve adoption of planned CWM. For effective management, regulatory arrangements are required to include access entitlements and powers to place restrictions on the timing and volume of water abstraction.

A number of researchers have assessed and confirmed the gains to be made from CWM (for instance, see Shah *et al.*, 2006). Where economic gains have been assessed in investigations and research studies, all show positive results. This knowledge has become a major piece of evidence used to promote the implementation of CWM. However, the extent of the analysis of socio-economic benefits is limited and mainly held in unpublished reports. It is rare to see detailed analyses of the benefits and costs of CWM; rather the data shows the incremental economic benefits when CWM is retrofitted to unplanned irrigation commands. In particular, it is rare to see an analysis of benefit and cost associated with planned CWM, and even rarer to see a discussion of the policy and institutional approaches supporting planned CWM.

17.4 REQUIRED INSTITUTIONAL STRUCTURES FOR EFFECTIVE CONJUNCTIVE WATER MANAGEMENT

CWM is not constrained mostly by a lack of technical understanding but rather by ineffective and incompatible institutional structures, with separate management arrangements, almost always established and operated by different institutions. As well, water resources at the sovereign level are often managed by a dedicated water resource agency, whilst irrigation commands are often managed by agricultural agencies or dedicated irrigation-command authorities. Overall water resource policy may be set at a jurisdictional scale with the irrigation sector required to operate under the authority of a regulatory agency. This results in a complex mosaic of planning and decision pathways that are not easily overcome in the pursuit of a planned conjunctive management model.

Foster and Van Steenbergen (2011) acknowledge that: "In many alluvial systems, the authority and capacity for water-resources management are mainly retained in surface-water-oriented agencies, because of the historical relationship with the development of irrigated agriculture (from impounded reservoirs or river intakes and major irrigation canals). This has led to little interest in complementary and conjunctive groundwater management. Some significant reform of this situation is essential – such as strengthening the groundwater-resource management function and/or creating an overarching and authoritative 'apex' agency". Similarly Shah *et al.* (2006) recognize that: "Water resources are typically managed by irrigation departments and groundwater departments. There is rarely any coordination between these ...".

Foster *et al.* (2010) also emphasize that: "The promotion of improved conjunctive use and management of groundwater and surface water resources will often require significant strengthening (or some reform) of the institutional arrangements for water resource administration, enhanced coordination among the usually split irrigation, surface water and groundwater management agencies, and gradual institutional reform learning from carefully monitored pilot projects." Bredehoeft (2011) emphasizes that effective management of conjunctive use "requires integrated institutions that can plan and sustain the management of the system for long periods". This is because it typically "takes more than a decade for significant changes in groundwater pumping... to have their full impact on the river" (as seen in the USA case he studied). Bredehoeft also stresses that, in much of the USA, the water management legal system based on prior appropriation fundamentally works against CWM: "Effective conjunctive management can probably only be accomplished by an approach that integrates the groundwater and surface water into a single institutional framework; they must be managed together to be efficient. Current institutions based upon the present application of the rules of prior appropriation make conjunctive management not practical." CWM will require major organizational change in water agencies. Furthermore reformed institutions need structures that can operate at the multiple scales with which groundwater, especially, requires.

Garduño *et al.* (2011) emphasised that "The promotion of more planned and integrated conjunctive use has to overcome significant socio-economic impediments through institutional reforms, public investments, and practical measures, including: (a) the introduction of a new overarching government agency for water resources, because existing agencies tended to rigidly follow historical sectoral boundaries and thus tend to perpetuate separation rather than the integration needed for conjunctive use; (b) gradual institutional reform learning from carefully monitored pilot projects; and (c) a long-term campaign to educate farmers through water user associations on the benefits of conjunctive use of both canal water and groundwater, crop diversification, and land micro-management according to prevailing hydrogeologic conditions." These commentators reflect a view that institutional strengthening is probably the most important challenge to CWM, especially in already developed irrigation systems where a more optimized management approach needs to be retrofitted.

17.5 MANAGEMENT OF STORAGE: MANAGED AQUIFER RECHARGE

In areas where there are seasonal excesses of surface water, supply side measures such as managed aquifer recharge (MAR) can protect, prolong, sustain or augment groundwater supplies. As one of a suite of integrated water resources management strategies, this expands local water resources, reduces evaporation losses, and assists with replenishing depleted aquifers. The amount of recharge that is economically or technically achievable is generally less than the annual groundwater deficit and a combination of demand management and recharge enhancement is essential to restore a groundwater system to equilibrium (Dillon *et al.*, 2009b).

There are many methods for recharging aquifers (*e.g.* Dillon *et al.*, 2009a) and these are selected based on the local hydrogeological characteristics, sources and quality of water available to be harvested. Importantly cost per unit volume needs to be competitive with the foregone net benefits of demand reduction, taking into account the costs of managing demand and supply.

As an alternative to recharging the aquifer, groundwater supplies can be augmented or replaced by surface water supplies, such as canals and pipelines. This has the effect of reducing demand on the aquifer, but is commonly perceived by groundwater users as a supply augmentation.

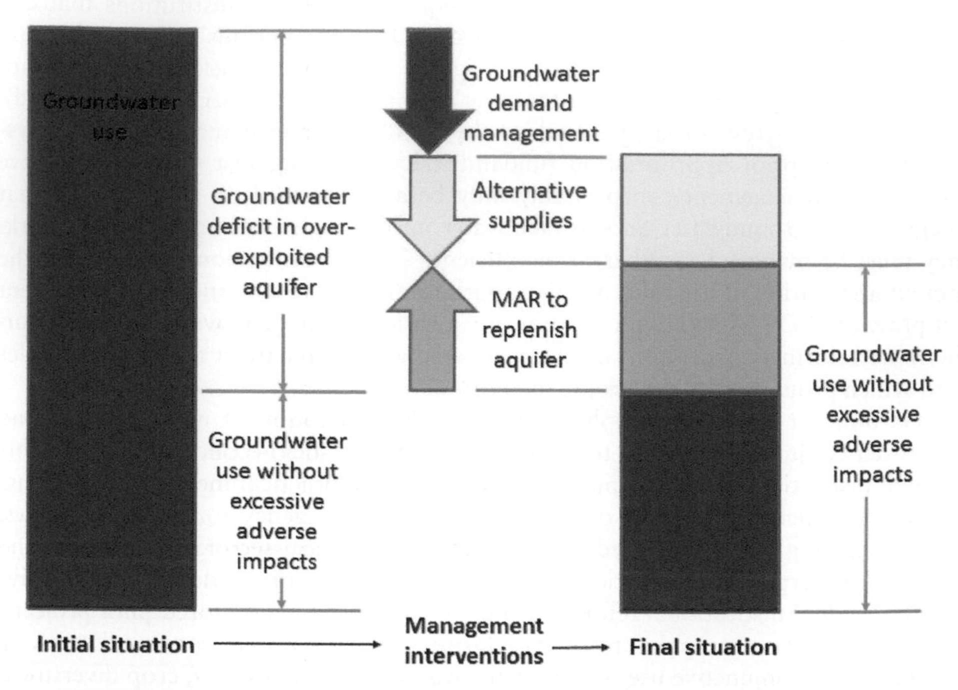

Figure 17.3 An aquifer can be brought into hydrologic equilibrium by either reducing extraction, or augmenting supplies, either through groundwater replenishment or providing alternative supplies (conjunctive use) (from Dillon *et al.*, 2012).

The complementary roles of demand management and expanding supplies, either via managed aquifer recharge or by providing alternative supplies are graphically depicted in Figure 17.3 for an over-allocated aquifer. Where surface water is in public ownership and groundwater in private ownership, the act of managed aquifer recharge effectively privatises a public good, so MAR is best implemented where water entitlements are divorced from land ownership. The synergistic effect of managed aquifer recharge on implementing demand management has much potential but is yet to be exploited systematically. However, MAR combined with conjunctive use is widely applied. For example, in a 445 km^2 groundwater irrigation area of Central Valley of California, USA, a combination of managed aquifer recharge (2.6 km^3) and conjunctive use (6.4 km^3) over a period of 46 years has stabilised falling groundwater levels 90 m above their projected levels (Scanlon *et al.*, 2016). Both measures are undertaken in California and Arizona and contribute substantially to supply security and to sustaining groundwater levels (Scanlon *et al.*, 2016) and at a cost substantially less than for constructing surface water storages (Perrone and Rohde 2016).

In arid climates the lack of availability of a water source constrains the opportunities for aquifer replenishment. Runoff is so infrequent in arid areas that assets need to be cost efficient as they are actively utilised only infrequently. Managed aquifer recharge is primarily for inter-year storage to increase long term yield. However alternative supplies such as treated wastewaters and desalinated water as a by-product of energy production are being adopted increasingly. In semi-arid and Mediterranean climates, water availability is a smaller constraint and seasonal demand for water can be high, meaning that inter-season storage has high value in addition to inter-year storage. Inter-season storage can have immediate commercial benefits, and water banking for buffering against drought and climate change can have higher value but needs institutional support. In humid climates, opportunities for natural recharge are greater and the demand for storage is less, so managed aquifer recharge is expected to have a minor or niche role.

Van Steenbergen and Tuinhof (2009) and Van Steenbergen *et al.* (2011) have reported a wide range of watershed interventions that enhance groundwater recharge, retain soil moisture, and reuse water, which they term the 3R concept for climate change adaption, food security and environmental enhancement. These have been applied in relatively small scale projects in arid and semi-arid areas of Africa, Asia and South America with startling results for improving the capability of land and farm income. They may be applied from land-holder scale up to sub-catchment and catchment scale and typically at very low cost and with active stakeholder participation and ownership by the community (*e.g.* in Rajasthan and Gujarat, India; Maheshwari *et al.*, 2014).

MAR can increase the value of water resources by transferring surface water in times of abundance to add to groundwater storage and thereby conserve water. This replenishes depleted groundwater and avoids evaporative losses, salinity increase and possibilities for blue green algal blooms if the water had been retained in surface reservoirs. The surface waters used for managed aquifer recharge may include natural waters from catchments, urban stormwater, water recycled from treated sewage effluent, desalinated water from brackish aquifers or the sea, and suitably treated industrial effluents.

For any aquifer, there will be a range of recharge options that can be ranked in order of increasing unit cost of supply. Similarly foregoing extraction for each use of groundwater will have a range of unit costs that can be ranked in increasing order. Each element of these lists has an associated volume and unit cost and the two lists may then be merged to identify the cheapest option and the volume of demand reduction or supply enhancement expected if that option were implemented. Depending on the degree of over-exploitation, a series of options may be required to achieve hydrologic equilibrium, or at least to extend the effective lifetime of the groundwater resource.

Improving water use efficiency and water productivity is generally the cheapest option, followed by MAR and finally reducing the area of high valued crops. Integrated management of surface water and groundwater would help ensure that the benefits of recharge upstream outweigh any valued decline in surface water availability downstream. Flood mitigation may in fact be a recharge benefit, as currently explored in Thailand and India (Pavelic *et al.*, 2012, 2015).

17.6 PROSPECTS FOR IMPROVING GOVERNANCE

17.6.1 Governance approaches

Effective CWM must involve the governance for both surface water and groundwater. Good governance principles associated with groundwater alone still apply, but they must be made to fit a broader governance paradigm – so-called Integrated Water Resource Management (IWRM). General water governance principles cover a number of main areas – authority, accountability, transparency, stakeholder participation and institutional integration. Authority relates to the policy and statutory powers vested in the government or delegated to an agency to administer and regulate on behalf of the government. The associated 'authority' becomes the decision-maker who must be held accountable for operationalizing policy and legislative instruments. Such an authority must be accountable for its decisions, with appropriate mechanisms in place, supportive of natural justice by enabling appeals against decisions to be independently reviewed. Such authorities typically operate at basin scale, which does raise boundary issues when, for example, river basin boundaries do not match the underlying aquifer system.

Transparency is required to demystify the decision-making process, support stakeholder confidence in the management process and provide the grounds for any appeal. Participation is required to ensure that there is ownership of the process by all stakeholders; this goes a long way to achieving planned outcomes. Finally, integration (both institutional and technical) is required to ensure that all aspects of water tenure are subject to a single basic water resource regime. Water is a single resource and should be managed accordingly.

Optimum water-resource use will be significantly advanced through planned management of conjunctive use. There are a number of areas where the governance model is crucial to the adoption of this planned management approach. However, it is useful to note that there is no single governance model that can be applied universally; rather elements of different approaches may need to be chosen depending on the specific circumstances for each case.

A recent example emerging from India shows the power of participatory management of groundwater based on groundwater level observations by farmers to influence cooperative decision making on dry season cropping, water use efficiency measures, drilling restrictions and development and maintenance of recharge structures in ephemeral streams (Maheshwari *et al.*, 2014).

Effective governance arrangements to underpin a CWM strategy are deemed to be the most significant challenge. Danton and Marr (2007) in a discussion of the governance arrangements associated with the Uttar Pradesh conjunctive use example, make the point that "multi-faceted governance arrangements are necessary for successful management of smallholder surface water irrigation systems. In managing conjunctive use ... these arrangements become more complex.... The greater complexity in management arises from the need for coordinated management of the two resources through greater participation and networking of stakeholders at each stage of water allocation, use and management." Further, Livingston (2005; as referenced in Danton and Marr, 2007) subdivides water governance models for water supply systems into three types: bureaucracy, community and market. Governance approaches may favour one model, but will ultimately include elements of all three.

Garduño and Foster (2010) listed a number of challenges when considering the governance of CWM. They reported that: "Serious impediments have to be overcome to realize such water resource management policies. They are primarily institutional in character, given that the structure of provincial government organizations often simply mirrors current water-use realities and tends to perpetuate the status quo, rather than offering a platform for the promotion of conjunctive management."

In summary, the governance model needs to address four areas of endeavour: legislative, organizational, capacity and socio-political. In many countries, the organizational aspect may require the most significant changes to be made.

17.6.1.1 Institutional strengthening

Institutions that manage water, at both the national and regional scale, need to be strengthened to remove impediments. This requires the adoption of frameworks that promote IWRM where surface water and groundwater functions operate collectively towards a single overarching objective, and the function of water and agriculture ministries are also aligned for this purpose. Institutions need to be clear on who operates and manages both physical infrastructure and the different parts of the hydrologic cycle. These arrangements may be inclusive of either the public or private sphere, or a combination of both. The resolution of chain of command issues across various levels of government also needs to be reviewed. That is, each level of government must understand its role in implementing national water resource policy and be effective in enacting that role. Any activities that undermine CWM must be confronted and remedies provided. Institutions need to have a strong compliance culture to ensure that outcomes are achieved.

17.6.1.2 Policy and legislation

In many instances, there is a need to understand and review the current approaches to allocating rights in water, and the form and attributes of those rights. In many situations, policies and regulations may be poorly formulated and hence not operating efficiently to achieve the intended outcomes. Effective water allocation planning is paramount. Such planning needs to be supported by strong national policy and to occur within a framework that ensures sustainable levels of take and use of the resource. This requires significant technical input, especially within the context of the need to assess the available consumptive pool. CWM relies on water policies and regulations that are efficient at promoting movement of water or access between the two resources when required and appropriate. Legal and market powers and mechanisms must be aligned to achieve this goal.

17.6.1.3 Planning

By its very nature, planned CWM requires a strong management platform. There is a need to clearly define objectives, outcomes, activities and performance measurement and compliance arrangements. Such plans should be based around water allocation mechanisms and have regard to the technical understanding of the total consumptive water available. Implementation planning requires definition of investment requirements and decisions about who will make those investments, and who will ultimately pay. Ideally, planning should incorporate the triple-bottom-line notions of achieving environmental, economic and social objectives. CWM also requires consideration of land use policy changes so that groundwater protection outcomes can be achieved.

This is not a usual set of policy decisions in most developing and developed countries and may not only require considerable input, but also political support.

17.6.1.4 Market and pricing approaches

Surface water and groundwater always have differential cost structures that apply to users. In centralized government systems, these cost structures may be heavily subsidized as a result of related policy decisions (for instance, those for food and energy) and there may be unwanted outcomes as a result; usually, these relate to poor water use efficiency outcomes. In general, groundwater users fully finance their associated infrastructure whereas surface water infrastructure has been either wholly or partly subsidized by the State. The different ownership models contribute to differential cost impacts for irrigators, leading to decisions that are inconsistent with optimized planning objectives. CWM needs to understand and remove these impediments. State-sponsored groundwater development is an area where investment may be required. There are also differences in economic approaches at the macro and micro scale, and any activity to enhance the water market needs to acknowledge the two different scales of benefits. This is also true where economic incentives are implemented.

17.6.1.5 On-the-ground implementation

Planned CWM will benefit strongly from, and possibly require, strong ownership especially by the irrigated farming sector. This can be achieved by building strong local water user groups through targeted education and enabling actions. This is simplest when drivers for change are inescapable, such as declining or erratic farm incomes or declining resource base. In the past, communities have been focused upon single issues (either surface water or groundwater) and there has been a reluctance to engage in management issues associated with the other side of the resource picture that would require reorganization to better reflect the distribution of users. Overcoming this issue is exacerbated by a number of factors including the absence of a revenue base for cost recovery and the politicization of the user groups towards maintaining subsidized surface water supplies. There needs to be a participatory culture of education, demonstration and capacity building between governments and the irrigation farming community and its key stakeholders.

17.6.1.6 Knowledge generation

To facilitate CWM, knowledge is required in two key areas – technical understanding of the spatial and temporal distribution of the total consumptive available water and support for planning through the capability to provide future impact scenarios. The latter may be in the form of a complex numerical model of aquifer-river basin performance or at time simple analytical approaches. CWM also requires the establishment or improvement of monitoring programmes so that the quantity and quality impacts of the use of surface water on groundwater and vice versa can be demonstrated, and so that the beneficial impacts of water management actions can be seen by all stakeholders.

17.6.2 Use of financial and market-based instruments to promote planned conjunctive water management

Financial and market-based instruments (FMBI) are a range of financial and economic measures that can be used to encourage specific actions and trends. In the context of water resource planning, FMBI can consist in direct financial incentives (*e.g.* taxation reduction, subsidies to lower electricity prices) or disincentives (*e.g.* taxation increases) or alternatively indirect trade-offs or offsets (*e.g.* pollution reduction schemes) and the introduction of water trading.

Some countries have favoured a regulatory approach to bring about various water resource outcomes, while other countries have tended to favour economic instruments, in the belief that clear financial signals are a strong lever to active policy objectives. In the case of CWM, in many countries subsidies that distort the true cost of water delivery (surface water and groundwater) bias irrigator behaviours and hence retard the potential for planned CWM to contribute to optimal water use outcomes.

Conversely other FMBI (*i.e.* those not aligned with subsidies) can be a very powerful tool to encourage the adoption of optimal CWM. The range of options tends to be very location- and culture-specific. Nonetheless schemes that provide both financial incentives (*e.g.* through taxation decreases), when a defined minimum volume of water is used conjunctively, and indirect economic offsets (*e.g.* for salinity control) are considered the most effective. These should generally be used to 'kick start' planned CWM and should not be seen as permanent measures.

The introduction of clearly defined water rights, the application of well-defined caps (*i.e.* maximum limits of use of groundwater and surface water) and the introduction of a water trading regime can operate to strongly facilitate more efficient total water use. Surface-water trading regimes currently operate in many countries, however groundwater-trading regimes are not so common. Surface water to groundwater (and vice versa) trading regimes are rare. Nonetheless water trading can represent a strong market instrument to encourage conjunctive use, if it is managed appropriately. There are, however, few examples in the world (*e.g.* Namoi Valley, Australia (described in Dillon *et al.*, 2012)) where this has occurred. This is especially an issue where market mechanisms are not designed to account for environmental impacts (*e.g.* salinity effects).

FMBIs are not readily recognizable where governments exercise centralized control as opposed to a market-based approach. However, in such centralized governance approaches, positive benefit-cost outcomes through similar initiatives as FMBIs can still be achieved in terms of measures of 'national good', that is, national gross production from irrigated agriculture, poverty alleviation, etc. The issue here is about applying the most appropriate reward and compliance signals to the water/irrigated agriculture sector.

This discussion also indicates that water management policy – and its role in planned CWM – is part of a larger policy position by governments that involves national food policy, poverty alleviation, economic growth, sustainability, climate change and energy considerations. Good governance is more likely to ensue once the impact on national water use policy of policy decisions (including subsidies) in these related areas is considered.

17.6.3 A suggested set of conjunctive water management principles for consideration within a governance approach

The following is a suggested set of principles for the implementation of CWM, where infrastructure and historical governance arrangements are in place:

- Planning should be undertaken with full and detailed knowledge of the characteristics of both the surface water and groundwater systems, of existing system operations and of the demands of the cropping systems/urban water needs.
- Goals should be established that are intended to optimize the water supply/demand balance, irrespective of existing institutional, governance and regulatory models.
- Opportunities for MAR are assessed to expand and secure beneficial use of the total resource and mitigate threats.
- A strong policy and legislative base that supports CWM.
- Institutional arrangements that enable the implementation of CWM.
- The combined surface water/groundwater system and their use should be managed so as to optimize net economic, social and environmental benefits, taking into account national energy, food security, population and poverty reduction, sustainability and climate change policies and programs.
- Stakeholder participation should be encouraged.

From an operational point of view, some key guidelines to implementing CWM include:

- a technically robust understanding of stream-catchment-aquifer interactions; water balance that is inclusive of connectivity between the surface and groundwater systems;
- technical assessment techniques commensurate with the understanding of the hydrological system and with explicit recognition as to the limitations to the validity and applicability of information;
- technical assessment of impacts of existing MAR infrastructure on local groundwater resources and downstream impacts on surface water and riparian ecosystems, and of MAR maintenance requirements;
- a strategic monitoring programme for the catchment, including the alignment of groundwater and surface water monitoring.

In summary, CWM planning is the structured water-planning process whereby the different characteristics (technical, economic, social and institutional) of groundwater and surface water are compared and weighed against each other so that the optimum use of the two water sources is achieved. The fact that this rarely occurs throughout the world is testament to the entrenched water institutional structures and the narrow understanding of fundamental technical processes.

17.6.4 Scope and potential for managing and improving groundwater storage and recovery

It is evident that entitlement to groundwater through land ownership or prior right does not work and cannot work in slowing or reversing groundwater depletion (Dillon *et al.*, 2012). These methods have conspicuously failed to allow volumetric allocations to be modified equitably as the environmentally protective allocatable resource pool becomes better defined. Furthermore, they oppose maximisation of the utility of the groundwater resource.

Improving irrigation efficiency and agronomic methods can reduce water use while sustaining or enhancing production. This should be considered in every portfolio of groundwater management policies along with MAR. Improved awareness of the magnitude and degree of resilience of the groundwater resources will help communities understand that the resource is finite, shared and there are severe consequences to all groundwater users and to connected streams and ecosystems if too much groundwater is extracted.

The key issue is for all groundwater users to understand they are sharing a common good, and that there is a finite limit to the total that can be shared. Two distinct and quite separate processes are required:

1 a scientific assessment of the magnitude of the allocatable resource, repeated periodically, based on credible monitoring of the groundwater storage, MAR and water use
2 a socially acceptable way for shares (entitlements) in that allocatable resource to be allocated to groundwater users, taking account of social, environmental and economic factors.

Allocations are made for a period based on multiplying the currently determined allocatable resource by the share of each groundwater user. Shares and allocations should be transferable, and should be registered as a property right, and traded in an open market subject to rules to protect the environment and other groundwater users. Where there is a plethora of small users of groundwater some form of participatory process needs to be adopted to allow a manageable number of aggregates to deal with internal and external trades. It is recognised that many countries lack the institutional basis for the establishment of individual tradeable water rights. In which case a first step is the establishment of individual surface water and groundwater licences which account for the impacts on the other resource.

Legislation may be required to vest the groundwater resource in the ownership of the State. Groundwater users recognise they have only an ambit claim to a continued right to the volume of groundwater previously attached to land ownership, as that volume will not be available unless total demand on the system was to reduce. However these users are taken into account in assigning shares of the allocatable resource.

In the event that there is disagreement among users, historical uses only should be taken into account, the shares of an individual should be based on the ratio of their historical use to the sum of historical uses of all individuals over a period concluding before share apportionment is calculated. Intended new users of groundwater would need to buy their allocation from a willing seller on the market at the price they agree. Similarly MAR operations would require entitlement to surface water in order to operate and a means of distributing or selling shares in recharged water that protected all existing users.

Demand management is a key element for sustaining groundwater supplies, and where other water resources are available, this can be assisted by managed aquifer recharge and supply augmentation. These additional measures can be applied most effectively where there is an entitlement system for groundwater use. For example new or existing groundwater users may be able to pay for managed aquifer recharge systems through the sale of some of the allocations that MAR may yield. Similarly, if supply augmentation with surface water systems occur, entitlement to access this water may require foregoing groundwater entitlements so as to ensure there is a benefit to the aquifer.

There are potentially significant benefits in incorporating managed aquifer recharge and/or supply augmentation where the costs of these options in monetary units per volume of water are less than the equivalent cost of reducing production, including management costs. There may be additional benefits where otherwise wasted water from urban areas or industries is harvested and treated to make it compatible with the aquifer and the existing uses of groundwater. Development of expertise has advanced to capture these opportunities and significant growth is anticipated, particularly where groundwater storage is in decline.

A framework for incorporating managed aquifer recharge into water resources management policies is presented by Ward and Dillon (2011 and 2012). It consists of applying the three instruments; entitlements, allocations and use conditions to each of the four key elements of managed aquifer recharge: access to recharge water, recharge, recovery and end use. It includes a recommended practical procedure, including constraints, on trading of recovery credits. This may be used to facilitate groundwater users associations, and provide a way of sourcing investment in managed aquifer recharge by beneficiaries across the groundwater basin. The entitlement to recover a volume of water that relates to water that has been recharged to an aquifer in general should be tradeable, but with constraints on trading entitlements into drawdown cones or trading into parts of aquifers that are fresher than the water being recharged.

17.7 CONCLUSIONS

a) There is a range of settings within which CWM can occur, and there do not appear to be any situation where CWM should not be practiced.
b) Planned CWM can be far better than spontaneous conjunctive use in terms of deriving a better set of outcomes from the point of view of the resource, of national good and economic return.
c) Most water resources have already been developed and few new 'greenfield' irrigation developments are likely at a significant scale, thus most implementation of CWM will be by retrofitting management arrangements to already existing systems.
d) There are major economic and social reasons to encourage planned CWM, an opportunity the world cannot afford to continue ignoring.
e) Poverty reduction in irrigation areas is closely linked to water supply efficiency and hence to CWM.
f) Institutional, economic, social and technical challenges will need to be addressed, probably in that order, at the sovereign scale.
g) The regulatory settings for water management for different sovereign States will be the most important setting for management approaches. Any institutional strengthening will need to be supported by strong policy and possible legislative changes.

h) CWM will be linked to sovereign policies related to energy, climate change adaption and to food security and hence a broader governmental approach will need to occur.
i) An important part of planned CWM is the identification of the true total cost of water resources and the separate cost to individual users (for example, electricity subsidies are very common), which can greatly differ.
j) The degree of connectivity of surface water and groundwater is an important technical consideration, but not one that will greatly influence whether CWM is successful.
k) Institutional strengthening around groundwater management and a fully integrated water agency will be a major challenge in most areas.
l) Institutional arrangements are required that address depletion and degradation of surface water resources owing to groundwater use, and vice versa.
m) Public education and supporting technical assessments are an important part of CWM.
n) Managed aquifer recharge can have a vital role to play in a planned CWM framework, particularly where groundwater storage is in decline, where aquifers are brackish, or where connectivity between surface water and groundwater is weak.

ACKNOWLEDGEMENT

This chapter is partially based on thematic papers written by the authors and others for the GEF- FAO Groundwater Governance Initiative.

REFERENCES

Bredehoeft, J.D. (2011) Hydrologic Trade-Offs in Conjunctive Use Management. *Groundwater*, 49 (4) 468–475.

Danton, D. & Marr, A.J. (2007) *Conjunctive use of surface water and groundwater in a water abundant river basin.* Australia, River symposium.

Dillon, P., Pavelic, P., Page, D., Beringen H. & Ward J. (2009a) *Managed Aquifer Recharge: An Introduction, Waterlines Report.* No: 13, [Online] Available from https://recharge.iah.org/files/2016/11/MAR_Intro-Waterlines-2009.pdf (accessed 9 Mar 2017)

Dillon, P., Gale, I., Contreras, S., Pavelic, P., Evans, R. & Ward, J. (2009b). Managing aquifer recharge and discharge to sustain irrigation livelihoods under water scarcity and climate change. *Improving Integrated Surface and Groundwater Resources Management in a Vulnerable and Changing World, IAHS Publ. 330*, 1–12.

Dillon, P., Fernandez, E.E. and Tuinhof, A. (2012). *Management of aquifer recharge and discharge processes and aquifer storage equilibrium.* IAH contribution to GEF-FAO Groundwater Governance Thematic Paper No; 4, [Online] Available from www.groundwatergovernance.org/resources/thematic-papers/en/

Dillon, P. and Arshad, M. (2016). Managed Aquifer Recharge in Integrated Water Resource Management. In Jakeman, A., Barreteau, O., Hunt, R., Rinaudo, J.D. and Ross, A. Springer (eds.) *Integrated Groundwater Management; Concepts, Approaches and Challenges.* [Online] Springer International Publishing. pp 435-452. Available from: http://link.springer.com/book/10.1007%2F978-3-319-23576-9

Evans, W.R., Evans, R. S. & Holland, G.F. (2012) *Conjunctive use and management of groundwater and surface water within existing irrigation comments: the need for a new focus on an old paradigm*. Thematic Paper 2. Groundwater Governance: a Global framework for Country Action.

FAO (1995) Land and water integration and river basin management. *Proceedings of a FAO informal workshop, Rome, Italy, 31 January - 2 February 1993*. Food and Agriculture Organization of the United Nations.

Foster, S. & Steenbergen, F. van (2011) Conjunctive groundwater use: a 'lost opportunity' for water management in the developing world? *Hydrogeology Journal* DOI 10.1007/s10040-011-0734-1.

Foster, S., Steenbergen, F. van, Zuleta, J. & Garduño, H. (2010) *Conjunctive use of groundwater and surface water: from spontaneous coping strategy to adaptive resource management*. GW-MATE Strategic Overview Series 2, [Online] Washington DC, The World Bank. Available from: www.worldbank.org/gwmate.

Garduño, H., Romani, S., Sengupta, B., Tuinhof, A. & Davis, R. (2011) *India Groundwater Governance: Case Study*. Water Papers, Water Partnership Program, The World Bank.

Garduño, H. & Foster, S.(2010) *Sustainable Groundwater Irrigation: approaches to reconciling demand with resources*. GW-MATE Strategic Overview Series, No 4, [Online] Washington DC The World Bank. http://documents.worldbank.org/curated/en/603961468331259686/Sustainable-groundwater-irrigation-approaches-to-reconciling-demand-with-resources (accessed 2 Nov 2016)

Maheshwari, B., M. Varua, J. Ward, R. Packham, P. Chinnasamy, Y. Dashora, S. Dave, P. Soni, P. Dillon, R. Purohit, Hakimuddin, T. Shah, S. Oza, P. Singh, S. Prathapar , A. Patel, Y. Jadeja, B. Thaker, R. Kookana, H. Grewal, K. Yadav, H. Mittal, M. Chew, P. Rao (2014). The role of transdisciplinary approach and community participation in village scale groundwater management: Insights from Gujarat and Rajasthan, India. *Water* Open Access Journal [Online] 6(6) 3386-3408. Available from: http://www.mdpi.com/journal/water/special_issues/MAR

Pavelic P, Srisuk K, Saraphirom P, Nadee S, Pholkern K, Chusanathas S, Munyou S, Tangsutthinon T, Intarasut T, Smakhtin V (2012) Balancing-out floods and droughts: opportunities to utilize floodwater harvesting and groundwater storage for agricultural development in Thailand. *Journal of Hydrology*, [Online] 470–471, 55–64. Available from: doi:10.1016/j.jhydrol.2012.08.007

Pavelic, P., Karthikeyan, B., Amarnath, G., Eriyagama, N., Muthuwatta, L., Smakhtin, V., Gangopadhyay, PK., Malik, R.P.S, Mishra, A., Sharma, B.R., Hanjra, M.A., Reddy, R.V., Mishra, V.K., Verma, C.L. and Kant, L. (2015). *Controlling Floods and Droughts through Underground Storage: From Concept to Pilot Implementation in the Ganges River Basin*. Intl Water Mgmt. Institute, Research Report No 165 http://www.iwmi.cgiar.org/Publications/IWMI_Research_Reports/PDF/pub165/rr165.pdf (Accessed 2 Nov 2016)

Perrone, D. and Rohde, M.M. (2016). Benefits and economic costs of managed aquifer recharge in California. *San Francisco Estuary and Watershed Science* 14 (2) Article 4. dx.doi.org/10.15447/sfews.2016v14iss2art4 http://escholarship.org/uc/item/7sb7440w (accessed 10 Nov 2016)

Scanlon, B.R., Reedy, R.C., Faunt, C.C., Pool, D. and Uhlman, K. (2016). Enhancing drought resilience with conjunctive use and managed aquifer recharge in California and Arizona. *Environmental Research Letters*, Vol 11, No. 3, 15p. http://iopscience.iop.org/article/10.1088/1748-9326/11/3/035013

Shah, T., Dargouth, S. & Dinar, A. (2006) *Conjunctive use of groundwater and surface water*. Agricultural and Rural Development Notes. Issue No 6, The World Bank.

Steenbergen van, F. & Tuinhof. A. (2009) *Managing the Water Buffer for Development and Climate Change Adaptation. Groundwater recharge, retention, reuse and rainwater storage*. [Online] Wageningen, The Netherlands, MetaMeta Communications. http://www.bebuffered.com/3rbook.htm (accessed 2 Nov 2016)

Steenbergen van, F., Tuinhof, A. & Knoop, L. (2011). *Transforming Landscapes Transforming Lives – The business of sustainable water buffer management.* [Online] Wageningen, The Netherlands, 3R Water Secretariat. http://www.bebuffered.com/downloads/transforming_land_3108_LQ_total.pdf (accessed 2 Nov 2016)

Tsur, Y. (1990) The stabilisation role of groundwater when surface water supplies are uncertain: the implications for groundwater development. *Water resources Research*, 26 (5).

Ward, J. & Dillon, P. (2011) *Robust policy design for managed aquifer recharge.* Waterlines Report Series. 38, 28p. Available from http://apo.org.au/files/Resource/waterlines_38.pdf (Accessed 2 Nov 2016)

Ward, J. & Dillon, P. (2012) Principles to coordinate managed aquifer recharge with natural resource management policies in Australia. *Hydrogeology Journal* 20 (5) 943–956.

Chapter 18

Global food and trade dimensions of groundwater governance

Arjen Y. Hoekstra[1,2]
[1]*Twente Water Centre, University of Twente, Enschede, The Netherlands*
[2]*Institute of Water Policy, Lee Kuan Yew School of Public Policy, National University of Singapore, Singapore*

ABSTRACT

About 22% of water use in the world is for producing export products. The overdraft of many aquifers is partially related to incentives to produce for the world market and the fact that water scarcity is not properly priced and therefore not reflected in the price of traded commodities. This chapter aims to discuss the relation between the global economy and unsustainable water use, with a focus on groundwater and global food demand. Unsustainable groundwater use isn't a local problem only, because increasingly global markets and companies and consumers worldwide depend on the products derived from unsustainable water supplies. There is a need to collaborate internationally in introducing forces or incentives to drive towards more sustainable groundwater use. Possible arrangements discussed are an international water pricing protocol, but also the institutionalization of groundwater footprint caps that reflect maximum sustainable abstraction levels and an international product label reflecting sustainability of water use.

18.1 INTRODUCTION

Groundwater is generally used locally. Water is too bulky to make transport over large distances economically feasible. There are exceptions, like for instance the Great Man-Made River project in Libya, designed to transfer fossil groundwater pumped from the Nubian Sandstone Aquifer beneath the desert in the south of the country to the northern coastal strip, for both urban and agricultural water supply, using a pipeline network of thousands of kilometres with a planned transport capacity of 6.5 million m^3 per day when fully completed (Sternberg, 2016). This, however, is an exception and luckily so given its unsustainable character. River water is generally used locally as well, although the number of examples of large-scale long-distance inter-basin water transfers of river water is bigger than for groundwater (Ghassemi and White, 2007). A disadvantage of groundwater is that it has to be pumped from the beginning, while surface water at least comes at surface level, with the possibility to use gravity to bring water from locations at higher to locations at lower altitude. In the end, however, inter-basin transfers of river water are generally accompanied with larger energy costs for pumping as well – the California State Water Project for instance uses 2 to 3% of all electricity consumed in the state (Cohen *et al.*, 2004). Water thus largely remains a resource available for local use only, say within a radius of ten to hundred kilometres. Probably more accurately stated, given the natural containment of freshwater within aquifers and river basins, one could argue that the use and availability of deep groundwater resources can best be studied at the geographic level of an aquifer, while the use and availability of interlinked groundwater – river water resources can best be studied at the geographic level of a river basin. But whatever is the precise scale used – that of an aquifer or river basin – the geographic scale remains far below the continental scale. Despite this, water is nowadays increasingly regarded as a *global* resource. How can that be? In addition, proposals are made to come to global arrangements around sustainable water use (Hoekstra, 2011). Why would a resource available for local use only need to be governed at a level beyond the local level?

Water is a global resource since an estimated 22% of water use in the world is for producing export products (Hoekstra and Mekonnen, 2012). There are many specific locations were close to 100% of the water is used for making export products. Consider for instance the use of groundwater from the Ica-Villacurí aquifer beneath the desert of the Ica Valley in Peru to produce asparagus for export to Europe (Hepworth *et al.*, 2010). The farmers in Ica can be characterized as sophisticated players in the global agriculture industry. The majority of land in the Ica Valley is farmed by large agro exporters, with more than 150 hectares per plot of land each, whose crops are primarily destined for export markets. They use sophisticated irrigation techniques and are at the forefront of agricultural innovation in Peru (Bullock, 2015). But growers are pumping water from the aquifer at a much faster rate than the recharge rate: according to the Local Water Authority, water withdrawals in 2013 exceeded recharge by a factor 1.8 for the Ica aquifer and by a factor 3.6 for the Villacurí aquifer.

Local water pollution for producing export products is as common as local water consumption for making export products. Take the example of vast pollution of both groundwater and surface water resources in India in relation to fertilizer and pesticides use in cotton growing for export (Chapagain *et al.*, 2006; Safaya *et al.*, 2016) or the water pollution with toxic effluents from the cotton processing industry in Bangladesh (Islam *et al.*, 2011). These examples show that if we try to understand local patterns of water use and pollution, we will often learn that they are driven by factors in the global economy. The rules of the global economy are apparently such that there is a lack of incentives to prevent overconsumption and pollution of water. Indeed, water scarcity generally remains unpriced – users generally pay for the energy, labour, etcetera to supply the water, but not for the water itself – and regulations or enforcement to ban unsustainable water use are generally lacking or insufficient.

In this chapter, I aim to show the relation between the global economy and unsustainable water use, with a focus on groundwater and global food demand, and I will point at the need and opportunities to collaborate internationally in introducing forces or incentives to drive towards more sustainable groundwater use. In the next section I will go a bit more in depth into the phenomenon of virtual water trade – the phenomenon that water resources used in the production of commodities for export are virtually traded as well. In the third section, I discuss the issue of unsustainable groundwater use in relation to trade. The fourth section addresses the implications of differences in national water endowments for international trade. In the fifth section, I argue that international cooperation in groundwater governance seems necessary to break through the lock-in situation whereby countries keep failing to properly translate

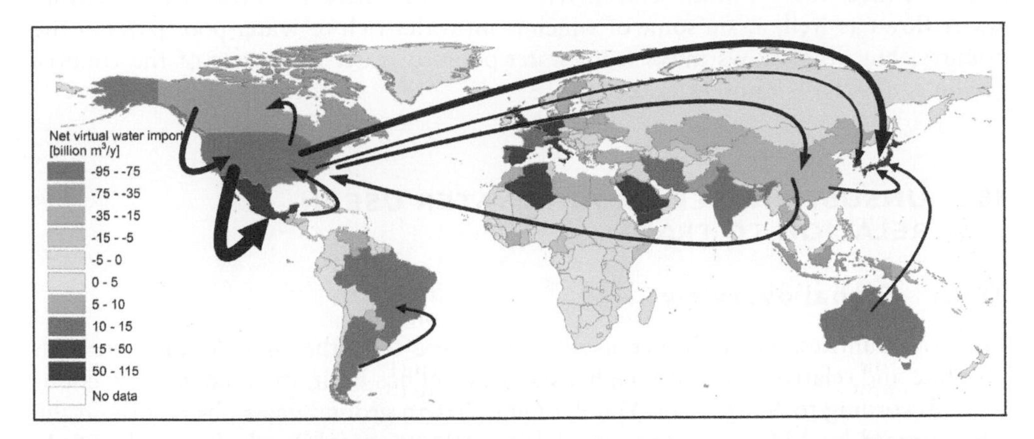

Figure 18.1 Virtual water balance per country and direction of gross virtual water flows related to trade in agricultural and industrial products over the period 1996–2005. Only the biggest gross flows (>15 billion m^3/y) are shown. Source: Hoekstra and Mekonnen (2012).

water scarcity into a price or put serious limitations to growing water use even in the most obvious cases of water depletion.

18.2 INTERNATIONAL AND INTERREGIONAL VIRTUAL WATER TRADE

The total amount of international virtual water flows has been estimated to amount 2320 billion m^3/y, of which about 300 billion m^3/y were blue water resources (Hoekstra and Mekonnen, 2012). Figure 18.1 shows the virtual water trade balance per country as well as the largest international virtual water flows.

The biggest gross virtual water exporters – using substantial portions of domestic water resources for producing export products – are the US, China, India, Brazil, Argentina, Canada and Australia. The degree to which the export commodities from these countries depend on groundwater differs. In most crop production in the world, rain is the primary source of water. Whether crops are irrigated with groundwater or surface water varies between and within countries, and similarly for water use in industries and for municipal water supply. Whether the water use for producing export products depends on renewable or non-renewable water supplies and whether they impact on flows necessary to maintain the functioning of ecosystems differs from place to place as well. The opposite of the virtual water exporters are the importers: Europe, North and Southern Africa, the Middle East, Mexico and Japan. In these regions, consumption partly depends on water resources use elsewhere. About forty percent of Europe's water footprint, the water use associated with the production of all commodities consumed in Europe, lies outside the continent, in countries like Brazil, Argentina, the US, China, India, Indonesia, and Turkey. Some of the international virtual water flows move from water-rich to water-poor countries, but other flows move from water-poor to water-rich countries. Northern Europe, for example, is relatively water abundant, but its consumption relies on substantial volumes of water use in regions where water is much scarcer. Within countries there are intra-national virtual water flows as well, again some of which from water-rich to water-poor parts of the country, but other flows move from water-poor to water-rich parts of the country (Dalin *et al.*, 2014; Zhuo *et al.*, 2016).

18.3 UNSUSTAINABLE GROUNDWATER USE IN RELATION TO TRADE

18.3.1 Global overview

In most countries, groundwater use has increased over the past decades, in both absolute and relative sense, although a stabilization has been observed in a few countries. According to Margat and Van der Gun (2013), groundwater abstraction in the US increased by 144% over the period 1950–1980 but stabilized afterwards. Wada *et al.* (2014) estimate that global groundwater withdrawals increased over the period 1979–2010 from 650 to 1200 billion m^3/y, an increase of 85%, while the ratio of global groundwater withdrawal to overall water withdrawals increased from 32.5% to 36.4% over the same period. During the period 1979–1990, global groundwater withdrawal increased by about 1% per year, but during the more recent period 1990–2010, groundwater withdrawals annually increased by about 3%. The growing importance of groundwater can possibly be explained as the result of increasing surface-water scarcity and the slowdown in the construction of new dams and reservoirs. Regional differences are large, as shown in Figure 18.2. In Europe, groundwater withdrawals account for about 30% of the total water withdrawal and has not increased substantially over the past decades. In North and Central America, however, groundwater withdrawal increased by more than 40% over the period 1979–2010, reaching about 60% of the total in 2010. In West Asia, groundwater withdrawal tripled, getting about 70% of the total in 2010. In South and East Asia, groundwater withdrawal nearly doubled over the period 1979–2010. In North Africa, groundwater withdrawal is about 30%. Over the other regions, like Southeast Asia and South America, groundwater withdrawals are less than 20% of the total.

Gleeson *et al.* (2012) estimate that groundwater withdrawals exceed groundwater availability – defined as groundwater recharges minus groundwater contributions to environmental stream flows – in 20% of the globe's aquifers. About 1.7 billion people live in these areas where abstractions exceed availability, where groundwater availability and/or groundwater-dependent surface water and ecosystems are thus at risk. The places with greatest levels of unsustainable groundwater use are: India, Pakistan, Saudi Arabia, Iran, Mexico, the USA, Northern Africa, China, and Central-Eastern Europe. In the Upper Ganges and Lower Indus Aquifers in India-Pakistan, the ratios of groundwater abstraction to availability average 54 and 18, respectively. In the North and South Arabian Aquifers in Saudi Arabia the ratios are 48 and 39, respectively, and in the Persian and South Caspian Aquifers in Iran the ratios are 20 and 98. In the Western and Central Mexico Aquifers the ratios are 27 and 9.1. In the High Plains and Central Valley Aquifers in the USA the ratios are 9.0 and 6.4. In the Nile Delta Aquifer in Egypt the ratio is 32, while the North Africa Aquifer shared by Algeria, Tunisia and Libya has a ratio of 2.6. In the North China Plain and Northern China

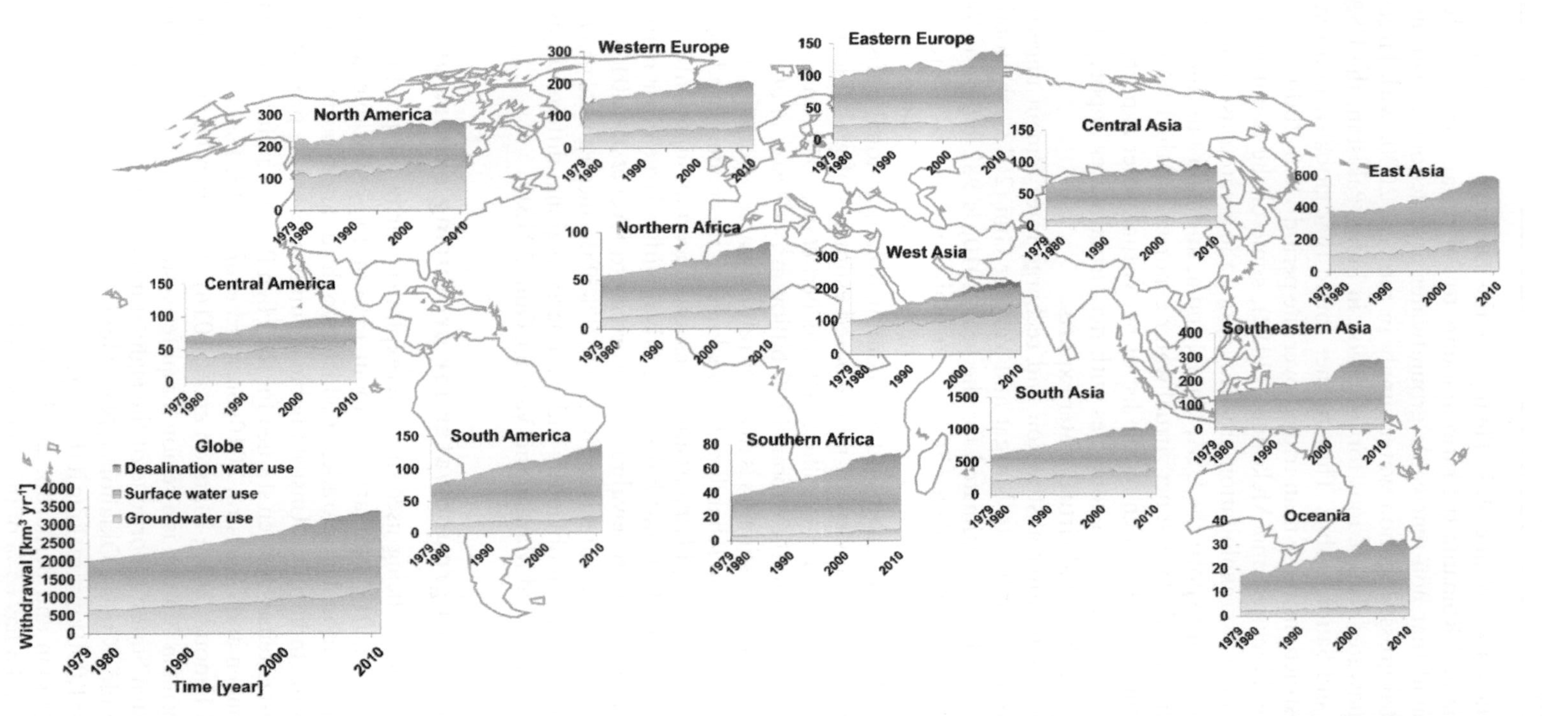

Figure 18.2 Regional trends of water withdrawal per source (groundwater, surface water and desalination water) over the period 1979–2010. The global figure is shown at the bottom left. Source: Wada *et al.* (2014).

Aquifers the ratios are 7.9 and 4.5. Finally, in the Danube Basin Aquifer below Hungary, Austria and Romania the ratio amounts to 7.4. In an earlier study, Wada *et al.* (2012) found that non-renewable groundwater abstractions constitute about 20% of the global gross irrigation water supply in the year 2000, with largest non-renewable groundwater abstractions for irrigation in India, Pakistan, the USA, Iran, China, Mexico and Saudi Arabia. They further found that globally, non-renewable groundwater abstractions more than tripled over the period 1960–2000.

As shown by Hoekstra and Mekonnen (2012), some of the countries with the greatest groundwater overexploitation are also among the greatest users of water for producing products for export: the USA, China, India, and Pakistan. India has been estimated to be the world's largest net virtual water exporter and the USA ranks third. Pakistan and China also rank high on the list of net virtual water exporters (seventh and eleventh, respectively). Even countries with net virtual water import, like Egypt and Iran, still have substantial virtual water exports.

Estimates of groundwater abstraction and recharge rates remain uncertain and vary across sources. According to Margat and Van der Gun (2013), the total global withdrawal of groundwater, estimated for the year 2010, is 982 billion m^3/y. They find that agriculture is responsible for 70% of the global withdrawal of groundwater, domestic water supply for 21% and industry for 9%. These estimates, adopted for example by FAO *et al.* (2016), are based on a compilation of national statistics. Wada *et al.* (2014) provides an overview of model-based estimates of global groundwater withdrawal, which range from 545 billion m^3/y (Siebert *et al.*, 2010; only considering groundwater use for irrigation) to about 1700 billion m^3/y (Wisser *et al.*, 2010). The weakness of statistics-based estimates is that they rely on scarce reported national statistics of unknown accuracy and reliability; the weakness of the model-based estimates is that they rely on various simplifying assumptions and uncertain input data. Presented groundwater abstraction data thus have to be taken with a large error margin. The same holds for groundwater renewal rates. According to FAO *et al.* (2016), the total global withdrawal of groundwater is equivalent to 8% of the mean global groundwater renewal. This global fraction obviously hides the large regional differences as shown by Gleeson *et al.* (2012) and Margat and Van der Gun (2013).

18.3.2 Overdraft of groundwater for producing export products

Worldwide, aquifers are being used and overexploited partly for producing export products. Dalin *et al.* (2017) estimate that about 11% of non-renewable groundwater is embedded in international crop trade, of which two thirds are exported by Pakistan, the United States and India alone. Unsustainable groundwater use for producing export products has most extensively been studied for the High Plains Aquifer in the Midwest of the US (Mekonnen and Hoekstra, 2010; Steward *et al.*, 2013; Esnault *et al.*, 2014; Williams and Al-Hmoud, 2015; Marston *et al.*, 2016). The High Plains Aquifer, also known as the Ogallala Aquifer, is a regional aquifer system located beneath the Great Plains in the United States in portions of the eight states of South Dakota, Nebraska, Wyoming, Colorado, Kansas, Oklahoma, New Mexico, and Texas. It covers an area of approximately 451,000 km^2, making it the largest area of irrigation-sustained cropland in the world (Peterson and Bernardo, 2003). Most of the aquifer underlies parts of three states: Nebraska has 65% of the aquifer's volume, Texas 12% and Kansas 10%

(Peck, 2007). About 27% of the irrigated land in the United States overlies this aquifer system, which supplies about 30% of the nation's groundwater used for irrigation (Dennehy, 2000). In 1995, the High Plains Aquifer contributed about 81% of the water supply in the High Plains area while the remainder was withdrawn from rivers and streams, most of it from the Platte River in Nebraska. Outside of the Platte River Valley, 92% of water used in the High Plains area is supplied by groundwater (Dennehy, 2000). Since the beginning of extensive irrigation using groundwater, the water level of the aquifer has dropped by 3 to 15 meters in most part of the aquifer (McGuire, 2007).

Major export products produced with water from the High Plains Aquifer include wheat, maize and cotton. Within the High Plains area, Kansas takes the largest share in wheat production (51%), followed by Texas and Nebraska (16% and 15% respectively). In Kansas, 84% of the wheat production comes from rain-fed areas. In Nebraska, this is 86% and in Texas 47%. The High Plains area accounts for about 14% of the total wheat production in the USA. Mekonnen and Hoekstra (2010) show that about 19% of the blue water footprint of wheat production in the USA lies in the High Plains area. They found a total blue water footprint of wheat production in the High Plains area of 1.1 billion m^3/y over the period 1996–2005. Texas takes the largest share (39%) in the blue water footprint of wheat production in the High Plains area, followed by Kansas (35%). There is a considerable variation in the blue water footprint per kilogram of wheat within the High Plains states, from 76 litre/kg in Kansas and 115 litre/kg in Nebraska to 304 litre/kg in Texas. Overall, the average blue water footprint per kilogram of wheat in the High Plains area is relatively large if compared to the average in the USA, which is 92 litre/kg.

In the period 1996–2005, the virtual water export related to export of wheat products from the USA was 57 billion m^3/y (Mekonnen and Hoekstra, 2010). About 98% of this virtual water export comes from domestic water resources and the remaining 2% is from re-export of imported virtual water related to import of wheat products. Taking the wheat consumption in the USA of about 88 kg/y per capita and a population in the High Plains area of 2.4 million, we find that only 2% of the wheat produced is consumed within the High Plains area and the surplus (98%) is exported out of the High Plains area to other areas in the USA or exported to other countries. This surplus of wheat constitutes 33% of the domestic wheat export from the USA. Figure 18.3 shows the major foreign destinations of wheat-related virtual water exports from the area of the High Plains Aquifer. Visualizing the hidden link between the wheat consumer elsewhere and the impact of wheat production on the depletion of water resources of the High Plains Aquifer is quite relevant in policy aimed at internalizing the negative externalities of wheat production and passing on those costs to consumers elsewhere.

18.3.3 Imported water risk

For importing countries it is increasingly important to understand their so-called 'imported water risk'. The water footprint of Jordanian consumption, for example, lies 86% outside its own territory and is largely located in the US, partially depending on unsustainable water supplies for wheat production and thus putting Jordan's food security at risk (Schyns *et al.*, 2015). Even countries that are relatively well endowed with freshwater resources may thus have a substantial imported water risk. About

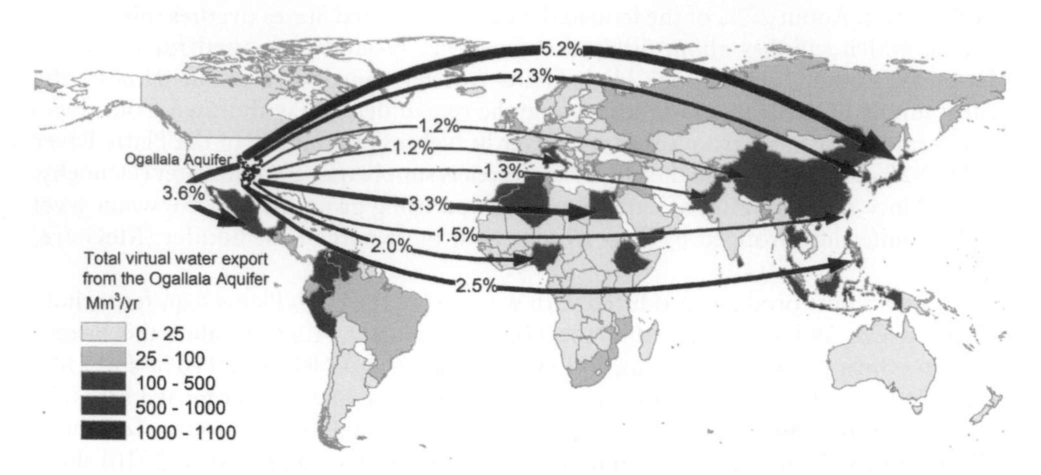

Figure 18.3 Major destinations of wheat-related virtual water exports from the area of the High Plains (Ogallala) Aquifer in the USA (1996–2005). About 58% of the total water footprint of wheat production in the area is for wheat consumption in the USA and 42% is for export to other nations. Only the largest exports (>1%) are shown. Source: Mekonnen and Hoekstra (2010).

75% of the total water footprint of UK consumption, for example, lies outside the UK and about half of UK's blue water footprint lies in areas where water use exceeds sustainable levels, for instance in Spain, the US, Pakistan, India, Iran and South Africa (Hoekstra and Mekonnen, 2016). This study for the UK shows the potential of establishing a relation between consumed products and underlying (remote) unsustainable water use, but tracing the water consumption and assessing the sustainability of that water consumption behind traded products remains a great challenge. There are relatively good data on international trade flows, but it's very difficult to estimate where precisely within an exporting country the export products have been produced. Besides, exports may be re-export of imported products as well, which requires further tracing. Another data challenge is still to distinguish the source of water being used in the places of origin, particularly to differentiate between surface water use and groundwater use. An increasing number of water footprint and virtual water trade studies explicitly differentiate between groundwater and surface water footprints (see for example Aldaya and Hoekstra, 2010; Dumont *et al.*, 2013), but this has to become standard practice.

18.4 THE IMPLICATIONS OF NATIONAL WATER ENDOWMENTS FOR INTERNATIONAL TRADE

There is an immense body of literature about international trade, but there are only few scholars who address the question to which extent international trade is influenced by regional differences in water availability or productivity. International trade is rather explained in terms of differences in labour productivities, availability of land, domestic subsidies to agriculture, import taxes, production surpluses and associated export subsidies, etcetera. According to international trade theory, which goes back to Ricardo

(1821), nations can gain from trade if they specialize in the production of goods and services for which they have a comparative advantage, while importing goods and services for which they have a comparative disadvantage. According to the Ricardian model of international trade, countries can best specialize in producing goods in which they have a relatively high productivity. In more precise, technical terms, economists say: countries have a comparative advantage in producing a particular good if they have a relatively high 'total factor productivity' for that good, whereby total factor productivity is a measure that relates output to all input factors (like labour, land, water). An alternative model of comparative advantage is the Heckscher-Ohlin model, which was formulated in the first half of the previous century. This model does not look at differences in factor productivity across countries, but at differences in factor abundance and in the factor intensity of goods. According to the Heckscher-Ohlin model, countries can best specialize in goods that use their relatively abundant factors relatively intensively. Neither model is comprehensive: whereas the Heckscher-Ohlin theory states that a country can best specialize in producing and exporting products that use the factors that are most abundant, Ricardo's theory says that a country can best focus on producing goods for which they have a relatively high productivity (output per input). But in any case, the rough idea is clear: production circumstances differ across countries, which gives some countries an opportunity in certain products, while it gives other countries an opportunity for other products, thus constituting mutual gains in trade. From the perspective of water, countries with either relative water abundance or relatively high water productivity (value of output per unit of water input), or a combination of both, will have a comparative advantage in producing and exporting commodities that are relatively water intensive.

An important question is what counts as 'water endowment' of a country. Groundwater resources are by far the most important freshwater resource in terms of stock (Gleeson *et al.*, 2016), but it would be a mistake to consider relative occurrence of water stocks across countries to estimate comparative water availabilities. Instead we have to consider freshwater renewal rates, *i.e.* the available flow of freshwater. The annual available freshwater flow is given by the precipitation over land, which splits up into usable green and blue water flows (evaporation and runoff). The blue water flow partly refers to direct surface runoff (precipitation directly going to small streams and rivers) and partly to indirect runoff (precipitation recharging groundwater and subsequently forming the base flow of rivers). Endowments in terms of groundwater flows and river water flows are both strongly related to the amount of rain. Hence, comparative advantages of countries in terms of water availability more or less follow relative rainfall rates. The advantage of groundwater is that it smooths out water availability within the year and between years, just like natural or artificial surface water reservoirs. For a measure of blue water availability, we have to consider the groundwater – surface water system as a whole, because groundwater and surface water resources are not additive. Groundwater recharge later forms the base flow of the river, so if we abstract water from the groundwater we cannot take it anymore from the river, and vice versa, if we want to take some base flow from the river, we shouldn't before already take it from the groundwater.

The relative abundance of green and blue water resources as a total – thus the rain as a total – and water productivity are the best measures of a region's relative comparative advantage in producing water-intensive commodities. The Midwest of the USA with its 400 to 600 mm of annual rainfall is thus not the place in the world most suitable for producing food for export, and the same holds for the North of China. The fact that those regions overexploit groundwater resources for producing export products shows the inadequacy of the global economy to take water scarcity properly into account, possible reasons being inadequate pricing and insufficient regulation. Although pricing and regulation can be implemented at national level, this isn't easily implemented because farmers living in countries that would apply more reasonable water pricing and protective regulations would be put in a disadvantageous position. Thus, implementing better pricing and regulation is politically difficult.

18.5 THE NEED FOR INTERNATIONAL COOPERATION IN GROUNDWATER GOVERNANCE

There is nothing more obvious than that overexploitation of aquifers can be regulated by setting a groundwater footprint cap for each aquifer, *i.e.* a maximum to the net water abstraction from an aquifer, and making sure that no more groundwater footprint permits are issued than available within the cap. These permits can be made tradable or issued based on some priority system or historical rights or whatever, but most important is to start with an agreement on a maximum abstraction rate. It is often assumed that the rate of groundwater withdrawal is "safe" or "sustainable" if it does not exceed the natural rate of recharge, but this is not correct (Alley *et al.*, 2002). The maximum sustainable level of the groundwater footprint on an aquifer depends on two variables: the groundwater recharge rate and which fraction of that can be sustainably withdrawn on annual basis without unacceptably affecting the base flow of the river that is fed with the groundwater outflow from the aquifer (Hoekstra *et al.*, 2011). What is unacceptable depends on the minimum flows to be maintained in the river system to preserve ecosystems and people that depend on this base flow. In practice this may imply that only 20–40% of the groundwater recharge rate is actually available for abstraction. Deep aquifers that are not or hardly recharged are not suitable for supplying water sustainably, so governments should be extremely careful in supplying abstraction permits in such cases. Even though it may take decades or centuries before aquifers get depleted, the risk of using them is that societies are built on a non-sustainable resource. The argument of possible substitution by another source in the long run generally doesn't hold, because the only reason to abstract fossil groundwater in the first place is that such alternative source is not available.

Agreeing on groundwater footprint caps for transboundary aquifers implies some form of international cooperation between the countries that share the aquifer. But probably another much larger form of international cooperation is needed to establish the whole idea of groundwater footprint caps. The reason is that setting such caps will influence the global food production pattern as a whole, with significant economic, social and political implications. Major aquifers with substantial importance in producing food for the global market are not slightly overcharged, but by factors of a few to fifty times the sustainable abstraction rate. Reducing those abstraction rates will inevitably reduce export from those production regions, and also have great consequences for the farmers involved. These consequences have to be mitigated, which will be easier, particularly in developing countries, when this is put in the context of international cooperation. When water and resultant food prices go up as a result of water use restrictions, there needs to be some international agreement to keep a level playing field.

A second incentive that governments could employ, together with institutionalizing groundwater footprint caps, is to introduce water pricing depending on local water scarcity. A differentiation between groundwater and surface water prices should be implemented where relevant. Prices will vary from catchment to catchment and vary within the year depending on water demand versus supply throughout the year. Pricing water based on its economic value was already agreed upon by the international community at the International Conference on Water and the Environment held in Dublin in 1992 (ICWE, 1992), but there has never been a serious follow-up. Earlier I have proposed that national governments start negotiations on an international Water Pricing Protocol as a way to jointly agree on a way forward (Hoekstra, 2011), because it is in the interest of everyone that our global food supply system is based on sustainable water use, but it is apparently difficult for individual countries to make substantial steps one-sided. An international agreement on water pricing should ensure that users pay the full cost of water use, including investment costs, operational and maintenance costs, a water scarcity rent and the cost of negative externalities of water use. Such an agreement would need to include all water-using sectors, including agriculture.

A very different form of international effort could be the introduction of sustainable water use criteria in existing environmental labels to products or a dedicated water label. This could be done by the public sector – for instance in Europe where environmental labelling of products is most advanced – or through private sector initiatives as well. Suppose that a brand like Coca Cola decides to avoid sugar in its drinks that is grown with unsustainable groundwater reserves or that brands like Unilever and Nestlé decide to apply such criterion across their whole product portfolio, this may drive change by itself and provide incentives to governments to respond – they will have to because such market changes affect local employment, economy and citizens. Finally, the international financial sector should agree on including sustainable water use criteria in investment decisions, comparable to taking into account the impact of investments on greenhouse gas emissions, which is increasingly done.

All of the above options for better groundwater governance – institutionalizing groundwater footprint caps and environmental base flow requirements, more sensible water pricing, incorporating water sustainability in environmental labels for products, applying sustainable water use criteria in investment decisions, and international collaboration in all these matters – face serious economic, social and political obstacles. However, business as usual will result in continued groundwater overexploitation rates. The problem that our global food economy is partly built on unsustainable water supplies can only temporarily be ignored.

18.6 CONCLUSION

Worldwide, aquifers are overexploited, partly in relation to international trade. The international trade system does not have any mechanism to reduce the problem. On the opposite, overexploitation remains attractive in the short term. Good governance of the world's groundwater resources requires shared rules on sustainable use. National governments need to agree keeping groundwater footprints below aquifer-specific maximum sustainable levels, and to issue limited groundwater footprint permits within these maximum levels. Further we need to move to a global economy that incorporates water scarcity in prices of water-based products. Consumers, companies and investors should prioritize products that are based on sustainable water use and avoid products that contribute to overexploitation of aquifers or rivers. This requires a greater transparency on how specific products relate to groundwater overexploitation.

ACKNOWLEDGEMENTS

The sections on the High Plains Aquifer and comparative advantage are based on Hoekstra (2013).

REFERENCES

Aldaya, M.M., Hoekstra, A.Y. (2010) The water needed for Italians to eat pasta and pizza, *Agricultural Systems*, 103(6): 351–360.

Alley, W.M., Healy, R.W., LaBaugh, J.W., Reilly, T.E. (2002) Flow and storage in groundwater systems, *Science*, 296(5575): 1985–1990.

Bullock, J. (2015) *Development of Peru's asparagus industry*, LAD case study, Stanford University, Stanford, UK.

Chapagain, A.K., Hoekstra, A.Y., Savenije, H.H.G., Gautam, R. (2006) The water footprint of cotton consumption: An assessment of the impact of worldwide consumption of cotton products on the water resources in the cotton producing countries, *Ecological Economics*, 60(1): 186–203.

Cohen, R., Nelson, B., Wolff, G. (2004) *Energy down the drain: The hidden costs of California's water supply*, Natural Resources Defense Council, New York, USA.

Dalin, C., Hanasaki, N., Qiu, H., Mauzerall, D.L., Rodriguez-Iturbe, I. (2014) Water resources transfers through Chinese interprovincial and foreign food trade, *Proceedings of the National Academy of Sciences*, 111(27): 9774–9779.

Dalin, C., Wada, Y., Kastener, T., Puma, M.J. (2017) Groundwater depletion embedded in international food trade, *Nature*, 543: 700–704.

Dennehy, K.F. (2000) *High Plains regional ground-water study*, USGS Fact Sheet FS-091-00, United States Geological Survey, Denver, Colorado, USA.

Dumont, A., Salmoral, G., Llamas, M.R. (2013) The water footprint of a river basin with a special focus on groundwater: The case of Guadalquivir basin (Spain), *Water Resources and Industry* 1(1–2): 60–76.

Esnault, L., Gleeson, T., Wada, Y., *et al.* (2014) Linking groundwater use and stress to specific crops using the groundwater footprint in the Central Valley and High Plains aquifer systems, USA, *Water Resources Research*, 50: 4953–4973.

FAO, UNESCO, IAH, World Bank Group, GEF (2016) *Global diagnostic on groundwater governance*, Food and Agriculture Organization, Rome, Italy.

Ghassemi, F., White, I. (2007) *Inter-basin water transfer: Case studies from Australia, United States, Canada, China and India*, Cambridge University Press, Cambridge, UK.

Gleeson, T., Befus, K.M., Jasechko, S., *et al.* (2016) The global volume and distribution of modern groundwater, *Nature Geoscience*, 9(2): 161–167.

Gleeson, T., Wada, Y., Bierkens, M.F.P., *et al.* (2012) Water balance of global aquifers revealed by groundwater footprint, *Nature*, 488(7410): 197–200.

Hepworth, N.D., Postigo, J.C., Güemes Delgado, B., Kjell, P. (2010) *Drop by drop: Understanding the impacts of the UK's water footprint through a case study of Peruvian asparagus*, Progressio, London, UK.

Hoekstra, A.Y. (2011) The global dimension of water governance: Why the river basin approach is no longer sufficient and why cooperative action at global level is needed, *Water*, 3(1): 21–46.

Hoekstra, A.Y. (2013) *The water footprint of modern consumer society*, Routledge, London, UK.

Hoekstra, A.Y., Chapagain, A.K., Aldaya, M.M., Mekonnen, M.M. (2011) *The water footprint assessment manual: Setting the global standard*, Earthscan, London, UK.

Hoekstra, A.Y., Mekonnen, M.M. (2012) The water footprint of humanity, *Proceedings of the National Academy of Sciences*, 109(9): 3232–3237.

Hoekstra, A.Y., Mekonnen, M.M. (2016) Imported water risk: the case of the UK, *Environmental Research Letters*, 11(5): 055002.

ICWE (1992) The Dublin statement on water and sustainable development, *International Conference on Water and the Environment*, Dublin, Ireland.

Islam, M.M., Mahmud, K., Faruk, O., Billah, M.S. (2011) Textile dyeing industries in Bangladesh for sustainable development, *International Journal of Environmental Science and Development*, 2(6): 428–436.

Margat, J., Van der Gun, J. (2013) *Groundwater around the world: A geographic synopsis*, CRC Press, Leiden, the Netherlands.

Marston, L., Konar, M., Cai, X., Troy, T.J. (2015) Virtual groundwater transfers from overexploited aquifers in the United States, *Proceedings of the National Academy of Sciences*, 112(28): 8561–8566.

McGuire, V.L. (2007) *Water-level changes in the High Plains Aquifer, predevelopment to 2005 and 2003 to 2005*, Scientific Investigations Report 2006–5324, United States Geological Survey, Reston, Virginia, USA.

Mekonnen, M.M., Hoekstra, A.Y. (2010) A global and high-resolution assessment of the green, blue and grey water footprint of wheat, *Hydrology and Earth System Sciences*, 14(7): 1259–1276.

Peck, J.C. (2007) Groundwater management in the High Plains Aquifer in the USA: Legal problems and innovations, In: Giordano, M. and Villholth, K.G. (eds.) *The agricultural groundwater revolution: Opportunities and threats to development*, CAB International, Wallingford, UK.

Peterson, J., Bernardo, D. (2003) High Plains regional aquifer study revisited: A 20 year retrospective for Western Kansas, *Great Plains Research*, 13(2): 179–197.

Safaya, S., Zhang, G., Mathews, R. (2016) *Toward sustainable water use in the cotton supply chain: A comparative assessment of the water footprint of agricultural practices in India*. Water Footprint Network, The Hague, Netherlands/C&A Foundation, Switzerland.

Schyns, J.F., Hamaideh, A., Hoekstra, A.Y., Mekonnen, M.M., Schyns, M. (2015) Mitigating the risk of extreme water scarcity and dependency: The case of Jordan, *Water*, 7(10): 5705–5730.

Siebert, S., Burke, J., Faures, J.M., Frenken, K., Hoogeveen, J., Döll, P., Portmann, F.T. (2010) Groundwater use for irrigation: A global inventory, *Hydrology and Earth System Sciences*, 14: 1863–1880.

Sternberg, T. (2016) Water megaprojects in deserts and drylands, *International Journal of Water Resources Development*, 32(2): 301–320.

Steward, D.R., Bruss, P.J., Yang, X., *et al.* (2013) Tapping unsustainable groundwater stores for agricultural production in the High Plains Aquifer of Kansas, projections to 2110, *Proceedings of the National Academy of Sciences*, 110: E3477–E3486.

Wada, Y., Van Beek, L.P.H., Bierkens M.F.P. (2012) Nonsustainable groundwater sustaining irrigation: A global assessment, *Water Resources Research*, 48: W00L06.

Wada, Y., Wisser, D., Bierkens, M.F.P. (2014) Global modeling of withdrawal, allocation and consumptive use of surface water and groundwater resources, *Earth System Dynamics*, 5: 15–40.

Williams, R., Al-Hmoud, R. (2015) Virtual Water on the Southern High Plains of Texas: The case of a nonrenewable blue water resource, *Natural Resources*, 6: 27–36.

Wisser, D., Fekete, B.M., Vörösmarty, C.J., Schumann, A.H. (2010) Reconstructing 20th century global hydrography: A contribution to the Global Terrestrial Network-Hydrology (GTN-H), *Hydrology and Earth System Sciences*, 14: 1–24.

Zhuo, L., Mekonnen, M.M., Hoekstra, A.Y. (2016) The effect of inter-annual variability of consumption, production, trade and climate on crop-related green and blue water footprints and inter-regional virtual water trade: A study for China (1978–2008), *Water Research*, 94: 73–85.

Chapter 19

Governance and management of transboundary aquifers

Shaminder Puri[1] *& Karen G. Villholth*[2]
[1]*International Association of Hydrogeologists (IAH), UK*
[2]*International Water Management Institute (IWMI), South Africa*

ABSTRACT

Transboundary aquifers are an increasingly important water resource around the world. The chapter gives a historic overview of the steps and progress achieved in terms of enhancing the recognition, assessment, mapping, and governance of these aquifers. The last two decades of pioneering work, including concrete experiences from joint management and governance of key transboundary aquifers around the world and the tabling of the UN International Law Commission Draft Articles on Transboundary Aquifers at the UN General Assembly, have been instrumental in bringing the issue of internationally shared aquifers on the political and scientific agenda as well as in building a competent community of practice. Future improvement of governance of these aquifers hinges on further awareness raising and capacity development at multiple levels, and critically, the finalisation of the status of the Draft Articles in a shape that becomes tractable with aquifer states, amenable to application in concrete cases taking into account issues of sovereignty, and also enabling the implementation of more integrated and conjunctive (surface water and aquifer) approaches.

19.1 INTRODUCTION

It may be confidently stated that addressing sound governance of transboundary aquifers, from near complete obscurity prior to 2000 to globally prominent recognition, has come of age in 2016. Examples of this evidence come from many sources, as further elaborated in this chapter, but the recent Earth Security Group's CEO briefing is indicative, because a business leadership role for transboundary aquifer governance is envisaged for national and multinational corporations (ESG, 2016). This is a strong signal of the maturation of the subject of transboundary aquifers, its critical importance for sustainable and peaceful multinational development, and the issues associated with their proper governance. This demonstrates the evolution from past academic interest to a current investment-related reality. It is therefore appropriate that the topic appears in a contemporary discourse on groundwater and water governance.

The purpose of this chapter is to review the evolution and take stock of the progress of governance of transboundary aquifers, underway since around the year 2000, to highlight some important milestones that have been reached and to draw key lessons learned during this period. The authors also give their assessment of desirable developments and processes when looking ahead to the next ten years and beyond.

Importantly, the chapter identifies the 'elephants in the room' in relation to transboundary aquifers – *i.e.* some critical issues that have been overlooked or deemed too sensitive for debate, *e.g.* the hydrohegemony and sovereignty aspects related to transboundary aquifers – but that need careful attention and emphasis going forward.

19.2 A HISTORIC REVIEW

19.2.1 Focus on domestic aquifers

To set the scene for the discussion that follows, a brief historical lookback is provided. At the outset, it is noted that a transboundary aquifer is a continuation of adjacent domestic aquifers demarcated by national borders, but nevertheless constituting a single aquifer or system of aquifers. Scientific studies on the evaluation and assessment of domestic aquifers can be traced back to the late 1950s in practically all countries of the world. The UN's Economic and Social Council started a global inventory to increase the scientific knowledge of hydrogeology in the interest of the growing attention to water, and pioneered countrywide 'groundwater status reports' for most of the UN member states (*e.g.* UN-DTCD, 1988, endnote 1). Looking through those reports now, with the benefit of hindsight, it is clear that in many parts of the world, two or more neighbouring countries shared (transboundary) aquifers. This is clear from the common geological formations, zones of recharge and discharge, and flow directions. However, due to the use of different geological nomenclature and hydrogeological terminology, their transboundary nature and significance were often obscured. Examples may be found in the ISARM Americas Reports (Villar, 2016). Certain international projects did, however, not stop at the border, *e.g.* the EC project 'Groundwater Resources of the EC' (1978–1982) (Fried, 1982) and may have influenced early thinking globally on more integrated mapping.

19.2.2 From national to international aquifers

Later, once the UN-DTCD's Country Reports were merged into continental hydrogeological maps, *e.g.* Groundwater in Africa (UN, 1973), the configuration of contiguous formations crossing national boundaries became clearer. A significant joint effort was expended by UN and national experts in ensuring the geological/hydrogeological consistency across national borders (endnote 2 and 3). A good summary of the progress from 'national' to 'collaborative international' evolution of the mapping is given by Gilbrich *et al.* (2014).

19.2.3 From regional aquifers to transboundary aquifers

Up to the mid-1970s, due to the post World War II tensions, the east–west confrontations and the strains of the 'iron curtain', as well as the progressing de-colonisations in Asia and Africa, reference to 'transboundary aquifers' was considered too politically sensitive. Rather, the preferred term was 'regional aquifers'. Discussion of scientific and other (*e.g.* socio-economic) conditions and aquifer management in a neighbouring country were a matter of contention, as they were considered to be 'national' issues

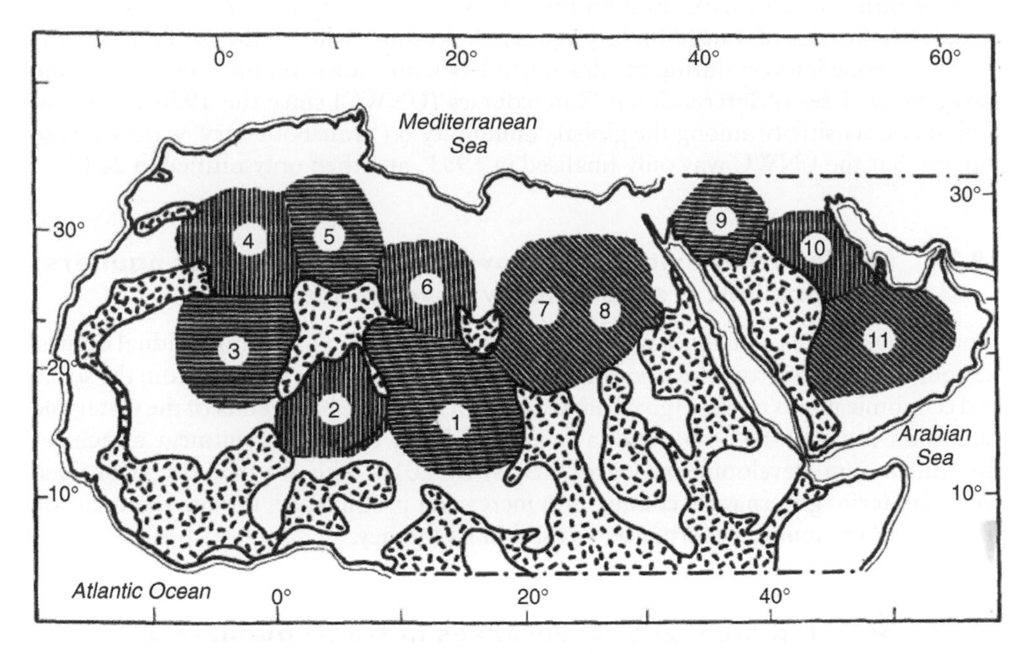

Figure 19.1 Burdon's (1977) representation of regional aquifers in North Africa without reference to international boundaries.

not to be interfered with by others. Given the socio-political stance at the time, the term transboundary governance was rarely raised. Even in the early pioneering work of Burdon (1977), Lloyd (1969) and others on 'regional aquifers', transboundary governance was not alluded to due to diplomatic sensitivities, and it remained difficult to compare national conditions in a transboundary context. As an example and resulting from this, regional maps of aquifers did not include indications of national borders (Figure 19.1).

By 1995, references to these transboundary resources were more explicit, *e.g.* Margat (1995) and Puri (1997), leading to the first international references to 'shared aquifer resources'. The Tripoli Statement of 1999 (UNESCO-IHP, 2001), following the international conference on 'Regional Aquifer Systems in Arid Zones', included the following recognition that "... some countries share aquifer systems; international law does not provide comprehensive rules for the management of such systems as yet, and clearly groundwater mining could have implications for shared water bodies". This Statement and associated discourse provided the impetus for the launch of the IAH-UNESCO ISARM (Internationally Shared Aquifer Resources Management Programme) (Puri and Aureli, 2005), which laid the foundations of modern assessment and joint management of these aquifers.

This was the scene at the time of the Hague World Water Forum in 2000, when the issue of transboundary aquifers was first raised in a global international forum, and then further explicitly discussed at the Kyoto World Water Forum in 2003. In contrast, the discussion on governance of transboundary river basins had already made its appearance much earlier, around the 1970s. From today's perspective, it may be stated that the issue of transboundary aquifers was omitted during the early and lengthy debates among lawyers during the drafting a UN Convention on the Law of the Non-navigational Uses of International Watercourses (UNWC) since the 1950s. Still, the diplomatic sensitivity among the global community on 'transboundary' waters was so intense that the UNWC was only finalised in 1991, and then only ratified in 2014.

19.2.4 Moving to individualised governance of domestic aquifers and incipient transboundary governance

From the late 1970s, countries were starting to pay attention to their individual domestic aquifers, not just from the scientific assessment perspectives, but also from the social and economic points of view (governance), driven by the global agenda of the water and sanitation decade (1981–90), and later in the 2000s, by the commitment to achieve the Millennium Development Goals (MDGs) by 2015. This was the origin of what we today term 'governance' coming into increasing prominence, though the focus on transboundary aquifer governance was still in its infancy.

19.2.5 Selected highlights of advances in transboundary aquifer management in the 21st century

While not exhaustive, Table 19.1 gives a listing of the key milestones of early advances made in transboundary groundwater governance.

Among the most noteworthy of the milestones is the conceptualisation, formulation and adoption in 2008 of the United Nations International Law Commission's (UNILC) Draft Articles on the Law of Transboundary Aquifers (Stephan, 2009). These were developed at a record pace, and were subsequently placed for the UN General Assembly's consideration in 2010. There is some speculation on the reason that these Draft Articles were concluded in record time (3 years), while the UNWC took 18 years to negotiate. It would seem that, while the UNWC was formulated almost exclusively by lawyers, in the case of the Draft Articles, from day 1, there was full engagement with practitioners of hydrogeology, thus continuously providing the legal drafters with the necessary science and the practice underlying such a complex issue (McCaffrey, 2009).

The publication of the WHYMAP (World-wide Hydrogeological Mapping and Assessment Programme) of transboundary aquifers of the world (Struckmeier *et al.*, 2006) in 2006 was another major step forward. This was presented at the 4th World Water Forum, held in Mexico City and demonstrated for the first time the global distribution of aquifers that are shared between countries.

Other noteworthy milestones are the completion of the European inventory (ECE, 1999), and the first global inventory of transboundary aquifers in 2009 (Puri and Aureli, 2009), the compilation of updates to the global map of transboundary aquifers by IGRAC (International Groundwater Resources Assessment Centre) in 2009 (IGRAC, 2009) and later (IGRAC, 2015, 2014, 2012), the UNECE adoption of Model Provisions on Transboundary Groundwaters (see Chapter 6 of this book) and the adoption by the Global Environment Facility (GEF) of several ISARM case studies under their International Waters Programme in 2008, strongly supporting and encouraging inter-country collaboration (see https://www.thegef.org/topics/international-waters).

Table 19.1 Summary of milestones in governance of transboundary aquifers, 1997–2017, with emphasis on early achievements.

1997 – IAH Commission on Transboundary Aquifer Management	The Executive of the International Association of Hydrogeologists (IAH) followed up suggestions made at their general assembly to focus on transboundary aquifers by setting up a Commission on Transboundary Aquifer Resource Management (IAH-TARM).
1999 – UNESCO Conference on Regional Aquifers, Tripoli, Libya	UNESCO took up the challenge and with the Govt. of Libya called for an international conference, whose outcome was the Tripoli Statement dedicated to managing non-renewable resources that were often shared between countries.
2000 – UNESCO-IHP resolution on establishing ISARM	The UNESCO-International Hydrological Programme (IHP)'s Intergovernmental Council took action on the Tripoli Statement and resolved to establish an IHP programme on shared aquifers, ISARM (Internationally Shared Aquifer Resources Management).
2001 – ISARM Framework Document	A framework document was developed by UNESCO-IHP ISARM and IAH-TARM Commission, setting out guidance for the development of the global inventory and approaches to the necessary data collation on transboundary aquifers.
2002 – IAH Congress, Mar del Plata, Argentina – Latin America's 1st Consultative Meeting on Transboundary Aquifers and launch of the ISARM, Americas	The 1st Latin American meeting of the ISARM-Americas (a pioneering initiative) was held in the course of the IAH Congress where it was resolved to pursue the issue of a continental inventory of transboundary aquifers and to scope the factors that need consideration by governments and other concerned stakeholders.
2003 – 3rd World Water Forum, Kyoto, Japan	UNESCO-IHP and IAH made significant contributions to the Forum by setting out the issue of groundwater governance and stressing the need for attention to transboundary aquifers – drawing on the experiences arising from the then ongoing Guaraní transboundary aquifer collaboration in Latin America.
2003 – UNILC's 1st Report on Shared Natural Resources	In 2002–03, the UN's ILC (International Law Commission) took shared natural resources as its topic for the progressive development of customary laws and decided to address transboundary aquifers. The first report (Yamada, 2003; UNILC, 2003) arose from significant science-based advice provided by technical experts fielded by UNESCO and IAH.
2004 – GEF STAP Consultative meeting on the relevance of groundwater in GEF operations	The GEF STAP (Scientific and Technical Advisory Panel) called for an advisory meeting of practitioners of hydrogeology to scope the issue of groundwater as a part of the global environmental assessments and in particular in the area of international waters (IAH, 2005).
2006 – Publication of the world map of Transboundary Aquifers by WHYMAP (scale 1:50M)	With the incoming data/information from the global TARM-ISARM inventory, WHYMAP (World-wide Hydrogeological Mapping and Assessment Programme) developed the first compilation of a global map of transboundary aquifers of the World, issued as a special publication for the 4th World Water Forum, Mexico City (Struckmeier *et al.*, 2006).
2008 – Final version of UNILC Draft Articles submitted to UNGA	The final version of the Draft Articles on the Law of Transboundary Aquifers developed by the UNILC (Stephan, 2009) were presented to the UNGA (UN General Assembly) for their consideration and eventual adoption.

(Continued)

Table 19.1 Continued.

2008 – Adoption by the GEF of the issue of transboundary aquifers	The GEF adopts transboundary aquifers as an important issue and, following the Guaraní financing, also finances projects on the Iullemeden transboundary aquifer, the NSAS (Nubian Sandstone Aquifer System), and the Groundwater and Drought Management in SADC project, among others.
2009 – First global inventory completed and the global map updated	Completion of the first global inventory of transboundary aquifers (Puri and Aureli, 2009) and updated global map of transboundary aquifers by IGRAC (International Groundwater Resources Assessment Centre) (IGRAC, 2009).
2010 – Transboundary Aquifers – Challenges and new directions, UNESCO Paris	UNESCO-IHP holds a major international conference devoted to ISARM – over 300 delegates participate and present cases from all over the world (UNESCO-IHP, 2010).
2012 – UN ECE's development on transboundary aquifers	Adoption of the Model Provisions on Transboundary Groundwaters, by the Parties to the UNECE Convention on the Protection and Use of Transboundary Watercourses and International Lakes.
2010–2017	This period marks a significant broadening of events and activities related to transboundary aquifers, demonstrating the evolving recognition of their importance, including the global GEF-funded TWAP (Transboundary Waters Assessment Programme) (UNESCO-IHP and UNEP, 2016) that set out to further understand transboundary aquifers around the world, including detailed case studies of particular aquifers.

19.3 CHALLENGES OF TRANSBOUNDARY AQUIFER MANAGEMENT

19.3.1 Asymmetric governance in domestic parts of transboundary aquifers

Irrespective of incipient governance of groundwater in the national context as alluded to in Section 19.2.4, the often-encountered context is one of a degree of asymmetry in the level and approach to the governance of groundwater in each domestic part of a transboundary aquifer. This reflects different trajectories of national development and little explicit regard for the subsurface extension or the natural resources on the other side of the border. The degree of asymmetry is governed by a number of socio-economic and institutional factors (Table 19.2) in addition to the climatic, hydrogeological and environmental features. Examples of asymmetric approaches to transboundary management include the Rum-Saq (also called the Saq-Ram) aquifer between Jordan and Saudi Arabia (Lloyd and Pim, 1990) and the shared aquifers between Mexico and USA (Varady *et al.*, 2014; Alley, 2013; Varady *et al.*, 2010).

19.3.2 Unknowns and information gaps complicate transboundary aquifer management

The inherent invisibility of aquifer systems and the associated typical lack of data and knowledge related to transboundary aquifers implies that impacts of groundwater

Table 19.2 Socio-economic and institutional characteristics defining the degree of asymmetry in transboundary aquifer management (from Varady *et al.*, 2014).

- Culture/language/educational system
- Colonial legacy
- Economy and prosperity levels
- Taxation and revenue-generation
- Legal framework
- Regulation and enforcement
- Administration: federal vs. decentralized
- Human resources and expertise
- Physical infrastructure, nexus sectors
- Robustness of institutions and civil society

development or various land uses are likely to take place without the explicit knowledge of all parties involved. This situation took place in the Rum-Saq aquifer from the 1970s through to the late 1980s, where unilateral developments took place in Jordan and Saudi Arabia for the large-scale extraction of groundwater (Puri and Elnaser, 2003). In analogous conditions, for river basins crossing international boundaries, cross-border impacts rapidly become visible and explicit. Examples include the transboundary Tigris-Euphrates Rivers, the Syr Darya–Amu Darya Rivers (Munia, 2016). In the case of aquifers, in contrast, the impacts are most often invisible, have a slow onset, and are often rather uncertain and fuzzy, in the absence of a significant amount of good aquifer systems data. Such context is exacerbated in situations with asymmetric governance. Countries with low level of management and knowledge may, despite good level of understanding in the neighbouring country, receive a downstream negative impact of development. Because of knowledge discrepancy, the hegemon may take advantage of the situation and only control in-country impacts. Table 19.3 summarises two conditions – one where aquifer or river system states[1] follow asymmetric governance; the other, where countries seek to adopt consistent joint governance approaches, even though the comprehensive data and aquifer characterisation may not yet be available. The table illustrates the added complexity of transboundary management of aquifers due to the invisibility and typical unknowns of aquifer systems and how it may jeopardize against the country with lower level of management.

19.3.3 Gaining the attention of decision makers and impacts on vulnerable populations

One observation to be made is that with inadequate data and documentation, scientists or stakeholders wishing to transmit their unilateral or bilateral concerns of transboundary interactions to possibly sceptical decision makers, have found it difficult to gain constructive audience. A good discussion of this is given by Villar (2016), whose description of the efforts by the scientific community to elevate the transboundary aspects of the Guaraní aquifer were only accelerated when the international community

[1]The terminology of aquifer system state is as given in the UNILC's Draft Articles on the Law of Transboundary Aquifers.

Table 19.3 The difference in impacts of asymmetric vs. symmetric governance between aquifers and river basin sharing countries

	Asymmetric governance in domestic portions of a transboundary water system	*Symmetric governance across domestic portions of a transboundary water system*
Aquifer systems	Transboundary impacts, and visibility of same, may be contrasting in neighbouring countries due to exploitation and management following separate trajectories. A hegemon may be able to use this to its advantage, paying unilateral attention to national concerns and impacts.	Impacts in neighbouring countries may be anticipated, actions can be planned to jointly mitigate adverse impacts, and benefits can be shared. Even with sparse data, some science-to-policy traction can be achieved.
River systems	Transboundary impacts are more obvious and more easily detected, implying that transparent and multi-lateral solutions can easier be negotiated and rationalized.	As above.

(through the GEF's support) entered onto the stage. Poor grasp of such concerns by decisions makers means that vulnerable human populations and groundwater-dependent ecosystems, mostly in countries with immature scientific, legal and technical capacity, are the ones to potentially suffer the consequences of such asymmetry (examples are transboundary aquifer areas in the Aral Sea Basin and aquifers in the Brahmaputra delta areas).

19.3.4 International versus domestic transboundary aquifers

Apart from 'international' transboundary aquifers, national aquifers shared between federal states (or other national sub-jurisdictional units) are also sometimes subject to similar 'diplomatic' constraints, *e.g.* in countries like Brazil (Villar, 2016), Australia and USA with federal constitutions (Varady *et al.*, 2014), where neighbouring states manage their affairs without reference to their neighbour. Some illustrative governance issues experienced in USA (over the Ogallala aquifer) are given in Box 19.1.

Box 19.1: Transboundary governance within a federal state: The Ogallala Aquifer, USA

- The Ogallala aquifer (approximately 583,000 km^2) is located in the High Plains of Colorado, Kansas, Nebraska, New Mexico, Oklahoma, and Texas. The depth and thickness vary across the region. From the 1900s, it has been a major water source for agriculture, municipal and industrial needs.
- Over time, the abstractions have exceeded the annual recharge. Recharge is non-uniform, consequently not all users within a state or all states face the same degree of crisis. Since Texas is more concerned than *e.g.* Nebraska,

Texas may develop a southern High Plains strategy as opposed to a six-state High Plains common strategy.

- Asymmetric development and governance of the hydrologically contiguous aquifer have been pursued by the federal states.
- While today, other factors have slowed down the rate of drawdown, including relatively higher energy prices, low crop prices, rain water harvesting, national/inter-state transboundary aquifer governance has yet to come together.

Source: www.meteor.iastate.edu/gccourse/issues/society/ogallala/ogallala.html

19.3.5 'Aquifer' vs. 'groundwater' in transboundary systems

There is copious literature, guidance and recommendations on the sound governance of water (*e.g.*, OECD, 2011). However, much of this relates to the *utilisation* of water in connection with anthropogenic needs. For example, the OECD guidelines cover: 1) water supply (domestic, industrial and agricultural); 2) pollution control and water treatment, both involving the establishment, monitoring and enforcement of standards, regulations and incentives. Where such guidelines are extended to include 'groundwater', they are focussed primarily on its *utilisation*, after it has been *withdrawn* from the aquifer, thus providing limited guidance for the governance of the aquifer *resource, i.e.* the water and the medium it is found in. Because of this, in the view of hydrogeologists, there is insufficient consideration of the subsurface space, which is the integral constituent of an aquifer.

Following the above line of thought, practitioners of transboundary hydrogeology consider that in using the term 'groundwater', it applies merely to the water and does not include the medium of its occurrence. In terms of developing sound governance of the aquifer resource, however, the important additional element, the host for the water – the rock or sediment medium – is implicitly discarded from the above-mentioned guidance. The adverse consequences, as well as potential missed opportunities, of this omission could be multiple. For instance, if the rock host is not considered to be an element or media for management, it thus misses out key benefits of aquifer utilisation, *e.g.* in conjunctive use, managed aquifer recharge, waste disposal or sub-surface infrastructure, mining, as well as for the avoidance of subsidence and aquifer contamination from mineral releases (see Chapter 20 of this book). It is for this reason that we argue and emphasize that the appropriate terminology in transboundary aquifer management is 'aquifer'.

In the debate on international transboundary groundwater law, there has been much discussion on sovereignty (McCaffrey, 2011), probably as a reflection of sovereignty over similar shared natural resources, such as petroleum (Zorn, 1983). Through the use of the term 'aquifer', which stands for the 'container and the water', it becomes clear that countries do have claim to absolute sovereignty over the immobile 'container', but may only exercise relative sovereignty over the mobile part. Such a definition clarifies the scope of responsibility of aquifer system states – viz. that the moving water is held in trust as a common and shared good, while the immobile part remains under sovereign jurisdiction. So, while an integrated systemic approach is advocated for the management, it is important to keep in mind the relative remit of the transboundary management. Such an explanation also serves to clarify that, despite the notion of sovereignty of states regarding the 'aquifer', the land surface on to which groundwater recharge or discharge occurs is to be managed and protected under the approach of wise stewardship of land in each country sharing the common aquifer.

19.3.6 Impediments arising from evolving pressures and geopolitics

Intuitively and from experience, asymmetric governance involving pressures (*e.g.* abstraction/contamination) occurring in one domestic part of a transboundary aquifer can become significantly magnified in time. The impacts of the pressures may not be noted as an aggravation until they have reached the neighbouring country. Furthermore, the lack of communication between the users and the regulators on one side of the border with those of the other side could lead to international diplomatic tensions (such as may be found in the Indus Valley Aquifers and in the Fergana Valley of Central Asia (de Bruijn, 2011)). It could also lead to a series of related side effects (such as restrained or uncontrolled cross border movements of people and local trade, *e.g.* nomadic peoples of Karamoja region on northeast Africa (FEWS-NET, 2005)). Finally, efforts at 'good governance' in the domestic aquifer on one side, with no parallel efforts on the other side may lead to limited, or even no benefit, to either party. As an example the case of the Mesozoic Aquifer in the coalfields of the Lublin region of Poland and Ukraine can be cited (Puri *et al.*, 2010)

Analysis of the practice of water governance reveals that many of the provisions of the developing international water law remain difficult to translate into actions (Puri, 2003). This difficulty is exacerbated in circumstances associated with political changes and unrest, such as those that took place following the collapse of the Soviet Union. This legacy continues to haunt the transboundary water relations between the recently independent countries. Examples may be found in relation to the aquifers of the Fergana Valley (Tajikistan, Kyrgyzstan, Uzbekistan – ECE and OSCE Transboundary waters evaluations (de Bruijn, 2011)) and the aquifers of the Dinaric Karst of the Balkan region (Puri, 2014). Further examples can be cited with reference to the extensive aquifers of the Indo-Gangetic Plains. The governance of the domestic parts of the aquifers, as practiced within India, Pakistan and Bangladesh, is much at odds with an ideal joint and integrated governance of the 'transboundary' whole. Recent work in the Mekong Basin on domestic aquifer assessments has also found contrasting practices of governance between countries, which must somehow be adapted for consideration of the transboundary whole and some joint form of governance mechanisms and approaches adopted (Ha *et al.*, 2015).

19.4 EASING IMPEDIMENTS OF TRANSBOUNDARY AQUIFER MANAGEMENT AND GOVERNANCE

Given that in the past decade and a half, governance of transboundary aquifers has matured from obscurity to general recognition in the transboundary waters arena, what can be done for further promoting the proper development and governance of these resources? What measures might be needed to accelerate the process of making effective processes, institutions and agreements between states sharing an aquifer? In addition, whose role is it to seek such progress? This will be discussed in the following sections.

19.4.1 Symmetric governance across domestic parts of transboundary aquifers

The transition from asymmetric governance in domestic parts of transboundary aquifers to consistent and joint governance across borders in recognition of transboundary continuity is a process and constitutes early steps towards good transboundary aquifer governance. Table 19.4 summarises a number of significant transboundary aquifers globally, selected based primarily on the degree of human dependence on them and lists a couple of possible indicators of the degree of governance related to them. These indicators are: 1. the question of whether the aquifer states agree on the nature of any transboundary issues that are of concern to them; and 2. whether there is a dispute resolution mechanism presently in place. It also lists 3. the corruption perception index; and 4. the government effectiveness score[2] for the aquifer states in question, as defined and assessed by the Transparency International and the World Bank, respectively. The latter two could give an indication of the degree of asymmetry of governance, even though it may be quite subjective. It would seem from the discrepancy of the indices between countries overlying the same transboundary aquifer, that a large number of aquifer-sharing countries conduct asymmetric governance for a plethora of reasons. In addition, it appears that a minority of sharing countries where to date some form of agreement related to the joint management of the aquifers has been reached (*e.g.* North West Saharan Aquifer System, Nubian Sandstone Aquifer System, Iullemeden, Guaraní, Rum-Saq), are in transition towards a more symmetric governance approach.

A hypothesis can be brought forward as follows: in countries with symmetrical, and positive overall governance and corruption perception scores, disputes may be settled without resort to jurisprudence, while under asymmetry and negative governance scores, there is a need to establish a dispute resolution mechanism. On the back of that, it is interesting to note that in the Genovese and the Guaraní cases, where the symmetry is more evident, and despite the fact that the nature of the 'transboundary problem' has not being stated explicitly, the countries have nevertheless opted to set up a dispute resolution mechanism, possibly being wary of conducting negotiations in the absence of an arbitrator. In contrast, for the Rum-Saq and the Nubian Sandstone Aquifer Systems, with disparate corruption and governance scores among countries, no dispute resolution mechanisms have been set up. It can be counter-hypothesised then that a symmetric context lends itself better to, and favours, agreements being reached through negotiations. However, relationships between symmetry and the fact of whether a dispute resolution agreement is established or not, cannot be clearly discerned due to the relatively small sample of aquifers where inter-state collaboration is underway. The overall conclusion is that no consistency or clear pattern emerges from the few transboundary aquifers with governance efforts presently underway.

[2]From World Bank – Government effectiveness captures perceptions of the quality of public services, the quality of the civil service and the degree of its independence from political pressures, the quality of policy formulation and implementation, and the credibility of the government's commitment to such policies.

Table 19.4 Key transboundary aquifers and parameters related to governance and asymmetry (partly after Velis, 2016).

Transboundary Aquifer System	*Consensus on nature of problem?*	*Dispute resolution mechanism?*	*Corruption perception index (2016)*[(a)]	*Government effectiveness score (2015)*[(b)]
Abbotsford-Sumas Aquifer	Yes	No	CA: 82 USA: 74	CA 1.8; US 1.5
Genovese Aquifer	No	Yes	FR 69; CH 86	FR 1.4 ; CH 2.0
Guaraní Aquifer System	No	Yes	BR 40; AR 36; PY 30; UY 71	BR −0.2; AR −0.1; PY −0.9; UY 0.5
Hueco-Bolsón Aquifer	Yes	No	US 82; MX 30	US 1.5; MX 0.2
Iullumeden Aquifer System	Yes	Yes	NE 35; NG 28; TD 20	NE −0.6; NG −1.0; TD −1.5
North-western Sahara Aquifer System	No	No	DZ 34; MA 37; TN 41	DZ −0.5; MA −0.1; TN −0.1
Nubian Sandstone Aquifer System	Yes	No	EG 34; LY 14; SD 14; TD 20	EG −0.8; LY −1.7; SD = 1.5; TD −1.5
Rum-Saq Aquifer	No	No	JO 48; SA 46	JO 0.1; SA 0.2
Stampriet Aquifer System[(c)]	Under discussion	Developing	ZA 45; NA 52; BW 60	ZA 0.3; NA 0.3; BW 0.5
PreTashkent Aquifer System[(c)]	Not discussed	Stalled	UZ 21; KZ 29	UZ −0.7; KZ −0.1
Trifinio Aquifer System[(c)]	Under discussion	Developing	HN 30; SV 36	HN −0.8; SV −0.2
Ramotswa Aquifer System[(c)]	Under discussion	Developing	ZA 45; BW 60	ZA 0.3; BW 0.5

Notes (acc. to Velis, 2016):
Dispute resolution – Does groundwater governance contain a dispute resolution mechanism?
Nature of problem – The nature of the problem is identified as overexploitation, pollution, underdevelopment, and/or lack of data.
(a) Source: http://www.transparency.org/news/feature/corruption_perceptions_index_2016 – the higher the score (range 0–100), the lower the corruption perceived
(b) Source: Kaufmann *et al.*, 2010 – the higher score (range −2.5 to +2.5), the better the governance systems in place
(c) Currently ongoing projects – not included in the analysis of Velis (2016)
Country abbreviations follow the ISO standard

19.4.2 International treaties and guidelines to support emerging symmetric domestic governance into sound transboundary governance

When symmetrical good governance is developed and practiced on both sides, and ideally efforts are coordinated, the issue of tensions in transboundary governance declines and may even become insignificant (examples include the US and Canada transboundary aquifers (Rivera, 2015) and the Swiss – French Genovese Aquifers (de los Cobos, 2010). It also potentially fosters and strengthens international relations and regional integration. However, this cooperation may not develop spontaneously, and there is a need for a 'glue' to join up separate domestic good governance efforts. This glue may be obtained through the provisions of the developing international water law

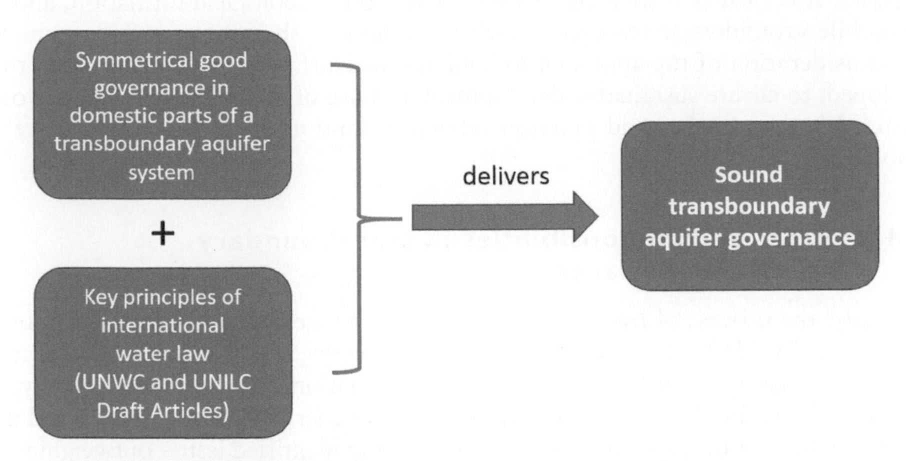

Figure 19.2 Combining domestic good governance with international water law for effective transboundary aquifer governance.

(Figure 19.2). In addition to symmetrical domestic governance, the overarching principles of 'reasonable and equitable utilisation', 'causing no appreciable harm', 'regular exchange of data and information', and 'protection of recharge and discharge zones' as encapsulated in the UNILC Draft Articles coupled with the UNWC, are essential to support onward progress on governance of transboundary aquifers.

The European Union has developed a comprehensive and integrated approach to water management through the European Water Framework Directive and its daughter Groundwater Directive (see Chapter 23 of this book). This implies that EU countries and especially the accession states have been required to adopt the EU environmental acquis[3] and to 'approximate' or harmonize their rules, regulations and approaches to water management (as well as to other sectors) resulting in consistencies across their national borders, though they may differ in the application. In addition, this requirement has been part of a legally binding agreement as a condition of admittance of new countries to the Union. Such efforts at 'approximation' and uniformity of approach remains a challenge in other parts of the world, and the approach to their transboundary groundwater management will benefit from the refinement and experience in application of universally accepted rules or guidelines. The UNWC provides one such global governance framework, though it must be adopted in close consideration of the Draft Articles, since the former remains unsuited for many aquifer configurations (Altchenko and Villholth, 2013).

19.4.3 Getting the 'aquifer' vs. 'groundwater' in transboundary systems right

The unclear distinction between 'aquifer' and 'groundwater' in the international water law instruments (UNWC and UNILC Draft Articles) continues to create confusion and lack of clarity in interpretation, adoption, adaptation and application of the principles. Efforts need to go into developing experience, guidance and tools to serve such

[3]Body of law accumulated by the European Union.

purposes. It is clear that sovereignty only relates to the geological formation, and not the mobile groundwater resource, which is a common shared good. Still, some further consideration of the approach to land and subsurface management needs to be developed, to ensure sustainable development and use of groundwater, which is often impacted by approaches and practices related to land use and subsurface space use (Gupta, 2016).

19.4.4 Roles and responsibilities in transboundary aquifer governance

Generally, the process of transboundary aquifer management rests with the national entities responsible for their territorial water management and for multilateral cooperation, with support of local, national and international stakeholders. Ideally, the process is driven by the demands of the sharing countries, based on a recognized need and assessment of the positive benefits of addressing identified issues outweighing the costs associated with the interventions. In practice, attention to transboundary aquifer management is influenced by promotion of various partners, such as UN Agencies, donors, NGO's and professional associations as part of a broader global agenda on water and the sustainable development goals (SDGs) as well as national and human security needs (Gupta, 2016). It has also evidently become a matter for multinational business corporations to take an interest as well (ESG, 2016). To the extent that this could work to enhance the outcomes as well as their corporate social responsibility, this may make a welcome contribution.

19.4.5 A generic approach to sound transboundary aquifer governance

Early work under ISARM helped set up processes for the delineation, inventory, characterisation and documentation of shared aquifers (Puri, 2001). In continuation, a semi-generic methodology for transboundary aquifer assessment and management is evolving under the GGRETA project managed by UNESCO and supported by IGRAC (GGRETA, 2016), and other similar efforts are underway in the RAMOTSWA project led by IWMI (International Water Management Institute)[4]. These efforts provide a good operational framework. Other efforts are underway for the US-Mexico transboundary aquifers (Varady *et al.*, 2014).

The experiences gained from activities on several transboundary aquifers suggest that the following elements are essential for progressing sound governance (Figure 19.3). A first requirement is to establish a firm science-based knowledge base of the transboundary aquifer – basically the hydrogeology, current and future demands and the socio-economic characteristics related to the use of the water. A second requirement is to conduct an analysis of the domestic and international laws that apply in the particular case – basically the features of the domestic laws that lend themselves to good national housekeeping and provisions contained in them for taking on the responsibility under international water law. The third requirement is an assessment

[4]http://ramotswa.iwmi.org/

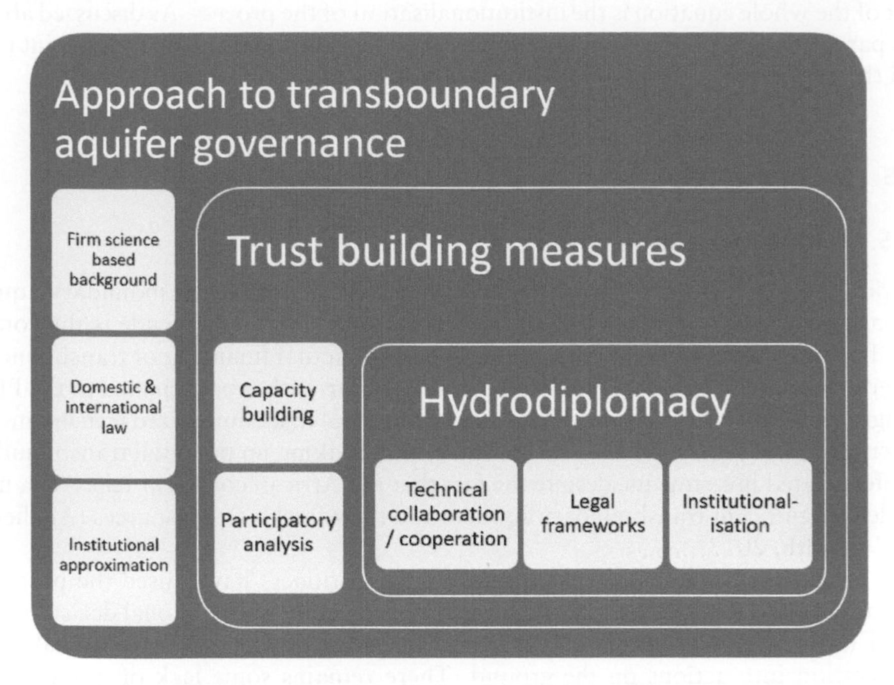

Figure 19.3 Framework for promoting and supporting sound transboundary aquifer governance.

of the domestic institutional frameworks in place in the aquifer states and how they allow for formal as well as informal links with their counterparts in the sharing countries. Clearly, if in one country the institution assigned for transboundary aquifers is *e.g.* the department for irrigation, and its primary focus is enhancement of agricultural output, while in the partner country, the responsibility lies with a geological survey, there will be a need to "approximate" or adapt their roles to address common issues.

19.4.6 Trust-building measures and hydrodiplomacy

Joint governance fundamentally arises through trust-building measures, which of necessity takes time to evolve and materialise. The extensive experience of the ISARM Programme shows that comprehensive and joint capacity building is the most effective pathway, including training, field surveys and case studies. Needless to add is a wide stakeholder participatory analysis and engagement.

The above measures have been found to be the basic building blocks of hydrodiplomacy in the transboundary context. Unlike diplomacy related to international trade between countries, the scope here is a common good that occurs in nature and is driven by not only anthropogenic forces. Thus, the diplomacy has to be structured around science-based technical cooperation and support from national scientists and practitioners of hydrogeology and related sciences (*e.g.* political economy, environment sciences, law). In addition, there may be a need for adjustments of the domestic legal frameworks and adoption of the basic norms of international water law – addressing reasonable and equitable use of resources, causing no appreciable harm, and the protection of the recharge and the discharge zones of the aquifer system. The most critical part of the whole equation is the institutionalisation of the process. As discussed above, this part remains a conundrum since there is no ideal and single format for a joint body and the design has to be adopted on a case-by-case basis.

19.5 DISCUSSION

19.5.1 Key lessons learned

Despite the increased recognition and evolving experience of transboundary aquifer management, one of the key lessons of the past one and a half decade is the continued 'back seat' that aquifer systems still take in the actual financing of transboundary water resources investments. For example, the countries that participated in the PIDA (Programme for Infrastructure Development in Africa) that emphasized transboundary water resources projects (PIDA, 2016) still omitted taking up potential transboundary aquifer related investments, despite the fact that the African continent relies on a multitude of significant transboundary aquifers and their freshwater resources (Altchenko and Villholth, 2013).

The science community of hydrogeology practitioners have raised the profile of groundwater governance, but may not yet have fully convinced national decision makers of the values of converting science and its policy related to international water cooperation into actions on the ground. There remains some lack of interest from countries in bringing what is perceived to be national groundwater matters to the international level, and there may also be aspects of unawareness, apprehension and ignorance due to the complexities of aquifer systems.

Another key lesson is that harmonising asymmetrical domestic governance systems into structured and systematic transboundary governance still remains a challenge as it is not quite clear how to proceed in terms of formulating functional institutional arrangements, especially on the back of already established river basin organizations conceptualised primarily on the basis of surface water concerns (Schmeier, 2010).

Finally, a general lesson is that formalisation of agreements on transboundary aquifers are understandably slow to develop. The transition from scientific understanding to policy development, to the diplomatic negotiation seems to take prolonged periods of time (sometimes decades), in contrast to other collaborative agreements such as those over trade or the development of transboundary petroleum resources (Zorn, 1983).

19.5.2 Sensitive aspects

Among the issues that have been too sensitive to be discussed openly in almost every circle – be it scientific, diplomatic or commerce – is the issue of hydro-hegemony (Zeitoun and Warner, 2006; Menga, 2015). Thus, this 'elephant in the room' is understated in most discourses on international water generally, and likely also in the transboundary aquifer debate, though no dedicated quantitative research on this exists. Countries that appear to hold the hydro-hegemony are secretive and reluctant to discuss the issue of their 'command' over their less powerful neighbours. However, aquifer management still remaining the 'distant cousin' of surface water, it does not receive effective exposure. Furthermore, due to the invisible and somewhat unpredictable character of groundwater and aquifer systems and the common localised character of aquifer impacts, it may be less clear who is impacting who and hence, who is the hegemon or the potential holder of the control of impacts and interventions. This is an area requiring future research.

Another primary area of controversy in legal circles relates to the application of the principle of sovereignty set out in the UNILC Draft Articles. Some scholars have suggested that this makes the Draft Articles flawed (McCaffrey, 2009), despite the fact that concerted efforts and consultations with a massive number of specialists (hydrologists as well as legal specialists) were exercised. Two issues may be relevant here from the perspective of aquifer states: the misconception or poor appreciation of sovereignty aspects in terms of aquifer systems set out in the UNILC Draft Articles as well as the difficulty of practically reconciling sovereignty related to the mobile and immobile parts of the aquifer (Section 19.3.5). The lesson learned from this is that further clarification of the concept, and associated principles, in legal terms as well as discussion of particular application cases is needed to identify concrete concerns and solutions.

19.5.3 Looking ahead

Among the noteworthy under-achievements to date is the continued uncertainty on the status of the UNILC Draft Articles. Either in relation to the controversy of the sovereignty issue discussed above, or due to lack of awareness, the status of the Draft Articles in the UNGA remains in hiatus. The report of the Oct 2016 General Assembly discussions indicates that many UN member states seem to be only partially aware of the relevance and effectiveness of the Draft Articles, making them take the position of 'wait and see'.

The following excerpt is taken from the UNGA report on their consideration of the status of the Draft Articles and has been reproduced verbatim, as an indication of the 'look ahead'[5].

"On the 20 October 2016: The UN General Assembly's (UNGA) Sixth Committee (Legal) discussed draft articles of a proposed law on transboundary aquifers, as UN and European Union (EU) leaders put forward examples of good practice. The draft articles provide guidance for countries entering into international agreements and arrangements for the proper management of transboundary aquifers. In the Sixth Committee discussion, speakers welcomed the draft articles that have been developed by the International Law Commission as a basis for countries to develop their own agreements with neighbors, and some called for the provision of further scientific information. Some countries foreshadowed the possibility of the draft articles forming the basis of an international convention. Countries also noted the importance of proper management of transboundary aquifers, which are said to provide 65% of the earth's drinking water. The Russian Federation stated that it was "too soon" to raise the issue of developing a convention, but that States should take the draft articles into consideration in their practice. Israel called for flexible guidelines, while observing that the topic was "not suitable" for codification in a binding form. Viet Nam believed the

[5] http://sdg.iisd.org/news/sixth-committee-discusses-transboundary-aquifer-management/

draft articles could become a convention, which would however, require a thorough review by States beforehand. Japan highlighted the importance placed on ensuring the sustainability of water resources in the 2030 Agenda for Sustainable Development, and Venezuela urged countries to prioritize Sustainable Development Goal 6 (SDG 6) on water and sanitation."

In view of some of the statements made by member states at the UNGA, suggesting that the matter of transboundary aquifers is premature, there is clearly an urgent requirement for the practitioner community to continue raising awareness and to seek agreement on adopting a final form of the Draft Articles. The Global Groundwater Governance project financed under the GEF (http://www.groundwatergovernance.org/) strongly advocated for increasing groundwater governance at national and transboundary level to support institutional as well as implementation processes.

The global inventory of transboundary aquifers has continued, and the current number of transboundary aquifers stands at over 600 (IGRAC, 2015), some of them aquifers within a sovereign country. However, in the coming years, there is a need to analyse the extent to which these aquifers match with the definition of transboundary systems as set out in the Draft Articles of the UNILC. To do this, additional knowledge may well be required – such as the more detailed hydrogeological information that will help determine the domestic and the international scope of obligations.

There is a need for conducting more analysis on new and previously studied transboundary aquifers, while also carrying out capacity strengthening at multiple levels. As has been noted above, the form of the most appropriate, new or existing, institutions that can address the relevant transboundary governance remains open and is subject to critical assessment in specific cases.

Finally, there is an emerging understanding that the transboundary water issues increasingly need a more integrated approach. With growing water stress, climate change, environmental and social adverse impacts, understanding the challenges more broadly, physically as in looking at conjunctive (groundwater and surface water) management and governance (Lautze *et al.*, 2017), is important. Additionally, bringing in cross-sectoral aspects, environmental and socio-economics, as well as addressing the increasing focus on national securitization, which could counter trends and processes towards cooperation, (Gupta, 2016) will be increasingly relevant.

19.6 CONCLUSIONS

The chapter gives a historic overview of progress towards managing and governing transboundary aquifers globally. This is a relatively recent endeavour and a growing field of research and assessments over the last couple of decades. Yet, it is an increasingly critical challenge for the achievement of water security, international cooperation, and sustainable development, because current trends indicate a growing reliance on groundwater due to global climatic and anthropogenic changes. It is becoming clearer that aquifers and groundwater are equally important in getting the solutions to sustainable transboundary water management right, and that they hold potential for both integrated and more robust solutions.

Ground breaking work in terms of mapping transboundary aquifers globally and drafting a set of generic articles on the law of transboundary aquifers have provided a solid basis for the governance of these systems, while it is also recognized that both mapping and finalizing the UNILC Draft Articles is an evolving and essential process in continually taking the work forward. As growing experience from building cooperation on specific transboundary aquifers is maturing in both emerging and developed economies, there is still a broad need to further develop awareness, tools, capacities, and political recognition.

While structured approaches and guidelines support concrete processes, it is also clear that because aquifer systems and the socio-economic and cultural context in which they exist, and the asymmetry of these factors across borders, are unique, each approach to governance and the mechanisms to drive it will also need to be customised. The complexity of aquifers, the partial inertia in lifting it in the political agenda, and the variable importance of groundwater across the globe may result in somewhat slower progress in setting up formal agreements and treaties on transboundary aquifers, as either separate from surface water or somehow in conjunction, and ultimately as part of broader cooperation efforts related to climate change adaptation, resilience building, and common resource management.

NOTES

1 The UN Department for Technical Cooperation for Development (UN-DTCD) was charged by the UN Secretary General in May 1958 to "promote coordinated efforts for the development of water resources." This resulted in a first report "Large Scale Groundwater Development" published in 1960 (UN, 1960). The UN Advisory Committee on Application of Science and Technology to Development gave priority to groundwater exploration and development. More than 100 national studies were conducted, entirely or partially devoted to groundwater prospecting.

2 The material and data in the series of UN-DTCD reports were prepared by hydrogeologists and groundwater engineers and specialists of the concerned countries. Examples of compilation of continental hydrogeological maps by merging national maps that illustrate the transboundary context include:

   1. Algeria Cartes hydrogéologiques, *e.g.* 1:200,000 and 1:1,000,000, Algiers, 1973
   2. Morocco Cartes hydrogéologiques, 1:200,000, 1:500,000 and 1:1,000,000, Rabat, 1960
   3. Tunesia Cartes hydrogéologiques des eaux souterraines, 1:200,000, Tunis, 1971

3 Efforts to bring consistency across national borders were made by the IAH Commission on Hydrogeological Maps. Struckmeier and Margat (1995) report that for the compilation of the European Hydrogeological Map, hydrogeologists from Austria, Czechoslovakia, the Federal Republic of Germany, France, Italy, Switzerland and Yugoslavia were involved during 1962 to 1964.

REFERENCES

Alley, W.M. (ed.), 2013. Five-year interim report of the United States – Mexico Transboundary Aquifer Assessment Program: 2007–2012: U.S. Geological Survey Open-File Report 2013–1059, 31 pp.

Altchenko, Y. and K.G. Villholth, 2013. Transboundary aquifer mapping and management in Africa: a harmonised approach. Hydrogeol. J. 21(7), 1497-1517. DOI 10.1007/s10040-013-1002-3.

Burdon, D.J., 1977. Flow of fossil groundwater. Quart. J. Eng. Geol. 10, 97-125.

De Bruijn, S., 2011. The Ferghana Valley: A Ticking Bomb? Conflict Potential and the Role of Youth in the Ferghana Valley. MSc thesis, Utrecht, 2011. gpm.ruhosting.nl/mt/2011MASG06BruijnSimonede.pdf (see also http://www.osce.org/eea/47095)

de los Cobos, G., 2010. The transboundary aquifer of the Geneva region (Switzerland and France): successfully managed for 30 years by the State of Geneva and French border communities. In UNESCO-IHP (2010). (http://www.siagua.org/sites/default/files/documentos/documentos/geneva.pdf.)

Earth Security Group, 2016. CEO Briefing: Global Depletion of Aquifers. Global companies must take an active role in groundwater governance to avoid existential risks. Accessible at *www.earthsecuritygroup.com* (Authors include Alejandro Litovsky (lead author), Margot Hill Clarvis, Caroline Hambloch, Pablo Orvananos, Rupert Bassett (information designer), Orlaith Delargy and Alvise Lisca.

Famine Early Warning Systems Network (FEWS-NET), 2005. Conflict Early Warning and Mitigation of Resource Based Conflicts in the Greater Horn of Africa: Conflict Baseline Study Report Conducted in the Karamajong Cluster of Kenya and Uganda (USAID). 60 pp.

Fried, J., 1982. Groundwater Resources of the European Community, Synthetical Report. Commission of the European Communities, Brussels, Belgium, published by Th. Schäfer GmbH, Hannover, Germany.

Gilbrich, W.H. and Struckmeier, W.F., 2014. Fifty years of hydro(geol)logical mapping. Published by UNESCO (http://unesdoc.unesco.org/images/0023/002310/231066e.pdf).

Governance of Groundwater Resources in Transboundary Aquifers (GGRETA), 2016. Governance of Groundwater Resources in Transboundary Aquifers (GGRETA) project – Overview and results of the assessment phase (2013–2015). UNESCO, Paris, 15 pp.

Gupta, J., 2016. The watercourses convention, hydro-hegemony and transboundary water issues. The Int. Spectator, 51, 3, 118–131, 6 http://dx.doi.org/10.1080/03932729.2016.1198558.

Ha, K., Minh Ngoc, N.T., Lee, E. and Jayakumar, R., 2015. Current Status and Issues of Groundwater in the Mekong River Basin. Korea Institute of Geoscience and Mineral Resources (KIGAM), CCOP Technical Secretariat, UNESCO Bangkok Office.

IGRAC, 2015. Transboundary Aquifers of the World [map]. Edition 2015. Scale 1: 50 000 000. IGRAC, Delft, Netherlands. https://www.un-igrac.org/sites/default/files/resources/files/TBAmap_2015.pdf.

IGRAC, 2014. Transboundary Aquifers of the World [map]. Edition 2014. Scale 1:50 000 000. IGRAC, Delft, Netherlands.

IGRAC, 2012. Transboundary Aquifers of the World [map]. Edition 2012. Scale 1:50 000 000. IGRAC, Delft, Netherlands.

IGRAC, 2009. Transboundary Aquifers of the World [map]. Edition 2009. Scale 1:50 000 000.

International Association of Hydrogeologists (IAH), 2005. Groundwater Policy Initiative. IAH Position Paper – issued to the IAH Council in preparation for the development of a GEF-funded programme on groundwater governance. Summary of the GEF STAP Workshop held at UNESCO, Paris, in March 2004.

Kaufmann, D.; Kraay, A.; Mastruzzi, M., 2010. The Worldwide Governance Indicators: Methodology and Analytical Issues. World Bank Policy Research Working Paper No. 5430. 31 pp.

Lautze, J., Holmatov, B., Saruchera, D. and Villholth, K.G., 2017. Conjunctive management of surface and groundwater in transboundary watercourses: A first assessment. Water Policy (in print).

Lloyd, J.W. and Pim, R.H., 1990. The hydrogeology and groundwater resources development of the Cambro-Ordovician Sandstone Aquifer in Saudi Arabia and Jordan. J. Hyd., 121, 1–20.

Lloyd, J. W., 1969. The Hydrogeology of the Southern Desert of Jordan. UNDP/FAO Publ. Tech. Rep. 1., Spec. Fund 212, Rome.

Margat, J., 1995. Les ressources en eau des pays de l'OSS – Évaluation, utilisation et gestion. UNESCO/OSS Publication. 80 pp. http://www.ircwash.org/sites/default/files/824-AFW95-13545.pdf.

McCaffrey, S.C., 2011. The International Law Commission's flawed Draft Articles on the Law of Transboundary Aquifers: the way forward. Water Int., 36, 5, 566–572.

McCaffrey, S.C., 2009. The International Law Commission adopts draft articles on transboundary aquifers. The American Journal of International Law, 103: 271–293.

Menga, F., 2015. Reconceptualizing hegemony: the circle of hydro-hegemony. Water Policy, 18, 2, 401-418. DOI: 10.2166/wp.2015.063.

Munia, H., Guillaume, J.H.A., Mirumachi, H., Porkka, H., Wada, Y. and Kummu, M., 2016. Water stress in global transboundary river basins: significance of upstream water use on downstream stress. Environ. Res. Lett., 11(1).

Organisation for Economic Co-operation and Development (OECD), 2011. Water Governance in OECD Countries, A Multi-level Approach. OECD, Paris. 244 pp. ISBN: 9789264119277.

Programme for Infrastructure Development in Africa (PIDA), 2016. PIDA Progress Report 2016. 52 pp. http://www.africa-platform.org/resources/pida-progress-report-2016.

Puri, S., 2014. Aquifer system TDA and legal frameworks – What lesson for DIKTAS? Proceedings of the International Conference on Karst Without Boundaries, Trebinje June 2014, pp. 256–263.

Puri, S., 2010. Education for hydrogeologists of the future: serving to ensure an environmentally secure world. Przeglad Geologiczny, 58, 9/1, 737-745.

Puri, S., 2003. Transboundary aquifer resources – International water law and hydrogeological uncertainty. Water Int., 28, 2, 276–279.

Puri, S. 2001. The challenge of managing transboundary aquifers – multidisciplinary and multifunctional approaches. In: Proceedings of the International Conference on Hydrological Challenges in Transboundary Water Resources Management, Koblenz, Germany, September 2001, 31-37.

Puri, S., 1997. Aquifers know no boundaries. ... but farmers do. So, who should care?!. Guest commentary in J. Int. Groundwater Tech., April/May 1997, p. 6.

Puri, S. and Aureli, A. (eds.). 2009. Atlas of Transboundary Aquifers – Global Maps, Regional Cooperation and Local Inventories. Paris; UNESCO Division of Water Sciences/International Hydrological Programme. 326 pp.

Puri, S. and Aureli, A., 2005. Transboundary Aquifers: a global programme to assess, evaluate and develop policy. Groundwater, 43, 661–668.

Puri, S. and Elnaser, H., 2003. Intensive use of groundwater in transboundary aquifers. In: Llamas, R. and E. Custodio (Eds.), 2003. Intensive Use of Groundwater: Challenges and Opportunities. A.A. Balkema Publishers, the Netherlands, pp. 415–439. ISBN 9789058093905.

Rivera, A., 2015. Transboundary aquifers along the Canada–USA border: Science, policy and social issues. J. Hydrol.: Regional Studies, 4B, 623–643.

Schmeier, 2010. Effective governance of transboundary aquifers through institutions – Lessons learned from river basin organizations. International Conference "Transboundary Aquifers: Challenges and New Directions" (ISARM2010).

Stephan, R.M. (ed.), 2009. Transboundary aquifers: managing a vital resource, the UNILC Draft Articles on the Law of Transboundary Aquifers, SC-2008/WS/35, UNESCO, Paris, 24 pp.

Struckmeier, W.F., Gilbrich, W.H., van der Gun, J., Maurer, T., Puri, S., Richts, A., Winter, P. and Zaepke, M., 2006. Groundwater Resources of the World. Transboundary Aquifer

Systems, scale 1:50M. WHYMAP and the World Map of Transboundary Aquifers. Special edition for the 4th World Water Forum, Mexico.

Struckmeier, W.F. and Margat, J., 1995. Hydrogeological Maps – a Guide and a Standard Legend. International Contributions to Hydrogeology. IAH, Vol. 17.

United Nations (UN), 1973. Groundwater in Africa, 1:17,000,000, U.N., New York.

United Nations (UN), 1960. Large-Scale Groundwater Development. Water Resources Development Centre, New York.

UN Department for Technical Cooperation for Development (UN-DTCD), 1988. Natural Resources/Water Series No. 18. Groundwater in North and West Africa. Dept. for Technical Cooperation for Development and Economic Commission for Africa.

UNECE, 1999. Inventory of Transboundary Groundwaters. Lelystad. Working Programme 1996/1999, Vol. 1. ISBN 9036952743, 283pp.

UNESCO-IHP, 2010. Transboundary Aquifers. Challenges and new directions. ISARM 2010 International Conference, 06–08 December 2010, UNESCO, Paris. Abstracts. UNESCO-IHP ISARM and PCCP Programmes. 188pp.

UNESCO-IHP and UNEP, 2016. Transboundary Aquifers and Groundwater Systems of Small Island Developing States: Status and Trends. UNEP, Nairobi.

UNESCO-IHP, 2001. Tripoli Statement. International Conference on 'Regional Aquifer Systems in Arid Zones – Managing Non-Renewable Resources', Tripoli, November 1999. http://unesdoc.unesco.org/images/0012/001270/127080e.pdf.

UN International Law Commission (UNILC), 2003. Shared Natural Resources: first report on outlines. By Mr. Chusei Yamada, Special Rapporteur. 55th Session, 30 June, 2003. http://www.internationalwaterlaw.org/bibliography/UN/UNILC/Groundwater/Yamada%201st%20Report-Addendum.pdf.

Varady, R.G., Scott, C.A., Megdal, S.B. and Wilder, M.O., 2014. Transboundary Groundwater Governance in the Western U.S.-Mexico (AZ-Sonora) Border Region Piecemeal Pragmatism vs. Comprehensive Idealism (http://www.watersecuritynetwork.org/wp-content/uploads/2014/08/Transboundary-Groundwater-Governance.pdf.)

Varady, R.G., Scott, C.A., Megdal, S.B. and McEvoy, J.P., 2010. Transboundary Aquifer Institutions, Policies, and Governance: A Preliminary Inquiry. In UNESCO-IHP (2010).

Velis, M., 2016. Understanding the Effectiveness of the Governance of Transboundary Aquifers: A Framework for Analysis. MSc thesis, Faculty of Geosciences, Utrecht University, The Netherlands. 121 pp.

Yamada, C. (UNILC Special Rapporteur), 2003. Shared Natural Resources: First Report on Outlines. U.N. Doc. A/CN.4/533 and Add.1. http://legal.un.org/ilc/documentation/english/a_cn4_533.pdf.

Villar, P.C., 2016. International cooperation on transboundary aquifers in South America and the Guaraní Aquifer case. Rev. Bras. Polit. Int., 59, 1 DOI: http://dx.doi.org/10.1590/0034-7329201600107.

Zeitoun, M. and Warner, J., 2006. Hydro-hegemony – a framework for analysis of transboundary water conflicts. Water Policy, 8, 435–460.

Zorn, S., 1983. Permanent sovereignty over natural resources – Recent developments in the petroleum sector. Nat. Resour. Forum. DOI: 10.1111/j.1477-8947.1983.tb00276.x.

Chapter 20

Governing extractable subsurface resources and subsurface space

Jac van der Gun[1] *& Emilio Custodio*[2]
[1]*Van der Gun Hydro-Consulting, Schalkhaar, The Netherlands*
[2]*Technical University of Catalonia (UPC), Barcelona, Spain*

ABSTRACT

This chapter briefly overviews governance of the main categories of subsurface resources: groundwater, geo-energy, mineral materials and space. Exploitation and use of these resources are highly diverse, and so is their governance. In spite of very significant efforts being spent, governance flaws and deficiencies can be observed virtually anywhere around the world. Increasing intensity and diversity of human subsurface activities are making the subsurface steadily more crowded and potentially lead to ever-expanding interferences and external impacts. More than any of the other subsurface resources are groundwater resources vulnerable to hazards produced by the extraction and use of other subsurface resources. Current governance of the subsurface resources is still highly fragmented; more coordination, especially by making linkages and creating synergy and transparency, will certainly bear fruit. A few options for moving towards coordinated governance of the subsurface are presented.

20.1 SUBSURFACE RESOURCES AND RELATED HUMAN ACTIVITIES

The upper part of the Earth crust – down to several thousands of metres deep – offers *four types of resources: water, geo-energy, mineral materials and space.* Table 20.1 lists the main related human activities: extraction of resources (items 1–3) and the use of the subsurface space (items 4–7).

Some of these activities have been practiced from time immemorial. Other ones emerged and developed only in more recent years. Anyway, in many areas around the world, the subsurface has become much more crowded than ever before, due to a greater diversity and intensity of human activities, encroaching to increasingly greater depths. As a result, the interferences between different subsurface activities have become stronger, as well as their individual and combined external impacts on the environmental conditions, especially in densely urbanized areas. It has become necessary to assess the potential external impacts and to implement mitigating measures through good governance of the subsurface. This chapter intends to give an overview of current governance of the different subsurface resources. Particular attention is paid to how the different practices affect groundwater and its use. In addition, some ideas are presented on how to move from currently fragmented governance, focused on

Table 20.1 Main subsurface human activities (modified after Van der Gun *et al.*, 2016).

Category	*Type of activity*
1. Groundwater development and management	Groundwater withdrawal for different uses Drainage of excess shallow groundwater
2. Geo-energy development (fluids as carrier of energy)	Oil and gas development High-enthalpy geothermal energy development Low-enthalpy geothermal resources development
3. Mining	Extraction of minerals, coal, lignite, building materials, etc.
4. Disposal and storage of hazardous waste	Waste disposal by deep well injection Carbon sequestration Subsurface storage of radioactive and dangerous chemical waste Waste from nuclear weapons testing and nuclear power accidents
5. Injection and recovery	Wet mining (*e.g.* using acids and other lixiviants) Injecting residual geothermal fluids Temporary storage of heat Storage of hydrocarbons and fluids associated with oil and natural gas production Hydraulic fracturing or 'fracking' for enhanced gas and oil production Managed aquifer recharge
6. Construction into the underground space	Pipelines, sewerage systems and cables Tunnels and underground railways Underground car parks and other underground constructions
7. Using the subsurface as a scientific archive	Exploring, visiting and conserving archaeological sites, fossil sites, protected karst sites (caves) and other subsurface natural or cultural heritage objects

separate resources and related activities, towards more coordinated governance of the subsurface. Much of the chapter's content has been taken from earlier publications, notably from Van der Gun *et al.* (2012; 2016) and GEF (2016a, 2016b).

20.2 DEVELOPING, GOVERNING AND MANAGING GROUNDWATER

20.2.1 Groundwater as a resource and as an environmental factor

Groundwater is a valuable resource that can be abstracted from the subsurface for domestic water supply, irrigation, industrial water supply or any other purpose. The global abstraction of fresh groundwater during 2010 is estimated at 982 km^3, of which around 70% was for irrigation, 21% for domestic use and 9% for industrial and other purposes (Margat and Van der Gun, 2013). The share of groundwater in total freshwater abstraction is approximately 26%, but its relative importance is likely to be higher than suggested by this percentage, because on average groundwater has some comparative advantages over surface water: a large buffer capacity (which makes it a reliable source of water, especially during droughts), lower needs for treatment in order to meet water quality standards, and its presence often at closer distance to where water is needed.

Groundwater is also an important environmental component. It plays a unique role as a buffer in the hydrological cycle, by transforming irregular recharge fluxes into slowly varying spring flows and baseflows, and by keeping water available during extended periods without rain. Shallow groundwater levels contribute to the sustainability of wetlands and to agricultural productivity.

Good groundwater governance is needed to ensure sustainability and wise use of groundwater as an extractable resource, as well as to preserve its in-situ functions. Many aspects of groundwater governance have been addressed in detail by the GEF Groundwater Governance project (GEF, 2016c). A selection is briefly summarised or commented below, mainly to enable comparison with governance of the other subsurface resources and to identify needs and scope for improved co-ordination.

20.2.2 Main actors: their roles, behaviour and interaction

The actors involved in groundwater governance are extremely numerous and diverse. Three main categories can be distinguished:

(a) Those who abstract groundwater: myriads of self-supplying individuals and smaller but still considerable numbers of self-supplying communities, water supply companies and self-supplying industries. Their behaviour is primarily focused on getting sufficient water of the desired quality at lowest possible cost and it is uncoordinated as long as groundwater management control is absent.

(b) Organisations in charge of groundwater resources management, usually government organisations with a special mandate for enforcing regulations and implementing measures. Their behaviour is ruled by prevailing policies and strategies on groundwater, intended to balance conflicts of interest, achieve maximum social benefit and ensure groundwater and environmental sustainability.

(c) Professionals assisting in groundwater development and management, including hydrogeologists, geophysicists, hydrologists, drillers, water supply engineers, economists, social scientists, layers, and the companies or agencies they belong to. Their role is to contribute to effective and efficient groundwater exploration, assessment, development and management; their behaviour is governed by a mix of professional, ideological, financial and ethical motives, depending on their affiliation.

Beyond these three categories there are still other types of actors, usually with less direct involvement in groundwater governance, *e.g.* the population supplied by water supply companies, commercial companies benefiting from productive uses of groundwater, potential groundwater polluters, and environmentalists concerned about water-related aspects.

Good groundwater governance requires effective coordination and co-operation between the main actors. Common obstacles to this difficult task are the very large number of stakeholders spread over the area, conflicts of interest arising from the 'common pool' nature of groundwater, poor understanding of groundwater's role in nature and the associated services, lack of leadership and vision, insufficient local information and knowledge, lack of awareness on groundwater, poor legislation and fuzzy mandates, lack of trust between users and managers of groundwater, insufficient capacity and funding of groundwater management agencies, lack of understanding between decision-makers and professionals involved in analysis, policy development and planning, and lack of stakeholder and civil society involvement. Countries that are most successful in groundwater management have found ways to remove, circumvent, or address these obstacles. In many countries, however, there is no effective cooperation yet between government agencies, the private sector and other stakeholders regarding groundwater, and involvement in sharing costs is usually limited. Good governance needs specialists able to highlight benefits, finding win-win solutions and forging consensus, while settling disputes.

20.2.3 Area-specific data, information and knowledge

Area-specific data, information and knowledge on groundwater are essential for effective groundwater management. The data and information should not be limited to the physical groundwater systems, but also relate to the connected socio-economic and environmental systems.

Outcomes of groundwater exploration, mapping and assessment are in many countries rather fragmentary, biased, and often not easily accessible. Few countries invest significantly in monitoring changes over time; lack of monitoring data of the relevant variables is therefore in most countries a major obstacle to effective groundwater management.

Structural provisions for the acquisition of data and information are needed, in particular for monitoring. Specialised scientific agencies should take care of information management and of transforming the information into area-specific knowledge. Modern technologies (telemetry, satellite based remote sensing, information and communication technology, etc.), national or international co-operation projects, and public-private partnerships may contribute to the development and efficiency of these groundwater governance provisions.

In spite of positive exceptions, sharing groundwater-related data and information widely is not yet common practice in the majority of the countries. These data and information, developed and tailored according to the envisaged target groups and the specific context, are not only crucial for studies, decision-making and planning, but also for raising awareness on groundwater issues among current and potential groundwater governance actors. Awareness is essential to create motivation for acting and to suggest what to do. Groundwater users and other local stakeholders should understand basic cause-and-effect relationships in groundwater. This is a precondition for a positive and cooperative attitude towards groundwater management interventions.

20.2.4 Policy and planning

Groundwater policies define goals related to groundwater and generic approaches on how to achieve these. Area-specific details and practical steps to be taken are subsequently elaborated in groundwater management plans. Countries and states may or may not have a specific groundwater policy. The existing ones show a large diversity in objectives, focus, coherence with closely related policy fields, principles adopted, categories of interventions to be considered, etc. Principles that have become widely adopted in recent years include the Integrated Water Resources Management (IWRM) concept, the sustainability principle and the 'polluter pays' principle. Area-specific programmes of groundwater management planning have been established only in a limited number of countries so far. Numerical model simulations are often used during plan development to provide guidance towards defining strategies optimally tuned to the adopted goals. After selecting a preferred strategy, a management plan has to specify the measures to be implemented in order to achieve these goals. Such measures may include physical interventions, enforcement of regulations, charging financial contributions for management, the provision of incentives for desired behaviour, and corrective measures in the case of deviations.

Most commonly observed deficiencies in existing groundwater policies include a limited scope, insufficient attention for linkages with other resources, lack of vision (especially for the long-term), unrealistic goals and a time mismatch between political cycles and hydrology, including the often long-delayed response of groundwater systems to external action. Furthermore, groundwater policies frequently rely on ineffective instruments or measures, or they ignore the need to involve stakeholders for effective implementation of measures. The fate of many groundwater plans is that they fail to survive beyond analysis and recommendations, so they do not get formal approval and implementation.

20.2.5 Legal and regulatory framework

Laws, regulations and other legal instruments related to groundwater exist in some way in virtually all countries. Nevertheless, they are often affected by significant shortcomings, such as excessive fragmentation, out-dated content, incompleteness, inconsistencies and insufficient compatibility with customary rights or local realities. Groundwater quantity and quality are addressed in separate laws and by different administrations in almost all countries and coherence with laws on related policy fields is rare. Groundwater quality issues are often more important than quantity aspects (Custodio, 2013), although this not always recognized due to pressure for quantity, especially in water-scarce areas, and the delayed detection of polluted groundwater. Individuals have often perceptions on groundwater abstraction rights that diverge from the formal legal status. In recent years, many countries have improved their legal frameworks related to groundwater. Aquifer-specific legal instruments are still largely missing for managing transboundary aquifers, crossed by either international boundaries or boundaries between autonomous provinces or regions within a given country.

Although there is still ample scope for improving these instruments, the most relevant bottle-neck is in most countries formed by poor enforcement of the law and lack of compliance with regulations. This is mainly a consequence of the earlier mentioned obstacles to effective co-operation between the main actors in groundwater governance, in particular by groundwater management agencies lacking the capacity needed to control compliance by usually very large numbers of stakeholders. In fact, in most cases, control cannot be effectively achieved without stakeholder involvement and the pressure from a well-organized and informed civil society. Good governance should favour these inputs. In this respect, bottom-up born and government-fostered groundwater associations concerned about resource management and protection besides taking care of their own specific water related affairs – that have developed and are still expanding in Mexico, India and Spain – are effective groundwater management institutions (Custodio, 2010). Aquifer contracts, as pioneered in France and Morocco, may be another helpful instrument, in particular to strengthen local commitment to avoiding over-exploitation (EA-SAC, 2007; Closas & Villholth, 2016).

20.2.6 Leadership, political support and finances

Given the many current and potential actors, often with conflicting interests, leadership is essential to develop good groundwater governance. Usually a government agency with a special mandate for groundwater is entrusted with this role. Leadership in groundwater governance has many facets and is very demanding. The leading agency should be aware and knowledgeable about groundwater in the area concerned, have a vision on its potential, its behaviour and on current and future threats, and be capable and motivated to address these effectively. It should have the capability to raise political support for groundwater governance and management, to put groundwater management on the agenda, to organize the overall groundwater management process with involvement of relevant stakeholders and it should be able to raise sufficient funding for carrying out the management tasks envisaged. This has to be secured regardless of political changes.

In many countries there is a need for redirecting finances, bringing public financing in line with policy priorities (for instance by cutting on energy subsidies and by recovering costs of management interventions), and for developing new financial incentives to encourage private commitment to sustainable groundwater management. Given the value of groundwater for the economy and society, sufficient and regular financing for the basic functions of groundwater governance should be secured, including monitoring, administration, regulation, capacity building and innovation.

20.2.7 Special challenges in groundwater resources management

The main issues to be addressed in groundwater management vary from one area to another. Preventing progressive declines of groundwater levels is commonly a main issue in arid and semi-arid regions (Custodio, 2002), while control of groundwater pollution has become a key issue in virtually all areas that are densely populated or intensively used for agricultural, mining or industrial purposes. Control of water levels may also be a priority in coastal areas in order to prevent undesired seawater intrusion, in areas where shallow groundwater tables support wet ecosystems or agricultural productivity, and in zones where land subsidence risks exist due to the presence of compressible formations. Allocation of exploitable groundwater becomes an issue where it is scarce. In water-scarce areas, considerable groundwater storage depletion has been produced, which has enabled in some cases impressive economic and social development, but in the long-term requires a fundamental change in the water use paradigm.

Diagnostic analysis has to reveal the main groundwater management issues, as well as local opportunities and constraints for each particular area. On a global scale, controlling groundwater quality against pollution and managing intensively exploited non- or weakly recharged aquifers are key challenges. Climate change and other global changes are nowadays considered important new challenges, in addition to the drivers of change traditionally taken into account in planning, like population growth and economic development.

20.2.8 Unconventional groundwater resources

Groundwater for ordinary uses is normally abstracted from shallow or medium-deep fresh-water aquifers. Increasing water scarcity triggers in some regions interest in unconventional groundwater resources as an additional source of water.

Considerable volumes of fresh groundwater are likely to be found beyond depths currently tapped by ordinary water wells. Development of these deep-seated groundwater resources may seem attractive, but comes with some disadvantages. In the first place, information on deep-seated aquifers tends to be scarce, fragmented and in most cases lacking the detail required to make a reliable judgement on the suitability of individual deep aquifers for freshwater withdrawal. Hence, expensive exploration and assessment activities covering large areas are needed to identify suitable deep-seated aquifers, with a significant risk of encountering saline or brackish water. Second, deep wells are expensive and so is the cost of water pumped from them. Furthermore, most of the deep-seated aquifers currently identified contain non-renewable resources (Foster and Loucks, 2006), while the remaining ones are only weakly recharged. Therefore, permanent exploitation of these deep groundwater resources is often not sustainable; rather they may serve as temporary supplies during and after emergencies (Vrba and Reza Salamat, 2007) or to boost local socio-economic development.

Another unconventional groundwater resource is brackish groundwater (*i.e.* water containing between 1 and 10 g/L of total dissolved solids (TDS). Brackish groundwater at shallow depths is abundant in many countries and can easily be withdrawn at low cost. It can be used either directly for purposes such as cooling water, aquaculture and a variety of uses in the oil and gas industry, and for many more purposes after treatment. Recent advances in membrane technology have reduced the cost and energy requirements of salinity reduction, thus making treated brackish groundwater a viable option for drinking-water supplies and for irrigating highly productive cash crops. Brackish groundwater is already used on a significant scale in several states of the USA (USGS, 2013) and in Eastern Spain and the Canary Islands, where several thousand small units supply touristic areas and mainly irrigation. There, farmers often use a mix of cheaper, relatively saline groundwater with more expensive imported water, local groundwater of artificially reduced mineral content, or desalinated sea water. Reducing the mineral content of water has the associated problem of disposing the residual saline water or brine. This often becomes a source of pollution when a network of collectors and safe disposal are not available.

A third unconventional resource, albeit a potential one so far, is submarine fresh or brackish groundwater. The presence of such groundwater offshore has been known since antiquity. Post *et al.* (2013) report on 33 large fresh to brackish offshore groundwater zones around the world and estimate the submarine groundwater reserves with TDS below 10 g/L to have a global volume of 0.5 million km^3, of which 0.3 million km^3 has a TDS lower than 1 g/L. Perspectives for economically feasible exploitation of this exhaustible and difficult to catch paleo-groundwater are not yet clear, but it might become a future strategic temporary resource in water-scarce coastal areas. Martin-Nagle (2016) and Kimani (2016) discuss governance aspects of such offshore fresh groundwater resources.

20.2.9 Interactions and important external impacts

Groundwater abstraction does not only produce the intended benefits of its use, but also unintended side-effects. Where abstraction sites are located close to each other within a single groundwater system, pumping likely interacts by reducing mutually the hydraulic head at each other's location, which leads to higher pumping cost and in some cases to springs, qanats, galleries and shallow wells running dry completely. Intensive abstraction by large numbers of wells may lead to depletion of aquifer storage, sea water intrusion, saltwater upconing and invasion of other poor-quality water. In some cases declines of the groundwater levels have a positive effect, for instance by inducing groundwater recharge from good-quality freshwater sources. Beside these internal side-effects that affect withdrawal conditions and the state of the aquifer resources, groundwater abstraction does also produce external environmental impacts, such as the reduction of the baseflow of streams, degradation of wetlands, loss of groundwater-related ecological services, yield reduction in water-table dependent agriculture, and land subsidence in zones where unconsolidated geological formations occur. It is difficult to control these interactions and impacts because decisions about groundwater abstraction are made in an extremely decentralised way, usually independently by myriads of individuals.

20.3 EXPLOITING AND GOVERNING OTHER EXTRACTABLE SUBSURFACE RESOURCES

20.3.1 Geothermal energy development

The geothermal energy of the Earth's crust originates from the formation of the planet and from decay of natural radio-active isotopes. The corresponding upward heat flow manifests as hot springs which have been used for millennia for bathing purposes and currently for heating purposes. The heat of hot groundwater (and vapour) tapped at depths where high temperatures prevail can be used for power generation. The zones with the highest subsurface temperatures at a given depth are in regions with active or geologically young volcanic activity. The typical subsurface arrangement for geothermal power development is a well doublet, consisting of an abstraction well and an injection well. Hot water or steam ejected or pumped from the abstraction well, after vaporization, is passed through turbines to generate electricity; the condensate of lower temperature is returned through the injection well to the subsurface.

Geothermal energy is still an underdeveloped source of energy. According to the International Geothermal Association, it produced during 2013 some 68 TWh of electricity (UCS, 2016), equivalent to 0.04% of the global primary energy demand (Table 20.2). The global technical geothermal potential, however, is comparable to the current global primary energy supply (Goldstein *et al.*, 2011). By January 2017, the top-10 countries in terms of installed geothermal power generation capacity included

Table 20.2 Global primary energy demand in 2013, by type of energy (after IEA, 2015).

Type of energy	*Global primary energy demand (2013)*			
	Mt_{oe} *(million tonnes* of oil equivalent)	*TWh* *(10^9 kWh)*	*EJ* *(10^{18} Joules)*	%
Coal	3 973	46 206	166	29
Oil	4 235	49 235	177	31
Gas	2 880	33 494	121	21
Nuclear	646	7 513	27	5
Hydropower	320	3 722	14	2
Bio-energy	1 366	15 887	57	12
Other renewable energy *(wind, solar, geothermal, marine)*	159	1 849	7	1
Total	13 579	157 924	569	100

the USA, Philippines, Indonesia, New Zealand, Italy, Mexico, Turkey, Kenya, Iceland and Japan (ThinkGeoenergy, 2017). The expectations are that geothermal energy will gain importance during the 21th century and potentially will cover more than 10% of the energy demand by 2100 (Goldstein *et al.*, 2011).

Geothermal energy is considered a renewable source of energy, because the tapped heat is continuously restored by natural heat production and by conduction and convection of heat from surrounding hotter zones. Nevertheless, in the long-term, cooling can be expected. It has only a minor impact on the groundwater resources, because most of the water used as a carrier of energy is re-injected into the subsurface. Geothermal fluids drawn from great depth carry a mixture of gases that, besides being a polluting hazard for local environment and shallow aquifers, may contribute to global warming and acid rain. Nevertheless, the environmental footprint remains comparatively low when action is correctly done. Existing geothermal electric plants emit an average of 122 kg/MWhe of CO_2, which is about one tenth of the emission intensity of conventional fossil fuel power plants (EIA, 2016).

20.3.2 Oil and gas exploration and production

Oil and natural gas derive from organic matter-rich sediments (source rock) located deep enough to get this material under pressure and heat transformed into fluids. Buoyancy forces cause the oil and gas to migrate from the source rock upward and eventually escape by seeps at the land surface. However, they may accumulate in porous and permeable geological formations (reservoirs) overlain by a poorly permeable formation (cap rock or seal) in a structural setting (*e.g.* an anticline or fault) that traps them and prevents the fluids from moving upward, thus forming conventional crude oil and natural gas fields. Natural gas may be found either in association with oil – floating on top of it – or in separate gas fields. There are also non-conventional oil resources: oil sands with partially biodegraded oil (crude bitumen or extra heavy crude oil) and oil shales (source rocks not yet long enough exposed to heat or pressure to convert trapped hydrocarbons into crude oil). There is also shale gas, trapped in poorly permeable source rock. Oil and gas exploration is based on geological studies, geophysical surveys (gravity, magnetic and seismic) and expensive, high-risk exploratory drilling. The production is by means of wells. Initially, the natural reservoir pressure tends to be sufficient to force the oil to the surface. For the development of shale gas, hydraulic fracturing ('fracking') is needed (see section 20.4.2).

Oil (petroleum) has been used since ancient times. It gained importance after the first oil refinery was built in 1856. Nowadays, the oil and gas industry represents one of the most important sectors of the world's economy. It meets more than 50% of the global primary energy demand (see Table 20.2) and supplies still around 90% of all vehicular fuel. In spite of a growing preference in many countries for replacing fossil energy by renewable energy, the global consumption of oil and gas is still increasing. By 2014, world oil reserves were estimated at 1659 billion (109) barrels and the production at 89 million barrels/day, resulting in a reserves-to-production ratio of 51 years. The corresponding figures for natural gas are 201,771 billion m^3, 3474 billion m^3/year and 58 years, respectively (Eni, 2015).

Oil and gas are non-renewable resources with considerable impact on groundwater and the environment. Large quantities of water – including fresh groundwater – are used during exploration, production and subsequent activities in the industry. As stated by IPIECA (2005) "we actually handle more water than we do oil". Even if the used waters are carefully treated and recycled, this has a significant impact on the water resources and their use, especially in water-scarce areas. Blow-outs, leakages, spillages and other accidents form additional threats. Oil and gas have a large share in global carbon dioxide emission. Depletion of oil and gas fields is likely to produce land subsidence and sometimes light tremors and fracturing.

20.3.3 Mining

The extractable resources targeted in the mining sector can be subdivided into (i) nonmetallic mineral deposits (industrial minerals; precious stones; construction materials); (ii) metallic mineral deposits (ferrous metals; non-ferrous or base metals; precious metals); and (iii) energy resources: fossil fuels (coal, lignite, etc.) and radioactive minerals (uranium). The extracted resource is non-renewable. Reconnaissance is based on a wide range of geological, geochemical and geophysical methods, depending on the type of mineral resource. Exploratory drilling is usually an additional component during the subsequent exploration stage. Exploitation is done either by surface mining (quarries; open pit mining; strip mining) or by underground mining, unsupported or supported by pillars and stoping (Lacy, 2015). Except for materials such as sand, gravel and building blocks, most extracted mining products require some form of processing (crushing, grinding, flotation or hydrochemical leaching, smelting, chemical processing, etc.).

Mining is indispensable to meet the needs of current and future populations, because its products have encroached in all facets of human life and human activities. It is estimated that the mining sector currently employs approximately 30 million people or approximately 1% of the world's economically active population (Kalindekafe, 2013). For many countries, the mining sector is important for creating wealth.

The mining industry is interacting intensely with aquifers and groundwater. Many mining projects – both surface and underground – require dewatering by artificial drainage to create adequate working conditions, which may have a profound impact on the regional hydrological and hydrogeological regimes, depending on the volumes pumped. The operations may result in permanent changes in aquifer conditions, such as by removing part of the aquifer, or soil and confining layers (leading to higher vulnerability to pollution). Underground mines tend to trigger fracturing, terrain collapses and land subsidence. After mine abandonment groundwater levels usually rise again. In open pit mining, large, deep lakes then may form, while in warm climates evaporation losses may become significant, thus affecting groundwater resources. More than 70% of all material excavated in mining operations is waste rock, which is often stored on spoil heaps. Together with tailings, usually deposited from suspension in water, they are substantial pollution sources. In addition, mining water often contains harmful dissolved solids and gases (Umweltbundesamt, 2003). Mining activities in the past have left a legacy of environmental problems.

20.3.4 Governance aspects

The exploitation of geothermal energy, oil and gas, and all types of mining resources, forms altogether a very heterogeneous set of activities, with operations that vary from relatively simple (quarries, sand withdrawal) to very complex (oil and gas industry). Due to the huge diversity of activities and conditions, governance is also highly variable. Therefore, only a few typical features and trends will be described, without claiming general validity.

Regarding actors in governance, there is a striking difference with the groundwater sector. Whereas the exploitation of groundwater is mainly in hands of myriads of individuals, the lion's share of the exploitation activities in the geo-energy and mining sectors is dominated by companies, often large companies, sometimes multinationals. This means that communication between only a few actors is in principle sufficient to negotiate and agree on key decisions and a modus operandi that strike a balance between public and private interests. For satisfactory deals, however, it is necessary that the government agencies in charge (and in some cases the private owners of an extractable resource) have sufficient capacity, support, knowledge and expertise to negotiate with specialized and powerful companies, and to monitor compliance with regulations and contractual obligations. The concentration of decision-making to a few key players has the disadvantage that it may lead to exclusion of other potential actors, in particular civil society. Only in recent decades have ordinary citizens become visible as stakeholders and actors, usually motivated by observed or feared harmful environmental impacts of resource exploitation and often acting as pressure groups.

The acquisition of local information and knowledge on the exploitable resource requires often huge investments in exploration activities. The outcomes are usually uncertain, with high financial risks. This explains the predominance of large companies. In general, exploration data and derived information are of high quality. Although governments usually stipulate this information to be made public, companies often are reluctant to share it with potential competitors and the general public.

Oil and gas resources, and the majority of other mining resources, are in most countries owned by the state. Dedicated mining laws are usually available. The government issues licenses to explore and exploit the resources. Such licenses may include a tax-and-royalty agreement, a product-sharing contract or a service contract. In any case, governments have to ensure that the operations are carried out in full compliance with the legal requirements on environmental protection.

Noteworthy is the unitization approach widely used in the oil and gas industry (Jarvis, 2011). It implies the joint operation of oil and gas reservoirs by all owners of rights in the separate tracts overlying these reservoirs, thus enhancing overall benefit and promoting a prominent role of the private sector in governance. In principle, unitization could also be considered for several other sectors of subsurface resource management.

Oil and gas companies, mining companies and other resource exploiters are driven by commercial goals (making a profit), while governments have to pursue that wealth is created for the society as a whole, that benefits are properly distributed, and that only acceptable environmental damage is produced. Governments should be aware of these conflicting goals and design their policies in such a way that they are adequately balanced. The role of the civil society is becoming steadily more pronounced and puts both the private sector and the government agencies under pressure. This is resulting steadily in shifts in governance practices towards an improved environmental and social behaviour of the private sector. Companies understand that complying with legal environmental regulations is not sufficient anymore, but that they also should get the image of contributing to sustainability and acquire a 'social license to operate', in order to avoid potentially costly conflicts and exposure to social risks (Prno and Slocombe, 2012). Unlike in the past, many projects nowadays include a final phase of 'reclamation' to restore land, water and environmental values as much as possible and have budgets allocated for this from the onset (Lacy, 2015). The role of the public opinion is particularly evident in the energy sector. In the first place, it triggers and catalyses government efforts in many countries to gradually phase out fossil energy sources and replace them by renewable ones – at least partly – in order to reduce carbon dioxide emissions, but also to reduce environmental impacts such as induced seismicity. Strong opposition of the civil society against hydraulic fracturing has in several countries resulted in a ban on the exploration and exploitation of shale gas. Disasters with nuclear power plants, like in Chernobyl (in 1986) and Fukushima (in 2011), mobilize public pressure for increased security but also for reducing or abandoning the exploitation of this energy source, although not always fully supported by well-documented evidence.

20.4 USING AND GOVERNING THE SUBSURFACE SPACE

20.4.1 Disposal and storage of hazardous waste

The disposal of hazardous waste is challenging. Burying it safely in the subsurface seems often an attractive option. The deep-well waste disposal technique is suitable for disposal of liquid waste or solid waste reworked to slurry. It requires a permeable geological formation with ample storage capacity to receive and store the waste, covered by an impermeable formation to ensure that the waste remains isolated for millions of years from the overlaying geological formations, the biosphere and the atmosphere. Typical categories of waste stored in this way are waste produced in the oil and gas industry (oil field brines, cuttings, drilling mud), liquid waste from solution mining operations and all kinds of industrial and municipal liquid waste. In some cases, the liquid waste may contain high concentrations of heavy metals, including mercury, arsenic and cadmium compounds.

Carbon capture and sequestration (or storage) – known as CCS – is a measure proposed for mitigating global climate change by capturing man-made carbon dioxide and storing it permanently in order to prevent it from being emitted to the atmosphere. Although deep ocean storage is theoretically an option, underground storage is generally preferred, for instance in depleted gas and oil reservoirs, or in deep saline reservoirs (Folger, 2013). The majority of the eleven large-scale CCS projects in operation by 2015 are located in North America. Together with another 34 projects in a planning stage around the world, they have a total CO_2 capture capacity of 80 million tonnes per annum; which is dwarfed by the approximately 4000 million tonnes of CO_2 to be captured and stored annually by 2040, as required to meet climate change targets (GCCSI, 2015).

The safe disposal of radioactive waste is still a controversial and unresolved issue. The problem of long-lived nuclear waste lies not only in the devastating impact of potentially released radiation on the exposed biosphere (including humans), but also in the enormous persistence of its hazardous properties due to the very long half-lives of some radioisotopes. Often a distinction is made between low-level waste (containing small amounts of mostly short-lived radioactivity) and high-level waste (as produced by nuclear reactors). Current practices in managing high-level radioactive waste, mostly from the front end and back end nuclear fuel cycle, and from nuclear weapons, are a combination of temporary storage and investigating options for permanent subsurface storage. Examples of potential sites studied in detail are in volcanic rock high above the water table (Yucca Mountain, USA, currently discontinued), in crystalline rock (Finland, Sweden and India), in thick impermeable clay (Boom Clay Formation, Belgium), and in salt domes (Gorleben, Germany). The disposal of short- and medium-lived radioisotopes, often of medical and industrial origin, is preferably done in isolated surface or small-depth repositories that are permanently monitored.

Special types of human activities that bring hazardous radioactive substances into the subsurface are nuclear weapons testing and nuclear power accidents. They form a significant and rather unpredictable risk to groundwater and the environment. Around 2010 underground nuclear tests have been carried out worldwide. Most terrestrial underground test sites are located in arid or semiarid areas with deep groundwater levels (>200 m). Apart from the Nevada Death Valley Test Site (NTS) in western USA, only limited data related to the impacts on groundwater has been made public.

If the subsurface disposal and storage of hazardous substances are properly planned and implemented, then the interaction with groundwater resources and the environment is often expected to be minimal. However, assumptions on the isolation and long-term stability of the chosen repositories may be too optimistic. Furthermore, past experience has shown that human errors and mechanical failures exist: they are hard to legislate against and to rule out. Given the hazardous properties of the stored substances, there is considerable risk for groundwater resources and the human environment.

20.4.2 Injection and recovery

Subsurface injection and recovery of fluids is being applied for a variety of purposes. Solution mining of salt by injecting water or steam became established in the second half of the 17th century. Acid and other chemical leaching processes are applied for wet mining of copper, gold, sulphur and uranium. The lixiviants are hazardous to groundwater quality in aquifers used for drinking water purposes.

Reinjection of geothermal residual fluids started purely as a disposal method, but has more recently been recognized as an essential and important part of geothermal reservoir management. Reinjection not only maintains reservoir pressure, but also increases energy extraction efficiency over the life of the resource. The location of injection wells in relation to production wells influences the ratio of injected fluid recovered in the production wells.

Temporal storage of heat is applied in shallow aquifers, using surplus heat available during low-demand periods of the year in order to recover it during periods when the energy demand is higher. A common application is heating of buildings during winter.

Injection and recovery of hydrocarbons and fluids associated with oil and natural gas production is used to create short-term buffer storage between production and the demand for hydrocarbons. Crude oil and natural gas may be injected into subsurface reservoirs or caverns in salt rock, using wells designed for both injection and recovery. Injection is also used to increase production and prolong the life of oil fields. This includes secondary recovery (water flooding) by reinjecting the saline water that was co-produced with oil and gas into the oil-producing formation. If needed, tertiary recovery methods may be applied, injecting gas (*e.g.* CO_2), water with special additives, and steam.

Exploiting the substantial quantities of gas located in the original source shale formations ('shale gas') scattered around the world requires a technique called hydraulic fracturing (or "fracking") to increase porosity and permeability of the formations to a level that makes exploitation of shale gas technically and economically feasible. Hydraulic fracturing is based on pumping a fracturing fluid by means of a horizontally extending wellbore at a rate sufficient to increase the pressure downhole to a value in excess of the fracture gradient of the formation rock. The pressure causes the formation to crack, allowing the fracturing fluid to enter and extend the crack farther into the formation. To keep this fracture open after the injection stops, a solid 'proppant' – commonly sieved round sand – is added to the fracture fluid. The propped hydraulic fracture then becomes a high-permeability conduit through which the formation fluids can flow to the well. The fluid injected into the rock is typically a slurry made of water, proppants and chemical additives. Additionally, gels, foams and compressed gases, including nitrogen, carbon dioxide and air, can be injected. Fracking is a highly controversial technique, because of its significant and irreversible impacts on the subsurface.

Managed Aquifer Recharge (MAR) includes artificial groundwater recharge and the subsequent recovery of the stored water at a moment when it is needed. It is a well-known technical groundwater management intervention to augment available water resources. It is applied in particular in arid and semi-arid regions, but also in zones where saltwater intrusion or saline water upconing has to be controlled, in which case part of the recharged water is to complement aquifer outflow to the sea (seawater barrier). A range of artificial recharge techniques is available, including spreading methods (ponding infiltration basins), injection wells, river bed management, recharge dams and other in-channel structures.

20.4.3 Construction into the underground space

This category of subsurface use is limited to the very shallow part of the subsurface domain. It includes accommodating at shallow depth – above the water table – pipelines (for water, gas and oil), sewerage systems and cables (for electricity and communication, including radio, television and internet) in densely populated areas. Leaks in sewerage systems and pipelines for oil and gas may constitute potential pollution risks for groundwater. Concerns are rising on groundwater pollution by leaks of emerging contaminants from household products, pharmaceuticals and cattle raising activities. Their behaviour and persistence in groundwater is still a subject of research.

Tunnels and underground railways go to tens of metres deep, occasionally more than one hundred metres. If they are located below the groundwater table – which is often the case – then the groundwater regime may be significantly disturbed during construction (by artificial drainage) and compaction of compressible layers may occur. Modified hydraulic properties of these layers, in combination with constructed artificial obstacles and drainage provisions, may have a permanent influence on the shallow groundwater regimes. Some tunnels form preferential entry paths for pollutants into groundwater systems.

Traditionally, buildings use the subsurface for foundations and cellars. The scarcity of space in urban agglomerations triggers a more intensive use of underground space for construction purposes. Underground car parks are mushrooming, while underground shopping centres, cinemas and commodity stores are likely to become more common in the near future. Their interaction with groundwater is comparable with that of tunnels and underground railways. In many urban areas initially supplied with local groundwater, the water table depletion due to groundwater abstraction allowed constructing underground infrastructures in the unsaturated zone. After abandoning these wells due to pollution or seawater intrusion – among other causes – and bringing surface water and groundwater from outside, the recovery of the water table, enhanced by leakages from distribution networks, may create water-logging, inundation and corrosion problems, requiring management.

20.4.4 Using the subsurface as a scientific archive

Using the subsurface as a scientific archive, includes exploring, visiting, protecting and conserving archaeological sites, special fossil sites, karst sites (caves) or stratigraphic type locations and other subsurface cultural or natural heritage objects. Apart from the exploration activities, these uses cause little or no interference with groundwater systems, but they may impose restrictions or delays on other uses of subsurface space or on the exploitation of subsurface resources.

20.4.5 Governance

Permanent storage of hazardous waste and most of the described injection and recovery activities represent considerable risks to groundwater and the environment. Therefore, these activities need to be carried out by parties having specialised technical expertise and on the basis of solid information on the subsurface conditions. These parties should obtain formal approval and licences from the mandated government agencies, and strictly comply with dedicated legislation, regulations and protocols. Careful supervision and monitoring are essential, not only to verify compliance, but also to detect as early as possible any unforeseen and undesired negative impacts – in case these would occur – such as leaking wells, failures of isolations, or subsurface layers being less impermeable than assumed.

In most countries, there is limited information in the public domain on these activities and on how decisions are taken. This lack of transparency often feeds distrust among the civil society and strengthens environmental activists in their opposition against certain controversial activities, such as carbon capture and sequestration, storage of nuclear waste or the application of fracking techniques. In several cases this has led to a moratorium on such subsurface activities until the most important pending questions have been answered, following the precautionary principle, but this may delay solutions while damage continues. Nuclear testing represents an extreme case of lack of transparency, because it is militarily and politically sensitive. Governments are reluctant to share information on it, even though some cases have been intensively studied. Among the injection and recovery interventions, MAR and temporary storage of heat score most favourably in terms of low risk to groundwater; the former even has an overall positive effect on groundwater resources and may offer good opportunities for active stakeholder involvement if the related projects are small. MAR has advanced to the point in which reclaimed waste water begins to be accepted as a source of water, if secure and well-monitored procedures are followed.

Regulations regarding different types of underground constructions are likely to exist in most countries, but probably a much smaller number of countries have adequate regulations on the protection and conservation of subsurface heritage objects. Weak enforcement capacity, budgetary problems, political opportunism and unexpected problems sometimes lead to poor compliance with existing regulations. Several classes of underground constructions require permanent monitoring to ensure that interactions with subsurface geological formations and groundwater remain within safety limits.

20.5 FROM FRAGMENTED TOWARDS COORDINATED GOVERNANCE OF THE SUBSURFACE RESOURCES

20.5.1 Overall diagnostic

The previous sections have presented in a nutshell an impression of the diversity of subsurface resources and their use. Each of these resources and the activities or projects intended for their exploitation and use is facing its own particular challenges and each of them has its own specific governance setting and characteristics. In addition, there are enormous geographic variations in the types and quantity of resources exploited and used, the approaches to their development, the benefits obtained, and the quality of governance. It is beyond the scope of this chapter to depict these variations.

In spite of huge investments in governance provisions, significant flaws and deficiencies in governance still prevail in many countries: limitations in data, information and knowledge; lack of consensus on facts and on predictions; poor consideration of groundwater impacts developing much more slowly than people imagine and beyond the time politicians remain in charge; inadequate legal and regulatory frameworks;

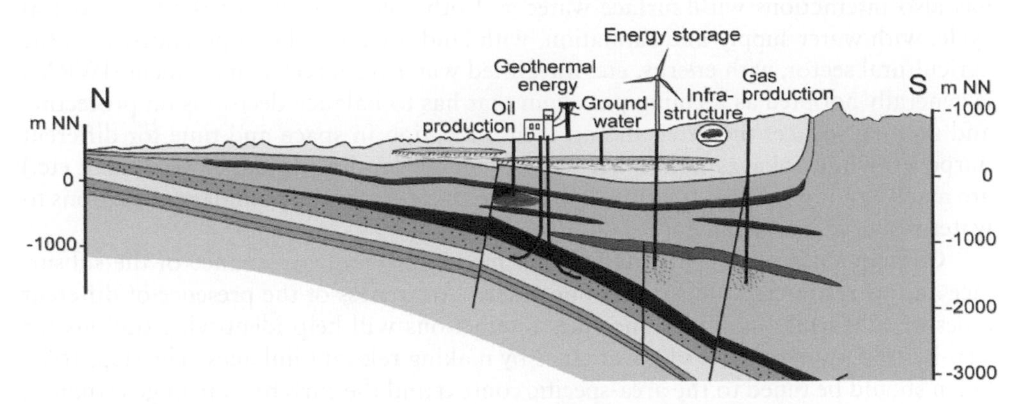

Figure 20.1 Cross-section showing the concurrence of different subsurface resources activities within a single area, at different depths (Source: Geomol, www.geomol.eu).

ineffective policies and poor government power; insufficient institutional capacity, mandate and funding; insufficient stakeholder involvement; lack of trust between key actors; lack of transparency and communication on planning, decision-making and implementation; lack of government capacity to supervise and monitor compliance with laws, regulations and contracts; vested interests and cavalier behaviour of companies. Good governance is comparatively easy in sectors or projects where financial profits are high and the number of key players in decision-making is relatively limited (oil, gas and mining sectors), but it is very difficult to achieve in opposite cases, like the groundwater sector.

The different uses of subsurface resources do interact. In particular groundwater is very vulnerable and at risk when exposed to either interferences between different groundwater exploiters or to the side-effects of the exploitation of other subsurface resources. As the multiple uses of the subsurface resources are becoming steadily more intensive, the interactions get stronger as well, producing mainly negative impacts. Governance, however, is currently still highly fragmented and focused on the direct objective of each subsurface activity separately, without much attention for other subsurface resources and activities. It is worthwhile to search for options to reduce this fragmentation and to establish and assure coordination and synergy.

20.5.2 Options for improving governance of the subsurface

Establishing fully integrated management and governance of the subsurface and all its resources seems at first glance an obvious and attractive ambition. However, if several types of resource exploitation and subsurface use overlap within one single area (see Figure 20.1), then this ambition seems to be unrealistic almost anywhere in the world, under current conditions. In the first place because such an integrated approach is operationally very complex and it requires a degree of maturity in governing the individual subsurface resources that only in rare cases has been reached. In the second place because current and potential interactions are not limited to the subsurface domain only; each of the subsurface resources and its exploitation interacts with systems, processes and other factors beyond the subsurface domain. Groundwater, for instance, has also interactions with surface water and other components of the hydrological cycle, with water supply and sanitation, with land use and related practices, with the agricultural sector, with energy, etc. Integrated water resources management (IWRM) is generally adopted as a ruling paradigm that has to balance decisions on protecting and tapping sources of water and on their allocation in space and time for different purposes; while linkages with other sectors (energy, land planning, employment, etc.) are made at a lower level, treating decisions in these sectors as boundary conditions to water resources management.

Consequently, an incremental approach to improving governance of the subsurface and its resources seems most appropriate. Awareness of the presence of different types of subsurface activities and their interactions will help identifying options for step-by-step improvement of governance by making relevant linkages. The steps to be taken should be tuned to the area-specific context and the current state of governance, including any adopted overarching principles.

Typical 'low-hanging fruit' options include: (a) creating synergy by promoting all data on the subsurface to be shared between those who could benefit from it; (b) creating transparency on all subsurface activities in the area concerned, *e.g.* by raising awareness and public information programs, web portals and publicly accessible databases; (c) announcing to the general public all significant planned new projects related to the subsurface and permitting stakeholders to appeal against these or propose amendments; (d) introducing obligatory 'groundwater impact assessments' for new projects aimed at any use of subsurface space or subsurface resources (Van der Gun *et al.*, 2016).

Other options that are likely to require more time before becoming effective include: (a) reforms of legal and institutional frameworks; (b) establishing public-private partnerships to widen the private sector's commitment and responsibility for subsurface governance, and to better profit from its unique expertise; (c) getting governance of the subsurface higher on the political agenda and on the agendas of international agencies (resulting in higher budgets, in cross-cutting consultations and in catalysing projects); (d) investing in research and development of technologies that reduce negative impacts of interactions; (e) educating a new generation of geoscientists familiar with both the technical and the management aspects of all kinds of exploitation and use of subsurface resources.

ACKNOWLEDGEMENTS

This chapter is partially based on a thematic paper written by the first author and others for the GEF-FAO-World Bank-UNESCO-IAH Groundwater Governance project. The authors would like to thank Todd Jarvis and Kirstin I. Conti for reviewing the first draft of this chapter and providing valuable comments.

REFERENCES

Closas, A., & Villholth, K. G. (2016). *Aquifer contracts: a means to solving groundwater over-exploitation in Morocco?* Colombo, Sri Lanka: International Water Management Institute

(IWMI). 20p. (Groundwater Solutions Initiative for Policy and Practice (GRIPP) Case Study Series 01). doi: 10.5337/2016.211.

Custodio, E. (2002). Aquifer overexploitation: What does it mean? In: *Hydrogeology. J.*, 10(2): 254–77.

Custodio, E. (2010). Intensive groundwater development: A water cycle transformation, a social revolution, a management challenge. In: *L. Martínez–Cortina, A. Garrido, E. López–Gunn (eds.) Rethinking Water and Food Security.* Botín Foundation/CRC Press: 259–298. http://www.rac.es/ficheros/doc/00734.pdf.

Custodio, E. (2013). *Trends in groundwater pollution: loss of groundwater quality and related services.* Groundwater Governance: A global Framework for Country Action. Thematic Paper no 1. http://www.groundwatergovernance.org/fileadmin/user_upload/groundwater-governance/docs/Thematic_papers/GWG_Thematic_Paper_1.pdf.

EA-SAC (2007). *Groundwater in the Southern Member States of the European Union: an assessment of current knowledge and future prospects. Country report for France.* European Academies, Science Advisory Council.

EIA (2016). *Environment, carbon dioxide emissions coefficients.* https://www.eia.gov/environment/emissions/co2_vol_mass.cfm

Eni (2015). *World Oil and Gas Review 2015*, 14th edition.

Folger, P. (2013). *Carbon capture and sequestration (CCS): a primer.* CRS Report for Congress, USA, https://www.fas.org/sgp/crs/misc/R42532.pdf

Foster, S.S.D., & P. Loucks (eds.) (2006). *Non-renewable groundwater resources. A guidebook on socially-sustainable management for water policy-makers.* Paris, UNESCO-IHP, 103 p.

GEF (2016a). *Global Diagnostic on Groundwater Governance.* Groundwater Governance – a Global Framework for Action. http://www.fao.org/3/a-i5706e.pdf.

GEF (2016b). *Global Framework for Action to achieve the Vision on Groundwater Governance.* Groundwater Governance – a Global Framework for Action. http://www.fao.org/3/a-i5705e.pdf.

GEF (2016c). *Project website.* Groundwater Governance – a Global Framework for Action. http://www.groundwatergovernance.org/home/en.

GCCSI (2015). *The Global Status of CCS.* 2015. Global CCS Institute Summary Report. http://hub.globalccsinstitute.com/sites/default/files/publications/196843/global-status-ccs-2015-summary.pdf.

Goldstein, B, Hiriart, G., Tester, J., *et al* (2011). Great expectations for geothermal energy to 2100. *Proc. 36th Workshop on Geothermal Reservoir Engineering*, Stanford University, Stanford, California, Jan–Feb 2011.

IEA (2015). *Energy and climate change.* World Energy Outlook Special Report, International Energy Agency, Paris, France.

IPIECA (2005). *Water resource management in the petroleum industry.* International Petroleum Industry Environmental Conservation Association, London, UK.

Jarvis, W.T. (2011). Unitization: a lesson in collective action from the oil industry for aquifer governance, *Water International*, 36:5, 619–630.

Kalindekafe, L. (2013). Enhancing the capacity of good governance and sustainable development in the mining sector. *UNCTAD Global Commodities Forum 2013*, 18–19 March 2013, Geneva, Switzerland.

Kimani, N. N. (2016). *The African Union's Role in the Governance of Offshore Freshwater Aquifers. The International Journal of Marine and Coastal Law* (Vol. 31). 620–651.

Lacy, W. (2015). *An introduction to geology and hard rock mining.* Rocky Mountain Mineral Law Foundation, Science and Technology Series.

Margat, J., & Van der Gun, J. (2013). *Groundwater around the World: A geographic synopsis.* Leiden, The Netherlands, CRC Press/Balkema.

Martin-Nagle, R. (2016). *Transboundary Offshore Aquifers: A Search for a Governance Regime*. http://www.brill.com/products/book/governance-regime-transboundary-offshore-aquifers.

OSWER (2012). *Introduction to mining and mineral processing*. PowerPoint presentation.

Prno, J. & Slocombe, D.S. (2012). Exploring the origins of 'social license to operate' in the mining sector: Perspectives from governance and sustainability theories. *Resources Policy* 37: 346–357.

ThinkGeoenergy (2017). *Overview on installed geothermal power generation capacity worldwide*. http://www.thinkgeoenergy.com/overview-on-installed-geothermal-power-generation-capacity-worldwide/ (Accessed 3 October 2017)

Umweltbundesamt (2003). Groundwater management in mining areas. *Proc. 2nd Image Train Advanced Study Course, Pécs, Hungary*. European Commission, Universität Karlsruhe, University of Newcastle, UBA-Vienna, http://www.umweltbundesamt.at/fileadmin/site/publikationen/CP035.pdf

UCS (2016). *Website of the Union of Concerned Scientists*. Accessed 22 July 2016. http://www.ucsusa.org/clean_energy/our- energy-choices/renewable-energy/how-geothermal-energy-works.html#.V45JoeTr274.)

USGS (2013). *National Brackish Groundwater Assessment*, USGS Info Sheet, U.S. Geological Survey.

Van der Gun, J., Merla, A., Jones, M., & Burke, J. (2012). Governance of the subsurface and groundwater frontier. *Thematic Paper 10*, Groundwater Governance, A Global Framework for Action. GEF, World Bank, UNESCO-IHP, FAO and IAH.

Van der Gun, J., Aureli, A., & Merla, A. (2016). Enhancing groundwater governance by making the linkage with multiple uses of the subsurface space and other resources. *Water* 2006, 8, 222; doi:10.3390/w8060222.

Vrba, J., & Reza Salamat, A. (2007). *Groundwater for emergency situations*. Proc. Intern. Workshop Tehran, 29–31 October 2006 UNESCO, IHP-VI, Series on Groundwater 15 (2007), Paris.

Part 4

Cases

Chapter 21

Groundwater governance in the Great Artesian Basin, Australia

Rien A. Habermehl[1,2]
[1]*(Previously) Geoscience, Australia*
[2]*(Previously) Bureau of Rural Sciences, Canberra, ACT, Australia*

ABSTRACT

The Great Artesian Basin, Australia's largest groundwater resource, is a multi-layered confined aquifer system, with artesian aquifers in Triassic, Jurassic and Cretaceous continental quartzose sandstones. It underlies three States and one Territory within Australia's Rangelands with low rainfall in arid and semi-arid regions. Pastoral activity, homestead, town water supplies, mining and petroleum ventures are all totally dependent on artesian groundwater from the basin. State Government investigations and legislation dominated during the period from the 1890s to 1960s, but Commonwealth Government funding and scientific investigations assisted since the 1960s. The States have the primary constitutional powers and responsibilities for land and water resources management under the Federal Australian Constitution. State (and Territory) Water Acts vest groundwater (and surface water) resources in the Crown, and provide for rights in water, control and management of works with respect to water conservation and protection. All artesian (flowing) bores, and sub-artesian (non-flowing) bores are required to be licensed, and construction and purpose of the bore are specified, as well as use, volumetric allocation and distribution works. Bores are to be constructed by suitably licensed (artesian) drillers. Bore owners are responsible for the maintenance of all bores, including casing, headworks and distribution networks. Combined Commonwealth and State projects and funding since 1989 for rehabilitation of bores alleviates the serious drawdown of artesian pressures. The *Environment Protection and Biodiversity Conservation Act 1999* (the EPBC Act) is the Australian Government's central piece of environmental legislation. It provides a legal framework to protect and manage nationally and internationally important flora, fauna, ecological communities (groundwater dependent ecosystems) and heritage places, defined in the EPBC Act as matters of national environmental significance, and is applicable to the Great Artesian Basin area.

21.1 GREAT ARTESIAN BASIN, AUSTRALIA – AUSTRALIA'S LARGEST GROUNDWATER RESOURCE, GEOLOGY AND HYDROGEOLOGY

The Great Artesian Basin is a confined groundwater basin, which underlies arid and semi-arid regions across 1.7 million km^2 or one-fifth of Australia, within the States of New South Wales, Queensland and South Australia and the Northern Territory (Figure 21.1). Most of the Basin area is within Australia's Rangelands with low rainfall in arid and semi-arid regions, except for the most northern parts which are in tropical seasonal high rainfall areas. The basin's groundwater resources were discovered around 1880, and their development allowed the establishment of an important pastoral industry. Pastoral activity, mainly sheep and cattle farming, homestead, town water supplies, mining and petroleum ventures are all totally dependent on artesian groundwater from the basin.

The Great Artesian Basin is a multi-layered confined aquifer system, with artesian aquifers in Triassic, Jurassic and Cretaceous continental quartzose sandstones. Intervening confining beds (aquitards) consist of siltstone and mudstone; Cretaceous marine sediments form the main confining unit (Habermehl, 1980, 1983, 2001a, b; Habermehl & Lau, 1997; Smerdon *et al.*, 2012; Ransley *et al.*, 2015, Figure 21.2). The Basin is up to 3000 m thick, and is a large synclinal structure, uplifted and exposed along its eastern margin and tilted southwest.

Recharge to the aquifers occurs in the relative high rainfall eastern margin, and the western margin in the arid centre of the continent receives minor recharge (Kellett *et al.*, 2003; Habermehl *et al.*, 2009, Keppel *et al.*, 2013). Regional groundwater movement in the aquifers is towards the southern, southwestern, western and northern margins, where artesian springs discharge (Habermehl, 1982; Keppel *et al.*, 2013; Love *et al.*, 2013) and produce sedimentary spring deposits, in the form of carbonate mounds. Many springs contain unique flora and fauna (Fensham *et al.*, 2010) and were favoured by Aboriginals as assured sources of water.

Residence or travel times of the artesian groundwater range from almost recent in the recharge areas to more than one million years near the centre of the Basin and the artesian groundwater source and origin is meteoric water, which recharged the basin's aquifers (Airey *et al.*, 1979, 1983, Calf & Habermehl, 1984, Bentley *et al.*, 1986, Torgersen *et al.*, 1991, 1993, Love *et al.*, 2000, Mahara *et al.*, 2007, 2009, Hasegawa *et al.*, 2010, 2016).

About 4700 flowing artesian waterbores were drilled into the main Lower Cretaceous-Jurassic (Cadna-owie Formation and Hooray Sandstone) aquifers, underlying the Cretaceous Rolling Downs Group (Figures 21.2 and 21.3) at depths of up to 2000 m, but average 500 m. Potentiometric surfaces (artesian pressure levels) of the Triassic, Jurassic and Early Cretaceous aquifers are still above ground level in most of the basin area, but pressure drawdowns of up to 100 m were recorded from the 1880s to the 1990s and 2000s in some relatively closely developed areas (Habermehl, 1980, 2001, Figure 21.4). Artesian groundwater extraction by the pastoral industry and for homestead and town water supplies, which peaked at about 2000 ML/day around 1918, caused this drawdown. About 3100 controlled and uncontrolled artesian waterbores remain flowing with an accumulated discharge of 1500 ML/day (Figure 21.5). About 25 000 non-flowing artesian waterbores, generally using windmill operated pumps, tap shallower Cretaceous aquifers. Waterbore development and the resultant lowered artesian pressures in many parts of the basin during the last 130 years caused some flowing artesian waterbores to cease flowing, necessitating groundwater to be pumped. As a result of the drawdown, most spring discharges declined, and in some areas springs have ceased to flow.

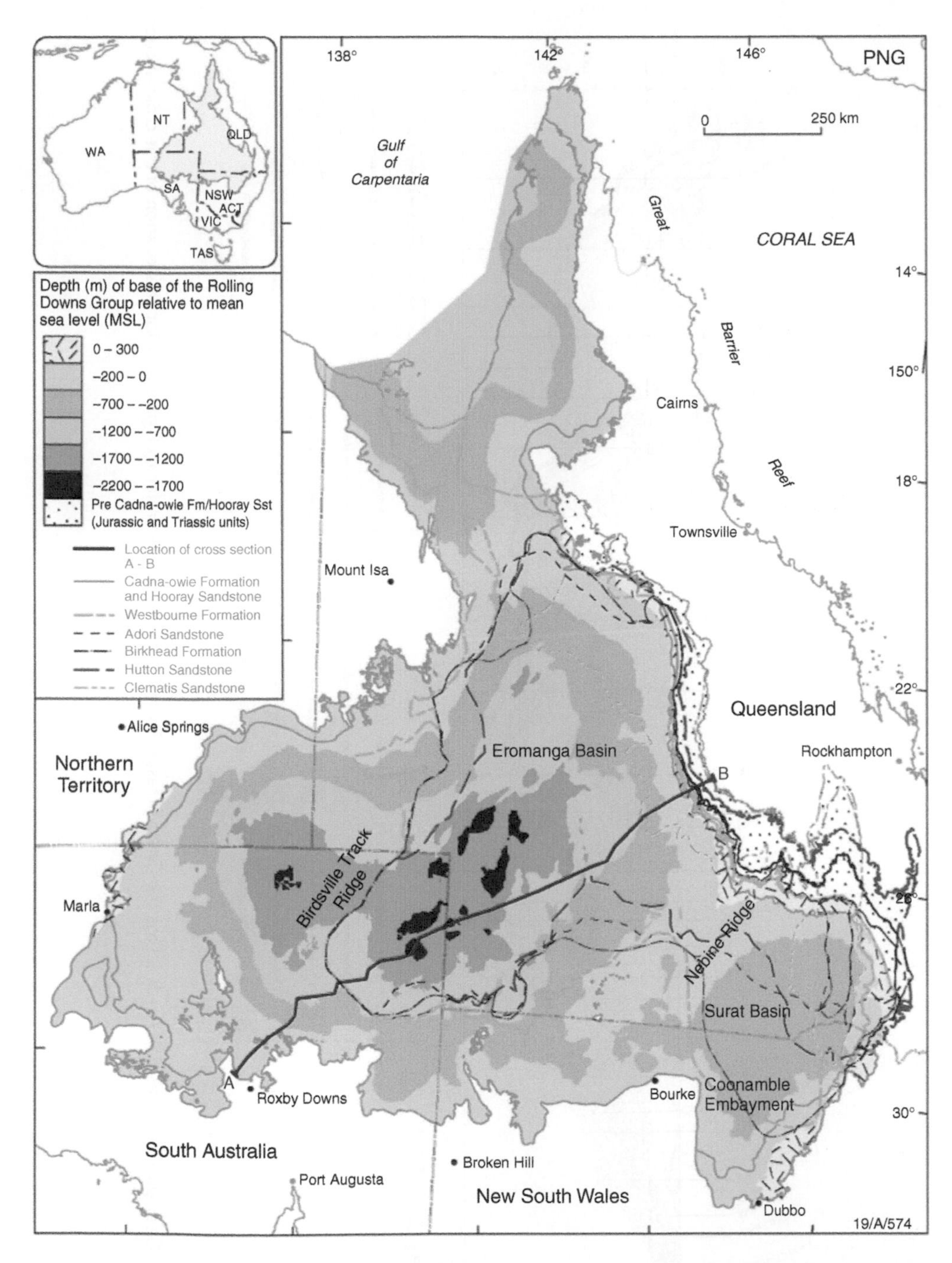

Figure 21.1 Extent of aquifers and depth to the Cadna-owie Formation – Hooray Sandstone aquifer, the main aquifer tapped by bores, and location of the cross-section A-B (Habermehl, 2001).

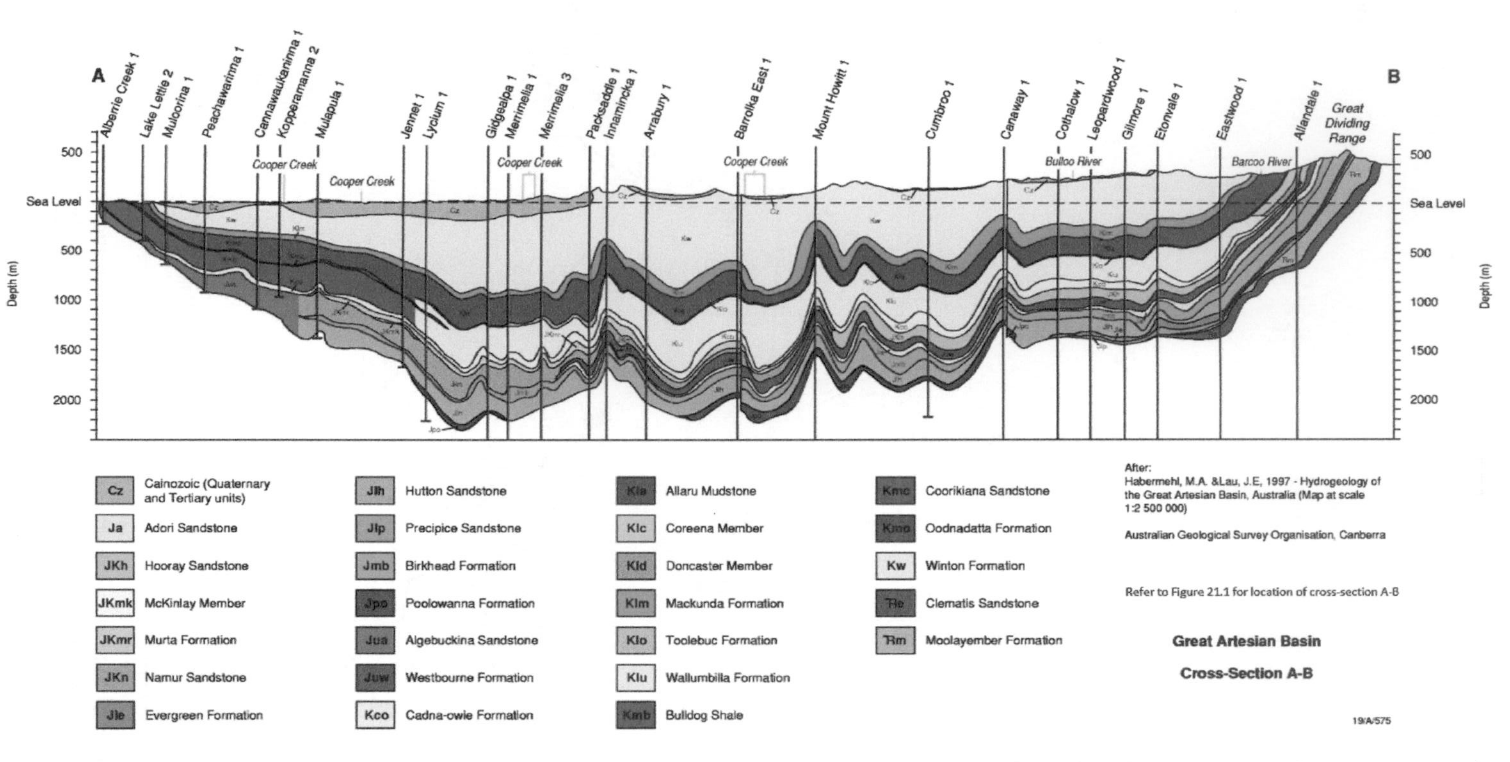

Figure 21.2 Geological cross-section A – B across the Great Artesian Basin (Habermehl, 2001; refer to Figure 21.1 for the location of this cross-section).

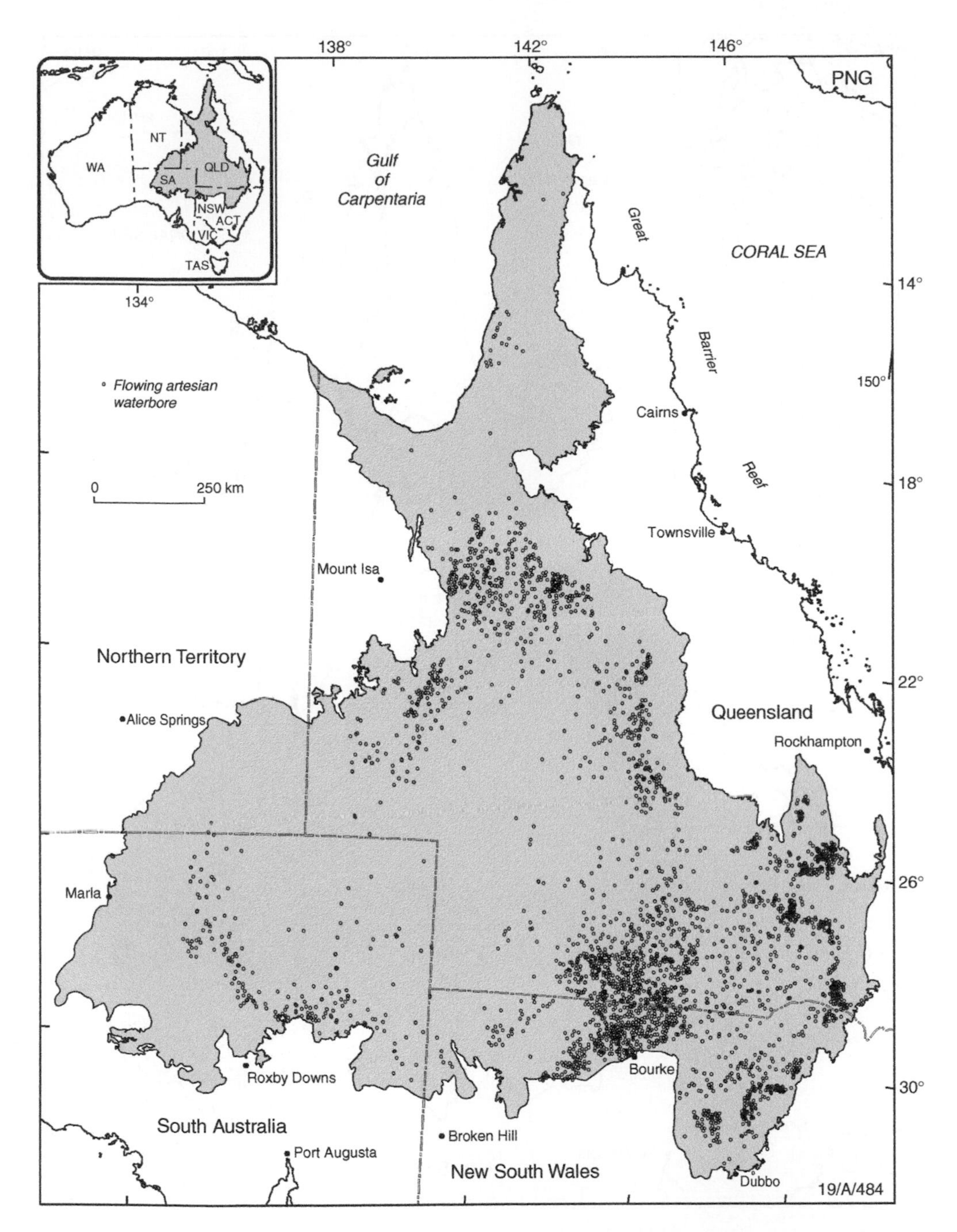

Figure 21.3 Location of flowing artesian bores (Habermehl, 1980, 2001; Habermehl & Lau, 1997).

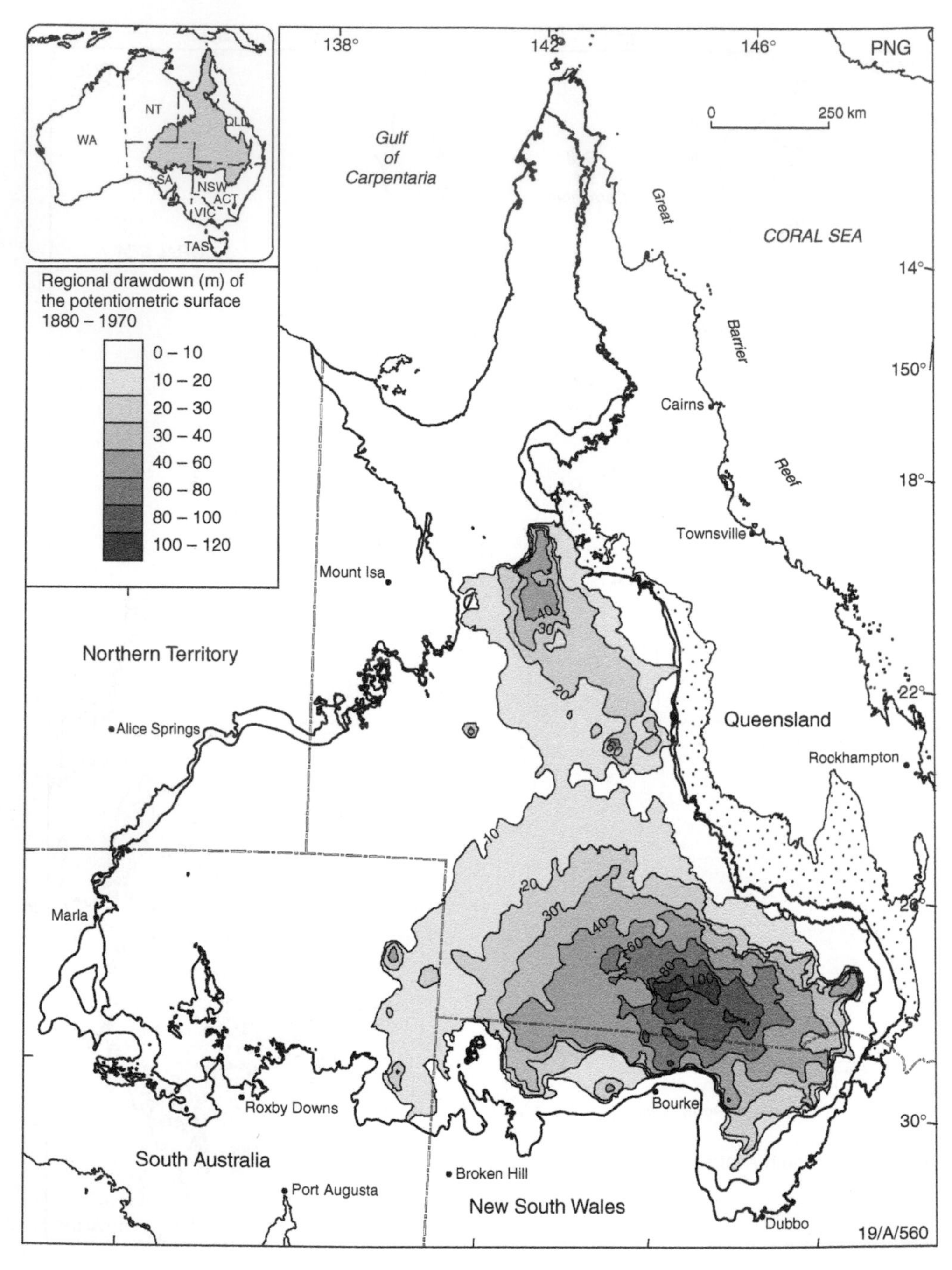

Figure 21.4 Regional drawdown of the potentiometric surface of the main aquifer, 1880–2000 (Habermehl, 1980, 2001; Habermehl & Lau, 1997).

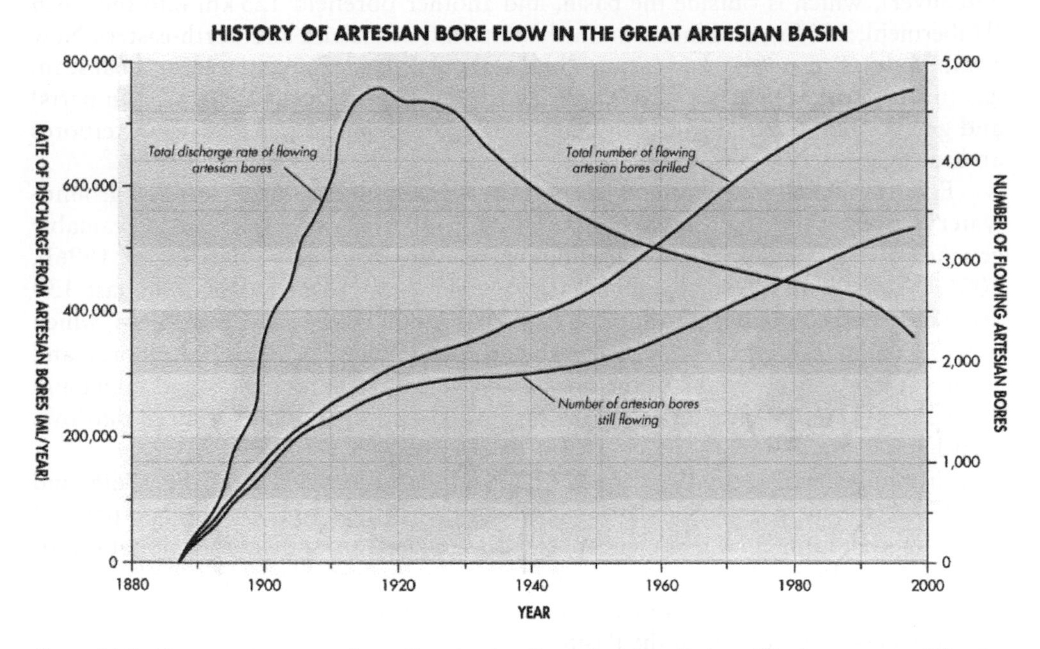

Figure 21.5 History of artesian bore flow in the Great Artesian Basin – Discharge rate of flowing artesian bores, total number of initial flowing artesian bores drilled and total number of bores still flowing (Habermehl, 1980; Reyenga *et al.*, 1998; GABCC, 1998).

Groundwater quality of the Lower Cretaceous-Jurassic aquifers is good at 500 to 1500 mg/L total dissolved solids (Habermehl, 1983, 1996a, Herczeg *et al.*, 1987, 1991; Radke *et al.*, 2000). Groundwater in most of the basin area is of the Na-HCO_3 type and is suitable for domestic, town water supply and stock use, though unsuitable for irrigation in most areas, as the hydrochemistry of the groundwater and the montmorillonite clays at the ground surface in many parts of the basin are incompatible. The southwestern part of the basin contains Na-Cl-SO_4 type water (Habermehl, 1986). Water quality is better in the lower aquifers than in the higher aquifers in the Lower Cretaceous-Jurassic sequence, which contains the main artesian aquifers. Groundwater temperatures at the bore heads range from 30° to 100°C, and are a potential geothermal energy source (Habermehl & Pestov, 2002). Spring temperatures range from 20° to 45°C.

Prior to the 1960s, artesian groundwater use was mainly for pastoral, homestead and town water supplies, but since then the development of petroleum resources in the basin area also used its groundwater. The basin (and underlying rocks, including those of the Late Carboniferous to Triassic rocks of the underlying Cooper Basin) comprises abundant hydrocarbon reservoir and source rocks, and commercial and sub-commercial oil and gas is produced from the Cooper Basin and from the Jurassic and Cretaceous sandstones, contradicting earlier beliefs that the basin-wide groundwater throughflow had flushed hydrocarbons out of the Jurassic and Cretaceous rocks. Since the 1980s, mineral mining in and outside the basin started using the basin's artesian groundwater, in particular from the Olympic Dam borefields in South Australia with borefields at the basin margin, 125 km from the mine site (copper, uranium, gold and silver), which is outside the basin, and another borefield 125 km into the basin (Habermehl, 2001a), and mines in north-western Queensland and north-eastern New South Wales. Present day and future additional oil and gas, mineral mining, coal seam gas (mainly from the lower parts of the Jurassic sequence in the eastern basin parts) and geothermal developments could affect artesian groundwater flow to waterbores and springs.

Environmental geology issues relate to the development of the artesian groundwater resources of the Great Artesian Basin and include aspects of sustainable groundwater use and groundwater and rangelands management (Habermehl, 1996b, 2006; Noble *et al.*, 1998). Extraction of artesian groundwater during the last 130 years has affected the Basin to varying degrees through large scale drawdowns, which reduced artesian pressures and reduced discharges from artesian waterbores and springs (Figure 21.4). Many of the artesian springs have unique flora and fauna and are easily affected by drawdown of the potentiometric surface and reduced outflows (Fensham *et al.*, 2010).

Groundwater use by the petroleum and mining industries since the 1960s and 1980s affect some parts of the Basin. The South Australian and Queensland parts of the Basin are examples where there are combinations of groundwater exploitation for the pastoral industry, town and homestead water supplies and petroleum and mining industries impacts on the Basin's groundwater conditions and on the artesian springs, the natural outflow points of the Basin.

Springs and areas of seepage are abundant in the marginal areas of the Basin, particularly in the southern, southwestern and northern areas, as well as in the recharge areas. The largest concentration of springs and their sedimentary deposits, mainly carbonates, forming conical mounds, occur in the southwestern margins (Habermehl, 1980, 2001a; Keppel *et al.*, 2011). Spring complexes have developed over several climatic cycles and age dating of spring deposits show age ranges up to 740 000 ± 120 000 years, with some spring deposits probably being older (Prescott & Habermehl, 2008).

21.2 DISCOVERY AND EARLY DEVELOPMENT 1880S–1900

Early work around 1880 suggested that meteoric water infiltrated the western slopes of the Great Dividing Range on the eastern margin of the Great Artesian Basin, and that artesian springs near Lake Eyre in the southwestern part indicated artesian water could be found across most of central and inland Australia. Geological investigations led to conclude in 1881 and 1885 that a large synclinal trough existed in western Queensland, in which pervious beds occurred with artesian water.

Artesian water in the Great Artesian Basin was first discovered in a shallow flowing waterbore southwest of Bourke (New South Wales) in 1878. Many shallow and deep bores were soon drilled, first near artesian springs at the margins of the basin, but later further basinwards. Droughts and poor water supplies led to the drilling of the first deep artesian bore near a town in central Queensland in 1887, and was followed by many others for stock, domestic and town water supplies. Drilling and geological investigations led to the discovery of artesian water nearly everywhere in the arid interior of the basin, and the extent of the artesian conditions was defined before the end of the nineteenth century (Habermehl, 1980).

Geological mapping by State Geological Surveys of the intake beds along the Great Dividing Range in Queensland and New South Wales and along the northern, north-western and southern margins, and waterbore information helped to outline the shape and size of the basin in the late 19th and early 20th century, so that by the end of the nineteenth century the basin was accepted as a classic artesian basin. Pressure data from field surveys confirmed the theory, though controversy remained for years as to whether the pressure head was hydraulic or hydrostatic. The hydraulic theory gained ground and became generally accepted, though the controversy lasted almost 20 years during the first years of the 20th century, but it inspired increased research in the basin.

Diminishing flows and pressures in artesian bores increasingly alarmed bore owners, and ultimately State Governments became involved, whose attention had already been drawn to the wastage of groundwater from many of the privately drilled bores. Once State legislation was passed in the early 1900s to control the use of sub-surface water, bores had to be licensed, detailed information provided and bores completed according to prescribed standards. State Water Authorities commenced systematic and periodical measurements which replaced the earlier intermittent data collection; and information on bores has been collected since that time, and extensive surveys carried out (Habermehl, 1980).

21.3 EARLY LEGISLATION AND CONTROL LATE 1800S–EARLY 1900S

The individual (sovereign) Colonies changed to that of States in Australia when they federated under the name of the "Commonwealth of Australia" in 1901. The Northern Territory was transferred from South Australia to the Commonwealth in 1911. However, land and other resources, including surface water and groundwater remained under State control. In terms of the Commonwealth Constitution primary responsibility for control and conservation of water rests with the individual State Governments. All minerals in Australia are vested in the Crown (Esmaeili, 2016).

In the States, sovereign rights are held by the State Governments with respect to mineral resources within their boundaries. Each State or Territory in the Commonwealth has its own Acts and regulations, and these Acts are similar in principle, but differ in detail.

The Commonwealth has only been involved in the development of water resources in cooperation with the States since the Australian Water Resources Council was established in 1962. The Commonwealth's role has expanded since, in cooperation with the States, particularly as a funding agent based primarily upon the financial assistance provisions of Section 96 of the Australian Constitution.

In Queensland an attempt was made in 1891 to introduce government control on groundwater, but the legislation was rejected and not passed until 1910. The delay was the result of insufficient knowledge whether the resource was limited and would last or would be abundant, and the initial private endeavours of developing bores by pastoralists in the far western undeveloped regions, as against the proposed introduction of stringent provisions for water conservation by government authorities and politicians on both sides of the arguments. It was eventually recognised that the State was the "owner" of all resources, and that much water was running to waste (Powell, 1991).

In New South Wales control was attempted in Acts introduced in 1894, 1897 and 1906, but was not achieved until 1912. Once legislation became effective, control was enforced and systematic measurements made by existing and newly formed State Water Authorities in Queensland and New South Wales, yet in South Australia no real control was exercised until 1976 (GABCC, 1998). In the late 1800s to early 1900s, information about artesian bores, especially concerning the diminishing flows, had already been gathered intermittently, but was incomplete, and only after legislation had been passed did official regular inspection, and collection of data begin, which continues to the present day. The new legislation provided for bores to be licensed, and required all relevant details to be submitted to the State Authority, and the casing and headworks to be completed according to prescribed standards, and construction and purpose of the bore are specified, as well as use, volumetric allocation and distribution works. Bores were to be constructed by suitably licensed (artesian) drillers. Bore owners were responsible for the maintenance of all bores, including casing, headworks and distribution networks. Nearly all bores in the Basin are lined with steel casing.

Problems related to the diminishing and ceasing flows, corrosion of bore casings, the extent of the basin, and the origin and movement of the artesian groundwater required an explanation.

A proposal was made by the New South Wales government in 1908 to the other Australian states to "form a consultative Board consisting of representatives from the various states, to take into consideration the question of whether the Artesian Water Supply of Australia was in danger of being seriously diminished, and, if necessary to advise as to the best means of combating the contingency". No such board was established, but in 1911 the matter was revived, and in 1912 the First Interstate Conference on Artesian Water was held in Sydney, with representatives attending from all mainland States. Though other artesian basins in Australia were also discussed, the emphasis was on the Great Artesian Basin, not only during this conference, but also during the following Interstate Conferences on Artesian Water.

21.4 FIVE INTERSTATE CONFERENCES ON ARTESIAN WATER 1912–1939

Systematic investigations of the groundwater conditions in the basin increased markedly as a result of the five Interstate Conferences on Artesian Water, held in 1912, 1914, 1921, 1924 and 1928 (ICAW, 1913, 1914, 1922, 1925 and 1929). No further meetings were held until 1939 (ICWCI, 1939). The five Interstate Conferences on Artesian Water were originally called to study the serious reduction in pressure and diminution or cessation of artesian flows, the extent of the basin, the origin and movement of the artesian water, the corrosion problems of bore casings and to inquire into better utilisation of the groundwater.

The reports of the first and second conferences are mainly concerned with the collection of information and the verbal presentations of the (State Authorities) delegates and pastoralists, bore owners, drillers, engineers, chemists and administrators. The reports of the third, fourth and fifth conferences record submissions by the State Authorities representatives (geologists and engineers), who presented detailed data about the basin, the bores, and progress in local (State) investigations. The conferences provided a better scientific understanding and led towards issues of management and governance of the Basin's groundwater resources (Habermehl, 1980).

No further meetings were held until 1939, when an Interstate Conference on Water Conservation and Irrigation (ICWCI, 1939) met in Sydney, which was concerned with most aspects of ground- and surface water, and not confined to artesian water in Australia. Discussion concentrated on the diminution and control of flow from bores and the improvement in distribution. It was recognised that the wastage of water from flowing artesian bores was the main problem in the Great Artesian Basin.

21.5 GEOLOGY AND GROUNDWATER HYDROLOGY INVESTIGATIONS AND REPORTS BY THE STATES 1920S–1950S

Systematic surveys of the geology and of investigations of hydraulic characteristics continued to be carried out at the State basis. State Water Authorities and State Geological Surveys participated, though the State Water Authorities were primarily dealing with Control and Administration of water resources and aspects of supply and conservation. Reports and publications contain geology, hydrogeology, groundwater hydrology and chemical characteristics, together with geological maps, sections, potentiometric maps, graphs and diagrams, calculations and interpretations on bore discharges, aquifer permeability, recharge and flowrates. Newer theories, which became available during this period, including elastic storage, were applied. Recommendations were made on water conservation by the partially closing of bores and improved distribution methods.

These activities were carried out in New South Wales and South Australia, and in Queensland (which contains the largest part of the Basin), where a major investigation by geologists and hydraulic engineers started in 1939 on the nature and structure of the Great Artesian Basin.

A First Interim Report was presented to the Queensland Government in 1945 (Queensland Government, 1945). Another major report by this Artesian Water Investigation Committee was produced in 1954 and presented to the Queensland Government, and formed the most comprehensive description of detailed investigations on the hydrogeology and bore data of the Queensland part (the largest part) of the Great Artesian Basin. An assessment of recharge and discharge by bores showed that the artesian diminution was not considered a threat, only a disability, as recharge occurred and groundwater would always be available at pumping depths where bores ceased flowing. The report also contains a summary of legislative history relating to artesian water in Queensland (Queensland Government, 1955; Ogilvie, 1955).

Recommendations concerned the proper use, regulation and control of water from bores, and the suggestion not to enter in the stringent conservation of flows from existing bores. Predictions were made about the number of bores which would cease flowing during the 50 to 60 years following the report date in addition to bores which already had stopped flowing, and the number of bores which would continue to flow forever with reduced yields. Restrictions on bore discharges, and the use of efficient distribution systems had already led to conservation of water, which would be improved even further by flow restrictions and bore re-conditioning, and by the use of piping.

The report states that the quality of the artesian groundwater (with its predominance of $NaHCO_3$) is unsuitable for prolonged irrigation on most of the (montmorillonite) soils. It was considered that domestic and stock requirements should have priority, and that no artesian groundwater should be used for irrigation unless surplus water was available. The report contents formed a basis upon which a State policy for the use and conservation of groundwater was defined, and led to recommendations stating the principles and methods by which the maximum benefits could be obtained from artesian groundwater supplies. However, a minority report criticised some aspects of the majority report, and expressed doubts about the inadequate premises and conclusions reached. It also recommended studying the Great Artesian Basin as an entity, in combination with the other States involved.

21.6 AUSTRALIAN WATER RESOURCES COUNCIL (COMMONWEALTH AND STATES) AND SUBSEQUENT AUTHORITIES, KEY LEGISLATION AND ADMINISTRATION, EDUCATION AND COOPERATION, 1960S ONWARDS

The Australian Water Resources Council (AWRC) was established by joint action of the Commonwealth and State Governments in 1962. The Council comprised Commonwealth and State Ministers primarily responsible for water resources, with the Commonwealth Minister as Chairman, and was serviced by a Standing Committee consisting mainly of heads of Departments responsible to these Ministers, and by a number of technical committees, including the Technical Committee on Underground Water (TCUW).

The Commonwealth interests in water and the Council and its successors were successively administered by a succession of Commonwealth Departments during the period from the 1960s to the present (2017), with the present one being the Department of Agriculture and Water Resources.

The primary objective of the Council was the provision of a comprehensive assessment on a continuing basis of Australia's water resources and the extension of measurement and research to provide a sound basis for the planning of future development. In terms of its objectives and functions, the Council has dealt with a wide range of topics, making recommendations and stimulating action by appropriate bodies. An accelerated water resource measure programme involved many more new and improved surface water gauging stations and groundwater investigations by Commonwealth and State Government authorities and began in 1964–1965 and was extended many times over the years and funding increased substantially. In addition to its own commitments in the Territories, the Commonwealth assisted the States with their programmes of water research investigations. A Water Research Fund administered by the Council approved new research programmes and aimed to improve the efficiency of water management in Australia, and complement research already undertaken by Commonwealth agencies, universities and other organisations.

The Council initiated amongst many programs a national water resources assessment program, a research program for a national approach to water resources management and preparation of guidelines for the quality of drinking water in Australia in conjunction with the National Health and Medical Research Council.

The Council published in 1965 a Review of Australia's Water Resources (Stream Flow and Underground Resources) 1963, and many other publications followed including Review of Australia's Water Resources 1975, (and others at 10-yearly intervals), Monthly Rainfall and Evaporation, Newsletter, Hydrological Series, Water Resources Research Inventory, Stream Gauging Information Catalogue and miscellaneous publications.

The Council also supported Australian participation in the scientific programmes of the International Hydrological Decade and studies on representative basins in Australia and an Australian Groundwater School, which was held every few years and later annually and attended mainly by staff from State Water Authorities, Commonwealth and State Geological Surveys, research organisations and universities, consultants, etc. Further, a range of national and international conferences, seminars and workshops on groundwater aspects were held during the 1970s to 1990s, which were supported and the papers published by the Council and various specialist committees, including the Technical Committee on Underground Water, which later became the Groundwater Committee.

The Australian Water Resources Council was followed and replaced by the Agriculture and Resource Management Council of Australia and New Zealand (ARMCANZ). Many of the programs and functions of the Council were subsequently taken up by the respective Australian Government Departments and the National Water Commission (2004–2015), which played an important role in the monitoring and auditing of water reform policy implementation and management nationally since the National Water Initiative (NWI) was agreed to in 2004.

The Council of Australian Governments (COAG, the peak intergovernmental forum of Commonwealth and State Governments) initiates, develops and monitors policy reforms that are of national significance and which require cooperative action by Australian governments. COAG's principal water policy agreement is the National Water Initiative (NWI, 2004) – Australia's enduring blueprint for water reform. Through it, governments across Australia agreed on actions to achieve a more cohesive national approach to the way Australia manages, measures, plans for, prices, and trades water.

The NWI was signed at the COAG meeting on 25 June 2004, leading to the establishment of the National Water Commission (NWC). The NWI continued a water reform process that had started in 1994 with the development by COAG of a water reform framework. This intergovernmental agreement set out to achieve a nationally compatible market, regulatory and planning based system – one that manages surface and groundwater resources for rural and urban use, and optimises economic, social and environmental outcomes. It represents a shared commitment by governments to increase the efficiency of Australia's water use, leading to greater certainty for investment and productivity, for rural and urban communities, and for the environment. The National Water Commission was abolished in 2015.

Responsibility for surface and groundwater policy at the Commonwealth Government level moved during the late 1990s/early 2000s from the Department of Primary Industries and Energy to the Department of Environment, Water, Heritage and Arts, Department of the Environment and more recently to the Department of Agriculture and Water Resources. State Water Authorities similarly changed names repeatedly and were moved from one to another State Government Department during their existence.

The Great Artesian Basin is partially overlain by surface water basins, including for a large part by the river systems and lakes of the Murray-Darling Basin in Queensland and New South Wales and the river systems and lakes of the Lake Eyre Basin in Queensland, New South Wales, South Australia and the Northern Territory. The Commonwealth Water Act 2007 provides the legislative framework for ensuring that Australia's largest (surface) water resource – the Murray-Darling Basin – is managed in the national interest. In doing so the Water Act recognises that Australian states (Queensland, New South Wales, Victoria, and the Australian Capital Territory) in the Murray-Darling Basin continue to manage Murray-Darling Basin (surface) water resources within their jurisdictions.

21.7 HYDROGEOLOGICAL STUDY OF THE GREAT ARTESIAN BASIN BY THE FEDERAL GEOLOGICAL SURVEY, 1971–EARLY 2000S BY BMR, AGSO, BRS AND GA

The AWRC Standing Committee resolved in 1964 that a TCUW Sub-Committee of officers from the relevant State and Commonwealth authorities should prepare a suitable hydrogeological programme for the study of the Great Artesian Basin and a statement of how the programme was the be carried out, bearing in mind that it would be desirable for the collation of results to be undertaken by one authority, the Commonwealth Bureau of Mineral Resources, Geology and Geophysics (BMR – the federal geological survey). The AWRC-TCUW (which included a representative from BMR for the Commonwealth) meeting in 1965 nominated a Sub-Committee of BMR and State representatives to prepare a suitable hydrogeological programme for the Great Artesian Basin study and a statement of how the programme should be carried out. Seven reasons were given to advance the study, as well as five programme items, including – geological mapping, stratigraphic study of the basin, study of water quality, study of aquifer characteristics and review and assessment of behaviour of the basin.

The State and BMR representatives prepared a more detailed proposal in 1966, but deferred the study until geological mapping of the basin was further advanced. State Geological Surveys were involved in the geological mapping of their States, and BMR carried out the geological mapping (at scale 1:250 000) of the GAB part in the Northern Territory and jointly with the Geological Survey of Queensland the geological mapping of Queensland. The Queensland part of the Great Artesian Basin represents the largest part of the Basin.

In 1969 it was agreed that BMR would be the authority to collate and interpret available data on the Basin. The BMR proposed in 1969 a study programme based on the AWRC-TCUW Sub-Committee programme recommended earlier in 1966, but with more emphasis on the use of computers in the storage and retrieval of data and modelling of the flow system. Progress since 1966 also made it possible to put less emphasis on stratigraphic synthesis, as had been proposed for the hydrogeological study at that time.

The Bureau of Mineral Resources (BMR) commenced the Great Artesian Basin hydrogeological study in 1971, with the main objectives being the review of the geological and hydrogeological data of the multi-layered confined aquifer system and to develop and apply a digital computer based model to simulate the groundwater hydrodynamics.

Basic data for the hydrogeological and model study included the results of geological surface mapping by BMR and State Geological Surveys obtained during the 1950s to early 1970s. Geological maps at scale 1:250 000 and their accompanying Explanatory Notes and also geological maps at scale 1:1 000 000 covered most of the Great Artesian Basin. BMR and petroleum company geophysical data covered large parts of the Basin. Subsurface information was obtained from hundreds of petroleum exploration and production wells. Stratigraphic drill-holes, both shallow and deep, drilled by BMR and the State Geological Surveys provided additional data, as did driller's logs of waterbores, which are available for most flowing artesian waterbores and for some non-flowing artesian (sub-artesian) waterbores.

Identification and correlation of hydrogeological units was further possible using geophysical (wire-line) logs of waterbores in the Basin. BMR wire-line logged about 1250 waterbores throughout the Basin in the period 1960 to 1975 (Habermehl, 2001b). State Geological Surveys also wire-logged a limited number of waterbores in their States. As almost all existing waterbores are lined with metal (steel) casings, the types of wire-line logs which can be run in the bores are restricted, mainly to natural gamma-ray, neutron, flow and temperature logs, as well as casing collar locator logs.

Hydrological data and detailed bore data, including driller's logs, casing data and records of depths of groundwater intersected, as well as the elevations to which the water rose and the size of the flows, were obtained from the files and records of the Water Authorities in New South Wales and Queensland and the Geological Surveys in South Australia and the Northern Territory. Records of the flowing artesian waterbores contain the bulk of the available hydrogeological data, as these bores represent the most significant artificial discharge points in the Basin and regular periodic (on intervals of a few years) measurements on discharge, pressure, temperature and chemistry have been carried out since the early development of the Basin. Non-flowing artesian waterbores (generally called sub-artesian), which are pumped, produce much smaller discharges and information on these bores is generally restricted to basic data obtained on completion of the bores. Most or almost all data in the State organisations was only available in paper files, as the State organisations had no computers and automatic data base systems in the early 1970s. After transcription, the data was entered into an automatic data-processing, storage and retrieval system designed by the BMR Great Artesian Basin Project.

The results of the compilation and interpretation of the geology and hydrogeology of the Great Artesian Basin study during the 1970s is summarised in Habermehl (1980), Habermehl & Seidel (1979) and the modelling aspects by Seidel (1980).

From the early 1970s onwards, BMR (and subsequently its successors the Australian Geological Survey Organisation – AGSO and later Geoscience Australia – GA) carried out isotope hydrology studies of the Great Artesian Basin, initially jointly with the Australian Atomic Energy Commission (subsequently the Australian Nuclear Science and Technology Organisation – ANSTO) and later with the Australian National University (ANU), and with overseas collaborators. Stable and radio-active isotopes and gases from the artesian groundwater were sampled throughout the Basin and used to date the groundwater and interpret groundwater movement and recharge to aquifers (Airey *et al.*, 1979, 1983; Calf & Habermehl, 1984; Bentley *et al.*, 1986; Torgersen *et al.*, 1991, 1993; Love *et al.*, 2000; Mahara *et al.*, 2009; Hasegawa *et al.*, 2010, 2016). Hydrochemistry sampling by BMR/AGSO/GA (and Bureau of Rural Sciences – BRS from 1998 to 2009, when the Great Artesian Basin Project and its staff were located in the Commonwealth Bureau of Rural Sciences) of waterbores across the Basin since the early 1970s and interpretation of the analyses results provided an understanding of the hydrochemistry of the artesian groundwater and chemical processes within the aquifer system (Habermehl, 1983, 1986, 1996a; Herczeg *et al.*, 1987, 1991; Radke *et al.*, 2000).

The Basin-wide geological, hydrogeological, hydrochemistry and isotope hydrology studies, together with the modelling studies since the 1970s and subsequent digital models and applications during the 1970s, 1980s, 1990s and 2000s (Seidel, 1980; Welsh, 2000, 2006, 2007) provided a detailed understanding of the Great Artesian Basin. Throughout the period from the 1970s to the present (2017) the results were communicated with Commonwealth and State organisations and Commonwealth and State Government Departments and published in BMR, AGSO and GA and (from 1998 until 2009 – BRS) publications and Australian and international conferences and journals. Hydrogeological aspects, related to environmental impact assessments for larger scale projects such as mining projects within and near the Basin were also dealt with for the Commonwealth Department of Environment.

In particular the modelling studies and results were of interest to the Commonwealth, State and Territory Water Resources Departments when a greater emphasis on management and regulation of the Basin's artesian groundwater resources appeared since the 1990s. The Great Artesian Basin groundwater modelling studies carried out by BMR, AGSO, BRS and GA and their results were used in a number of management applications and reviews by Commonwealth and State organisations and in particular in relation to the Great Artesian Basin Sustainability Initiative (GABSI) and other management activities (Seidel, 1980, Welsh, 2000, 2006, 2007).

21.8 INTERSTATE COMMITTEE ON THE GREAT ARTESIAN BASIN – DIAGNOSTICS AND RELATED GOVERNANCE PROVISIONS – 1980S

The Australian Water Resources Council – Groundwater Committee (AWRC-GC), the successor of the Technical Committee on Underground Water (AWRC-TCUW), also consisted of Commonwealth and State Water Authorities. It initiated several hydrogeological and groundwater studies in the Great Artesian Basin. In 1985, the Great Artesian Basin – Sub-Committee was established to review matters concerning the Basin. In its final report, the Sub-Committee, identified considerable wastage of artesian water by the main users, the pastoral industry, mainly because of the distribution of artesian water from flowing bores by open earth channels or drains (which can be up to 100 km in length), owing to seepage, transpiration and evaporation, causing wastage of more than 95% of the groundwater produced, and that a large number of waterbores were unserviceable. The report, prepared by State and BMR hydrogeologists recommended better management on a basin-wide basis, and proposed the Great Artesian Basin Bore Rehabilitation Program (Woolley *et al.*, 1987).

An Interstate Meeting of Chief Executives of the State Water Authorities requested State and Commonwealth hydrogeologists in 1987 to prepare a report on "Improved Efficiency of Artesian Groundwater Use, Rehabilitation of Bores and Conversion from Drains to Piping."

The report from the Great Artesian Basin Sub-Committee (Woolley *et al.*, 1987) was directed to the AWRC-GC. The Interstate Meeting of Chief Executives of the State Water Authorities confirmed the Sub-committee conclusions and recommendations. These were endorsed by the AWRC. In response to these recommendations, the Interstate Working Group on the Great Artesian Basin was established in 1987. The group consisted of representatives from New South Wales, Queensland, South Australia and the Commonwealth (BMR). The Terms of Reference for the Interstate Working Group on the Great Artesian Basin were to: 1 – Examine and report on the future use of water from the Basin; 2 – Study the performance of the Basin to enable it to be better understood; 3 – Study the environmental consequences of current or proposed management; 4 – Review relevant legislation; 5 – Examine the benefits of better water management.

The Interstate Working Group met annually to review monitoring results and developments in the Great Artesian Basin. It acted as an avenue for cooperation and coordination in the Great Artesian Basin, and reported annually to the Water Resources Management Committee of ARMCANZ. To clearly identify its role in relation to the newly established Great Artesian Basin Consultative Council, the Great Artesian Basin Interstate Working Group changed its name to the Great Artesian Basin Technical Working Group in 1997.

21.9 ADDITIONAL GREAT ARTESIAN BASIN GOVERNANCE PROVISIONS SINCE 1997 – GABCC (1997–2004) – GAB MANAGEMENT PLAN (2000) AND GABCC (2004–PRESENT/2017)

The Great Artesian Basin Consultative Council (GABCC) was established in 1997 by the stakeholders in the Great Artesian Basin. Council had 17 representatives from federal, state and local governments, pastoral, petroleum and mining industries, traditional landholders, bore owners, community and conservation groups including representatives of local communities (State Advisory Bodies), industry (National Farmers Federation, Minerals Council of Australia, Australian Petroleum Production and Exploration Association), traditional landholders (Aboriginal and Torres Strait Islander Commission), conservation (Australian Conservation Foundation), local, State and Commonwealth governments (Australian Local Government Association, State Water Agencies, Commonwealth Department of Primary Industries and Energy, and its successors, and Environment Australia). The GAB Technical Working Group consisting of hydrogeologists and groundwater hydrologists from the State Government Water Authorities and the Australian Government – Australian Geological Survey Organisation subsequently Geoscience Australia, was also represented, and provides technical support to the Council.

GABCC prepared a Great Artesian Basin Resource Study (GABCC, 1998, Cox & Barron, 1998), which dealt with most aspects of the Basin, including physical aspects, hydrogeology, traditional use, water extraction, values associated with the groundwater resource, sustainability and environmental issues, policy framework for basin management, current management and research.

The initial role of the Council was to coordinate Great Artesian Basin management policy between States and provide advice to State and Commonwealth Governments on a strategic framework for sustainable use of the Basin for all users and on water resource management issues in relation to the Basin as a whole. The Council developed a Great Artesian Basin Strategic Management Plan in 2000 (GABCC, 2000), to address basin-wide management issues, sustainable development and use of the artesian groundwater. State Advisory Committees were established and were an important link between water users, State Governments and the Council, with a key role in advising on implementation and helping facilitate cooperative arrangements between all partners and stakeholders.

State and Territory Governments have a primary responsibility for water resource management, policy legislation and operational delivery of programs. The Commonwealth Government has a range of national and Basin-wide interests in implementation. It has policy and program functions to provide leadership and strategic investment sufficient to trigger investment by others in partnership, in particular, to provide leadership in developing Basin-wide natural resource management policy; make strategic investments and provide leadership in negotiation of shared investments; provide technical support; ensure national and international obligations are met; and maintain momentum and commitment.

The Great Artesian Basin Technical Working Group consisting of State and Commonwealth Government hydrogeologists was established with a primary role of providing technical support to Council.

The Great Artesian Basin Coordinating Committee replaced the Great Artesian Basin Consultative Council in 2004, with as its primary role to provide advice to relevant Australian and State and Territory Ministers on the sustainable whole-of-basin resource management, and to coordinate activity between stakeholders. Specific responsibilities of the Committee include: reviewing the progress of programmes; reporting on the implementation of the Strategic Management Plan; promotion/publicity and communication about whole-of-basin values against an approved work plan; objective analysis of policy issues and provision of advice to Ministers; coordination of policy/management across sectors where appropriate; coordination of technical activity (*e.g.* standards) and research, and examination and preparation of recommendations to Ministers on cross-border issues. The Great Artesian Basin Coordinating Committee has four Sub-Committees: Strategic Focus; Communications; Research and Development, and Finance.

The Australian government facilitates the cooperative management of water resources. Federal environment laws apply when actions may affect nationally listed threatened species or ecological communities such as the community dependent on Basin's natural discharge springs (EPBC, 1999). There are 19 Natural Resource Management boards across the Basin. They address some of the broader natural resource issues such as weeds and pest animals, arising from the use and management of Basin waters. Water users, land managers and communities contribute to Basin management by participating in the Great Artesian Basin Sustainability Initiative, adopting new technologies, improving land management and industry practices and participating in water planning.

Strategic Management Plan Reviews by outside independent consultants took place in 2006 and 2015, with an analysis of the 2000 Strategic Management Plan and changes since 2000 and defining a sharper focus for implementing the Strategic Management Plan at State, Territory and regional level, including national policy principles on groundwater management, sustainability and biodiversity, and complement State and Territory water resource legislation. A new Strategic Management Plan has been developed (2017).

21.10 NATIONAL POLICIES AND GROUNDWATER MANAGEMENT – RECENT GAB PROJECTS

21.10.1 Commonwealth and State activities under National Water Reforms

During the 1980s and 1990s the Commonwealth Government increased its involvement by introducing policy matters dealing with surface water and groundwater, and cooperation with the State Governments. Several Federal programs provided funding for additional investigations, research and works by Commonwealth and State organisations, including through the Federal Water Resources Assistance Programme (FWRAP), the Natural Heritage Trust (NHT) and the National Landcare Programme (NLP), the latter a key part of the Commonwealth commitment to natural resource management. Funding supports local and environmental agriculture projects and stewardship of the environment through practical action in urban, rural and regional communities. The National Landcare Programme still continues through an investment by the Australian Government of $A 1 billion through the Programme over four years from 2014–2015.

Management of groundwater in the Great Artesian Basin occurs at a variety of scales, on the farm, through local government, regional boards and water supply authorities, by State and Territory governments, and through the Commonwealth Government (Habermehl, 1996b). Groundwater policies, institutional arrangements and practices by State and Territory governments are guided by a suite of natural resource management policies and strategies at both State and Territory and National levels.

The national policy context for natural resource management in the Great Artesian Basin is derived from three key Policy Statements: – National Strategy for Ecologically Sustainable Development; – Decade of Landcare Plan; – National Strategy for the Conservation of Australia's Biological Diversity (GABCC, 1998). These are supported by more detailed statements on specific issues on natural resource management, particularly rangelands management, wetland conservation and management, and indigenous requirement, and the Council of Australian Government's National Water Reform Framework.

21.10.2 State legislation, initial and recent

State legislation regulating the access to groundwater and spring water resources of the Great Artesian Basin originate from the 1890s, with the New South Wales Artesian Well Act 1897 initiating sharing of artesian supplies through Bore Trusts, which promoted both development and sharing of artesian supplies with government assistance to allow groups of settlers to construct bores that serviced their collective properties with artesian water distributed via open bore drains. The Water Act 1912 required all bores greater than 30 metres in depth to be licensed and to be completed to a prescribed standard. Later changes to the legislation and new Acts added all bores to be licensed, be constructed by a suitably licensed 'artesian driller' to a minimum bore specification, and all artesian bores constructed to be piped. Bore owners are primarily responsible for the maintenance of all artesian bores, including casing, headworks and distribution networks. The Water Management Act 2000 provides better ways for the equitable sharing and management of the State's water resources, with the Great Artesian Basin water resources in New South Wales at high risk in relation to the decline in artesian pressure because of over extraction and free-flowing of groundwater, stress to groundwater dependent ecosystems (springs) due to high levels of extraction and threats to the grazing industries due to a reduction in access for both domestic and stock users and licensed users.

In Queensland, where the first artesian bores were drilled in the 1880s, attempts to introduce legislation in the early 1890s faltered because of discord whether the artesian pressures and flows were inexhaustible or would diminish in time and could lead to mining of the artesian groundwater resources. Once legislation was passed in the early 1900s to control the use of artesian groundwater, bores had to be licensed, after diminishing flows and pressures increasingly alarmed bore owners, and attention had already been drawn to the wastage of artesian water from many privately drilled bores. Detailed information had to be provided to State Water Authorities and bores completed according to prescribed standards. State Water Authorities commenced, as in New South Wales, systematic and periodical measurements on pressures and flows, temperatures and hydrochemistry from bores, which have been collected since that time, and extensive surveys were carried out during the following decades. Legislation was changed and updated throughout the 1900s and early 2000s. The Water Resource (Great Artesian Basin) Operations Plan 2006 contains the most up-to-date arrangements for the sustainable management for the benefit of future generations and provides enhanced certainty and security for water users and the natural environment.

In South Australia the first artesian bores were also drilled in the 1880s, and State legislation to control the use of water came into operation in 1967. The Water Resources Act 1976 vested all water in South Australia in the Crown, including groundwater, and was designed to protect groundwater resources from over-exploitation. The Water Resources Act 1997 and the Natural Resources Management Act 2004 provided the authority for actions to manage the Great Artesian Basin. Exceptions are two Indenture Acts for the petroleum operation at Moomba, where Basin groundwater is a by-product of its operations, and the Olympic Dam mine and the adjoining town of Roxby Downs, which uses artesian groundwater from two borefields in the Great Artesian Basin and is piped more than 200 km to the mine and town. The Indenture Act permits the extraction of groundwater subject to certain conditions, including drawdown limits. The legislation is to achieve responsible use of underground water, eliminate wasteful practices, ensure ecosystem health and to clarify the rights and responsibilities of users of the Great Artesian Basin resources in South Australia. These objectives are consistent with those of the Great Artesian Basin Strategic Management Plan, the Commonwealth Environment Protection and Biodiversity Conservation Act 1999 and the 2004 Intergovernmental Agreement on a National Water Initiative.

The many Great Artesian Basin springs in this area, essentially surface discharge points of the Basin's aquifers (groundwater dependent ecosystems), support populations of unique and threatened fauna and flora (Fensham *et al.*, 2010) and are of immense cultural and ecological importance, in particular to the Aboriginal people, who have a strong connection with their land and the associated resources. The Great Artesian Basin is critical to the health of ecological communities and the viability of the pastoral, mining and tourist industries in this region. Two critical components associated with the Basin's artesian groundwater are its pressure and temperature, which provide environmental and economic benefits in their own right.

Clark (1979) deals with groundwater law and administration in Australia and reviews and comments in some detail on the early and subsequent laws enacted in New South Wales, Queensland and South Australia and the Northern Territory dealing with the Great Artesian Basin since the discovery of the Basin until the late 1970s. It includes the development of common law, determination of groundwater resources, regulation of construction, maintenance and abandonment of bores, regulation of extraction of groundwater, licensing of drillers, regulation of groundwater quality and techniques of introducing controls, rights of appeal and surviving private rights of action.

Nelson (2010, 2016) and Nelson & Casey (2013) cover more recent developments in groundwater law in Australia and Nelson & Quevauviller (2016) review and compare legal principles in the United States of America, Australia and the European Union.

21.10.3 Commonwealth legislation (Environmental Protection and Biodiversity Act 1999)

Significant for the environmental aspects of natural resource management is the *Environment Protection and Biodiversity Conservation Act 1999* (the EPBC Act). It is the Australian Government's central piece of environmental legislation. It provides a legal framework to protect and manage nationally and internationally important flora, fauna, ecological communities and heritage places – defined in the EPBC Act as matters of national environmental significance. It is important for the protection and management of the artesian springs in the Great Artesian Basin and the fauna and flora associated with the springs.

Developments within the Great Artesian Basin and developments outside the Basin, using or influencing the artesian groundwater resources are required to consider the preparation of Environmental Impact Statements, to be approved by the Australian Government (Minister of Environment). These are sometimes assessed by State Governments under accreditation agreements with the Australian Government. The EPBC Act enables the Australian Government to join with the States and Territories in providing a truly national scheme of environment and heritage protection and biodiversity conservation. The EPBC Act focuses Australian Government interests on the protection of matters of national environmental significance, with the states and territories having responsibility for matters of state and local significance.

The EPBC Act comes into play when a proposal has the potential to have a significant impact on a matter of national environmental significance, and an application is required for approval to proceed under the EPBC Act. This approval process under the EPBC Act would be in addition to any state or local government approval that might be required. Proposals must be referred to the Department of Environment and Energy. This 'referral' is then released to the public, as well as relevant State, Territory and Commonwealth ministers, for comment on whether the project is likely to have a significant impact on matters of national environmental significance. The minister will then decide whether the likely environmental impacts of the project are such that it should be assessed under the EPBC Act. Any relevant public comments are taken into consideration in making that decision. Five different levels of assessment are involved, depending on the significance of the project and how much information is already available, with each level considering technical information assembled by the proponent and comments made by the public.

21.10.4 New matter of national environmental significance – water trigger

Amendments to the EPBC Act became law on 22 June 2013, making water resources a matter of national environmental significance, in relation to coal seam gas and large coal mining development. This very significantly increases the influence of the EPBC Act (and the federal government) on groundwater (Nelson, 2016). The nine Matters of National Environmental Significance (MNES) are: – world heritage properties; – national heritage places; – wetlands of international importance (often called 'Ramsar' wetlands after the international treaty under which such wetlands are listed); – nationally threatened species and ecological communities; – migratory species; – Commonwealth marine areas; – the Great Barrier Reef Marine Park; – nuclear actions (including uranium mining); – a water resource, in relation to coal seam gas development and large coal mining development.

The EPBC Act affects any group or individual (including companies) whose actions may have a significant impact on a matter of national environmental significance. This includes: landowners; developers; industry; farmers; councils; State and Territory agencies and Commonwealth agencies. Great Artesian Basin springs usually are of national environmental significance, as 'native species dependent on natural discharge of groundwater from the Great Artesian Basin' are a listed ecological community under the EPBC Act. These populations of unique and threatened flora and fauna are of immense cultural and ecological importance.

21.10.5 Great Artesian Basin Bore Rehabilitation Project (GABBRP) 1989–1999 and Great Artesian Basin Sustainability Initiative (GABSI) 1999–2017, – Commonwealth and State activities under National Water Reforms

The Great Artesian Basin Bore Rehabilitation Program (GABBRP) was a joint initiative of the Commonwealth and State governments, carried out by the State government authorities of Queensland, New South Wales and South Australia (Reyenga *et al.*, 1998). The Program commenced in 1989, though in South Australia rehabilitation of uncontrolled waterbores had already been underway since 1977, mainly because many of the bores in South Australia are government property. In New South Wales rehabilitation of flowing bores took place from 1952 to 1976.

The GABBRP Program has been partly funded by the Commonwealth Government through the Commonwealth Department of Primary Industries and Energy and its successors, under the Federal Water Resources Assistance Program, and subsequently by the National Landcare Program. Funding for rehabilitation was based on a 40:40:20 split of the costs by the Commonwealth Government, the respective State Government and the bore-owner or landholder. Participation by the landholder is voluntary. In South Australia, the funding was on a 50:50 basis, with the Commonwealth Government and State Government providing equal shares, with no landholder contribution, as most waterbores are government owned.

State and Territory legislation requires all waterbores to be licensed. Flow rates allowed are based on the area to be watered and number of sheep or cattle which can be carried on a property. Regular measurements of artesian waterbores for flow, pressure, temperature and chemistry at several year intervals have been performed throughout the 20th and early 21th century by the State Water Authorities. The regular inspections and measurements of the artesian waterbores and the original drilling and casing data provide an extensive data base of the bore characteristics. Most waterbores are privately owned, and generally used for sheep and beef cattle grazing in the arid and semi-arid rangeland parts of the Great Artesian Basin, with some crop growing in the south-eastern areas. Much of the land within the Basin area is lease-hold from State Governments, with free-hold land mainly located in the eastern marginal areas.

Given the magnitude of the problems of wastage in the Basin, the Great Artesian Basin Bore Rehabilitation Program (GABBRP) aimed to rehabilitate artesian waterbores to eliminate or at least reduce the large number of uncontrolled flowing artesian waterbores and reduce the wastage of water. The rehabilitation or reconditioning of artesian waterbores involved the installation of proper headworks, including control valves or the repair or replacement of the current headworks. The repair or replacement of broken or corroded borehole casings was also included in the Program. As a result of these measures, the pressures in the groundwater system were expected to increase or be partially restored, and the Basin-wide diminution in artesian pressures reduced. Artesian pressure increases have been confirmed since the implementation of the Bore Rehabilitation Program, particularly in the eastern half of the Basin in Queensland.

The costs of reconditioning waterbores vary considerably. In some cases the work is very difficult and quite expensive. Under these circumstances, consideration may be given to plugging the bore with cement and drilling a new adjoining bore. This is determined in consultation with the landholder, taking into account the benefit likely to be gained for the costs involved. However, a new bore will not be considered if the old bore cannot be plugged.

Replacement of the open earth bore drains for distribution of water from bores is being encouraged in the interests of more efficient use of water. The past and present distribution of artesian groundwater by the main users, the pastoral industry, from flowing artesian bores by open earth drains is extremely wasteful, owing to seepage, transpiration and evaporation of water. Bore drains have lengths of many tens of kilometres, with lengths of up to 100 km. This causes wastage of more than 95 percent of the groundwater produced. Introduction of polythene piping to replace the earth drain reticulation system will significantly reduce the demand on flowing artesian waterbores for groundwater and could almost eliminate wastage of water if piping is combined with float-valve-controlled tanks and trough systems. Piping of the water will also reduce the environmental effects caused by the introduction of large amounts of water and watering points in the semi-arid and arid landscape. The availability of water in these areas has resulted in land degradation, the spread of introduced weeds, shrubs and animals attracted by the water, and affected the biodiversity around waterbores and bore drains, and near artesian springs, with reduced outflows (Noble *et al.*, 1998).

Reduced demand also provides resources for alternative industries, and mining and oil and gas production have become significant users and producers of artesian groundwater. Groundwater extraction by these industries has caused results similar to effects of groundwater development for the pastoral industry, *i.e.* large drawdowns of the potentiometric surfaces, which affect other users and naturally occurring flowing artesian springs (Habermehl, 2001a). Other environmental issues include the establishment of large scale borefields for the production of coal seam gas in the eastern parts of the Great Artesian Basin.

In Queensland, New South Wales and South Australia the State Water Authorities carry out most or all of the work related to the Bore Rehabilitation Program, including evaluation and assessment of the possibility of the rehabilitation and the estimates of costs. Few activities are carried out by private industry, because not all waterbore drillers have the capacity, or are prepared, to undertake the work, which in some cases include a high risk factor, because of badly corroded headworks, casings and/or high temperature and pressure outflows from large holes, where the steel casing has largely corroded and disappeared. In addition, the costs related to the work on isolated waterbores requires high mobilisation components, and without long-term activities on a series of bores in a region, private contractors are reluctant to take on some of this work many hundreds or more than a thousand kilometres from their base.

In most circumstances, related activities undertaken by the State government authorities, at least in Queensland, includes the preparation, together with the property owner, of a Property Management Plan. It is the total overview of a property, and its characteristics, which provides a good basis for planning the physical and economic operating conditions of the property, including the provision and distribution of water.

Participation in the Great Artesian Basin Bore Rehabilitation Program in Queensland and New South Wales is based on the applications by landholders to the State government water authorities. The State government authorities use guidelines to determine priorities for waterbores to be rehabilitated. These include reduction in flow and wastage, security of supply, potential to reduce land degradation, physical condition of the waterbore and willingness to pipe water from the bore. Work only commenced when the landholder has agreed to the terms and conditions of the work.

An important criterion for acceptance and incorporation in the Program was the capability of the landholder to contribute 20 percent of the cost of the works. The depressed economic circumstances of many landholders during parts of the Program meant that many were not able to contribute the required funding, and as a result many waterbores needing rehabilitation did not get on the priority list. State government authorities had lists of waterbores which required rehabilitation, but unless the landholder contributed 20 percent of the costs, the waterbores were not rehabilitated.

In the period 1989 to 1997, 40 percent of 1192 uncontrolled artesian waterbores, or waterbores with corrosion or maintenance problems in Queensland, New South Wales and South Australia were rehabilitated, representing a water "saving" of around 46 000 ML per year. Much of the water which was previously wasted is now conserved in the aquifer. However, the continued use of bore drains means that a significant proportion of the groundwater was still wasted.

Changes in artesian pressures or water levels as a result of rehabilitation of waterbores have been significant. Artesian pressures have risen in several areas where large numbers of waterbore were rehabilitated, to the extent that previous flowing artesian waterbores which had ceased flowing, have begun flowing again. Major improvements in pressure heads of several metres have been recorded in several large regions as result of rehabilitation (GABCC, 1998; Macaulay *et al.*, 2009; Smerdon *et al.*, 2012). The replacement of bore drains by piping following rehabilitation led to the largest changes in artesian pressures, as less groundwater is extracted from the bores.

A limited number of bore drain replacements to piping was carried out in an area in Queensland, with a 80 percent subsidy (40 percent each from Commonwealth and State governments) and in New South Wales with a 20 percent State – Commonwealth subsidy for piping.

Up until the 1950s, artesian groundwater that came to the ground surface under natural artesian pressure by the sinking of waterbores had been allowed to flow uncontrolled into open earth drains and creeks for distribution to stock. However, even in well maintained drains, up to 95 percent of this water was being wasted through evaporation and seepage. Such uncontrolled flow of artesian groundwater from bores and open earth drains in the Great Artesian Basin area threatened the continued access to artesian groundwater by pastoralists and the health of important groundwater dependent ecosystems. In addition, it had become difficult for new water users in the Great Artesian Basin area to obtain access to groundwater resources.

The Great Artesian Basin Sustainability Initiative (GABSI) Program followed the Great Artesian Basin Bore Rehabilitation Program (GABBRP) in 1999 to accelerate bore rehabilitation and bore drain replacement by piping.

Under the Great Artesian Basin Sustainability Initiative (GABSI) the Australian Government invested in the order of $A 115 million over 15 years (1999–2014) to accelerate work on the repair of uncontrolled artesian bores and the replacement of open earth bore drains with piped water reticulation systems. The Australian Government provided $A 112 million to the New South Wales, Queensland and South Australia for works under GABSI from 1999 to 2014. Under GABSI State and Territory expenditure matched Australian Government expenditure.

Under GABSI 695 bores were controlled for the period 1999–2014 and 20,736 km of open bore drains were deleted for the period 1999–2014, and 30,672 km of piping was installed for the period 1999–2014. The estimated artesian groundwater savings achieved under GABSI works carried out over the period 1999–2014 was 219 872 ML/year (megalitres per annum).

In late 2014 the Australian Government announced further funding of up to $15.9 million over three years to undertake further GABSI work. The GABSI Initiative continues to be delivered through State agencies and the Australian Government makes its contributions jointly with State governments and bore owners.

The Great Artesian Basin Sustainability Initiative (GABSI) was reviewed several times, the first time after the first three years (1999 to 2002), independent consultants were engaged to review the achievements and performance (Hassall & Associates, 2003), and the findings were that GABSI achieved its targets for pressure recovery and water savings and was regarded as a successful program. The GABSI Initiative was again reviewed in late 2007, during the second five-year phase of GABSI (GABSI Mid-Term Review of Phase 2 (SKM, 2008). The mid-term review for GABSI 3 was completed in early 2013 (GABSI Mid-Term Review of Phase 3, 2014). The review found that significant achievements had been accomplished and that future groundwater management in the Great Artesian Basin should be based on a thorough assessment of priorities, eligibility criteria, and funding arrangements. It also recommended conducting a value for money review of GABSI across the three phases, and the development of objective metrics to determine whether continued government funding could be justified beyond GABSI 3.

As recommended in the mid-term review of GABSI Phase 3, a Value for Money Review of GABSI using a total economic valuation (TEV) approach was completed in early 2014. The objectives of the Review were to objectively measure trends in the return on investment achieved by the respective governments (Commonwealth and States) over the three phases of GABSI and assess whether the completion of the GABSI 3 represents value for money (GABSI Value for Money Review, SKM, 2014).

The GABSI Initiative has assisted with the implementation of key actions of the Strategic Management Plan 2000 (SMP 2000) prepared by the Great Artesian Basin Consultative Council (GABCC, 2000). The Strategic Management Plan 2000 is a long term strategic framework for responsible groundwater and related natural resource management in the Great Artesian Basin, and the use of the Great Artesian Basin groundwater resources, including springs (GABCC, 2000; Fensham *et al.*, 2010).

The Great Artesian Basin Bore Rehabilitation Program 1989–1999 (GABBRP), (Reyenga *et al.*, 1998) and the subsequent Great Artesian Basin Sustainability Initiative (GABSI, 1999–2018) provides a substantial level of Australian Government and State Government assistance to landholders (most properties are leasehold from State Governments) to rehabilitate uncontrolled and/or corroded flowing artesian bores (Habermehl, 2009) and replace the open earth bore drains, with lengths of up to 100 km, which causes up to 95 percent wastage of the water, with piped water distribution systems using polythene pipes within the Basin. GABSI assists the implementation of the Great Artesian Basin Strategic Management Plan prepared by the Great Artesian Basin Coordinating Committee (GABCC, 2000). The Plan provides for the restoration of the environmental assets of the Great Artesian Basin with an emphasis on springs (GABCC, 2000; Fensham *et al.*, 2010).

These programs aim to rehabilitate waterbores in poor condition (Habermehl, 2009) and equip bores with control valves. The replacement of the inefficient open earth drain distribution system, with a piping system is encouraged, and in more recent years also included in the GABSI Program. These measures will benefit groundwater management and rangeland management and assist to alleviate land degradation and plant and animal pest problems.

In addition, the Environment Protection and Biodiversity Act (EPBC) 1999 provides environmental protection for the GAB springs, and is the main regulation tool for (larger) developments by the States and Commonwealth, in particular for mining and petroleum developments.

21.10.6 Great Artesian Basin Water Resources Assessment 2012–2014

A scientific assessment of the Great Artesian Basin during 2012–2014, the first since 1980, showed the advance in the understanding of the Basin (Smerdon *et al.*, 2012; Ransley *et al.*, 2015). The Great Artesian Basin Water Resource Assessment by the CSIRO (Commonwealth Scientific and Industrial Research Organisation), Geoscience Australia and Consultants was requested by the Australian Government and provides an analytical framework to assist water managers in the Basin to meet National Water Initiative commitments and communicates the best available science to the Australian Government. The assessment was a review of existing information and an update on the geology and hydrogeology, groundwater conditions and groundwater modelling, with in addition evaluation of the potential impacts of climate change and groundwater development by modelling from 2010 to 2070.

Assessment of groundwater level maps for 20-year intervals beginning in 1900 clearly illustrates the decline in groundwater levels in the early part of the last century, but in the most recent decade (circa 2000–2010) an increase (recovery) of groundwater levels is evident from bore capping under the Great Artesian Basin Initiative (GABSI) and previous government programmes.

21.11 CONCLUSIONS

State and Commonwealth legislation and governance guided the development of the artesian groundwater resources in the Great Artesian Basin since the early 1900s. Extensive development and wastage of groundwater from free flowing artesian waterbores caused significant drawdowns and reductions in artesian flows from the late 1800s to the early 2000s.

Following the implementation of GABBRP and GABSI and rehabilitation of bores and reduction of their outflows, potentiometric surfaces have become stable and increased in some areas and may increase flows from springs. Rehabilitated bores and headworks, abandonment of open earth channel distribution of the groundwater and the installation of piping and float-valve controlled tanks and troughs will be of benefit to the landholders and the managers of the artesian groundwater resources. Securing groundwater flows to springs, particularly those with high conservation values, could be an addition to the criteria used to select bores for rehabilitation and piping to maximise opportunities to sustain or re-activate springs.

REFERENCES

Airey, P.L, Bentley, H., Calf, G.E., Davis, S.N., Elmore, D., Gove, H., Habermehl, M.A., Phillips, F., Smith, J., & Torgersen, T. (1983) – *Isotope hydrology of the Great Artesian Basin, Australia, Papers of the International Conference on Groundwater and Man.* Australian Government Publishing Service, Canberra, 1983, Sydney, 1–11.

Airey, P.L, Calf, G.E., Campbell, B.L., Habermehl, M.A., Hartley, P.E., & Roman D. (1979) – *Aspects of the isotope hydrology of the Great Artesian Basin, Australia, International Symposium on Isotope Hydrology – International Atomic Energy Agency and United Nations*

Educational, Scientific and Cultural Organisation, Neuherberg, Fed. Rep. Germany. Isotope Hydrology 1978. International Atomic Energy Agency, Vienna, 1979, Neuherberg, Germany, 205–219.

Bentley, H.W., Phillips, F.M., Davis S.N., Habermehl, M.A., Airey, P.L., Calf, G.E., Elmore, D., Govem, H.E. & Torgersen, T. (1986) – Cl-36 Dating of Very Old Groundwater. 1. The Great Artesian Basin, Australia. *Water Resources Research 22*, 1991–2001.

Calf, G.E. & Habermehl, M.A. (1984) – Isotope hydrology and hydrochemistry of the Great Artesian Basin, Australia. Paper for International Atomic Energy Agency (IAEA) & United Nations Educational Scientific and Cultural Organization (UNESCO), *International Symposium on Isotope Hydrology in Water Resources Development, Vienna, Austria, 12–16 September 1983*. In Isotope Hydrology 1983, IAEA, Vienna, 397–413.

Clark, S.D. (1979) – Groundwater law and administration in Australia. Department of National Development (Australian Water Resources Council Research Project No. 73/50*) Australian Water Resources Council Technical Paper* No. *44,* Australian Government Publishing Service, Canberra, 379 p.

Cox, R. & Barron, A. (Editors), (1998) – *Great Artesian Basin Resource Study*, p. 33–43 & p. 44–59. Great Artesian Basin Consultative Council (GABCC), Brisbane, 235 p.

Environment Protection and Biodiversity Act (EPBC) (1999) Commonwealth of Australia.

Esmaeili, H. (2016) – Boundaries of land, fixtures and ownership of minerals and resources: The search for certainty. In: *The boundaries of Australian property law, Esmaeili, H. & Grigg B. (Editors),* Cambridge University Press, Port Melbourne, 293 p.

Fensham, R.J., Ponder, W.F. & Fairfax, R.J. (2010) – *Recovery plan for the community of native species dependent on natural discharge of groundwater from the Great Artesian Basin.* Queensland Department of Environment and Resource Management, Brisbane.

GABCC Great Artesian Basin Consultative Council (1998) – *Great Artesian Basin Resource Study*, Cox, R. & Barron, A. (Editors), 235 p.

GABCC Great Artesian Basin Consultative Council (2000) – *Great Artesian Basin Strategic Management Plan.* Great Artesian Basin Consultative Council for the Australian Government Department of Agriculture, Fisheries and Forestry, Canberra, 60 p.

Habermehl, M.A. (1980) – The Great Artesian Basin, Australia. *BMR Journal of Australian Geology and Geophysics 5*, 9–38.

Habermehl, M.A. (1982) Springs in the Great Artesian Basin, Australia – their origin and nature. Bureau of Mineral Resources, Australia, Report 235.

Habermehl M.A. (1983) – *Hydrogeology and hydrochemistry of the Great Artesian Basin, Australia. Invited Keynote Paper for International Conference on Groundwater and Man, Australian Water Resources Council (AWRC), International Association of Hydrological Sciences (IAHS), International Association of Hydrogeologists (IAH), Australian Academy of Science (AAS)*. Papers of the International Conference on Groundwater and Man, Sydney, 1983. Australian Water Resources Council Conference Series Australian Government Publishing Service, Canberra, Sydney, 83–98.

Habermehl, M.A. (1986) – Regional groundwater movement, hydrochemistry and hydrocarbon migration in the Eromanga Basin. In: D.I. Gravestock, P.S. Moore & Pitt, G.M. (eds) *Contributions to the geology and hydrocarbon potential of the Eromanga Basin.* Geological Society of Australia Special Publication No. 12, 353–376.

Habermehl, M.A. (1996a) – Sources of fluoride in groundwater in North Queensland, Australia. *Paper for 13th Australian Geological Convention, Geological Society of Australia Abstracts No. 41*, Canberra, 176.

Habermehl, M.A. (1996b) – Water allocation in the Great Artesian Basin. In: *Managing Australia's Inland Waters – Roles for Science and Technology. Prime Minister's Science and Engineering Council 14th Meeting, 46–53*. Commonwealth Government Department of Industry, Science and Tourism, Canberra.

Habermehl, M.A. (2001a) – Hydrogeology and environmental geology of the Great Artesian Basin, Australia. in: V.A. Gostin (ed) *Gondwana to greenhouse – Australian environmental geoscience.* Geological Society of Australia Inc., Special Publication 21, 127–143, 344–346.

Habermehl, M.A. (2001b) – *Wire-line logged waterbores in the Great Artesian Basin, Australia – Digital data of logs and waterbore data acquired by AGSO.* (Report + data CD + 3 maps at scale 1:2 500 000) Bureau of Rural Sciences, Canberra , 98 p., & Australian Geological Survey Organisation Bulletin 245, 98 p.

Habermehl, M.A. (2006) – The Great Artesian Basin, Australia. In: Foster, S. & Loucks, D.P. (Editors), 2006 – *Non-Renewable Groundwater Resources – A guidebook on socially-sustainable management for water-policy makers.* United Nations Educational, Scientific and Cultural Organization – International Hydrological Programme (UNESCO-IHP), Paris, France, UNESCO-IHP-VI, Series on Groundwater No. 10, 82–88.

Habermehl, M.A. (2009) – *Inter-Aquifer Leakage in the Queensland and New South Wales parts of the Great Artesian Basin.* Final Report for the Australian Government Department of Environment and Water Resources, Heritage and Arts. Bureau of Rural Sciences, Canberra. 33 p. + 22 appendices

Habermehl, M.A., Devonshire, J. & Magee, J. (2009) – *Sustainable Groundwater Allocations in the Intake Beds of the Great Artesian Basin in New South Wales (Recharge to the New South Wales part of the Great Artesian Basin)* – Final Report for the National Water Commission, Australian Government. Bureau of Rural Sciences, Canberra, 44 p. + 5 appendices

Habermehl, M.A., & Lau, J.E. (1997) – *Hydrogeology of the Great Artesian Basin, Australia* (Map at scale 1:2 500 000) Australian Geological Survey Organisation, Canberra. ISBN 0 642 27305 7

Habermehl, M.A., & Pestov, I. (2002) – Geothermal resources of the Great Artesian Basin, *Australia. Geo-Heat Center Quarterly Bulletin, 23 (2)*, 20–26.

Habermehl, M.A. & Seidel, G.E. (1979) – Groundwater resources of the Great Artesian Basin. in: *e.g.* Hallsworth and J.T. Woodcock (eds). *Proceedings of the Second Invitation Symposium Land and Water Resources of Australia – Dynamics of Utilisation,* Australian Academy of Technological Sciences, Sydney, Australian Academy of Technological Sciences, Melbourne, 71–93.

Hasegawa, T., Mahara, Y., Nakata, K., and Habermehl, M.A. (2010) – Verification of ^{4}He and ^{36}Cl dating of very old groundwater in the Great Artesian Basin, Australia. in: M. Taniguchi and I.P. Holman (eds). *Groundwater response to changing climate. Selected Papers on Hydrogeology 16, International Association of Hydrogeologists.* CRC Press/Balkema Taylor & Francis Group, London/Leiden. Chapter 9, 99–111.

Hasegawa, T., Nakata, K., Mahara, Y., Habermehl, M.A., Oyama, T. & Higashihara, T. (2016) – Characterization of a diffusion-dominant system using chloride and chlorine isotopes (^{36}Cl, ^{37}Cl) for the confining layer of the Great Artesian Basin, Australia. *Geochimica et Cosmochimica Acta 192*, 279–294.

Hassall & Associates Pty Ltd (2003) – *Review of the Great Artesian Basin Sustainability Initiative, Mid Term Review Phase 1*, 141 p.

Herczeg, A.L., Torgersen, T., Chivas. A.R., & Habermehl, M.A. (1987) – Geochemical evolution of groundwaters from the Great Artesian Basin, Australia. *EOS Transactions American Geophysical Union 68*, 1275–1276.

Herczeg, A.L., Torgersen, T., Chivas, A.R. & Habermehl, M.A. (1991) – Geochemistry of ground waters from the Great Artesian Basin, Australia. *Journal of Hydrology* 126, 225–245.

ICAW (1913) – Report on the Interstate conference on artesian water, Sydney, 1912. *Interstate Conference on Artesian Water*, Government Printer, Sydney.

ICAW (1914) – Report on the Second Interstate conference on artesian water, Brisbane, 1914. *Interstate Conference on Artesian Water*, Government Printer, Brisbane.

ICAW (1922) – Report on the Third Interstate conference on artesian water, Adelaide, 1921. *Interstate Conference on Artesian Water,* Government Printer, Adelaide.

ICAW (1925) – Report on the Fourth Interstate conference on artesian water, Perth, 1924. *Interstate Conference on Artesian Water,* Government Printer, Sydney.

ICAW (1929) – Report on the Fifth Interstate conference on artesian water, Sydney, 1928. *Interstate Conference on Artesian Water*, Government Printer, Sydney.

Interstate Conference on Water Conservation and Irrigation (1939) – *Interstate Conference on Water Conservation and Irrigation*, Sydney.

Kellett, J.R., Ransley, T.R., Coram, J., Jaycock, J., Barclay, D.F., McMahon, G.A., Foster, L.M. & Hillier, J.R. (2003) – *Groundwater recharge in the Great Artesian Basin intake beds, Queensland.* Queensland Department of Natural Resources and Mines, Brisbane, Queensland.

Keppel, M., Clarke, J.A., Halihan, T., Love, A.J. & Werner, A.D. (2011) – Mound springs in the arid Lake Eyre South region of South Australia: A new depositional tufa model and its controls. *Sedimentary Geology* 240, 55–70.

Keppel, M., Karlstrom, K.E., Love, A.J., Priestley, S., Wohling, D., & De Ritter, S., (2013) – *Allocating water and maintaining springs in the Great Artesian Basin,. Volume 1 – Hydrogeological framework of the western Great Artesian Basin.* National Water Commission, Canberra, 7 Volumes.

Love, A.J., Herczeg, A.L., Sampson, L., Cresswell, R.G. & Fifield, L.K. (2000) – Sources of chloride and implications for ^{36}Cl dating of old groundwater, southwestern Great Artesian Basin, Australia. *Water Resources Research* 36, 1561–1574.

Macaulay, S., Carey, H. & Habermehl, M.A. (2009) – *Artesian pressure trends within the Great Artesian Basin.* Technical report. Bureau of Rural Sciences, Canberra, 123 p.

Mahara, Y., Habermehl, M.A., Miyakawa, K., Shimada, J., & Mizuochi, Y. (2007) – Can the ^{4}He clock be calibrated by ^{36}Cl for groundwater dating? *Nuclear Instruments and Methods in Physics Research* B 259 (2007), Proceedings AMS 10, p. 536–546.

Mahara, Y., Habermehl, M.A., Hasegawa, T., Nakata, K., Ransley, T.R., Hatano, T., Mizuochi, Y., Kobayashi, H., Ninomiya, A., Senior, B.R., Yasuda, H. & Ohta, T. (2009) – Groundwater dating by estimation of groundwater flow velocity and dissolved ^{4}He accumulation rate calibrated by ^{36}Cl in the Great Artesian Basin, Australia. Earth and Planetary Science Letters, 287, 43–56.

Nelson, R. (2010) – *The guidelines for groundwater protection in Australia: regulatory review* (prepared for Geoscience Australia) Geoscience Australia

Nelson, R. (2016) Broadening Regulatory Concepts and Responses to Cumulative Impacts: Considering the Trajectory and Future of Groundwater Law and Policy. *Environmental and Planning Law Journal*, 33, 356–371.

Nelson, R. & Casey, M. (2013) Taking Policy from Paper to the Pump: Lessons on Effective and Flexible Groundwater Policy and Management from the Western U.S. and Australia *Water in the West Working Paper*, Stanford University, September 2013

Nelson, R. & Quevauviller, P. (2016) Groundwater law In: Jaleman, T., Barreteau, Hunt, R., Rinaudo & Ross, A. (2016) *Integrated Groundwater Management. Concepts, approaches and challenges.* Springer Publishing. P. 173–196.

Noble, J.C., Habermehl, M.A., James, C.D., Landsberg, J., Langston, A.C., & Morton, S.R. (1998) – Biodiversity implications of water management in the Great Artesian Basin. *Rangeland Journal*, 20 (2), 275–300.

Ogilvie, C. (1955) – The hydrology of the Queensland portion of the Great Australian Artesian Basin. Appendix H In: Dept. Co-ord Gen. Public Works, Brisbane, *Artesian water supplies in Queensland. Parliamentary Paper* A, 56-1955, 21–61.

Powell, J.M. (1991) – *Plains of promise, rivers of destiny: water management and development of Queensland 1824 – 1990.* Boolarong Publications, Brisbane. 395 p.

Prescott, J.R. & Habermehl, M.A. (2008) – Luminescence dating of spring mound deposits in the southwestern Great Artesian Basin, northern South Australia. *Australian Journal of Earth Sciences* 55, 167–181.

Queensland Government (1945) – *Artesian water supplies in Queensland. First Interim Report (1945) of Committee appointed by the Queensland Government to investigate certain aspects relating to the Great Artesian basin (Queensland Section) with particular reference to the problem of diminishing supply*. Government Printer, Brisbane, Parliamentary Paper A1–1945.

Queensland Government (1955) – *Artesian water supplies in Queensland. Report following First Interim Report (1945) of Committee appointed by the Queensland Government to investigate certain aspects relating to the Great Artesian basin (Queensland portion) with particular reference to the problem of diminishing supply.* Department of Co-ordinator- General of Public Works Queensland, Parliamentary Paper A56-1955.

Radke, B.M., Ferguson, J., Cresswell, .RG., Ransley, T.R. & Habermehl, M.A. (2000) – *Hydrochemistry and implied hydrodynamics of the Cadna-owie – Hooray Aquifer, Great Artesian Basin, Australia.* Bureau of Rural Sciences, Canberra, 229 p.

Ransley, T.R., Radke, B.R., Feitz, A.J., Kellett, J.R., Owens, R., Bell, J., Stewart, G. & Carey, H. (2015) – *Hydrogeological Atlas of the Great Artesian Basin. Geoscience Australia*, Canberra. 134 p.

Reyenga, P.J., Habermehl, M.A., & Howden, S.M. (1998) – *The Great Artesian Basin – Bore rehabilitation, rangelands and groundwater management.* Bureau of Resource Sciences, Canberra, 76 p.

Seidel, G.E. (1980) – Application of the GABHYD groundwater model of the Great Artesian Basin, Australia. *BMR Journal of Australian Geology and Geophysics* 5, 39–45.

SKM (2008) – *Great Artesian Basin Sustainability Initiative – Mid-Term Review of Phase 2*, 129 p.

SKM (2014) – *Great Artesian Basin Sustainability Initiative (GABSI) Value for Money Review Final Report*, 52 p.

Smerdon, B.D., Ransley, T.R., Radke, B.M., & Kellett, J.R. (2012) – *Water resource assessment for the Great Artesian Basin* – A report to the Australian Government from the CSIRO Great Artesian Basin Water Resource Assessment. CSIRO Water for a Healthy Counrty Flagship, Australia. (Australian Government Department of Sustainability, Environment, Water, Population and Communities & National Water Commission)

Torgersen, T., Habermehl, M.A., Phillips, F.M., Elmore, D., Kubik, P., Jones, B.G., Hemmick, T. & Gove, H.E. (1991) – Chlorine-36 dating of very old groundwater 3. Further studies in the Great Artesian Basin, Australia. *Water Resources Research* 27, 3201–3213.

Torgersen, T. & Phillips, F.M. (1993) – Reply to "Comment on 'Chlorine-36 Dating of very old groundwater 3. Further results on the Great Artesian Basin, Australia' by T. Torgersen, *et al.*" by J.N. Andrews and J.C. Fontes. *Water Resources Research* 29, 1875–1877.

Welsh, W.D. (2000) – *GABFLOW: A steady state groundwater flow model of the Great Artesian Basin.* Bureau of Rural Sciences, Canberra.

Welsh, W.D. (2006) – *Great Artesian Basin transient groundwater model.* Bureau of Rural Sciences, Canberra.

Welsh, W.D. (2007) – *Groundwater Balance Modelling with Darcy's Law.* PhD thesis, Fenner School of Environment and Society, The Australian National University. *http://thesis.anu.edu.au/public/adt-ANU20070703.165654/index.html.*

Woolley, D.R., Habermehl, M.A., Palmer, J., Sibenaler, X., & O'Neill, D., (1987) - Report to the Australian Water Resources Council Groundwater Committee (AWRC-GC) by the Great Artesian Basin Sub-Committee – *Improved Efficiency of Artesian Groundwater Use, Rehabilitation of Bores and Conversion from Drains to Piping.* 18 p.

Chapter 22

Institutions and policies governing groundwater development, use and management in the Indo-Gangetic Plains of India

M. Dinesh Kumar
Institute for Resource Analysis and Policy, Hyderabad, India

ABSTRACT

The Indian part of the Indo-Gangetic Plains extending from the north-western side to the north-eastern side of the sub-continent is the "cereal-bowl" of the country and has a history of settled agriculture, with irrigation from wells, river diversion systems and storage reservoirs and canal networks. In the past five decades, the region has experienced a virtual explosion in well irrigation. The various parts of the plains display distinctly different physical, ecological and socio-economic characteristics. Hence, the groundwater problems and management challenges are also different. While farmers in the Indus plains experience secular declines in groundwater levels, soil salinization and rising cost of well drilling, their counterparts in the eastern Gangetic plains are faced with problems of rising cost of energy and high inequity in access to groundwater. This chapter provides an overview of the groundwater issues in the Indus and Ganges basins within India and the governance and management challenges they pose. It analyses the performance of the existing formal and informal institutions concerned with groundwater development for irrigation and resource management. It also reviews various policies related to groundwater, energy and food security in the basins. Various recommendations are put forward to move from a primary development mentality to a management and governance system that takes equity and resource sustainability as overriding goals.

22.1 INTRODUCTION

Since the late 1960s, the Indo-Gangetic Plains (IGP), which are historically known for settled agriculture, have experienced a virtual explosion in well irrigation (Pandey, 2014; Mukherji *et al.*, 2009; Pant, 2004 & 2005). There has been a manifold increase in number of groundwater abstraction structures in the form of wells and pump sets in this heavily populated vast region (Ojha *et al.*, 2012) spread over an area of 2.55 million km^2. The alluvial plains of the two basins, viz., the Indus and the Ganges, which fall within the Indian part of the sub-continent, (hereafter called IGPI) however, display distinct characteristics, in terms of hydrogeological regime, agro climate and socio-economic characteristics, which largely determine the access to and use of groundwater (Kumar *et al.*, 2012).

The Indus basin part of the IGP has low rainfall and high aridity (Ullah *et al.*, 2001). But the region is economically prosperous; cropping and irrigation intensities are very high; farm sizes are large, and the agricultural productivity is one of the highest in the world, with very high yield of cereals such as wheat and paddy (rice) (Zwart and Bastiaanssen, 2004). Due to high intensity of cropping and low rainfall, the groundwater use dependence is very high (Pandey, 2014), with problems of over-exploitation. The major issues here are falling water levels, increasing cost of construction of wells and constantly rising cost of pumping water from greater depths.

Contrasting conditions exist in the eastern part, in the Gangetic plains. Though groundwater resources are abundant, the aggregate demand for water for irrigation is relatively lower, owing to high to very high rainfall and sub-humid to humid climate, and abundant surface water resources available (rivers, wetlands, ponds and lakes). Yet, where it makes sense to use groundwater for irrigation, only a small percentage of the farm households enjoy direct access to it, owing to poor access to electricity and rising cost of diesel, lack of access to finance for well construction, poor economic conditions of the marginal holders who dominate the agricultural landscape and high degree of land fragmentation. Millions of farmers depend on informal groundwater markets to access water for irrigation.

It is argued that while sustainability of resource use is an issue in the Indus basin part of the IGPI, inequity in access and economic scarcity of the resource are the key issues in the eastern part of IGPI. Hence, the governance and management challenges are different in the two situations and so are the institutional and policy measures required. In the former case, the institutional and policy interventions should promote efficient and sustainable use, whereas in the latter case, they should be designed to promote water security and distributional equity.

22.2 INSTITUTIONAL LANDSCAPE AND EXAMPLES OF DIRECT STATE INTERVENTIONS FOR GROUNDWATER DEVELOPMENT, USE AND MANAGEMENT IN THE IGPI

22.2.1 Intervention by Uttar Pradesh State agencies for developing groundwater irrigation

A national programme of constructing public tube wells for marginal farmers, who did not own modern groundwater abstraction devices, was a major intervention in the groundwater irrigation sector in the 1980s in Uttar Pradesh (UP). This public policy of supporting state-owned tube wells for irrigation received support from intellectuals who believed that the private tube owners could create short run diseconomies on traditional open dug wells, owned by marginal farmers (Pant, 2005). In the 1980s, there was evidence of correlation between caste and ownership of water abstraction devices as well as landholding size in the country. However, by the mid-2000s, supremacy of high caste people in owning modern water abstraction devices started waning (Pant, 2005) with well drilling and pumps becoming more affordable.

Pant (2004) reported that a higher proportion of farm households depended on public irrigation schemes, such as canals and public tube wells, in 2002 as compared to early 1980s, indicating increasing equity in access. Private water markets, where well-owning farmers sell water to other farmers, had also become a dominant factor in irrigation amongst smallholders, with an increase in the percentage of farmers depending on them over the two-decade period (Pant, 2004), again indicating more equitable access. However, another explanation could be the fact that many farmers having their own wells also depend on purchased water due to increasing degree of land fragmentation. Though researchers have downplayed the impact of public tube well irrigation programmes on improving equity in access to groundwater in eastern IGPI and eulogized water markets as the main explanation (Mukherji, 2008; Pant, 2004 & 2005; Shah, 2001), it is probably fair to say, that these schemes did play a role.

22.2.2 Groundwater markets for irrigation, eastern IGPI

Informal groundwater markets exist in the IGPI, but are most prevalent in the eastern part (Pant, 2004 & 2005). The access to groundwater for irrigation and the degree of equity in access are heavily dependent on the way these markets function. A large proportion of the rural people in the eastern IGPI, especially in eastern UP, Bihar and West Bengal, are very poor, with very small and fragmented landholdings. These factors reduce their ability to invest in individual wells and pump sets for smaller fragments of land. In the western IGPI, especially in Punjab, there is no widespread incidence of water markets, though kinship partners owning wells jointly is reported by researchers (Tiwari and Sabatier, 2009).

22.2.3 Legal and regulatory instruments for checking over-development in West Bengal and Punjab

State governments have made sporadic attempts to control and regulate groundwater use through top-down legislative measures. In 1993, the government of West Bengal in eastern IGPI imposed restrictions on drilling of low duty tube wells in some districts through an executive order. However, the Minor Irrigation Census of 2002 showed a vast increase in number of tube wells in the state, indicating that this step was not effective in controlling groundwater development.

In 2005, the state government further introduced checks on issuing of new electricity connections for farmers who wanted to install electrified pump sets for their wells. The state groundwater department also started issuing licenses for drilling agro wells in order to regulate groundwater development and control threat to groundwater sustainability (Mukherji, 2006)[1]. Some researchers criticized this as a 'knee jerk' reaction from officialdom to the perceived threat of groundwater mining in the state, a way to demonstrate action against arsenic problems (by some associated with groundwater irrigation), and also an avenue for rent seeking by collecting a payment for issuing such licenses (Mukherji, 2006). They also strongly argued for deregulating the process of

[1]In 2011, the State Water Investigation Directorate (SWID) has changed a provision of the Groundwater Act of 2005. Farmers located in the 'safe' groundwater blocks and owning pumps of less than 5 HP and tube wells with discharge less than 30 m^3/hour will no longer need a permit from SWID before applying for electricity connection from the West Bengal State Electricity Distribution Corporation Ltd (IWMI, 2012).

drilling of wells in the state on the basis of lack of evidence of impacts on the resource (Mukherji *et al.*, 2012). However, an issue, which is largely ignored, is that the current groundwater resource assessment methodology is not robust enough to capture all negative consequences of groundwater intensive use, such as its impact on wetlands, a factor very crucial for states like West Bengal (see Section 22.2.5 for details). It is probably safe to say that a proper understanding of the resource capacity is missing, while various perceptions are used to drive different political agendas (IWMI, 2012).

In Punjab in western IGPI, the state government requested farmers to delay the sowing time for *kharif* (crops grown during the monsoon season) paddy to reduce the dependence on groundwater through an ordinance in 2008, which became a law in 2009, called the 'Punjab Preservation of Sub-soil Water Act 2009'. The act prohibits the farmers from raising paddy nursery before 10th May and transplanting of paddy saplings before 10th June, and those who are found violating the act would be fined (150 US$/ha), and their crops would be destroyed in the field at their own expense.

The farmers in the region traditionally do nursery and transplantation of monsoon paddy early (by early April and early May, respectively, which is the hottest part of summer for north-western India) in order to obtain the best yield, and use water from their wells to raise the paddy nursery and also give 1–2 watering to the transplanted paddy saplings. The penalty is used to discourage farmers from pre-monsoon sowing, which increases the crop water and irrigation requirement. Some reports and studies have suggested large-scale adoption of this practice by Punjab farmers and positive impact on groundwater conservation (Singh, 2009; Tripathy *et al.*, 2016), electricity consumption and the environment, while reducing the revenue losses from subsidized electricity supply for the Punjab State Electricity Board (Singh, 2009). In addition, there are no reported incidences of defaulting by farmers. However, the effect of delayed transplanting on paddy yields needs to be studied, though Perry and Steduto (2017) argue that this can be a good strategy for enhancing crop water productivity and achieve real groundwater savings in the region.

22.2.4 The mechanism of minimum support price for cereals

It has long been argued that the agricultural procurement policy adopted in Punjab and Haryana in western IGPI of offering a minimum support price (higher than the market price) for cereals (paddy and wheat), combined with free electricity for agricultural groundwater pumping and the storage facilities offered by FCI (Food Corporation of India), keeps the rice-wheat cropping system popular among farmers, while depleting the groundwater resources in the region (Shergill, 2007). A logical response has been to argue for removing the minimum support price for wheat and paddy in the states. However, the negative impact of paddy-wheat cropping system on the water ecology of Punjab is largely a perception, which has originated from the fact that irrigation water requirements to *kharif* paddy in a semi-arid region like Punjab with sandy and sandy loam soils are very high. However, what is equally important, but often ignored, is the indication that a large fraction of the water applied in paddy fields for inundation percolates down to recharge the groundwater. There are no proper water balance studies done in Punjab to substantiate this claim. Studies in the Indus basin in Pakistan and the Indian part of the Indus basin have shown that the evapotranspiration (ET) from the paddy-wheat system is much less than the total amount of water applied in the form of irrigation and rainfall (Ahmad *et al.*, 2004; Singh, 2005). Therefore, more than the water intensity of the crops, what matters is the high intensity and extent of irrigated crop production in the state, with 95% of the land under cultivation and 90% of the same under irrigation. This is what drives the high aggregated water demand. The fact is also that the two states have contributed significantly to the nation's food basket, with large stocks of wheat and rice from these states going to the buffer storage of grains for extended times.

Shift in cropping pattern toward low water consuming crops is naturally debated in the context of Punjab (Singh, 2008; Pandey, 2014). However, having a minimum support price and assured purchase programme introduced for other crops, such as groundnut and maize, will be infeasible for Punjab given the WTO restrictions and budgetary compulsions of avoiding a fiscal crisis (Shergill, 2007). In addition, a large-scale shift in cropping pattern from cereal to oil seeds and other cash crops with less water demand could pose threat to national food security (Shergill, 2007).

However, these constraints may be partly out-weighted, as states such as Madhya Pradesh, which have large amounts of cultivable land, have recently become a major supplier of wheat and paddy to the nation's buffer stock, with its irrigated area expanding remarkably during the past 10–12 years (Ninan, 2017) because of development of large irrigation schemes based on surface water. Similarly, Gujarat, which also has large amounts of arable land, has recently seen remarkable increase in crop production, with quantum jump in production of wheat and paddy, as a result of expansion of its cropped area owing to irrigation water from the recent Sardar Sarovar Project (Jagadeesan and Kumar, 2015).

In spite of the fact that the cereal demand in India continues to rise as a result of population pressure and rise in per capita income, irrigation development happening in land-rich regions would surely provide a new window of opportunity to relieve the pressure on Punjab's water resources. This opportunity is still not fully realized, as the yield levels for cereals are still quite low in states such as Gujarat and Madhya Pradesh, unlike Punjab and Haryana, which have achieved peak yield levels.

22.2.5 Resource evaluation methodologies and planning approaches

The state and central agencies concerned with groundwater are engaged in groundwater resource investigation, resource evaluation and planning. Each state falling in the IGPI has a state level agency for promoting groundwater development and is capable of carrying out surveys and investigations, resource monitoring and assessment. The central agency, the Central Ground Water Board (CGWB) is also engaged in these activities and provides handholding to the state agencies in the sense that new groundwater survey investigation techniques and new resource assessment methodologies are often developed by them. In order to avoid duplication, the central agency works in coordination with the state agencies, validates and uses the data available from their well monitoring network for broader resource evaluation and planning.

However, the current resource evaluation methodology currently used by CGWB has some limitations. The assessment is based on estimates of annual recharge from rainfall and induced recharge from other sources such as canals, irrigated fields, ponds, tanks and reservoirs (Chatterjee and Ray, 2014). Recharge from rainfall is estimated using the water level fluctuation approach. However, the assessment does not take into account factors such as lateral flow and groundwater discharge to streams (base flow) that determine the utilizable recharge available in the aquifer, especially in regions characterized by highly-undulating and hilly terrains where these become major components of the water balance and are important for environmental flows.

Further, the assessment of the stage of groundwater development is based on water balance considerations of the annual recharge relative to abstraction during the previous 5-years (CGWB, 2012)[2]. It does not integrate the wide range of physical, social, economic and environmental considerations that together determine the acceptable level of groundwater exploitation in an area (Kumar and Singh, 2008). Some of the factors, which need to be considered are: longer-term changes in water levels in an area, in addition to the short-term changes in water balance; groundwater stock in the area, over and above the dynamic component; characteristics of the geological formations (Custodio, 2000); contribution of surface water irrigation systems in changing the groundwater balance (Kumar, 2007); and depth to groundwater levels in the area and its social, economic and environmental implications (Custodio, 2000). Poor resource assessment methodology leads to poor management decisions. For instance, required groundwater discharge to streams (base flow) is not estimated while assessing utilizable groundwater recharge. This leads to under-estimation of the stage of groundwater development in highly undulating and hilly terrains with the result that many such regions are still classified as 'safe', in spite of experiencing problems of water level drawdown and widespread well failure (Kumar *et al.*, 2012). A fallout of this is that institutional financing for groundwater development continues in such areas.

In the case of western IGPI, the discussion on arresting groundwater depletion has largely focused on reducing water withdrawal for paddy production and implementing artificial recharge of groundwater. There is little recognition of the fact that a significant proportion of the water applied in paddy is available for recharge through irrigation return flows (Tripathy *et al.*, 2016). To improve the groundwater balance in the areas with negative groundwater balance, the contribution for meeting the crop ET requirements from surface water and other sources, like treated wastewater, needs to be enhanced (Kumar, 2007). But, policy makers and practitioners push for water harvesting and groundwater recharge schemes for arresting groundwater depletion in the region, while the fact remains that the region does not have significant surplus runoff (Prathapar *et al.*, 2012). A poor understanding of the region's hydrology is one factor, which encourages such management decisions (Kumar *et al.*, 2008). The larger political economy in the country, which is more favourable to government taking up water harvesting schemes than implementing large-scale water transfer projects, is another factor.

In the case of eastern IGPI, understanding groundwater-surface water interactions is important to ascertain as to what extent the groundwater use could be intensified in

[2]In India, stage of groundwater development is assessed for administrative units and is estimated as the ratio of the average annual groundwater abstraction and average annual recharge, expressed in percentage terms. The annual recharge estimates consider the natural recharge and the recharge from ponds, tanks, lakes, irrigation return flows and seepage from canals (Chatterjee and Ray, 2014). The estimate does not consider static groundwater resources.

the region. This is urgent, as the proposal for flood water harvesting and recharging combined with groundwater pumping for reducing flood risk in the eastern IGPI has gained momentum in the recent past (Amarasinghe *et al.*, 2016). Blanket use of the concept of intensive groundwater pumping during summer to reduce monsoon flooding in eastern IGPI can cause havoc in areas that are not flood-prone, by drying up the existing wetlands. Therefore, refining the current resource evaluation methodologies to suite the unique hydrological conditions is required, and specific development strategies need to be developed and institutionalised.

In sum, the methodology for assessing the status of groundwater development needs to be refined by integrating multiple dimensions of sustainability. Environmental criteria in the decision making should also consider the negative impacts of groundwater over-exploitation on water quality, such as dissolution of arsenic in groundwater.

Finally, there are no regulatory instruments or incentive mechanisms in place presently to bring about behavioural change among farmers to prevent the widespread use of nitrogenous fertilizers in irrigated paddy fields in intensively cropped and irrigated areas, which cause groundwater pollution. Rules will have to be framed to affect changes in agricultural practices that can lead to reduced use of nitrogenous fertilizers, which are built on either economic incentives for reducing fertilizer use or other disincentives for excessive use.

22.3 ANALYSIS OF POLICIES RELATED TO ENERGY INFLUENCING GROUNDWATER DEVELOPMENT AND USE IN THE INDO-GANGETIC PLAINS

22.3.1 Policies for providing power connections in the farm sector

All Indian states have adopted friendly policies towards the farm sector, when it comes to issuing power connections to farmers. However, the outcomes of this policy have not been uniform across states. States such as Punjab and Haryana made early progress in providing universal agricultural power connections, and therefore water markets are not common. This rapid pace in rural electrification[3], which happened in these states to support tube well irrigation, was the backbone of the Green Revolution in the early 1970s (Randhawa, 1971). This change was enabled by the strong demand from the rural sector (consisting mostly of large farmers who are highly enterprising, efficient, literate, knowledgeable and conversant with mechanized farming (Randhawa, 1971), good state finances owing to good economic growth resulting from industrialization, especially the agro-based industries, and good overall working of the power sector in the states.

Contrary to this, in states such as UP, Bihar and West Bengal in central and eastern IGPI, the pace of rural electrification has been slow, owing to the poor state finances,

[3]While the number of electric tube wells in Punjab has increased manifold from 0.28 million (in 1980–81) to 1.1 million in 2010–11, the number of diesel tube wells has reduced from 0.28 million to 0.27 million during the same period (Singh, 2012).

poor overall governance and management of the power sector, and poor economic condition of the population in rural areas, which suppressed the demand for electricity. However, these states are increasingly power-starved, especially in the agriculture sector, with growing gap between demand and supplies.

In UP and Bihar, owing to the poor quality of power supply (*i.e.* limited hours of supply, voltage fluctuations and poor reliability) de-electrification (gradual replacement of electric pumps by diesel pumps) has been happening in rural areas for some time, with some improvements in the recent past (Pant, 2004 & 2005).

22.3.2 Electricity supply policies for the farm sector

Most states in the IGPI provide free power to the farmers for agricultural use, which is mainly used for pumping of groundwater. The only exceptions are Uttarakhand and West Bengal, where groundwater pumping is charged. While Uttarakhand introduced the policy of metering agricultural power connections in 2007 and later in 2010 revoked, West Bengal has continued the policy of metering power supply to agriculture and charging it on a pro rata basis. For the other states, and since the state subsidy on electricity to the farm sector is very high, and most of the revenue of the utilities comes from commercial and industrial sectors, the utilities have not prioritized providing quality power supply to the agricultural sector. They have tried to restrict electricity use in the sector through rationing of power supply in order to reduce the subsidy burden (Shah and Verma, 2008). In none of the states, power supply today is round-the-clock, and in most cases, it is now available for a maximum of six hours daily. This affects the quality of irrigation. The resource rich farmers try to overcome this constraint by installing higher capacity pump sets.

The recent introduction of heavy capital subsidy for irrigation pumps based on solar photovoltaic power by many state governments, including Punjab and Rajasthan in western IGPI, is likely to create further incentive for groundwater over-abstraction. Farmers who were using diesel engines and incurring heavy energy costs might switch over to solar pumps if they can afford the investment and wherever the physical condition is favourable (Bassi, 2015). Since these systems will be highly decentralized, this can create new governance challenges for groundwater in the IGPI. New research in participation with farmers are now testing various models of farmers selling back the solar power to the utilities at favourable prices, counter-acting incentives to use the power for irrigation expansion (Shah *et al.*, 2016). However, evidence shows that for farmers to sell the electricity they produce, the purchase price offered by the utilities will have to be much higher than the average price at which they currently sell electricity to consumers. Since the utilities are unlikely to agree to such models, the solar PV adopters might end up selling water (Nair, 2016).

22.3.3 Electricity pricing policies and diesel subsidies

It has long been argued by scholars in economics and public policy that free electricity for groundwater pumping, which results in almost zero marginal cost of pumping, has resulted in over-exploitation of groundwater in semi-arid and arid regions of India (Saleth, 1997). Empirical studies done in alluvial north Gujarat and eastern IGPI (south Bihar and eastern UP) validate the arguments in favour of pricing of electricity on the basis of consumption, which show that pro rata pricing promotes efficiency, equity and sustainability in groundwater use (Kumar, 2005; Kumar *et al.*, 2011, 2013a)[4]. These studies have shown that the energy prices, at which the farmers start responding to tariff changes in terms of reducing the demand for these inputs would be socio-economically viable (Kumar *et al.*, 2011, 2013a). This is because they are able to improve the efficiency of use of irrigation water and other farm inputs and modify their farming systems, when confronted with positive marginal cost of electricity and higher unit tariff. The result is that they get not only higher water productivity in physical and economic terms (kg/m^3 of water and Rs/m^3 of water), but also higher net income per unit area of land (Kumar *et al.*, 2010, 2011).

Not-withstanding, the electricity utilities in many Indian states, including UP, Bihar and Punjab, continue the practice of providing free electricity to the farm sector for groundwater pumping (sometimes combined with rationing), or they charge a flat rate on the basis of the connected load (capacity of the pump set). One major justification offered by the utilities for this has been that the transaction cost of metering electricity use in the farm sector would be higher than the revenue from sale of electricity to that sector. This is due to the fact that the wells are located in remote rural areas and that electricity anyway will have to be supplied at heavily subsidized rates to the farmers given the poor affordability among the farming community. However, the transaction cost argument is often used as an excuse by field level officials of the state power utilities. Under flat rate system, they seek 'rent' from farmers who rampantly do under-reporting of the connected load of their wells. Importantly, there are other reasons for the lack of interest in metering farm power consumption and charging for it. They include the influence of the powerful farm lobby on the political class, who use the favourable policy as a vote catcher, arguing that free electricity or the subsidized flat rate based electricity pricing benefits all well owners (see also Section 22.4).

22.3.4 Provision of subsidized diesel engines

Millions of farmers in the IGPI, especially in Bihar, UP, west Bengal and Assam in eastern IGPI, use diesel engines for pumping groundwater in lieu of the fact that obtaining power connections for running electric pumps is extremely difficult in rural areas of blocks where groundwater development has reached a 'critical stage'. These states were also offering capital subsidies for purchase of diesel engines in the early 1980s (Shah, 2001). More importantly, diesel is still subsidized in India, though the market price has increased significantly in recent years (Shah, 2009). Kumar *et al.* (2010) showed that given the shallow water table conditions, under a price, which is still moderate, the diesel well-owning farmers in eastern IGPI are not very cautious about diesel use with the result that the energy use efficiency is quite low (when compared to farmers

[4]The empirical study involved: a] survey of large number of farmers who pay for irrigation water from diesel and electric well owners on hourly basis, diesel well-owning farmers who incur marginal cost of pumping groundwater, and electric well-owning farmers who do not incur marginal cost of pumping groundwater in south Bihar and eastern UP; and, b] farmers who are paying for electricity on the basis of metered consumption and farmers who pay for electricity on flat rate basis in north Gujarat (for details, see Kumar *et al.*, 2011).

who buy water from them), though higher as compared to electric well commands (Kumar *et al.*, 2010).

The study also showed that there is substantial scope for increasing diesel prices in the farm sector, which can result in improved energy use efficiency through improved water use efficiency in irrigated crop production, without adversely affecting farm incomes. The empirical analysis, which supported this inference, indicated that farmers who purchased water from diesel well owners and paid high price for water were found to be having higher water use efficiency in their cropping system as compared to the diesel well-owning counterparts, who incurred much lower cost for pumping water from their own wells. More importantly, despite the high cost for irrigation water, the water buyers were getting almost the same net income per unit area of land as the diesel well owners, because they allocated more water and land for high-value crops (especially vegetables), which give higher return per unit volume of water and area of land (Kumar *et al.*, 2010).

22.4 IMPACT OF INSTITUTIONAL AND POLICY MEASURES IN THE GROUNDWATER SECTOR ON IRRIGATION, WITH PARTICULAR REFERENCE TO ACCESS EQUITY

There are two important factors, which determine access equity in groundwater. The first one concerns the legal right to use groundwater. In India, right to use groundwater is attached to land ownership rights, which follows English Common Law. Any person having a piece of land has the right to use groundwater underlying it. Hence, landowners can access groundwater whenever he/she has the resource to invest in a well, and can pump out as much water as he/she deems fit (Saleth, 1996) in areas where there are no government regulations on well drilling. But millions of farmers in the rural areas of the IGPI do not have the wherewithal to invest in wells and pump sets. In spite of the fact that the investment in well construction in the eastern IGPI is relatively low, high degree of land fragmentation and poor economic conditions make it an unattractive proposition for the marginal farmers to invest in individual wells (Kishore, 2004; Kumar, 2007). It is mostly the large and medium farmers, whose numbers run in a few hundred thousand, and a very small fraction of the tens of millions of marginal farmers who own wells (Kumar *et al.*, 2013b). Furthermore, lowering water table conditions have a negative impact on access equity in groundwater, as the cost of construction or deepening of wells increases and hits disproportionally the poor. This is an important concern for western IGPI, especially in Punjab, where groundwater levels are falling in large areas.

The other important factor influencing access equity is the cost of pumping groundwater. For a long time, the public policy debate on groundwater, especially for eastern IGPI, revolved around access equity and how energy pricing and energy subsidies could promote access equity by influencing the functioning of water markets. It was argued that the free electricity or flat rate system of pricing electricity in the farm sector encourage electric well owners to pump more groundwater and offer to their neighbours who do not own wells. It was further argued that since there would be competition amongst well owners to pump out water and sell, the price of water would drop in the markets and the non-well owning farmers would consequently benefit, improving access equity in groundwater. Conversely, they argued that when electricity is metered and charged on a pro rata basis, the pump owners would not have incentive to sell water. And even when they decided to sell water, they would pass on the cost burden induced by the marginal cost of pumping water to the buyers, and with that the price of water would go up (Mukherji, 2008).

Field evidence questions these arguments. A survey in West Bengal showed that the diesel well owners were found to be selling water more aggressively to make profits than those who had electric submersible pumps and paying for electricity on the basis of connected load. Against an average area of 22.8 ha irrigated by diesel well owners, the total area of water buyers was 19.2 ha, *i.e.*, nearly 84% of the pumping was done to provide irrigation service to neighbouring farmers. Conversely, in the case of electric well owners, against a total area of 27.0 ha irrigated by the tube well, the area irrigated by water buyers was 22.3 ha (82%). Hence, the argument that under metered tariff (or in this case, non-freely energy-supplied farmers), well owners would have less incentive to sell water may not hold (Source: based on data provided in Mukherji, 2008).

Further, such arguments missed the point that under flat rate pricing of electricity, the large well owners, whose implicit cost of irrigation is very low and return from crop production high, may enjoy monopoly power and decide the price at which water should be sold in the market. Conversely, the smallholders owing wells, whose implicit cost of irrigating own farm is high, are left without much choice but to look for buyers to whom they could sell water to earn extra income, at a price decided by the market as they will have limited bargaining power. Such prices obviously offer high profit margins for the large farmers, but not for the small farmers.

It was found that the Monopoly Price Ratio (MPR) – the ratio of the price at which a commodity is sold and the actual cost of production – charged by electric well owners, who incur zero marginal cost and very low implicit cost of pumping water, is far higher than that charged by the diesel well owners who incur high marginal cost of pumping (Kumar *et al.*, 2013a)[5].

Overall, water markets under flat rate system of electricity pricing are inequitable in the sense that the opportunity cost of not having own source of irrigation is high for the water buyers (with the market price of water far greater than the cost of pumping groundwater) and that they are forced to grow high value crops that involve high production and market risk to recover the high input costs. At the same time, the well owners earn substantial revenue from water sales and do not pass on the subsidy benefits to the buyers (Kumar *et al.*, 2014).

If electricity is charged on pro rata basis, both small and large farmers will incur the same unit cost of pumping water, and the large farmers will not enjoy a comparative advantage over small farmers. In view of the fact that there are no fixed costs of keeping pump sets, the small landholders are not under pressure to sell water at a very low price

[5]Studies in eastern UP, West Bengal and south Bihar show that these markets can be highly monopolistic in the sense that the price, which well owners charge from the buyers, is much higher than the actual cost incurred by them for abstracting water, depending on supply and demand in the market (Kumar *et al.*, 2013a, 2014). Electric well owners who typically incur very low cost for abstracting water, owing to very low cost of energy, charge as much as what a diesel pump owner, who incur much higher energy cost, charges and therefore the monopoly price of electric well owners was found to be very high (Kumar *et al.*, 2013a). Over a period of three decades or so, the monopoly prices charged by these well owners have however dropped, with greater proportion of farmers owing wells and pump sets (Kishore, 2004).

to stay in the market. Hence, both large and small farmers would have equal incentive to invest in wells and sell water, and have equal opportunities to make profits in the shallow groundwater areas. This will lower the price at which water would be sold in the market. Empirical evidence from West Bengal substantiates this view[6] (Kumar *et al.*, 2014).

22.5 INSTITUTIONAL PROPOSITIONS FOR SUSTAINABLE GROUNDWATER MANAGEMENT

22.5.1 Correct pricing of electricity and diesel in the farm sector

The optimum pro rata energy tariff, which would be most efficient in terms of demand reduction, affordability for farmers and improvement of the viability of the power sector, will have to be a function of the depth to groundwater table. For shallow water table areas, the unit energy price should be higher, while it should be lower for deep water table areas, in order to make sure that the total cost of abstracting a unit volume of groundwater remains more or less the same across geo-hydrological environments, while also reflecting the scarcity of water by adjusting the price in depleting areas relatively upwards. There is a clear trade-off between assuring sustainability and social acceptability. Such approach will ensure inter-regional parity in well irrigation costs, as there is significant variation in depth to water table across regions. However, the price for electricity would also have to be a function of what farmers are willing to pay. The willingness to pay for water would be a function of the economic surplus generated from the use of water for irrigation. This is a function of the opportunities, which is decided by the cropping system that is feasible under the given agro climate, the demand for the crop in the market and the access to the market. These factors need to be considered when fixing the price for electricity.

Consumption-based pricing of electricity is followed in West Bengal and Gujarat. While West Bengal with shallow water table conditions charges a rate of 3.75 Rs/kWh (0.06 US$/kWh) of electricity across tube well owners, Gujarat charges a rate of 0.60 Rs/kWh (0.001 US$/kWh) for those who have metered power connections. While 100% of the agricultural tube well connections in West Bengal are metered, 80% of the power connections to agro wells in Gujarat are metered (source: based on fieldwork carried out in both the locations by the author). The government of West Bengal could introduce it because of the strong social base of the party in power (Communist Party of India-Marxist). The political stability of the government in Gujarat, run by a party, which came to power for the 3rd consecutive term, gave them the leverage to introduce power sector reform including metering of agro wells. Empirical study in north Gujarat, which compared farming enterprise of well owners under flat rate system of electricity pricing and well owners under metered tariff showed that pro rata pricing had resulted in improved efficiency of groundwater use (Kumar *et al.*, 2011).

---

[6]Analysis of primary data from villages in WB (Source: based on Mukherji, 2008) shows that under the flat rate system, the monopoly power enjoyed by diesel well owners (estimated in terms of monopoly price ratio, MPR = 1.90), who are confronted with positive marginal cost of pumping groundwater, was much lower than that of electric well owners (MPR = 16.70).

22.5.2 Metering of farm power supply and groundwater withdrawal and improving energy efficiency and viability of the power sector

The state electricity utilities and policy makers in the Indian government recognize the importance of metering farm electricity use from the point of both cost recovery and improving energy efficiency. They have also been toiling with the idea of carrying out metering of farm power in a way that makes it fool proof as well as cost-effective. Today, technologies exist and have been attempted, not only for metering but also for controlling energy consumption by farmers (Aarnoudse *et al.*, 2016; Kumar and Amarasinghe, 2009; Zekri, 2008). The pre-paid electronic meters[7], which are typically operated through scratch cards and can work on satellite and internet technology, are fit for remote areas to monitor energy use and control groundwater use online from a centralized station. The technology builds on services provided by internet and mobile (satellite) phone services, which have improved remarkably in India over the last 15 years, especially in the rural areas, with a phenomenal increase in the number of consumers (Kumar *et al.*, 2011).

Pilferage of electricity is prevalent under the current flat rate system of pricing. It is also prevalent with conventional energy metering through tampering of meters. Pre-paid meters prevent electricity pilferage happening through manipulation of pump capacity, as the pumping equipment will stop running once the fixed quota of energy that is paid for is used up. They can be operated through tokens, scratch cards, magnetic cards or recharged digitally through internet and SMS. They help electricity utilities restrict the use of electricity. The utility can decide on the "energy quota" for each farmer on the basis of reported connected load, total hours of power supply, or estimated sustainable abstraction levels per unit of irrigated land. For operationalizing this, information for every agricultural consumer on connected load, coordinates, and field data to assess sustainable withdrawal levels, among other data, are required.

These measures can bring about larger societal benefits by checking groundwater over-draft, improving financial viability of state electricity utilities and raising agricultural productivity. However, politicians generally believe that such measures are risky as some researchers argue that metering and pricing electricity and reduction in energy subsidies would adversely affect the income and livelihoods of millions of marginal farmers, though studies show that under the present structure power subsidies largely benefit the large and medium farmers. On the other hand, the bureaucrats and technocrats are not keen to change the 'status quo', which requires adoption of new work culture, technologies and practices.

22.5.3 Improving the direct access of marginal farmers to groundwater for irrigation in eastern IGPI

We have earlier argued that pro rata pricing would benefit the poor non-well owning farmers in eastern IGPI by reducing the monopoly price ratio for water sold in the market, as it allows more smallholder farmers to become well-owners and also

[7]Meters that are connected to electrical equipment that works only for a limited period on the basis of energy charges paid in advance for a certain quota of energy from the electricity utility.

encourage them to actively participate in the market as they would enjoy opportunities equal to that of large farmers. Though the marginal cost of pumping groundwater would go up under the metered tariff system, the price at which water is traded in the market wouldn't increase as the monopoly power of the well owner, which determines the market conditions, will reduce, as a result of increase in number of sellers against buyers.

In addition, the state governments in eastern IGPI states can think of formulating schemes that offer micro diesel engines (fuel-efficient and with low operating heads) to marginal farmers at subsidized rates. The financial implications of this for the state exchequer will be lower than that of offering free power connections. It would not only be economically viable, but also financially feasible for the resource-poor farmers, with very small landholdings, who at present are not able to invest in wells in this shallow groundwater region due to high cost of the pumping devise. The financial burden on the states resulting from the provision of free micro diesel engines would also be much less as compared to that resulting from the new policy of giving free power connections to farmers (Kumar *et al.*, 2014). However, proper targeting of the subsidy would be essential to prevent pilferage and misappropriation by the rural elite.

22.5.4 Evolving a functional water rights system and its enforcement in the 'over-exploited' areas

Researchers have long argued for tradable water rights for ensuring sustainability of water use in developing countries (Rosegrant and Binswanger, 1994; Saleth, 1996). Tradable property rights for groundwater exist in many countries, including United States (Kansas, Texas and California), Australia (Murray Darling Basin), Mexico and Chile. Analyses of experiences with tradable water rights in California, Mexico, Chile (Rosegrant and Gazmuri S., 1995) and Murray Darling basin (NWC, 2010) show that it can serve as a viable instrument for (re)allocation of water to improve allocative efficiency in water use. However, there are legal, institutional and technology challenges that need to be overcome for establishing and enforcing property rights system for groundwater, given the invisible nature of the resource, the pattern of its use and the legal status vis-à-vis ownership of the resource. The proposed water rights system, if designed to address concerns of the current inequity in access to the resource, can rebalance access. With many millions of farmers in remote rural areas accessing groundwater, monitoring volumetric resource use by individual farmers, required for tracking water allocations, however, would require new institutions with the associated transaction costs.

The process of establishing water rights[8] and restricting water use by farmers is feasible in areas where farmers use electricity supplied by the state electricity utility, as the energy consumption is used as proxy for groundwater abstraction. Restricting farmers' energy use for pumping groundwater is analogous to rationing groundwater withdrawal for irrigation volumetrically, and hence energy rationing can be equivalent to a functional water rights system. This can be done through pre-paid electricity

[8]These are not absolute ownership rights over water, but rights to use water defined in volumetric terms, and can also be called 'water entitlements' or 'quotas'.

meters as previously described. As shown by Kumar (2005) through a comparative analysis of well owners who are confronted with zero marginal cost of using electricity and water and farmers who have fixed water entitlements, farmers would allocate a greater proportion of the available water to economically more efficient crops besides improving the physical efficiency of water use when confronted with volumetric water rationing. Restricting energy use will have positive impact on efficiency of groundwater use by all categories of farmers. More importantly, it will help achieve sustainable groundwater use, if the energy quota is fixed by taking into account the sustainable yield of the aquifer. In such cases, it is important that the consumers are informed about their energy quota, and the approximate number of hours for which they could pump water from their wells well in advance of the agricultural season. Such information helps them choose the crops depending on the availability of power (and hence water) over the cropping season(s) of the year. Here again, the energy quota will have to be decided on the basis of the geo-hydrological environment prevailing in the area and the lower optimum irrigation requirements.

However, there are governance challenges in implementing this idea. First, the utility has to frame rules/norms regarding allocation of energy quota amongst the farmers, which will have to be based on groundwater rights/entitlements. The decision on these entitlements should use sound criteria based on principles of access equity and resource sustainability. Framing such rules can be politically sensitive because the decisions would ideally result in allocation of water rights/entitlements to many farmers who currently do not enjoy direct access to groundwater, and limits to the use of groundwater by many large and medium farmers. Politicians and bureaucrats are generally averse to taking decisions, which have potentially large social ramifications.

22.6 CONCLUSIONS

The western and eastern parts of the IGP in India characterise two distinctly different situations with respect to groundwater environment and development, and hence the associated differences in governance and management challenges. Public policies in water, agriculture, food security and energy, as part of a broader water-food-energy nexus, have profoundly influenced the way groundwater resources are developed, used and managed in both parts. In the western part of IGPI, the early and rapid rural electrification, free or subsidized power to the farm sector, large productive farmers, and attractive procurement prices for major cereals led to intensive use of groundwater, with low water and energy use efficiency and unsustainable resource use emerging as major challenges. The very low opportunity cost of using groundwater is the major factor driving over-exploitation and inefficient use of the resource.

In the eastern IGPI, groundwater is relatively abundant, and demand for the resource for irrigation is comparatively lower. Here, inequity in access to groundwater is a major challenge in the region, with high monopoly price for the water traded in the market for irrigation. At the same time, resource use efficiency is higher among the millions of small diesel well owners and water buyers, as indicated by the high water productivity in their farming systems. An analysis presented in the chapter showed that free electric power connections or flat rate power are unlikely to result in improved access equity in groundwater or intensification of well irrigation, as the benefits of energy subsidy are not passed on to the water buyers (Kumar *et al.*, 2014). So, maintaining high diesel prices in this region is recommended. In sum, to improve access equity in groundwater in eastern IGPI, the following measures are required: a) provision of subsidies for purchase of micro diesel engines for marginal farmers; b) introduction of pro rata pricing of electricity; and, c) rationalising the price of diesel used in the farm sector.

To improve the sustainability of groundwater use in the western IGPI, two policy measures are recommended. They are pro rata pricing of electricity and a water rights system, both working in conjunction. Pre-paid energy meters can be used to introduce a pro rata pricing system and monitor energy use of individual farmers by the electricity utility. The cost of these meters will have to be borne by farmers. Metered pricing can be combined with energy rationing (quotas). However, the electricity prices that would incentivize efficient use of water, but also be affordable to the farmers need to be worked out for different socio-economic and environmental conditions. Energy rationing can also become a proxy for a functional water rights system, with energy quotas of individual farmers decided on the basis of their new volumetric water rights or entitlements.

On the knowledge front, it is important to develop methodologies for assessing the stage of groundwater development that integrate a variety of environmental considerations such as protection of wetlands and preservation of groundwater quality. Also, rules and incentives need to be framed on use of nitrogenous fertilizers in intensively cropped regions like Punjab.

To conclude, poor resource assessment methodologies and inefficient pricing of electricity & diesel pose major governance challenges for groundwater resources in the IGPI. To improve the governance of groundwater, the goal for deciding on the mode of energy pricing in both parts of IGPI should be the long-term sustainability of resource use and access equity rather than short-term income gain for the farmers and transaction cost reduction. To begin with, the agriculturally more prosperous region of IGPI, *i.e.*, western IGPI, needs to be chosen for introducing metering and pro rata pricing of electricity, with a low unit tariff. This should be combined with the strategy of improving the quality of power supply in terms of duration and voltage stability to enhance the acceptability of this reform. The tariff can be raised over a period of time to improve the financial working of the utility.

Similarly, the norms for fixing the prices of electricity and diesel should be based on consideration of water/energy use efficiency, cost recovery, farmer affordability and agricultural productivity enhancement. In the groundwater scarce regions, which experience problems of overdraft (primarily western IGPI), the energy supply policy (in terms of duration and quantum) should be driven by considerations of equity and sustainability of resource base along with agricultural productivity. Once such a policy is introduced, farmers on their own will adopt measures such as late planting of paddy etc. to conserve groundwater and to reduce their input costs. To make this happen, the first step is education of the politicians on the trade-off between short term gains and long term effects of following populist policies. As the next step, the farmers need to be educated about the fact that switching over to metered tariff is a 'win-win' situation for them and the electricity utility. Interventions like the solar PV systems for farmers, which are now becoming popular in the IGPI, can pose new governance challenges for groundwater, if farmers have more incentive to sell water than electricity.

REFERENCES

Ahmad, M. D., I., Masih and H. Turral (2004) Diagnostic analysis of spatial and temporal variations in crop water productivity: A field scale analysis of the rice-wheat cropping system of Punjab, Pakistan, *Journal of Applied Irrigation Science*, 39 (10).

Amarasinghe, U., Muthuwatta, L., Surinaidu, L., Anand, S., and Jain, S.K. (2016) Reviving the Ganges Water Machine: Potential, *Hydrology and Earth Systems Sciences*, 20, 1085–1101.

Ameren Illinois (2012) Advanced Metering Infrastructure (AMI) Cost/Benefit Analysis, July 2012.

Aarnoudse, E., Wei Qu, B. Bluemling & T. Herzfeld (2016) Groundwater quota versus tiered groundwater pricing: two cases of groundwater management in north-west China, International Journal of Water Resources Development, DOI: 10.1080/07900627.2016.1240069

Bassi, Nitin (2015) Irrigation and Energy Nexus: Solar Pumps are Not Viable, *Economic and Political Weekly*, L (10): 63–66.

Central Ground Water Board (2014) *Ground Water Year Book 2013-14*, Central Ground Water Board, Ministry of Water Resources, Government of India, Faridabad.

Chatterjee, R., and Ray, R.K. (2014) *Assessment of Groundwater Resources: A Review of International Practices*, Central Ground Water Board, Ministry of Water Resources, Govt. of India, Faridabad.

International Water Management Institute (2012) *Agricultural Water Management Learning and Discussion Brief*, AGWAT Solutions, improved livelihood for small holder farmers, April 2012.

Jagadeesan, S. and M.D. Kumar (2015) *The Sardar Sarovar Project: Assessing Economic and Social Impacts*, Sage Publications, New Delhi.

Kishore, Avinash (2004) Understanding Agrarian Impasse in Bihar, *Economic and Political Weekly*, Review of Agriculture, XXXIX (31): 3484–3491.

Kumar, M. Dinesh (2007) *Groundwater Management in India: Physical, Institutional and Policy Alternatives*, New Delhi: Sage Publications.

Kumar, M. Dinesh and O. P. Singh (2008) How Serious Are Groundwater Over-exploitation Problems in India? A Fresh Investigation into an Old Issue, in Kumar, M. Dinesh (Ed) *Managing Water in the Face of Growing Scarcity, Inequity and Declining Returns: Exploring Fresh Approaches*, 7th Annual Partners' meet of IWMI-Tata water policy research program, ICRISAT, Patancheru, AP, 2–4 April, 2008.

Kumar, M. Dinesh and Upali Amarasinghe (Eds) (2009) *Water Productivity Improvements in Indian Agriculture: Potentials, Constraints and Prospects*, Strategic Analysis of the National River Linking Project (NRLP) of India: Series 4, Colombo: International Water Management Institute.

Kumar, M. D., Singh, O.P., and Sivamohan, M.V.K. (2010) Have Diesel Price Hikes Actually Led to Farmer Distress in India? *Water International*, 35(3):270–284.

Kumar, M. Dinesh, Christopher Scott and O.P. Singh (2011) Inducing the Shift from Flat-Rate or Free Agricultural Power to Metered Supply: Implications for Groundwater Depletion and Power Sector Viability in India, *Journal of Hydrology*, 409 (2011): 382–394.

Kumar, M. Dinesh, M.V.K. Sivamohan and A. Narayanamoorthy (2012) The Food Security Challenge of the Food-Land-Water Nexus in India, *Food Security*, published on 3rd August, 2012.

Kumar, M. Dinesh, Christopher Scott and O.P. Singh (2013a) Can India raise agricultural productivity while reducing groundwater and energy use? *Int. Journal of Water Resources Development*, 29 (4): 557–573.

Kumar, M. Dinesh, V. Ratna Reddy, A. Narayanamoorthy and M.V.K. Sivamohan (2013b) Analysis of India's Minor Irrigation Statistics: Faulty Analysis, Wrong Inferences, Economic and Political Weekly, XLVIII (45 & 46), November 16, 2013.

Kumar M.D., N. Bassi, M.V.K. Sivamohan and L. Venkatachalam (2014) Breaking the agrarian crisis in eastern India. In: Kumar M.D., Bassi N., Narayanamoorthy A, Sivamohan MVK (eds) Water, energy and food security nexus: lessons from India for development. Routledge, London, United Kingdom, pp. 143–159.

Mukherji, A. (2006) Political Ecology of Groundwater: The Contrasting Case of Water Abundant West Bengal and Water Scarce Gujarat, India, *Hydrogeology Journal* 14(3): 392–406.

Mukherji, Aditi (2008) "The paradox of groundwater scarcity amidst plenty and its implications for food security and poverty alleviation in West Bengal, India: What can be done to ameliorate the crisis?," Paper presented at 9th Annual Global Development Network Conference, Brisbane, Australia, 29–31 January, 2008.

Mukherji, A., K.G. Villholth, B.R. Sharma and J. Wang (Eds.) (2009) Groundwater Governance in the Indo-Gangetic and Yellow River Basins: Realities and Challenges. Series: IAH – Selected Papers on Hydrogeology, Volume 15. CRC Press, Taylor & Francis Group. 325 pp. ISBN 978-0-415-46580-9

Mukherji, Aditi, Tushaar Shah and Parthasarathy Banerjee (2012) Kick-starting a Second Green Revolution in Bengal, Commentary, *Economic and Political Weekly*, May 05, 2012.

Nair, Avinash (2017) Gujarat: Solar co-operative at Dhundi village sells water instead of electricity, The Indian Express, August 14, 2016.

National Water Commission (2010). The impacts of water trading in the southern Murray–Darling Basin: an economic, social and environmental assessment, NWC, Canberra.

Ninan, T.N. (2017) How Madhya Pradesh Worked a Miracle in Agriculture, *Business Standard*, May 09, 2017.

Pandey, Rita (2014) *Groundwater Irrigation in Punjab: Some Issues and Way Forward*, Working Paper 2014-140, National Institute of Public Finance and Policy, New Delhi, August.

Pant, Niranjan (2004) Trends in Groundwater Irrigation in Eastern and Western UP, *Economic and Political Weekly*, Review of Agriculture, XXXIX (31): 3663–3468.

Pant, Niranjan (2005) Control of and Access to Groundwater in UP, Economic and Political Weekly, 40 (26), pp. 2672–2680.

Perry, Chris and P. Steduto (2017) Does Improved Irrigation Technology Save Water? A Review of the Evidence, Discussion paper on irrigation and sustainable water resources management in the Near East and North Africa, Food and Agriculture Organization of the United Nations, Cairo, Egypt.

Prathapar, S. A., B. R. Sharma and P.K. Aggarwal (2012) 'Hydro, Hydrogeological Constraints to Managed Aquifer Recharge', Water Policy Highlight 40, IWMI-Tata Water Policy Research Program.

Randhawa, M. S. (1971) *Green Revolution in Punjab*, Punjab Agricultural University Press, Ludhiana, July 03, 1971.

Rosegrant, M. and Binswanger, H. P. (1994) Markets in Tradable Water Rights: Potential for Efficiency Gains in Developing Country Water Resource Allocation, *World Development*, 22 (11): 1613–1625.

Rosegrant, M. and Gazmuri S. Renato (1994) Reforming Water Allocation Policy through Markets in Tradable Water Rights: Experience from Chile, Mexico and California, paper presented at the DSE/IFPRI/ISISI workshop on Agricultural Sustainability, Growth and Poverty alleviation in East and South East Asia, October 3–6, 1994.

Saleth, R. Maria (1996) *Water Institutions in India: Economics, Law and Policy*. New Delhi: Commonwealth Publishers, 299 pp.

Saleth, R. Maria (1997) Power Tariff Policy for Groundwater Regulation: Efficiency, Equity and Sustainability. *Artha Vijnana*, XXXIX (3): 312–322.

Shah, T., N. Durga, S. Verma and R. Rathod (2016) Solar power as remunerative crop. IWMI-Tata Water Policy Research Highlight no. 10.

Shah, T. and S. Verma (2008) Co-management of electricity and groundwater: an assessment of Gujarat's Jyotigram Scheme. Economic and Political Weekly, 43(7): 59–66.

Shah, Tushaar (2001) *Wells and Welfare in Ganga Basin: Public Policy and Private Initiative in Eastern Uttar Pradesh, India*, Research Report 54, International Water Management Institute, Colombo, Sri Lanka.

Shah, T. (2009) Crop per Drop of Diesel? Energy Squeeze on India's Smallholder Irrigation, Economic and Political Weekly, 42 (39): 4002–4009.

Shergill, H. (2007) Wheat and paddy cultivation and the question of optimal cropping pattern for Punjab. Journal of Political Science, 12: 239–250. http://www.global.ucsb.edu/punjab/12.2_Shergill.pdf.

Singh, Karam (2008) Water table behaviour in Punjab: Issues and Policy Options, in Kumar, M. D. (Ed) in "Managing water in the face of growing scarcity, inequity and declining returns: exploring fresh approaches," International Water Management Institute, Hyderabad, 2–4 April 2008.

Singh, Karam (2009) Act to Save Groundwater in Punjab: Its Impact on Water Table, Electricity Subsidy and Environment, Agricultural Economics Research Review Vol. 22 (Conference Number) 2009, pp. 365–386.

Singh, Karan (2012) Electricity Subsidy in Punjab Agriculture: The Extent and Impact, *Indian Journal of Agricultural Economics*, 67 (4): 617–632.

Singh, Ranvir (2005) *Water productivity analysis from field to regional scale: integration of crop and soil modeling, remote sensing and geographic information*, doctoral thesis, Wageningen University, Wageningen, The Netherlands.

Tiwari, R. and J.L. Sabatier (2009) Anthropological perspectives on groundwater irrigation: Ethnographic evidence from a village in Bist Doab, Punjab, in Mukherji, A., K.G. Villholth, B.R. Sharma and J. Wang (Eds) Groundwater governance in the Indo Gangetic and Yellow River Basins: Realities and Challenges, CRC Press.

Tripathy, A., Mishra, A. K. and Verma, G. (2016) Impact of Preservation of Sub-Soil Water on Groundwater Depletion: The Case of Punjab, India, *Environmental Management*, 58 (1): 48–59.

Ullah, M. K., Z. Habib, S. Muhammad (2001) *Spatial distribution of reference and potential evapotranspiration across the Indus Basin Irrigation Systems*, Working Paper 24, Lahore, Pakistan: International Water Management Institute.

Zekri, S. (2008) Using Economic Incentives and Regulations to reduce Seawater Intrusion in the Batinah Coastal area of Oman, *Agricultural Water Management*, 95 (3), March.

Chapter 23

Groundwater governance in the European Union, its history and its legislation: an enlightening example of groundwater governance

Jean Fried[1], *Philippe Quevauviller*[2] & *Elisa Vargas Amelin*[3]

[1]*Urban Planning and Public Policy, School of Social Ecology, University of California, Irvine, USA*

[2]*Department of Hydrology and Hydrological Engineering, Vrije Universiteit Brussel, Brussels, Belgium*

[3]*European Commission, DG Environment, Brussels, Belgium*

ABSTRACT

This Chapter provides a unique historical perspective of the way groundwater governance has evolved in the EU within the past three decades, based on a dialogue science-policy and a change of priorities from purely economic considerations to environmental preservation. First, the 1976 "Dangerous Substances" Directive and the drafting of the first EU groundwater directive (1980) are described, with the groundwater survey and mapping carried out in the early 1980s, followed by the Ministerial Seminars of Frankfurt in 1988 on the Community Water Policy and The Hague in 1991 on Groundwater. The main orientations of the 1996 Groundwater Action Plan are then introduced, which have been taken aboard in the policy discussions that led to the adoption of the 2000 *European Union (EU) Water Framework Directive* (WFD). In this context, the call for a new groundwater regulation (under Article 17 of the WFD) is explained, as well as the consultation process and the related scientific support that enabled the adoption of the 2006 "Daughter" Groundwater Directive and its implication for groundwater governance at the EU level. Besides, the Chapter insists on an essential dimension of Groundwater Governance considering River Basin Management, introduced and developed by the WFD, providing a coherent planning and management system integrating surface water and groundwater, water quantity and quality, land-use planning and the interactions with the sectoral economic policies.

23.1 INTRODUCTION

Although groundwater and surface water are two components of the same hydraulic system and have to be managed conjunctively, the need for specific groundwater governance has been generally acknowledged. Due among others, to the invisibility of groundwater, which makes it hard to measure and to study, or its complex interactions with the geological domain and with human activities in the subsurface. Furthermore, although it represents almost 99% of all fresh water stored on Earth, it is often neglected by the policy-makers and unknown from the public. This specificity has been recognized by the European Union water legislation as soon as 1976.

An essential dimension of groundwater governance is the consideration of river basin management principles and this has been introduced and developed by the EU *Water Framework Directive* – WFD – (2000). It defines a geographical area determined by a river or stream catchment limits, including surface water and groundwater, taking into account interactions between groundwater and surface water in the basin, between water quantity and quality and between land and water, upstream and downstream, which turns river basins from a geographical area into a coherent system. In terms of planning, the river basin management approach is much broader than traditional water management as it includes land-use planning, agricultural policy, environmental management and other policy areas.

This chapter will provide a unique historical perspective of the way groundwater governance has evolved in the EU within the past three decades, based on a dialogue science-policy and a change of priorities from purely economic considerations to environmental preservation. First, the 1976 "Dangerous Substances" Directive and the drafting of the first EU groundwater directive (1980) will be described, with the groundwater mapping carried out in the early 1980s, followed by the Ministerial Seminars of Frankfurt in 1988 and The Hague in 1991. The main orientations of the 1996 Groundwater Action Plan will then be introduced, which have been taken aboard in the policy discussions that led to the adoption of the 2000 WFD. In this context, the call for a new groundwater regulation (under Article 17 of the WFD) will be explained, as well as the consultation process and the related scientific support that enabled the adoption of the 2006 "Daughter" Groundwater Directive and its implication for groundwater governance at the EU level. This will be continued by an update of the latest developments of the groundwater law (from the period of its adoption to date).

23.2 1976 TO 1996: FROM PREVENTING THE DISTORTION OF COMPETITION TO PROTECTING THE ENVIRONMENT, BIRTH AND EVOLUTION OF A EU GROUNDWATER POLICY[1]

23.2.1 Three paradoxes

The EU Groundwater Policy is built on two paradoxes concerning groundwater in general and a third paradox typical of the EU:

- Although surface water and groundwater are physically strongly related and, therefore, should theoretically be managed in an integrated way, it is usual to separate governance, policy and management of groundwater from surface water, for practical reasons due to the specific characteristics of groundwater, such as its invisibility and its strong local character. It happened in the EU, with the introduction of a specific groundwater directive from the very beginning.
- Although groundwater is the major fresh water resource worldwide, constituting up to 99% of all available fresh water, and about 75% of EU residents depend on

[1] Although the concept of European Union has been created in 1992 only, by the Treaty of Maastricht, we will constantly use the denomination "European Union" or "EU" for the sake of simplicity.

it for water supply, it is typically undervalued by governments. It is often weakly governed and underfunded and underrepresented in water policy discourse, largely neglected by policy-makers and even by hydrologists, as compared to surface water. Why? It is essentially due to the fact that it is an invisible resource and, therefore, unknown. This physical invisibility has resulted in a political invisibility. Therefore, it is rather interesting to see that the EU has made a kind of revolution as it conceived and developed a genuine groundwater legislation, and stimulated a better knowledge of that water resource by, among others, ordering a descriptive survey of the European aquifers, as we will see hereafter.

- The EU water policy and its related groundwater policy have started on a third specific paradox, which concerns the European environmental policy as a whole: the idea of protecting the environment at European level did not stem from the ecological necessity of protecting human health and life, but from the political-economic principle of *preventing the distortion of competition.* Some Member States have adopted environmental protection laws stricter than some others and, for clear economic reasons, industries could choose to invest in the countries with the least environmental constraints, which could be economically detrimental and constitute a distortion of competition. We will see how the EU philosophy in that matter has evolved with time, adapting to the changing needs of its populations.

23.2.2 Main steps and dates in the conception and elaboration of the first Groundwater Directive (1980) and its follow-up towards a specific Groundwater Directive (2006) jointly with the Water Framework Directive (2000)

Box 23.1: Summary Timeline

1976 Dangerous Substances Directive
1980 Groundwater Directive
1978–1986 Survey and mapping of the European Aquifers
1988 Frankfurt Ministerial Seminar on Water Policy
1991 The Hague Ministerial Seminar on Groundwater Policy
1996 A Groundwater Action Programme
2000 Water Framework Directive
2006 Groundwater Directive
2012 Repeal of the 1980 Directive under the WFD

The first wave of legislation concerning water took place from 1975 to 1980, resulting in a number of directives and decisions which either laid down *Water Quality Objectives (WQO)*[2] for specific types of water, *e.g.* the 1975 Drinking Water

[2]The Water Quality Objective (WQO) approach defines the minimum quality requirements of water to limit the cumulative impact of emissions, both from point sources and diffuse sources. The WQO's outline a desirable quality level of the water resource.

Abstraction Directive, or established *Emission Limit Values (ELV)*[3] for specific water uses, like the 1976 Dangerous Substances Directive and its Daughter Directives on various individual substances (1982–1986). Since then, the question of which approach was most appropriate has been the subject of long scientific and political debates. Subsequently, more recent legislation both at European and Member State levels, has been based on a *'combined approach'* where ELVs and WQOs are used to mutually reinforce each other. In any particular situation, the more rigorous approach will apply. The new European Water Policy, and its operative tool, the WFD (2000/60/EC), are based on this 'combined approach' (see, in particular, Article 10 of the latter) (Kallis and Nijkamp, 1999).

The 1976 *Directive 76/464/EEC **on pollution caused by certain dangerous substances discharged into the aquatic environment of the Community***, or "dangerous substances directive" as it is usually called, is at the origin of the EU groundwater legislation because its Article 4 explicitly referred to the drafting of a specific groundwater directive. This 1976 directive in itself has been an important component of EU water legislation as it provided the framework for subsequent regulation to control the discharge of specific dangerous substances, classified in two lists according to the danger they present (see hereafter).

The drafting of a groundwater directive, *Directive 80/68/EEC on the protection of groundwater against pollution caused by certain dangerous substances*, started at the end of 1976 and it took three years to complete. Finished in December 1979, it was adopted in January 1980. It provided a protection framework for groundwater that required prevention of the (direct or indirect) introduction of highly dangerous pollutants into groundwater (list I)[4] and limiting the introduction into groundwater of other pollutants considered as less dangerous (list II)[5] to avoid pollution of this water by these substances. It was replaced by a directive on the protection of groundwater against pollution and deterioration, adopted in 2006, jointly with the more general Water Framework Directive, or WFD, of 2000 (European Commission, 2008).

The directive also defined the concepts of direct and indirect discharges of pollution into groundwater: *'direct discharge'* means the introduction into groundwater of substances in lists I or II without percolation through the ground or subsoil and *'indirect discharge'* means the introduction into groundwater of substances in lists I or II after percolation through the ground or subsoil. To support the elaboration of the groundwater directive, a *survey and a mapping of the geographical extension, quantity, quality and vulnerability of the aquifers* were undertaken, concerning the nine EU Member States of the time, extending to Greece in 1981, and finally completed when Spain and Portugal joined the Community in 1986.

---

[3]The Emission Limit Value approach (ELV) focuses on the maximum allowed quantities of pollutants that may be discharged from a particular source into the aquatic environment.
[4]List I: List of certain individual substances selected mainly based on their toxicity, persistence, and bioaccumulation, except for those which are biologically harmless or which are rapidly converted into substances which are biologically harmless.
[5]List II: List containing substances which have a deleterious effect on the aquatic environment, which can, however, be confined to a given area and which depend on the characteristics and location of the water into which they are discharged.

In June 1988 a *ministerial seminar on the Community Water Policy*, organized in Frankfurt, reviewed the existing legislation and identified a number of improvements that could be made and gaps that could be filled. The conclusions of the Ministerial Seminar highlighted the need for Community legislation covering ecological quality of surface water. The Council in its resolution of 28 June 1988, immediately following the ministerial seminar, asked the Commission to submit proposals to improve ecological quality in Community surface waters. Although mentioned in the discussions, groundwater was not specifically debated.

This resulted in a second wave of water legislation, whose first results were, in 1991, the adoption of the *Urban Waste Water Treatment Directive*, providing for secondary (biological) waste water treatment, and even more stringent treatment where necessary, and the *Nitrates Directive* (European Commission, 1991a), addressing water pollution by nitrates from agriculture. Although not groundwater specific, both directives may have consequences on groundwater, through artificial recharge of aquifers by treated waste water for the first one, and pollution of aquifers by fertilizers for the second one.

Because of the original characteristics of groundwater, a specific groundwater Ministerial Seminar took place in November 1991 in The Hague (Netherlands), along the same methodological lines. It defined a European Union groundwater policy, identifying the threats to groundwater and the corrective and preventive measures to be taken to ensure sustainable groundwater management. It also stressed the necessary policy, governance and management integration of surface water and groundwater as a whole, paying equal attention to both quantity and quality aspects. It called for a program of actions to be implemented by the year 2000 aiming at sustainable management and protection of freshwater resources.

In its resolutions of 25 February 1992, and 20 February 1995, the Council requested an action program for groundwater and a revision of the 1980 Groundwater Directive as part of an overall policy on freshwater protection. In 1996 such a *Groundwater Action Programme* was adopted (see section 23.3). Pressure for a fundamental rethink of Community water policy came to a head in mid-1995: the Commission, which had already been considering the need for a more global approach to water policy, accepted requests from the European Parliament's Environment Committee and from the Council of Environment ministers. It corresponded to an increasing awareness of citizens and other involved parties for their water. At the same time water policy and water management had to address problems in a coherent way. Therefore, a new European Water Policy was developed in an open consultation process involving all interested parties.

A Commission Communication was formally addressed to the Council and the European Parliament, but at the same time invited comment from all interested parties, such as local and regional authorities, water users and non-governmental organizations (NGOs). A score of organizations and individuals responded in writing, most of the comments welcoming the broad outline given by the Commission.

As the culmination of this open process a two-day Water Conference was hosted in May 1996. This Conference was attended by some 250 delegates including representatives of Member States, regional and local authorities, enforcement agencies, water providers, industry, agriculture and, not least, consumers and environmentalists.

The outcome of this consultation process was a widespread consensus that, while considerable progress had been made in tackling individual issues, the current water policy was fragmented, both in terms of objectives and means. All parties agreed on the need for *a single piece of framework legislation* to resolve these problems. In response to this, the Commission presented a Proposal for a *Water Framework Directive (WFD)* with the following key aims:

- expanding the scope of water protection to all waters, inland surface waters, transitional waters (bodies of water near river mouths, partly saline but substantially influenced by freshwater flows), coastal waters and groundwater
- achieving "good status" for all waters by a set deadline
- water management based on river basins
- "combined approach" of emission limit values and quality standards
- getting water-pricing right
- getting the citizen involved more closely
- streamlining legislation

Following the introduction of this WFD in 2000, a specific directive on the protection of groundwater against pollution and deterioration was adopted in 2006. It should be stressed that, for the first time in the EU water policy history, purely environmental concerns were introduced (Quevauviller *et al.*, 2011).

23.2.3 The conception, design, draft and adoption of the European Groundwater Policy and Governance, a model of Science-Policy dialogue and a Governance instrument

The elaboration and drafting of the European Groundwater Policy and Governance can be considered as a model of dialogue between Policy and Science, and used as such in the education and training of policy- and decision-makers who will be in charge of water resources policy, governance and management (Fried, 2016).

Starting in 1976, the Commission established a working group made up of scientists of various disciplines, lawyers and policy-makers, originating from the Member States, who used to meet several times a month in Brussels at the Commission. With the exception of the hydrogeologists of the group, nobody knew what groundwater was and, during the first discussions, the group drafted and adopted basic definitions that are now part of the groundwater directive, *e.g.* groundwater or pollution:

- 'groundwater' means all water which is below the surface of the ground in the saturation zone and in direct contact with the ground or subsoil;
- 'pollution' means the anthropogenic discharge, directly or indirectly, of substances or energy into groundwater, the results of which are such as to endanger human health or water supplies, harm living resources and the aquatic ecosystem or interfere with other legitimate uses of water.

The scientists of the group spent quite some time explaining to their colleagues, political scientists and lawyers, what they meant by groundwater or pollution, which also proved beneficial for them. They worked in English but had to assist the translators to choose the right words in every European Union language and reach the right juridical meaning. Of course the texts had to go to *lawyers-linguists* at the end of the process.

Another important feature of the dialogue between the scientists and their legal and policy counterparts was to understand that the protection of the environment was not in itself a motivation for their work but that the policy people were guided by the *political-economic principle of preventing the distortion of competition*, as explained above in the preliminary notices. And this is where *the dialogue acquired its real value*: the scientists, in a few years, managed to convince the political people of the need of protecting the water resources for the sake of the populations on the one hand, and, on the other hand, for the sake of the environment per se, and to open their minds to a more ecological approach, which in the late 1980s gave way to the concept of *sustainable development*. To support the preparation of this specific groundwater directive, it was decided, at Commission level, to do the assessment of the European groundwater resources. This assessment consisted in the survey and mapping of the aquifers of the Member States of the time. It was comprised of the general mapping, country by country, of all known aquifers, showing their geographic positions and whether confined or unconfined, their storage coefficients and their transmissivities, and an evaluation of the available water quantity as the balance between abstraction and recharge. It was completed by a mapping of their vulnerabilities, measured in residence times (the time it takes for an ideal pollutant to migrate from the ground surface to the water table). A general synthetic report and the country reports were produced and the whole was published in 1982. It was completed by the mappings and reports concerning Greece (1981) and Spain and Portugal (1986).

With the preparation and drafting of other directives, such as the "nitrates from agricultural sources" directive or the "wastewater treatment" directive that were adopted in 1991, the science-policy dialogue did take new dimensions as it was extended to exchanges with various political and economic pressure groups such as farmers' unions, ecological associations or representatives of the industrial sector. Meetings were often informal and person to person to prepare for the Commission formal meetings with the representatives of the Member States, policy-makers, lawyers and scientists, which provided an opportunity to the scientists to explain science to their counterparts and to be informed about the political and legal constraints.

At the same time, a small group, consisting of about four or five Commission administrators, essentially lawyers and political scientists, and an external expert-consultant, was analysing the European Union environmental legal realizations (Fried, 2016). They came to the conclusion that the existing water legislation, dealing essentially with water quality, was a good start but far from enough, and a more comprehensive legislation was needed, meaning that a political framework had to be designed and adopted at European level, a genuine European Water Policy. They decided that they would use *the power of initiative of the Commission* to have a policy adopted and had the idea of proposing the organization of a *ministerial seminar* for that purpose. According to this concept, a ministerial seminar was a way of having the top decision-makers, the ministers, discuss as informally as possible the various sensitive aspects of a water policy. A working group of national experts prepared the agenda and drafted a policy document and an action plan to be discussed during the seminar and adopted at the end, being confirmed by an official council of ministers the very next day. The ministerial seminar was held on June 27 and 28, 1988 in Frankfurt and the council of ministers on June 28, 1988 in Brussels. And the result proved successful as the European Union adopted a water policy for the first time of its history[6], whose results have been briefly presented in §2.2. Also presented in §2.2, the 1991 Ministerial Seminar on Groundwater of The Hague served as the basis for the presentation by the Commission of a proposal for a Decision of the European Parliament and the Council on an *Action Programme for Integrated Protection and Management of Groundwater* which was adopted on 25th November 1996 and in which the Commission pointed to the need to establish procedures for the regulation of abstraction of freshwater and for the monitoring of freshwater quality and quantity (see section 23.3).

These considerations coincided with the request made by the European institutions to the Commission to come forward with a proposal for a Directive establishing a framework for a European water policy. They were hence naturally embedded into the development of this large policy framework development, which resulted in the adoption of the *WFD2000/60/EC* on 23rd October 2000, and, in 2006, of the *Directive 2006/118/EC on the protection of groundwater against pollution and deterioration* (see sections 23.4 and 23.5).

From a science-policy and a methodology of action perspective, the very preparation of these ministerial conferences and the elaboration and adoption of their conclusions proved to be an interesting case study with regard to communication and exchanges between environmental scientists and policy-makers with the following objectives:

- Facilitate the scientific exchanges between scientists and technicians of different cultures, trying to elaborate common approaches to similar problems;
- Express the scientific problems and solutions in a language which could be understood by the policy-makers and, conversely, bring the political and economic preoccupations and constraints of the policy-makers to the scientists and technicians, in order for them to include them in their assumptions and guide them in their proposals of guidelines for the policy document and the action plan, which are part of the traditional output of a ministerial conference.

23.3 THE 1996 GROUNDWATER ACTION PROGRAMME

As stressed above, the The Hague declaration (1991) underlined the need to "establish a programme of actions to be implemented by the year 2000 at national and Community level aiming at sustainable management and protection of water resources". It also stressed that "the objective of sustainability should be implemented through an integrated approach, which means that groundwater and surface water should be managed as a whole, paying equal attention to both quantity and quality aspects; that all interactions with soil and atmosphere should be duly taken into account;

[6]A document of conclusions in English, German and French has been issued at the end of the Seminar. It served as the basis for the Council of Ministers that followed the next day. It should still be in the Commission archives. I have a copy. The English name is "The Community Water Policy".

and that water management policies should be integrated within the wider environmental framework as well as with other policies dealing with human activities such as agriculture, industry, energy, transport and tourism". The Groundwater Action Programme (GWAP) (European Commission, 1996a) called for Community actions and requested that a detailed action programme be drawn up for comprehensive protection and management of groundwater as part of an overall policy on water protection.

The GWAP stressed that Community groundwater policy had suffered from a lack of overall planning and a lack of concerted actions across the Community. In particular, the directive on groundwater protection from 1980 (European Commission, 1980) was said to "rather narrowly focus on control of emissions of substances from industrial and urban sources while provisions for control of diffuse sources from agriculture, forestry and other sources only recently have been added to Community environment legislation[7]". The absence of provisions on groundwater quantity and abstraction of fresh water was also noted. Following the analysis of existing legislation, the Commission adopted a Communication on European Water Policy (European Commission, 1996b), which constituted the foundation of the future WFD (see section 23.2). It stipulated that basic provisions on groundwater management would be incorporated, including provisions of the Groundwater Directive described in section 23.2 of this chapter (European Commission, 1980). Further integration of water policy into other Community policies would also be ensured, in particular regarding agriculture and regional development, and the importance of research was emphasized (Schmitz *et al.*, 1994). The communication highlighted the need to closely follow-up progress in implementing Community water legislation, notably the *Nitrates Directive* (European Commission, 1991a) and the *Urban Waste Water Treatment Directive* (European Commission, 1991b).

It should be stressed that the benefits of the programme were merely turned towards the achievement of a stable and sufficient supply of high quality fresh water from groundwater for domestic as well as for industrial or other uses. This aim was related to the need to reduce costs for expensive heavy infrastructure for water supply and the reduction of treatment and purification costs. Identified costs were related to treatment of waste water, control of nitrates, and environmental impact assessments (which were actually not incurred to the action programme), as well as other costs linked to the directives on integrated pollution prevention and control (European Commission, 1996c), drinking water quality (European Commission, 1998), landfills (European Commission, 1999) and the WFD (European Commission, 2000).

From this on, groundwater features were actually discussed in the framework of the negotiation of the WFD, and the GWAP never really entered into force before the adoption of the WFD and the new Groundwater Directive which are briefly described in sections 23.4 and 23.5. Discussions about whether GWAP recommendations were fully taken on board by the regulatory regime now in place are developed elsewhere (Quevauviller, 2008).

[7]This refers to the Nitrates, Plant Protection Products and Biocides Directives which are mentioned below.

Table 23.1 WFD definitions relevant to groundwater.

Ref. WFD	*Good status*
Good quantitative status (Annex V.2.1.2)	The level of groundwater in the groundwater body is such that the available groundwater resource is not exceeded by the long-term annual average rate of abstraction. Accordingly, the level of groundwater is not subject to anthropogenic alteration such as would result in: (a) failure to achieve the WFD environmental objectives for associated surface waters, (b) any significant diminution in the status of such waters, and (c) any significant damage to terrestrial ecosystems which depend directly on the groundwater body. Alterations to flow direction resulting from level changes may occur temporarily, or continuously in a spatially limited area, but such reversals do not cause saltwater or other intrusion, and do not indicate a sustained and clearly identified anthropogenically induced trend in flow direction likely to result in such intrusions.
Good chemical status (Annex V.2.3.2)	The chemical composition of the groundwater body is such that the concentration of pollutants do not exhibit the effects of saline or other intrusions (as determined by changes in conductivity) into the groundwater body, do not exceed the quality standards applicable under other relevant Community legislation in accordance with Article 17 of the WFD, and are not such as would result in failure to achieve the WFD environmental objectives for associated surface waters not any significant diminution of the ecological or chemical quality of such bodies nor in any significant damage to terrestrial ecosystems which depend directly on the groundwater body.

23.4 THE GROUNDWATER POLICY FRAMEWORK UNDER THE WFD

The WFD (Directive 2000/60/EC) is considered as a quite advanced regulatory framework for the protection of all (surface and ground) waters in order to achieve 'good status' objectives. As described below, it is based on specific milestones and operational steps which have to be undertaken by Member States. With regard to groundwater, the directive stipulates that Member States have to implement measures necessary to prevent or limit the input of pollutants into groundwater and to prevent the deterioration of the status of all groundwater bodies. In this context, Member States have to protect, enhance and restore all groundwater bodies, ensure a balance between abstraction and recharge, with the aim to achieve good groundwater (chemical and quantitative) status by 2015, following the definitions given in Table 23.1. A range of derogation clauses are in place regarding force majeure cases, technical feasibility etc. (Article 4 of the directive).

The Directive also requires the implementation of measures necessary to reverse any significant and sustained upward trend in the concentration of any pollutant resulting from the impact of human activity in order to progressively reduce groundwater pollution.

Under this Directive, the framework for groundwater protection imposes on Member States to:

- Delineate groundwater bodies within River Basin Districts and characterise them through an analysis of pressures and impacts of human activity on the status of groundwater in order to identify groundwater bodies presenting a risk of not achieving WFD environmental objectives (following vulnerability studies).
- Establish registers of protected areas within each river basin districts for those groundwater areas or habitats and species directly depending on water, including all bodies of water used for the abstraction of water intended for human consumption (European Commission, 1998) and all protected areas covered by the *Bathing Water Directive* 76/160/EEC (European Commission, 1976), vulnerable zones under the *Nitrates Directive* 91/676/EEC (European Commission, 1991a) and sensitive areas under the *Urban Wastewater Treatment Directive* 91/271/EEC (European Commission, 1991b), as well as areas designated for the protection of habitats and species including relevant Natura 2000 sites designated under Directives 92/43/EEC (European Commission, 1992) and 79/409/EEC (European Commission, 1979). In this context, vulnerable zones are defined as "*all known areas of land in Member States territories which drain into the waters affected by pollution and waters which could be affected by pollution and which contribute to pollution*". For these vulnerable zones, action programmes are required under the *Nitrates Directive* to reduce pollution caused or induced by nitrates and prevent further pollution.
- Based on the results of the characterisation phase (resulting from the above mentioned pressures and impacts analysis, enabling to identify groundwater bodies "at risk" of not achieving WFD objectives), establish a groundwater monitoring network providing a comprehensive overview of groundwater chemical and quantitative status (Quevauviller, 2005), and design a monitoring programme that has to be operational by the end of 2006. Monitoring will have to be reported, following requirements summarised in a guidance document (European Commission, 2007).
- Set up a river basin management plan (RBMP) for each river basin district, including a summary of pressures and impact of human activity on the groundwater status, a presentation in map form of monitoring results, a summary of the economic analysis of water use, a summary of the programme(s) of protection, control or remediation measures, etc. The first RBMPs were to be provided for the period 2009–2015 by Member States and their assessment was published by the European Commission in several reports. The second RBMPs cycle (2016–2021) is now on-going.
- Establish a programme of measures for achieving WFD environmental objectives (*e.g.* abstraction control, measures to prevent or control pollution). Basic measures include, in particular, controls over the abstraction of groundwater, controls (with prior authorisation) of artificial recharge or augmentation of groundwater bodies (providing that it does not compromise the achievement of environmental objectives). Point source discharges and diffuse sources liable to cause pollution are also regulated under the basic measures. Direct discharges of pollutants into groundwater are prohibited, subject to a range of provisions.

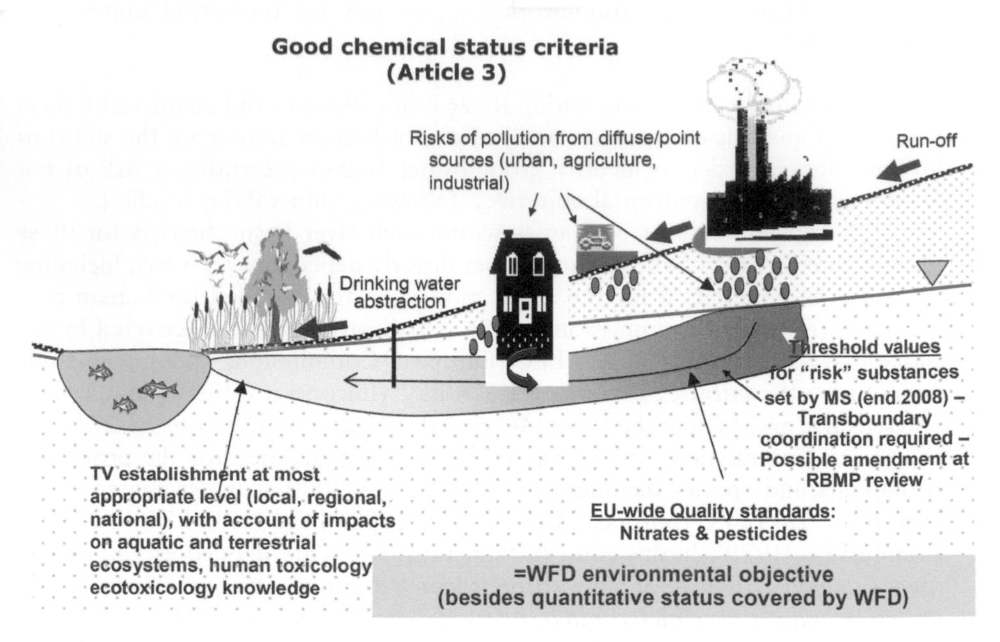

Figure 23.1 Criteria for good groundwater chemical status (TV stands for "ThesholdValues"; MS refers to Member States), from Quevauviller (2008).

23.5 THE 'DAUGHTER' GROUNDWATER DIRECTIVE 2006/118/EC

While quantitative status requirements are clearly covered by the WFD, it did not include, however, specific provisions on chemical status, *i.e.* the different conceptual approaches to groundwater protection did not allow achieving an agreement on detailed provisions within the WFD at the time of its adoption. Therefore, a new proposal was developed by the Commission, setting specific measures to prevent and control groundwater pollution, which led to the adoption of the "daughter" directive 2006/118/EC at the end of 2006 (European Commission, 2006). This directive is based on three main pillars, namely:

- Criteria linked to good chemical status evaluation, which are based on compliance to EU existing environmental quality standards (nitrates, plant protection products and biocides) and to "threshold values" (playing the same role as EQS) for pollutants representing a risk to groundwater bodies (Figure 23.1). The latter category of standards had to be established by Member States, using common methodological criteria, at the most appropriate scale (national, regional or local), taking account of hydrogeological conditions, soil vulnerability, types of pressures etc.
- Identification of sustained upward pollution trends and their reversal (Figure 23.2). Under this pillar, Member States (MS) need to identify trends for any pollutant characterising groundwater as being at risk, which is directly linked to the analysis of pressures and impacts carried out under the WFD. In this context, the obligation to apply trend reversal is established at 75% of values of EU-wide groundwater

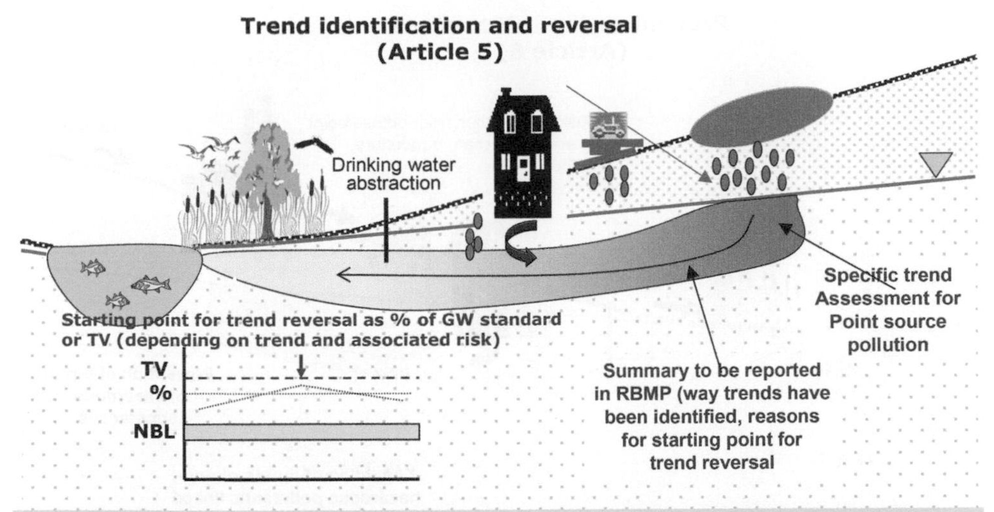

Figure 23.2 Criteria for trend identification and reversal (TV stands for "Theshold Values"; NBL refers to Natural Background Levels; RBMP stands for River Basin Management Planning), from Quevauviller (2008).

quality standards or threshold values (a rule that might be adapted according to local justified circumstances). Trend reversal should take place through the programme of measures of the RBMPs.

– Requirements on the prevention/limitation of pollutant inputs to groundwater, which ensure a continuity of the 80/68/EEC Directive (European Commission, 1980) regime after its repeal in 2013, *i.e.* the same principle of prevention of hazardous substances introduction and limitation of other pollutants so as to avoid pollution still apply (Figure 23.3).

Other elements concern clarifications about the groundwater use as drinking water (albeit this is well covered by Article 7 of the WFD) and its relation with the present directive, which relates to WFD environmental objectives. Recommendations to undertake research on groundwater ecosystems were also expressed in a recital, illustrating the awareness for a needed scientific integration. Finally, review of technical annexes of the directive (in particular concerning the establishment of groundwater threshold values and methods for identifying and reversing pollution trends) were requested, taking into account scientific progress, following "comitology" rules, *i.e.* adoption of possible technical amendments of annexes by a regulatory committee composed of Member States.

As said above, it should be kept in mind that related directives are part of the 'basic measures' listed in Annex VI of the WFD. In other words, they have to be efficiently implemented to allow for the environmental objectives to be attained. This concerns in particular the *Nitrates Directive* which aims to reduce water pollution caused or

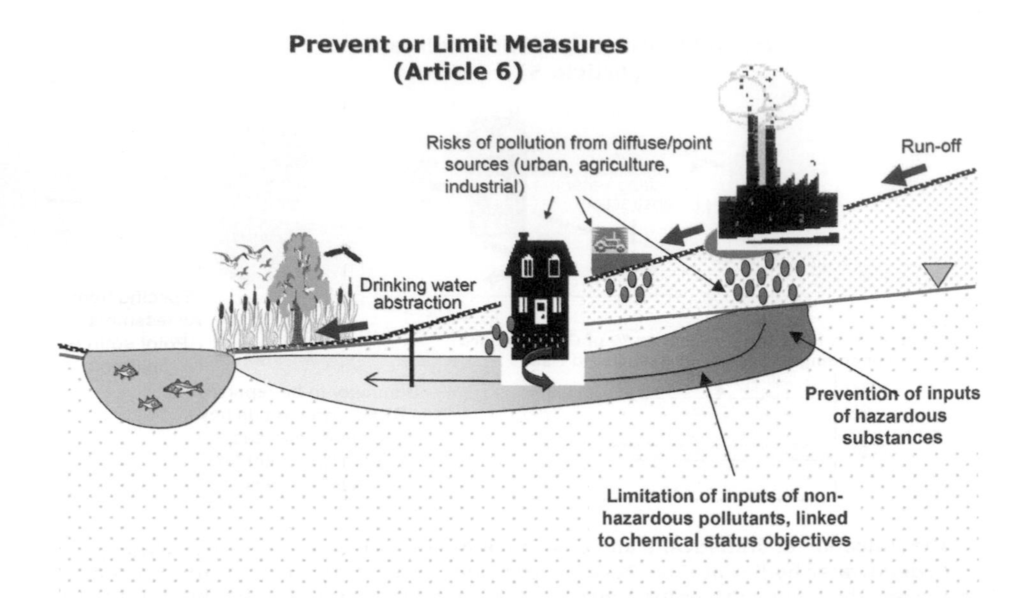

Figure 23.3 Requirements for prevention or limitation of pollutant inputs, from Quevauviller (2008).

induced by nitrates from agricultural sources and to prevent further such pollution. The GWD, *i.e.* the daughter Directive of 2006, reinforces this directive in imposing that nitrate contamination levels should not be over the trigger value (value triggering actions to reduce nitrate contamination but not set a standard *sensu stricto*) set at 50 mg/l (this argument has been used for proposing this value as an EU groundwater quality standard for groundwater in the new Groundwater Directive). The *Plant Protection Products Directive* (European Commission, 1991b) concerns the authorisation, placing on the market, use and control within the Community of plant protection products in commercial form. Regarding groundwater, authorisations are only granted whether plant protection products have no harmful effect on human or human health, directly or indirectly, or on groundwater, and they have no unacceptable influence on the environment, particularly contamination of water including drinking water and groundwater. The directive makes a direct reference to groundwater contamination (with drinking water standards not allowed to be exceeded), which therefore requires to be monitored. Similarly to nitrates, the pesticides values found in this directive have been used to set up EU-wide groundwater quality standards that appear in Annex I of the new Groundwater Directive. The trend reversal obligation also applies to pesticides. A similar approach has been followed regarding the *Biocides Directive* (European Commission, 1998b) for biocidal products and related authorisations, *i.e.* regarding EU-wide groundwater quality standards setting and trend reversal obligations.

23.6 THE NEW GROUNDWATER DIRECTIVE

To set quality standards Member States had to establish threshold values for a series of pollutants, and the 2006 Groundwater Directive (GWD) included a minimum list to consider. With the revision of the GWD Annexes in 2014 (European Commission, 2014), a few additional ones were included to the original list of the 2006 Directive, and the current list for pollutants is made of: Nitrates, Pesticides, Arsenic, Cadmium, Lead, Mercury, Ammonium, Chloride, Sulphate, Trichlorethylene, Tetrachlorethylene, Conductivity, Nitrites, and Phosphorus/Phosphates.

Of all these pollutants, there are only legal maximum values set at the EU level for Nitrates (50 milligrams/litre) and pesticides (individual 0.1 microgram/l and total 0.5 microgram/litre). The large differences in terms of geology and other factors across Europe, in addition to analytical difficulties and differences in views and political positions have made difficult the possibility of incorporating legal values for other pollutants.

With the completion of the first cycle and the assessment of the first EU RBMPs some preliminary conclusions were drawn for groundwater resources. In general, it was estimated that by 2015 about 90% of water bodies would reach good quantitative status and 77% good chemical status (European Commission, 2015), which in comparison to surface water bodies were quite positive figures. However, there were many gaps in status assessment and numerous groundwater bodies were classified as 'unknown status' by EU Member States due to lack of data and gaps in monitoring networks. Thus, it is still to be determined, through the assessment of the second RBMPs, what a more realistic picture of groundwater resources in the EU could be.

Furthermore, many of the main pressures (*e.g.* diffuse pollution from agricultural practices) are far from being reduced, and there is a huge variability in ranges of threshold values set across Europe due to different approaches followed and existing natural background levels of pollutants and their behaviour, all of which make the comparability process a quite complex exercise. This problem is currently being addressed by the existing technical working group on groundwater under the WFD Common Implementation Strategy (CIS), described in the following section.

23.7 SUPPORT TO IMPLEMENTATION: INVOLVEMENT OF STAKEHOLDERS

With the adoption of the WFD in 2000, many specific technical challenges and questions arose. Definitions and objectives were clearly stated in the Directive, but it was not always clear how to achieve the latter. Just to give an example, Member States were uncertain as how to set the boundaries of water bodies or how their aggregation should take place. To address such type of technical challenges and promote the Directive's implementation in a comparable and homogenous way, an informal and collaborative structure was created in 2001: the Common Implementation Strategy (CIS). This collaborative programme was developed to promote a joint work and understanding between the European Commission and Member States. Rapidly the CIS incorporated EFTA Countries[8], acceding and candidate countries, potential candidate countries, other stakeholders and NGOs to ensure that different views and positions could be

[8]The European Free Trade Association (EFTA) is an intergovernmental organisation set up for the promotion of free trade and economic integration to the benefit of its four Member States: Iceland, Liechtenstein, Norway and Switzerland.

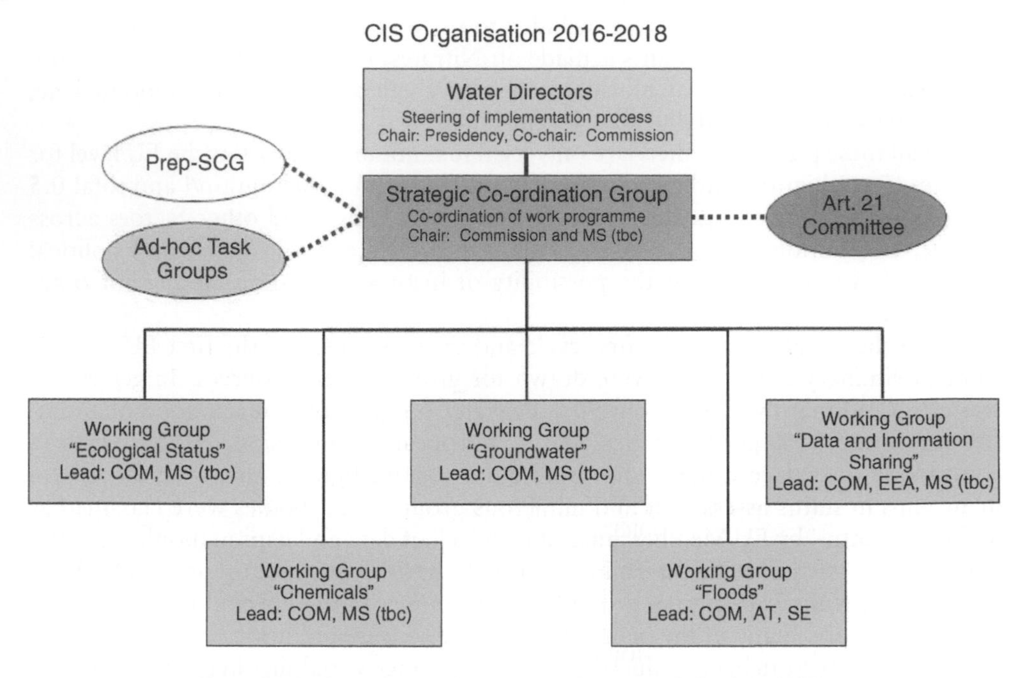

Figure 23.4 Common Implementation Strategy (CIS) of the WFD.

considered in technical works. The CIS has been considered a milestone in working together towards successful implementation of the core water policies at the EU level.

While the CIS structure has evolved and changed since its inception, its main scheme has remained with the following main structures: EU Water Directors as main political body to endorse decisions, the Strategic Coordination Group made up of representatives from EU Member States, other countries, NGOs and stakeholders, the regulatory committee according to article 21 of the WFD (only made up of Commission and MS representatives), in addition to working groups to address specific technical issues with their own working structure and mandate. The programme and structure of the CIS has been usually organised in a three-year long programme and a work plan developed accordingly and endorsed by Water Directors.

Under the framework of the CIS, 34 guidance documents[9] had been produced by the end of 2015, as well as numerous technical reports, which have been agreed and endorsed by Water Directors over the past years. All these documents are publically available, as well as minutes and presentations of all held meetings. While most are relevant to groundwater management, there are specific ones that have been very relevant for the first phase of implementing the WFD. It is the case for instance of the guidance document on groundwater monitoring (European Commission, 2007a), groundwater in drinking water protected areas (European Commission, 2007b), preventing or limiting direct and indirect inputs (European Commission, 2007c), status and trend assessment (European Commission, 2009) or risk assessment and conceptual models

[9]http://ec.europa.eu/environment/water/water-framework/facts_figures/guidance_docs_en.htm.

(European Commission, 2010). Some recent technical reports address challenges that were hardly addressed in the first cycle such as methodologies to assess groundwater dependent terrestrial and aquatic ecosystems (2014 and 2015).

Guidance documents are not legally binding, but they constitute the basis for applying technical specifications of EU water legislation and implementing agreed methodologies between the Commission and Member States and stakeholders. While some technical gaps still exist, the whole process has contributed to achieve more harmonised and comparable ways of implementing the WFD and linked Directives.

Figure 23.4 depicts the CIS structure for the period 2016-2018 (European Commission, 2015b). The main objectives for the 2016-2018 Work Programme are as follows:

1 Improving the implementation of the WFD and coordination with implementation of other water-related directives;
2 Increasing the integration of water and other environmental and sectoral policy objectives;
3 Contributing to fill in possible gaps and to identify potential improvements within the EU framework on water, including contributions toward the 2019 review of the WFD.

23.8 CONCLUSIONS

The historical perspective about EU groundwater policy presented in this chapter is intimately linked with interactions among policy-makers, the scientific community and a wide range of stakeholders. Science-policy interfacing has been largely explored in the past ten years, in particular in the light of the 2006 Groundwater Directive development (Quevauviller, 2010). At the time of the development of this daughter directive to the WFD, it was indeed recognised that research would be needed to develop a common methodology for the establishment of threshold values, and cooperation among different groundwater actors had hence to become effective to deliver scientific inputs to the political negotiation of the regulation. The difficulty, when dealing with groundwater science and policy, is to understand the interdependent nature of hydrologic, hydrogeologic and water-use systems, which makes the continuum between data, information and knowledge of particular importance. This understanding requires a flow of hydrological data of verified quality as well as data related to (ground)water uses. It also relies on a continuous refinement of the scientific foundation, both physical and social, upon which policy and management solutions are developed in context of the river basin management plans. Examples of projects funded by the European Commission have illustrated this type of progressive approach to groundwater information needs, with concrete interactions orchestrated within the Groundwater Working Group of the Common Implementation Strategy of the WFD (see section 23.7). The lack of data and scientific understanding of groundwater patterns often represents a critical gap undermining the development of groundwater policy and related management approaches, which may be a serious drawback in the implementation of the EU Groundwater policy framework described in this chapter.

The absence of data often limits the degree to which hydrogeologists are able to quantify and describe complex aquifer dynamics. Equally important are the ways in which raw data and information are treated, presented and used. Information is only useful if it is used. In many instances, however, the absence of information (or the non-use of essential information) creates situations in which emerging problems and management options are poorly understood. In other words, the push for a wide range of data to set sometimes difficult-to-interpret scenarios should possibly give more room to simpler approaches and improved communication of the knowledge base to policy-makers and the general public.

The challenge ahead of us is now to be able to reconcile necessary research on complex hydrogeological systems and their "translation" into outputs which may be usable by policy-makers in the context of groundwater policy implementation, as well as accessible to other non-technical stakeholders. It is only through a mediation mechanism among science and policy that the two worlds will cross-fertilise and be more intelligible to the general public.

REFERENCES

European Commission (1976) Council Directive 76/160/EEC of 8 December 1975 concerning the quality of bathing water, Official Journal of the European Communities, L 31, 5.2.1976, p.1.

European Commission (1979) Birds Directive 79/409/EEC, Official Journal of the European Communities, L 103, 25.4.1979, p.1.

European Commission (1980) Council Directive 80/68/EEC of 17 December 1979 on the protection of groundwater against pollution caused by certain dangerous substances, Official Journal of the European Communities, L 20, 26.1.1980

European Commission (1991a) Council Directive 91/676/EEC of 12 December 1991 concerning the protection of waters against pollution caused by nitrates from agricultural sources, Official Journal of the European Communities L 375, 31.12.1991, p.1.

European Commission (1991b) Council Directive 91/271/EEC of 21 May 1991 concerning urban waste treatment, Official Journal of the European Communities L 135, 30.5.1991, p.40.

European Commission (1991c) Council Directive of 15 July 1991 concerning the placing of plant protection products on the market, Official Journal of the European Communities, L 230, 19.8.1991, p.1.

European Commission (1992) Habitats Directive 92/43/EEC, Official Journal of the European Communities, L 206, 22.7.1992, p.7.

European Commission (1996a) Groundwater Action Programme, Official Journal of the European Communities, C 355, 25.11.1996, p.1.

European Commission (1996b) Communication from the Commission to the Council and European Parliament on European Community Water Policy, COM(96) 59 final, 21.2.1996.

European Commission (1996c) Council Directive 96/61/ECof 24 September 1996 concerning integrated pollution prevention and control, Official Journal of the European Communities, L 257, 10.10.1996, p. 26.

European Commission (1998) Council Directive 80/778/EEC of 15 July 1980 relating to the quality of water intended for human consumption, Official Journal of the European Communities L 229, 5.12.1998, p.32, as amended by Council Directive 98/83/EC of 3 November 1998, OJ L 330, 5.12.1998, p. 32.

European Commission (1998b) Directive 98/8/EC of the European Parliament and of the Council of 16 February 1998 concerning the placing of biocidal products on the market, Official Journal of the European Communities L 123, 24.4.1998, p.1.

European Commission (1999) Council Directive 1999/31/EC of 26 April 1999 on the landfill of waste, Official Journal of the European Communities, L 182, 16.07.1999, p.1.

European Commission (2000) Directive 2000/60/EC of the European Parliament and of the Council of 23 October 2000 establishing a framework for Community action in the field of water policy, Official Journal of the European Communities, L 327, 22.12.2000, p.1.

European Commission (2003) Common Implementation Strategy for the Water Framework Directive, European Communities, ISBN 92-894-2040-5, 2003.

European Commission (2006) Directive of the European Parliament and of the Council of 12 December 2006 on the protection of groundwater against pollution and deterioration, Official Journal of the European Communities, L 372, 12.12.2006, p. 19.

European Commission (2007a) Monitoring Guidance for Groundwater, CIS Guidance N° 15, Common Implementation Strategy of the WFD, European Commission.

European Commission (2007b) Guidance on Groundwater in Drinking Water Protected Areas, CIS Guidance N° 16, Common Implementation Strategy of the WFD, European Commission, 2007.

European Commission (2007c) Guidance on Direct and Indirect Inputs of Pollutants in the context of the Directive 2006/118/EC, CIS Guidance N° 17, Common Implementation Strategy of the WFD, European Commission, 2007.

European Commission DG Environment (2008) *Groundwater Protection in Europe, the new Groundwater Directive-Consolidating the EU Regulatory Framework,* http://bookshop.europa.eu.

European Commission (2014) Commission Directive 2014/80/EU of 20 June 2014 amending Annex II to Directive 2006/118/EC of the European Parliament and of the Council on the protection of groundwater against pollution and deterioration Text with EEA relevance.

European Commission (2015) Communication from the Commission to the European Parliament and the Council, the Water Framework Directive and the Floods Directive: Actions towards the 'good status' of EU water and to reduce flood risks. COM(2015) 120 final.

European Commission (2015b) Discussion paper on the CIS Work Programme 2016–2018. Water Directors – Riga 26–27 May 2015.

Fried J. (2016) *Notes for the Course "Groundwater Policy and Governance, the examples of the United States and the European Union"*, PPD 100, University of California, Irvine, (unpublished).

Grath J., Ward R., Scheidleder A. and Quevauviller Ph., *J. Environ. Monitor.*, 2007, 9, 1162–1175

Kallis G. & Nijkamp P. (1999) *Evolution of EU water policy: A critical assessment and a hopeful perspective*. Research Memorandum 1999-27, Serie Research Memoranda, Faculteit der Economische Wetenschappen en Econometrie, Vrije Universiteit Amsterdam

Quevauviller Ph. (2005) *J. Environ. Monitor.*, 2005, 7, 89–102.

Quevauviller Ph. (2008) From the 1996 Groundwater Action Programme to the 2006 Groundwater Directive What have we done, what have we learnt, what is the way ahead? *J. Environ. Monitor.*, 10, 408–421.

Quevauviller Ph. Ed. (2010) *Groundwater Science and Policy – An international overview*, The Royal Society of Chemistry, Cambridge, ISBN: 978-0-85404-294-4, 2007, p. 754.

Quevauviller Ph., Borchers U., Thompson K.C., Simonart T., Eds. (2011a) *The Water Framework Directive – Action Programmes and Adaptation to Climate Change*, RSC Publishing, Cambridge, ISBN: 978-1-84973-053-2, 214 p.

Schmitz B., Reiniger P., Pero H., Quevauviller Ph. and Warras M. (1994) Europe and scientific and technological cooperation on water, European Commission, EUR 15645 EN.

Struckmeier W. (2011) in *Water Status under the Water Framework Directive*, U. Borchers, K.C. Thompson, T. Simonart and Ph. Quevauviller Eds., John Wiley & Sons Ltd., Chichester.

The full text of the Directives mentioned in this chapter can be found on the website of the Official Journal of the European Communities

Chapter 24

Groundwater governance in the United States: a mosaic of approaches

Sharon B. Megdal[1], *Adriana Zuniga Teran*[2], *Robert G. Varady*[2], *Nathaniel Delano*[3], *Andrea K. Gerlak*[2] & *Ethan T. Vimont*[1]
[1]*Water Resources Research Center, The University of Arizona, Tucson, AZ, USA*
[2]*Udall Center for Studies in Public Policy, The University of Arizona, Tucson, AZ, USA*
[3]*U.S. Environmental Protection Agency Region 2, New York, NY, USA*

ABSTRACT

Groundwater is an important water supply for meeting municipal, industrial, and agricultural water demands and for supporting riparian and other ecological systems in the United States (U.S.). Effective groundwater governance is therefore crucial to the wise use of this largely non-renewable resource (recharge rates are slower than extraction rates). While minimum, federally-established drinking-water quality and water-discharge regulations do exist, the framework of the laws and regulations governing groundwater use in this country is highly decentralized. Each state determines its own groundwater priorities and governance approaches, with the further potential for states to delegate significant responsibilities to sub-state jurisdictions. Painting the groundwater governance picture in the U.S. with a single brushstroke is therefore impossible; a more refined analysis is required to characterize the mosaic of groundwater governance priorities and approaches. In this chapter, we report on findings of a nationwide survey on groundwater governance. We address the variation in circumstances and approaches across the country, and provide insights into some challenges identified by survey respondents. To demonstrate the changing nature of groundwater governance and groundwater debates, we consider two Western U.S. states, California and Arizona. Arizona has long practiced groundwater management in designated parts of the state, while California only recently adopted a comprehensive approach to groundwater management. We conclude with a synopsis of ongoing national-scale research and a look to the future of groundwater governance in the U.S.

LIST OF ACRONYMS

ADWR – Arizona Department of Water Resources
AM – Adaptive Management
AMAs – Active Management Areas
CAP – Central Arizona Project
DWR – Department of Water Resources
GMA – Groundwater Management Act
IWRM – Integrated Water Resources Management
SGMA – Sustainable Groundwater Management Act

UA – University of Arizona
WRRC – Water Resources Research Center

24.1 INTRODUCTION

Like elsewhere in the world, water users in the United States (U.S.) are increasing their reliance on groundwater (Maupin *et al.*, 2015). Since the invention of the centrifugal pump after World War II, extraction of groundwater in this country has grown exponentially. Six decades later groundwater is the source for drinking water for about half of the population (Leshy, 2008). However, there are significant variations to this trend, depending on the water-use sector and region of the county. Irrigated agriculture is a major user of groundwater, consuming approximately two-thirds of the groundwater withdrawn in this country, mostly by the mostly arid Western States (Maupin *et al.*, 2015). In general, human needs met by groundwater varies widely by state; groundwater constitutes 80% of Kansas' fresh water supply, whereas Virginia relies on groundwater for less than 5% of its water (Figure 24.1). Human needs include public supply (including irrigation of greenspace), self-supplied domestic, irrigation, livestock, aquaculture, self-supplied industrial, mining, and the generation of thermoelectric power (our percentages are based on the most recently published data from the U.S. Geological Survey; Maupin *et al.*, 2015).

In contrast to governance in most other nations, water governance in the U.S. is highly decentralized. While the federal (*i.e.*, central) government establishes minimum drinking-water and water-discharge standards, individual states are largely responsible for managing their surface-water and groundwater supplies (Leshy, 2008; Megdal *et al.*, 2015). This state-level authority yields a mosaic of priorities for and approaches to both surface-water and groundwater governance. Moreover, many states do not recognize the connection between surface-water and groundwater, leading to different sets of laws and regulations for what is, actually, a single resource (Gerlak *et al.*, 2013). As a result, water governance is rendered particularly complex because of variability in reliance on groundwater versus surface water, and due to a near-total lack of uniformity in state-level statutory approaches to implementing and enforcing regulations (Megdal *et al.*, 2015).

Although western states rely on groundwater resources to accommodate growing demands more than states in the rest of the country, groundwater allocation does not follow such an east-west split. Each state has a particular set of laws and regulations regarding groundwater, and that kind of patchwork approach can even vary within the same state, as it does in Arizona, for example. The Groundwater Management Act of 1980 identified areas with heavy use of groundwater, known as Active Management Areas (AMAs). Although all AMAs have to comply with regulations stated in the Groundwater Code, certain regulations are AMA-specific in order to recognize differences in groundwater use and conditions. It is not difficult to imagine, then, particular governance structures and institutions developing as expressions of each state's individual laws and regulations. This is particularly problematic considering the paucity of scholarly focus on this topic until recent years.

But notwithstanding this lack of uniformity, a groundwater-governance framework is very important to managing groundwater over time. Typically, physical limitations are not the major barrier to sustainable groundwater use. Rather, it is

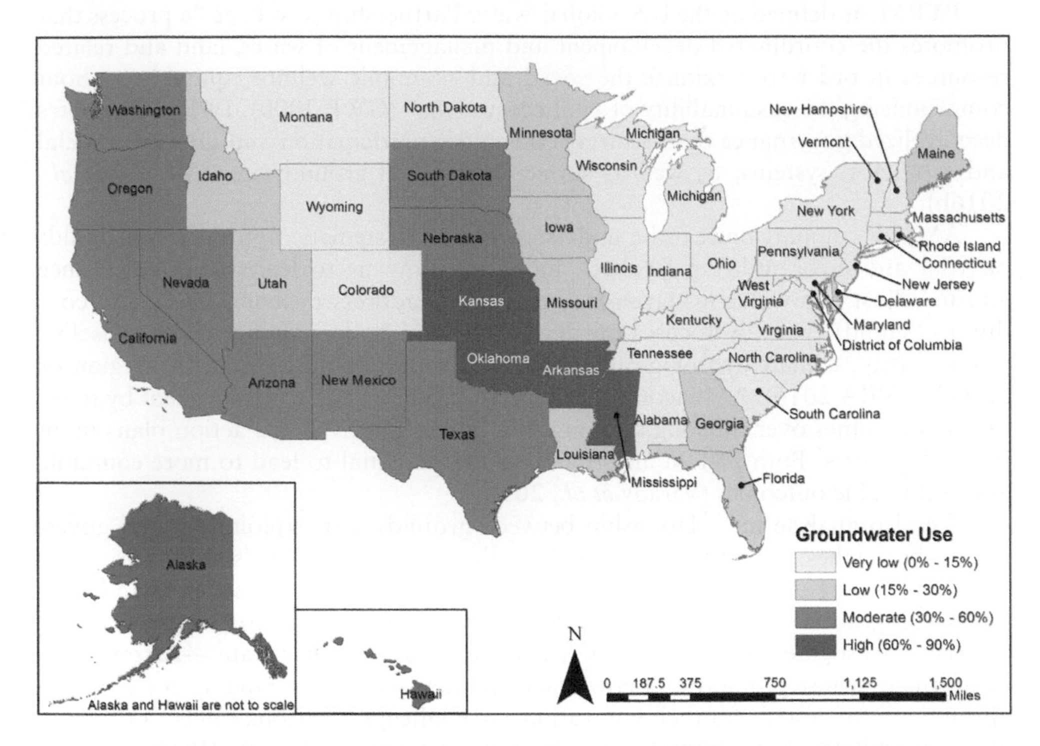

Figure 24.1 Percent of human needs met by groundwater using USGS data for 2010 (Maupin *et al.*, 2015).

ineffective governance regimes (*e.g.*, when some stakeholders are not considered in the decision-making process, or when extractions are not monitored) that often lead to overuse of the resource (Mukherji & Shah, 2005; Varady *et al.*, 2016a). In the past groundwater overexploitation has been exacerbated by a lack of critical evaluation of groundwater-governance structures (GEF *et al.*, 2016). In recent years, however, observers have begun to delve into the complexities associated with creating effective groundwater-governance regimes around the world (Varady *et al.*, 2016a). At the University of Arizona (UA), researchers at the Water Resources Research Center (WRRC) and the Udall Center for Studies in Public Policy have engaged in a multi-phased effort to understand groundwater governance in the country. The work began with a 2013 initial national-scale survey and continues with case-study analysis and a second survey (Gerlak *et al.*, 2013; Megdal *et al.*, 2014; Megdal *et al.*, 2015; Megdal *et al.*, 2016).

The present chapter is designed to improve knowledge about the variation of groundwater-governance challenges and approaches in a single (in this case, large) nation and to explore in greater depth the approaches of two states for which groundwater is a crucial water resource for all water-using sectors. Specifically, we use the most recent national data available to provide a picture of state-by-state dependence on groundwater across the nation and provide additional analysis of our 2013 survey results, in which we assessed groundwater governance across the country. We report those survey findings regarding the adoption of effective groundwater-governance approaches in each state, including Integrated Water Resources Management (IWRM) and Adaptive Management (AM).

IWRM, as defined by the U.S. Global Water Partnership (GWP) as "a process that promotes the coordinated development and management of water, land and related resources in order to maximize the social and economic welfare equitably without compromising the sustainability of vital ecosystems" (GWP 2000). IWRM promotes decentralized governance that ensures stakeholder participation and integrates social and ecological systems, as well as surface water and groundwater (Varady *et al.*, 2016b).

AM, by comparison, can be understood as a "systematic approach that builds on trial and error utilizing feedback loops to allow us to learn from experience and to adjust our water management practices to address evolving issues and conditions ... (and) typically focuses on developing and understanding of the baseline physical (*e.g.*, climate/hydrological), legal and socioeconomic aspects of a region or basin" (AWRA 2016). AM incorporates uncertainty into water management by monitoring outcomes over the long term and revisiting objectives and action plans in an iterative process. Both IWRM and AM have the potential to lead to more equitable and sustainable outcomes (Varady *et al.*, 2016b).

We also analyse the relationship between groundwater exploitation and governance approaches, identifying states with declining subsurface levels as a priority. We overlay survey results with state-level data on groundwater use released in 2015 by the U.S. Geological Survey, and we also include new analysis of the 2013 survey results. We then investigate groundwater governance in two particular states—Arizona and California—whose groundwater governance frameworks are commonly cited nationally and internationally for their innovation and challenges. Arizona can be considered an "early adopter" of comprehensive groundwater governance. In 1980 it enacted arguably the most rigorous and forward-looking groundwater regulations and governance structure in the United States. California, by contrast, can be seen as a "late adopter". It took until 2014 for that state to legislate comprehensive groundwater governance. Arizona was an early adopter of a comprehensive groundwater governance approach because its long history of groundwater overdraft caused land subsidence. Groundwater regulation was a pre-condition to obtaining federal funding to convey Colorado River water into the region. Recently, severe drought conditions and curtailment of surface water supplies have resulted in significant overdrafting of groundwater, which in turn led to state adoption of groundwater use regulations. The summaries are evidence of the complexities and challenges associated with groundwater governance and how it has been customized to address state-level conditions.

We conclude with a synopsis of ongoing research and speak to the future of groundwater governance in the U.S.

24.2 EXPLORING VARIATION IN GROUNDWATER GOVERNANCE IN THE U.S.

24.2.1 Methodology and approach to the 2013 survey

At the time the survey was conducted in 2013, there existed no compendium or analysis of groundwater-governance or management practices in the United States. A large body of work examined legal regimes in specific cases (*e.g.*, Bryner & Purcell, 2003; Torres,

2012; Dellapenna, 2012; Stevens, 2013), but few analysts had examined trends and commonalities in governance across the country. The 2013 UA study was designed to begin to address that knowledge gap. This preliminary study had three goals: (1) to understand each state's broad groundwater-governance framework, (2) to clarify the factors driving changes to groundwater governance, and (3) to assess state-by-state priority for groundwater governance.

While this book defines *groundwater governance* elsewhere (and includes variations among chapter authors), it is helpful to understand the definition we used to guide our design of the 2013 survey. We defined groundwater governance as "the overarching framework of groundwater use laws, regulations, and customs, as well as the process of engaging the public sector, and civil society" (p. 678) (Megdal *et al.*, 2015). It is important to differentiate water governance from water management. In the same study, we defined groundwater management as "the actions to implement those laws, policies, and decisions. It consists of the routine, practical, and effective ways that enable us to achieve predetermined goals and objectives" (p. 678) (Megdal *et al.*, 2015).

Groundwater governance includes coordinating administrative actions and decision-making among different jurisdictions, which may include: federal, state, tribal, county, or local (*i.e.*, municipal) governments; private business, individuals, and non-governmental institutions. In addition, there are other jurisdictional units that directly or indirectly affect groundwater governance such as Active Management areas (AMAs), U.S. Forest Service, Bureau of Reclamation, Bureau of Land Management, U.S. Fish and Wildlife Service, National Park Service, U.S. Amy Corps of Engineers special regions and protected areas, and other units of administration. A groundwater governance framework typically includes four dimensions: political, socio-cultural, economic, and ecological (Varady *et al.*, 2013). Also important to the survey design was consideration of *groundwater management*. While governance largely consists of the sets of laws, regulations, and customs in each state, groundwater management refers to the expression of governance in the routine practice of groundwater administration (Varady *et al.*, 2013). The 2013 survey was designed with these definitions in mind, and contained questions addressing the *de-jure* groundwater governance in each state, as well as the *de-facto* groundwater management conducted in the state on a day-to-day basis.

Specifically, the survey, employing the online service Survey Monkey, featured questions regarding the extent and scope of groundwater use, groundwater laws and regulations, and groundwater tools and strategies. The targeted respondents were largely identified through the network of federally authorized Water Resources Research Institutes at universities across the country (See NIWR.info). A single state-level respondent filled out the survey for each of the 50 states and the District of Columbia, for a total of 51 responses. The state agencies represented included: 22 water-quality agencies, 19 water-quantity agencies, and seven that managed both water quality and quantity in their state. In three states a representative could not be identified, so instead a researcher from the state's Water Resources Research Institute responded. Of the respondents: 8 identified as mid-level manager, 20 as manager, 5 as director/political appointee, 12 as engineer, 4 as hydrologist/geologist, 3 as planner, and one each identified as lawyer, researcher, or public-relations specialist.

The question-by-question results were posted on-line in "Groundwater Governance in the U.S.: Summary of Initial Survey Results" (Gerlak *et al.*, 2013), and

Table 24.1 Summary of key survey responses.

Questions	*Yes*	*No*	*No resp.*
States have groundwater laws (formal or informal)	50	0	1
State law recognizes the connection between surface-water & groundwater	25	23	3
State law recognizes groundwater quality	43	5	3
State law recognizes groundwater conservation	36	12	3
State law recognizes groundwater-dependent ecosystems	25	21	5
State agencies have groundwater oversight & enforcement authority	48	0	3
Local agencies have groundwater oversight & enforcement authority	31	1	19
Different state agencies oversee water quantity & water quality	36	15	0
State agencies have sufficient capacity to carry out responsibilities	25	23	3
Respondents have observed substantial changes in groundwater management	35	15	1

the survey questions are shown in Appendix A. The 2015 article, "Groundwater Governance in the United States: Common Priorities and Challenges" (Megdal *et al.*, 2015), expounded upon the fragmented nature of groundwater governance identified in the original survey, and identified three common priorities for groundwater governance: water quality and contamination, conflicts between users, and declining groundwater levels.

In this chapter, we offer summary results along with more in-depth and additional analysis of the 2013 survey results. We report findings regarding how the states categorize their groundwater-governance approaches, including their deployment of IWRM and AM. We also analyze the relationship between groundwater exploitation and governance approaches, identifying states with declining groundwater levels as a priority.

24.2.2 Key findings of the 2013 survey

Results of the survey show the decentralized nature of groundwater governance in this country, where most states have groundwater laws and agencies that oversee and enforce these laws (Table 24.1).

24.2.2.1 Groundwater use

As expected, the survey results show wide variation in groundwater use across and within states. Respondents offered a broad spectrum when asked to estimate the percentage of human demands met by groundwater in their state. "Human demands", as defined in the survey, include domestic, commercial, industrial, and agricultural – all uses except groundwater dependent ecosystems (see Appendix A). The percentages provided by these experts often did not coincide with the numbers developed by the U.S. Geological Survey (whose results can be seen in Figure 24.1). This discrepancy suggests a data-collection issue or respondents' incorrect perceptions of their state's reliance on groundwater. There was, however, strong agreement that groundwater reliance varied within each state. Two-thirds of respondents noted that reliance on groundwater varies by region within their state, and 88% said that the proportion of groundwater use by

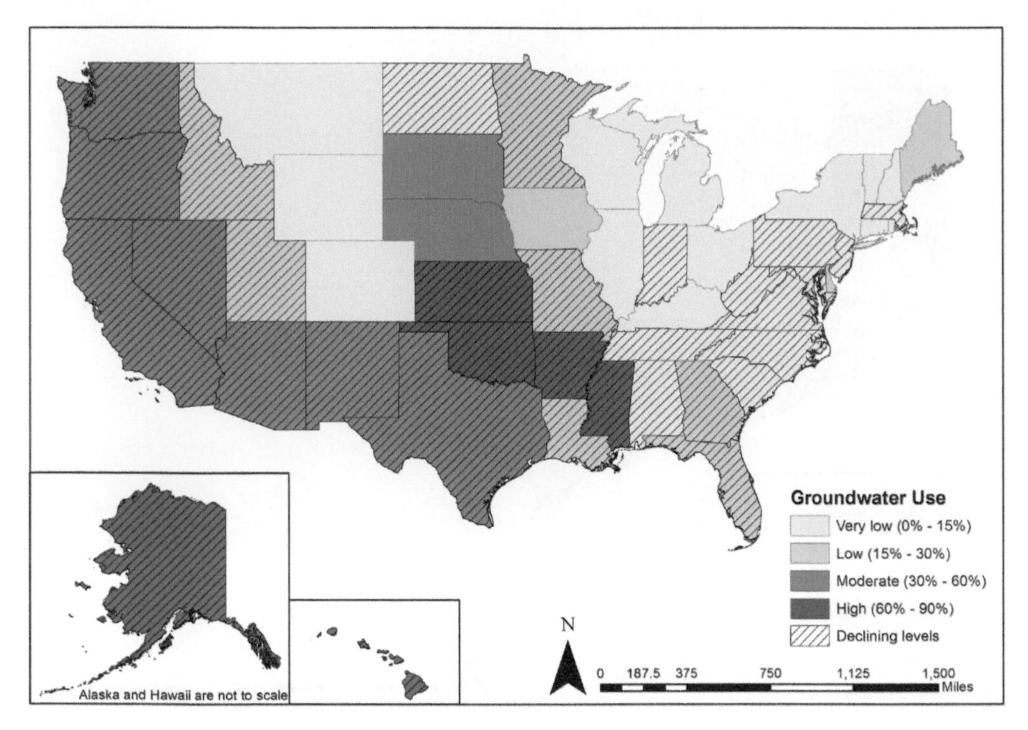

Figure 24.2 Percent of withdrawals from groundwater (USGS, 2010) combined with reported declining levels in aquifers (data from survey).

each major water-use sector (municipal, industrial, and agricultural) varies spatially within their states.

But not all states use groundwater resources in the same proportion. Our results are based on expert opinion and show that the 16 states that use groundwater for more than 30 percent of their human needs are located in both the eastern and western parts of the U.S (see "moderate" and "high" in Figure 24.1). Responses to our survey also revealed 32 states that have declared maintaining groundwater levels as a governance priority. Reliance on groundwater for their human needs combined with declining levels of groundwater suggests a "potential critical condition" of groundwater resources—that is, high reliance and decreasing supply (Figure 24.2).

Our analysis identified 13 states in this potential critical condition (Figure 24.3). Four of these (Arkansas, Kansas, Mississippi, and Oklahoma) rely on groundwater for more than 60 percent of their human uses (high) and they reported declining levels of groundwater. The remaining nine states in potential critical condition (Alaska, Arizona, California, Hawaii, Nevada, New Mexico, Oregon, Texas, and Washington) depend on groundwater for between 30 and 60 percent (moderate) of their human uses and report declining groundwater levels as a priority concern. Most of these are western states. It is important to acknowledge that this "critical" condition is likely to vary inside each state. Some regions within a state, for instance, may not be moderately or highly reliant on groundwater, or it may be that not all aquifer levels are declining.

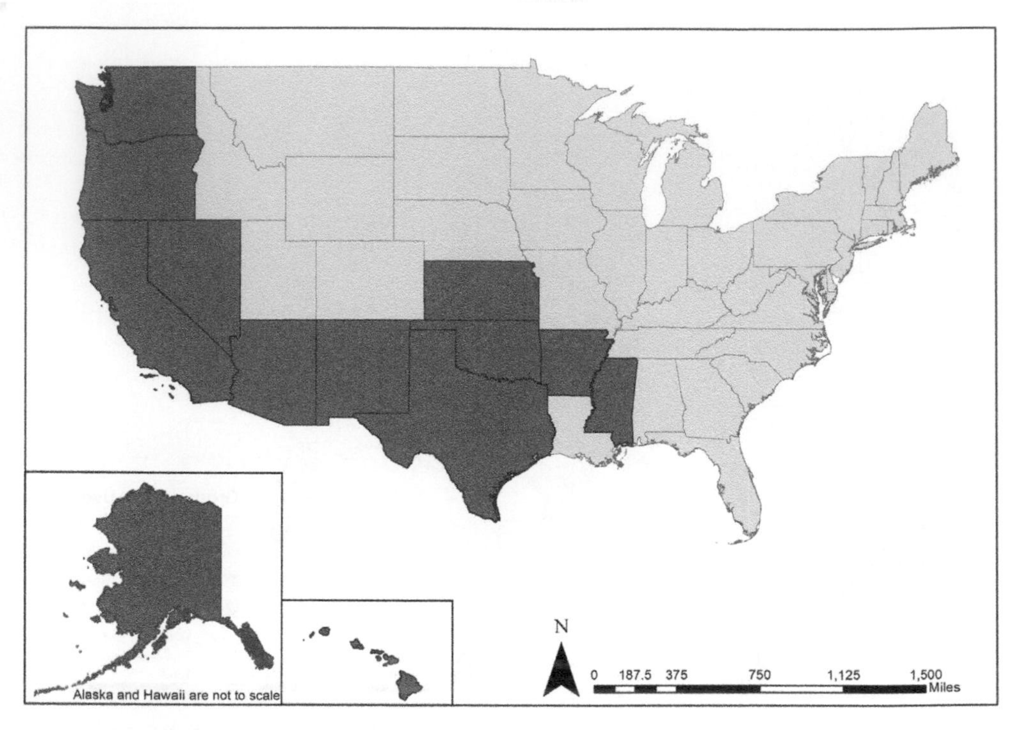

Figure 24.3 States that face a potential critical condition in their groundwater resources—more than 30 percent of human needs are met by groundwater (USGS, 2010) and reported declining aquifer levels (data from survey).

24.2.2.2 Groundwater-governance framework

As expected, the survey responses revealed considerable variation in existence and types of groundwater laws and regulations, though some uniformity did emerge. Ninety-six percent of respondents indicated that there exist formal groundwater policies, rules, or regulations in their state. While nearly all states have formal groundwater-governance structures in place, the structure can range from formal and explicit to a patchwork that is only indirectly applied to protect groundwater. Illinois and Nebraska, for example, have explicit laws protecting groundwater quality and quantity, while Pennsylvania does not have policies designed explicitly to protect groundwater; it relies instead on a piecemeal approach, using specific statutes or regulations for various activities to work for groundwater-quality protection.

Whatever the structure, the respondents generally agreed that this state of affairs had developed during the last few decades. Fully 70% of respondents had observed changes in the way groundwater is managed in the state, with the most common being: the passage of new legislation, changes in permitting processes, aquifer modelling, aquifer storage and recovery, and greater integration in management of groundwater and surface water.

While the responses showed significant attention to groundwater priorities, the survey also revealed areas of need. One of the biggest impediments to effective, sustainable groundwater governance is likely the continued lack of explicit recognition

Table 24.2 State groundwater governance priorities (N = 48).

Priorities	*Count of states*
Water quality/Contamination	45
Conflicts between water users (e.g., well interference)	36
Declining groundwater levels	32
Quantification of water rights	20
Regulatory disputes	12
Access	9
Other	8
Inter-agency jurisdictional conflict	3
There have been no clearly articulated priorities	2

of the inescapable connection between surface and groundwater in state water laws. Only half the respondents indicated that their state unequivocally recognized this connection. Additionally, 49% said that the courts are active in groundwater issues in their states, and 29% indicate existence of programs or settlements addressing Native American groundwater. Just more than a majority of respondents (54%) note that state law considers the water needs of groundwater dependent ecosystems.

The survey disclosed widely differing priorities regarding successful groundwater governance and management. Top priorities included: water quality/contamination, conflicts between users—*e.g.*, well interference, and declining groundwater levels (Table 24.2).

The survey showed that the application of groundwater regulation varied among water user groups. Twenty-nine respondents (59%) said that regulation applied to industrial users and publicly-owned community water systems; 28 (57%) indicated that it applied to privately-owned community water systems; 31 (63%) believed that it applied to all user groups. Of course, the nature of the regulation differed according to the type of water user. Some are exempt (*e.g.*, domestic household users), some have fewer restrictions (*e.g.*, agriculture users), while others face more stringent regulations (*e.g.*, users of public, industrial drinking water).

The survey also showed frequent delegation of groundwater-governance authority from state to local government levels. While nearly all respondents identified state agencies as responsible for oversight and enforcement, fully two-thirds mentioned that local agencies also have oversight and enforcement responsibilities. Groundwater oversight typically rests with environmental and natural-resources agencies and departments, water-resource boards and departments, and health departments—at both state and local governance levels.

24.2.2.3 *Groundwater management*

In addition to explaining groundwater-governance frameworks in each state, the survey considered variations across the nation. Common focal areas for the agencies surveyed included: permitting (88% of respondents), monitoring (80%), planning (70%), and protected areas (54%). Monitoring, both of groundwater quality and quantity, is widely practiced across water-use sectors, including municipal, industrial, and agricultural. Tools that are widely utilized by the agencies include public education programs, and increasing public access to data on water rights, groundwater use, and groundwater supply.

Holding governance power and having the ability to use it were seen to be very different attributes in the survey. While many state and local agencies theoretically hold groundwater-enforcement power, only half the respondents said that agencies had sufficient capacity to carry out mandated responsibilities. Insufficient staffing and/or programmatic money were the most commonly cited reasons for insufficient capacity to enforce laws and regulations. Where capacity fails, many states place their trust in voluntary measures to address groundwater issues. Eighty-eight percent of respondents say their state encourages the use of such measures for such actions as contamination clean-up, technical guidance, and information and education.

24.2.3 IWRM and AM in the U.S. states

Even though IWRM and AM are widely acclaimed by scholars, and in theory are portrayed as effective governance approaches that can be implemented to groundwater (Lemos, 2015; Varady *et al.*, 2016b), U.S. practice shows a big gap. Survey results suggest that neither IWRM nor AM governance approaches are commonly employed. Sixteen states reported using IWRM, while 14 states claim they are adhering to an AM approach. From these states, only nine reported using both IWRM and AM (California, Colorado, Connecticut, Hawaii, Massachusetts, Nebraska, Nevada, New Mexico, Oregon, Washington and Wyoming; Figure 24.4).

It is possible, however, that states may in fact be implementing principles from both IWRM and AM without referring to these practices with these terms; alternatively, it may be that some state representatives who responded did not recognize one or both of the terms. We asked the question *"In your state, are any of these groundwater management strategies in use? (check all that apply)."* Answer options included both IWRM and AM—along with other strategies such as aquifer recharge and storage programs, engagement of regional-planning or management organizations, economic incentives, public education programs, and "other." Interestingly, not all the states that reported using the IWRM approach (16 out of 51 or 31.3 percent) also reported having a law that recognized explicitly the surface-water/groundwater connection, an essential element of IWRM. Survey results show the 10 states (California, Colorado, Connecticut, Hawaii, Massachusetts, New Mexico, Oregon, Washington, West Virginia, and Wyoming) that reported having this law reportedly use the IWRM approach. Six states (Florida, Louisiana, Nebraska, Nevada, and New Jersey) reported using IWRM in spite of lacking a law that recognizes the surface-water/groundwater connection. It is important to note that to fully validate and update our reported findings; there needs to be substantially more research on the issue of recognizing this connection.

As with IWRM, a minority of states reported using an AM approach to groundwater governance (14 out of 51, or 27.4%) (Figure 24.4). Essential elements of AM are monitoring and flexible planning, so that learning can occur even in the face of uncertainty. However, only five of the 14 agencies that reported using AM (representing Connecticut, Hawaii, Nebraska, Nevada, and North Carolina) also stated that they plan and monitor both groundwater quality and quantity. The other nine agencies (from California, Colorado, Idaho, Kansas, Massachusetts, New Mexico, Oregon,

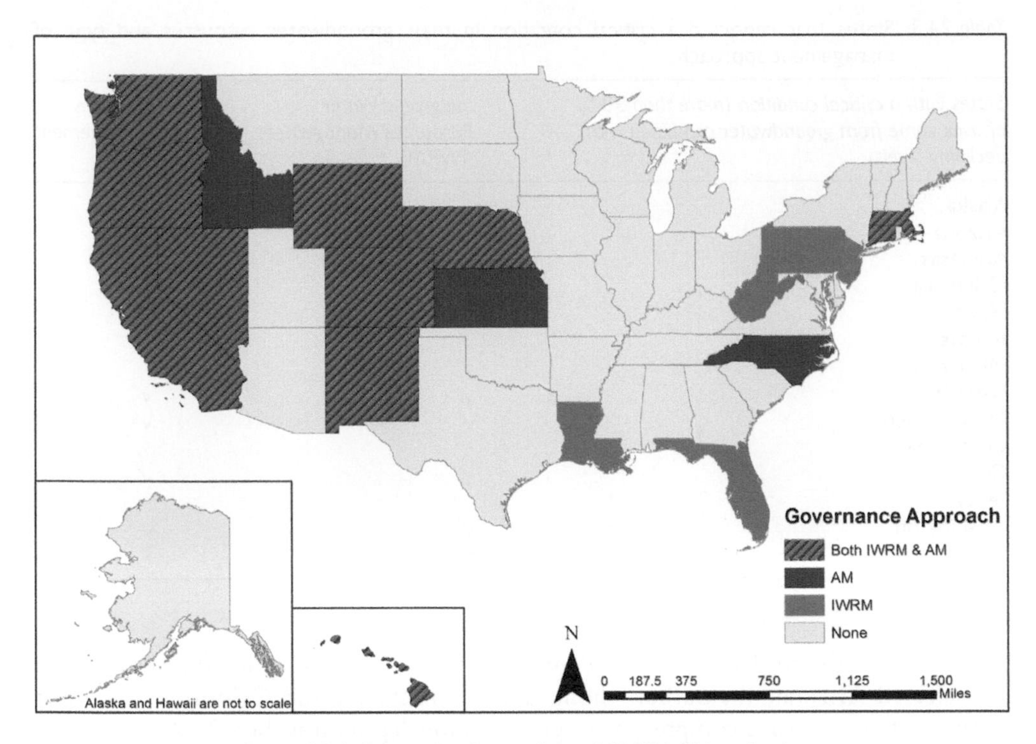

Figure 24.4 States that have adopted IWRM, AM, or both.

Washington and Wyoming) state that they employ AM, but do not plan or monitor groundwater quality and quantity.

Obviously, given the vast geographic diversity of the country, not all states face the same water-resources challenges. To examine more closely the espousal of IWRM and AM, we identified which of the states reporting a "potential critical condition in the groundwater resources" (*i.e.*, depending on groundwater for more than 30% of human use and experiencing declining aquifer levels) have actually implemented either IWRM or AM, or both approaches (Table 24.3). At the time of our survey, of the 13 states with critical conditions, six (California, Hawaii, Nevada, New Mexico, Oregon, and Washington) adopted both IWRM and AM. One additional state (Kansas; using AM), as identified and based on the survey, has carried out one of the effective governance approaches (IWRM or AM). Alaska, Arizona, Arkansas, Mississippi, Oklahoma, and Texas reported a potential critical condition of groundwater resources, yet did not report adopting either IWRM or AM.

24.3 RECENT DEVELOPMENTS IN GROUNDWATER GOVERNANCE IN ARIZONA AND CALIFORNIA

As is apparent from the survey results, groundwater governance and management can vary greatly across state lines. Transboundary groundwater governance issues between states and neighbouring countries are critical but were not within the scope of this

Table 24.3 States that reported a critical condition in their groundwater resources and type of management approach.

States with a critical condition (more than 30 % of uses come from groundwater combined with declining levels)	*Integrated Water Resources Management (IWRM)*	*Adaptive management (AM)*
Alaska		
Arizona		
Arkansas		
California	✓	✓
Hawaii	✓	✓
Kansas		✓
Mississippi		
Nevada	✓	✓
New Mexico	✓	✓
Oklahoma		
Oregon	✓	✓
Texas		
Washington	✓	✓
Total	6	7

analysis. In order to delve a bit more into detail for a few states, for further analysis we selected Arizona, the authors' home state (Box 24.1), and California, the most recent state to approve comprehensive groundwater legislation (Box 24.2). While the two states share a border and some similarities in climate and geology, their respective approaches to groundwater governance have been very different.

Arizona, an arid, inland state, has long relied on groundwater and imported surface water. As such, it has often taken a comprehensive approach to groundwater regulation in designated areas of the state, passing in 1980 the Arizona Groundwater Management Act (GMA), widely regarded as one of the nation's most progressive groundwater-governance laws, where within the five AMAs there is a mandate that developers and water providers demonstrate a 100-year assured water supply to support community growth and there exist other regulations designed to reduce or eliminate groundwater overdraft. While groundwater stresses continue to emerge, large-scale, progressive programs such as groundwater-recharge/recovery and water reuse have allowed Arizona to limit overdraft problems and address groundwater level declines for regulated areas of the state.

Box 24.1 Arizona: early adopter of comprehensive groundwater governance—but only in parts of the State

As Arizona's population grew from less than 2 million in 1970 to 7 million in 2017 (http://worldpopulationreview.com/states/arizona-population/), the state became increasingly reliant on groundwater to meet its human needs (WRRC, 2007). As in California, Arizona afforded landowners the right to pump essentially unlimited groundwater. This led to widespread extraction, subsidence, and the dubious distinction of featuring the nation's largest metropolitan area that was solely reliant on groundwater (Tucson). These facts led Arizona leaders to attempt to bring surface water from the Colorado River—which forms Arizona's western boundary with California—to the heavily populated central portion of the state. The process of obtaining approval for planning and building the US$4.7 billion 541 km (336 mi) long Central Arizona Project (CAP) canal is an entire book in itself. One of the most important points of Arizona's water history is the passage of the 1980 Arizona Groundwater Management Act (GMA). Prior to funding the CAP, the federal government insisted that Arizona control its groundwater-overdraft problem, and the GMA was the result (Colby and Jacobs, 2007; Megdal, 2012; Ferris *et al.*, 2015).

The GMA's goal was to eliminate severe overdraft in areas where this was a problem. The Act created four Active Management Areas (AMAs), which were populated areas of the state where groundwater overdraft was to be managed or halted (Figure 24.5). Later legislation split the one of the AMAs (Tucson) into two parts. Management goals were specified in legislation for each of the AMAs, and a new state agency, the Arizona Department of Water Resources (ADWR), was to implement and enforce the new law's provisions. For four of the five AMAs, safe-yield, which requires AMA-wide balance between groundwater withdrawals and natural and artificial recharge, is the goal. Only the management goals for the largely agricultural Pinal AMA (see Figure 24.5) allows for declining groundwater levels.

The GMA mandated conservation by the agricultural, industrial, and municipal sectors in AMAs. Every 10 years, the ADWR is to adopt a management plan for each AMA that codifies these mandatory conservation regulations and measures progress toward meeting the AMA's management goal. Farmers were not permitted to increase their irrigated area, as measured in acres (0.4 hectare per acre), and they are expected to adhere to irrigation efficiency requirements and/or implement best management practices. Municipal providers were also required to adopt conservation practices. Early management plans established target per-capita volumes (in gallons) per day for users of municipal water, targets that have largely been replaced by best management practices.

Another innovative component of the GMA designed to reduce municipal water use is the Assured Water Supply Program. This rules-based program requires new residential developments to demonstrate that there is sufficient water to meet the need of existing and new residents for 100 years. Water supplies relied upon to serve new community needs must meet water-quality standards and be physically, continuously, and legally available for 100 years, and water utilization must be consistent with the statutory water management goal for the AMA. The management plans assess progress made toward meeting GMA management goals and these provisions have reduced groundwater mining considerably in the state, leading to forward-thinking programs designed to increase Arizona's water sustainability.

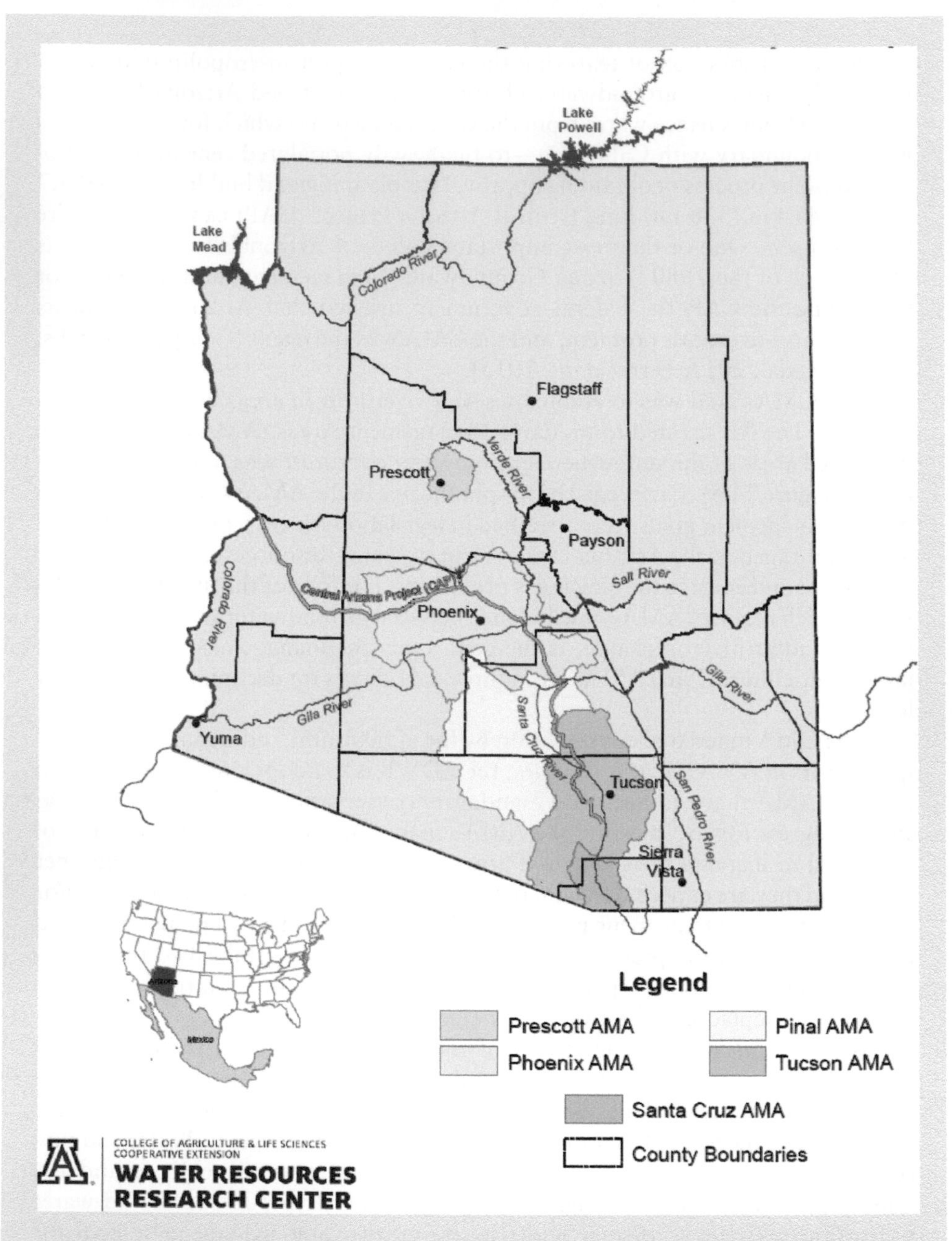

Figure 24.5 Active Management Areas (AMAs) in Arizona (map created by the WRRC).

The delivery of Colorado River water through the CAP and restrictions on groundwater withdrawals prompted development of a statewide recharge and recovery program. The statutory framework, which was last revised in 1994, requires permitting of recharge facilities, storage activities, and the recovery of stored water. Significant annual storage capacity has been permitted in the region served by the CAP. Utilization of the storage and recovery framework has provided flexibility in how water providers have met the requirements of the GMA. The Arizona legislature allowed water suppliers to use water recharged in one place within an AMA as the fulfilment of the Assured Water Supply requirements related to use of renewable supplies, lending flexibility to water providers who do not have direct access to renewable supplies such as the CAP. The state legislature created the Central Arizona Groundwater Replenishment District as a vehicle to facilitate meeting the Assured Water Supply Rules' requirement that water use be consistent with the management goal of the AMA. Additionally, the Arizona Water Banking Authority was created to utilize on an annual basis any unused Colorado River water entitlement (Megdal, 2012).

Arizona has a long history of progressive groundwater management, and continues to be a leader in development of innovative ways to accommodate growth while limiting overdraft of its groundwater resources.

In contrast, for much of its history California largely allowed unregulated groundwater pumping. This decision permitted massive agricultural development in the Central Valley—a vast region stretching some 350 km (220 mi) northward and southward from Fresno—and contributed to municipal growth. But the practice led to land subsidence and ongoing concerns about the additional impacts of drought and climate

Box 24.2 California: late adopter of comprehensive groundwater governance—but too early to assess success

Historically, groundwater in California was managed under the correlative rights doctrine, which requires equitable sharing of all groundwater users overlying an aquifer by comparing demands of other overlying rights. Each person with land overlying an aquifer was allowed a "reasonable amount" for his/her use; however, California had no comprehensive statutory framework to regulate groundwater, so much of the correlative rights doctrine was open to interpretation (Thompson Jr. *et al.*, 2012).

Beginning in 2012 drought hit California and much of the state has remained under various levels of drought conditions. These conditions prompted increased reliance on groundwater that exacerbated overdraft and subsidence in many areas. For many in the state, it became clear that something had to be done to better manage groundwater. The Sustainable Groundwater Management Act (SGMA) is the amalgamation of three legislative bills (SB 1168, SB 1319, and AB 1739), and was approved by Governor Jerry Brown on September 16, 2014 (California Department of Water Resources 2016a).

The SGMA authorizes that groundwater resources in California will be managed for sustainable, long-term reliability and for economic, social, and environmental benefits for current and future beneficial uses. In order to

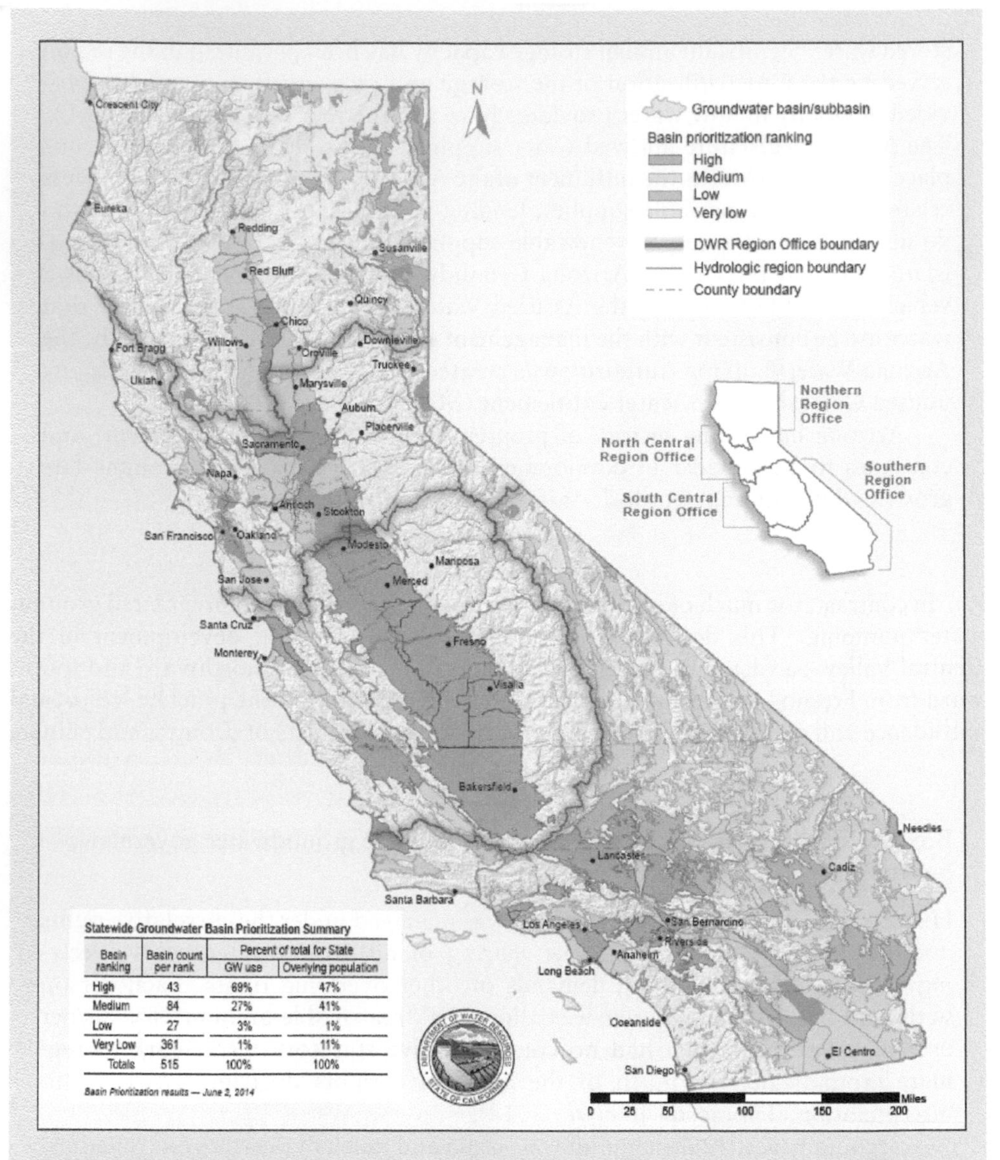

Statewide Groundwater Basin Prioritization Summary

Basin ranking	Basin count per rank	Percent of total for State	
		GW use	Overlying population
High	43	69%	47%
Medium	84	27%	41%
Low	27	3%	1%
Very Low	361	1%	11%
Totals	515	100%	100%

Basin Prioritization results — June 2, 2014

Figure 24.6 California Statewide Groundwater Basin Prioritization (California Department of Water Resources 2016a).

accomplish this, it requires the California Department of Water Resources (CDWR) to assign each groundwater basin to one of four priority categories based on condition of long-term overdraft and inadequate groundwater sustainability plan, among other conditions: high, medium, low, and very low (Figure 24.6). A timeline is assigned to each basin based on these priorities and whether the basin has been determined to be in critical overdraft by

CDWR. Higher priority basins with critical overdraft must be managed under a groundwater-sustainability plan or coordinated groundwater sustainability plan by 2020 (California Assembly 2014; California Senate 2014a, California Senate 2014b). Due to the regional management structure of groundwater in California, each groundwater basin must create its own plan to be reviewed by CDWR (Nishikawa, 2016). The coordinated groundwater sustainability plan concept reflects the idea that groundwater does not necessarily follow political boundaries, so multiple agencies can agree upon a single sustainability plan. High- and medium-priority basins that do not have critical overdraft must be managed under one of these plans by 2022. Recognizing that having groundwater basins create their own sustainability plans could be like having the fox in charge of the hen house, these plans must be reviewed by CDWR, which will determine if the plans will be able to achieve groundwater sustainability as per the specified timelines. Sustainability is achieved when withdrawals are equal to recharge, and must be achieved within 20 years of the plans' acceptance (California Assembly 2014; California Senate, 2014a, 2014b).

The act places further restrictions on groundwater basins, and provides a state "backstop" if basins are unable to meet requirements. Starting in 2025, groundwater basins that are designated high or medium priority can be placed in a probationary status by the State Water Resources Control Board (Board) if groundwater extractions result in significant depletions of interconnected surface waters. Areas in a basin where this occurs can be exempt if they meet the sustainability goal. If a basin is found deficient at creating an acceptable sustainability plan or following the approved one, the Board is also authorized to make interim plans for probationary basins in coordination with CDWR. If actions are not taken to rectify the problem, the interim plan may be adopted by the Board after notice and a public hearing (California Assembly 2014; California Senate 2014a, California Senate 2014b).

Groundwater management is highly decentralized in California (Nishikawa, 2016). In order to bring the various facets of management under one roof, the SGMA authorizes local agencies to elect a groundwater sustainability agency. Areas that are not managed by a groundwater sustainability agency will be managed by the encompassing county. These agencies empowered by the act to: (1) require registration of a groundwater extraction facility, (2) require an extraction facility to be measured with a water-measuring device, and (3) regulate groundwater extraction. The agency may also obtain inspection permits, conduct inspections, and impose fees. These powers come with a significant caveat: they can only be applied in consultation with the cities and counties. Extraction regulations must be consistent with the general plan of cities and counties unless there is insufficient sustainable yield to meet those purposes. Obversely, however, planning agencies for cities and counties must take groundwater sustainability plans into account when making changes to general plans (California Assembly 2014; California Senate 2014a, 2014b). These stipulations ensure that agencies do not regulate groundwater without good reason, and that entities within their jurisdiction do not make plans that are not in line with the goal of sustainability.

The Act is paid for with money from the Water Rights Fund after appropriation by Legislature, but all of the money expended must be eventually recovered from fees. Some of this money will be recovered by levying fines for violations of cease and desist orders that are issued by the Board (California Assembly 2014; California Senate 2014a, 2014b). The Act is California's first attempt at regulating groundwater, and its degree of success remains to be seen; even the highest priority basins are not required to meet groundwater sustainability goals for nearly 25 years.

change. In 2014 California passed the Sustainable Groundwater Management Act. This law is designed to limit groundwater mining (*i.e.*, non-replenished extraction) and return groundwater withdrawals to sustainable levels across the state.

In the present section, to better understand the contexts in which states are governing groundwater use, we examine in greater detail recent developments in governance in these two states.

Discussion of two state-level approaches shows the importance of drivers for enactment of major groundwater legislation. Arizona experienced groundwater challenges over several decades. Population growth combined with increasing reliance on groundwater resources led to subsidence and aquifer depletion. Arizona leaders worked hard to bring water from the Colorado River via the CAP canal. Securing funding to construct the CAP led to adoption of the GMA. The 1980 law created AMAs that included the most populated areas in the state. The GMA also includes an Assured Water Supply Program, which requires that the water supply new development have enough water supply for its residents for the next 100 years in a manner consistent with statutorily established groundwater management goals.

Multiyear drought in California combined with diminishing surface water supplies exacerbated groundwater overdraft and subsidence in the state. Public attention focused on the rapidly increasing reliance on aquifers to supplement diminished surface water supplies. As a consequence, groundwater governance in California shifted from the correlative rights doctrine to the 2014 approval of the SGMA. The law's intent is for long-term sustainable management of aquifers. The overall sustainability goal is to have equal amount of withdrawals and recharge in all basins by 2022. Local agencies are authorized to select a groundwater sustainability agency; areas not managed by one of these agencies are managed by the county.

24.4 GROUNDWATER GOVERNANCE: LESSONS FROM THE U.S.

Governance and management of groundwater across the U.S. is decentralized, uneven, complex, and dynamic. From the completeness of groundwater laws and regulations, to the most basic recognition of the connection of surface and groundwater, to the recognition of the water needs of the environment, governance structures differ radically across state lines. While some similar priorities exist across states, the differences and gaps in governance frameworks are apparent through even a cursory examination of the survey results.

States have different mixtures of regulatory authority over groundwater, with some states delegating more to local governments than others. On the one hand, some respondents indicated that different agencies have regulatory authority over different portions of water-use sectors in different regions of a state, further fragmenting governance and management. For a common resource such as groundwater, fragmentation can lead to gaps in regulation, overlapping and sometimes contradictory rules, and lack of attention to larger environmental problems (Doremus, 2009). On the other hand, some degree of fragmentation can be beneficial, because it can lead to responsive, local governance that can increase efficiency (Blomquist & Schlager, 2005). Determining a proper balance between central and local authority is an ongoing challenge for many states.

The difficulty in establishing appropriate roles for state and local regulatory bodies is exacerbated by state groundwater laws that do not codify basic hydrologic facts. Many states fail to recognize the surface-water/groundwater connection (n = 24), the interplay between groundwater quality and quantity (n = 13), or the water needs of the environmental sector (n = 26). Most state laws do not acknowledge that surface waters infiltrate and feed aquifers, and that groundwater basins often provide base flow to rivers. And typically, quantity and quality are overseen by different state agencies, complicating coordination of data and information. Groundwater-dependent ecosystems such as springs preserve biodiversity and are important to tourism and quality of life. Yet many legal and regulatory frameworks fail to recognize these linkages, making it difficult to implement sustainable groundwater management structures.

We found that groundwater challenges span the entire nation. We identified 13 states having what we termed critical condition in their groundwater resources (listed in Table 24.3). These are places where groundwater reliance is considerable (more than 30 percent of human uses) while supply is decreasing (declining levels are a national priority as reported by 32 states). From these states, six have adopted what are generally considered effective governance approaches (IWRM and AM)—most of these in the continental western part of the country (except Hawaii).

In spite of this lack of integrated management, many states aim to use groundwater sustainably by maintaining levels and limiting contamination. States also largely use complementary management tools—such as permitting, monitoring, and encouragement of voluntary activities by water users—to try to achieve those goals. Employing such tools to address common goals is sure to become more important as human population growth and climate change continue to alter the hydrologic cycle (Georgakakos *et al.*, 2014). Perhaps most encouraging is the fact that many survey respondents felt their state's groundwater governance had evolved over the last few decades. If that trend continues, states may be able to develop and apply appropriate tools and techniques to meet critical groundwater-governance priorities.

Our analysis shows that both Arizona and California face potential critical conditions in their groundwater resources. However, Arizona did not report adopting either IWRM or AM, while California reported adopting both. But regardless of how these states refer to their management approaches, both states *have* addressed their challenges by significantly changing their groundwater governance frameworks.

On the one hand, Arizona is recognized as an early adopter of a comprehensive groundwater management approach by securing water from the Colorado River for aquifer recharge, and by creating active management areas in the most populated regions (Colby & Jacobs, 2007; Megdal, 2012). California, on the other hand, has been a late adopter of a comprehensive groundwater management approach, but it has done so throughout the entire state. Both states have realized the importance of managing their groundwater resources sustainably and have taken clear steps toward this direction. In time we will learn the results of California's actions.

24.5 NEXT STEPS

Additional research on the very complex topic of groundwater governance in the U.S. is needed. Survey responses can fill in only part of the mosaic. Surveying a larger sample of representatives of each state and delving more deeply into actual governance practices can assist in developing a more complete and more refined picture. Updates are needed because the mosaic changes over time, as is evident by the discussion of California.

UA researchers at the Water Resources Research Center and Udall Center for Studies in Public Policy have engaged in more in-depth case-study analysis, as well as additional survey work. A second survey, largely aimed at groundwater quality, is expected to yield a much more nuanced understanding of variations in groundwater-quality governance and management. It will build on initial results by probing more closely themes of groundwater concerns and use, such as quality management and monitoring, and regulation; quality-quantity connections; and extent of resources, research, and collaboration in each state.

Our research to now has built on vital knowledge gained from our 2013 survey. But much remains to be done. In-depth analysis of groundwater-quantity laws and regulations, further investigation into centralization vs. fragmentation (including the relative value of each type of governance), and regulatory adaptation to growing threats such as population growth, urbanization, and climate change are all areas of groundwater governance and management research that require further investigation.

Similarly, groundwater governance across state and international borders has been inadequately studied and remains poorly understood, with the exception perhaps of the Memphis Sand Aquifer underlying Arkansas, Mississippi, Tennessee and Kentucky (Fried and Ganoulis 2016). Aquifer-assessment work at the border between the U.S. and Mexico is one effort that has been proceeding (Callegary *et al.*, 2016) and we believe that analogous initiatives would be a welcome addition to the literature and to the arsenal of practical applications.

In this essay, we have documented that groundwater governance across the United States is fragmented, uneven, non-static, and almost certainly ripe for change. Further research along the lines we have suggested would go a long way to enhancing understanding of promising approaches. These would identify best practices that could be shared to promote conservation and effective governance of groundwater—not just in the U.S., but in other areas of the world facing groundwater challenges.

ACKNOWLEDGMENTS

The authors thank Jacob Petersen-Perlman for his review of this chapter. Funding for the WRRC's 2013 survey effort was provided by the University of Arizona Technology and Research Initiative Fund (TRIF), which also supported WRRC staff efforts associated with writing this article. Additional support was provided by: the International Water Security Network, funded by Lloyd's Register Foundation (LRF); the Inter-American Institute for Global Change Research (IAI) Project SGP-CRA005; the Morris K. Udall and Stewart L. Udall Foundation in Tucson, Arizona.

REFERENCES

American Water Resources Association (AWRA). (2009) Adaptive management and water resources II. [Online] Available from: http://www.awra.org/meetings/SnowBird2009/ [Accessed 8 Nov. 2016]

Blomquist, W. & Schlager, E. (2005) Political Pitfalls of Integrated Watershed Management. *Soc. Natur. Resour.* [Online] 18 (2), 101–17. Available from: 10.1080/08941920590894435 [Accessed 11 Nov. 2016]

Bryner, G. C. & Purcell, E. (2003) *Groundwater Law Sourcebook of the Western United States*, [Online]

Boulder, CO, Natural Resources Law Center University of Colorado School of Law. Available from: http://scholar.law.colorado.edu/books_reports_studies/74/ [Accessed 11 Nov. 2016]

California Assembly. (2014) *Assembly Bill No. 1739.* [Online] Available from: http://leginfo.legislature.ca.gov/faces/billNavClient.xhtml?bill_id=201320140AB1739 [Accessed 11 Nov. 2016]

California Department of Water Resources. (2016) *CASGEM Basin Prioritization.* [Online] Available from: http://www.water.ca.gov/groundwater/casgem/basin_prioritization.cfm [Accessed: 4 Nov. 2016]

California Department of Water Resources (2016) *Groundwater Information Center.* [Online] Available from: http://www.water.ca.gov/groundwater/groundwater_management/legislation.cfm

California Senate. (2014) *Senate Bill No. 1168.* [Online] Available from: http://leginfo.legislature.ca.gov/faces/billTextClient.xhtml?bill_id=201320140SB1168 [Accessed 11 Nov. 2016]

California Senate (2014) *Senate Bill No. 1319.* [Online] Available from: http://leginfo.legislature.ca.gov/faces/billNavClient.xhtml?bill_id=201320140SB1319 [Accessed 11 Nov. 2016]

Callegary, J.B., Minjárez Sosa, I., Tapia Villaseñor, E.M., dos Santos, P., Monreal Saavedra, R., Grijalva

Noriega, F.J., Huth, A.K., Gray, F., Scott, C.A., Megdal, S.B., Oroz Ramos, L.A., Rangel Medina, M., & Leenhouts, J.M. (2016) *San Pedro River Aquifer Binational Report.* International Boundary and Water Commission

Colby, B. & Jacobs, K. (Eds.), (2007) *Arizona Water Policy: Management Innovations in an Urbanizing, Arid Region.* Washington, DC, RFF Press.

Dellapenna, J. W. (2012) A Primer on Groundwater Law. *Idaho L. Rev.* 49, 265.

Doremus, H. (2009) CALFED and the Quest for Optimal Institutional Fragmentation. *Environ. Sci. Policy.* [Online] 12 (6), 729–32. Available from: http://www.sciencedirect.com/science/article/pii/S1462901109000793 [Accessed 11 Nov. 2016]

Ferris, K., Porter, S., Springer, A., Eden, S., & Megdal, S. (2015) *Keeping Arizona's Water Glass Full.* [Online] 107th Arizona Town Hall. Available from: http://www.

aztownhall.org/resources/Documents/107%20Background%20Report%20web.pdf [Accessed 11 Nov. 2016]

Fried, J. and Ganoulis, J. (Eds.) (2016). Transboundary Groundwater Resources – Sustainable Management and Conflict Resolution. Lambert Academic Publishing.

Georgakakos, A., Fleming, P., Dettinger, M., Peters-Lidard, C., Richmond, T.C., Reckhow, K., White, K., & Yates., D. (2014) Water Resources. In: *Climate Change Impacts in the United States: The Third National Climate Assessment.* U.S. Global Change Research Program.

Gerlak, A.K., Megdal, S.B., Varady, R.G., & Richards, H. (2013) *Groundwater Governance in the U.S.: Summary of Initial Survey Results.* [Online] University of Arizona Water Resources Research Center. Available from: https://wrrc.arizona.edu/sites/wrrc.arizona.edu/files/pdfs/GroundwaterGovernanceReport-FINALMay2013.pdf [Accessed 11 Nov. 2016]

Global Environment Facility (GEF), Food and Agriculture Organization, UNESCO-IHP, International Association of Hydrogeologists, & World Bank. (2016) *Groundwater Governance - A Global Framework for Action.* [Online] Available from: http://www.groundwatergovernance.org/home/en/ [Accessed 11 Nov. 2016]

Global Water Partnership (GWP) (2000). Global Water Partnership – Technical Advisory Committee (TAC) Background Paper No. 4. Integrated Water Resources Management. Stockholm, Sweden.

Lemos, M.C. (2015) Usable Climate Knowledge for Adaptive and Co-Managed Water Governance. *Curr. Opin. Environ. Sustain.* 12, 48–52.

Leshy, J.D. (2008) Interstate Groundwater Resources: The Federal Role. *Hastings W.-Nw. J. Envt'l L. & Pol'y* 14 (Summer), 1475–98.

Maupin, M. A., Kenny, J.F., Hutson, S. S., Lovelace, J. K., Barber, N. L., & Linsey. K. S. (2015) *Estimated Use of Water in the United States in 2010.* [Online] U.S. Geological Survey. Report number: 1405. Available from: http://pubs.usgs.gov/circ/1405/ [Accessed 11 Nov. 2016]

Megdal, S.B. (2012) *Arizona Groundwater Management.* [Online] The Water Report. Report number: 104. Available from: http://www.thewaterreport.com/Issues%20101%20to%20104.html [Accessed 11 Nov. 2016]

Megdal, S. B., Gerlak, A.K., Varady, R.G., & Huang, L.-Y. (2015) Groundwater Governance in the United States: Common Priorities and Challenges. *Groundwater* 52 (1), 622–84.

Megdal., S. B., Gerlak, A.K., Huang, L.-Y., Delano, N., Varady, R. G., & Petersen-Perlman, J.D. (2016) Innovative Approaches to Collaborative Groundwater Governance in the United States: Case Studies from Three High-growth Regions in the Sun Belt. *Environ. Manage.* (under review).

Mukherji, A. & Shah, T. (2005) Groundwater Socio-Ecology and Governance: A Review of Institutions and Policies in Selected Countries. *Hydrogeol. J.* 13 (1), 328–45.

Nishikawa, K. (2016) The End of an Era: California's First Attempt to Manage Its Groundwater Resources Through Its Sustainable Groundwater Management Act and Its Impact on Almond Farmers. *Environ. Claims J.* [Online] 28 (3), 206-222. Available from: http://www.tandfonline.com/doi/abs/10.1080/10406026.2016.1129294?journalCode=becj20 [Accessed 11 Nov. 2016]

Stevens, I. (2013) California's Groundwater: A Legally Neglected Resource. *Hastings W.-Nw. J. Envt'l L. & Pol'y* 19, 3.

Torres, G. (2012) Liquid Assets: Groundwater in Texas. *Yale LJ Online* 122, 143–837.

Thompson Jr., B., Leshy, J., Abrams., R. (2012) Legal Control of Water Resources. 5th Edition. St. Paul, West Academic Publishing.

University of Arizona Water Resources Research Center (WRRC). (2007) *Layperson's Guide to Arizona Water.* [Online] Water Education Foundation. Available from: http://www.azwater.gov/AzDWR/IT/documents/Layperson's_Guide_to_Arizona_Water.pdf [Accessed 11 Nov. 2016]

U.S. Army Corps of Engineers (2010) *National Report: Responding to National Water Resources Challenges: Building Strong Collaborative Relationships for a Sustainable Water Resources Future.* [Online] Civil Works Directorate. Available from: http://www.building-collaboration-for-water.org/documents/nationalreport_final.pdf [Accessed 11 Nov. 2016]

Varady, R. G., van Weert, F., Megdal, S.B., Gerlak, A., Abdalla Iskandar, C., & House-Peters, L. (2013) *GROUNDWATER GOVERNANCE: A Global Framework for Country Action GEF ID 3726.* [Online] Groundwater Governance. Available from: http://www.yemenwater.org/wp-content/uploads/2015/04/GWG_Thematic5_8June2012.pdf [Accessed 11 Nov. 2016]

Varady, R., Zuniga-Teran, A., Gerlak, A., & Megdal, S. (2016a) Modes and Approaches of Groundwater Governance: A Survey of Lessons Learned from Selected Cases across the Globe. *Water* [Online] 8 (10), 417. Available from: doi:10.3390/w8100417 [Accessed 11 Nov. 2016]

Varady, R. G., Zuniga-Teran A. A., Garfin, G.M., Martin, F., & Vicuña, S. (2016b) Adaptive Management and Water Security in a Global Context: Definitions, Concepts, and Examples. *Curr. Opin. Environ. Sustain.* (in press).

APPENDIX A: 2013 SURVEY

Introduction: This is a short questionnaire intended to acquire first-hand knowledge from state agency personnel about your state's groundwater governance practices, including the institutions and laws involved.

The study aims to describe the state of the practice in the U.S. and produce a national-scale report identifying the range of approaches to groundwater governance.

This questionnaire should take approximately 15 minutes to complete.

Please note: Data or comments obtained in this survey project will not be attributed to particular individuals. Respondents may skip questions, as necessary.

I. Basic Information

State you represent: ____________________

Name of the agency you represent: ____________________

Part I: Groundwater Use

1 In an average year, what approximate percentage of total human demands (*i.e.* domestic, commercial, industrial, and agricultural) are met through use of groundwater supplies in this state?

% of all water withdrawn for human demands in the state that comes from groundwater: ____________________

2 Is the importance of groundwater use consistent throughout the state or does it vary by region in terms of relative reliance on groundwater to supply human demands? Check only one.

__Reliance is consistent throughout the state

__Reliance varies by region

3 Is the proportion of groundwater use by each major groundwater-using water sector consistent throughout the state or does it vary by region? Check only one.
__Proportions of use by each sector consistent throughout the state
__Proportions of use by each sector varies by region

Part II: Groundwater Laws and Regulations

1 Are there formal groundwater policies, rules, or regulations in the state?
__Yes
__No
Please provide names and dates of relevant statute/rule(s):
____

2 Have you observed substantial changes in how groundwater is managed in the state over the past few decades?
__Yes
__No
Please explain: ____

3 In what agencies do authorities for groundwater oversight/enforcement reside? Please list all.
Local agencies: ____
State agencies: ____

4 Do separate agencies deal with water quantity and water quality? Yes____No____

5 What are the state's groundwater governance priorities? Check all that apply.
__Declining groundwater levels
__Conflicts between water users (*e.g.* well interference)
__Access
__Quantification of water rights
__Water quality/Contamination
__Regulatory disputes
__Inter-agency jurisdictional conflict
__There have been no clearly articulated priorities
__Other

6 Are there programs or settlements addressing international, interstate or Native American groundwater issues in the state?
__Yes
__No
Please explain: ____

7 Are there programs or settlements addressing Native American groundwater issues in the state?
__Yes
__No
Please explain: ____

8 Are there water conservation regulations applicable to groundwater use in the state law?
__Yes

__No
Please provide names and dates of relevant statute(s): ____________________

9 Does state law explicitly recognize or address the connection between surface water and groundwater?
__Yes
__No
If yes, how? Please provide names and dates of relevant statute(s):
____

10 Does state law explicitly address groundwater quality?
__Yes
__No
If yes, how? Please provide names and dates of relevant statute(s):
____

11 Does state law consider the water needs of groundwater dependent ecosystems?
__Yes
__No
If yes, how? Please provide names and dates of relevant statute(s):

12 Do enforcement agencies have sufficient capacity to carry out policies and responsibilities?
__Yes
__No
Comments: ____

13 Are the courts active in groundwater issues in the state?
__Yes
__No
Please list relevant court decisions: ______________________________

14 To which of the following user groups do groundwater regulations apply? Check all that apply.
__Household/domestic wells
__Industrial Users
__Privately owned community water systems
__Publicly owned community water systems
__Irrigation associations
__All of these
__Other

15 Do regulations differ for each water user types listed above (*e.g.* municipal use vs. irrigation)?
__Yes
__No
If yes, please explain: ______________________________________

16 Does your state encourage the use of voluntary measures for addressing groundwater issues?
__Yes
__No
If yes, please explain: ______________________________________

Part III: Groundwater Tools and Strategies

1 Which tools do the state use to manage groundwater use/quantity? Check all that apply.
__Permits
__Planning
__Land use development laws/regulations
__Protected areas
__Pricing
__Extraction fees
__Monitoring
__Other

2 For which water sectors is groundwater use metered or monitored? Check all that apply.
__All water sectors
__Municipal
__Industrial
__Agricultural
__Other

3 What aspects of groundwater are monitored? Check all that apply.
__Groundwater levels
__Groundwater abstractions
__Amount in storage
__Conductivity properties
__Groundwater quality
__Rates of recharge
__Rates of discharge
__Other
__None

4 Which tools does the state use to manage groundwater quality? Check all that apply.
__Permits
__Planning
__Land use development laws/regulations
__Protected areas
__Pricing
__Extraction fees
__Monitoring
__Other

5 How are the activities of groundwater management agencies (*e.g.* permit reviews, monitoring) funded? Check all that apply.
__User fees
__Taxes
__State general fund
__Mitigation fees
__Other

6 How widely is information about groundwater resources and rights reported? Check all that apply.
__Information about groundwater supplies is publicly available
__Information about groundwater use is publicly available
__Information about water rights of all users is publicly available
__Information about groundwater supplies is provided directly to water users
__Information about groundwater use is provided directly to water users
__Information about water rights of all users is provided directly to water users
__Information about groundwater resources and water rights including access to water rights registers is not reported
7 To what extent is groundwater information publicly accessible?
__Extremely accessible
__Somewhat accessible
__Not publicly accessible at all
8 In your state, are any of these groundwater management strategies in use? Check all that apply.
__Integrated Water Resources Management (IWRM)
__Aquifer recharge and storage programs
__Regional planning or management organizations
__Economic incentives
__Adaptive management
__Public education programs
__Other

Part IV: Future Research and Contacts

1 Contact information (optional)
Your name (optional) ____________________
Telephone (optional) ____________________
Email (optional) ________________________
2 May we contact you with additional questions in the future?
Yes___
No___

3 Please indicate which of the following categories best describes your professional title.
__Engineer
__Economist
__Planner
__Mid-level administrator
__Manager
__Political appointee/Director
__Researcher/Academic
__Lawyer
__Other (please specify)

Thank you for taking our survey!

Chapter 25

Turning the tide – curbing groundwater over-abstraction in the Tosca-Molopo area, South Africa

Paul Seward[1] *& Gabriel Stephanus du Toit van Dyk*[2]

[1]*Independent Groundwater Consultant, Cape Town, South Africa*

[2]*Department of Water and Sanitation, Kimberly, South Africa*

ABSTRACT

The Tosca-Molopo case presents a unique example of an effective groundwater management intervention in South Africa, where groundwater governance has otherwise been described as weak to non-existent. Prior to 1998, groundwater was a private resource used and exploited by landowners. The National Water Act of 1998 defined groundwater as a national resource, which provided the basis for management intervention by the government. The circumstances precipitating the particular interventions in Tosca-Molopo over the period 1993 to 2011, their nature, and the consequences are described, and general lessons are sought. Rapid and uncontrolled expansion of irrigation use lead to over-abstraction of a dolomite and sedimentary aquifer system. Groundwater levels declined, shallow boreholes dried up, livestock and domestic water supply systems failed, leaving land without water. The existing water use entitlements were verified and validated, unauthorised use was stopped, a further reduction above a threshold of maximum irrigation area per farmer was implemented, and subsequently an even stricter compulsory licensing was initiated, though providing for fair and equitable entitlements. A local water user association was established for water use and groundwater level monitoring as well as general support of the process. Through these management measures, the water levels stabilised over a number of years to prevent further failure of the water supply systems. The case illustrates that a combination of conducive factors within a properly framed legal system, including political will, mutual perception of the problems among stakeholders, sufficient technical capacity for implementing, monitoring and enforcing tightened regulations, and a functioning collaboration between the national and local level, can lead to reversing trends in groundwater over-abstraction.

25.1 INTRODUCTION

Groundwater governance in South Africa has been assessed as weak to non-existent, based on results of in-depth interviews with key stakeholders (Knüppe, 2011), case studies (Pietersen *et al.*, 2011) and government employee experience (Seward, 2015). The weak groundwater governance stands in stark contrast to South Africa's water legislation, which has been described as 'progressive', 'advanced', 'forward-looking' and even 'revolutionary' (Burns *et al.*, 2006). This incongruity between forward-thinking legislation and retrograde implementation of that legislation is common for many other countries where groundwater is important (Mukherji & Shah, 2005; López-Gunn & Cortina, 2006). However, anomalous cases of effective groundwater regulations, whether direct or indirect, in regions dominated by ineffective governance can be found (Mukherji and Shah, 2005). In South Africa, one such anomaly can be found at Tosca-Molopo, where a strong groundwater management intervention was implemented in response to conflicts sparked by excessive groundwater use and its consequences. Within ten years (1993-2000), groundwater abstraction for irrigation purposes increased from 100 ha (0.77 Mm^3/a) to 2000 ha (18.9 Mm^3/a). Groundwater levels had declined between 20 to 60 m, shallow boreholes dried up with failure of supplies for livestock and domestic use. The subsequent "land without water" created a major conflict between landowners.

The causes, nature, and results of the intervention at Tosca-Molopo are described and discussed, followed by a general discussion of the effectiveness of groundwater governance to ensure socially accepted regulations and environmentally sustainable water use. For the purposes of this chapter, groundwater governance is taken to mean the process by which rules about groundwater are made (or not made) and implemented (or not implemented). Groundwater management is taken to be a subset of groundwater governance related to technical implementation. Good groundwater governance will be taken to mean a process that is: (1) 'democratic'; (2) efficient and effective—the decisions made are implemented and impacts are positive and significant; and (3) sustained—the process is ongoing.

25.2 THE TOSCA-MOLOPO AREA

The Tosca-Molopo geographic area (Figure 25.1) is located in the North West Province in northern South Africa, close to the border with Botswana. The small settlement known as Tosca is situated approximately 150 km north of Vryburg – the largest and closest significant town and an industrial and agricultural centre of the region. The Tosca-Molopo area, constituting the settlements of Tosca and Vergelee (Vergelegen in English), is rural, covers an area of 1625 km^2 and has a population of approximately 4500 people.

The area is predominantly flat, with an elevation declining from 1210 m in the western watershed area to 1070 m in the east of the Molopo river valley (Figure 25.2). Geographical features include non-perennial rivers and the Waterberge hills that rise 50 m above the plain in the north. Rivers are important for intermittent and focused groundwater recharge.

The development of the Disaneng, Setumo and Modimola Dams about 100 km upstream on the Molopo river has impeded flow in the ephemeral Molopo river that once flowed frequently after heavy rainfall. Reports suggest that before 1980, the Molopo river flowed every 2 to 3 years. However, since 1990 (after the dams were constructed) the Molopo river was reported to flow less frequently.

The area is semi-arid with an average annual rainfall of 399 mm. Evaporation in the area is high, between 2050 and 2250 mm/a, indicating that only a small percentage of rainwater is available to recharge groundwater. Recharge estimates range from 0.5% to 3% of the annual precipitation (van Dyk, 2005). Groundwater is the only water source

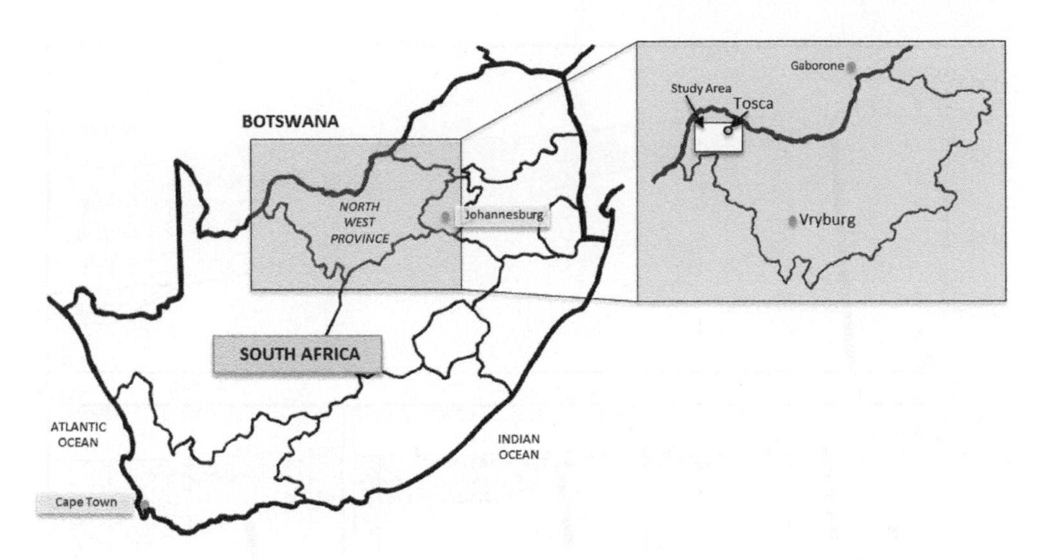

Figure 25.1 Map of the Tosca-Molopo area in South Africa.

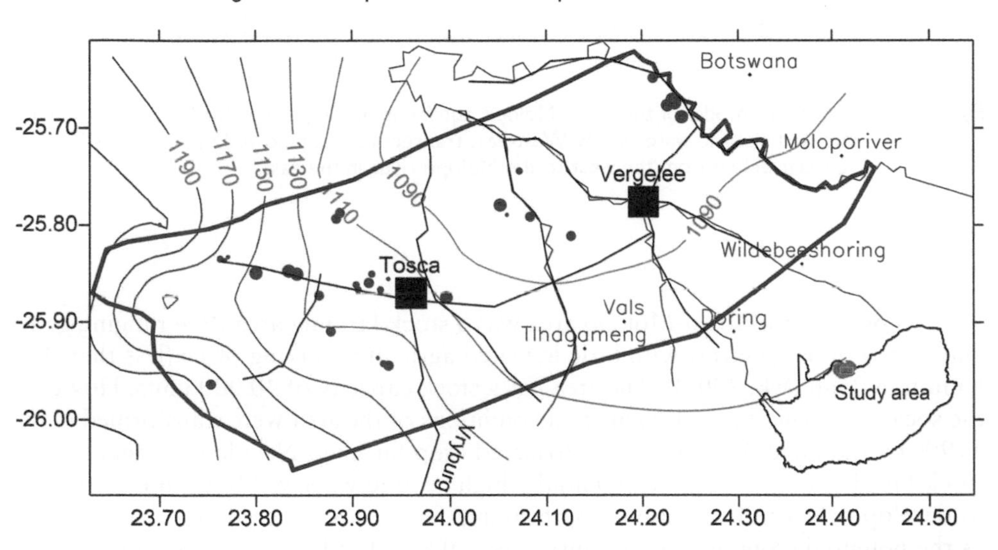

Figure 25.2 Map of the Tosca-Molopo area with surface elevation, rivers and irrigation areas in blue.

for all users in the Tosca-Molopo area. Two distinctive superimposed and hydraulically interconnected aquifers are present in the area: a porous intergranular aquifer formed by fine-grained sediments of the Kalahari Group on top of a high-yielding fractured karstic dolomite aquifer of the Ghaap Plateau formation (Figure 25.3). The sediments contribute largely towards the storage of the aquifer system, while the fractures of the dolomite contribute to high yielding wells. Numerous semi-pervious dolerite dykes are present. Duvenhage and Meyer (1991) characterised the groundwater resources as high-yielding, although isotope and water quality samples showed that some of the groundwater was practically non-renewable.

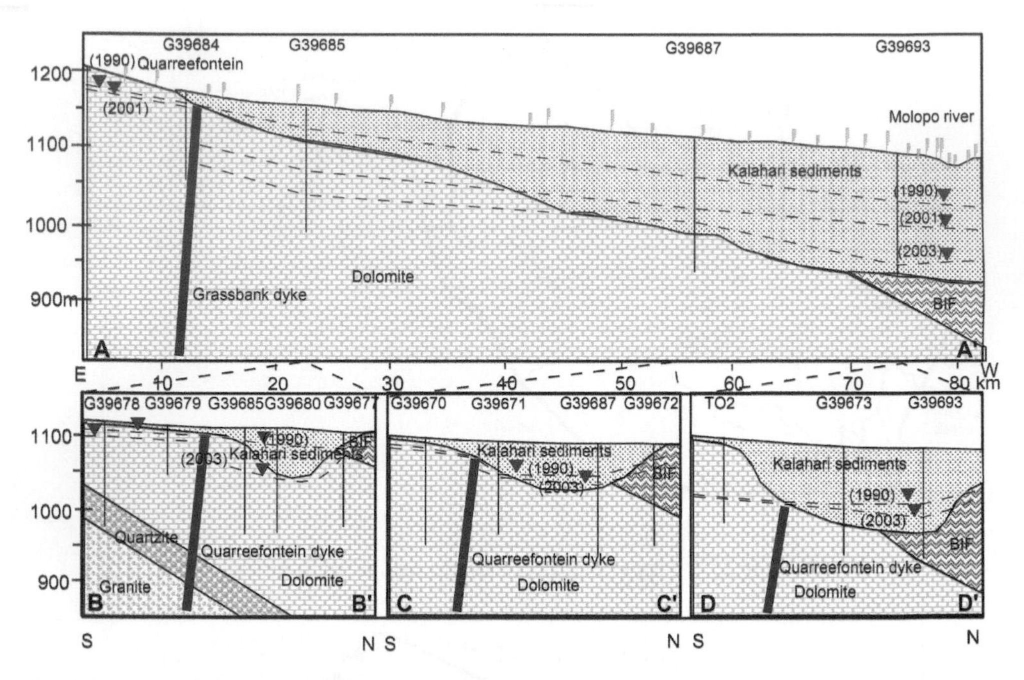

Figure 25.3 Conceptual model of the Tosca-Molopo aquifer, indicating the major boundaries, aquifer units and historic water levels. West-east transect at top, followed by three north-south transects below, from the west to the Molopo river in the east.

Historically, the Tosca-Molopo area was a stock farming area (free-ranging game and cattle) on large farms with little irrigated agriculture taking place (less than 100 ha in total (van Dyk, 2005)). The area has a stock capacity of 40 000 units. However, the socio-economic and environmental conditions of the area were transformed in the 1990s by the rapid development of irrigated crop farming. New landowners bought stock farming land seeing the potentially high groundwater yields as an opportunity to develop or expand irrigation crop farming. Soon, land prices more than doubled as the potential economic opportunity was realized. Further factors encouraging the expansion of irrigation were: (1) the growing of crops had become more profitable; (2) dryland cultivation had become problematic as a result of less favourable climatic and economic conditions; (3) efficient pumps and irrigation systems had become easily available to farmers; (4) electricity from the national grid had become available from 1995; (5) completion of a tarred road from Vryburg to Tosca in 1994 had facilitated the marketing of crop produce; and (6) the knowledge and technology to drill deep boreholes in the dolomites for irrigation purposes had advanced.

It was estimated that in 2002, 2000 ha was under irrigation, abstracting 18.9 Mm^3 of water annually, compared to 100 ha before 1990s (van Dyk, 2005). The crops irrigated were: corn (41% of irrigation area), paprika (19%), peanuts and wheat (30%), and potatoes and alfalfa (10%). There were approximately fifty landowners, each irrigating on average 40 ha (van Dyk, 2005). The majority irrigated less than 20 ha while a few individuals irrigated more than 100 ha.

25.3 THE PROBLEM: WATER USE CONFLICT

The initial indicator of a groundwater problem was declining groundwater levels. Water level drops of 20 to 60 m caused many relatively shallow stock water boreholes to dry up, which in turn required significant investment to replace them by deeper boreholes. The declining water levels were brought to the attention of the Department of Water Affairs and Forestry (DWAF)[1]. Their analysis indicated that the resource was being over-utilised with possible short-term depletion of the water resource and water crisis for the area. Some irrigation farmers adapted through drilling deeper boreholes equipped with high yielding pumps funded through irrigation revenues. The stock farmers, however, were mostly dependent on old shallow boreholes with wind pumps and could not afford the replacement of boreholes and deeper pumping.

However, simply stating that the problem was declining water levels and possible over-abstraction does not fully describe the problem. Stock farming (or ranching) requires minimal amounts of water but large areas of land to be economically viable in the Tosca-Molopo area. Cultivation of crops using irrigation water, on the other hand, requires large amounts of water but relatively small amounts of land to be economically viable. The profits, in 2003 prices, from 1 ha of cultivated land under irrigation was estimated to be R2 887, while 1 ha of land yielded only R114 in income for stock farming. Thus, crop irrigation is clearly far more profitable per ha than stock farming. However, total profits from stock farming, estimated to be R45 684 640 in 2003, were still substantially higher than the R4 828 074 profits from crop irrigation in the same year. These higher total profits from stock farming were, however, at risk from boreholes drying up as a result of the quest for higher per ha returns from crop irrigation. Therefore, inherent to the conflict, was a discussion of what was the most beneficial use of the groundwater.

The authorised use of groundwater needs to address the capital investments made by both crop irrigators and stock farmers to ensure their livelihoods. In the case of stock farming, this includes ensuring that crop irrigation does not cause their water levels to drop to such an extent that the drilling of replacement deep boreholes cannot be justified by stock-farming incomes.

25.4 THE MANAGEMENT RESPONSE

25.4.1 Water use authorisation types in South Africa

To better understand the options and mechanisms available in South Africa for making and implementing decisions about groundwater use, it is necessary to understand the

[1]The Department of Water Affairs and Forestry (DWAF) is now the Department of Water and Sanitation (DWS). For many years in between, it was called the Department of Water Affairs (DWA).

basic types of water use authorisation in place since the implementation of the National Water Act (NWA) (Republic of South Africa, 1998). Basically, there are four main types of water use authorisation, of which only licensing (the fourth) requires written permission: (1) **Schedule 1 use:** small volumes of water for household use and livestock watering where impacts on the resource can reasonably be expected to be negligible; (2) **General Authorisations:** this applies to larger volumes of water than Schedule 1 use, but use is well within the limits of the resource for limited irrigation and industrial use.; (3) **Existing Lawful Use:** this allows water use that was existing before the 1998 NWA came into effect to continue; and (4) **Licensed Water Use:** Large scale water use for mining, irrigation, industrial and domestic users are required to apply to DWAF for licences if their planned use exceeds that of the previous three types of authorisation. The aim of the license is to ensure that the authorised use is issued with conditions that would manage the impact on the water quantity and quality and not be detrimental to the resource and existing users. Irrespective of the above authorizations, if a resource is already over-utilised, a compulsory licensing process may be initiated for all users of that resource with the intention of reducing total use to sustainable levels. The aim of compulsory licensing is to equitably and sustainably re-allocate water from a resource or a sector.

25.4.2 Initial reduction of water use authorizations by implementing existing rules

A water problem was identified by the community dependent on the Tosca-Molopo groundwater resource. To address this, the average yield of the groundwater resource was estimated by DWAF. According to the reserve determination (Godfrey & van Dyk, 2002), the average sustainable abstraction rate for the resource was 11.1 Mm^3/a. In comparison, groundwater use for crop irrigation was estimated at 18.1 Mm^3/a, based on crop water requirement data (Table 25.1). To help reduce crop irrigation to a sustainable level, a number of initial steps were taken by DWAF during 1999 to 2003 that involved the implementation of and adherence to existing rules and thus did not require extensive discussions and consensus building. The steps implemented were:

- 1. **Termination of unused water rights:** A number of users had not yet developed the groundwater they were legally entitled to. These unused rights were withdrawn and total rights for these farmers were reduced to the general authorisation for the area. This reduced potential development of legal use by 2.0 Mm^3/a (from 2.6 to 0.6 Mm^3/a).
- 2. **Termination of unauthorized water use:** Potential unauthorised groundwater use for irrigation between February 1999 and March 2002 was identified using satellite imagery. Fifteen potential unauthorised users were sent written requests to document the legality of their water use. Their responses were evaluated by a DWAF Regional Water Use Authorisation Committee in December 2002. Six users provided evidence that their water use was authorised. The remaining nine were sent directives to curtail their use. Following the issue of these directives, additional two users provided sufficient information that resulted in their directives being cancelled. The reduction in groundwater use from the remaining seven unlawful

Table 25.1 Initial groundwater abstraction reduction measures implemented.

Type of use/intervention	*Amount (Mm^3/a)*
Existing registered use	18.1
Termination of unused water rights (would not save water, but avoid new demand)	−2.0
Termination of unauthorized use	−3.7
New water use authorisation	0.98
Use after water authorisation controls implemented	15.38

users was 3.7 Mm^3/a, estimated from their cropping areas, crops and crop water requirements over the previous three years.

– **3. Restrictions for new water use applications:** To ensure equitable access to the resource while taking cognizance of its depleted state, DWAF decided that all new water use applications would be limited to the applicable general authorisation. This would limit increase in water use from new applications to 0.98 Mm^3/a.

25.4.3 Further water use reduction through the implementation of irrigation reduction

The implementation of the administrative steps above led to a reduction in demand from the resource by 3–4 Mm^3/a. This equates to a total demand of 14–15 Mm^3/a (Table 25.1). However, according to groundwater model predictions (Van Dyk, 2005), confirming the reserve determination, the system would only stabilise at a long-term average abstraction rate of 11 Mm^3/a. Thus, a further reduction of approximately 3.7 Mm^3/a, or 30%, was needed to ensure sustainable use of the resource.

DWAF discussed the need for a further reduction with the water users, who agreed that water restrictions could relieve the stress on the aquifer. However, the water users requested that, to ensure the economic viability of their farming, the first 10 ha of land or 75 000 m^3/a per farmer should be exempt from restrictions, and that the 30% reduction should only apply to use exceeding that volume. DWAF agreed to this request. For the restrictions to be effective, and to assess their effectiveness, DWAF proposed that: (1) The restrictions would initially apply for five years; (2) The total volume of water abstracted per borehole and/or property should be measured using flow meters; (3) The water user would be responsible for recording accumulated water use on a monthly basis and forwarding this information to DWAF; (4) DWAF would carry out annual inspections of cropping areas (with remote sensing) and metering devices; and (5) DWAF had the right to abolish the authorization of and terminate the abstraction of any user not complying with the restrictions. The proposed restrictions were approved by the Director General of DWAF and published in 2004 (Republic of South Africa, 2004b).

25.4.4 Enforcement of water abstraction restrictions

DWAF officials noted that while all groundwater users agreed to the regulations, most did not believe there would be serious actions taken to enforce compliance with the water abstraction regulations. This was indicated by the fact that, as of December 2004, six water users were documented who exceeded their allocated water use. After discussions between DWAF Kimberly (Regional Office) and DWAF Legal Division (National Office), it was decided that two cases could be exempted for various reasons (*e.g.*, miscalculation of water abstraction or error in authorized use granted). Hence, enforcement actions were taken against the remaining four users in 2005. These actions were taken at the end of the irrigation season to minimise loss of income. The enforcement actions taken by DWAF were: (1) removal of pumps from boreholes used for irrigation; (2) sealing of boreholes used for irrigation; (3) recovery of the costs of these actions from the water users in question; and (4) access and entitlement to water use were restored after the offending users had paid costs and signed an agreement with the water user association (WUA) to comply with existing regulations regarding water use. In the process, the demand of approximately 1 Mm^3/a equivalent to 100 ha irrigation was removed from the system.

25.4.5 Establishment of a water users association

The establishment of a WUA was a process that ran in parallel to the enforcement of rules by DWAF rather than a process that was central to resolving the over-abstraction issues. The steering committee of the WUA was central to enforcement actions, was consulted by DWAF, and influenced decisions within a legal and socio-economic context. In South Africa, WUAs are seen as support institutions. They may police allocations made by higher institutions and they may help ensure that their members use their allocations wisely, but they are not regarded as a body that makes water allocation decisions. This role is enshrined in the NWA:

> *Although water user associations are water management institutions, their primary purpose, unlike catchment management agencies, is not water management. They operate at a restricted localised level, and are in effect co-operative associations of individual water users who wish to undertake water-related activities for their mutual benefit.*

In South Africa, any interested party can motivate for the establishment of a WUA. In the Tosca-Molopo case, it was the water users in the community that initiated the establishment of the WUA. The water conflict, water depletion, competition for water entitlements, economic stakes and environmental damages were the major drivers. An informal steering committee was established in January 2001 to compile a draft WUA constitution for submission to the DWAF Minister, who must approve or reject the formation of the WUA. Even at this stage, there were disputes, with the stock farmers requesting that the steering committee be re-elected on the grounds that the committee was biased in favour of the irrigation farmers. However, DWAF continued to support this committee on the grounds that its role was merely to draw up a constitution and not to make decisions regarding water management. To facilitate the establishment of the WUA, DWAF appointed a consultant to assist in the drafting of the WUA constitution. After a series of consultative meetings, a draft constitution was finalised in December 2002 and submitted to the DWAF Minister. A number of improvements to the constitution were required before the Minister approved it in July 2004 (Republic of South Africa, 2004a). The resulting WUA oversees an area of approximately 162 500 ha containing 53 registered irrigation water users, a bulk water domestic supplier, and 200 stock water users. The area of responsibility was delineated on the boundaries of the dolomite aquifer and included all irrigation water users.

25.4.6 Establishing compulsory licensing

The third management intervention round was the implementation of compulsory licensing in order to arrive at the predetermined total sustainable groundwater abstraction threshold of 11.1 Mm^3/a. The term compulsory licensing is used because every water user within the area affected *must* apply for their use to be licensed at a reduced level. With an individual licence, a prospective user is only required to apply for a licence *if* their prospective use exceeds certain volume criteria. In the compulsory licensing process, the total limits of the resource are evaluated and water is allocated among users within these limits. The Tosca-Molopo compulsory licensing process involved the following:

- A verification process was carried out by DWAF in 2008. The process involved determining actual use and legally authorised use (before compulsory licensing)
- The DWAF Regional Office (Kimberly) explained the compulsory licensing process to the users involved (2009–2010). DWAF National Office (Pretoria) assisted the groundwater users to complete the application forms for compulsory licensing
- The WUA held meetings to discuss and facilitate the compulsory licence application process and kept their members updated on developments using SMS. The WUA also participated in an interview with the state radio station Radio Sonder Grense (Radio Without Borders) on the compulsory licensing process (Grobbelaar, 2016)
- After evaluation by DWAF of the applications, the allocation for compulsory licensing was published in the Government Gazette in 2011 (Republic of South Africa, 2011). Volumes allocated were: 598 722 m^3/a for the reserve (*i.e.* the environment and basic human water needs) and international allocations[2] and 10.3 Mm^3/a for consumptive use (the residual of the estimated sustainable yield, after satisfying other needs). Most consumptive use was allocated for agricultural purposes, except for an allocation of 153 870 m^3/a to the Molopo Local Municipality. The total allocations amounted to 10.9 Mm^3/a.
- A 30-day time period was provided for objections to be submitted. The WUA assisted the users who wished to make objections

The compulsory licensing process was driven and implemented by DWAF. The WUA's contribution to the compulsory licensing process was that of support. It did not decide on the water allocations to be made, but supported and facilitated the process.

A chronology of the groundwater events in the area is given in Table 25.2.

[2]The dolomite aquifer is transboundary and extends into Botswana (Turton *et al.*, 2006)

Table 25.2 Events related to groundwater management in the Tosca-Molopo case.

Date	*Event*
1990	Exploration and assessment of the resource to identify potential groundwater resources, prompted by WAF (Duvenhage *et al.*, 1991)
1993	First concerns voiced by individual farmers in the community (letter to the DWAF Minister) of declining water levels and failing water resources
1994	Information session with the community. Groundwater was private resource prior to the 1998 National Water Act (NWA) (Republic of South Africa, 1998), implying that legal solutions were limited
1996	First estimation of extent of irrigation in the area (surface area irrigated, volume abstracted)
1998	Promulgation of the NWA — providing an enabling frame to regulate groundwater use legally Registration of water use by property owner, with the volume of use required to establish entitlement
2001 (Jan)	Establishment of a steering committee to form a water users association (WUA) in order to ensure cooperation on water management through a local structure, as required by the NWA
2001	Individuals and groups, broadly representing water users in the area, voiced unified concerns (letters to the DWAF Minister)
2001 (April)	Commence with annual monitoring of groundwater levels by DWAF to establish the status, trends and dynamics of the resource
2001 (Aug)	The depleted status of the resource discussed with water users in a public meeting called by DWAF and steering committee of the WUA
2001 (Sept)	Larger community expresses concerns regarding the resource – DWAF commitment to cooperate to address the problem
2002	A number of public meetings called by DWAF and steering committee of the WUA to discuss and apply for establishment of WUA
2002 (Jun)	A technical report (Godfrey and van Dyk, 2002) suggests that the resource is over-allocated by more than 100% considering all water uses
2002 (Aug)	Verification of users by satellite imagery – 15 potentially unauthorised irrigation water users identified
2002 (Oct)	Fifteen potentially unauthorised irrigation users are given opportunity to prove that they are legal water users. Eight comply and their water use is recognised, while the remaining seven water user are declared illegal.
2002 (Dec)	A draft WUA constitution is sent to DWAF head office for approval, is approved, and the WUA is established in 2004 in accordance with procedures
2003 (Jan)	Directives issued against seven illegal users: stop or reduce irrigation activities by March 2003
2003 (Mar)	Field inspections and communications confirm limited cooperation of seven illegal water users
2003 (Mar)	DWAF and user discussions regarding a 30% reduction in water use: not all water users in favour of the restrictions
2003 (Sep)	Commencement of restrictions. Authorised users reluctant to comply. Confirmation of cooperation of seven unauthorised water users. Contravening directives issued against only three of them for non-compliance
2003 (Oct)	Three users charged with unauthorised water use. However, the South African Police Department shows little interest in the prosecution
2004 Oct	Approval for 30% restriction on irrigation water use and publication of 30% water restrictions in Government Gazette
2004 Dec	Establishment meeting and election of WUA management committee
2004 Dec	On request from WUA, DWAF issue six directives for unauthorized use
2005 Apr	Letters to gain access to properties week of 16 May 2005 issued
2005 May	Enforce water restriction compliance by removing pumps of four non-compliant water users
2011	Compulsory licensing of 59 users to the composite volume of 10.3 Mm^3/a

25.5 ASSESSMENT OF GROUNDWATER GOVERNANCE AS APPLIED IN THE TOSCA-MOLOPO CASE

25.5.1 Assessment criteria

Our working definition of good groundwater governance is that: (1) it involves a fair, transparent and democratic decision-making process; (2) the decisions made must lead to effective results; and (3) the support is sustained. This section addresses whether these three criteria were met.

25.5.2 Fair, transparent and democratic decision-making

The decision-making behind the intervention was well-documented through minutes of public meetings, letters to the WUA and individual users, reports, and media coverage in papers and magazines. Responses from community members in meetings and letters were evaluated and responded to. User groups were also represented on the WUA and their concerns were attended to through the WUA or elevated to DWAF for consideration in decision-making. The documentation also makes it clear that: (1) DWAF responded appropriately to the concerns of the water users; (2) DWAF used the powers available to it in the NWA to implement and sanction solutions that had the broad support of the users. Thus, it can be concluded that fair and transparent decision-making took place. It can also be concluded that the NWA contains the necessary tools for operational groundwater governance, and they are more than sufficient when there is the political will, cooperation and an ability to apply the existing legal framework. The relative success of this groundwater governance case is measurable in the implementation of legal and regulatory measures to prevent the depletion of a groundwater resource and ensure water available for competing water uses (*i.e.*, domestic, stock farming and irrigation). The WUA is still operational after more than fifteen years providing water management functions. Water users comply with the requirements of allocation, water use and resource monitoring. With integrated water management principles, it is possible that environmental integrity and economic benefit to all water users can co-exist if the needed cooperation, structures and decision making converge in managing the water resource.

25.5.3 Effective outcomes

One of the indicators that the sequence of interventions was effective is that the rate of decline in groundwater levels have decreased and stabilization appears to be happening, but at various rates across the area. The broad pattern is that water levels stabilised some years after 2005. Representative water level graphs are shown in Figure 25.4. However, recent water level elevations are well below the 1990 water level elevations before irrigation commenced, indicating that a new equilibrium situation is being established.

The water level elevations stabilised due to a combination of interventions to reduce the demand. The water use data collected by the WUA from crop production and other uses are graphically represented in Figure 25.5. From 2001, there was a significant decline in water abstraction due to the described intervention actions and the economic conditions (*e.g.* escalation of energy cost). The abstraction from the

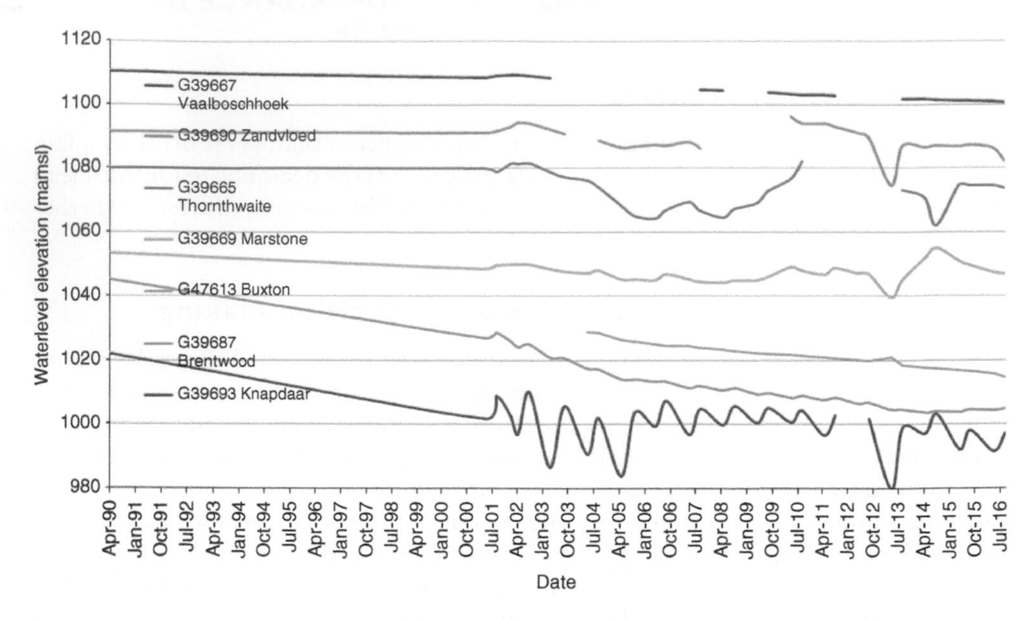

Figure 25.4 Characteristic water level fluctuations, 1990–2016 (note: from 1990 to 2000 limited monitoring).

resource declined to between 5 and 6 Mm^3/a after 2008 – well below the estimated resource potential of 11.1 Mm^3/a.

25.5.4 Sustainability of governance

An indication of the sustained efforts of groundwater governance at Tosca-Molopo is the long-term engagement over the period from 1999 to today.

Overall, the compulsory licensing process was successfully implemented, thus indicating successful governance. However, groundwater management at Tosca-Molopo is clearly driven by a national government department supported by regional offices (DWAF) and is, therefore, strongly dependent on the effective functioning of this department and its regional offices in cooperation with local structures like the WUA. The WUA was instrumental in assisting the farmers in utilising and administering their allocations by crop planning; general administrative issues related to water allocations; and handling of complaints. The WUA was, however, unable or unwilling to deal with many of these complaints, and consequently complaints were often escalated to DWAF. This suggests that the WUA, rather than enforcing rules such as metering of water levels and volumes of water extracted, is more concerned with maintaining good relations with its members.

Of the complaints received, subsequent to the third round of interventions, very few have been from stock farmers, whose initial concerns were one of the catalysts for this governance intervention. Most of the shallow boreholes have also been replaced by deeper ones, and with many of the stock-farmers also practising irrigation, there are opportunities to cross-subsidise the costs of replacing shallow boreholes that have dried up.

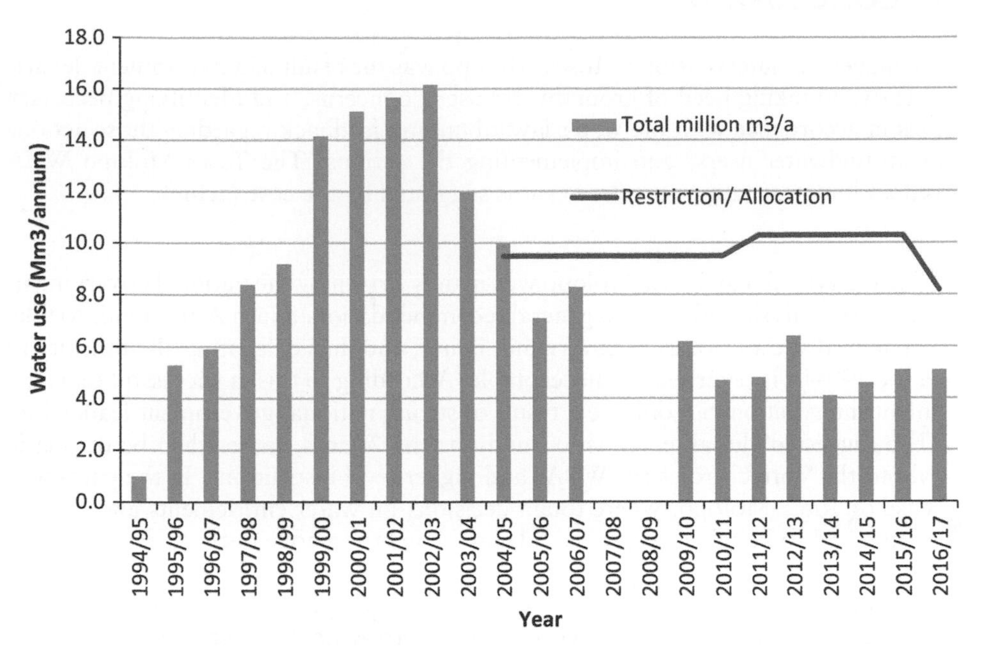

Figure 25.5 Estimated water use (from crop and demand data collected by WUA) from the resource 1994–2016.

25.5.5 Remaining challenges

However, some facts point to outstanding challenges. Volumes of groundwater abstracted have not been measured by the users as stipulated in the water use authorisations. Although total crop water use was estimated using remote sensing, this shortcoming has been addressed from 2016 by enforcing previous regulations for irrigation water users to install water flow meters and measure local water levels. However, climate change and variability may aggravate the precarious equilibrium achieved. The recent (2015–2016) drought experienced in the area forced water users to adapt to crops that use less water and crops that grow during the rainy season from January to April, when frequent rainfall events reduce the dependency on water pumped from the subsurface. The drought forced water users to change irrigation crops from the water-intensive crops like paprika and maize to potatoes and other tubers. The deep sandy virgin disease-free soils of the area are now sought for the production of potatoes from seed. It is estimated that more than 50% of potato tubers grown in South Africa is produced on approximately 500 ha in the area. The crop water demand associated with potato tubers is less than half of the demand of water-thirsty crops. Renewed complaints of failing water systems by stock and game farmers during the drought of 2015/16 has led to new water restrictions of 20% on irrigation water use implemented in 2016. In order to achieve and maintain sustainable water use, continuous decisions on available resources and use need to be made and implemented.

25.6 CONCLUSION

The management intervention at Tosca-Molopo was the result of a government department (DWAF) taking heed of groundwater users' concerns, and identifying necessary actions in accordance with the water law, obtaining feedback regarding those actions from groundwater users, and implementing the actions. The Tosca-Molopo WUA played an important support role. Lessons suggested by the case include:

- The intervention at Tosca-Molopo was mainly driven by the national government. However, this should not be a general recommendation. South Africa tends to take a paternalistic approach to governance issues, and thus delegating all authority to a local WUA is regarded as unacceptable. According to this, a successful management intervention can only be a result of strong national government leadership. The suggested global lesson is to build on what works, rather than be dogmatic about the 'correct' roles for WUAs and higher-level institutions. In the particular case of Tosca-Molopo, where tough decisions on water entitlements and curtailments had to be taken, it proved advantageous to remove decision making from the local level to maintain relations at local level. The WUA, however, was needed to support, coordinate, implement and monitor compliance to decisions.
- There is nothing significantly wrong with South Africa's water laws. These laws contain many powerful tools for good groundwater management. Revising them is essentially a distraction from implementing good governance rather than an aid. The implementation of, and compliance with legal requirements and regulations of water and the structures for implementation should be the focus.
- It is not the laws that require attention, but their implementation. The laws were used effectively at Tosca-Molopo. That they are not being used effectively in the rest of South Africa is the result of a lack of institutional capacity (Seward, 2015) rather than ineffective laws. Knüppe (2011) suggested that lack of institutional capacity may be more to do with mind-sets and attitude, and not necessarily shortage of staff, skills and funding.
- The conditions for good groundwater governance are probabilistic, not deterministic. The more a certain number of required coincident properties are present, the higher the likelihood that good groundwater governance will happen. These properties are contained in tools for assessing the effectiveness of groundwater governance, such as the Foster *et al.* (2010) 20 benchmarking criteria (Pietersen *et al.*, 2011) and the Ostrom design principles (Cox *et al.*, 2010; Seward, 2015). It is suggested that these tools could, and should, provide the framework for improving groundwater governance. Once, an area has been assessed using these criteria, the missing factors or particular challenges can be prioritised for implementation based on an understanding of the local institutional strengths and weaknesses. Pilot studies probing the effectiveness of interventions are likely to yield far more useful data than observation of passive groundwater governance cases. Wherever effective groundwater governance *is* taking place, *has* taken place, or *could* take place, scientific-based interventions should be made using an adaptive iterative and participatory management approach to actively probe the criteria needed for sustainable groundwater management and good groundwater governance.

REFERENCES

Burns, M., Audouin, M., Weaver, A. (2006) Advancing sustainability science in South Africa. *South African Journal of Science, 102*, 379–374.

Cox, M., Arnold, G., Villamayor-Tomas S. (2010) A review and reassessment of design principles for community-based natural resource management. *Ecology and Society.* 15 (4):38 [online]. Available at http://www.ecology and society.org/vol15/iss4/art38/ Accessed 15 January 2013.

Duvenhage, A.W.A., Meyer, R. (1991) *Geohydrological and Geophysical Investigation in the Tosca-Vergelee area.* Report no. EMI-C 91179. CSIR, Pretoria, South Africa.

Foster, S., Garduno, H., Tuinhof, A., Tovey. (2010) *Groundwater governance–conceptual framework for assessment of provisions and needs.* GWMate Strategic Overview Series No. 1, World Bank. Washington, D.C.

Godfrey, L., van Dyk, G.S. du T. (2002) *Reserve determination for the Pomfret-Vergelegen Dolomitic aquifer.* CSIR. Report no. ENV-P-C 2002-031.

Grobbelaar, M. (agrimolopobu@vodamail.co.za) (30 July 2016) Questions for Maryke. E-mail to: Seward, P. (sewardp@vodamail.co.za).

Knüppe, K. (2011) The challenges facing sustainable and adaptive groundwater management in South Africa. *Water SA.* 37(1):71–79.

López-Gunn, E., Cortina, L.M. (2006) Is self-regulation a myth? Case study on Spanish groundwater user associations and the role of higher-level authorities. *Hydrogeology Journal.* 14:361–379.

Mukherji, A., Shah. T. (2005) Groundwater socio-ecology and governance: a review of institutions and policies in selected countries. *Hydrogeology Journal.* 13:328–345.

Pietersen, K., Beekman, H.E., Holland, M. (2011) South African Groundwater Governance Case Study. WRC Report No. KV 273/11, ISBN 978-1-4312-0122-8.

Republic of South Africa (1998) *National Water Act No. 36 of 1998.* Government Printer. Pretoria, South Africa.

Republic of South Africa (2004a) *Establishment of Tosca-Molopo Water User Association.* Government Gazette 26552, Notice 828, 16 July 2004, Government Printer, Pretoria, South Africa.

Republic of South Africa (2004b) *Restrictions on taking water from Tosca-Molopo Dolomite Aquifer.* Government Gazette 26848, Notice 1138, 8 October 2004, Government Printer, Pretoria, South Africa.

Republic of South Africa (2011) *Final Allocation Schedule in terms of Section 47 of the National Water Act, 1998 for the Tosca-Molopo geographic area.* Government Gazette 34474, Notice 606, 22 July 2011. Government Printer, Pretoria, South Africa.

Seward, P. (2015) *Rethinking groundwater governance in South Africa.* PhD Thesis. Department of Earth Sciences. University of the Western Cape. Bellville. South Africa.

Turton A, Godfrey L, Julien F, Hattingh H (2006) *Unpacking groundwater governance through the lens of a trialogue: a Southern African case study.* Proc. International Symposium on Groundwater Sustainability (ISGWAS), Alicante, Spain, 24–27. January 2006.

van Dyk, G. S. du T. (2005) *Managing the impact of irrigation on the Tosca Molopo Groundwater Resource.* DWAF. Report number: GH4023.

Chapter 26

Governing groundwater in the Middle East and North Africa Region

François Molle[1,2], *Alvar Closas*[2] & *Waleed Al-Zubari*[3]

[1]*Institut de Recherche pour le Développement (IRD), France*
[2]*International Water Management Institute (IWMI), Cairo, Egypt*
[3]*Water Resources Management Program, College of Graduate Studies, Arabian Gulf University, Manama, Kingdom of Bahrain*

ABSTRACT

Groundwater is a key resource in the Middle East and Northern Africa region for both water supply and agriculture. While farmers in irrigated areas have tapped aquifers as a response to dwindling and uncertain surface water supply, investors have drilled wells and expanded agriculture into arid or desert areas. Most countries in the region have adopted standard water regulations, with emphasis on zoning (of stressed areas), licensing, metering and sometimes pricing, but enforcement has been problematic, and results on the ground modest to non-existent. This reflects a lack of material and financial means to handle tens and sometimes hundreds of thousands of dispersed wells but also a lack of political will. This can be explained by the state's lack of appetite for curtailing access to a resource that substantially contributes to the rural economy, a drop in state authority after the Arab Spring, inter-sectoral policy contradictions, the interest of politically well-connected investors, and sometimes the complexity of tribal politics. While state-centred governance has been largely ineffective, attempts at co-management or instilling a degree of participation from users have also been limited and unconvincing, reflecting wider governance systems in the region. The worrying depletion of many vital aquifers in the region is increasing the cost of abstraction, heightening social differentiation as smaller farmers are pumped out, and foreshadows a gradual collapse of the groundwater economy.

26.1 INTRODUCTION

Poor surface water endowment and rapid population growth rates since the 1970s (the region's population doubled between 1980 and 2011 from 170 million to over 350 million) have reduced the Middle East and North Africa (MENA) region's[1] per-capita freshwater share – from around 3,500 m^3 per year in 1960 to 700 m^3 per year in 2011 (Al-Zubari, 2012). Groundwater is the region's second conventional water resource after surface water, and in countries such as Jordan, Oman, Saudi Arabia and Yemen, it accounts for half the total water withdrawals (UNDP, 2013; Wada *et al.*,

[1]Includes Morocco, Algeria, Tunisia, Libya, Egypt, Jordan, Palestine, Israel, Lebanon, Syria, Iraq and the Arabian Peninsula.

2012). Worryingly, most groundwater-based agriculture and domestic and urban use are unsustainable. It has been estimated that the value of national GDP contributed by groundwater over-abstraction could be equivalent to 2 percent in the case of Jordan and around 1.5 percent in Yemen (World Bank, 2007). Yet these figures do not convey the strategic importance of the resource for rural livelihoods, food production or supply for cities. Neither do they reflect the associated environmental impact.

Groundwater quality also presents challenges for the region. With many large urban areas located near the coast, untreated sewage from these centres, refuse sites or septic tanks leak into and pollute coastal aquifers (often used for drinking-water supply) (*e.g.* in Lebanon, the West Bank and Egypt) (El-Fadel *et al.*, 2002). Industrial effluents and pollution from agricultural sources (fertilisers and pesticides) are also a cause for concern, especially in shallow aquifers (*e.g.* Ras al Jabal region in Tunisia and the Nile Delta in Egypt) (UNDP, 2013).

As a semi-arid to arid region, MENA has long depended on its groundwater resources. Historically, groundwater was accessed through shallow dug wells (mostly in alluvial formations) and natural springs, as well as 'horizontal wells' called *qanats* (or various other names across the region: *e.g. khettaras*, *fogara*, *aflaj*). Groundwater sourced from springs and qanats has been managed through customary rights and rules, generally articulated around private property rights associated with land access and the initial investment of capital and labour by shareholders. Many such common property resources have been successfully managed for centuries (and some are still in place in countries such as Algeria and Oman), indicating substantial, but dwindling, social capital around the collective management of groundwater resources.

As in many other countries in the world, tube wells (often deep), new drilling techniques, submersible pumps and electrification have contributed to an unsettling of the historic balance between the available resource and its use, quickly leading to an overall situation of overexploitation in many aquifers. What is special, therefore, with the MENA region regarding groundwater? First, MENA is facing acute water scarcity, and the overall reliance on groundwater is higher than in most regions. Second, and perhaps correspondingly, many aquifers are barely- or non-renewable. Third, the region displays a contrast between the age-tested, community-based management of springs, wells or qanats and the current, state-centred modes of governance, which are proving to be largely ineffective. Fourth, large-scale agribusiness companies are increasingly prominent in the expansion of groundwater-based agriculture.

This chapter first describes the unprecedented boom in groundwater use in the MENA region and then investigates in more detail the responses of the states and users. The last section analyses the major causes of those dynamics and the implications for groundwater governance. We focus on water depletion issues and allude to the contamination of aquifers only in passing, since this – locally serious – problem is not well documented and is still largely unaddressed by public policies.

26.2 THE GROUNDWATER BOOM

In line with what has been observed worldwide, groundwater development in the MENA region has boomed over the past 40 years based on the initiative of both the state and individual users. The surge in the exploitation of groundwater, mostly for agriculture, has been observed in three different situations, as examined here.

26.2.1 Conjunctive use in irrigation schemes

Iconic large-scale public irrigation schemes were developed by most states between the 1960s and 1980s. This was notably the case in Morocco, where King Hassan II's 'hydraulic policy' sought to develop one million hectares of irrigated land, but was also witnessed in northern Tunisia, Syria (Orontes and Euphrates River Basins) and Iraq, not to mention the continued expansion of the Nile Delta/Valley system. Supplied by diverted surface water, with time these schemes faced occasional to persistent shortages, due to upstream development (*e.g.* Euphrates), irrigation 'overbuilding' (*e.g.* Morocco), competition with domestic or other uses, or increasing climatic variability. As a result, farmers turned to conjunctive use, tapping water available in both nearby drains and streams as well as underlying aquifers, which to a large extent are fed by irrigation from surface water (Van Steenbergen and El Haouari, 2010). In some areas, wells pepper the landscape and provide more water than is distributed through public canals (see Kuper *et al.*, 2012 for a graphic example of the Tadla scheme in Morocco).

On the positive side, the overall efficiency of the schemes is greatly enhanced as return flows are recaptured and farmers enjoy higher flexibility and security in their water supply. This is especially important for capital-intensive crops – fruit trees in particular. On the negative side, conjunctive use incurs additional running costs for farmers, with successive technological adjustments being required by the continual drop in the water table.

The intensity of conjunctive use can be considered as a good indicator of the adequacy of irrigation water supply. In schemes like Tadla in Morocco a significant number of individual wells have been drilled since the 1980s, when the Oum Er-Rbia basin started to be overexploited (Kuper *et al.*, 2012). A similar situation can be observed in the Maghreb region as a whole and in the Ghab Valley, Syria. Elsewhere, the phenomenon is more recent, such as in the Nile Delta, where intensive well drilling has been observed over the last 10 to 15 years (El-Agha *et al.*, 2017). Importantly, the phenomenon is also observed in small-scale communal schemes based on river diversion, springs or qanats, notably in oases.

26.2.2 Supplemental irrigation in rainfed areas

A second situation where wells have proliferated is that of rainfed agriculture, where the supply of groundwater has allowed farmers to increase, and more importantly secure, their yields in times of drought. This has been observed in Morocco, Tunisia and Yemen, but most prominently in Lebanon, where much of the Bekaa plain came to be irrigated from different surface and groundwater sources, and Syria, where the plateau between the Orontes River and the eastern deserts has massively resorted to groundwater.

Securing rainfed agriculture, as well as capitalizing on this secure source of water to diversify into cash crops, has generally been a result of individual investment. There are cases, however, where the state has attempted to establish collective distribution networks based on public wells, as in many parts of Tunisia (Hamdane, 2014) or in the Souss, Morocco. A key example of the state's direct role in conveying groundwater to rainfed areas for irrigation, and also for cities, is Libya's 'Great Man-Made River' project, which aimed to transfer 6.5 Mm^3 of groundwater per day from large aquifers in the desert in the south of the country to the coastal area.

26.2.3 Expanding the frontier into deserts

Groundwater has also been mobilized to expand agriculture into arid or desert areas. Large agribusiness companies in, for example, Sudan, Egypt, Jordan, Morocco and Saudi Arabia have invested in tube wells to tap (often fossil) groundwater at great depths. In Egypt 75 percent of all desert reclamation (for agriculture and urban projects) has relied on private investors and corporate modes of farming (Sims, 2015). In general, such capitalist ventures in agriculture have benefitted from generous government subsidies and financial incentives and/or been encouraged by permissive state regulations and a lack of control of groundwater abstraction. In many cases the state has mediated such investments, providing long-term leases for the land (and well licenses), for instance to the companies initially allowed to exploit the Disi aquifer in southern Jordan.

Egypt's desert development projects since the 1950s have included well fields in various oases as part of the 'New Valley' project in the western desert. Since the 1990s the government has reclaimed thousands of hectares for irrigation in oases such as Kharga and Dakhla further south near the border with Sudan (East and West Oweinat). Some public wells have been handed over to groups of farmers to manage them. The state may be the indirect beneficiary through parastatal companies or, more directly, it could be the army (Sims, 2015). Although state-sponsored or private agribusiness dominates such kinds of enterprise across the region, there are also cases where small-scale investors are able to develop small to medium-scale projects in the desert or arid lands, based on one or more wells. This is the case in the highlands of Jordan, in Morocco and on the fringes of the Nile Delta. Conducive conditions for this generally include road accessibility and the relative proximity of urban centres or export facilities, modes of land tenure that make claims/access to land possible, and the ability to obtain well licenses (or authorities turning a blind eye to groundwater development).

26.2.4 Main drivers and consequences

The initial boom in groundwater-based agriculture has largely been fuelled by direct or indirect subsidies by the states or donor-funded programs. Tapping groundwater was seen as an easy and decentralized means of supporting rural development and poverty alleviation (Allan, 2007), of compensating for uncertain supply in public schemes, and sometimes of improving food security in the country (Al-Zubari, 2014). The role of the state in driving groundwater development in the MENA region is ubiquitous and can be illustrated by a few cases.

In Tunisia, there are over 1,000 user groups managing irrigation schemes of between 30 and 300 ha, most of which were established by the state and based on groundwater resources (Elloumi, 2016). Encouraging individual well development through various types of incentive, for example drilling wells, buying pumps or, indirectly, buying micro-irrigation equipment, has been a central component of agricultural policies aimed at supporting rural livelihoods. This has also been the case in Algeria, Morocco and the United Arab Emirates (UAE).

The state has often played an active role in the development of groundwater resources as part of a campaign to settle tribes and nomads in rural areas, in some cases as part of a wider program to ensure rural livelihoods or support the development of rural water supply via wells (*e.g.* Jordan, Saudi Arabia, UAE). Since the early 1980s UAE citizens have received plots of between 2 and 10 ha, ready for cultivation with one well/ha, and interest-free loans to purchase pumps and other equipment (Fragaszy and McDonnell, 2016). Consequently, farmland in the Liwa region, which was approximately 1,000 ha in 1987, peaked at around 21,000 ha in 2002, leaving the oasis peppered with 35,000 wells (many of which are now defunct due to falling groundwater levels).

The development of groundwater can also be linked to the maintaining of patronage relationships within the state apparatus and the state's relationship with local leaders (*e.g.* Yemen) (Zeitoun *et al.*, 2012). Community leaders and specific groups or elites benefit from permissive regulation and/or political connections in their development of agricultural ventures and in drilling wells. Jordan is a case in point and Egypt even more so (notably on the western fringe of the Nile Delta), where the army has received licenses to develop large tracts of desert land (Sims, 2015).

Either directly or indirectly, many state policies have encouraged well drilling and groundwater-based agriculture. In Syria, for example, groundwater-based agriculture accounts for 53 percent of the total irrigated land. During the 1980s and 1990s the development of groundwater abstraction wells and irrigation (from 53,000 wells in 1988 to 124,000 in 1994) was fuelled by input subsidies sponsored by the government, such as a diesel fuel subsidy and crop procurement price support (Aw-Hassan *et al.*, 2014). Donors have also been supportive of groundwater development. In Yemen money from international development agencies, such as the World Bank in the 1970s, or German or Dutch bilateral funds, has been used to develop the country's water-management infrastructure and institutions.

This 'good' idea of harnessing groundwater resources to achieve all these desirable development policy goals eventually turned sour. Uncontrolled well drilling gradually resulted in declining water tables (with typical observed drawdowns of 1 m/year), which made abstraction costlier (either due to increasing fuel costs or the expense of deepening wells) and sometimes pushed smaller farms out of business. Over-abstraction also affected the systems fed by the discharge of the aquifers: springs dried up everywhere; wetlands such as the Azraq oasis in Jordan virtually disappeared; qanats – and the huge cumulated investments as well as social capital that sustained them – fell apart (Al-Zubari, 2014). In coastal areas (*e.g.* Lebanon, Tunisia, Morocco, Oman) lower aquifer levels led to the intrusion of seawater, causing land salinisation and the destruction of agriculture. With the use of deeper layers of aquifer, changes in water quality were frequently observed, particularly in terms of increased salt content (*e.g.* Jordan), forcing farmers to adapt their crops or discontinue agriculture. In other cases a race to the bottom unfolded on both sides of a transboundary aquifer, raising tensions between neighbouring countries (*e.g.* the Saq-Ram aquifer between Jordan and Saudi Arabia).

26.3 RESPONSE OPTIONS

Because many countries in the MENA region have declared water a state or public resource, and because groundwater is intertwined with many key social and economic issues, widespread groundwater over-abstraction readily puts the onus on the state. International and development agencies have contributed to the dissemination of standard policies and laws considered to be internationally sanctioned best practices. The form of the responses to the degradation of groundwater quantity and quality therefore bears similarities across the region. This section reviews and categorizes the different types of response, and briefly reports the outcome of their implementation. For analytical purposes we distinguish here between three types of policy objective:

1 How to control (limit or reduce) the *number* of existing/active wells;
2 How to control (limit or reduce) the *amount of groundwater* abstracted by existing/active wells;
3 How to 'bring additional water in' to reduce the degree of groundwater overexploitation.

26.3.1 Controlling well expansion

26.3.1.1 Registration of wells

Most countries in the MENA region have put in place a system of licenses or permits for groundwater use (World Bank, 2007), with varying procedures. In some instances (*e.g.* Jordan, Lebanon, Tunisia) a distinction is made between the drilling license and the subsequent exploitation license, both of which must be obtained.

In most cases, the majority of wells were drilled at a time when licensing was either loosely enforced or non-existent. This means that when stricter legislation is passed, making licensing mandatory, the question of the legality of existing wells comes to the fore. In general, a grace period is granted for regularization. Farmers rarely comply for a number of reasons. These include reluctance to follow a burdensome procedure, the (sometimes annual) fee to be paid, the fear of being charged for water in the future, a lack of trust of the authorities' intentions, and the ingrained belief that groundwater is not the government's business. As a result, deadlines are often extended. In Morocco a regularization period for wells dug before 1995 was open until the end of 1998; in 2009 wells dug before 2009 were to be registered within a three-year period, later extended to 2015. In Jordan the 2002 Groundwater Bylaw gave one year to illegal well owners to regularize their wells; an amendment in 2003 gave another six months, and eventually all wells older than 2005 were considered illegal and liable to be backfilled (Al-Naber and Molle, 2017).

In Morocco the authorization of new wells requires the submission of a detailed file, which includes situation maps as well as a technical study for the well and its impacts on neighbouring wells and local water resources (Del Vecchio, 2013). Registration is also burdensome and costly in Egypt, where a large majority of farmers in the Nile Delta have not bothered to register their wells, especially after the 2011 revolution (the Arab Spring) and the undermining of state and police authority (El-Agha *et al.*, 2017). In Lebanon, further to the recent decision to use private companies for the revision of the technical aspects of permit applications, their number has fallen from 2,000 to 500 per year on account of the increasingly costly (US$960 per well) and tedious process (Nassif, 2016).

In most countries, therefore, the percentage of unregistered wells (whether or not they are known to the administration) remains high (Kuper *et al.*, 2016), with certain exceptions, such as Bahrain and Abu Dhabi (which is currently concluding a national inventory). In Jordan, illegal (known) agricultural wells number around 1,268 out of 3,721 registered wells (in 2011), but there are also unknown illegal wells, with some still being drilled in regions such as Azraq and Jafr. In Yemen, licensed wells have permits for the abstraction of specific quantities of groundwater per year although no monitoring takes place and the majority of wells do not have permits (Morill and Simas, 2009). Information from a recent UNDP-funded project in Lebanon estimated that there are around 59,124 unregistered (yet inventoried) private wells and 20,537 registered private wells in the country (UNDP, 2014).

26.3.1.2 Controlling and banning well drilling

Limiting the drilling of new wells can be achieved by controlling the drillers rather than the farmers. In Yemen, all heavy drilling rigs and metal well casing must meet technical specifications issued by the National Water Resources Agency (NWRA) (Morill and Simas, 2009). However, while 125 drilling contractors were licensed by the end of 2006, some estimates suggest there could be a total of 400, or even as many as 900, drilling rigs in the country (NWSSIP, 2008; FMWEY, 2015). Despite the use of sophisticated technology (*e.g.* GPS tracking and satellite imagery in the hands of a 'rig tracking unit'), illegal drilling has continued, with blatant violations of regulations carried out by influential people in plain sight (Ward, 2015; Van Steenbergen *et al.*, 2015).

In Oman, only government-registered contractors may carry out well construction and maintenance, yield testing and pump installation (Morill and Simas, 2009). In the UAE, only one company is licensed by the government. The 2002 Groundwater Bylaw in Jordan requires licenses and authorizations for drilling equipment and drillers. Since 2013, the Ministry of Water and Irrigation (and Jordanian security forces) have confiscated drilling rigs (up to 159 rigs by April 2015) (Jordan Times, 2015). However, drillers have become more creative and it was reported that they now manufacture rigs that can be stealthily loaded on pick-ups (Al-Naber and Molle, 2017). In 2005, in the Souss-Massa in Morocco, the River Basin Agency launched an initiative to control borehole drilling and, by 2010, 190 drilling machines had been seized (BRLI and Agro-Concept, 2012). The Agency is also trying to create a professional association for the drilling companies, which could potentially become a management actor.

Drilling bans have been implemented in various countries, such as in critical zones in Algeria, Bahrain and Tunisia (Faysse *et al.*, 2011) and more broadly in Jordan (agricultural wells, since 2002). On balance, it has proved difficult to control both legal and illegal drillers, and the level of implementation of these bans has been weak. From Jordan to Morocco, fingers are in particular being pointed at Syrian operators, notorious for specializing in legal/illegal well drilling across the region.

Well expansion or existing abstraction levels can be monitored indirectly with the use of technology such as satellite imagery. In Jordan the Ministry of Water Resources and Irrigation uses this technology to identify land use with irrigation and locate illegal wells (Herald Globe, 2014). The control of irrigation surface in dry areas with this type of technology is supported through different donor programs in several countries in the MENA region (*e.g.* the MAWRED project funded by NASA and USAID), but its use is incipient.

26.3.1.3 Backfilling of illegal wells

The most radical measure in dealing with illegal wells is identifying and backfilling them. Jordan's Ministry of Water Resources and Irrigation launched a campaign in 2013 aimed at sealing and backfilling illegal wells. According to newspaper sources, the authorities (in tandem with Jordanian security forces) sealed 644 illegal wells from 2013 to mid-2015 (Jordan Times, 2015). The task is made difficult in some cases by resistance on the part of the owners, who deploy guards and dogs at the property gates and deny entry to the inspectors – a situation found in Balqa (Jordan Times, 2015), Azraq and other places. However, off-the-record information indicates that the wells that have been sealed were either already dry or unused. This still clearly heralds a tightening of regulation by the government (Al-Naber and Molle, 2017). In 2017, Environment Abu Dhabi (EAD) launched an awareness campaign on the negative impacts of well drilling, encouraging people to report illegal wells as a first step towards their filling.

In the Souss-Massa, Morocco, the policy of removing unauthorized wells or pumps (for example around 70 percent of the pumps found in the Souss area) was never fully implemented (BRLI and Agro-Concept, 2012). Illegal wells identified through campaigns carried out in some Algerian provinces, such as in Oran in 2006, were also meant to be sealed, but there is no sign that anything effectively happened on the ground (Amichi, 2015). In general, it seems that cases where authorities have cracked down are linked to over-abstraction threatening domestic use, or to very specific situations where the owner of an illegal well had conflicting relationships with the authorities or the police (*e.g.* refused to pay bribes, etc.) (Molle and Closas, 2017).

26.3.2 Controlling and reducing abstraction in existing wells

Once the expansion of wells has been addressed, the second main policy objective is to control (limit/reduce) *actual* groundwater use in *existing* wells. Metering, pricing, abstraction quotas and technology improvements for irrigation are some of the main instruments used by governments to control the abstraction of groundwater in the MENA region.

26.3.2.1 Groundwater metering

Even though groundwater metering is not a tool in itself to reduce groundwater abstraction, it is often seen as a prerequisite for the implementation of pricing instruments and abstraction quotas. Hardly any country in the MENA region has been able to ensure that licensed agricultural wells are fitted with working meters. Jordan prescribed the obligatory use of meters in the mid-1990s and started to equip wells with the support of USAID (Venot and Molle, 2008). However, while in the Amman-Zarqa Basin, 90 percent of wells were equipped with meters, only 61 percent of those meters were found to be working in 2004 (Chebaane *et al.*, 2004). A recent survey in Azraq found that of a total of 334 wells surveyed only 192 water meters were working (IRG, 2014). Besides technical issues, meter tampering and vandalism are common (Chebaane *et al.*, 2004).

In Syria, despite the fact that current regulations require groundwater wells to have meters installed, the lack of training and technical skills by engineers and public officials affects the implementation of this measure (Albarazi, 2014). This, however, is a typical official explanation and obscures the likely lack of political will to enforce such regulations. In Tunisia, meters are not required, even for wells deeper than 50 meters, on account of the admitted incapacity of the state to monitor them (Hamdane, 2014). There is also recognition that meters are costly and this raises the question of who has to pay for them. One option is to shift the cost onto farmers' shoulders but this may make licensing even less attractive to them. In other countries, such as Yemen and Egypt, the requirement to install water meters is nullified by the fact that wells are not registered in the first place.

In Abu Dhabi, well metering (and pricing) are floated as possible measures but (so far) are stiffly resisted by farmers. In Oman, a law in 1990 established that wells be licensed and metered (with penalties for tampering), but it has not yet been effectively applied (McDonnell, 2016). In Bahrein, the metering of all wells was initiated in 1997 and applied for some time (Al-Zubari and Lori, 2006). After a long period of ups and downs, most groundwater wells are now metered and their production is read on a monthly basis. This is made possible by the fact that most agricultural wells are within a one-hour drive in the same (north-west) region and are limited in number (around one thousand). Bahrain passed groundwater regulation as early as the 1960s and enforcement has probably been facilitated by the non-tribal nature of the population. In the Souss, Morocco, the installation of meters was a key measure of the aquifer contract (*contrat de nappe*) but was postponed when it became clear that other clauses had not been implemented (BRLI and Agro-Concept, 2012).

26.3.2.2 Abstraction quotas

Where wells are licensed, it is generally the rule that a maximum annual abstraction volume (or average discharge) is granted, as in Lebanon, Bahrain, Tunisia and Morocco. However, it is often unclear what would happen if the limit were exceeded. In any case, there is no evidence of any serious volumetric monitoring of agricultural groundwater use in the MENA region to verify whether quotas are respected.

Quotas can be associated with a block-tariff, where the user is allowed to pump water beyond a minimum quota (which is in general granted for free) but at a cost, with the price of water increasing with the volume abstracted. In Jordan, abstraction licenses granted between 1962 and 1992 generally specified the amount allowed to be pumped, most commonly 50,000 or 75,000 m^3 per year per well (Venot and Molle, 2008), although these limits were not enforced. The Groundwater Bylaw of 2002 established a new quota system based on block-tariffs (see below).

The application and enforcement of quotas require abstraction to be measured and are therefore difficult and prone to corrupt practices as in Syria, or users tampering with meters as in Jordan for example. Jordan, however, is now increasingly using proxies, such as electricity consumption and the cropping area (with the help of remote sensing), to estimate and charge for groundwater abstraction (Al-Naber and Molle, 2017). This is certainly an option that is worth exploring in the region, especially since aridity facilitates the identification of irrigated crops by remote sensing.

26.3.2.3 Groundwater pricing

Volumetric tariffs have the potential to encourage users to abstract less groundwater yet have barely been used in the region (despite the emphasis on pricing instruments in state and donor policy documents). One reason stems from the evidence discussed above that metering in agriculture is rare and often dysfunctional. A second reason is that it is close to impossible for governments to tax legal groundwater users volumetrically at a level that is sufficient to elicit conservation measures without impacting income. The situation is different, however, with groundwater for domestic or industrial uses, where the potential for volumetric metering and pricing is higher (Molle and Berkoff, 2007).

In Tunisia, groundwater users in small, publicly managed irrigated areas, covering around 24 percent of the irrigated area in the country, have to pay for groundwater supplied through pressurized networks via state-established but user-run decentralized associations (GDAs) (Frija *et al.,* 2014). However, prices are not high enough to affect behaviour, and, if they rise, farmers tend to shift to individual unmetered and uncharged wells (Ghazouani and Mekki, 2015). In Bahrein, a decree issued in 1997 on groundwater pricing for all purposes could not be implemented due to socioeconomic and political constraints (Al-Zubari and Lori, 2006). In Lebanon, users abstracting over 100 m^3/day need an authorization by decree and are supposed to pay for water, but monitoring, and payment, are virtually non-existent (Nassif, 2016).

In Jordan, groundwater pricing is in place via a system of block-tariffs. This system was introduced in 2002 by the Groundwater Bylaw and charged any water use over a threshold of 150,000 m^3 per year per well at a level of 0.007 USD per m^3 for volumes between 150,000 and 200,000 m^3 per year and 0.085 USD per m^3 beyond. Such tariffs, and the generous free-block, were found to be ineffective with regard to water conservation, especially in fruit-tree farms with high income (Venot and Molle, 2008; Demilecamps, 2010). It proved difficult for field staff to collect water fees and hardly anyone paid. In 2009, the ministry sent bills with cumulated arrears and since then has stepped up its pressure on farmers to elicit payment. Yet, there are clear limits to the taxation that farmers with legal wells are ready to accept since they already pay to access groundwater and do not accept government intrusion in such matters. In 2010, the Jordanian government moved to using tariffs to hit illegal wells, which are now given no or much smaller free blocks and whose taxation can be justified not because they are using water but because they are illegal (Al-Naber and Molle, 2017). People with unpaid bills are now being barred from accessing other state services or obtaining official documents. This double pressure is starting to be felt and sends a strong signal about the ministry's growing resolve.

26.3.2.4 Buying out wells

The buying out of wells by the state, a last resort in controlling abstraction and the number of wells, has been considered by the government of Jordan. Chebaane *et al.* (2004) found that 50 percent of farmers were in principle in favour of such an option.

The successful implementation of this measure requires substantial funds, as legal well owners are likely to expect significant compensation reflecting the profitable use of water for cash crops. On the other hand, the government rules out paying for illegal wells, which would be antithetic to the law. Zekri (2007) notes that groundwater gains achieved by this option could be pumped again by the active farmers in the area if no further abstraction limits were imposed and controlled. Control of expansion is therefore essential before existing wells are tackled. This combined approach has not yet been implemented in the MENA region.

26.3.2.5 Technology fixes to reduce groundwater use

Technology is invariably cast as central to the strategy of reducing water use in agriculture. This includes solutions such as the use of remote sensing, high-tech meters (with automatic data transfer) and, most commonly, micro-irrigation to 'reduce water losses' and improve irrigation efficiency. While Morocco, Saudi Arabia and Tunisia have extended subsidies to farmers for micro-irrigation, other countries such as Lebanon, Jordan and Egypt have been more conservative, largely leaving it to the market. In Oman, between 2000 and 2016, farmers (with a licensed well) buying modern irrigation technologies received 100 percent subsidies (McDonnell, 2016).

In Tunisia, the National Programme for Water Savings, established in 1995, offers subsidies to farmers of 40 to 60 percent of investment costs for water-saving irrigation technologies. Yet, national data show no decrease in water use per ha, just the expansion of irrigated areas (Frija *et al.,* 2014). The Green Morocco Plan (*Plan Maroc Vert*), Morocco's national strategy to improve the country's agricultural efficiency and output, can fund up to 100 percent of drip irrigation and 80 percent of the cost of drilling a well (Molle and Tanouti, 2017). Although micro-irrigation generally reduces the quantity of water applied, its impact is mostly neutral, and occasionally negative, with regard to plot-level water consumption (Molle and Tanouti, 2017; Perry and Steduto, 2017).

26.3.2.6 Indirect incentives

Groundwater-based agriculture can potentially be shaped and regulated indirectly, by modifying the prices of input and output factors. While it is next to impossible to impose administered volumetric taxes on well abstraction to encourage water conservation, it could be easier to influence use through the cost of the energy needed to pump water or through targeted crop input/output subsidies.

Indirect incentives through electricity supply and pricing have been extensively studied in the case of India by scholars such as Shah (2009). With regard to the MENA region, there has been debate in Tunisia about the possibility of establishing an electricity-based pricing system for private bore-wells – an idea negotiated between the Ministry of Agriculture and the Tunisian Society of Electricity and Gas (Frija *et al.,* 2014). In Jordan, a new law has made it impossible to obtain a connection to the grid without a valid well license, condemning illegal wells to using diesel at a much higher cost; with the unintended adverse effect, however, of pushing them towards solar energy, which results in even more groundwater over-exploitation as pumping costs are generally drastically reduced.

In Syria, the cancellation of diesel and fertilizer subsidies for farmers in 2008 and 2009 (with the intention of integrating the country into the global trade system and joining the World Trade Organisation) pushed prices up and forced many farmers either to revert to rain-fed agriculture or stop agriculture altogether when their wells dried up (de Châtel, 2014; Wendle, 2016). The combination of a severe and lasting drought, which resulted in many aquifers being depleted in northern Syria, and high energy prices fuelled population displacement (Wendle, 2016), which may later have contributed indirectly to the popular uprising. In Saudi Arabia, demand for water decreased as a result of policies with intended as well as unintended consequences. In 1994 the government reduced the price of wheat (from SRI 2000/ton to SRI 1500/ton). More importantly, the price of the main input for pumping water – diesel – was tripled, making it far more expensive to pump groundwater (Al-Sheikh, 1997). Saudi Arabia has also recently reduced input subsidies as part of its program to reduce agricultural groundwater use in the 2000s and phase out wheat cultivation by 2018; it has banned fodder exports and recently taken steps to limit alfalfa production significantly.

Water use can also be modified by subsidizing certain crops or paying farmers for certain behaviour. In Abu Dhabi, all farm owners receive 90,000 AED/year not to grow Rhodes grass or alfalfa on more than 10 percent of their farmland (if they do not have more than 120,000 AED/year of property income). An additional 10,000 AED/year are provided to all farm owners with 60 or more date palms (low water users) on their farm (Fragaszy and McDonnell, 2016).

26.3.3 Supply-side measures

Groundwater over-abstraction can be reduced via supply augmentation of either surface water (transfers) or groundwater (through aquifer recharge), and also by seawater desalination. The use of technology and the provision of additional resources are seen as the most 'conflict-free' management options. Bringing more water, especially if it is cheaper and/or of better quality, helps decrease the pressure on groundwater and potentially reduce its use. However, in the mid to long term, more water being available at the dwelling or farm plot could well result in expanded use and consumption.

26.3.3.1 *Bringing more surface water in*

Several countries in the MENA region have created infrastructure projects to 'bring more water in' and partly replace the existing demand for groundwater with surface water. Morocco's Water Strategy for 2030 includes a 'water highway', from the north of the country to the south, expected to transfer around 800 Mm^3 of water per year. New water could relieve pressure on the Haouz aquifer in particular by supplying Marrakech with drinking water. The Souss-Massa basin in the south is home to the Guerdane irrigated area (10,000 hectares of citrus plantation), which now receives surface water derived from a reservoir upstream in the basin in order to compensate for declining groundwater resources (Houdret, 2012). The Saïss Plain will benefit from a water transfer from the Sebou river with the aim of reducing aquifer abstraction and ending groundwater pumping for drinking water entirely by 2030 (Del Vecchio, 2013).

In Egypt, new investors in the West Delta area have been expanding groundwater-based, high-value agribusinesses since the 1990s. The West Delta Canal project was conceived to complement or replace dwindling groundwater stocks by surface water from the Nile via a water transfer (Barnes, 2012). However, it has yet to be implemented due to bidding and procurement issues and, more recently, the 2011 revolution. In Tunisia, approximately 30,000 km of networks transfer water from the north and aquifers in the west towards the Cap Bon and the coast, including Tunis (INECO, 2007). The transfer to the Cap Bon Peninsula in north-east Tunisia provides water for cities as well as agriculture and serves the western (Grombalia) and southern parts of the Cap Bon region and its 15,000 hectares of citrus.

26.3.3.2 Aquifer recharge

Managed aquifer recharge (MAR) and storage can be used especially where groundwater is a strategic reserve for emergency situations. The injection and enhanced infiltration of water can serve as a way of storing excess rainfall and storm runoff, which otherwise could turn into flooding, or as intermediate storage of treated effluent for later agriculture and industrial use.

In Tunisia, wastewater represents around 30 percent of the country's agricultural water supply, and aquifer recharge with wastewater has helped prevent coastal aquifer salinisation caused by groundwater over-abstraction (*e.g.* in the Korba plain in the Cap Bon) (Ouelhazi *et al.*, 2013). Bahrain also applies aquifer-recharge technology, using gravity-fed aquifer recharge through gulleys, pits, chambers and recharge wells to direct urban runoff from storms to the Khobar aquifer (Klingbeil, 2014). In Abu Dhabi, the Liwa aquifer is to be used to store excess desalinated water to increase the Emirate's freshwater reserves from 3 to 90 days (Fragazy and McDonnel, 2016). In the Salalah coastal aquifer, Oman, an artificial recharge scheme of 40 tube wells installed in 2003 has been effective in pushing back the saline zone front (Shammas, 2008).

26.3.3.3 Desalination

Half of the world's desalination capacity is found in the Arab world, with Saudi Arabia and the UAE jointly producing more than 30 percent of the world's desalinated water (UNDP, 2013). More than 55 percent of water supplied to urban areas in the Gulf comes from desalinated water, used directly or mixed with groundwater, contributing around 1.8 percent of the region's total water supply (UNDP, 2013). Countries with less intensive use of desalination, such as Tunisia, are considering expanding the sector. In Algeria, desalination plants have been ensuring the supply of drinking water in coastal areas since the inauguration in 2008 of the Hamma desalination plant near Algiers (a second plant is in operation near Oran). The Red Sea-Dead Sea project in Jordan, in whatever form it will eventually take, intends to desalinate all of the water it brings in.

While the increased use of desalination has meant a reduction in the demand for groundwater for urban supply, this has been a response to increasing urban demand rather than a means of addressing the pressure on groundwater resources. Desalination is energy-intensive, produces costly water (between 0.50 and 4 USD per m^3 in the Arab region depending on the level of subsidies) (UNDP, 2013) and has environmental impacts (disposal of brine and emission of harmful oxides from burning fossil fuels).

26.3.4 Attempts at participatory management of groundwater

Some governments have experimented with participatory management with users, either having realised they cannot manage groundwater alone or having been persuaded by donors who routinely promote participatory approaches as a means of enhancing water governance.

In Bsissi, Gabès Governorate, Tunisia, a decree prohibiting well drilling in 1987, and subsequent decisions by the Ministry of Agriculture to discontinue agricultural subsidies and close illegal wells, triggered farmer protests at the highest level. The regional office of the Ministry of Agriculture (CRDA: Commissariat Régional au Développement Agricole) proposed a negotiation and prompted the formation in 2000 of an association of 103 farmers. Farmers and the CRDA agreed upon a series of give-and-take measures to control both the drilling of new wells and abstraction from existing wells (quota, micro-irrigation). In exchange, the wells of members would be regularized and connected to the electricity grid (lowering pumping costs); and subsidies would be extended for shifting to micro-irrigation and other needs. Farmers committed to assist the CRDA in closing disused or abandoned wells. Despite promising results, it seems that the association has now been affected by the challenge to state authority that came with the 2011 revolution (Frija *et al.*, 2014; Leghrissi, 2012; Hamdane, 2015).

The experience in Morocco with aquifer contracts started in 2004 when the Souss-Massa River Basin Agency carried out an awareness campaign about the new water law and proceeded to close illegal wells. This triggered social unrest, and the governor of the region suspended the decision and decided to approach the problem by creating a commission with representatives from 20 institutional partners (BRLI and Agro-Concept, 2012). An agreement was signed in 2007, which included 22 small dams and 5 large dams to be constructed by the state and the regularization of 'illegal' wells, against a freezing of the expansion of irrigated areas for citrus and vegetables, a (subsidized) shift to drip irrigation, an increase of groundwater user fees and reinforcement of the water police (Closas and Villholth, 2016). Although that contract was never implemented, the government put aquifer contracts in the limelight again in 2014 with a policy that they be established in all (later reduced to three) major aquifers in Morocco by 2016 (L'Economiste, 2014).

In the highlands of Jordan, with financial support from the German Cooperation Agency (GIZ), the Ministry of Water and Irrigation attempted to bring together 60 stakeholders (agricultural water users, government institutions, NGOs and research institutions) in a 'Highland Water Forum', a multi-organisational dialogue mechanism for the area (Mesnil and Habjoka, 2012). The Forum's Secretariat is now based at the Ministry of Water and Irrigation, which saw the Forum as a means of mediating its reforms, while farmers largely took it as a means of claiming benefits. No clear measures to curb abstraction have been agreed.

Although the region is rich in traditional community-managed systems of water use, as recalled in the introduction, these new approaches have largely been disappointing. This has to do in part with the lack of trust between the state and citizens and the fact that state officials are generally unwilling to share power or support co-management. Ultimately, rural livelihoods are precarious, and the effort needed to curb abstraction are often unmanageable given the level of overexploitation that has hitherto been allowed.

26.3.5 The case of non-renewable aquifers

The MENA region has considerable regional non-renewable groundwater systems that extend between neighbouring Arab countries and across the border of the region. These are contained within relatively deep geological formations and store significant amounts of water, but this water has a finite lifespan at present exploitation levels as well as quality limitations (UNESCO, 2012; LAS *et al.*, 2010).

Being non-renewable or 'fossil' groundwater, it is impossible to achieve sustainable development in absolute terms, assuming that managed aquifer recharge cannot make up for the amounts withdrawn. Therefore, sustainability of non-renewable groundwater needs to be interpreted in a social and economic, rather than merely physical context, implying that full consideration must be given not only to the immediate benefits but also to the 'negative impacts' of development, the 'what comes after?' question and, thus, to long-term horizons (World Bank, 2003).

In general, there are two ways in which non-renewable groundwater resources are being utilized in the region. The first is through 'planned schemes', where the mining of aquifer reserves is calculated from the outset to be for a limited time (*e.g.* the Libyan Sarir Basin; the North Western Sahara Aquifer in Libya, Tunisia and Algeria; and Al-Sharqiyah Sand and Al-Massarat Basin in Oman). The second type of utilization is unplanned, leading to the fast depletion of aquifer reserves and the deterioration of its water quality. Unfortunately, this is the case in most Arab countries (*e.g.* Saq-Ram aquifer, Tawilah aquifer in Yemen, Sana'a basin and the Palaeogene [Rus-Umm er Radhuma-Dammam] aquifer in the Arabian Peninsula) (LAS *et al.*, 2010).

The unplanned depletion of non-renewable groundwater reserves can undermine, and potentially erode, the economic and social vitality of the traditional groundwater-dependent community, and instances of the collapse of such rural communities are known (UNESCO, 2006). In the 'planned scheme', the management goal is the orderly utilization of aquifer reserves and appropriate 'exit strategies', developed and implemented before the aquifer is seriously depleted. This scheme includes balanced socioeconomic and strategic choices on the use of aquifer storage reserves and on the transition to a less water-dependent economy and modern agricultural technologies. Since agriculture is the main user of these waters in the MENA region it is vital that such reserves are used with maximum hydraulic efficiency and economic productivity; but it is even more important to be able to limit and control expansion. A key consideration in defining the 'exit strategy' will be the identification of appropriate water resource substitution options, such as the desalination of seawater or brackish groundwater. In terms of governance, public awareness campaigns on the nature, uniqueness and value of non-renewable groundwater are essential to create social conditions conducive to aquifer conservation and management, including, wherever possible, full user participation.

26.3.6 Transboundary groundwater resources

While the governance and management of groundwater resources are challenging and complex nationally, they become all the more intense and convoluted in the case of transboundary resources. In some MENA countries the groundwater dependency ratio is extremely high, and almost every country depends, to a varying degree, on aquifers that are non-renewable and shared with neighbouring countries. For example, the non-renewable Rus-Umm er Radhuma-Dammam aquifer system, which extends from the north to the south of the Arabian Peninsula, is shared between the majority of the Peninsula countries and, in most of those, plays a key role in meeting their water needs.

Despite such high dependency, most transboundary groundwater resources are managed unilaterally, without comprehensive international agreements (for example building on the 2008 UN GA resolution A/RES/63/124 on Transboundary Aquifers). In the absence of such agreements, uncontrolled development practices have had detrimental effects, such as high depletion rates, increased pumping costs, the deterioration of water quality due to the mixing of water between multi-layered aquifers, and the reversal of flow direction in some locations and across international boundaries. The development of groundwater to meet increasing water demand is expected to lead to further mining of shared aquifers, with the potential for regional disputes (LAS *et al.*, 2010; ACSAD, 2009).

However, some modest but encouraging steps have been taken. For example, a technical MOU was signed in 2007 between Saudi Arabia and Jordan, who share the Saq-Ram aquifer system. The agreement prohibited the drilling of new production wells and the expansion of agricultural activities within an area of 10 km along both sides of the border (UN-ESCWA and BGR, 2013). With the support of OSS, UNESCO and others, scientists from Algeria, Libya and Tunisia have been working together since the 1960s to develop a common database on their shared North Western Sahara Aquifer System (NWSAS). Having agreed on the impact of different resource use scenarios, their example illustrates the benefits of finding consensus on datasets and building trust (Benblidia, 2005). Likewise, a protocol was signed between Egypt, Libya, Sudan and Chad, which share the Nubian Sandstone Aquifer System (NSAS), providing a framework for scientific collaboration, data exchange, joint capacity development and aquifer development plans (Salem, 2007).

26.4 OVERALL MISMANAGEMENT?

In the previous sections, we reviewed the diversity of policy tools designed and implemented in the region and suggested that their efficacy has been generally limited. We turn here to analysing the reasons for this state of affairs and try to pinpoint regional commonalities and specificities, while recognising that countries elsewhere in the world fare little better.

26.4.1 The top-down enforcement of regulations with weak results

Although in theory virtually all countries issue permits for well drilling and exploitation, many prohibit the drilling of new wells and control drillers, and several have volumetric quotas and even pricing, the top-down application of these tools has proven difficult in practice. In Syria, Yemen, Jordan, Morocco, Oman and elsewhere, despite comprehensive regulatory frameworks, wells are still being drilled without permits; meters have not become the rule; and use is barely monitored by the state. Merely codifying such measures within the law makes little difference to farmers' choices and strategies. Although the situation is not radically different from other regions of the world, the seriousness of water scarcity in the MENA region makes this regulatory failure more worrying.

The top-down application of regulation neglects the fact that the power of the state dramatically dwindles as one penetrates the countryside. Locally, people find ways around regulations, and the reality on the ground is often very far from what governments like to proclaim. Legal pluralism, where the state's very control of water and land resources is contested (see Al-Naber and Molle, 2016 for the case of Jordan), is pervasive. In addition, when attempted, regulation is generally enforced 'without teeth'. In Yemen, as Alhamdi (2012) reflects, "we have a water law; we have good policies and reasonably good strategies; but the problem is their implementation and enforcement". There is no real political will to tackle the problem, and water management policies are not at the top of the political agenda. In the MENA region in general, it is perceived that "although policies exist, enforcement is lacking" (Tutundjian, 2012).

In the case of Oman, despite the issuing of strict provisions in 2009 and most aquifers being declared 'at risk' (having a negative water balance), drilling is not banned but still allowed with a permit. Likewise, Morocco has yet to define any 'zone of prohibition' as provided by the law, despite the catastrophic situation of many aquifers. This demonstrates limited political resolve to tackle overdraft for fear of curtailing the source of important social and political gains (Kuper *et al.*, 2017; Molle and Closas, 2017).

The organizational culture and professional background (mostly hydraulic engineering) of water agencies and departments means they attach great importance to data acquisition and modelling but less to understanding the agricultural sector and the diversity of actors (Faysse *et al.*, 2011). Yet, assessments of groundwater resources remain problematic (Leduc *et al.*, 2017) and technical parameters are often out of date. In Tunisia, the distinction between shallow and deep groundwater (the 50 m limit) appears to be very arbitrary (Hamdane, 2014). It does not take into account the aquifer's characteristics, such as possibly having interconnected layers. Similarly, in Lebanon, the age-old depth limit of 150 m, below which no drilling permit is required, is obsolete but has yet to be scrapped.

26.4.2 A constrained environment

Campaigns to regularise existing wells, the control and sanctioning of illegal wells and the monitoring of water use invariably cause logistical nightmares, not least since most authorities are short-staffed and lacking in funds. In Yemen, between 2003 (when the Water Law was passed) and 2007, the National Water Resource Authority received around 2,000 license applications, of which 47 percent were approved. But this must be contrasted with the close to 100,000 wells existing across the country (Redecker, 2007). In Abu Dhabi, installing meters in the 100,000 existing wells would require a great effort, not to mention that the equipment itself is expensive and must also be able to withstand the very harsh conditions of the UAE desert. The same applies to Oman's 130,000 wells. In Lebanon there is glaring understaffing at the department responsible for the licensing and monitoring of wells. An official at the Ministry of Energy and Water reported that around 100 officials are needed to conduct regular field visits across the country, while no more than 10 employees were currently working in the department (Nassif, 2016). The technical side of well-drilling projects and monitoring has been outsourced to private companies.

In the Souss-Massa region of Morocco the water police is short-staffed and users drill overnight and during holidays and weekends (BRLI and Agro-Concept, 2012). In Lebanon, officials indicate that while department officials can theoretically perform field visits without being accompanied by the Internal Security they prefer not to do so for safety reasons (Nassif, 2016). In Jordan, the number of field staff is insufficient (*e.g.* only three for the 400 registered wells of Azraq and as many illegal wells) and they are too close to farmers (Al-Naber and Molle, 2017). In Tunisia, the lack of authority and absence of a 'water police' monitoring compliance with rules and restrictions are major obstacles (Hamdane, 2014). In many countries, including Jordan, Yemen and Morocco, staff can face intimidation and violence, or simply do not have the power to risk antagonising powerful people.

A third limitation to state action is corruption and the abuse of political power. Lebanese officials are frequently reported to accept bribes to turn a blind eye to infringements (Nassif, 2016). In Aleppo, Syria, the lack of clearly defined administrative roles within the Ministry of Agriculture (responsible for the issuing of groundwater permits) has led farmers to seek 'informal permits' from the local police (Albarazi, 2014). The corruption of enforcement officers also undermines the application of rules on the ground. As described by Wendle (2016), well drillers had connections and trusted contacts to local government officials, on whom they could count "to look the other way" if they bent the rules.

When the question of groundwater control is embroiled in larger issues of land speculation it becomes even more intractable. The "disastrous management of public land" in Egypt, as bleakly described by Sims (2015), reveals dysfunctional property titling, registration and transfer systems. Since 2007, numerous land scams involving officials and influential people have been unearthed and made publicized. According to Sims (2015), the mess remains untidied for a purpose: "to keep the machine producing 'rents' and windfall profits for vested interests and space for pervasive corruption." Accessing public (desert) land, whether for productive or speculative purposes, is conditional upon the obtaining of licenses to drill wells or the capacity to ensure that authorities turn a blind eye to illegal occupation and drilling. A similar situation is found in certain parts of Jordan, such as Azraq, where the engine of groundwater depletion is partially linked to the preservation of the Bedouins' capacity to mediate investors' access to land against payment (Al-Naber and Molle, 2016).

26.4.3 Administrative organization of the groundwater sector

The administrative and organizational regulation of groundwater abstraction within state structures and governments in the Arab world can be problematic. In some cases the amalgamation of the regulator and operator roles within the same ministry carries the risk of confusing these roles and weakening the implementation of legislation. In Tunisia, the Ministry of Agriculture and Environment is responsible for surface and groundwater monitoring and control, as well as agriculture. Indeed, CRDAs are strong local authorities, which need to weigh resource conservation on one hand and development needs as expressed by the population and the governor on the other.

It is more common, however, to have splintered administrations, with a large number and diversity of state and sectoral actors with a stake in groundwater use and/or management. A first issue is that of the positioning of the 'regulator', responsible for 'resource management', which often generates turf battles for prerogatives and derived benefits. A second issue is linked to the inconsistencies and lack of coordination associated with an administrative organization in sectoral silos (Faysse *et al.*, 2011; Al-Zubari, 2014).

Where development and regulatory agencies are separate, it is generally the former that dominates over the latter. In Morocco, groundwater regulation is the responsibility of the River Basin Agencies, but these lack funds and are understaffed. They are confined to data collection, contracting out technical studies, planning tasks and the licensing of well drilling, and must compromise with the ORMVAs (Office de Mise en Valeur Agricole, representing agriculture) and the Ministry of Public Works, under which they are situated (Tanouti and Molle, 2013). In Yemen, "at least until the updated 2010 Sector Strategy [...], the Ministry of Agriculture and Irrigation looked upon the Ministry of Water and Environment 'as a menace to its power'" and resentment between these ministries is expressed through inequitable budgets (Alderwish *et al.*, 2014). The newly created NWRA (dependent upon the Ministry of Water and Environment) is seen by the Ministry of Agriculture as "ineffectual and hostile to the interests of farmers", with little ability to plan or act (Ward, 2015). In Algeria, the proliferation of illegal wells was partly due to a lack of coordination between those authorities responsible for water resources, with the division of administrative tasks between the *wilayas* (in charge of issuing drilling permits) and the National Agency of Hydrological Resources (in charge of studying and monitoring the resource) (FAO, 2008).

As described in the Introduction, groundwater development in the MENA region has often initially been part and parcel of plans aimed at raising rural incomes, ensuring national food self-sufficiency or the settlement of Bedouins. Government subsidies to develop agriculture or enhance its performance, such as domestic price support, barriers to imports or energy subsidies, have further fuelled groundwater abstraction and the expansion of irrigated agriculture, encouraging farmers to over-irrigate or use water for low-value crops (World Bank, 2007). While policies supporting irrigated agriculture were important historically, they have rarely been phased out once their impact on resources was recognised. A World Bank study (2007) identified that most countries in the region still had incentives for irrigation in place, notably subsidized credit for farmers. With few exceptions, such as the wheat policy in Saudi Arabia, policies by the ministries of energy, agriculture, urban development or tourism still frequently incentivise the use of (ground)water.

A textbook illustration of policy contradiction is provided by the current 'Plan Maroc Vert' – a major nationwide investment plan to boost agriculture. It makes light of environmental requirements and even provides a loophole for farmers to use illegal wells in overexploited aquifers and receive subsidies, which contradicts attempts by the River Basin Agencies to regulate (over)exploitation (Molle and Tanouti, 2017). In addition, the subsidised development of drip irrigation results in increased water consumption rather than 'savings'. Water consumption increases at the plot level (better uptake of water by the plants, densification of plantations, *e.g.* in Morocco or Tunisia, where olive groves with 200 trees/ha have been transformed into olive + fruit trees at 800 plants/ha), as well as due to the expansion of the irrigated area. Since an individual well's abstraction is more or less unchanged (and based on the capacity of the well), the water 'saved' can be reallocated to an expanded plot (see Molle and Closas, 2017). While both production and water productivity increase, benefiting farmers and the nation, the total water consumed also increases, to the detriment of return flows (which, in many cases, is the main recharge of the aquifer), thereby rendering it an unsustainable enterprise in the long term (Molle and Tanouti, 2017). In Jordan, the Ministry of Energy and Mineral Resources favours the development of solar energy for groundwater pumping (and even received subsidies from the EU for 200 units), while the Ministry of Water and Irrigation fiercely opposes the idea, as it leads to greater groundwater exploitation.

Finally, state capacity in managing groundwater can be confused by the contradictory influences donors may have on water management and water sector reforms. In Yemen, many projects have made the country's institutions dependent on foreign funds, such as the NWRA where, by 2009, more than a third of its staff was contractual and paid by donors (Ward, 2015). According to Alhamdi (2012), "Water is always perceived in Yemen as a donor sector, pushed by donors, with a weak role of the Yemeni side", which creates low levels of leadership and ownership by the Yemeni authorities.

26.4.4 The state, vested interests and groundwater governance

A last major difficulty for state-centred groundwater governance in the MENA region is the vulnerability of the state, associated with multiple structures of authority, and its tolerance of various types of rent-seeking strategies by actors both internal and external to the state.

In Yemen, tensions underlying water management and water politics arise from the contest between established traditional authorities and a relatively young Yemeni state attempting to establish itself and find its own legitimacy within the country's various tribal systems and allegiances. Forces external to the formal branches of government and various vested interests affect the development of national politics, such as families controlling the bulk of commerce and tacit coalitions between sheikhs and local security officials and representatives (Zeitoun, 2009). In such a political context, patronage and control of political life are pervasive (Van Steenbergen *et al.*, 2012). In the past, access to groundwater played a role in integrating tribal elites into the government's 'formal ruling establishment' (Moore, 2011). In many tribal rural areas, tube wells came to be seen as a sign of wealth and prestige, and financing groundwater abstraction amongst tribal elites became an effective patronage mechanism. Wells were used as political gifts, through which local leaders could be co-opted into power.

A similar situation can be found in Jordan where, as Richards wrote (1993), "all government decisions must be viewed through the lens of His Majesty who must balance contentious internal and external forces", and agriculture is viewed as "a source of patronage for key constituencies [in particular Bedouin tribes] whose support is essential to achieve domestic stability/foreign policy goals, or as a source of income for the population". Controlling access to and distribution of (state) land and water to specific constituencies is part of a key political balancing act (Al-Naber and Molle, 2016, 2017).

Corruption, in terms of mediating benefits that can be obtained from the state, is often, as in Yemen, considered to be "the main point of business rather than its murky illegal underside" (Bafana, 2012). It is part of a prevalent system of patronage, networks of influence, political, economic and social power where *wasta* (or connections) with people well positioned within the establishment is key (Robinson *et al.*, 2006). In the UAE, where the tribal community structure is considerable, landowners wield significant power through wasta in opposing the implementation of regulation. As one agricultural official in Abu Dhabi summed up: "here there isn't such a thing as 'illegal' for a local who has good connections" (Fragaszy and McDonnell, 2016, page 42). In Lebanon's Bekaa, local political leaders sometimes intervene at the Ministry of Energy and Water to issue permits for their electors (Nassif, 2016).

The power of particular interest groups can also be seen in their capacity to obtain land concessions or well licenses, to attract state investment and subsidies and the preferential allocation of water, as in the El Guerdane project in Morocco (Houdret, 2012) and the (stalled) West Delta Project in Egypt. The example of Israel shows that organized agricultural lobbies can deflect attempts by the treasury to curtail the abstraction of water and tax its use (Feitelson, 2005).

26.4.5 Coping strategies and induced social dynamics

The overexploitation and salinisation of groundwater resources have had negative environmental and social impacts. Facing groundwater depletion and the 'race to the bottom', farmers and other users have devised solutions to cope with the loss of the resource. Dropping aquifers have forced farmers to continuously deepen their wells, especially in agricultural frontier areas, such as the Jordanian highlands and the Souss-Massa in Morocco. In large plains, such as the Saïss and Haouz (Morocco) and Bekaa (Lebanon), as well as in large-scale public irrigation schemes, the combined use of surface and groundwater gives farmers more flexibility, as in the Tadla in Morocco and the Nile Delta in Egypt. Seawater intrusion has led to the destruction of agricultural areas and the loss of livelihoods (*e.g.* Oman, Morocco and Tunisia).

Farmers adapt to changing conditions by deepening their wells, investing in collective wells (*e.g.* Egypt, Morocco, Yemen), adjusting cropping patterns or even reverting to rain-fed agriculture, but many abandon agriculture altogether when they can no longer meet the rising costs. Access to capital is therefore essential in determining the type of strategy followed by groundwater users (Faysse *et al.*, 2011). It is apparent that small farmers are often displaced by wealthier users who can afford to invest and pump deeper (*e.g.* Souss-Massa, Morocco, Syria, Lebanon). Unmanaged aquifers therefore exacerbate social differentiation.

26.5 CONCLUSIONS

The groundwater situation in the MENA region is certainly worrisome when judged by the many key aquifers that are subject to overexploitation or degradation, including those that are non-renewable. On paper, most countries appear to have adequate formal institutions, especially at the national level, including strategies and laws. The shortcomings of their implementation on the ground are often ascribed to the fact that these institutions still need a lot of support (capacity building, strengthening of technical capacities of staff and budgetary support). Although this is certainly the case, such assessments can serve as a fig leaf for a deeper lack of political willingness and/or an ability to address the issue head on. This is illustrated by a number of countries, including the UAE, Oman, Lebanon, and Morocco, which have yet to apply drilling bans, despite the critical status of most of their aquifers. This lack of resolve has undermined the implementation on the ground of tools such as well registration, metering, pricing and quotas. One obvious conclusion is that it would be senseless to implement measures aimed at reducing abstraction while well expansion remains uncontrolled. Registration, not to mention metering, has proven to be a costly logistical nightmare and is rarely exhaustive (except in a few cases, including Bahrain and Abu Dhabi).

Policies 'without teeth' have been shown to relate, in some cases, to tribal politics, clientelistic practices and the pervasive use of wasta to circumvent or evade regulation. More profoundly, the economic and political benefits from groundwater use accruing to both farmers/users (in particular powerful investors with high-level connections) and the state are too great to allow the control of the growth (and bust) of the groundwater economy (Molle and Closas, 2017). Examples of the vulnerability of this economy to changes in the price of energy have sent a strong message to governments: in Yemen a reduction in diesel subsidies generated a crisis of unprecedented severity and exacerbated the political and social unrest that beset the country following the Arab Spring. In Syria the withdrawal of diesel and fertilizer subsidies, combined with consecutive years of drought, reduced crop yields and increased food prices, causing numerous farmers to abandon their land and fuelling a humanitarian crisis in and beyond the north-east of the country (Kelley *et al.,* 2015).

Current policies certainly overestimate the power of the state in such contexts and point to a deficit in participatory groundwater management. It is striking that the few attempts at such forms of management, as observed for example in Bsissi (Tunisia) and the Souss region of Morocco, originated in confrontations between groundwater users and state bent on enforcing regulations. In contrast, Jordan's Highland Water Forum can be seen as the culmination of 15 years of donor projects in the Azraq region rather than an endogenous state initiative. This helps explain the low level of participatory management in the MENA region, and reflects the overall 'top-heavy' system of governance (Al-Zubari, 2014).

In sum, the benefits provided by groundwater to rural communities, and thus its political role as an 'escape valve', work against regulation, as does its use in agribusiness ventures by investors often with political connections. The authority is hampered by a lack of political will (due to those benefits) as well as by the discouraging magnitude of a problem that has spiralled out of control, the weakening of state power, particularly in the wake of the Arab Spring, and the troubled and mistrustful relationship between the state and its citizens that runs counter to co-management solutions.

ACKNOWLEDGEMENTS

Background research for this chapter was developed by IWMI under USAID-grant AID-263-IO-13-00005 aimed at studying groundwater governance in the Arab world.

REFERENCES

Abderrahman, W.A. 2003. Should intensive use of non-renewable groundwater resources always be rejected? Intensive Use of Groundwater 191–203. Balkema Publishers. Lisse, The Netherlands.

Abderrahman, W.A. 2005. Groundwater management for sustainable development of urban and rural areas in extremely arid regions: a case study, *Water Resources Development*, 21(3), 403–412.

AbuZeid, K., Elrawady, M. 2008. Sustainable development of non-renewable groundwater. UNESCO Congress on water scarcity, University of Irvine, California.

ACSAD. 2009. Water Economy in the Arab region. Arab Centre for the Studies of Arid Zones and Dry Lands, Damascus.

Albarazi, G. 2014. Is Syria's water institution capable of addressing current and future challenges? An exploration of Syria's water policy, administration, and law, MSc Thesis, Unpublished, University of Oxford.

Alderwish, M.A., AlKhirbash, S.B., and Mushied, M.H. 2014. *Review of Yemen's control of groundwater extraction regime: situation and options*, International Research Journal of Earth Sciences, 2(3), 7–16.

Alhamdi, M. 2012. 'Water scarcity and the need for policy redirection in Yemen', in Yemen in Transition: Challenges and Opportunities, October 19–20, 2012, Harvard University, Video recording, http://vimeo.com/52884952 (Accessed 6th April 2014).

Allan, J.A.T. 2007. "Rural economic transitions: groundwater use in the Middle East and its environmental consequences", in Giordano, M., and K.G. Villholth (eds.) *The agricultural groundwater revolution: opportunities and threats to development*, Wallingford, UK: CABI, 63–78.

Al-Naber, M. and Molle, F. 2016. The politics of accessing desert land in Jordan. *Land Use Policy* 59: 492–503.

Al-Naber, M. and Molle, F. 2017. Controlling groundwater overabstraction: policies vs. local practices in Jordan Highlands. *Water Policy*.

Al-Sheikh, H.M.H. 1997. Water policy reform in Saudi Arabia. In: Proceedings of the Second Expert Consultation on National Water Policy Reform in the Near East, Cairo, 24–25 November, 1997. FAO, Regional Office for the Near East. http://www.fao.org/docrep/006/ad456e/ad456e00.htm

Al-Zubari, W. 2012. 'Groundwater governance in the Kingdom of Bahrain', PowerPoint Presentation, Regional Consultation on 'Groundwater governance in the Arab World', 8–10 October 2012, Amman.

Al-Zubari, W.K., and I.J. Lori 2006. Management and sustainability of groundwater resources in Bahrain, *Water Policy*, 8, 127–145.

Al-Zubari, W.K. 2014. Synthesis report on groundwater governance regional diagnosis in the Arab Region. Groundwater Governance – A Global Framework for Action. GEF and FAO.

Amichi, F. 2015. Personal communication.

Aw-Hassan, A., Rida, F., Telleria, R., and A. Bruggeman 2014. The impact of food and agricultural policies on groundwater use in Syria, *Journal of Hydrology*, 513, 204–215.

Bafana, H. 2012 "Case samples: corruption methodology in Yemen", http://blog.haykal.sg/the-yemen/13-case-samples-corruption-methodology-in-yemen (Accessed 6th August 2015).

Barnes, J. 2012. Pumping possibility: agricultural expansion through desert reclamation in Egypt, *Social Studies of Science*, 42(4), 517–538.

Benblidia, M. 2005. Les Agences de Bassin en Algérie. Background paper to Making the Most of Scarcity: Accountability for Better Water Management Results in the Middle East and North Africa. Washington, DC: World Bank.

BRLI, and Agro Concept 2012. Gestion de la demande en eau dans les pays méditerranéens: gestion de la demande en eau – étude de cas du Maroc, Décembre 2012.

Chebaane, M., El-Naser, H., Fitch, J., Hijazi, A., and A. Jabbarin 2004. Participatory groundwater management in Jordan: development and analysis of options, *Hydrogeology Journal*, 12, 14–32.

Closas, A. and K.G. Villholth 2016. Aquifer Contracts - A Means to Solving Groundwater Overexploitation in Morocco? Colombo, Sri Lanka: International Water Management Institute (IWMI). 20p. (Groundwater Solutions Initiative for Policy and Practice (GRIPP) Case Study Series 01). doi: 10.5337/2016.211.

de Châtel, F. 2014. The Role of Drought and Climate Change in the Syrian Uprising: Untangling the Triggers of the Revolution, *Middle Eastern Studies*, 50(4), 521–535.

Del Vecchio, K. 2013. Une politique contractuelle sans contrôle? La régulation des ressources en eau souterraine dans la plaine du Saïss au Maroc, Mémoire de Master 2, Université Lumière Lyon 2, Sciences Po Lyon.

Demilecamps, C. 2010. *Farming in the desert: analysis of the agricultural situation in Azraq Basin*, German-Jordanian Programme "Management of Water Resources", GIZ.

El-Agha, D., Closas, A., and F. Molle. 2017. Below the radar: the boom of groundwater use in the central part of the Nile Delta in Egypt, *Hydrogeology Journal*, doi:10.1007/s10040-017-1570-8.

El-Fadel, M., Bou-Zeid, E., and Chahine, W. 2002. Long term simulations of leachate generation and transport from solid waste disposal at a former quarry site, Journal of Solid Waste Technology and Management, 28(2), 60–70.

Elloumi, M. 2016. La gouvernance des eaux souterraines en Tunisie. IWMI Project Report, Groundwater governance in the Arab World, USAID. IWMI.

FAO 2008. *Rapport Algérie*, Etude sur la gestion des eaux souterraines dans les pays pilotes du Proche-Orient, Bureau régional de la FAO pour le Proche-Orient.

Faysse, N. Hartani, T., Frija, A., Marlet, S., Tazekrit, I., Zaïri, C., and A. Challouf 2011. Agricultural Use of Groundwater and Management Initiatives in the Maghreb: Challenges and Opportunities for Sustainable Aquifer Exploitation. Economic Brief. AfDB.

Feitelson, E. 2005. Political economy of groundwater exploitation: the Israeli case, *Water Resources Development*, 21(3), 413–423.

FMWEY, 2015. Former Minister of Water and Environment of Yemen (anonymous), personal communication, 4th August 2015.

Fragaszy, S. and McDonnell, R. 2016. Oasis at a crossroads: agriculture and groundwater in Liwa, United Arab Emirates. IWMI project publication – Groundwater governance in the Arab World – Taking stock and addressing the challenges, USAID. IWMI.

Frija, A., Chebil, A., Speelman, S., and N. Faysse 2014. A critical assessment of groundwater governance in Tunisia, *Water Policy*, 16, 358-373.

Ghazouani, W., and I. Mekki 2015. *Les ressources en eaux souterraines de la plaine de Haouaria : Etat fragile, acteurs multiples et nécessité d'un changement intégré*, "USAID Groundwater Governance in the Arab World " Internal Project Report, Cairo : IWMI.

Hamdane, A. 2014. Personal communication, Former Director General, Ministry of Agriculture, Tunisia.

Hamdane, A. 2015. Le contrôle de l'utilisation des eaux souterraines et la gestion participative des nappes. Report to FAO.

Herald Globe 2014. "Jordan examining satellite images to locate illegal water wells", 19th November 2014, www.heraldglobe.com/index.php/sid/227753817 (Accessed 17th August 2015).

Houdret, A. 2012. The water connection: irrigation, water grabbing and politics in Southern Morocco, *Water Alternatives*, 5(2), 284–303.

INECO 2007. *Governance and water management structures in the Mediterranean basin*, National Technical University of Athens, Sixth Framework Programme (2002–2006) European Commission, INCO –CT-2006-517673.

IRG 2014. *Analysis report: socio-economic survey of groundwater wells in Jordan*, Institutional Support and Strengthening Program (ISSP), USAID, Jordan.

James, I. 2015. Dry springs and dead orchards: Barren fields in Morocco reveal risks of severe depletion in North Africa. www.desertsun.com/story/news/environment/2015/12/10/morocco-groundwater-depletion-africa/76788024/

Jordan Times 2015. Cracking down on water theft. Jul 14, 2015. www.jordantimes.com/opinion/editorial/cracking-down-water-theft.

Kelley, C.P., Mohtadi, S., Cane, M.A., Seager, R., and Y. Kushnir 2015. Climate change in the Fertile Crescent and implications of the recent Syrian drought, *Proceedings of the National Academy of Sciences*, 112(11), 3241–3246.

Klingbeil, R. 2014. 'Managed aquifer recharge – aquifer storage and recovery: regional experiences and needs for further cooperation and knowledge exchanges', International Association of Hydrogeologists (IAH), 41st International Congress, 'Groundwater challenges and strategies', Marrakech, Morocco, 15–19 September.

Kuper, M., Hammani, A., Chohin, A, Garin, P., and M. Saaf 2012. When groundwater takes over: linking 40 years of agricultural and groundwater dynamics in a large-scale irrigation scheme in Morocco, *Irrigation and Drainage*, 61(S1), 45–53.

Kuper, M.; Faysse, N.; Hammani, A.; Hartani, T.; Marlet, S.; Hamamouche, M.F. and Ameur, F. 2016. Liberation or anarchy? The Janus nature of groundwater use on North Africa's new irrigation frontiers. In Jakeman, A.; Barreteau, O.; Hunt, R.J.; Rinaudo J.-D. and Ross, A. (eds.), Integrated groundwater management: concepts, approaches and challenges, pp. 583–615. Cham: Springer International Publishing.

LAS, UNEP, CEDARE, 2010. Environment Outlook of the Arab Region – Environment for development and human well-being (EOAR), Nairobi, Kenya: UNEP.

Lattemann, S., and T. Hopner 2008. Environmental impact and impact assessment of seawater desalination, *Desalination*, 220(1–3), 1–15.

L'Economiste 2014. *Eaux souterraines: les contrats de nappes bientôt généralisés*, Édition N° 4242 du 2014/03/27.

Leduc, C.; Pulido-Bosch, A. Remini, B. 2017. Anthropization of groundwater resources in the Mediterranean region: processes and challenges. *Hydrogeology Journal.*

Leghrissi, H. 2012. Importance des arrangements institutionnels dans la gestion des puits privatifs dans un périmètre irrigué : cas du périmètre Bsissi-Oued El Akarit en Tunisie, MSc Thesis, Supagro Montpellier.

McDonnell, R. 2016. Water and groundwater in Oman – resources and management. IWMI project publication – Groundwater governance in the Arab World – Taking stock and addressing the challenges, USAID. IWMI.

Mesnil, A., and N. Habjoka 2012. The Azraq Dilema: Past, Present and Future Groundwater Management. Amman: GIZ.

Molle, F. and Closas, A. 2017. Groundwater governance: a synthesis. Report submitted to USAID (Vol. 6). IWMI, Colombo.

Molle, F. and J. Berkoff (eds.) 2007. Water pricing in irrigation: the gap between theory and practice, Wallingford: CABI.

Molle, F. and Tanouti, O. 2017. Squaring the circle: impacts of irrigation intensification on water resources in Morocco, Agricultural water Management (forthcoming).

Moore, S. 2011. Parchedness, politics, and power: the state hydraulic in Yemen, *Journal of Political Ecology*, 18, 39–50.

Morill, J., and J. Simas 2009. "Comparative analysis of water laws in MNA countries", in Jagannathan, N.V., Mohamed, A.S., and A. Kremer (eds.) *Water in the Arab*

World: management perspectives and innovations, Washington DC: The World Bank, 285–334.

Nassif, M. 2016. Groundwater governance in the Central Bekaa, Lebanon. IWMI project publication – Groundwater governance in the Arab World – Taking stock and addressing the challenges, USAID. IWMI.

NWSSIP 2008. *Update of the National Water Sector Strategy and Investment Programme*, the NWSSIP Update, Republic of Yemen, December 17th, 2008.

Ouelhazi, H., Lachaal, F., Charef, A., Challouf, B., Chaieb, H., and F.J. Horriche 2013. Hydrogeological investigation of groundwater artificial recharge by treated wastewater in semi-arid regions: Korba aquifer (Cap-Bon Tunisia), *Arabian Journal of Geosciences*, September 2013.

Perry, C.J. and Steduto, P. 2017. Does hi tech irrigation save water? A review of the evidence. Regional Initiative Series No. 4. FAO, Regional Office for Near East and North Africa, Cairo, Egypt.

Redecker, G. 2007. *NWSSIP 2005-2009 Two years of achievements ... and an outlook – A donor's perspective*, KfW Office Sana'a, Yemen.

Richards, A. 1993. *Bananas and Bedouins: political economy issues in agricultural sector reform in Jordan*, Democratic Institutions Support (DIS) Project, Governance and Democracy Program, Near East Bureau, USAID.

Robinson, G.E., Wilcox, O., Carpenter, S., and A.G. Al-Iryani 2006. *Yemen corruption assessment*, USAID-Yemen.

Salem, O. 2007. Management of Shared Groundwater Basins in Libya. African Water journal, 1(1): 106-117. www.amcow-online.org/docs/journal/African%20Water%20Journal%20-%202007.pdf

Shah, T. 2009 Taming the anarchy: groundwater governance in South Asia, Washington DC: RFF Press.

Shammas, M. I. 2008. The effectiveness of artificial recharge in combating seawater intrusion in Salah Coastal Aquifer, Oman. *Environmental Geology*, 55: 191

Sims, D., 2015. Egypt's Desert Dreams: Development or Disaster? AUC Press: Cairo.

Tanouti, O., and F. Molle 2013. Réappropriations de l'eau dans les bassins versants surexploités: le cas du bassin du Tensift (Maroc), *Etudes Rurales*, 192(2), 79–96.

Tutundjian, S. 2012. Third regional consultation: Arab states region, Regional consultation report, Groundwater Governance: A global framework for country action, GEF ID 3726, GEF, The World Bank, UNESCO-IHP, FAO, IAH.

UNDP (United Nations Development Programme). 2013. Water governance in the Arab Region: managing scarcity and securing the future, New York, USA: UNDP.

UNDP. 2014. Groundwater Assessment and Database Project, Final Output. May 2014.

UNESCO. 2012. The United Nations World Water Development Report 4: Managing Water under Uncertainty and Risk. World Water Development Programme (WWAP). Paris, France.

UNESCO, 2006, Non-renewable groundwater resources, A guidebook on socially-sustainable management for water-policy makers. Foster, S. and Loucks, D. P (eds.). IHP-VI, Series on Groundwater no. 10. UNESCO, Paris. Available at: http://unesdoc.unesco.org/images/0014/001469/146997E.pdf

United Nations Economic and Social Commission for Western Asia [UN-ESCWA] and BGR (Bundesanstalt für Geowissenschaften und Rohstoffe) 2013. *Inventory of Shared Water Resources in Western Asia*. Beirut. E/ESCWA/SDPD/2013/Inventory.

Van Steenbergen, F., and N. El Haouari 2010 "The blind spot in water governance: conjunctive groundwater use in MENA countries", in Bogdanovich S., and L. Salame (eds.) *Water policy and law in the Mediterranean: an evolving nexus*, Serbia: Faculty of Law Business Academy Novi Sad, 171–189.

Van Steenbergen, F., Bamaga, O.A., and A.M. Al-Weshali 2012. *Groundwater security in Yemen: role and responsibilities of local communities in conserving groundwater*, Embassy of the Kingdom of the Netherlands, Water and Environment Centre, MetaMeta Research.

Van Steenbergen, F., Kumsa, A., and N. Al-Awlaki 2015. Understanding political will in groundwater management: comparing Yemen and Ethiopia, *Water Alternatives*, 8(1), 774–799.

Venot, J.P., and F. Molle 2008. Groundwater depletion in the Jordan Highlands: can pricing policies regulate irrigation water use?, *Water Resources Management*, 22(12), 1925–1941.

Wada, Y., van Beek, L.P.H., and M.F.P. Bierkens 2012. Nonsustainable groundwater sustaining irrigation: a global assessment, Water Resources Research, 48(6), 1–18.

Ward, C. 1998. Practical responses to extreme groundwater overdraft in Yemen, International Conference Yemen: the challenge of social, economic, and democratic development, April 1998, University of Exeter, Centre for Arab Gulf Studies.

Ward, C. 2015. *The water crisis in Yemen: managing extreme water scarcity in the Middle East*, London : IB Tauris.

Wendle, J. 2016. Syria's climate refugees, *Scientific American*, 314(3), 42–47.

World Bank. 2003. Utilization of Non-Renewable Groundwater a socially-sustainable approach to resource management. GWMATE Briefing Note 11. http://documents.worldbank.org/curated / en / 621881468137375750 / Utilization-of-non-renewable-groundwater-a-socially-sustainable-approach-to-resource-management.

World Bank 2007. *Making the most of scarcity: accountability for better water management in the Middle East and North Africa*, Washington DC: The World Bank.

Zeitoun, M. 2009. *The political economy of water demand management in Yemen and Jordan: a synthesis of findings*, Water Demand Management Series, Regional Water Demand Initiative in the Middle East and North Africa, IDRC-Canada, CIDA-Canada, IFAD.

Zeitoun, M., Allan, T., al Aulaqi, N., Jabarin, A., and H. Laamrani 2012. Water demand management in Yemen and Jordan: addressing power and interests, The Geographical Journal, 178(1), 54-66.

Zekri, S. 2007. Water use licensing versus electricity policy reform to stop seawater intrusion. In Lamaddalena N., Bogliotti C., Todorovic M., Scardigno A. (eds.) Water saving in Mediterranean agriculture and future research needs [Vol. 3]. Bari: CIHEAM.

Chapter 27

Perspectives on Guarani Aquifer Governance

Luiz Amore
Foreign Affairs Chief Adviser, National Water Agency – ANA, Brasilia, DF, Brazil

ABSTRACT

The Guarani Aquifer System is an important transboundary groundwater reservoir located in central-east part of South America and shared by Argentina, Brazil, Paraguay and Uruguay. Groundwater use in the region has increased very fast since the 1970s.

During the last decades, the four countries have developed laws, institutional frameworks and regulations related to water resources management in general, but gaps in groundwater issues still exist. After previous scientific studies showing the importance of the aquifer, these four countries have jointly prepared and implemented the Project on the Protection and Sustainable Development of the Guarani Aquifer System (PSAG, 2003–2009), with the support of the Organization of the American States and the World Bank, and funded by the Global Environment Facility and other international organizations. The main project objective was to support countries in establishing a management framework for the Guarani Aquifer. As a result of the scientific knowledge developments, the project appropriation process by country representatives and a structured project execution process carried out by the project team, a Strategic Action Plan (SAP) was developed. Additionally, the countries signed an Agreement on the Guarani Aquifer (2010), now under national parliament approval in Paraguay (Argentina Brazil and Uruguay have already approved it).

The objective of this chapter is to discuss the Guarani Aquifer Governance perspectives in practical terms, correlating the existence of applicable management instruments and the responsibilities of different institutional levels involved in groundwater regulation in the countries. Four institutional levels are distinguished: a) Regional or global; b) national; c) state or province; and d) municipal or local. Currently, the main management responsibilities are at the national level in the unitary countries and at state or province level in the federal countries. Besides that, according all knowledge developed by the Guarani Aquifer Project the weakest and most crucial level to foster groundwater governance is the local or municipal level, because it is at this level that all contamination and overexploitation problems of the aquifer really occur. Many expectations are supposed to be resolved after the Guarani Aquifer Agreement's enforcement; one of them is how regional and national level can effectively support the local level, a critical dimension to mitigate impacts and develop protection strategies to the Guarani Aquifer.

Figure 27.1 Location of the GAS.

27.1 GUARANI AQUIFER CHARACTERISTICS

27.1.1 General features

The transboundary Guarani Aquifer System (GAS) is an important groundwater reservoir located in the southern portion of the South-American continent (between 16–32° South and 47–60° West). It has an area of approximately 1.1 million km^2 spread over territories of Argentina, Brazil, Paraguay and Uruguay (Figure 27.1). It is associated with a series of rocks originating from accumulation of sediments (gravel, sand, silt and clay) of the Paraná Basin (Brazil and Paraguay), the Chaco-Paraná Basin (Argentina) and the North Basin (Uruguay). The Guarani sandstones are composed of round-shaped sand grains, produced by eolian deposition in a huge Paleo-Cretaceous desert. The architectural aspects and compartments of this extensive basin and corresponding sediment layers are encompassed by huge structural elements, known as arcs and synclines.

Water is stored in the sandstone pores, reaching large storage volumes and moving at an extremely slow rate in some portions of the aquifer. The reservoir is protected by a thick layer of non-permeable basaltic rock that can reach up to 1.5 km. Recharge and discharge areas are located at sandstone outcrops, bordering the geological basin. At the western border, discharge waters feed some rivers that go to the Pantanal wetland. The water is in general of good quality and accessible through deep wells, mainly used for the urban supply and tourism.

The total area of the GAS is 1,087,879 km^2 and its distribution by country is shown in Table 27.1. The aquifer's outcrop area is 124,650 km^2, of which 83,500 km^2 (67%) is effectively considered as recharge area (aquifer replenishment area).

Table 27.1 Total estimated area and distribution of the GAS by country.

Current Project Estimation	*Argentina*	*Brazil*	*Paraguay*	*Uruguay*	*GAS Area*
Area (km²)	228,255	735,918	87,536	36,170	1,087,879
% of the total	20.98	61.65	8.05	3.32	100
% related to country area	8.1	8.7	21.5	19.5	–

Source: Organization of American States, 2009.

Total water availability in the GAS was assessed on technical and economic aspects of groundwater use basis. Estimates for static reserves, drainable reserves above 400 m deep and elastic reserves are 29,551 km^3 (± 4000 km^3), 2,014 km^3 (± 270 km^3) and 25 km^3, respectively. According to studies on current use of the Guarani Aquifer's resources, the total rate of groundwater abstraction equals approximately the average rate of natural recharge (aquifer replenishment).

27.1.2 Hydrodynamic Characterization

The Hydrogeological Map of the GAS (Organization of American States, 2009) shows the main groundwater flow directions based on water levels of wells drilled in the aquifer. This information was obtained using the hydrogeological database (HDB), including inventory and sampling of almost 1,800 wells from a total of 8,000 available and registered wells in the region. Full information was found on 1,348 wells. Such information encompasses static and dynamic levels and data on flows. The results show that most of the well discharges are less than 50 m^3/h. Some wells, however, produce discharges of between 150 and 300 m^3/h, and are used for public water supply for large and medium-size towns, especially in the State of São Paulo. With respect to specific discharges (*i.e.* discharges per unit of water level drawdown) the majority of wells yielded less than 6 m^3/h/m, and most had specific discharges of less than 2 m^3/h/m. In general water quality is very good (conductivity $< 1,000$ μS/cm), and potable without any treatment. But in some restricted areas, GAS water contains some natural elements in high concentrations like arsenic, cadmium, zinc, boron and nickel. Local inorganic anomalies need to be assessed in greater detail.

Four large hydrodynamic domains (northeast, east, west and south) can be identified. The Arc of Ponta Grossa divides the aquifer into two potentiometric domains and the Arc of Rio Grande–Asunción is responsible for an important reduction in the sandstone thickness in the subsoil and for a major regional directional change in GAS groundwater flows.

27.1.3 Guarani Aquifer System (GAS) management zones

Based on previous scientific studies and a first schematic management map of the Guarani Aquifer (Amore *et al.*, 2001), many developments were implemented by the Guarani Aquifer Project. Based on existing and new information on hydrogeology, geometry and hydrodynamic, hydrochemical, isotopic and hydrothermal characteristics of the GAS it was possible to identify distinct zones with similar groundwater flow and time-of-residence characteristics. The knowledge on these zones could provide inputs for management of the GAS. Such zoning should be interpreted as a starting point and a reference to the management to be developed by groundwater management institutions. The principal zones defined are described as follows:

(1) **Outcrop zones (ZA):** The outcrop zone (ZA) serves as the aquifer's regional discharge and replenishment zones (Figure 27.2, areas I). There are interactions between surface water and groundwater and, generally, base flows of rivers and other surface water bodies are derived from GAS discharges. In such areas, the aquifer is generally unconfined and rainfall is the main origin of replenishment water. Water is young, circulation is fast, and it is generally potable. It is predominantly of calcium bicarbonate or – secondarily – of calcium-magnesium and calcium-sodium bicarbonate. These areas are highly vulnerable to anthropogenic pollution.

(2) **Confinement/transition zones (ZC):** They represent fringes immediately adjacent to the outcrop areas (10 to 50 km wide) where fractured basalt up to 100 m thick overlies the GAS (Figure 27.2, areas II). Basalts may act as aquitards in certain cases and water availability depends upon their vertical hydraulic conductivity. There are evidences of existing hydraulic connections between the GAS and post-GAS units through geological faults and fractures of basalts. In basalt windows, the GAS presents an unconfined aquifer regime. It is an important theme, and object of specific investigations to be carried out on the future. In these areas, waters are of calcium-bicarbonate and sodium-bicarbonate type. The aquifer in these areas is slightly vulnerable, in view of the fractured basalts, their proximity to the replenishment zone, and of the advection movement of pollutants through the GAS itself.

(3) **Highly confined zones (ZFC):** These correspond to areas marked with III, IVa, IVb, V, VI, and VII in Figure 27.2 and represent all GAS areas confined by basalts of more than 100 m in thicknesses, where abstracted water stems from mechanical decompression of the aquifer. Rates of circulation are extremely slow, leading to very low replenishment rates. Waters are predominantly sodium-bicarbonate and sulphate. They present heterogeneities and, therefore, are subdivided into different areas as showed in the map of Figure 27.2:

    (a) *Confined area III (northwest):* Located in the western region of the GAS. Water circulation is more dynamic in this zone if compared with eastern and central GAS confined sectors. Its waters have a specific isotopic signature of Oxygen 18 (^{18}O), probably associated with inflows from outcrop zones to the north (Goias and Mato Grosso) and from basalt windows that serve as replenishment zones. It is not at all vulnerable.

    (b) *Confined area IVa (northeast):* The area is in the north-eastern region of the GAS. Its waters are old, according ^{18}O isotopic signature. It is not vulnerable, and the IVa confinement area presents the highest natural groundwater temperatures of the GAS.

    (c) *Confined area IVb (to the north of Asuncion-Rio Grande Arc)*: Located in the northern part of the central region of the GAS, along the thickest stretch of the GAS, and the thickest stretch of the post-GAS cover. Its southern limit is associated with the Asunción-Rio Grande rise (São Gabriel Arc). Like those

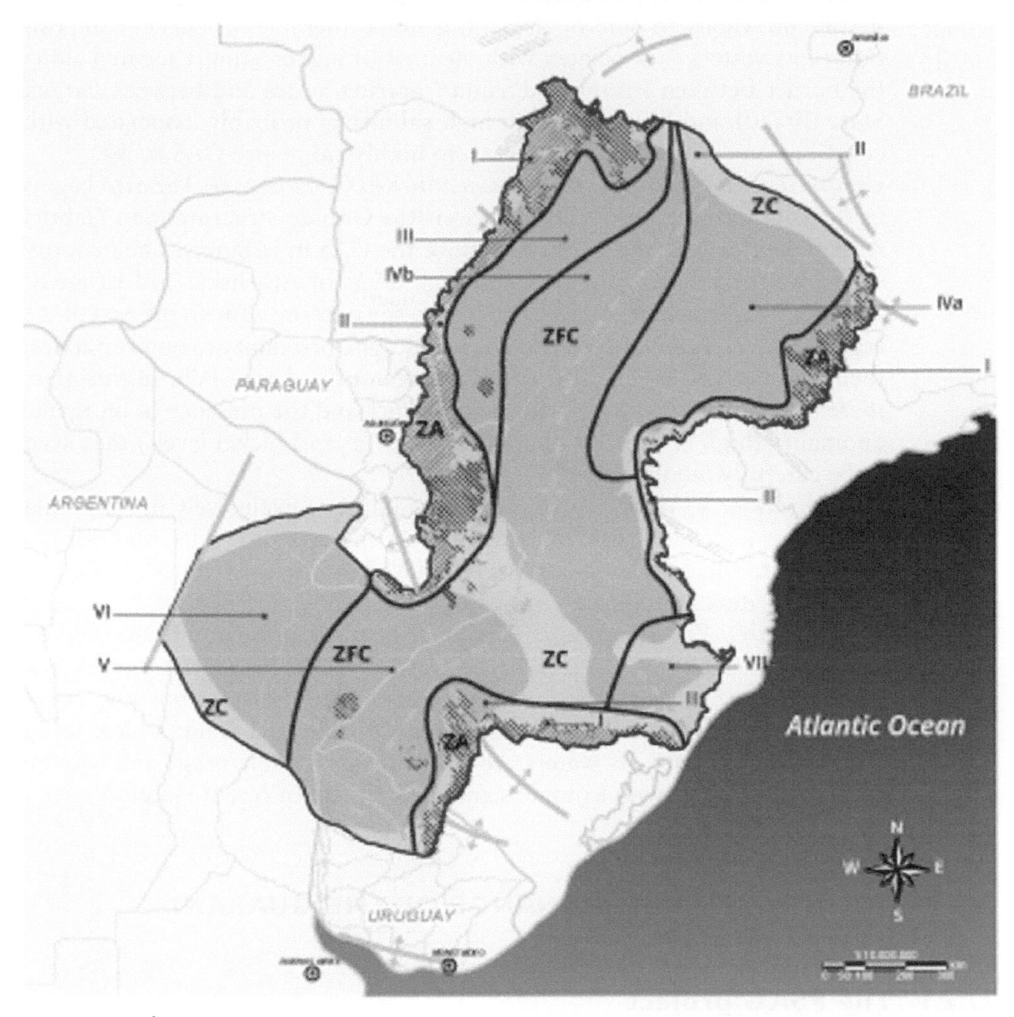

Legend

ZFC Confined/ strongly confined zone

ZC Transition/confined zone

ZA GAS outcrop zone

ZA Replenishment area

ZA Discharge area

⊙ National capital

I Recharge or discharge areas

II Area of fractured basalt overlying the GAS up to 100 m thick

III Northwest confined areas by basalts more than 100 m thick

IVa Northeast confined area

IVb Confined area to the north of Asunción–Rio Grande arc

V Confined area to the south of Asunción–Rio Grande arc

VI Southwest confined area

VII Southeast confined area

Figure 27.2 Proposed delimitation of GAS management areas Source: Strategic Action Plan in Organization of American States, 2009.

of confined zone IVa, its waters are of high temperature and relatively old, despite proximity to outcrop areas. It is not vulnerable. Chemical composition of its waters is associated with the area of higher salinity located along the border between Paraná and Santa Catarina States and between Paraná State (Brazil) and Paraguay. Such high salinity is probably associated with contributions of interstitial waters from highly-saline pre-GAS rocks.

(d) *Confined area V (to the south of Asuncion-Rio Grande Arc):* The area begins in the southern region of the Asunción-Rio Grande structure (São Gabriel Arc) and extends to the southern limit of the GAS in Uruguay. The majority of the wells of this region are in thermal areas of Argentina and Uruguay. The area's replenishment zone is related to the outcrop zone to the east of the region. Its waters tend to be very old, and despite proximity to outcrop zones, their age tends to be like that of waters of confined area IVb. In this area, there are also thermal wells with high salinity and the presence of inorganic anomalies (high arsenic, cadmium, zinc, boron and nickel levels) that need to be carefully monitored.

(e) *Confined area VI (southwest):* It is located almost exclusively in Argentina (except for a small portion in Paraguay). As there are no wells, no information of water quality is available. It is believed that to the west of the Paraná River its waters are saline.

(f) *Confined area VII (southeast):* This small portion of the aquifer located on the eastern limit of the GAS, is characterized by local flows, waters with low potentiometric levels and little replenishment. In the eastern portion, water level is over 400 meters deep, and thus unfeasible as a water source using current technologies. Its waters are calcium bicarbonate (east) and sodium bicarbonate (west) with isotopic signatures typical of recent waters.

27.2 GROUNDWATER GOVERNANCE OF THE GUARANI AQUIFER SYSTEM

27.2.1 The PSAG project

The Project on the Protection and Sustainable Development of the Guarani Aquifer System (PSAG, 2003–2009) has been a milestone in the development of groundwater management and governance regarding the Guaraní Aquifer. In line with the GEF's methodology, the main product was the Strategic Action Program (SAP) which was developed based on a Transboundary Diagnostic Analysis (TDA). The SAP presented a synthesis of all achievements from different studies developed by international and local consulting companies and experts, the management strategies developed during a participatory process coordinated by national institutions and the main priorities to be implemented in the future, jointly defined by the main stakeholders related to groundwater management in the region.

The project execution process was a very intense process in all different phases. Experts on groundwater shared their knowledge with national technicians, politicians and diplomats, from federal to state/province or local levels. The Project Execution Council (CSDP) met twice a year to discuss technical and management aspects of the project, creating confidence among countries and trust on developed knowledge. The national coordination level in each country was very important to internalize all knowledge, to support regional agreements and to get local institutional involvement and counterparts. The countries invested 25 additional million dollars during project execution, such as for national counterparts, exceeding previous estimation.

Local and regional processes had different, but complementary dynamics. Four pilot areas were initially selected to try out local developments: Concordia-Salto (Argentina and Uruguay transboundary local area), Itapúa (Paraguay), Ribeirão Preto (São Paulo State, Brazil) and Rivera-Santana do Livramento (Uruguay and Brazil transboundary cities). In the four pilot areas local people and involved institutions started some important developments that need to be continued. In other areas of potential conflict as Ponta Porã-Pedro Juan Caballero (Brazil and Paraguay transboundary cities) similar approaches and methodologies could be applied.

27.2.2 Guarani Aquifer Management Instruments

In the Strategic Action Program (SAP) preparation process of the Guarani Aquifer System, four management instruments were prioritized for the protection and management of the aquifer. Two of them are technical with the objective to provide more accurate and integrated information to the management process. A third one was designed to involve local stakeholders and water managers at the problem level to finding integrated solutions on water management and aquifer protection. The fourth one is related to the importance of sharing all technical information developed by the project and to promote capacity building mechanisms for technical and institutional development. The four management instruments are briefly described below:

(1) **The Information System SISAG:** Its objective is to make relevant information on more than 8,000 wells available to stakeholders on a geographic platform. The system includes 32 workstations located in federal and state water management institutions. A single base map integrating geographical information from 191 different local maps was also developed. A Technical Advisory Commission with representatives from national institutions responsible for water information participated in SISAG elaboration and this Commission is supposed to support the required development in the future. Operational support, technical updating, performance supervision and maintenance of SISAG are provided by Argentina.

(2) **Monitoring Network and Mathematical Modelling:** Using information of 1,800 wells, the GAS was subdivided into seven main zones (see Figure 27.2), with different hydrogeological characteristics. Regional and local models were developed to predict groundwater flow, including interferences between wells in critical areas (pilots), in terms of groundwater levels, temperature and water quality. A monitoring network of 180 wells, distributed over the delineated aquifer zones, was designed using technical criteria. Each country is responsible for data collection in this network and may add new monitoring wells. A joint Monitoring and Mathematical Modelling Commission will address monitoring network needs and evaluate modelling activities, while overall support will be provided by Brazil.

(3) **Local Management Support Committees:** Local Management Support Commissions, with participation of municipalities, users, academic institutions and the

civil society, were established in four critical areas and achieved important results. At the end of project, country representatives proposed to share some supporting responsibilities (*indicates the country that will be responsible for supporting the transboundary local commission activities):

- *Concordia – Salto* (Argentina* – Uruguay): Involvement of local authorities from both countries in the Local Committee. Definition of minimum distance between wells and other measures to avoid water level interferences and temperature decrease, using mathematical models.
- *Ribeirão Preto* (Brazil): A landscape zoning act was developed and includes groundwater abstraction restrictions and protection of recharge areas to reduce a large observed groundwater level decline in Ribeirão Preto downtown (60 m during the last 30 years). The Local Committee was incorporated into the Pardo Watershed Basin Committee and the Act was approved by the Water Resources Council of São Paulo State.
- *Itapúa* (Paraguay): Creation of the local watershed committee, supported by a local university. Development of studies on water supply wells and agricultural uses and protection of the aquifer.
- *Rivera-Santana do Livramento* (Uruguay* – Brazil): A large educational campaign on the aquifer was developed with municipal schools. Based on previously integrated systems on energy supply, jointly solutions for the lack of sewage systems (almost 50% in both cities) and waste disposal were suggested to municipalities and governments.

(4) **Capacity Building for Groundwater Management and Knowledge Dissemination:** The four countries recognize that knowledge dissemination and empowerment of institutions are crucial for the management of the Guarani Aquifer. A permanent capacity building process of institutional press officers, national environmental journalists, NGOs and academic sectors should be established and strengthened. Paraguay will organize the process of supporting the implementation of planned activities.

27.3 GROUNDWATER GOVERNANCE AT COUNTRY LEVEL

Generally, water laws in all four countries have been developed based on surface water strategies. Fragmented approaches and institutional dispersion make all developments on groundwater management laws especially difficult. During project execution phase some laws and plans were developed based on the integrated management approach. Many developments on groundwater legal aspects occurred in parallel to or motivated by the Guarani Aquifer Project (preparation and execution phases), catalysed by the cooperation and knowledge environments, water managers, researchers and representatives from different governmental institutions, universities, private sectors and NGOs from all involved countries. Table 27.2 summarises the main legal and institutional provisions for groundwater governance in the four countries, while a description follows for each of the countries separately.

27.3.1 Groundwater Governance in Argentina

The Constitution of the Argentinian Confederation (1853) established that Provinces have full mandate on all matters not delegated to the Nation (Art. 101). Since that and in accordance with the National Constitution reform (1994), natural resources belong to provincial jurisdiction (Art 124). The Provinces in Argentina have their own water resources laws and regulation authorities.

Based on the National Constitution, the national government can dictate general guidelines for the environmental protection, and the provinces dictate complementary ones (Art. 41). The General Law on the Environment (Law 25.675/2002) established guidelines for environmental management. Based on those guidelines, the Environmental Management of Water (Law 25.688/2002) proposed other guidelines for water protection and creation of watershed committees in interjurisdictional rivers but this law is not fully operational. There are previous interjurisdictional agreements and committees such as on the Colorado river (1956), the Lower Atuel river (1983) and the Argentinian La Plata Watershed (1983). The Integration Committee of the La Plata River was based in Buenos Aires (1967), with the support of the La Plata River Treaty (Argentina, Bolivia, Brazil, Paraguay and Uruguay, 1969). There are 14 interjurisdictional watershed organizations (committees, authorities, or groups) and 11 transboundary initiatives.

The institutional framework on the environment and water management has been continuously modified since 1970. In 1992, water policy development issues were included in the National Water Resources Sub-Secretariat (SSRH) competencies, linked to the Ministry of Economy, Works and Public Services. The National Water Institute (INA) was created in 1973 and oriented to technical investigations. The Regional Centre on Groundwater (San Juan) was incorporated in the INA in 1998, and provides information to private and governmental institutions at national, provincial and municipal levels.

The Federal Council on Water (COHIFE) was created initially by some provinces in 2002, and was recognized by the Law 26.438/2008, with the objectives of coordinating water policies and promoting integrated water management to contribute with the sustainable development of Argentina.

During the Guarani Aquifer Project preparation and execution, the provinces developed their specific water laws (surface and groundwater) and provincial institutions responsible for the implementation process. During the Project execution period, some provinces established specific Water and/or Guarani Aquifer Laws such as: Misiones (Law XVI – n° 95/2007), Chaco (Law 5.446/2004), Formosa (Law 2.401/2006) and Corrientes (Law 5.641/2004). To all of them, water belongs to the public domain and some of them established Guarani Aquifer Councils in which all distinct institutions related to groundwater are represented. The Entre Rios Province Regulatory Agency (ERRTER, 2010) launched a Thermal Resources Exploitation Development Plan and set 10 kilometres as a minimum distance between wells.

In 2003, the COHIFE presented some Water Policy Basic Principles, a document with 49 agreements on a shared vision of an upcoming national water policy. Some provinces have adopted these principles, like Santa Fe province (Law 13132/2010) in the Guarani Aquifer area. A National Plan on Water Resources was proposed by the SSRH in 2006 to COHIFE and a first edition was published in 2008. A National Water

Table 27.2 Main legal and institutional provisions for groundwater governance in the four countries.

	Argentina	*Brazil*	*Paraguay*	*Uruguay*
Groundwater ownership	Public good	Public good	Public good	Public good
Main jurisdictional level for groundwater management	Provinces	States	Country	Country
Relevant laws at national level	National Constitution (1853; 1994) General Law on the Environment (2002)	Federal Constitution (1988); Environmental Policy Law (1981); Mining Code (1967)	Law on the Environment (2000); Water resources law (2007)	National Constitution (1997; 2004) Water Code (1978)
Relevant laws at state/ provincial level	Provincial water laws laws	State water resources management	–	–
Relevant institutions at national level	National Water Institute (INA) National Water Resources Sub-secretariat (SSRH); Federal Council on Water (COHIFE)	National Water Agency (ANA); Brazilian Geological Survey; National Water Council (CNRH); Technical Groundwater Chamber	National Environmental Council; Environmental Secretariat (SEAM); Water supply & Sanitation agencies	Ministry of the Housing, Territorial Management and Environment (MVOTMA); National Council on Water, Environment and Spatial Planning; National Directorate on Water (DINAGUA)
Relevant institutions at state/provincial level	Provincial implementing agencies	State implementing agencies State and watershed committees	–	Three regional Water Councils

	Argentina	*Brazil*	*Paraguay*	*Uruguay*
Water policy	COHIFE Water Policy principles (2003); Federal Plan on Groundwater (2008) National Water Plan (2016)	National Water Policy (1997); National Water Plan; State water policies; Sector policies	Ambition to establish a national water resources plan	National Water Policy (2009)
International Treaties and Agreements	La Plata River Treaty; Guarani Aquifer System Agreement	La Plata River Treaty; Guarani Aquifer System Agreement	La Plata River Treaty; Guarani Aquifer System Agreement (still to be approved by the parliament)	La Plata River Treaty; Guarani Aquifer System Agreement
Relevant committees/ institutions at basin/aquifer level	Colorado River; La Plata River	La Plata River; and some tributary rivers committees	La Plata River	La Plata River; Guaraní Aquifer Committee
Information Management	Groundwater Info System SIFAS	Groundwater Information System SIAGAS	–	–

Plan (2016–2019) was launched in 2016, including a specific Plan on Drinking Water and Sanitation.

With the support of the National Water Resources Sub-Secretariat (SSRH), the Council (COHIFE) proposed in 2008 a Federal Plan on Groundwater (PFAS). Based on a previous information system and the development of the Guarani Information System (SIAGAS), a recent product was the Groundwater Information System (SIFAS), with the Guarani Aquifer database incorporated in its structure. In fact, the SIFAS integrated all 23 provinces. At the time of the Guarani Aquifer Project, Argentina counted 200 wells in its national segment of the Guarani Aquifer, and SIFAS presents a total amount of 20,000 wells in the country. INA has been monitoring some wells in the Guarani Aquifer, and in the Concordia-Salto thermal area.

In the Concordia-Salto area (Argentina and Uruguay), municipal, provincial and national representatives are committed to the implementation of project monitoring and mathematical modelling running permanently in both countries, also promoting the dissemination of information to the local society for local institutional development. Specific meetings have been organized by the Entre Rios Province department to coordinate capacities and efforts among local and national representatives with the participation of both involved countries.

27.3.2 Groundwater Governance in Brazil

In accordance with the Brazilian Federal Constitution (1988), governments at federal, state and municipal levels are autonomous and have specific attributions in the republican context. Water is a public good under two different jurisdictions: interstate and transboundary rivers and lakes are national, while other rivers and groundwater belong to the state domain. Brazil has a National Water Resources Policy and specific state water policies in the states, all based on participatory and integrated management principles. There is no reference on transboundary aquifers, but there was a strong interaction process among institutions from national level and subnational states on the Guarani Aquifer Project, during both the preparation and the execution phases.

There are also different sectorial policies that intervene on water in general and on groundwater specifically, demanding from governmental institutions a frequent interaction process. The Environmental Policy (Law 6.938/1981) is oriented to environmental protection, licensing infrastructural works and water quality standards. Both urban and rural spatial planning are managed by municipalities. Mining, and thermal and bottled waters have specific regulations and are under a concession regime by Federal Government (Decree Law 227/1967). Municipalities are also responsible for urban drainage, garbage collection, water supply and waste water treatment and disposal; they can contract out the related services.

The current National Water Policy, established by the Law 9.433/1997, includes a bottom-up management system and management instruments, some of them designed for surface waters. Since 1997, the Brazilian Geological Survey is developing a Groundwater Information System (SIAGAS), based on information technology, with 282,563 wells (2016) in the whole country. The National Water Agency (ANA) was created (Law 9.984/2000) to implement the National Water Policy.

In the terms of the law, there are different technical instruments and an integrated management system. The main technical instruments are: watershed plan; water quality goals; information system; water permits; fees and compensation to municipalities (specifically the ANA's payment for the environmental/water services program). The management unit is the watershed, and the watershed committees are responsible for plan approval, fees definition and resources allocation with participation of government (max. 40%), multiple water users (30%) and civil society organizations (30%). Also, the management system is integrated by the committee's execution agencies; the state and national water resources councils; and water resources institutions at state and federal levels, responsible for policy definition (water secretariats) and its implementation (national and state agencies). There are nine federal committees and more than 250 state committees (ANA 2015), mainly in higher populated and conflictive areas. All states have their own water resources management laws and specific management strategies to implement the policy instruments and managerial system in water resources under their domain, including groundwater.

Since the Guarani Aquifer Project preparation, the creation of integrated management units in all involved States (Goias, Minas Gerais, Mato Grosso do Sul, Mato Grosso, Parana, Rio Grande do Sul, Santa Catarina and São Paulo) and a National Management Supporting Unit was very important. This was a key element to develop a coherent set of instruments and mechanisms to joint efforts from all responsible institutions, civil society and people interested in aquifer management and institutional development.

The National Water Council (CNRH) created a permanent Technical Groundwater Chamber (Act CNRH 9/2000) with the main objective of presenting proposals for the inclusion of groundwater management into the National Water Policy and suggesting institutional mechanisms to integrate groundwater and surface water management. The Groundwater Chamber approved many different acts related to integrated groundwater and surface water management development, for example: general principles for surface and groundwater management, including transboundary aquifers (Act CNRH 15/2001); groundwater and water permits (Act CNRH 16/2001 and 37/2004); groundwater aspects to be considered in water plans (Act CNRH 22/2001 and 48/2005); bottling and mineral water (Act CNRH 76/2007); water classification (Acts CNRH 91 and 92/2008 and Act CONAMA 396/2008); groundwater protection (Act CNRH 92/2008); approval of the National Groundwater Plan (Act CNRH 99/2009) in the National Water Plan; groundwater monitoring network (Act 107/2010); criteria for groundwater recharge (Act 153/2013). All these acts have been very helpful for developments on groundwater use and regulation in the states.

The National Water Plan was approved by the National Water Council for the period 2012–2016 and presents four main components, 13 programs and 30 subprograms. The National Groundwater Program is part of the National Water Plan and presents three subprograms, related respectively to (i) development of hydrogeological knowledge (oriented to interstate and transboundary aquifers and management instruments in general); (ii) development of legal and institutional aspects; and (iii) capacity building, communication and social mobilization. The National Plan established actions considering time frames of 2007, 2011, 2015 and 2020.

Considering full mandate on groundwater at subnational states level, the National Water Agency (ANA) is acting to foster integrated management of surface and groundwater. In addition, the National Groundwater Agenda (under implementation by ANA, since 2007) has supporting programs that helps states on different water management levels to implement their own water policies, based on goals achievement. ANA has also an important South-South Cooperation project portfolio to support countries on water management, capacity building and institutional development. Based on the Climate Convention ANA presented some guidelines to the National Plan for climate change adaptation. Coherently to their long boundaries, the discussions on climate change repercussions in Brazil should involve neighbour countries, regional and international water agencies too.

27.3.3 Groundwater Governance in Paraguay

The Law on the Environment (1.561/2000) has established the environmental policy and created the National Environmental System, which includes a National Council and an Environmental Secretariat (SEAM). The SEAM-Paraguay was created with a wide range of objectives on policy formulation, coordination and implementation, from territorial planning till environmental impacts, natural disasters, water resources, biodiversity, indigenous people and climate change. To support water management issues in the country, SEAM established a Directorate on Water Resources.

Water resources management in Paraguay was established by Law 3.239/2007. Among different principles, water is public property and river basin is the basic management unit. The establishment of a water resources plan is one of the policy objectives and – as a next step – the integration of water into the general national plan. Based on that law, water resources protection will be based on a system approach including the definition of aquifer recharge areas. By law, all users must request for water abstraction rights, while a small amount of water is a declared human right.

The law and some linked acts were developed during the Guarani Aquifer Project execution phase. Some technical instruments were established, including water user records (Act 2194/2007), the national water plan and an information system. National and watershed water balances must be calculated and all users should pay water fees. Spring protection zones must be defined, extending as far as 100 m from the springs. During the Guarani Aquifer Project, different acts were also developed at national level in Paraguay: the act 222/2002 adopting water quality standards; the act 2155/2005 presenting guidelines for drilling deep wells; the act 170/2006 on the creation of water resources councils.

There are some national institutions in charge of water supply and sanitation but few people are engaged in water resources management. A remarkable experience is related to the administration of wells by local communities. Since the creation of the Environmental Secretariat (SEAM-Paraguay) some institutional improvements have been implemented. Groundwater is part of the duties of small water management teams with a small budget. Important knowledge and developments on groundwater have been developed in the framework of special projects, such as the German (BGR)–Paraguayan (SEAM) cooperation that was discontinued after 2011.

There are important challenges on groundwater management in Paraguay. The Patiño Aquifer is in one of the most populated areas of the country and is exposed to high abstraction rates and some risks of overexploitation and pollution (sanitation and industries). There are also some interests on the development of a management strategy of the Yrendá-Toba-Tarijeño Aquifer (Paraguay-Argentina and Bolivia borders). The hot spot in the transboundary area of the Guarani Aquifer in Itapúa (agricultural impacts) was implemented during the project. The hot spot in Pedro Juan Caballero

(Paraguay) – Ponta Porã (Brazil), where wells interact in a transboundary thermal touristic area, is still to be implemented.

During the political conflict between Paraguay and Mercosur in 2012, different agreements were rejected, including the Agreement on the Guarani Aquifer under discussion in the Parliament. To face important issues on groundwater, Paraguay will open again discussions on the Agreement on the Guarani Aquifer.

In 2014, Paraguay approved a National Development Plan (2030). The plan recognizes the importance of implementing an efficient policy for the protection and conservation of the Guarani Aquifer considering climate change. As stated in the plan, 85% of the population has potable water supply and the plan recognizes problems on water resources, contamination and lack on management of waste waters, since only 11% of all wastewater is collected. Groundwater management in Paraguay has been related the Guarani Aquifer and some developments on water management in general have been implemented.

27.3.4 Groundwater Governance in Uruguay

The Law 14.859/1978 establishing a general water code in Uruguay is still valid. Water uses for irrigation are regulated through the Law 16.858/1997. After the concession of water supply services in some coastal cities (*e.g.* Maldonado) a plebiscite in 2004 modified the National Constitution (1997) and included in its Art. 47 that water is essential to life and the access to water supply and sanitation are considered fundamental human rights.

A general plan related to the management of thermal wells was included in that law. The Act 214/2000 established some specific technical rules: dynamic level less than 150 metres, maximum instantaneous abstraction rate of 150 m^3/h, minimum distance between wells of 2000 m, maximum daily period of pumping of 16 hours, and maximum user permit period of 10 years.

After 2005, the water authority moved from the Ministry of Works to the Ministry of the Housing, Territorial Management and Environment (MVOTMA), responsible for coordinating water use in the country with other involved ministries. The Law 18.610/2009 established the Basic Principles of the National Water Policy and the National Directorate of Water and Sanitation (or only National Directorate of Water after 2011; DINAGUA/MVOTMA) whose is responsible for policy implementation.

The Basic Principles of the National Water Policy (Law 18.610/2009) regulate the creation of Watershed and Aquifer Committees. Regional Water Councils were created in 2012 in three areas: Uruguay river, Merin Lagoon and La Plata river/estuary in accordance with a decentralization strategy. Each council is formed by 21 members (civil society, government and water users, equally represented). The objectives of the Regional Councils are sectorial coordination and formulation and monitoring of regional water plans. A National Council on Water, Environment and Territorial Planning was also created. The president of the council is the Minister of Housing, Territorial Planning and the Environment and water policy development is coordinated by the National Directorate on Water (DINAGUA). The Commission on the Guarani Aquifer was created in 2013 with 33 members (12 government, 11 civil society and 10 users), as stated in Act 183/2013. The authorization process for the construction of wells in confined areas of the Guarani Aquifer requires public hearings. Additionally, Guarani wells should be monitored once a year, in accordance with the Network established by the Guarani Aquifer Project.

DINAGUA, with support of the National Directorate of the Environment DINAMA/MVOTMA that is responsible for water quality management, elaborated in 2011 a document with the guidelines for a National Water Plan. The Water Resources Plan proposal was finally presented in 2016; it includes a diagnostic on water resources in the country and 11 prioritized programs and goals till 2030.

At the end of the Guarani Aquifer Project, Uruguay installed the Liaison Unit just as the countries' agreement in the Project Execution Council, but it is not yet operational. Uruguay and UNESCO installed a Regional Centre on Groundwater Management in Montevideo, the first in the Americas to support investigation at national and international levels on aquifer management, capacity building and knowledge dissemination.

27.4 THE AGREEMENT ON THE GUARANI AQUIFER

As recognized by the countries involved, the technical and diplomatic discussions that took place in the Guarani Aquifer Project Execution Council contributed to building confidence between the countries. Monitoring the technical results on scientific understanding of the aquifer's functioning and flow regime, the general aspects of water resources management and environmental protection in each country, and the basic principles of transboundary management were main ingredients of the envisaged agreement on the Guaraní Aquifer. It is important to underline that all cooperation practices existing in the region and in very different diplomatic sectors were the cements of the agreement designing process. On request of the countries, the General Secretariat presented a draft, a first proposal was consolidated by high-level Argentinian diplomats and an experienced diplomatic group from the four countries built the final version presented and signed by national presidents in 2010 in San Juan, Argentina.

Considering groundwater flow characteristics, the aquifer management framework and the types of users, groundwater management should be locally oriented and focused on critical issues, in consonance with the national priorities and harmonized by joint regional strategies. Each country has its own groundwater management strategies and needs to consider and develop new efficient strategies based on its own policies and administrative sovereignty.

The Agreement presents guidelines on how to act in case of controversies among the countries and obliges countries to exchange technical information on studies, activities and works related to the GAS and to inform each other on activities that may have transboundary effects. An important discussion took place on the adoption of negotiation or arbitration mechanisms, on local or external frameworks. In fact, negotiation processes are the pillars in modern water management framework and the adoption of arbitration mechanisms in the Guarani Aquifer or their countries could be considered a step backwards on integrated management strategies and a first step to put water policy in courts. Considering future discussions related to the Art. 19, the Guarani countries will have the opportunity to contribute to the international law related to Groundwater Governance.

Naturally, some aspects included in the Strategic Action Plan (SAP) have not been included in the International Agreement. The countries are facing a fast expansion of groundwater exploitation, including in the outcrop zones and in important confined areas where exploitation can be just groundwater mining. The countries also agreed on sharing the coordination process of groundwater management instruments, such as an information system (Argentina); groundwater monitoring networks and mathematical modelling (Brazil); knowledge dissemination and capacity building (Paraguay); and the support of a Liaison Unit (Uruguay). The SAP defined some priorities that the parties need to develop in their own groundwater governance strategies since they are responsible for water and aquifer management.

The Guarani Aquifer Agreement was the best possible agreement considering the existing level of information, the existing institutional framework and the specific state of water management in the world and in the region. The experience in this region was considered also during the development of the Draft Articles on the Law of Transboundary Aquifers sponsored by UNESCO and other participants (UN General Assembly Resolution 63/124). The main challenge now is to put the agreement operational and implement the approved jointly management strategy. The approval of the Agreement will facilitate institutional development and society organization at the country level, necessary to face upcoming important management issues on the aquifer and related ecosystems under the pressure of climate change. The experience of National Coordination Units that functioned during project execution in each country was important for avoiding overlaps, integrating efforts, supporting high-level technical discussions and fostering groundwater governance at different levels.

The General Assembly of Uruguay was the first to approve the Guarani Aquifer Agreement in 2012. The Argentinian Congress approved the Agreement on the Guarani Aquifer in the same year 2012 (Ley 26.789). At the same year, Paraguayan Parliament rejected the Guarani Aquifer Agreement and all agreements from the Mercosur countries, due to political reasons unrelated to groundwater or the Guarani Aquifer. The parliament of Brazil has just approved the agreement (2017) and the parliament of Paraguay reassumed the discussions and an imminent approval is expected. In accordance with the agreement, a Commission integrated by the countries will coordinate the cooperation and will establish its own operational regulations.

Countries ought to acknowledge the fact that the Agreement amplifies the national competency to face groundwater management challenges, it is also a must that each country remains responsible for protecting and managing the water resources of its aquifers. Indeed, cooperation can help in harmonizing general strategies, in providing mutual technical support to maximize benefits and in dealing with transboundary and national hotspots calling for immediate and innovative solutions.

27.5 REGIONAL COOPERATION AND LOCAL MANAGEMENT

The development of the Guarani aquifer management tools depends on the joint efforts of the four countries. National and subnational governments need better information to establish consistent predictions and adjusted decision-making processes. the efforts of national and subnational technical commissions to the integration of all management instruments developed by the Project should help the technical institutions also in their own development (*e.g.* SIAFAS-SISAG integration in Argentina). The implementation of the Information System (SIAGAS), the constant operation of the monitoring network, the regional and local mathematical modelling (M&M) will benefit water authorities, economic users and the civil society. Capacity building, transparency and dissemination of information is a key element to the aquifer governance process.

Data availability beyond the reconnaissance level has been already needed by institutions and Local Committees in the pilot areas (*e.g.* Concordia-Salto to run a better prediction model) and will be equally important for predictions under climate change scenarios and the increase of groundwater uses in all defined management regions. Strengthening the local commissions in the pilot areas requires some autonomy and support from subnational, national and regional institutions. The pilot areas are facing conflicts between groundwater and urban land uses and problems caused by existing gaps in national and state/provincial laws. Local transboundary commissions' achievements will probably depend on specific agreements and technical support from cooperation projects.

Since groundwater is a hidden resource, a better use of an aquifer implies better knowledge and innovative research. The Strategic Action Program of the Guarani Aquifer has established different priorities for national, state and local management, and improved mechanisms for regional cooperation and for adequate use of the management tools developed. The Guarani Project has demonstrated that countries and international agencies can mobilize resources, technical companies, and the academic sector for the common objective of developing knowledge. In this respect, interruptions in research and multilateral management consensus-building projects mean loss of resources and time.

The SAP defined ten priority lines of action of the countries, with a set of detailed actions to improve transboundary groundwater management in the four countries, also at subnational levels since the provinces and states are responsible for groundwater management in Argentina and Brazil (federal countries). The countries are currently addressing different priorities as defined in SAP. The National Water Agency in Brazil developed a detailed study in a recharge area, while Uruguay and Paraguay showed their interest in applying the same methodology in their areas. The Water Resources Secretariat in Argentina implemented the Guarani Aquifer Information System; both its provinces and the other countries could use that experience for implementing their own information systems. Considering the number of institutions related to groundwater in each country and province/state, the restructuration of national and subnational coordination units can optimize local efforts. At regional level, the technical commissions with national experts could support countries in the implementation and coordination of each management instrument.

The cooperation process among the countries was a catalytic element in developing groundwater managerial tools and criteria. Locally in Concordia-Salto, in the border area of Argentina and Uruguay, a minimum distance between deep thermal wells was applied to avoid interference (10 km and 2 km respectively). In Ribeirão Preto (Brazil), where 100% of the inhabitants are supplied by groundwater only, a land management zoning act was approved by the São Paulo State Water Resources Council. In Santana do Livramento (Brazil) and Rivera (Uruguay), the discussions were focused on the delineation of areas for concentrations of exploitation wells far from sanitary contamination sources. In Itapúa (Paraguay), where the local committee cooperated with the watershed management institution, the representatives dealt with conflicts related to deforestation and soy beans cultivation, to avoid future overexploitation and contamination problems. Complementary, regional cooperation catalysed the development of complex numerical models to support local groundwater resource management in each of the pilot areas and countries.

All established priorities should be incorporated into the actions of water resources management institutions at different sectors and levels of public administration. Actions at country level must be integrated by a National Unit for Management Support to guide and integrate the efforts of all institutions, until now acting in a dispersed and fragmented manner in the use, management and protection of the GAS. In the case of federal countries such as Argentina and Brazil, Units of State management support should be created also, considering groundwater technical and policy integration requirements. It is also important to integrate representatives from different users and social sectors in the National/States units to construct a technically based dialog about wells, groundwater use and aquifer management to avoid different types of misconceptions (or hydromyths, as they are called by Prof. E. Custodio in 2003) on the Guarani Aquifer.

The creation of Local Management Support Committees must be encouraged to support the development of appropriate mechanisms for the protection of GAS in areas where significant impacts occur. The experience of the creation of local committees has been very positive in the pilot areas, as in Ribeirão Preto where the community proposed and the State Water Resources Council approved a zoning act with measures to protect areas, to control groundwater uses and to face water level deplaning of the aquifer in the municipality.

Considering the existing institutional fragmentation regarding groundwater in the four countries, a regional cooperation framework has been proposed (Figure 27.3), inspired by the project execution structure. National Units for Management Support should be installed by each country and Technical Advisory Committees (with experts from all four countries) for each of the managerial instruments, while a Liaison Unit integrated by technical experts could coordinate actions, facilitate communication and support decision-making by the GAS Regional Committee.

The Office of the Liaison Unit was installed in Montevideo with the support of Uruguay but no staff members have been designated by the countries yet. This Unit could be supported by the UNESCO Centre on Groundwater created in Uruguay and already functioning in Montevideo. Other successful Project initiatives as the Citizenship Fund, to support the participation of civil society organizations; the University Fund, to support academic investigations and specific events involving environmental journalist associations should be preserved by the countries. Unlike when the GAS project started, the regional population has now some information on the aquifer resources, mainly due to the participation of civil society organizations.

The implementation and development of the Guarani aquifer management instruments require substantial sub-national and regional co-operation efforts, supported at all levels by appropriate institutions. At the local level, institutional involvement needs to be promoted and supported by mentioned Local Management Support Committees. At the national and state/provincial level, Management Support Units shall be strengthened. In consonance with the Guarani Aquifer Agreement, the GAS Regional Committee with participation of the four countries representatives should coordinate

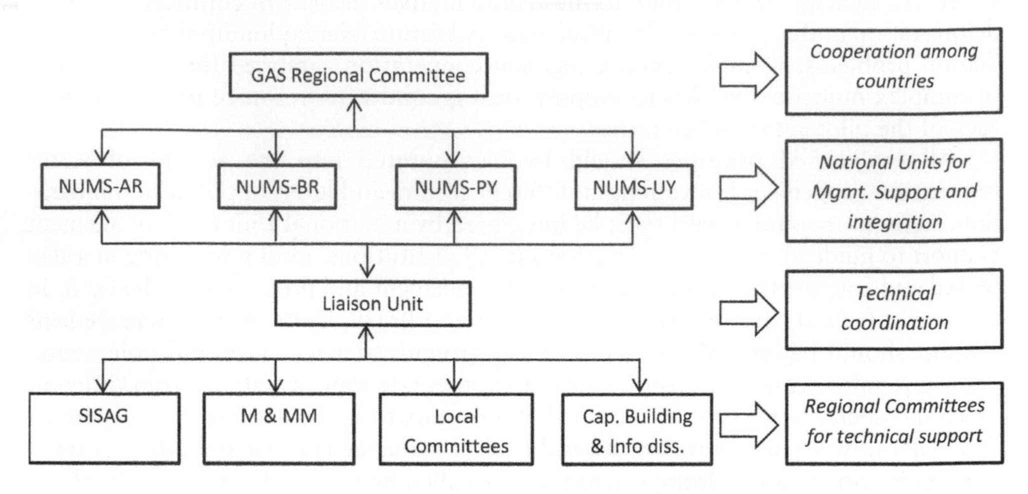

Figure 27.3 Regional cooperation framework.

all joint efforts. The Committee will be established based on the Project Execution Council's experience and under the umbrella of the Treaty of the River Plate Basin (1969).

27.6 CONCLUSIONS

The Guarani Aquifer and groundwater management in the region have been subject to an important development since the PSAG Project was prepared and implemented. A huge amount of information, studies, manuals and management proposals have been developed at local, subnational, national and regional levels. At the international level, regional experts participated in the elaboration of the Draft Articles on the Law of Transboundary Aquifers (UN Assembly, 2008). Groundwater management is mainly local; transboundary aquifer management must be coordinated by the countries; sovereignty and cooperation are complementary in national and regional contexts; and regional and international institutions should act as catalytic support.

Upon the end of the project, the local responsible institutions decided to start a new and challenging phase, based on their own capacity and resources only. Maybe the countries should have received additional support for the internalization of project results and the initial implementation of the instruments in the countries. As previously explained and due to reasons unrelated to groundwater, approval of the Guarani Aquifer Agreement requires more time than expected and it still needs to be approved by the Paraguayan Parliament.

Groundwater management is a locally based activity, that requires continuous national/regional support and cooperation to be sustainable. Conforming to investigations in the Guarani Aquifer pilot areas, problems emerge in restricted areas but they could spread spatially and be replicated in other areas, causing socioeconomic problems. In such conditions, lack of groundwater management could result in problems impossible to be solved and to be mitigated only at high cost.

There is clearly great potential for the development of the Guarani aquifer's functions and for increasing environmental resilience, for the benefit of the countries and their people. Aquifer management and protection are the pillars of future development. Effective protection needs to consider basic knowledge and the local dimension (where potential pollutants and overexploitation problems can present adverse effects), all levels of administration and all matrix dimensions of transboundary groundwater management. In consonance with the experience gained in the Guarani Aquifer Project, an adequate decision-making process should be technically and diplomatically based.

All major developments on groundwater management which occurred in recent years in the four countries are related to the Guarani Aquifer Project. The institutions in these countries can count on a great amount of knowledge in high-level studies, guidebooks, civil society experiments, management instruments and strategic planning; and they have an institutional framework capable of developing the Guarani Aquifer and groundwater governance in general. Furthermore, Argentina, Brazil, Paraguay and Uruguay have an important agreement signed by national representatives, which should be enforced as soon as it has been approved by all national parliaments. Aquifer governance is not a theoretical definition but a result of integrated knowledge, efficacious legislation, strengthened institutions and coordinated procedures that need to be continuously developed and renewed. All knowledge and both positive and negative experiences should be shared by our institutions and people of different expertise and levels. The experience acquired on the Guarani Aquifer also needs to be shared among other countries and some management actions can be replicated under similar and controlled conditions.

There are many uncertainties on water management in general, and specifically on groundwater under variability and climate change scenarios. There are also gaps in information and knowledge on the use of the Guarani Aquifer, but the main conditions related to its governance are established and its management will always be a result of both government decisions and social participation. Since the Guarani Aquifer Project products as national developments on groundwater governance resulted from a diligent work of experts and institutions, the maintenance of the same involvement at local, subnational, national and international levels is indispensable for aquifer governance and groundwater management.

REFERENCES

Amore, L. (2011) *The Guarani Aquifer: From Knowledge to Water Management*, International. Journal of Water Resources Development, vol. 27, no. 3, pp. 463–476.

Amore, L. Tröger, U. (2010) *Transboundary Guarani Aquifer System and Groundwater Management Mechanisms*. IN: UNESCO-IAH-UNEP Conference, Paris, France. Proceedings.

Amore, L., Vargas, F.P.H., & Oliveira, W.A. (2001) *The Schematic map of the Guarani Aquifer System: as water management support*. Annuals..., II Simposio Paraguayo de Geología y III Simposio Paraguayo de Aguas Subterráneas y Perforación de Pozos. Asunción, Paraguay. [CD-ROM]. [Portuguese Language]

Argentina, Brazil, Paraguay and Uruguay (2009) *The Guarani Aquifer Agreement*. Itamaraty. Brasilia. http://www.itamaraty.gov.br/ [Portuguese and Spanish]

Argentina Environmental Management of Waters Law (2002) *Ley 25688*. Available from: servicios.infoleg.gob.ar [accessed Jan 10th 2017]

Argentina Entre Rios Province Water Law (2002) *Ley 4 9172*. Available from: mininterior.gov.ar/provincias/entrerios/agua-entre-rios.pdf [accessed Jan 10th 2017]

Argentina Chaco Guarani Law (2004) *Ley 5446*. Available from: www.inti.gob.ar/salta/leyesAgua/chaco/ley_5446_acuiferoguarani.pdf. [accessed Jan 10th 2017]

Argentina Corrientes Guarani Aquifer Law (2004) *Ley 5641 de 2004*. Available from: www.citargentina.org/download/leyes/ley_5641.pdf [accessed Jan 10th 2017]

Argentina Corrientes Water Law (2001) *Decreto Ley 191/01*. Available from: www.icaa.gov.ar/documentos/ingenieria/codigo_aguas_ley191_01.pdf [accessed Jan 10th 2017]

Argentina Misiones Guarani and Groundwater Law (2007) *Ley XVI – n °95 (or Ley 4326)*. Available from: www.faolex.fao.org/docs/pdf/arg144337. pdf [accessed Jan 10th 2017]

Argentina General Environmental Law (2002) *Ley 25675*. Available from: servicios.infoleg.gob.ar [accessed Jan 10th 2017]

Argentina Santa Fe Water Principals approval (2010) *Ley 13132 de 2010*. Available from: www.santafe.gov.ar/normativa/item.php?id=109601&cod=2d9f2cd8692d10690d9c1ebc92eg2b9c [accessed Jan 10th 2017]

Argentinian Confederation Constitution, 1853 (as amended 1860, 1866, 1898, 1957 and 1994) *Ley 24.430*. Available from: servicios.infoleg.gob.ar [accessed Jan 10th 2017]

Brazil National Water Policy (1997) *Law 9433 of 1997*. Available from: www.planalto.gov.br/ccivil_03/leis/L9433.htm [accessed Jan 10th 2017]

Brazilian Geological Survey – CPRM (1997) *Brazil National Groundwater System – SIAGAS*. Available from: siagasweb.cprm.gov.br [accessed Dec. 20th, 2016]

Brazilian National Water Council Acts (Resoluções 15, 16, 22, 48, 76, 91, 92, 99, 107, 153) Available from: cnrh.gov.br/index.php?>option=com_content&view=article&id=14. [accessed Jan. 10th, 2017]

Brazilian National Water Plan (2006) *Plano Nacional de Recursos Hídricos*. Available from: cnrh.gov.br/index.php?>option=com_content&view=article&id=14. [accessed Jan. 10th, 2017]

DINAGUA/MVOTMA Uruguay (2016) *National Water Plan Proposal de 2016*. Available from: www.mvotma.gub.uy/ciudadania/item/10008231-propuesta-del-plan-nacional-de-aguas.html [accessed Jan 10, 2017]

ERRTER - Entre Rios Province Regulatory Agency (2010) *Plan Estratégico de Desarrollo de los Recursos Termales de la Provincia de Entre Ríos*. Available from: www.entrerios.gov.ar/termas/userfiles/files/otros_archivos/Informe%20Final%20Plan%20Estrategico%20TERMAS%202010.pdf [accessed Jan 10th 2017]

FAO, UNESCO,IAH,WB & GEF (2015) Groundwater Governance a call for action: A shared global vision for 2030. Avalilable from: http://www.groundwatergovernance.org/fileadmin/user_upload/groundwatergovernance/docs/GWG_VISION_EN.pdf [accessed Jan 10th 2017)

Federal Council on Water – COHIFE (2008) *Plan Hídrico Nacional Federal de la República Argentina*. Available from cohife.org [accessed Jan 10th 2017]

Kemper, K. Mestre, E. & Amore, L. (2003) Management of the Guarani Aquifer System. Water International, 28(2), 185–200. doi: 10.1080/02508060308691684

Organization of American States (2009) *Strategic Action Program – SAP*. Environmental Protection and Sustainable Development of the Guarani Aquifer System Project. Brasília. 409 p. [http://www.ana.gov.br/bibliotecavirtual/arquivos/20100223172013_PEA_GUARANI_Ing.pdf]

Organization of American States (2007) *Transboundary Diagnostic Analysis – TDA*. Environmental Protection and Sustainable Development of the Guarani Aquifer System Project. Montevideo. 249 p. [Spanish Language]

Organization of American States (2007) *University Fund: Advances in knowledge to sustainable management*. Environmental Protection and Sustainable Development of the Guarani Aquifer System Project. BNWPP, ALHSUD. Montevideo. 176 p. [Portuguese and Spanish Languages]

Organization of American States (2009) *Project on "Sustainable Management of the Water Resources of the La Plata Basin with respect to the Effects of Climate Variability and Change.* Buenos Aires. Project Document. 59 p. http://www.cicplata.org/marco/

Orme, M., Cuthbert, Z, Sindico, F. Gibson, J. & Bostic, R. (2015) *Good transboundary water governance in the 2015 Sustainable Development Goals: a legal perspective.* Water International, 40(7), 969-983. doi: 10.1080/02508060.2015.1099083

SEAM Paraguay (2007) *Resolución 2.194 de 2007.* Establece el Registro Nacional de Recursos Hídricos, los procedimientos para la inscripción en el mismo y para el otorgamiento del certificado de disponibilidad de recursos hídricos. Available from www.seam.com.py. [accessed Jan. 10th 2017]

Paraguay Law on Water Resources (2007) *Ley 3.239 de 2007.* Available from www.seam.com.py. [accessed Jan. 10th 2017]

Paraguay Law on Environment (2000) *Ley 1.561 de 2000.* Available from www.seam.com.py. [accessed Jan. 10th 2017]

Rebouças, A.C. and Amore, L. (2002) *The Guarani Aquifer System.* IN: Brazilian Groundwater Review, 16. Brazilian Groundwater Association – ABAS. p. 103–110. [Portuguese Language].

Santa Cruz, J. (2016) *Sistema Acuífero Guaraní: Nivel base de conocimiento en el Piloto Concordia-Salto* (2003-2012). 94p. Personal Communication Sep. 16th 2016.

Subsecretaria de Recursos Hídricos (2016) *Plan Nacional del Agua: Objetivos, Políticas, Estrategias e Acciones.* http://www.mininterior.gov.ar/plan/docs/plan-nacional-agua.pdf. 1st Version. [11/1/2016].

Van der Gun, J. Groundwater and global change: trends, opportunities and challenges. 2012.

Sugg, Z., Varady, R., Gerlak, A. & Grenade, R (2015) *Transboundary groundwater governance in the Guarani Aquifer System: reflections form a survey of global and regional experts.* Water International, 40(3), 377-400. doi: 10.1080/02508060.2015.1052939

UNEP – Paraguay (2006) *Usos y Gobernabilidad del Agua en el Paraguay.* PNUD ed. 97p.

Uruguay Water Code 1978 (as amended in 2014) *Ley14.859.* Código de Águas del Uruguay. www.parlamento.gub.uy/documentosyleyes/codigos [accessed Jan 10, 2017]

Uruguay Management Plan of the Guarani Aquifer (2000) *Decreto 214 de 2000 Aprobación del Plan de Gestión del Acuífero Infrabasaltico Guarani en Uruguay.* Available from: www.impo.com.uy/bases/decretos/214-2000 [accessed Jan 10, 2017]

Uruguay National Water Policy (2009) *Ley18.610 de 2009.* Available from: www.ose.com.uy [accessed Jan 10, 2017]

Chapter 28

Groundwater governance in São Paulo and Mexico metropolitan areas: some comparative lessons learnt

Ricardo Hirata[1] & *Oscar Escolero*[2]

[1]*Full Professor and Vice Director of Groundwater Research Center (CEPAS|USP), Institute of Geosciences, University of São Paulo, Brazil*

[2]*Principal Researcher, Department of Regional Geology, Institute of Geology, National Autonomous University of Mexico (UNAM), Mexico*

ABSTRACT

São Paulo and Mexico City are the two largest metropolitan areas in Latin America. Both cities have populations around 20 million inhabitants and GDP greater than US $ 400 billion/year. These cities also have recurrent problems of water shortage. Groundwater is a key resource in the two regions, however in Mexico City it is largely used for public water supply (80%), and in São Paulo for private use (more than 12000 wells that withdraw more than 10 m^3/s). In Mexico City groundwater extraction has caused extensive subsidence. Other problems consist of loss of well efficiency, anthropogenic contamination and water salinization. In São Paulo there are serious problems of contamination in previous and current industrialized areas, localized losses of well efficiency, and a large number of illegal wells (around 60%). On the other hand illegal wells in Mexico City are virtually non-existent. This contrast stems from the existence of the following regulation factors in Mexico: a) the Federal Government Finance Ministry collects fees on groundwater extraction, not the water sector; b) there is an effective market for buying and selling the water right; c) the water value is determined by an active and effective market, and increases yearly due to the strong competition among users; c) the industry or service owner has to obtain the right for extracting water from a legal well in order to have the electricity installation by the government. In both cities the legal regulatory system, governing the use of water resources, is complex and fragmented. Besides the São Paulo system is largely inefficient and the State government, responsible for the groundwater management, has little capacity to enforce the regulation on the groundwater extraction and the planning for the future is almost absent.

28.1 INTRODUCTION

The Metropolitan Area of Mexico City (Mexico City) and Metropolitan Region of São Paulo (São Paulo) are the two largest urban conglomerates in Latin America and have several similarities: (i) population around 20 million inhabitants; (ii) large urban-population growth in the decades 50–80, accompanied by lack of land use and urban planning; (iii) GDP greater than US$ 400 billion/year; (iv) significant proportion of the

Table 28.1 Characteristics São Paulo and Mexico City Metropolitan Regions.

	São Paulo	*Mexico City*
Population ($x10^6$)	21.1 (2016)	20.1 (2015)
Population increase rate (%/yr.)	1	1.2
Territorial area (km^2)	8051	7954
Urban area (km^2)	2139	1400
GDP (US$ x 10^9)	450	411 (2011)
Public water supply (m^3/s) (pop. served %)	68 (95%)	81.9
Public groundwater supply	<1%	>88%
% private water supply	>15%	>12%
Total number of wells	12000	3000 (1581 for drinking water supply)
Relative economic & social importance of groundwater	medium	high
Level of illegal wells	>60%	>10%
Main groundwater use	Municipal, industrial	Municipal, industrial

water public supply comes from watersheds outside the metropolitan regions, imposing high water transposition costs and complex institutional arrangements; and (v) recurrent problems of water shortage (Table 28.1).

Groundwater is a key resource in the two regions, however in Mexico City it is largely used for public water supply (80%), and in São Paulo for private use (more than 12000 wells that withdraw more than 10 m^3/s). In the latter, the average discharge is 5 m^3/h/well, which supplies 15% of the population, but during the last water crisis (2013–2015), the proportion reached 24%. In contrast, in Mexico there are only 3000 wells (600 public wells) extracting more than 50 m^3/s from a very productive sedimentary aquifer (average flow of 100 m^3/h/well).

The Mexico City's groundwater extraction has caused problems related to extensive land subsidence, well efficiency loss, anthropogenic contamination and salinization of water. In São Paulo there are serious problems of contamination in earlier and present industrialized areas, localized losses of well efficiency, and a large number of illegal wells (around 60%). On the other hand, there are not illegal wells in Mexico City practically (<10%). Although both cities have a rather complex legal regulatory system, there are still many problems of governance of the water resource.

In order to discuss these problems, this chapter aims at detailing the aspects mentioned above, presenting the main problems and their causes, related to groundwater resource status in these two metropolitan areas. This study addresses the following issues: a) legal and regulatory aspects; b) government institutional structure; and c) compensatory mechanisms and financial incentives associated with water use.

28.2 METROPOLITAN REGION OF SÃO PAULO

The Metropolitan Region of São Paulo is practically all inserted in the Upper Tiete River Basin (Tiete Basin), which occupies an area of 5720 km^2 and has a population of 21.2 million inhabitants (2016), distributed in 32 municipalities.

The public water supply system (SABESP) is made by a complex arrangement of 8 surface water sources. Overall, there is already a shortage of available water (67.90 m^3/s) to meet the demands (69.28 m^3/s in 2005) (FUSP, 2009), while the maximum treatment capacity is still sufficient, although close to its limit. This is especially noticeable in times of extended periods of low rainfall; further causing problems to water supply rationing. In addition, the 8 water-producing systems are not fully integrated. More recently, and in response to a major water crisis affecting the entire region, new networks have been built to interconnect the major producer systems with one another and to allow the waters of one system to reach the entire public water supply mains.

Although public water supply uses just over 0.7 m^3/s (1%) of groundwater, there are more than 12000 private wells in the region, which add more than 10 m^3/s, which complement the public system. Thus, groundwater accounts for 15% of the total water consumed in São Paulo (24% during the recent water crisis). The loss of this resource, either through contamination or overexploitation, can create a stress on the public water supply, because water companies cannot provide additional 10 m^3/s of water in a short time. The aggravating problem is that 60% of all wells in the Tiete Basin are illegal, *i.e.* they do not have an operating license. Therefore, the Tiete Basin's water security depends to a large extent on illegal private wells without state control (Hirata *et al.*, 2015) (Figure 28.1).

Groundwater also plays additional roles in the urban hydrological cycle that are likewise poorly understood by managers, planners and stakeholders of the Tiete Basin. Groundwater actively contributes to the base flow of rivers that drain into the basin, making the rivers and streams perennial during the dry months. This also continues its role of sustaining aquatic life, transporting sediments, diluting urban and agricultural contamination, and keeping the natural scenic beauty.

28.3 Hydrogeology of São Paulo

The Upper Tiete Basin is a hydrological unit that includes the domains of the São Paulo Sedimentary Basin (1452 km^2), and the Precambrian rocks of the crystalline basement (4323 km^2). This geological context defines the two main aquifer systems, the sedimentary and the crystalline (Hirata & Ferreira 2001) (Figure 28.2a,b).

The Crystalline Aquifer System (CAS) is composed of three aquifer units based on the description of the type of rock: (A) *granitoid*; (B) *metasedimentary*; and (C) *metacarbonatic*, and these rocks exhibit two different hydraulic behaviors from top to base due to weathering process. The first, related to portions of the weathered rock, forms a heterogeneous and unconfined aquifer, with average thicknesses of 30–40 m. Under the weathering mantle, and often hydraulically connected, a crystalline aquifer itself occurs, where the water circulates through the discontinuities (faults and fractures). This unit is unconfined to semi-confined, sometimes confined by overlaid sediments, and strongly heterogeneous and anisotropic (Hirata & Ferreira, 2001) (Table 28.2).

The Sedimentary Aquifer System (SAS) covers only 25% of the watershed area and is divided into two units: São Paulo and Resende formations. This aquifer system is unconfined to semi-confined, has primary porosity, and is heterogeneous.

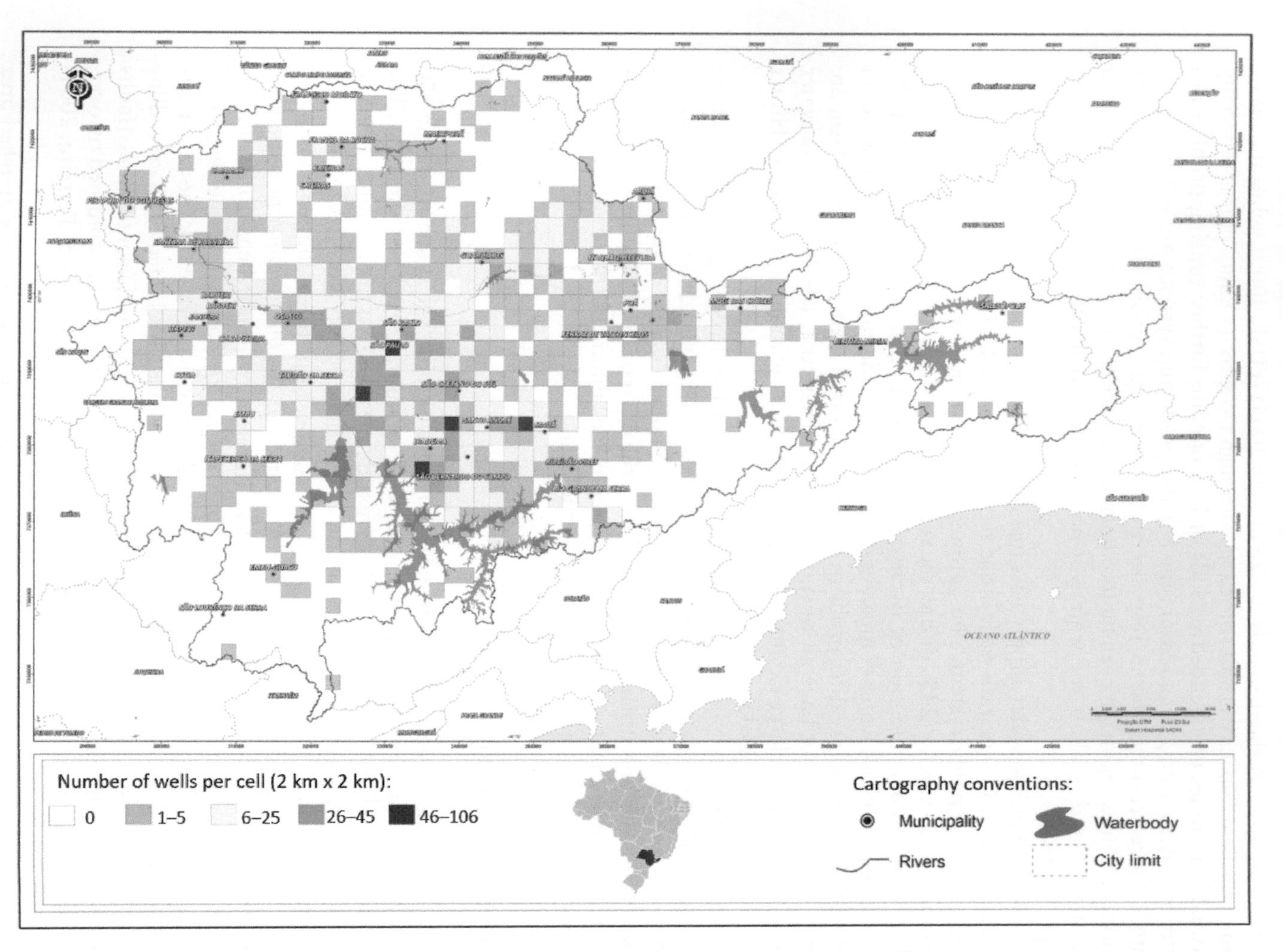

Figure 28.1 The distribution of the density of wells per cell (2 km × 2 km) in Upper Tiete Basin (only wells with license) (Hirata *et al.*, unpublished) (São Paulo location in the small map).

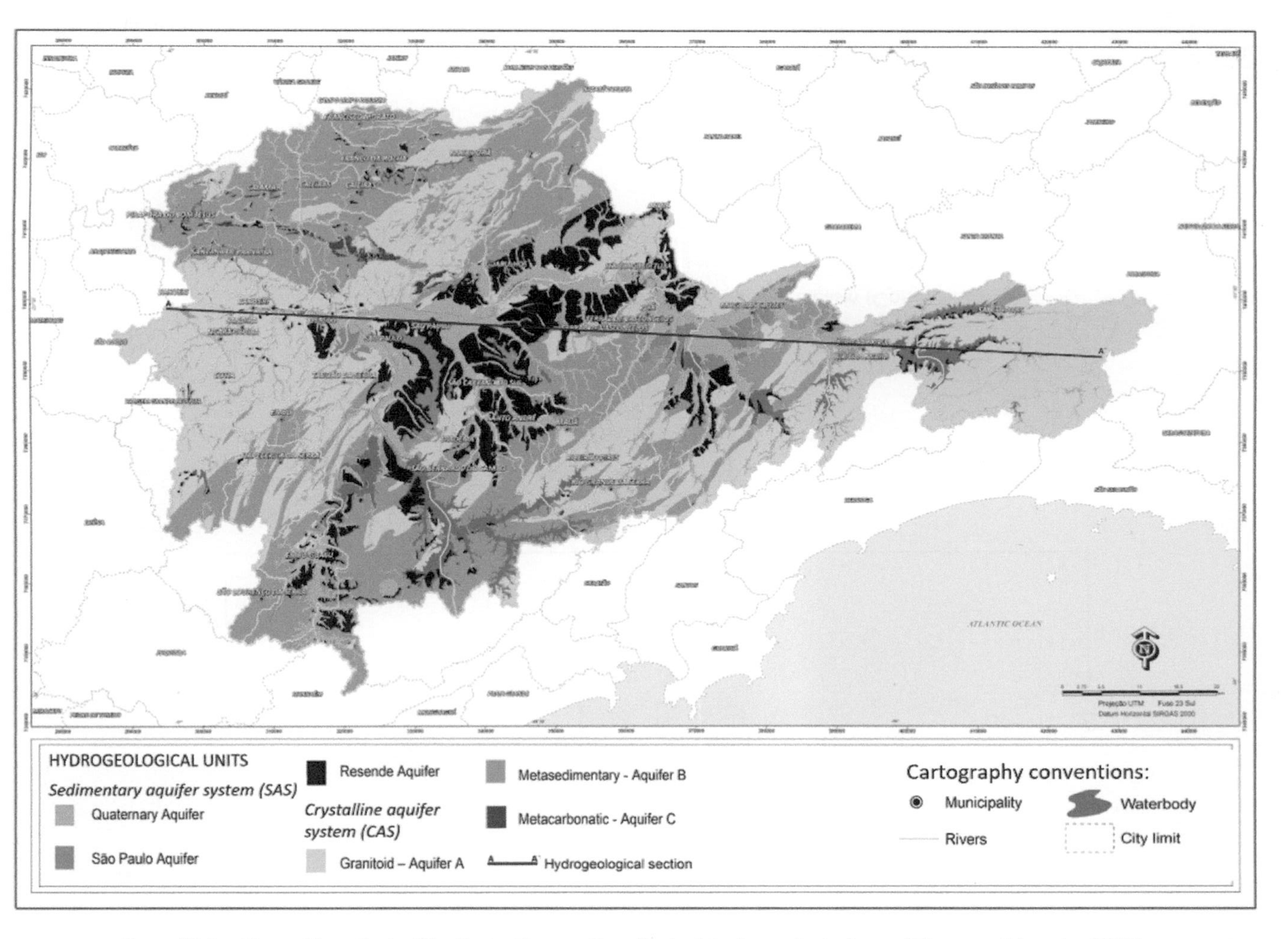

Figure 28.2a Upper Tiete Basin (São Paulo Metropolitan Region) hydrogeological map (Hirata and Ferreira, 2001).

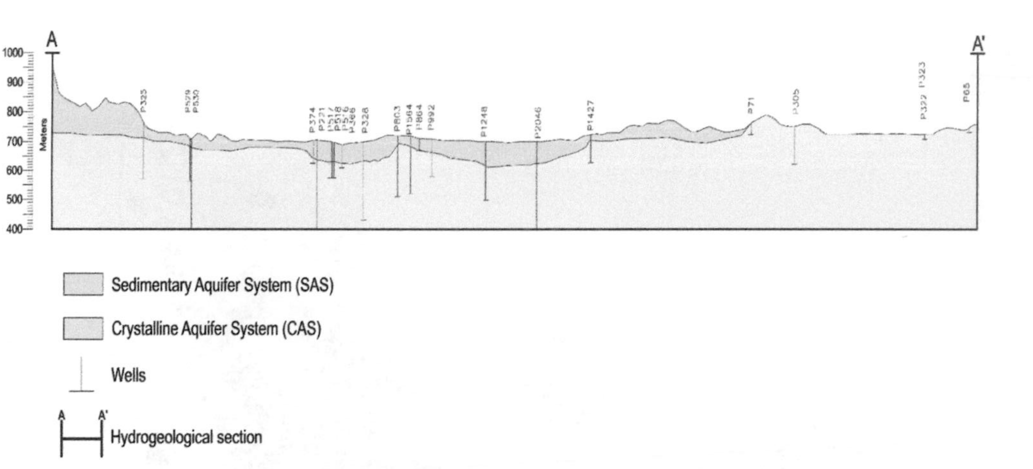

Figure 28.2b The hydrogeological transection of Upper Tiete Basin (Conicelli, 2014).

Table 28.2 Aquifers properties in São Paulo and Mexico City Metropolitan Regions.

Type of aquifer	*São Paulo* *Alluvial and pre-Cambrian fractured*	*Mexico City* *Alluvial and volcanic rocks*
Aquifer thickness (average)	120 m (SAS) and 300 (SAC)	1000
Aquifer water age	Young	8000
Cost of water extraction	Low	High
Well flow rate (average, m^3/h)	5	100
Depth of well (average, m)	150	300
Drilling well cost ($\times 10^3$ US$)	20	50
Balance & availability of groundwater = [extraction/availability] %	30	130

The conceptual model of groundwater circulation from Tiete Basin shows that the groundwater flows naturally, moving toward to major surface water bodies of the basin, with Tiete river as the main discharge area. Currently, well deepening and increase in the number of new wells have lowered the hydraulic heads, causing convergence zones of groundwater flow.

The recharge of the aquifers systems is associated to natural infiltration in part from water surplus, leakage of water public mains and, more restrictedly, sewer system and storm sewers. The recharge of unconfined aquifer from Tiete Basin ranges from <150 to 450 mm/year that sum than 30 m^3/s (407 mm/y) in the whole area, of which 50% comes from leaks from the water and sewage network (Hirata *et al.*, 2013; Bertolo *et al.*, 2015).

The groundwater flow in the Sedimentary Aquifer is part of a local to semi-local circulation system, especially because this unit is 100 m thick in average. The groundwater flow in the Crystalline Aquifer (Figure 28.2a,b) occurs through fractures, and it can be limited if fracture connectivity is absent. Due to the lower effective porosity, fractured rock aquifers present high groundwater flow velocity, which can increase its vulnerability.

Groundwater in the Tiete Basin has a good water quality and is suitable for human consumption and other uses. However, in very specific areas there are problems with anomalous concentrations of fluoride as well as iron and manganese (FUSP, 2009; Hirata *et al.*, 2013). More serious problems are related to imbalance between human activities and aquifer vulnerability, provoking local aquifer degradation in 2844 localities in the Tiete Basin, associated more to hydrocarbons; halogenated hydrocarbons; and heavy metal (CETESB, 2016).

On the other hand, overexploitation problems are still restricting to some areas in the Tiete Basin, although in some of them the lowering water level is higher than 100 m below natural levels.

28.3.1 Institutional arrangements for groundwater management in the Upper Tiete Basin

The state of São Paulo has a decentralized water resource management system considering of 22 water basin units (Unidade de Gerenciamento de Recursos Hídricos,

UGRHI). The Upper Tiete Basin is one of them. The Water Basin Committee (Comitê de Bacia Hidrográfica), comprising representatives of the state government, 36 municipalities and civil society – that have equal voting rights – is a deliberative forum that deals with matters related to water resource management and planning in the Tiete Basin and the Water Basin Agency, which is the executive office of Water Basin Committee, implements its policies and gives all necessary technical work support (FUSP, 2009). Despite the creation of the basin committee in 1991, its implementation is still very poor in practice. The Basin Committee is composed of representatives of the state government, municipal governments and civil society have equal voting rights. However, there is a clear imbalance among the members of civil society. Stronger economic sectors have better prepared members than those in the less capitalized civil society sectors, such as the environment or even small farmers. Likewise, the state government has a strong command of decisions, since it controls quite intensively the financing of projects. Actual water resource management is still centralized in state government institutions. Within São Paulo state government, the State Secretariat for Water Supply and Sanitation through e Department of Water and Electric Energy (DAEE) is in charge of water extraction permits (water rights) and water planning and sanitation policies; the Environmental Agency (CETESB) is responsible for controlling human activities that can contaminate soil, air and water; and Health Secretariat is in charge of sanitary surveys and control end-users water quality.

The Tiete Basin has a serious legal problem related to groundwater management: it is estimated that more than 60% of all wells are illegal (with no extraction permit). Although there are regulations controlling the drilling for and exploitation of groundwater in the State of São Paulo, there are serious issues with the enforcement of these regulations. The State government agencies claim that a lack of technical staff and administrative infrastructure is to blame. However, the problem is much more complex and relates to many other factors, including: (i) there is not much societal and government awareness of the true value of groundwater and their role for the society and the environment; (ii) groundwater regulation is flawed because it only penalizes illegal users (many thousands) and not the companies that drill the wells (less than a hundred); (iii) groundwater users see no benefit in having wells legalized, because of the bureaucratic hurdles and fee that must be paid to do so and there is virtually no oversight; and (iv) the groundwater overexploitation and contamination conflicts exist but these are not apparent (or perceived) by users; therefore, users who are affected by such problems do not recognize them nor pressure the state to implement control measures (Hirata *et al.,* 2014; Mahlknecht *et al.,* 2014).

Although there is a cooperation agreement among the Environmental Agency, the Secretariat of Water Resources and Sanitary Surveillance organism to assess whether permission to use groundwater will not cause quality problems, there is no integrated management and planning of the quality and quantity of water resources in the Tiete Basin. The Environmental Agency acts by evaluating and controlling potentially contaminating activities individually and does not analyse the water resource as a whole. The Sanitary Surveillance system mainly monitors the public water distribution systems, regardless of the origin of the water resource.

28.4 THE METROPOLITAN AREA OF MEXICO CITY

The Metropolitan Area of Mexico City includes the territory of the Federal District (Mexico City) and 38 municipalities of the states of Mexico and Hidalgo. This territory has been characterized by the constant urban-rural transformations since ancient times. In it settled diverse indigenous towns and was constituted in the center of the Aztec Empire, the New Spain and the present Mexican Republic. The Mexico City's population is around 20.1 million inhabitants today.

Currently Mexico City aquifer supplies 42 m^3/s equivalents to 66% of the drinking water supply. The import of surface water from the Cutzamala basin represents 24% of the water supply, the import of groundwater from the Lerma basin represents 8%, and the springs in the south of the city produce the remaining 2% of the drinking water supply. All industries, hotels, shopping malls and other economic activities that self-supply of water do so through wells drilled in the aquifer within Mexico City (Figure 28.3).

The intensive exploitation of groundwater in Mexico City has caused tremendous economic and environmental problems. Since the potentiometric levels have fallen to depths greater than 90 m (See Figure 28.4), from an economic point of view, there has been increase in the pumping, the construction and maintenance of wells costs. Many private shallow wells of little depth are inactive, and the migration of water with poor quality increased the costs for groundwater potabilization. The subsidence and fracturing of the land have caused economic losses in houses, buildings and urban infrastructure. The most important environmental impacts are the disappearance of springs and phreatophyte vegetation, the drying of the lakes, and the loss of flow base in rivers (Escolero *et al.*, 2009).

28.4.1 Hydrogeology in Mexico City

Mexico City refers to a valley located in the lower part of the closed Basin of Mexico. The Basin of Mexico is a graben structure that is closed hydrologically by mountains of the Mexican Trans-Volcanic Belt (Demant, 1978). It has a total area of 7740 km^2 of which 1507 km^2 is the central lacustrine plain at an elevation of 2236–2250 m above sea level (masl). The Valley floor has an elevation of about 2236 masl and the lowest part is still occupied by what remains of Lake Texcoco. The basin boundaries are the mountains that surround the Valley (frequently with more than 3000 masl) and the lowest pass across the mountains (about 2260 masl). Elevations in the Sierra de las Cruces to the west are commonly above 3000 masl. In the east, the volcanoes Iztaccihuatl and Popocatepetl rise to over 5000 masl and have permanent glacial ice (Durazo & Farvolden, 1989). The Sierra Chichinautzin, a broad ridge that rises some 700 m above the Valley floor, formed mostly of Quaternary basaltic flows, closes the Basin of Mexico to the south. Pyroclastic cones of various sizes are scattered widely across the mountain ridge (Durazo & Farvolden, 1989).

These lava flows form large perched aquifers that capture the infiltration of rainwater in the Chichinautzin Sierra, discharging into medium flow springs, which are collected for drinking water supplies to populations in the south of Mexico City (Escolero *et al.*, 2009). These springs have been used since the Aztecs' times.

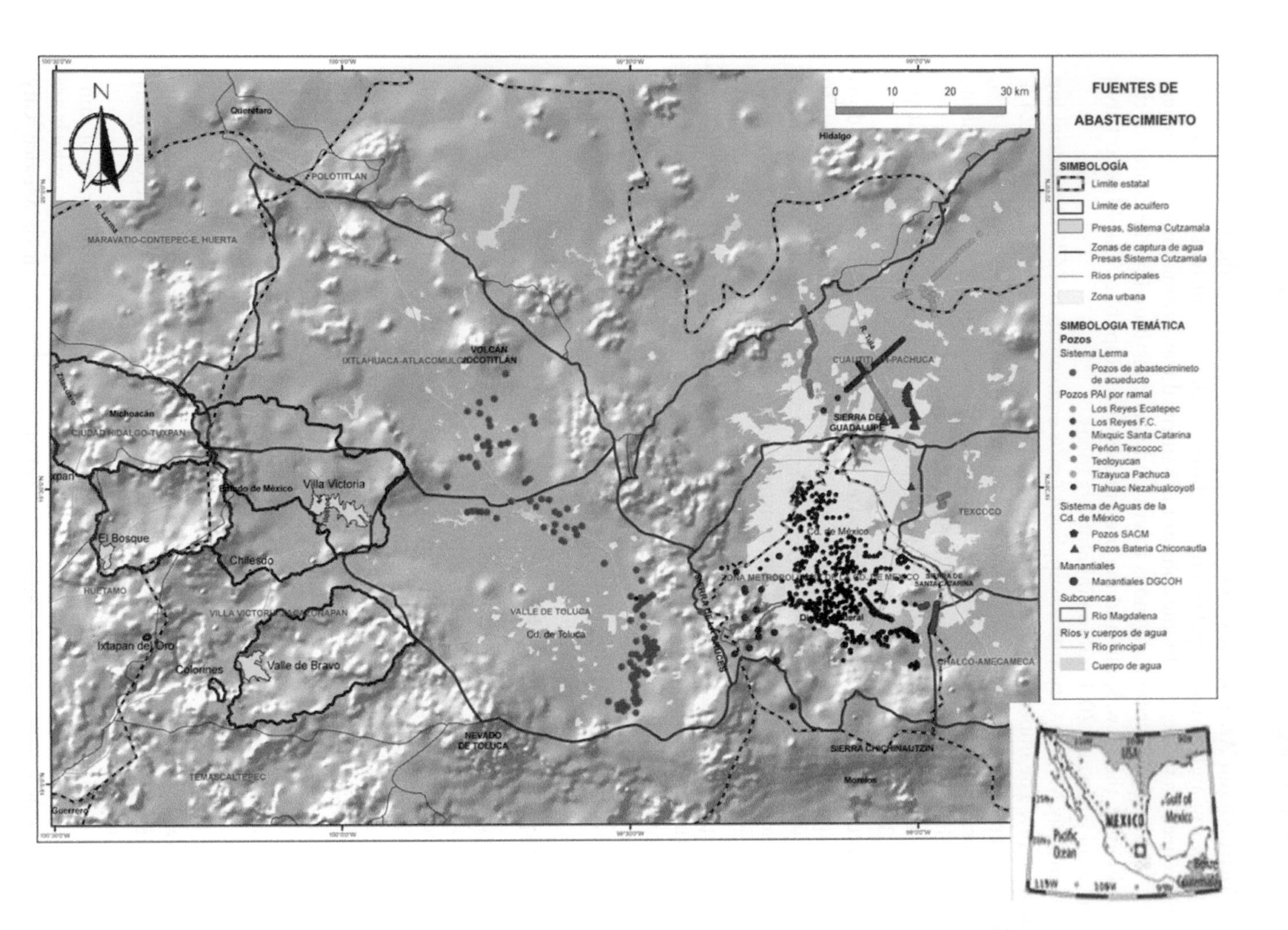

Figure 28.3 Drinking water sources for water supply to Mexico City (Escolero *et al*., 2016).

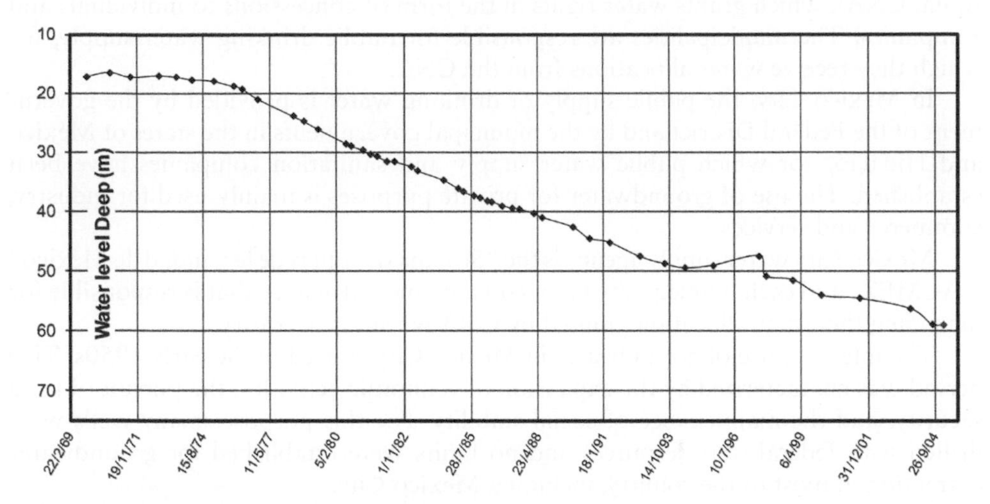

Figure 28.4 Decrease of water levels in the aquifer of Mexico City between 1969 and 2005, the data correspond to well 5311 and was elaborated with data of Escolero *et al.* (2009).

The sediments that fill up the Valley include thick, extensive layers of compressible lacustrine clays with a high-water content, which form shallow aquitards with a variable thickness between 30 and 300 m and which acts as an aquitard in the central part of the Valley (Table 28.2).

Under this shallow aquitard, there is the aquifer currently in operation, formed by granular material of pyroclastic and conglomerates of volcanic origin. Its thickness is variable (between 200 and 700 m) as well as its hydraulic properties. The wells for drinking water supply are drilled to depths between 300 and 700 m. Fractured volcanic rocks, whose top reach 3000 m deep in the center and increase towards the flanks of the Basin, underlie this aquifer. Water wells have recently been constructed that reach depths between 700 and 1000 m, capturing good quality groundwater in the volcanic rocks.

The three units present a wide range in their hydrodynamic parameters and are hydraulically connected especially the last two (volcanic rocks and granular material). The interrelation of the first (clay pack) with the second (granular material) has been demonstrated by the subsidence that affects Mexico City (Martinez *et al.*, 2015).

After an intense earthquake in 1985, a program of drilling five deep wells was started in Mexico City, reaching depths between 2000 and 3000 m, to know the regional basement of the Valley and limestone rocks were found at depths between 1500 and 3000 m. Another deep-water drilling program is currently underway at 2000 m depth and it has been possible to extract cores from the limestone rocks found at 1500 m depth and are under study to know their hydraulic properties and water quality and age (Escolero *et al.*, 2016).

28.4.2 Institutional arrangements for groundwater management in Mexico

In Mexico, groundwater is owned by the nation, and water is administered by the federal government, through the National Water Commission (Comisión Nacional del

Agua, CNA), which grants water rights in the form of concessions to individuals and companies. The municipalities are responsible for public drinking water supply, for which they receive water allocations from the CNA.

In Mexico City, the public supply of drinking water is provided by the government of the Federal District and by the municipal governments in the states of México and Hidalgo, for which public water supply and sanitation companies have been established. The use of groundwater for private purposes is mainly used for industry, commerce and services.

Mexico City water supply agency is the "Sistema de Aguas de la Ciudad de México" (SACMEX) and each municipality has its own municipal agency that is responsible for managing the water allocation granted by CNA within its territory.

The intensive use of groundwater in Mexico City started in the early 1950s. This period was characterized by the expansion of economic activities, the pursuit of food security, and the maintenance of social stability. For this purpose, many wells were drilled with federal-state resources and no limits were established for groundwater extraction in most of the country, including Mexico City.

The increase in the intensive exploitation of groundwater in many regions was intensified by the corruption of officials responsible for grant drilling permits, or by pressures from local politicians and governors to support electoral campaigns.

The second stage begins in the late 1980s, with the publication of a new water law and the establishment of a single federal agency responsible for all aspects of water (CNA). At this stage, controls and limitations were established to grant new permits to drill wells, reinforced the scheme of groundwater concessions, created the water rights market, and regulate the transfer of water rights between different water uses (Garduño, 2001; Escolero & Martinez, 2007).

To reinforce the concessions scheme, it was established that farmers and also municipal drinking water agencies to have access to the federal subsidies should have in place and comply with the volume of water concessions.

In the case of private companies, they must pay a right to the federal government per volume of groundwater used, which directly enters to the Ministry of Finance, carrying out the tax collection, so there is a strict monitoring of the volumes of water used by these private companies.

Due to the rapid expansion of urban areas in Mexico City, there has been a strong transfer of groundwater rights from agriculture to urban and industrial uses, so that water rights for agriculture have virtually disappeared. Due to the strong economic competition for groundwater rights, public drinking water agencies and municipalities have adopted the strategy of authorizing new private real estate developments in agricultural areas only when the acquisitions of land are accompanied by the purchase of the water rights, so that later the private developers transfer these water rights to urban public use in favor of drinking water agencies.

Because of this process of concentration of groundwater rights in the hands of municipal water agencies, 88% of the 507 Mm^3/y that is extracted from the aquifer in Mexico City is destined to supply drinking water, 10% for industrial use, 1% for commercial centers and the rest for other groundwater users.

Due to the concessions and permits granted, the authorized annual volume far exceeds the sustainable volumes in the long term, which represents a huge challenge since almost all the water extracted is destined for drinking water. Even if all water concessions granted for industry and commerce were cancelled, it would not be possible to reduce the extraction to match the sustainable volumes in the medium and long term, estimated at 279 Mm^3/y by CNA.

The water law published in 1989 establishes the forms of participation of users in decision-making on water management, but in the case of Mexico City, due to the high concentration of water rights in municipal water agencies, the municipal water companies prefer to negotiate subsidies directly through political agreements with the Federal Government, which is why they do not accept the formation of associations of water users. For this reason, there have been no associations of water users in this region yet, although there is a Basin Council integrated mainly by federal, state and municipal authorities, with a limited participation of other water users, designated by the same authorities. All decisions on water management in Mexico City are made from federal, state and municipal government structures.

28.5 COMPARISON BETWEEN THE TWO METROPOLITAN REGIONS

Groundwater plays an important and distinctive role for the metropolitan regions of São Paulo and Mexico. The Mexico City's public water supply is based on 88% of the groundwater resource. In contrast, in São Paulo, the public system uses almost exclusively surface water (99%), and groundwater has the role of complementary supply and restricted to private activities. Although this private use is important (15% of the total water supply in the Tiete Basin), it is distributed among thousands of small users across the watershed and with little State control.

The water management system presents differences as well. In Mexico, it is quite centralized in government agencies and user and society participations in the decisions are much more restricted than in São Paulo. Although water management in São Paulo is done through a basin committee, with the participation of the state, municipal and civil society, decisions on groundwater are not in its agenda. The importance of groundwater is hardly recognized for those who make the decisions about this resource in Tiete Basin. Thus, in practice, groundwater management is all centralized in government agencies. In contrast, the control of soil and groundwater contamination in São Paulo is more developed than in Mexico, and the few interventions to restrict groundwater use in Tiete Basin occur more because of contamination problems than due to problems of intense exploitation.

The groundwater development relies in different stages too. In Mexico, the level of irregularity in the wells is less than 10% and is decreasing drastically, in contrast to São Paulo, where illegality is greater than 60%. Success in controlling illegality occurs through the application of various financial and management mechanisms. However, this phenomenon is more complex and involves in addition the following variables/explanatory factors:

a) The number of wells in Mexico City is much lower than in São Paulo, making government inspection easier. In Mexico City there are just over 3000 wells, against 12000 wells in São Paulo. This is because the wells in Mexico are more expensive, deeper and much more productive. The stage of development of the aquifer is also different. With deeper potentiometric levels due to exploitation, the wells have to be deeper and more costly in Mexico.

b) The awareness level of the groundwater importance in Mexico is higher than in São Paulo. The conflicts over the groundwater user are also more evident: there are geotechnical problems associated with land subsidence, which are visible to the entire population. There is a need to increase the depth of wells more frequently in Mexico than in São Paulo, where the aquifer exploitation is still less developed and problematic.

c) Mexico has implemented some financial regulation that has controlled its water uses, which do not exist in São Paulo: (i) the Federal Government Finance Ministry collects fees on groundwater extraction; (ii) there is an effective market for buying and selling the water right; (iii) the water value is determined by an active and effective market, and increases yearly due to the strong competition among users; (iv) the industry or service owner has to obtain the right for extracting water from a legal well, in order to have the electricity installation by the government.

d) Due to the water shortage in Mexico, there is strong competition for water. In São Paulo the right to water use is given to almost every request. There are some limitations, but they are associated with problems of groundwater contamination, which are still very limited, or if the well is less than 200 m apart from a proven source of contamination.

28.6 CONCLUSIONS

The stage of aquifer exploitation development in Mexico City and São Paulo is quite different. In Mexico City there are several problems associated with the intensive exploitation of aquifers, in contrast to São Paulo, where the few restrictions in the exploitation license applications are associated to the contamination caused by anthropic activities. The high dependence on groundwater for public supply and the relationship between demand and availability in Mexico City lead to an increasing population awareness, and forced the State to have more control to water resource users. In São Paulo, an average citizen has little knowledge on the origin and status of groundwater, due to the fact that the conflicts among users, although existent are not visible to the population and government agencies. They do not understand the aquifer problems related to overexploitation and, consequently, do not claim their rights and are not aware about the importance of legalizing their well.

Financial incentive policies have apparently proved to be efficient in Mexico to reduce the numbers of illegal wells, where there is a market for water use rights that gives commercial value to water. The government subsidy to groundwater exploitation by farmers requires them to have their legalized wells, although create an artificial value to water. The same happens with the sector of public water supply, which leads to a waste and inefficient use of water resources. The strong concentration of the right to water use in the hands of public water government owner companies has reduced the imbalance in the unfair competition between small and large water users, although this is still a real problem in Mexico.

In both cases, decisions on groundwater management are centralized in government agencies, in the case of Mexico by federal government agencies and municipalities, and in the case of São Paulo by state government agencies. So, in both cases there is little concern of government agencies about the environmental impacts on groundwater. In both cases there is a lack of vision towards integrated water management including environmental, social and economic aspects.

In the case of São Paulo, participation in decision-making by groundwater users is limited by the large number of illegal wells, and is limited to the participation in the Basin Committee. In the case of Mexico, the participation of other users of groundwater is overshadowed by the presence of municipal water companies with high economic and political power, that control 88% of groundwater rights.

For Mexico, the construction of new deeper wells represents a temporary solution of a structural-type problem, which will only allow deferring in time the making of non-structural decisions that require the co-responsible participation of the inhabitants of Mexico City. In the meantime, it will be necessary to revise the institutional arrangements to allow greater involvement of civil society organizations. In the case of São Paulo, the solution for the increase of water production is based on the construction of large works of water transposition and integration of the various sources of surface water production. Although groundwater plays a significant role in balancing supply and demand, no strategic action is being planned.

Finally, there will only be an efficient and effective management of the groundwater resource with the real involvement of its users in this process. To do this, it is necessary to develop social communication schemes and awareness of the problems and conflicts associated with the exploitation of groundwater and to articulate alternative action in the face of possible problems, including those associated with global climate change.

REFERENCES

Bertolo, R; Hirata, R; Conicelli, B; Simonato, M; Pinhatti, A; and Fernandes, A. (2015) Água subterrânea para abastecimento público na Região Metropolitana de São Paulo: é possível utilizá-la em larga escala?. Revista DAE, v. 63, p. 6–17, 2015.

Conicelli, B. (2014) Gestão das águas subterraneas na Bacia Hidrográfica do Alto Tietê. Doctorate Dissertation. Institute of Geosciences. University of Sao Paulo.

CETESB. (2016) *Cadastro de áreas contaminadas.* [on line] Available from: http://areascontaminadas.cetesb.sp.gov.br/cadastro-de-acs/ [25 Dec 2016].

Demant, A. (1978) Caracteristicas del Eje Neovocanico Transmexicano y sus problems de interpretatcion. Rev. Inst. Geol., Univ. Nac. Auton. Mex., Mex., D.F. 2, pp. 172–178.

Durazo, J. and Farvolden, R.N. (1989) The groundwater regime of the Valley of Mexico from historic evidence and field observations. Journal of Hydrology 112:1–190.

Escolero O. and Martínez, S. (2007). The Mexican Experience with Groundwater management. In: The Global Importance of Groundwater in the 21st Century: Proceedings of the International Symposium on Groundwater Sustainability. National Ground Water Association, NGWA Press, ISBN 1-56034-131-9. p. 97–103.

Escolero, O., Martínez, S. E., Kralisch, S., Perevochtchikova and M. Delgado-Campo, J. (2009) *Vulnerabilidad de las fuentes de abastecimiento de agua potable de la Ciudad de México en el contexto de cambio climático.* Informe final. Centro Virtual de Cambio Climático. Available from: http://www.cvcccmatmosfera.unam.mx/sis_admin/archivos/agua_escolero__inffinal_org.pdf [25 Dec 2016].

Escolero, O., Kralisch, S., Martínez, S. E. and Perevochtchikova, M. (2016) Diagnóstico y análisis de los factores que influyen en la vulnerabilidad de las fuentes de abastecimiento de agua potable a la Ciudad de México, México. Boletín de la Sociedad Geológica Mexicana, octubre de 2016; 68(3), 409–427.

Fundação da Universidade de São Paulo – FUSP. (2009) Plano da Bacia Hidrográfica do Alto Tietê. Comitê de Bacia Hidrográfica do Alto Tietê. São Paulo. 4 vol.

Garduño, H. (2001) *Water Rights Administration: Experience, Issues and Guidelines*. UN-FAO Legislative Study 70: Rome, Italy.

Hirata, R. and Ferreira, L. (2001) Os aquíferos da bacia hidrográfica do Alto Tietê: disponibilidade hídrica e vulnerabilidade à poluição. Revista Brasileira de Geociências, São Paulo, v. 31, n.1, p. 43–50.

Hirata, R; Foster, S and Oliveira, F. (2015) A'guas Subterrâneas Urbanas no Brasil: avaliação para uma gestão sustentável. 1. ed. São Paulo: Instituto de Geociências, v. 1. 112p.

Hirata, R; Bertolo, R; Conicelli, B and Maldaner, C. (2013) Hydrogeology of the Upper Tiete Basin. In: Arsênio Negro; Makoto Namba; Vivian Sanches; Andrea Dyminski; Alessander Kormann. (Org.). Soils of the metropolitan regions of São Paulo e Curitiba. 1ed. São Paulo: D'Livros, 2013, v. 1, p. 67–88.

Mahlknecht, J; Hirata, R. and Ledesma-Ruiz, R. (2014) Urban groundwater supply in Latin American cities. In: Ismael Aguillar; Jurgen Mahlknecht; Jonathan Kaledin; Marianne Kjellen; Abel Mejía. (Org.). Water and cities in Latin America: challenges for sustainable development. 1ed.Florence, USA: Routledge; Francis Taylor Group, 2014, v. 1, p. 126–146.

Martinez, S; Kralisch, S; Escolero O. and Perevochtchikova M. (2015) Vulnerability of Mexico City's water supply sources in the context of climate change. Journal of Water and Climate Change. Vol. 6, No. 3, 518–533. doi:10.2166/wcc.2015.083

Wigle J. (2010) The 'Xochimilco model' for managing irregular settlements in conservation land in Mexico Cities. Vol. 27: 337–347.

Printed and bound by CPI Group (UK) Ltd, Croydon, CR0 4YY
01/11/2024
01782603-0001